Liquid Chromatography–Mass Spectrometry

Third Edition

CHROMATOGRAPHIC SCIENCE SERIES

A Series of Textbooks and Reference Books

Editor: JACK CAZES

Liquid Chromatography–Mass Spectrometry

Third Edition

Wilfried M.A. Niessen

hyphen MassSpec Consultancy
Leiden, The Netherlands

CRC Press
Taylor & Francis Group
Boca Raton London New York

CRC Press is an imprint of the
Taylor & Francis Group, an **informa** business

A TAYLOR & FRANCIS BOOK

CRC Press
Taylor & Francis Group
6000 Broken Sound Parkway NW, Suite 300
Boca Raton, FL 33487-2742

First issued in paperback 2020

© 2006 by Taylor and Francis Group, LLC
CRC Press is an imprint of Taylor & Francis Group, an Informa business

No claim to original U.S. Government works

ISBN-13: 978-0-367-57782-7 (pbk)
ISBN-13: 978-0-8247-4082-5 (hbk)

Library of Congress Card Number 2006013709

Library of Congress Cataloging-in-Publication Data

Niessen, W. M. A. (Wilfried M. A.), 1956-
    Liquid chromatography--mass spectrometry. -- 3rd ed. / Wilfried M.A. Niessen.
        p. cm. -- (Chromatographic science series ; 97)
    Includes bibliographical references and index.
    ISBN-13: 978-0-8247-4082-5 (acid-free paper)
    ISBN-10: 0-8247-4082-3 (acid-free paper)
    1. Liquid chromatography. 2. Mass spectrometry. I. Title. II. Series: Chromatographic science ; v. 97.

QD79.C454N54 2007
543'.84--dc22   2006013709

Visit the Taylor & Francis Web site at
http://www.taylorandfrancis.com

and the CRC Press Web site at
http://www.crcpress.com

PREFACE TO THE THIRD EDITION

Before one starts to write the preface to the third edition of one's book, one obviously rereads the prefaces to the previous two editions. This third edition significantly differs from the previous two editions. Most chapters are completely new or have been extensively rewritten. With the new text and the update to current developments, the orientation on technology and on the hyphenated character of LC–MS, nowadays also including sample pretreatment and data processing, was kept. In the first edition, the main focus was on (interface) technology. The second edition still paid considerable attention to interface technology, but the application section had grown to 200 pages. In this third edition, there are two application sections, covering more than two-thirds of the text (420 out of the 600 pages). The message that can be read from this is that the LC–MS technology has become established and mature, whereas still rapid and exciting developments occur in its many application areas.

This book provides a literature overview. The focus is on principles, technologies, and especially applications and analytical strategies. Contrary to the previous editions, I did not at all intend to achieve comprehensive literature coverage in this third edition. Between 1998 and today, more than 15,000 papers were published on the topics discussed in this book. It is impossible for me to read all these papers, due to time limitations, and certainly to give proper attention to their contents, due to space limitations. In each individual chapter, I have tried to tell a story relevant to the topic of the chapter, providing a reasonable complete account on LC–MS related developments in that field. The goal was to provide an introduction and overview of the strategies and technologies important in each of the selected application areas. Papers were more-or-less randomly selected to serve as illustrations to the story and to help me in telling the story. In most cases, attention is focussed on discussing the role of LC–MS in the selected application areas and to highlight important analytical strategies, and not so much on the actual results obtained. I have to apologize to the authors of so many excellent papers, that I could not cite in the present text. There are far more applications than I could cover in this edition of the book.

In the past years, LC–MS has definitively come out of the mass spectrometry specialist's laboratory to find its place in many chromatography laboratories. Small-molecule application areas in environmental, food safety, and clinical analysis are the clearest and most striking examples of this. Obviously, the huge impact of LC–MS in pharmaceutical drug discovery and development continued. At the same time, the proteomics field developed, and LC–MS contributes significantly to these developments.

This third edition is most likely also the last edition, at least in this form. The exciting and spectacular growth of LC–MS in the past years is such that it is no longer possible for one person to comprehensively cover and follow all relevant developments in the wide variety of application areas.

Finally, I have to thank the many people who have inspired me over the years to continue with my efforts in completing this book. This includes among others the many people I meet during my courses and consulting work in LC–MS, my colleagues and the Ph.D. students in my part-time job at the Free University in Amsterdam, my international collaboration partners. I thank my wife and family, who had to share me, because a large part of me was writing this book.

Wilfried Niessen
2006

PREFACE TO THE SECOND EDITION

When the first edition of this book was published early 1992, LC–MS could already be considered an important and mature analytical technique. However, at that time, the great impact on LC–MS that electrospray and atmospheric-pressure chemical ionization (APCI) would have could already be foreseen. Since then, the versatility and application of LC–MS really exploded. Numerous LC–MS systems have been sold in the past 6 years and have found their way into many different laboratories, although the pharmaceutical applications of LC–MS appear to be most important, at least in terms of instrument sales. LC–MS-MS in selective reaction monitoring mode has now become the method of choice in quantitative bioanalysis.

This second updated, revised and expanded edition of this book on LC–MS was written and finished in a period when interface innovations somewhat calmed down. Electrospray and APCI have become the interfaces of choice. At present, no major developments in interface technology can be foreseen that will lead to another breakthrough in LC–MS. In terms of applications and versatility, innovations continue to appear, *e.g.*, in the use of LC–MS in characterization of combinatorial libraries and in other phases of drug development, in the advent of electrospray time-of-flight instrumentation for impurity profiling, in applications in the field of biochemistry and biotechnology.

In view of these developments, older interfaces like thermospray, particle-beam and continuous-flow fast-atom bombardment appear to be obsolete. Nevertheless, it was decided to keep the second edition of this book as the comprehensive introduction and review of all important aspects of LC–MS interfacing and as a comprehensive guide through the complete field of LC–MS, covering all major interfaces and paying attention to the history of the technique as well. However, all chapters have been extensively revised and expanded. The discussions on interface technology and ionization methods have been integrated. Experimental parameters and optimization are covered in much more detail in the various interface-related chapters. Another major change concerns the attention paid to applications: instead of one 50-page chapter, like in the first edition, the major fields of application of LC–MS, *i.e.*, in environmental, pharmaceutical, biochemical and biotechnological analysis and in the analysis of natural products and endogenous compounds, are reviewed in five chapters, covering almost 200 pages, in this second edition.

The author would like to thank the people who reviewed some of the new chapters and whose valuable comments were used to enhance the quality of the text: Dr. Jaroslav Slobodník (Environmental Institute, Koš, Slovak Republic), Dr. Arjen Tinke (Yamanouchi Europe, Leiderdorp, the Netherlands), and Dr. Maarten Honing (AKZO-Nobel Organon, Oss, the Netherlands).

Wilfried Niessen
1998

PREFACE TO THE FIRST EDITION

In the early 1970s several groups started research projects aiming at the development of the on-line coupling of liquid chromatography and mass spectrometry (LC–MS). These research efforts were mainly inspired by the great success of combined capillary gas chromatography mass spectrometry (GC–MS) in solving analytical problems. However, the development of on-line LC–MS turned out to be a demanding and challenging task. In the past 20 years many approaches to LC–MS have been described. Some of these are successful and commercially available. LC–MS is no longer a highly sophisticated technique being used in laboratories of specialists only. LC–MS has grown to become a mature and routinely used technique in many areas of applications. LC–MS still is a rapidly developing technique, expanding its analytical power and attracting more and more users.

In a period of rapid developments, this book on LC–MS is written. The core of this book is therefore focussed more on principles and strategies than on reviewing applications. All aspects of LC–MS are covered in this comprehensive review, giving a survey of the field from various angles and both for newcomers and experienced users. For the newcomers, the text affords a comprehensive introduction and review of all important aspects in LC–MS interfacing. Experienced users will find an extensive review of the various aspects, and perhaps some new viewpoints and inspiration for new experiments to develop and optimize LC–MS. Since the field of LC–MS is moving extremely fast, some of the chapters will unfortunately need updating on appearance of this volume. This is certainly true for the Ch. 9 and 10. In principle, all literature available to us by the end of 1990 is incorporated in this text. In some chapters, some later appeared papers have been included, either by brief mention in the text or in the applications tables and review.

This text is written from the 'true hybrid' philosophy on LC–MS. For that reason, concise introductions in liquid chromatography (Ch. 1) as well as mass spectrometry (Ch. 2) precede a general discussion on interfacing chromatography and mass spectrometry (Ch. 3). Subsequently, the various interfaces for LC–MS are discussed from a technological point of view. After a historical overview, in which all approaches to on-line LC–MS are discussed (Ch. 4), the commercially available and therefore most widely applied LC–MS interfaces are discussed, i.e., the moving-belt interface (Ch. 5), direct liquid introduction (Ch. 6), thermospray (Ch. 7), continuous-flow fast atom bombardment (Ch. 8), particle-beam interfaces (Ch. 9) and electrospray and related methods (Ch. 10). Developments in combining supercritical fluid chromatography and capillary electrophoresis to mass spectrometry are reviewed as well (Ch. 11 and 12) to fit LC–MS in the whole analytical framework of separation methods coupled with mass spectrometry. Next, the field of LC–MS is approached from the ionization point of view. Attention is paid to specific aspects of ionization under LC–MS conditions. In this respect, attention is paid to electron impact ionization (Ch. 13), chemical ionization (Ch. 14), ion evaporation (Ch. 15), and fast atom bombardment (Ch. 16), while a chapter on various ways to induce fragmentation (Ch. 17) closes the section on ionization. The third angle on LC–MS is from the application point of view. Applications from the fields of environmental, pharmaceutical and biochemical analysis as well as the analysis of natural products are discussed (Ch. 18), not to provide in-depth information in that particular field of application, but from a general analytical point of view, allowing the comparison of the different interfaces and the assessment of applicability ranges of the various LC–MS interfaces. Finally, LC–MS is considered as a hybrid technique. First, some aspects related to mobile phase compatibility problems are reviewed (Ch. 19). Then, LC–MS is considered from a general point of view. The various experimental parameters related to the separation, the interface, the ionization, and the mass analysis as well as aspects related to data handling are considered from the hybrid point of view. Developments in the various fields, that are combined in LC–MS as a hybrid technique, are reviewed. Important areas of future research are indicated (Ch. 20). Each chapter is written as a separate unit, that can be read apart from the other chapters, while extensive cross-referencing is provided.

Finally, this text could not have been completed without the inspiration, research activities, help and advice from many of the people in our laboratories at the Leiden University (Center for Bio-Pharmaceutical Sciences) and the department of structure elucidation and instrumental analysis at TNO. We would like to thank especially U.R. Tjaden, C.E.M. Heeremans, E.R. Verheij, R.A.M. van der Hoeven, P.S. Kokkonen, A.C. Tas, G.F. La Vos, L.G. Gramberg, M.C. ten Noever de Brauw, A.P. Tinke, D.C. van Setten, J.J. Pot and his people at the photography and drawing department of the Gorlaeus Laboratories, Ms. P. Jousma-de Graaf and M. van der Ham-Meijer.

Wilfried Niessen
Jan van der Greef
August 1991

CONTENTS

ABBREVIATIONS

2D	two-dimensional
ADME	adsorption, distribution, metabolism and excretion
AES	atomic emission spectrometry
AfC	affinity chromatography
ALS	acid-labile surfactant
AmOAc	ammonium acetate
ANIS	analogue internal standard
APCI	atmospheric-pressure chemical ionization
API	atmospheric-pressure ionization
APPI	atmospheric-pressure photoionization
BIRD	black-body infrared radiative dissociation
BLAST	Basic Local Alignment Search Tool
BSA	bovine serum albumin
CE	capillary electrophoresis
Cf-FAB	continuous-flow fast-atom bombardment
CI	chemical ionization
CID	collision induced dissociation
CIEF	capillary isoelectric focussing
CYP	cytochrome P450 complex
DAD	photodiode array detection
DCI	direct chemical ionization
DDA	data-dependent acquisition
DLI	direct liquid introduction
ECD	electron-capture dissociation
ECNI	electron-capture negative ionization
EDC	endocrine disrupting compound
EHI	electrohydrodynamic ionization
EI	electron ionization
ELSD	evaporative light scattering detection
ESA	electrostatic analyser
ESI	electrospray ionization
FAB	fast-atom bombardment
FAC	frontal affinity chromatography
FAIMS	high-field asymmetric-waveform ion-mobility spectroscopy

FD	field desorption ionization
FT-ICR-MS	Fourier-transform ion-cyclotron resonance mass spectrometry
FWHM	full-width at half maximum
GC	gas chromatography
GE	gel electrophoresis
H/D	hydrogen/deuterium exchange
HFBA	heptafluorobutyric acid
HILIC	hydrophilic interaction chromatography
HPAEC	high-performance anion-exchange chromatography
IAC	immunoaffinity chromatography
ICAT	isotope-coded affinity tag
ICP	inductively coupled plasma
ID	internal diameter
IEC	ion-exchange chromatography
IEF	isoelectric focussing
IEV	ion evaporation ionization
ILIS	isotope-labelled internal standard
IMAC	immobilized metal-ion affinity chromatography
IMER	immobilized enzyme reactor
IRMPD	infrared multiphoton dissociation
IS	internal standard
LC	liquid chromatography
LCxLC	comprehensive liquid chromatography
LINAC	linear acceleration collision cell
LIT	linear ion trap
LLE	liquid-liquid extraction
LOQ	lower limit of quantification
MAGIC	monodisperse aerosol generation interface for chromatography
MALDI	matrix-assisted laser desorption ionization
MBI	moving-belt interface
MRL	maximum residue level
MS	mass spectrometry
MS-MS	tandem mass spectrometry
MSPD	matrix solid-phase dispersion
MTBE	methyl-t-butyl ether
MudPIT	multidimensional protein identification technology
MUX	multiplexed electrospray interface
NMR	nuclear magnetic resonance spectroscopy
PAGE	polyacrylamide gel electrophoresis
PBI	particle-beam interface
PBMC	peripheral blood mononuclear cells
PD	plasma desorption ionization
PEG	poly(ethylene) glycol
PFK	perfluorokerosene
PFTBA	perfluorotributylamine
PMF	peptide mass fingerprinting

PPG	poly(propylene) glycol
PSA	peptide sequence analysis
PS-DVB	poly(styrene–divinylbenzene)
PTM	post-translational modification
Q-LIT	quadrupole-linear-ion-trap hybrid
Q-TOF	quadrupole-time-of-flight hybrid
RAM	restricted-access material
RF	radiofrequency
RPLC	reversed-phase liquid chromatography
S/N	signal-to-noise ratio
SALSA	scoring algorithm for spectral analysis
SBSE	stir-bar sorptive extraction
SCX	strong cation-exchange chromatography
SDS	sodium dodecylsulfate
SEC	size exclusion chromatography
SFC	supercritical fluid chromatography
SILAC	stable isotope labelling with amino acids in cell cultures
SIM	selected-ion monitoring
SIMS	secondary-ion mass spectrometry
SNP	single nucleotide polymorphisms
SORI	sustained off-resonance irradiation
SPE	solid-phase extraction
SPME	solid-phase microextraction
SRM	selected-reaction monitoring
SS-LLE	solid-supported liquid-liquid extraction
STP	sewage treatment plant
TAG	triacylglycerides
TCA	trichloroacetic acid
TDM	therapeutic drug monitoring
TFA	trifluoroacetic acid
TFC	turbulent flow chromatography
TMT	tandem mass tags
TOF	time-of-flight mass analyser
TSP	thermospray ionization
UV	ultraviolet detection

INTRODUCTION

1

LIQUID CHROMATOGRAPHY
AND SAMPLE PRETREATMENT

1. Introduction

Chromatography is a physical separation method in which the components to be separated are selectively distributed between two immiscible phases: a mobile phase is flowing through a stationary phase bed. The technique is named after the mobile phase: gas chromatography (GC), liquid chromatography (LC), or supercritical fluid chromatography (SFC). The chromatographic process occurs as a result of repeated sorption/desorption steps during the movement of the analytes along the stationary phase. The separation is due to the differences in distribution coefficients of the individual analytes in the sample. Theoretical and practical aspects of LC have been covered in detail elsewhere [1-5].

This chapter is not meant to be a short course in LC. Some aspects of LC, important in relation to combined liquid chromatography–mass spectrometry (LC–MS), are discussed, e.g., column types and miniaturization, phase systems and separation mechanisms, and detection characteristics. In addition, important sample pretreatment techniques are discussed. Special attention is paid to new developments in LC and sample pretreatment.

3

2. Instrumentation for liquid chromatography

In LC, the sample is injected by means of an injection port into the mobile-phase stream delivered by the high-pressure pump and transported through the column where the separation takes place. The separation is monitored with a flow-through detector. In designing an LC system, one has to consider a variety of issues:

• The separation efficiency is related to the particle size of the stationary phase material. A higher pressure is required when the particle size is reduced. With a typical linear velocity in the range of 2–10 mm/s, a pressure drop over the column can exceed 10 MPa, obviously depending on the column length as well.

• In order to maintain the resolution achieved in the column, external peak broadening must be reduced and limited as much as possible. In general, a 5%-loss in resolution due to external peak broadening is acceptable. In practice, this means that with a 3–4.6-mm-ID column, a 20-μl injection volume and a 6–12 μl detector cell volume can be used in combination with short, small internal-diameter connecting tubes. Avoiding external peak broadening is especially important when the column internal diameter is reduced [6].

• The quality of the solvents used in the mobile phase is important in LC–MS. Phthalates and other solvent contaminants can cause problems [7]. Appropriate filtering of the solvents over a 0.2–0.4-μm filter is required. Degassing of the mobile phase is required to prevent air bubble formation in the pump heads, but also in interface capillaries.

Table 1.1: Characteristics of LC columns with various internal diameters					
Type	ID (mm)	F (μl/min)	V_{inj} (μl)	C_{max} at detector[1]	Relative loading capacity[2]
Conventional	4.6	1000	100	1	8333
Narrowbore	2.0	200	19	5.3	1583
Microbore	1.0	47	4.7	21.2	392
Microcapillary	0.32	4.9	0.49	207	41
Nano-LC	0.05	0.120	0.012	8464	1

[1] Based on column ID; [2] Based on given injection volume.

- High-throughput LC–MS analysis demand for high-pressure pumps capable of delivering an accurate, pulse-free, and reproducible and constant flow-rate. A small hold-up volume is needed for fast gradient analysis. High-pressure mixing devices are to be preferred. Modern LC pumps feature advanced electronic feedback systems to ensure proper functioning and to enable steep solvent gradients.
- Injection valves with an appropriate sample loop volume, mainly determined by the external peak broadening permitted, are used. Reduction of sample memory and carry-over is an important aspect, especially in quantitative analysis. Modern autosamplers allow a more versatile control over the injection volume by the application of partially filled loops and enable reduction of carry-over by needle wash steps.

2.1 The column

The column is the heart of the LC system. It requires appropriate care. Conventionally, LC columns are 100–300-mm long and have an internal diameter of 3–4.6 mm with an outer diameter of $^1/_4$ inch. In LC–MS, and especially in quantitative bioanalysis, shorter column are used, *e.g.*, 30–50 mm, and packed with 3–5 μm ID packing materials. A variety of other column types, differing in column inner diameter, are applied. Some characteristics of these columns are compared in Table 1.1.

The microcapillary packed and nano-LC columns are made of 0.05–0.5-mm-ID fused-silica tubes. The packing geometry of these columns differs from that of a larger bore column, resulting in relatively higher column efficiencies. These type of columns are frequently used in LC–MS applications with sample limitations, *e.g.*, in the characterization of proteins isolated from biological systems.

With respect to packing geometry and column efficiency, microbore columns are equivalent to conventional columns, except with respect to the internal diameter. Since most electrospray (ESI) interfaces are optimized for operation with flow-rates between 50 and 200 μl/min, the use of 1–3-mm-ID microbore columns is advantageous, because no post-column solvent splitting is required.

Asymmetric peaks can have a number of causes: overloading, insufficient resolution between analyte peaks, unwanted interactions between the analytes and the stationary phase, *e.g.*, residual silanol groups, voids in the column packing, and external peak broadening.

In most cases, the use of a guard column is advised. It is placed between the injector and the analytical column to protect the latter from damage due to the injection of crude samples, strong adsorbing compounds, or proteins in biological samples that might clog the column after denaturation. In this way, the performance and lifetime of the expensive analytical column can be prolonged. Guard columns inevitably result in a loss of efficiency.

2.2   General detector characteristics

The detector measures a physical parameter of the column effluent or of components in the column effluent and transforms it to an electrical signal. A universal detector measures a bulk property of the effluent, *e.g.*, the refractive index, while in a specific detector only particular compounds contribute to the detector signal.

A detector can be either a concentration sensitive device, which gives a signal that is a function of the concentration of an analyte in the effluent, or a mass-flow sensitive device, where the signal is proportional to the mass flow of analyte, *i.e.*, the concentration times the flow-rate.

The analyte concentration at the top of the chromatographic peak c_{max} is an important parameter, related to the dilution in the chromatographic column. It can be related to various chromatographic parameters:

$$c_{max} = \frac{4 \; M \; \sqrt{N}}{\pi \; d_c^2 \; \varepsilon \; L \; (1 + k') \; \sqrt{2\pi}}$$

where M is the injected amount, N is the plate number of the column with an internal diameter d_c, a length L, and a column porosity ε, and k' is the capacity ratio of the analyte. Guided by this equation, a particular detection problem can be approached by optimizing the separation parameters, *e.g.*, amount injected, column diameter, plate number, and capacity ratio. It also is an important equation in appreciating the use of miniaturized LC column.

Other important characteristics of a detector for LC are:
- The noise, which is the statistical fluctuation of the amplitude of the baseline envelope. It includes all random variations of the detector signal. Noise generally refers to electronic noise, and not to the so-called 'chemical noise', although the latter generally is far more important in solving real-life analytical problems.
- The detection and determination limits, which are generally defined in terms of signal-to-noise ratios (S/N), *e.g.*, an S/N of 3 for the detection limit and of 5–10 for the determination limit of lower limit of quantification.
- The linearity and linear dynamic range. A detector is linear over a limited range only. In ESI-MS, the linearity is limited inherent to the ionization process. A linear dynamic range of at least 2–3 order of magnitude is desirable.
- The detector time constant. The detector must respond sufficiently fast to the changes in concentration or mass flow in the effluent, otherwise the peaks are distorted.

2.3   Detectors for LC

Next to the mass spectrometer, which obviously is considered being the most important LC detector in this text, a number of other detectors [8] are used in various applications:

- The UV-absorbance detector is the most widely used detector in LC, which is a specific detector with a rather broad applicability range. The detection is based on the absorption of photons by a chromophore, *e.g.*, double bonds, aromatic rings, and some hetero-atoms. According to the equation of Lambert-Beer, the UV detector is a concentration-sensitive device.
- The fluorescence detector is a specific and concentration-sensitive detector. It is based on the emission of photons by electronically excited molecules. Fluorescence is especially observed for analytes with large conjugated ring systems, *e.g.*, polynuclear aromatic hydrocarbons and their derivatives. In order to extend its applicability range, pre-column or post-column derivatization strategies have been developed [9].
- Evaporative light-scattering detection (ELSD) is a universal detector based on the ability of particles to cause photon scattering when they traverse the path of a polychromatic beam of light. The liquid effluent from an LC is nebulized. The resulting aerosol is directed through a light beam. The ELSD is a mass-flow sensitive device, which provides a response directly proportional to the mass of the non-volatile analyte. Because it can detect compounds that are transparent to other detection techniques, the ELSD is frequently used in conjunction with LC–MS to obtain a complete analysis of the sample [10].
- Nuclear magnetic resonance spectroscopy (NMR) coupling to LC has seen significant progress in the past five years [11]. Continuous-flow NMR probes have been designed with a typical detection volume of 40–120 µl or smaller. The NMR spectrum is often recorded in stop-flow mode, although continuous-flow applications have been reported as well.
- An inductively-coupled plasma (ICP) is an effective spectroscopic excitation source, which in combination with atomic emission spectrometry (AES) is important in inorganic elemental analysis. ICP was also considered as an ion source for MS. An ICP-MS system is a special type of atmospheric-pressure ion source, where the liquid is nebulized into an atmospheric-pressure spray chamber. The larger droplets are separated from the smaller droplets and drained to waste. The aerosol of small droplets is transported by means of argon to the torch, where the ICP is generated and sustained. The analytes are atomized, and ionization of the elements takes place. Ions are sampled through an orifice into an atmospheric-pressure–vacuum interface, similar to an atmospheric-pressure ionization system for LC–MS. LC–ICP-MS is extensively reviewed, *e.g.*, [12].

Table 1.2: Separation mechanisms in LC	
adsorption	selective adsorption/desorption on a solid phase
partition	selective partition between two immiscible liquids
ion-exchange	differences in ion-exchange properties
ion-pair	formation of ion-pair and selective partition or sorption of these ion-pairs
gel permeation / size exclusion	differences in molecular size, or more explicitly the ability to diffuse into and out of the pore system

Table 1.3: Phase systems in various LC modes		
Mechanism	**Mobile phase**	**Stationary phase**
adsorption (normal-phase)	apolar organic solvent with organic modifier	silica gel, alumina bonded-phase material
adsorption (reversed-phase)	aqueous buffer with organic modifier, *e.g.*, CH_3OH or CH_3CN	bonded-phase material, *e.g.*, octadecyl-modified silica gel
ion-pair	aqueous buffer with organic modifier and ion-pairing agent	reversed-phase bonded-phase material
partition	liquid, mostly nonpolar	liquid, physically coated on porous solid support
ion exchange	aqueous buffers	cationic or anionic exchange resin or bonded-phase material
size exclusion	non-polar solvent	silica gel or polymeric material

3. Separation mechanisms

A useful classification of the various LC techniques is based on the type of distribution mechanism applied in the separation (see Table 1.2). In practice, most LC separations are the result of mixed mechanisms, *e.g.*, in partition chromatography in most cases contributions due to adsorption/desorption effects are observed. Most LC applications are done with reversed-phase LC, *i.e.*, a nonpolar stationary phase and a polar mobile phase. Reversed-phase LC is ideally suited for the analysis of polar and ionic analytes, which are not amenable to GC analysis. Important characteristics of LC phase systems are summarized in Table 1.3.

3.1 Intra- and intermolecular interactions

Various intra- and intermolecular interactions between analyte molecules and mobile and stationary phase are important in chromatography [5] (Figure 1.1):
- The covalent bond is the strongest molecular interaction (200–800 kJ/mol). It should not occur during chromatography, because irreversible adsorption and/or damage to the column packing material takes place.
- Ionic interactions between two oppositely charged ions is also quite strong (40–400 kJ/mol). Such interactions occur in ion-exchange chromatography, which explains the sometimes rigorous conditions required for eluting analytes from an ion-exchange column.
- Ions in solution will attract solvent molecules for solvation due to ion-dipole interactions (4–40 kJ/mol).

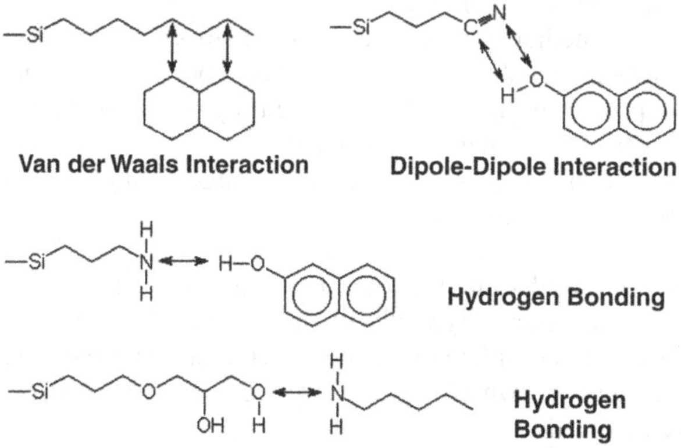

Figure 1.1: Intra- and intermolecular interactions important in chromatography. Based on [5].

Packing	Non-polar	Polar	Anion Exchange	Cation Exchange
Table 1.4: Interactions between analytes and stationary phase packing materials. (❶ Primary Interaction; ❷ Secondary Interaction; ✚ Silanol Activity)				
Octadecyl (C_{18}), octyl (C_8), and phenyl ($-C_6H_5$)	❶	❷		✚
Ethyl (C_2), cyano ($-C\equiv N$), and diol ($2\times -OH$)	❶	❶		✚
Silica ($-Si-OH$)		❶		✚
Amino ($-NH_2$), and diethylaminopropyl (DEA)	❷	❶	❶	✚
Quaternary Amine (SAX)	❷	❷	❶	✚
Carboxylic Acid (CBA)	❷	❷		❶
Benzenesulfonic (SCX)	❶	❷		❶

- The hydrogen atom can interact between two electronegative atoms, either within one or between two molecules. Hydrogen bonding can be considered as an important interaction (4–40 kJ/mol) between analyte molecules and both the mobile and the stationary phase in LC. In reversed-phase LC, both water and methanol can act both as acceptor and donor in hydrogen bonding, while acetonitrile can only accept, not donate.
- The third type of medium-strong interaction (4–40 kJ/mol) is the Van der Waals interaction, which are short-range interactions between permanent dipoles, a permanent dipole and the dipoles induced by it in another molecule, and dispersive forces between neutral molecules.
- Weaker interactions (0.4–4 kJ/mol) are longer range dipole–dipole and dipole–induced dipole interactions.

Alternatively, intermolecular interactions can be classified as:
- polar interactions, where hydrophilic groups like hydroxy, primary amine, carboxylic acid, amide, sulfate or quaternary ammonium groups are involved.
- nonpolar interactions, where hydrophobic groups like alkyl, alkylene, and aromates are involved.
- nonpolar interactions were carbonyl, ether, or cyano groups are involved.
- ionic interactions, *i.e.*, between cations and anions.

Along these lines, the interactions in various column packing materials can be classified (Table 1.4). The most important LC modes are briefly described below.

3.2 Reversed-phase chromatography

Reversed-phase LC is ideally suited for the analysis of polar and ionogenic analytes, and as such is ideally suited to be applied in LC–MS. Reversed-phase LC is the most widely used LC method. Probably, over 50% of the analytical applications are preformed by reversed-phase LC. Nonpolar, chemically-modified silica or other nonpolar packing materials, such as styrene-divinylbenzene copolymers (XAD, PRP) or hybrid silicon-carbon particles (XTerra), are used as stationary phases in combination with aqueous-organic solvent mixtures. Silica-based packing materials are used more frequently than polymeric packing materials.

Conventional chemically-modified silica materials are stable in organic and aqueous solvents in the pH range 2.5-8. The styrene-divinylbenzene copolymers and the XTerra material can be used in a wider pH-range.

Specific analyte-solvent interactions, *e.g.*, solubility effects, are most important in reversed-phase LC, because the interaction of the analyte with the bonded-phase material is a relatively weak, nonspecific Van der Waals interactions. The retention decreases with increasing polarity of the analyte. Mixtures of water or aqueous buffers and an organic modifier (methanol, acetonitrile, or tetrahydrofuran (THF)) are used as eluent. The percentage and type of organic modifier is the most important parameter in adjusting the retention of nonionic analytes. THF is generally not recommended for LC–MS applications, because of the possible formation of highly-reactive peroxide free radicals in the ion source. Because of the higher solvent strength and the lower viscosity in mixtures with water, acetonitrile is often preferred over methanol.

A buffer is frequently used in reversed-phase LC to reduce the protolysis of ionogenic analytes, which in ionic form show little retention. Phosphate buffers are widely applied for that purpose, since they span a wide pH range and show good buffer capacity. The use of buffers is obligatory in real world applications, *e.g.*, quantitative bioanalysis, where many of the matrix components are ionogenic. LC–MS puts constraints to the type of buffers that can be used in practice. Phosphate buffers must be replaced by volatile alternatives, *e.g.*, ammonium formate, acetate or carbonate.

3.3 Chromatography of ionic compounds

Ionic compounds often show little retention in reversed-phase LC. There are a number of ways to enhance the retention characteristics:
- Ion-suppressed chromatography, which means the analysis of acidic analytes under low pH conditions, thereby reducing the protolysis. The mode can be unfavourable for ESI-MS, which in principle is based on the formation of preformed ions in solution.
- Ion-pair chromatography, where a lipophilic ionic compound is added as a counter-ion. This results in the formation of ion-pairs, that are well retained on

the reversed-phase material. Widely used counter-ions are quaternary ammonium compounds and sulfonic acids with long alkyl chains for the analysis of organic acids and bases, respectively, cannot be used and must be replaced in LC–MS with shorter-chain ammonium salts or perfluoropropionic or -butyric acids. A column once used in ion-pair LC may continue to bleed ion-pairing agents for a very long time.

- Ion-exchange packing materials are chemically modified silica or styrene-divinylbenzene copolymers, modified with ionic functional groups, *e.g.*, n-propylamine, diethylaminopropyl, alkyl-$N^+(CH_3)_3$, carboxylic acid, or benzenesulfonic acid. The retention is primarily influenced by the type of counter-ion, the ionic strength, the pH and modifier content of the mobile phase, and the temperature.

- Ion chromatography is used for the separation of ionic solutes such as inorganic anions and cations, low molecular-mass water-soluble organic acids and bases as well as ionic chelates and organometallic compounds. The separation can be based on ion-exchange, ion-pair and/or ion-exclusion effects. Special detection techniques like ion-suppressed conductivity detection or indirect UV detection have to be used because most analytes are transparent to conventional UV detection.

4. Other modes of liquid chromatography

4.1   Perfusion chromatography

In order to improve the separation efficiency and speed in biopolymer analysis a variety of new packing materials have been developed. These developments aim at reducing the effect of slow diffusion between mobile and stationary phase, which is important in the analysis of macromolecules due to their slow diffusion properties. Perfusion phases [13] are produced from highly cross-linked styrene-divinylbenzene copolymers with two types of pores: through-pores with a diameter of 600–800 nm and diffusion pores of 80–150 nm. Both the internal and the external surface is covered with the chemically bonded stationary phase. The improved efficiency and separation speed result from the fact that the biopolymers do not have to enter the particles by diffusion only, but are transported into the through-pores by mobile-phase flow.

4.2   Immunoaffinity chromatography

Affinity chromatography [14] is a highly-specific separation method based on biochemical interactions such as between antigen and antibody. The specificity of the interaction is due to both spatial and electrostatic effects. One component of the interactive pair, the ligand, is chemically bonded to a solid support, while the other, the analyte, is reversibly adsorbed from the mobile phase. Only components that match the ligand properties are adsorbed. Elution is performed by the use of a mobile phase containing a component with a larger affinity to the ligand than the analyte, or by changes of pH or ionic strength of the mobile phase. Most stationary phases are based on diol- or amine-modified silica to which by means of a 'spacer' the ligand is bound. In this way, free accessibility of the bonding site of the ligand is achieved.

Sample pretreatment methods based on immunoaffinity interactions (IAC) have been developed for LC–MS. An aqueous sample or an extract is applied to a first column, packed with covalently-bound antibody. After loading, the IAC column is washed, and eluted onto a trapping column, which is then eluted in backflush mode onto a conventional analytical column for LC–MS analysis. Sample pretreatment by IAC was reviewed by Hennion and Pichon [15].

4.3   Chiral separation

The separation of enantiomers is especially important in the pharmaceutical field, because drug enantiomers may produce different effects in the body. Enantiomer separations by chromatography require one of the components of the phase system to be chiral. This can be achieved by: (a) the addition of a chiral compound to the mobile phase, which is then used in combination with a nonchiral stationary phase, or (b) the use of a chiral stationary phase in combination with a nonchiral mobile phase. The chiral phase can either be a solid support physically coated with a chiral stationary phase liquid or a chemically bonded chiral phase. For mobile-phase compatibility reasons, a chiral stationary phase is preferred in LC–MS. However, most chiral stationary phases have stringent demands with respect to mobile-phase composition, which in turn may lead to compatibility problems. Three types of phase systems are applied in LC–MS:

- Columns like Chiralpak AD, Chiralpak AS, and Chiralcel OJ-R are used with normal-phase mobile phases of an alkane, *e.g.*, hexane, *iso*-hexane, or heptane, and a small amount of alcohol, *e.g.*, methanol, ethanol, isopropanol. With ESI-MS, post-column addition of an alcohol in water or 5 mmol/l aqueous ammonium acetate must be performed.
- The Chirobiotic series of columns (T based on teicoplanin and V on vancomycin as immobilized chiral selector) can be used with polar-organic mobile phase of over 90% methanol and a small amount of aqueous acid or salt solution.
- Other chiral columns such as Chiral AGP, Chirex 3005, Cyclobond (based on β-cyclodextrin), and Bioptick AV-1 can be used with highly-aqueous mobile phases containing buffer and methanol or isopropanol.

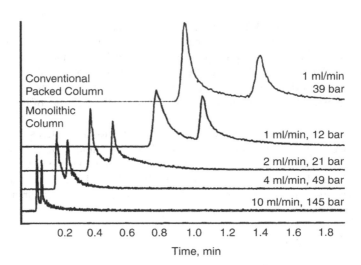

Figure 1.2: LC–MS chromatograms of a drug and its metabolites on conventional packed column and a monolithic silica rod at various flow-rates. Reprinted from [16] with permission, and adapted. ©2002, John Wiley and Sons Ltd.

4.4   Monolithic columns

Monolithic columns were introduced in the mid 1990's. Due to their biporous structure of small mesopores, providing a large surface area for sufficient analyte capacity, and larger through-pores, these columns can be operated at higher flow-rates with reasonable back-pressure. Various types of monoliths are produced: (modified) silica rods, polyacrylamide, polymethacrylate, and polystyrene–divinylbenzene polymers. Monolithic columns show an efficiency equivalent to 3–5-μm-ID silica particles, but with a 30–40% lower pressure drop [16]. Therefore, these columns can be applied in high-throughput analysis for proteomics (Ch. 17.5.2) or in quantitative bioanalysis (Ch. 11.7.2). The separation of a drug and its major metabolite on a conventional column and a monolithic column at various flow-rates is compared in Figure 1.2. The high-flow operation has distinct advantages in the removal of interference materials.

4.5   Hydrophilic interaction chromatography

In hydrophilic interaction chromatography (HILIC), LC is performed on a non-modified silica column, using an aqueous-organic mobile phase. Compared to reversed-phase LC, the retention order is reversed, *i.e.*, highly polar analytes are more strongly retained. For ESI-MS applications, basic compounds can be eluted with an acidic mobile phase and detected in the positive-ion mode, while acidic analytes are eluted at neutral pH and detected in the negative-ion mode. Analytes poorly retained in reversed-phase LC showed good retention in HILIC. Applications of HILIC in quantitative bioanalysis are discussed in Ch. 11.7.3.

4.6   Coupled-column chromatography

In coupled-column chromatography, two analytical columns are applied. A peak of interest is heartcut from the first dimension of LC and transferred to the second dimension of LC, often via a trapping column enabling intermediate concentration and mobile-phase switching. The power of coupled-column LC in significantly enhancing the selectivity of the LC separation and the reduction of interferences was demonstrated by Edlund *et al.* [17] already in 1989 for the analysis of methandrostenolone metabolites. The potential of coupled-column LC in the reduction of matrix effects was demonstrated by Sancho *et al.* [18] in the determination of the organophosphorous pesticide chlorpyrifos and its main metabolite 3,5,6-trichloro-2-pyridinol in human serum, and by Dijkman *et al.* [19] in a comparison of various methods of reducing matrix effect in the direct trace analysis of acidic pesticides in water.

Two-dimensional LC, based on a combination of ion-exchange and reversed-phase LC, is widely applied in the field of proteomics (Ch. 17.5.4 and Ch. 18.3.2).

5.   Sample pretreatment strategies

A wide variety of sample pretreatment methods have been used in combination with LC–MS. Some of the most important ones are briefly discussed here. A more general guide to sample pretreatment for LC and LC–MS can be found in a book by Wells [20]

5.1   Protein precipitation

Protein precipitation as a sample pretreatment method is very popular in quantitative bioanalysis, because it is a very fast and almost generic approach. First, the protein precipitation additive is added. After mixing and centrifugation, the supernatant can be directly injected into the LC–MS system. Typical additives are trichloroacetic acid (TCA), zinc sulfate, acetonitrile, ethanol, or methanol. The use of zinc sulfate in LC–MS requires a divert valve to avoid excessive source contamination. TCA might result in significant ion suppression. In some cases, poor analyte recovery is observed, probably due to inclusion of analytes in the precipitating proteins.

The effectiveness of various protein precipitation additives in terms of protein removal and matric effects was investigated by Polson *et al.* [21]. Acetonitrile, TCA, and zinc sulfate were found most effective in removing proteins (applied in a 2:1 additive-to-plasma ratio). By a post-column infusion setup (see Ch. 11.5.1 and Figure 11.6), these three methods were further evaluated for five different mobile-phase compositions with respect to matrix effect.

Protein precipitation was automated into a 96-well plate format by means of a robotic liquid handler by Watt *et al.* [22]. Plasma samples (50 μl) were transferred from a 96-rack of tubes to a 96-well plate by means of a single-dispense tool. Acetonitrile (200 μl) was added to the wells by means of an 8-channel tool. The plate was removed, heat sealed, vortex-mixed for 20 s, and centrifuged (2000g for 15 min). Using the 8 channel tool, the supernatant was transferred to a clean plate, to which first 50 μl of a 25 mmol/l ammonium formate buffer solution was added. The plate is then removed, heat-sealed, vortex-mixed, and transferred to the autosampler for LC–MS analysis. The procedure takes ~2 hr per plate. A fourfold improvement in sample throughput on the LC–MS instrument was achieved, compared to previous manual protein precipitation procedures.

5.2   Liquid extraction and liquid-liquid extraction

Liquid-liquid extraction (LLE) is a powerful sample pretreatment, based on the selective partitioning of analytes between two immiscible liquid phases. It is simple, fast, and efficient in the removal of nonvolatiles. Analytes are extracted from an aqueous biological fluid by means of an immiscible organic solvent, *e.g.*, dichloromethane, ethyl acetate, methyl *t*-butyl ether, or hexane. It enables analyte enrichment by solvent evaporation. It can be selective by means of a careful selection of extraction solvent and pH of the aqueous phase. Some method development and optimization is needed. Unless performed in an automatic, 96-well plate format, LLE can be time-consuming and labour-intensive. A critical step in the process is the phase separation. LLE may yield a significant amount of chemical waste of organic solvent.

LLE in 96-well plate format has been pioneered by the group of Henion [23]. As an example, the LLE procedure for methotrexate (MTX) and its 7-hydroxy metabolites is described here. In a 1.1-ml deep-well 96-well plate, 200 μl of plasma were pipetted. An aqueous standard solution (20 μl) was added. This resulted in plasma spiked at 0.1–500 ng/ml with MTX and at 0.25–100 ng/ml with the 7-hydroxy metabolite. Next, 500 μl of acetonitrile were added for protein precipitation. The acetonitrile added contained 10 ng/ml $[D_3]$-MTX and 20 ng/ml $[D_3]$-7-hydroxy-MTX as isotopically-labelled internal standard. The plates were sealed and mixed at 40 rpm for 10 min, and then centrifuged for 4 min at 2500 rpm. The supernatant was transferred into a second deep-well plate. Now, 500 μl of chloroform were added, the plate was sealed again, mixed and centrifuged at 2500 rpm. Next, the aqueous layer was transferred into a third 96-well plate. The plate was blown with N_2 to remove residual organic solvent and then sealed and stored at 4°C prior to LC–MS analysis. All liquid handling was performed using a Tomtec Quadra 96 sample preparation robot. With this approach, four 96-well plates could be prepared by one person in 90 min. Subsequently, it took ~11 hr to analyse these four plates with LC–MS, providing an analysis time of 1.2 min per sample. Limit of quantification was 0.5 ng/ml for MTX and 0.75 ng/ml for its metabolite. Intra-day and inter-day precision was better than 8%. LLE in 96-well plate format has become very popular, especially in quantitative bioanalysis (Ch. 11).

In solid-supported LLE (SS-LLE) or liquid-liquid cartridge extraction, the aqueous sample is applied on to a dry bed of inert diatomaceous earth particles in a flow-through tube or in 96-well plate format. After a short equilibration time (3–5 min), organic solvent is added. The organic eluate is collected, evaporated to dryness, and reconstituted in mobile phase. Compared to conventional LLE procedures, SS-LLE avoids the need for vortex-mixing, phase separation by centrifugation, and phase transfer by aqueous layer freezing.

For extraction from solid samples, e.g., biological materials and homogenates (plant, tissue, food), liquid extraction can be applied using for instance acetone, methanol, or acetonitrile. Often, extracts are filtered prior to injection to LC–MS.

Instead of a liquid, a supercritical fluid can be for the extraction of solid samples. Carbon dioxide is an ideal solvent. The solvation strength can be controlled via the pressure and temperature. The high volatility of CO_2 enables concentration of the sample and easy removal of the extraction liquid.

A number of alternatives to Soxhlet extraction have been described. By pressurized liquid or accelerated solvent extraction, the extraction efficiency can be enhanced. Superheated water extraction, taking advantages of the decreased polarity of water at higher temperature and pressure, has been used for liquid extraction of solid samples as well.

5.3   Solid-phase extraction

The general setup of any SPE procedure consists of four steps: (1) condition the SPE material by means of methanol or acetonitrile, followed by water, (2) apply the aqueous biological sample to the SPE material, (3) remove hydrophilic interferences by washing with water or 5% aqueous acetonitrile, and (4) elute the analytes.

SPE can be performed in a number of ways: in single cartridges for off-line use, in 96-well plate format, either using cartridges or extraction disks, and in on-line modes (Ch. 1.5.4), either on top of the analytical column, or preferably on a precolumn in a column-switching system. In addition, related procedures have been described such as solid-phase microextraction (SPME), microextraction in packed syringes, stir-bar sorptive extraction. SPE enables significant analyte enrichment, especially when large sample amounts are available, like in environmental analysis.

A wide variety of materials have been used in SPE. This can be considered both as a strong and as a weak point: appropriate material can be selected to achieve optimum performance, but the selection must be made from a large variety of packings. The most widely applied packings are based on silica or chemically-modified silica, *e.g.*, C_{18}- or C_8-material, but materials based on ethylbenzene–divinylbenzene or styrene–divinylbenzene copolymers, graphitized nonporous carbon ,and graphitized carbon black are available as well.

Procedures of SPE on a cartridge can be automated by means of a Gilson ASPEC or a Zymark RapidTrace robotic liquid handling system. Unfortunately, these ASPEC procedures are rather slow. Therefore, SPE procedures in 96-well plate format were developed [24-25]. Again, both cartridge and disk SPE systems have been used. As an example of a 96-well plate SPE procedure, the procedure for the determination of fentanyl in plasma [26] is briefly described here. Plasma sample vials were vortex-mixed, centrifuged at 2000 rpm for 10 min, and then 250 µl were transferred into 1-ml 96-deep-well plates using a Multiprobe II automated sample handler. The plasma was diluted with 250 µl water, containing the [D_5]-ILIS at 50 ng/ml. The plate is manually transferred to a Tomtec Quadra 96 robot. A 25-mg mixed-mode SPE cartridge plate (see below) was placed on a Tomtec vacuum manifold. The SPE plate was conditioned by 0.5 ml of methanol and 0.5 ml of water. To the sample plate, 250 µl of 5% aqueous acetic acid was added. After mixing by three cycles of sequential aspiration and dispensing, the samples were transferred to the SPE plate and drawn through it by weal vacuum. The SPE plate was washed with 0.5 ml of 5% aqueous acetic acid and 0.5 ml of methanol. After drying by vacuum for 3 min, a clean sample plate was positioned under the SPE plate. The analytes were eluted by two portions of 0.375 ml of 2% ammonium hydroxide in 80% chloroform in isopropanol. The samples were evaporated to dryness and reconstituted in 100 µl of 90% aqueous acetonitrile, containing 0.5% TFA. The plate was sealed and ready for LC–MS analysis. The use of the 96-well plate SPE procedure reduced sample work-up time from ~3.5 hr to ~2 hr. The 96-well plate SPE procedures have become very popular, especially in high-throughput quantitative bioanalysis (Ch. 11).

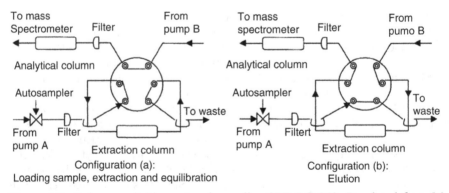

Figure 1.3: Column-switching setup for on-line SPE–LC–MS. Reprinted from M. Jemal, *High-throughput quantitative bioanalysis by LC–MS–MS*, Biomed. Chromatogr., 14 (2000) 422 with permission. ©2000, John Wiley & Sons, Ltd.

Next to SPE on silica-based C_{18}-materials with analyte retention based on hydrophobic interactions, mixed-mode materials like Oasis HLB, which is a divinylbenzene–*n*-vinylpyrrolidone copolymer, become more popular. In mixed-mode materials, the retention is based on combined hydrophobic interaction and ion-exchange interaction.

5.4   On-line SPE–LC

The typical column-switching setup for on-line SPE–LC–MS is shown in Figure 1.3. In a typical application, the sample is loaded by the autosampler onto a precolumn. The sample volume can be larger than the typical injection volume of an analytical column. Analytes are adsorbed onto the chosen stationary phase under weak solvent conditions, while more hydrophilic sample constituents are flushed through. A washing step of the SPE column may be included in the procedure. Next, the valves are switched from the load to the inject position. The SPE column is eluted, in most cases in backflush mode, and the analytes are transferred to the LC column for separation and subsequent LC–MS detection. Examples of on-line SPE–LC–MS are discussed in Ch. 7.3.2 in environmental analysis, in Ch. 11.6.4 for quantitative bioanalysis, and in Ch. 17.5.2 for peptide analysis.

Because often only limited resolution is required for adequate LC–MS determination of target compounds, an alternative approach to on-line SPE–LC–MS was explored: the single-short-column. A single-short-column is a short (10–20 mm) column, similar to the cartridge columns applied in on-line SPE, but high-pressure packed with 3–5-μm-ID particles instead of manually-packed with 20–60-μm-ID particles. The same column is used for both trace enrichment and separation. The approach was successfully applied in target-compound analysis for environmental analysis in combination with MS and MS–MS, both on quadrupole and ion-trap instruments [27] (see Ch. 7.3.2).

5.5   Turbulent-flow chromatography

In turbulent-flow chromatography (TFC), SPE is performed at very high flow-rates on either columns packed with 50 μm Cyclone HTLC particles or monolithic columns (Ch. 1.4.4). The high linear flow through the column results in a flat turbulent flow profile rather than the more common laminar flow profile. This results in a more efficient mass transfer between mobile and stationary phase, leading to a similar chromatographic efficiency in much shorter analysis time. There is no need for protein precipitation prior to the analysis: plasma samples are just centrifuged and then injected. The combination of the high linear speed of the aqueous mobile phase and the large particle size resulted in the rapid passage of the proteins and other large biomolecules through the column. TFC is performed in a one-column setup for TFC–MS or in a two-column setup for TFC–LC–MS. TFC is introduced as a tool for high-throughput quantitative bioanalysis by Cohesive

Technologies Inc. However, the approach of high flow-rate sample pretreatment is frequently applied with other instrumentation as well.

A typical two-column setup featuring two six-port switching valves was described by Herman [28] (Figure 1.4). The procedure consists of four steps: (1) the eluent loop is filled with 40% acetonitrile in 0.05% aqueous formic acid, (2) the sample is loaded onto the 50 × 1 mm ID Cyclone HTLC column (50 µm) at a flow-rate of 4 ml/min during 30 s, (3) the eluent loop is discharged at 0.3 ml/min for 90 s to transfer the analytes from the TFC column onto the 14 × 4.6 mm ID Eclipse C_{18} column (3 µm) and 0.05% aqueous formic acid at 1.2 ml/min in added post-column, and (4) LC–MS is performed using a ballistic gradient at 1 ml/min (5–95% acetonitrile in 0.1% aqueous formic acid in 2 min). Sample throughput can be further increased by applying two- or four-channel staggered parallel TFC.

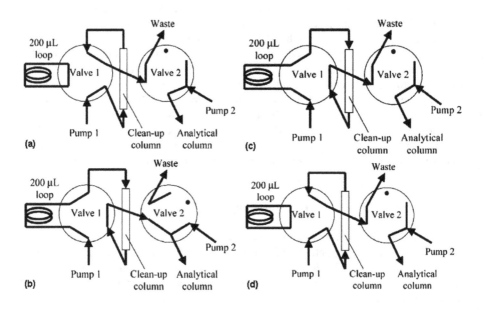

Figure 1.4: Valve-switching setup for two-column TFC–LC–MS. (a) Sample loading and clean-up, (b) sample transfer, (c) sample elution and loop fill, and (d) column equilibration. Reprinted from [28] with permission. ©2002, John Wiley and Sons Ltd.

5.6 Restricted-access stationary phases

Restricted-access material (RAM) columns combine the size-exclusion of proteins by the hydrophilic outer surface of the packing and the simultaneous enrichment by SPE of analytes that interact with hydrophobic groups at the inner surface of the packing. These columns allow the direct injection of plasma samples without protein precipitation. Often, on-line RAM–LC–MS is described, following a procedure identical to on-line SPE–LC–MS (Ch. 1.5.4). The use of RAM columns has been reviewed by Souverain *et al.* [29].

6. References

1. J.C. Giddings, *Unified Separation Science*, 1991, Wiley&Sons Ltd, New York, NY.
2. C.F. Poole, K. Poole, *Chromatography Today*, 1991, Elsevier, Amsterdam, The Netherlands.
3. V.R. Meyer, *Practical High-Performance Liquid Chromatography*, 2nd Ed., 1994, Wiley & Sons, New York, NY.
4. J.W. Dolan, L.R. Snyder, *Troubleshooting LC Systems*, 1989, Humana Press, Clifton, NJ.
5. R.F. Venn (Ed.), *Principles and practice of bioanalysis*, 2000, Taylor & Francis, London, UK.
6. J.C. Sternberg, in: J.C. Giddings, R.A. Keller (Ed.), *Advances in Chromatography*, Vol. 2, 1966, Marcel Dekker Inc., New York, NY, p. 205.
7. B.S. Middleditch, A. Zlatkis, *Artifacts in chromatography: an overview*, J. Chromatogr. Sci., 25 (1987) 547.
8. R.P.W. Scott, *Liquid Chromatography Detectors*, 1987, Elsevier, Amsterdam.
9. H. Lingeman, W.J.M. Underberg (Ed.), *Detection-Oriented Derivatization Techniques in Liquid Chromatography*, 1990, Marcel Dekker Inc., New York, NY.
10. S. Cardenas, M. Valcarcel, *ELSD: a new tool for screening purposes*, Anal. Chim. Acta, 402 (1999) 1.
11. K. Albert, *LC–NMR spectroscopy*, J. Chromatogr. A, 856 (1999) 199.
12. M. Montes-Bayón, K. DeNicola, J.A. Caruso, *LC–ICP-MS*, J. Chromatogr. A, 1000 (2003) 457.
13. N.B. Afeyan, S.P. Fulton, F.E. Regnier, *Perfusion chromatography packing materials for proteins and peptides*, J. Chromatogr.A, 544 (1991) 267.
14. M.M. Rhemrev-Boom, M. Yates, M. Rudolph, M. Raedts, *(I)AC: a versatile tool for fast and selective purification, concentration, isolation and analysis*, J. Pharm. Biomed. Anal., 24 (2001) 825.
15. M.-C. Hennion, V. Pichon, *Immuno-based sample preparation for trace analysis*, J. Chromatogr. A, 1000 (2003) 29.
16. Y. Hsieh, G. Wang, Y. Wang, S. Chackalamannil, J.-M. Brisson, K. Ng, W.A. Korfmacher, *Simultaneous determination of a drug candidate and its metabolite in rat plasma samples using ultrafast monolithic column LC–MS–MS*, Rapid Commun. Mass Spectrom., 16 (2002) 944.
17. P.O. Edlund, L. Bowers, J.D. Henion, *Determination of methandrostenolone and its*

metabolites in equine plasma and urine by coupled-column LC with UV detection and confirmation by MS–MS, J. Chromatogr., 487 (1989) 341.

18. J.V. Sancho, O.J. Pozo, F. Hernández, *Direct determination of chlorpyrifos and its main metabolite 3,5,6-trichloro-2-pyridinol in human serum and urine by coupled-column LC–ESI-MS–MS*, Rapid Commun. Mass Spectrom., 14 (2000) 1485.

19. E. Dijkman, D. Mooibroek, R. Hoogerbrugge, E. Hogendoorn, J.-V. Sancho, O. Pozo, F. Hernández, *Study of matrix effects on the direct trace analysis of acidic pesticides in water using various LC modes coupled to MS–MS detection*, J. Chromatogr. A, 926 (2001) 113.

20. D..A. Wells, *High throughput bioanalytical sample preparation. Methods and automation strategies*, 2003, Elsevier Science, Amsterdam, the Netherlands.

21. C. Polson, P. Sarkar, B. Incledon, V. Raguvaran, R. Grant, *Optimization of protein precipitation based upon effectiveness of protein removal and ionization effect in LC–MS–MS*, J. Chromatogr. B, 785 (2003) 263.

22. A.P. Watt, D.Morrison, K.L. Locker, D.C. Evans, *Higher throughput bioanalysis by automation of a protein precipitation assay using a 96-well format with detection by LC–MS–MS*, Anal. Chem., 72 (2000) 979.

23. S. Steinborner, J. Henion, *LLE in the 96-well plate format with SRM LC–MS quantitative determination of methotrexate and its major metabolite in human plasma*, Anal. Chem., 71 (1999) 2340.

24. B. Kaye, W.J. Heron, P.V. Mcrae, S. Robinson, D.A. Stopher, R.F. Venn, W. Wild, *Rapid SPE technique for the high-throughput assay of darifenacin in human plasma*, Anal. Chem., 68 (1996) 1658.

25. J.P. Allanson, R.A. Biddlecombe, A.E. Jones, S. Pleasance, *The use of automated SPE in the '96 well' format for high throughput bioanalysis using LC coupled to MS–MS*, Rapid Commun. Mass Spectrom., 10 (1996) 811.

26. W.Z. Shou, X. Jiang, B.D. Beato, W. Naidong, *A highly automated 96-well SPE and LC–MS–MS method for the determination of fentanyl in human plasma*, Rapid Commun. Mass Spectrom., 15 (2001) 466.

27. A.C. Hogenboom, P. Speksnijder, R.J. Vreeken, W.M.A. Niessen, U.A.Th. Brinkman, *Rapid target analysis of microcontaminants in water by on-line single-short-column LC combined with atmospheric pressure chemical ionization MS–MS*, J. Chromatogr. A, 777 (1997) 81.

28. J. L. Herman, *Generic method for on-line extraction of drug substances in the presence of biological matrices using TFC*, Rapid Commun. Mass Spectrom., 16 (2002) 421.

29. S. Souverain, S. Rudaz, J.-L. Veuthey, *RAM and large particle supports for on-line sample preparation: an attractive approach for biological fluids analysis*, J. Chromatogr. B, 801 (2004) 141.

2

MASS SPECTROMETRY

1. Introduction

Mass spectrometry (MS) is based on the production of ions, that are subsequently separated or filtered according to their mass-to-charge (m/z) ratio and detected. The resulting mass spectrum is a plot of the (relative) abundance of the generated ions as a function of the m/z. Excellent selectivity can be obtained, which is of utmost importance in quantitative trace analysis. This chapter is not a brief introduction in MS, but rather highlights important aspects for the discussions on liquid chromatography–mass spectrometry (LC–MS) to come. General discussion and tutorials in MS can be found elsewhere [1-2].

The mass spectrometer is a highly sophisticated and computerized instrument, which basically consists of five parts: sample introduction, ionization, mass analysis, ion detection, and data handling. In principle, liquid chromatography is just one of the possible analyte techniques, or the mass spectrometer just another detector for LC. However, on-line chromatography–MS systems offer additional value, especially in terms of selectivity.

2. Analyte ionization

A wide variety of ionization techniques is available for organic mass spectrometry [1-2]. Analyte-ionization techniques can be classified as 'hard' or 'soft', depending on the extent of fragmentation occurring during the ionization process. Electron ionization (EI) is a typical example of a hard ionization method, while the currently extensively applied electrospray ionization and matrix-assisted laser desorption ionization (MALDI) are soft ionization techniques.

2.1   Electron ionization

In EI, the analyte vapour is subjected to a bombardment by energetic electrons (typically 70 eV). Most electrons are elastically scattered, others cause electron excitation of the analyte molecules upon interaction, while a few excitations cause the complete removal of an electron from the molecule. The latter type of interactions generates a radical cation, generally denoted as $M^{+\bullet}$, and two electrons:

$$M + e^- \rightarrow M^{+\bullet} + 2\,e^-$$

The $M^{+\bullet}$ ion is called the molecular ion. It is an odd-electron ion ($OE^{+\bullet}$). Its m/z ratio corresponds to the molecular mass M of the analyte. The ions generated in EI are characterized by a distribution of internal energies, generally centred around 2–6 eV. The excess internal energy of the molecular ions can for different structures give rise to unique unimolecular dissociation reactions resulting in fragment ions, $i.e.$, the formation of an ionized fragment accompanied by the loss of either a radical $R^\bullet$ or a neutral N.

EI is performed in a high-vacuum ion source (typically $\leq 10^{-3}$ Pa); intermolecular collisions are avoided in this way. As a result, EI mass spectra are highly reproducible. Extensive collections of standardized EI mass spectra are available [3-4], also for on-line computer evaluation. An important limitation of EI is the necessity to present the analyte as a vapour, which excludes the use of EI in the study of nonvolatile and thermally labile compounds. EI is widely applied in GC–MS [5]. In LC–MS, its applicability is limited to the particle-beam interface and the moving-belt interface.

2.2   Chemical ionization

Chemical ionization (CI) is one of the most versatile ionization techniques as it relies on chemical reactions in the gas phase [6]. CI is an important ionization technique in LC–MS. It is based on ion-molecule reactions between reagent-gas ions and the analyte molecules. Gas-phase ion-molecule reactions comprise proton transfer, charge exchange, electrophilic addition, and anion abstraction in positive-ion CI and proton transfer (abstraction) in negative-ion CI. CI can be performed under various pressure conditions:
- Low-pressure CI (<0.1 Pa) can only be performed in systems that allow for an elongated sample residence time in the ion source, $e.g.$, in ion-cyclotron resonance and ion-trap cells. Low-pressure CI is hardly used in LC–MS.
- Medium-pressure CI at ion-source pressures between 1 and 2000 Pa is widely used. In LC–MS, it is important in particle-beam and thermospray interfacing. Either an externally-added reagent gas like methane, isobutane, or ammonia is

used (conventional CI) or solvent molecules from the LC mobile phase (solvent-mediated CI).

- Atmospheric-pressure CI can be performed in an atmospheric-pressure ion source, *i.e.*, atmospheric-pressure chemical ionization (APCI) (Ch. 6.4).

2.3   Electron-capture negative ionization

An alternative procedure for the production of negative ions is electron-capture negative ionization (ECNI). It is as a highly selective ionization method, as only a limited number of analytes are prone to efficient electron capture, *e.g.*, fluorinated compounds or derivatives. It takes place by capture by the analyte molecules of 'thermal' electrons, and results in the generation of radical anions. The process must be performed in a medium-pressure ion source in order to slow down the electrons and to remove excess energy from the radical anion formed upon electron attachment. The formation of negative ions by electron capture can occur by two mechanisms:

- Associative electron capture, where an intact molecular anion $M^{-\bullet}$ is generated.
- Dissociative electron capture, where the molecular anion generated immediately fragments into a fragment anion and a radical fragment

2.4   Energy-sudden or desorption ionization

A wide variety of desorption ionization methods is available [7]: desorption chemical ionization (DCI), secondary-ion mass spectrometry (SIMS), fast-atom bombardment (FAB), liquid-SIMS, plasma desorption (PD), matrix-assisted laser desorption ionization (MALDI), and field desorption (FD). Two processes are important in the ionization mechanism, *i.e.*, the formation of ions in the sample matrix prior to desorption, and rapid evaporation prior to ionization, which can be affected by very rapid heating or by sputtering by high-energy photons or particles. In addition, it is assumed that the energy deposited on the sample surface can cause (gas-phase) ionization reactions to occur near the interface of the solid or liquid and the vacuum (the so-called selvedge) or provide preformed ions in the condensed phase with sufficient kinetic energy to leave their environment.

Most desorption techniques have been applied in on-line coupling of LC and MS:

- A fractional sampling interface for LC–MS via ^{252}CF PD (Ch. 3.2.2, [8]).
- Ionization via SIMS and FAB from a moving-belt interface (Ch. 4.4.3, [9]).
- Continuous-flow FAB (Ch. 4.6, [10]). This was the most successful approach to an on-line combination of LC–MS via a desorption ionization technique.
- Continuous-flow MALDI and aerosol MALDI with liquid introduction (Ch. 5.9) [11].

Next to electrospray ionization (Ch. 6.3), MALDI is the most important ionization technique in the MS analysis of biomacromolecules. MALDI was introduced by Tanaka *et al.* [12] and Karas and Hillenkamp in 1988 [13]. Today, MALDI plays an important role in the characterization of proteins, oligonucleotides, sugars, and synthetic polymers. MALDI–time-of-flight MS is an essential tool in the current proteomics research (Ch. 18.2).

In MALDI, the sample is deposited on a target and co-crystallized with a solid matrix [14-15]. The target is transferred to vacuum and bombarded by photon pulses from a laser, in most cases a nitrogen laser (337 nm) nowadays. The ionization results from efficient electronic excitation of the matrix and subsequent transfer of the energy to the dissolved analyte molecules, which are desorbed and analysed as protonated or cationized molecules [7]. The ionization process is not fully understood. Extremely high molecular-mass compounds, *e.g.*, in excess of 200 kDa, can be analysed using the MALDI, if performed on a time-of-flight mass spectrometer (Ch. 2.4.3).

2.5 Nebulization ionization

Nebulization ionization is the process involved in the analyte ionization in thermospray [16] and electrospray [17] interfacing. No primary ionization, *i.e.*, a filament or a discharge electrode, is applied. The ionization mechanism is not fully understood (Ch. 6.3). The general understanding can be summarized as follows: Upon nebulization, charged droplets of a few μm ID are generated. The fate of these droplets is determined by a number of competing processes, the relative importance of which may dependent on the nature of the analyte:

- Charge-preserving solvent evaporation from the droplets, resulting in smaller droplets with a higher number of charges.
- When the repulsive forces due to the surface charges exceed the forces due to surface tension, the droplets explode as a result of field-induced Rayleigh or electrohydrodynamic instabilities. Multiple smaller droplets are generated.
- Due to repetitive electrohydrodynamic instabilities, microdroplets containing only one charged molecule may be generated. By soft desolvation of this droplet, the ions will be free in the gas phase and amenable to mass analysis.
- At a certain droplet-size/charge ratio, field-induced ion evaporation [18] of preformed ions in solution may take place. The resulting free ions in the gas phase are amenable to mass analysis.
- Gas-phase electrolyte ions, *e.g.*, NH_4^+, may be formed during these processes as well. These ions may act as reagent gas ions in ion-molecule reactions taking place in the gas-phase, *i.e.*, chemical ionization processes, or at the liquid-gas interface of the microdroplets.
- Furthermore, the ions generated may be collisionally activated due to the many ion-molecule collisions in the high-pressure ion source. This may result in the formation of fragment ions.

3. Information from mass spectrometry

All mass analysers perform a separation of ions according to their m/z. A singly-protonated molecule of a compound with molecular mass of 400 is observed at m/z 401, while a compound with a molecular mass of 16,000 carrying 40 charges due to protonation is also observed at m/z 401. Multiple charging enables the use of mass analysers with limited mass range in the analysis of high molecular-mass compounds.

A number of concepts important in mass spectrometry is summarized and defined in Table 2.1.

The term '*molecular ion*' by definition refers to a radical cation or anion of an intact molecule. Molecular ions are odd-electron ions, which may thus be generated by EI. Unfortunately, the term 'molecular ion' is also frequently used to indicate the even-electron ionic species produced by electrospray and APCI. This obviously is not correct. In the soft ionization techniques, predominantly even-electron *protonated molecules* are generated in positive-ion mode, and *deprotonated molecules* in negative ions. In addition, various adduct ions may be generated (Table 2.2). These all are even-electron ions, and should therefore not be referred to as molecular ions. Alternatively, the term 'protonated molecular ions' is used, which again is incorrect: one cannot protonate a radical cation!

In addition, the terms 'pseudomolecular ion' and 'quasimolecular ion' are frequently applied to indicated cationized or anionized molecules. Although the second one is actually recommended by the IUPAC [19], it will not be used in this text. Further information on nomenclature issues can be found elsewhere [19-21].

The molecular mass of the analyte can be determined from the m/z of the molecular ion in EI, if it is observed in the mass spectrum. With a soft-ionization method, next to, or even instead of, the protonated molecule in the positive-ion mode or the deprotonated molecule in the negative-ion mode, a variety of adducts ions may occur in the mass spectrum (Table 2.2). The presence of adduct peaks can be often helpful in assigning the correct molecular mass. The use of both the positive-ion mode, resulting in m/z of the $[M+H]^+$ ion, and the negative-ion mode, resulting in m/z of the $[M-H]^-$ ion, if applicable, also leads to an unambiguously molecular-mass determination for an unknown compound.

Because most elements consist of a mixture of stable isotopes (Table 2.3), isotope peaks are observed in mass spectra. For a given elemental composition, the isotope pattern can be predicted using a computer program. The equidistant peaks in an isotope pattern, *i.e.*, at one m/z unit for a single-charge ion, represent a series of ions with relative abundances, that should closely agree with the theoretically predicted values.

Table 2.1: Important concepts in mass spectrometry	
Average molecular mass (molecular weight)	Calculated from the elemental composition using average atomic weights (Table 2.2). It is the chemical mass. In MS, it can be important in the analysis of large molecules.
Monoisotopic molecular mass	Calculated from the elemental composition using the monoisotopic atomic masses of the most abundant isotopes in nature (Table 2.2). Depending on the mass resolution, it is expressed as a nominal mass, or as an exact mass, based on exact monoisotopic atomic masses.
Isomers	Isomers have an identical elemental composition, but a different structure. They have identical nominal and exact mass.
Isobars	Isobars have different elemental compositions. Isobars have identical nominal mass, but (slightly) different exact mass.
Mass defect	Difference between the monoisotopic exact mass and the monoisotopic nominal mass. Most compounds of biological origin show a positive mass defect.
Mass resolution	Indicates the ability to discriminate two ions with an m/z difference of Δm. Resolution is often expressed as the full width at half of the maximum height (FWHM) of the peak in the profile mass spectrum. For a single-charge ion, the resolution can be calculated from: $$R = (m/z) \, / \, FWHM$$ A quadrupole mass analyser is operated at unit-mass resolution with a FWHM of *ca.* 0.6 u for a single-charge ion.
Mass accuracy	Difference between the measured m/z and the calculated m/z. The mass accuracy can be expressed as an absolute error (in mDa), or as a relative error (given in ppm): $$\textit{Relative Error} = \frac{\textit{Measured} - \textit{Calculated}}{\textit{Calculated}} \times 10^6$$

Table 2.2: Common adducts ion observed under electrospray and APCI conditions			
Positive ions		**Negative ions**	
protonated *molecule*		*deprotonated* *molecule*	
$[M+H]^+$	M + 1	$[M–H]^-$	M – 1
$[M+nH]^{n+}$	(M + n) / n	$[M–nH]^{n-}$	(M – n) / n
cationized */adduct*		*anionized* */adduct*	
$[M+NH_4]^+$	M + 18	$[M+HCOO]^-$	M + 45
$[M+Na]^+$	M + 23	$[M+OAc]^-$	M + 59
$[M+CH_3OH+H]^+$	M + 33	$[M+TFA]^-$	M + 113
$[M+K]^+$	M + 39	$[M+Cl]^-$	M + 35 [iso 37]
$[M+CH_3CN+H]^+$	M + 42		
preformed ion		*preformed ion*	
M^+; $[M+H]^+$	M; M + 1	M^-; $[M–H]^-$	M; M – 1

In electrospray ionization, compounds with a mass in excess of 500 Da may be prone to the formation of multiple-charge ions, either in positive-ion or in negative-ion mode [22-23]. The averaging algorithm [23-24] can be used for molecular-weight determination from an ion envelope of multiple-charge ions (Ch. 16.2.2). Various automated computer-based procedures for the deconvolution or transformation of the electrospray mass spectra of proteins have been introduced.

While for a single-charge ion, the isotope peaks are one *m/z* unit apart, in a multiple-charge ion with *n* charges, the peaks in the isotope pattern are 1/*n* *m/z* units apart. Therefore, an instrument with unit-mass resolution cannot distinguish the individual isotope peaks in the pattern of the multiple-charge ion. As a result, the average molecular weight of the protein is determined rather than the monoisotopic molecular mass.

3.1 *m/z* Axis calibration

The mass spectrometer only provides accurate mass measurement, if the *m/z* axis is properly calibrated. Calibration can be performed using automated procedures available in the software of the instrument. During the procedure, a (mixture of) reference compound(s) is introduced and ionized. This provides ions of known *m/z* within the mass range of the measurement.

The two most widely used reference compounds for EI are heptacosafluoro-tributylamine (PFTBA, $(C_4F_9)_3N$) and perfluorokerosene (PFK, $CF_3–(CF_2)_n–CF_3$).

Table 2.3: Relative abundance, nominal mass, exact mass, and average mass of some common elements				
Element	Nominal Mass	Rel. Abund. (%)	Exact Mass	Average Mass
H	1	100	1.0078	1.008
	2	0.016	2.0141	
C	12	100	12.0000	12.011
	13	1.08	13.0034	
N	14	100	14.0031	14.007
	15	0.38	15.0001	
O	16	100	15.9949	15.999
	17	0.04	16.9991	
	18	0.2	17.9992	
F	19	100	18.9984	18.998
Na	23	100	22.9898	22.990
Si	28	100	27.9769	28.086
	29	5.1	28.9765	
	30	3.35	29.9738	
P	31	100	30.9738	30.974
S	32	100	31.9721	32.060
	33	0.78	32.9715	
	34	4.4	33.9679	
Cl	35	100	34.9689	35.453
	37	32.5	36.9659	
Br	79	100	78.9183	79.904
	81	98	80.9163	
I	127	100	126.9045	126.905

For soft ionization methods, the calibration method is less standardized. A wide variety of reference compounds have been proposed for use in electrospray LC–MS. Proper calibration of the m/z axis is especially important in the analysis of proteins, because the error in the mass measurement of a multiple-charge unknown is magnified by the number of charges at the unknown ion. Frequently used reference compounds for m/z calibration in electrospray MS are:
• Cesium iodide or cesium carbonate cluster ions were proposed [25–26]. A mixture of 2.5% cesium iodide in sodium iodide is recommended by Waters. The use of cluster ions of a variety of cesium salts for calibration up to m/z 10,000

was proposed [27]. Cesium salts of monobutyl phthalic acid, heptafluorobutyric acid, tridecafluoroheptanoic acid, and perfluorosebasic acid were used.

- Poly(ethylene glycol) (PEG) and poly(propylene glycol) (PPG) are frequently applied as reference compounds. PPG is recommended by Applied Biosystems MDS Sciex. PEG and PPG are only applicable in a limited mass range (up to *ca.* m/z 4000), because larger PEG or PPG molecules are prone to the formation of multiple-charge ions.

- Readily available proteins have been suggested as reference compounds as well. A mixture of the peptide MRFA and myoglobin has been used for calibration by Thermo Finnigan. The use of a protein solution rather than CsI/NaI or PPG as a reference compound is sometimes recommended for protein analysis, because the multiple-charge ions of proteins may behave differently from single-charge cluster ions [28].

  In this respect, the paper of Zaia *et al.* [29], entitled *"The correct molecular weight of myoglobin, a common calibrant for mass spectrometry"*, is also of interest. It shows that the correct average molecular mass of horse heart myoglobin is 16,951.49 Da rather than frequently given 16,950.5 Da.

- Ultramark 1621, a mixture of fluorinated phosphazenes, was proposed as a reference compound for calibration in the range of m/z 800 to 2000 [28]. It was also recommended for ion-trap instruments from Thermo Finnigan. Ultramark 1621 strongly adsorbs to surfaces and is difficult to remove due to its poor solubility in common solvents [30]. A proprietary mixture of fluorinated phosphazenes is recommended as reference compound by Agilent Technologies.

- Water cluster ions have been proposed for calibration up to m/z 3000 [31].

- The use of sodium trifluoroacetate cluster ions as an alternative to Ultramark 1621 were proposed by Moini *et al.* [30]. This can be applied up to m/z 4000 Da.

3.2 Full-spectrum analysis and selected-ion monitoring

Mass spectrometry can be performed in two general data-acquisition modes:

- Full-spectrum analysis, where a series of mass spectra is acquired. This mode is typically applied in qualitative analysis.

- Selected-ion monitoring (SIM), where the ion abundances of preselected ions are acquired. Selected-ion monitoring is applied in routine quantitative analysis. In quadrupole and magnetic sector instruments, acquisition in SIM mode provides a substantial gain in signal-to-noise ratio (S/N).

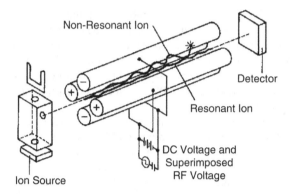

Figure 2.1: Linear quadrupole mass spectrometer.

4. Mass analysis

Mass analysis, *i.e.*, the separation of ions according to their *m/z* in either time or space, can be achieved in a number of ways. This section does not discuss the basic principles of the most important types of mass analysers [1-2], but focusses on aspects and developments important in relation to LC–MS. An excellent discussion on the various mass analysers and their advantages and limitations is given by Brunnee [32].

4.1   Linear quadrupole MS and MS–MS instruments

The linear quadrupole mass analyser (Figure 2.1) consists of four hyperbolic or cylindrical rods, that are placed parallel in a radial array. Opposite rods are charged by a positive or negative direct-current (DC) potential at which an oscillating radiofrequency alternating-current (RF) voltage is superimposed. The latter successively reinforces and overwhelms the DC field. Ions are introduced into the quadrupole field by means of a low accelerating potential, only a few Volts. The ions start to oscillate in a plane perpendicular to the rod length as they traverse through the quadrupole filter. At a given combination of DC and RF applied to the rods, the trajectories of the ions of one particular *m/z* are stable. These ions are transmitted towards the detector. Ions with other *m/z* have unstable trajectories and do not pass the mass filter, because the amplitude of their oscillations becomes infinite. They are discharged on the rods and/or lost in the vacuum system. Ions of increasing *m/z* can consecutively be transmitted towards the detector by sweeping the DC and AC potentials at a constant ratio. The resolution depends on the ratio of DC and RF potentials. Generally, the quadrupole is operated at unit-mass resolution.

The resolution of a quadrupole is not only determined by the ratio of DC and RF, but also by the quality and the alignment of the rods. The potential of enhanced resolution in a quadrupole mass filter was investigated by Tyler *et al.* [33]. Enhanced resolution leads to a significant loss in response. Recently, Thermo Finnigan introduced a triple-quadrupole instrument with improved quadrupole design, RF power supply stability, and temperature control (TSQ Quantum), enabling 0.1-unit-mass resolution without dramatic losses in response [34].

Another important application of a linear quadrupole is operation in the RF-only mode, *i.e.*, where the DC voltage is zero. In this mode, all ions with m/z between a low-mass and high-mass cut-off are transmitted. RF-only quadrupoles (or hexapole or octapole) are used as an ion guide between API sources and mass analysers (Ch. 5.4.5) and as collision cells in various MS-MS instruments.

The quadrupole is easy in use. The electric voltages can be easily and rapidly varied under computer control. The quadrupole mass filter is the most widely applied mass analyser in LC–MS.

Tandem mass spectrometry

In structure-elucidation problems, one often wants to obtain more information on the ions generated in the soft-ionization process. Increasing the internal energy of the even-electron ion is a way to induce fragmentation. This can be achieved by collisional activation via collisions of (selected) ions with neutral gas molecules (collision-induced dissociation, CID). CID is a two-step process: ion translational energy is converted into ion internal energy in the collision event, and subsequently unimolecular decomposition of the excited ion may yield various product ions. Energy redistribution within the ion may take place between the two steps. With most instruments, low-energy CID is performed, *i.e.*, the collision energy does not exceed 100 eV. In sector instruments, high-energy CID, *i.e.*, with collision energies in the low keV range, are also possible. This opens a variety of other fragmentation pathways that require higher ion internal energies.

In an API instrument, CID can be achieved in a number of ways:
- Fragmentation may be induced by increasing the potential difference between the ion-sampling orifice and the skimmer at the front-end of the instrument (Ch. 5.4.4). In this case of in-source CID, all ions entering this region through the ion-sampling orifice might be fragmented. No precursor selection is performed.
- A particular ion of interest may be selected by means of a mass analyser and be subjected to CID. In this case, the fragmentation of the accelerated ions will actually be induced in a argon-filled collision cell at a somewhat higher pressure. Two stages of mass analysis are required: one to select the precursor ion from other ions generated in the ion source, and one to analyse the product ions after the collisions. Therefore, this approach is called tandem mass spectrometry (MS–MS) [35-36]. Various instrumental setups are available for these experiments, as discussed below.

Table 2.4: Analysis modes in tandem mass spectrometry			
Mode	**MS1**	**MS2**	**Application**
product-ion (daughter ion)	selecting	scanning	to obtain structural information from ions produced in the ion source
precursor-ion (parent ion)	scanning	selecting	to monitor compounds which in CID give an identical fragment
neutral-loss	scanning	scanning	MS1 and MS2 are scanning at a fixed *m/z* difference: to monitor compounds that lose a common neutral species
selected reaction monitoring (SRM)	selecting	selecting	to monitor a specific CID reaction

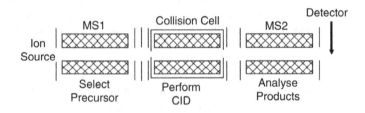

Figure 2.2: A triple-quadrupole instrument for MS–MS.

Triple-quadrupole instrument for MS–MS

The most versatile and most widely used MS–MS configuration is the triple quadrupole instrument, where mass analysis is performed in the first and third quadrupoles, while the second quadrupole is used as a collision cell in the RF-only mode, *i.e.*, in a Q–q$_{coll}$–Q configuration. Alternative collision cells have been implemented by various instrument manufacturers, *e.g.*, RF-only hexapoles, octapoles, or ion tunnels. In a linear-acceleration high-pressure collision cell (LINAC), as applied in recent instruments available from Applied Biosystems MDS Sciex, the rod distance at the ion-entrance side is larger than that at the ion-exit side [37]. By applying an axial DC voltage over the rods, the ions are accelerated through the LINAC. This significantly reduces cross-talk between components, for which the same product ion was selected, in ultra-fast multi-component quantitative bioanalysis.

Analysis modes in MS–MS

MS–MS was originally introduced as a way to achieve fragmentation of ions generated in the ion source, *e.g.*, by soft ionization methods. In such an experiment, the first MS selects a particular precursor ion, which is dissociated in the collision cell, and the product ions are analysed with the second MS. This is the product-ion analysis mode, which is widely used in structure elucidation. However, other scan modes are possible as well (Table 2.4). The monitoring of a selected collision-induced fragmentation reaction by selected reaction monitoring (SRM) significantly improves the selectivity, resulting in greatly improved S/N. While SRM is the method of choice in quantitative bioanalysis, the precursor-ion and neutral-loss analysis modes are particularly useful in qualitative analysis and screening of food or environmental samples for contaminants.

4.2   Quadrupole ion trap MS and MS–MS^n instruments

An important development in quadrupole technology is the three-dimensional ion trap [38-39]. A quadrupole ion trap consists of a cylindrical ring electrode to which the quadrupole field is applied, and two end-cap electrodes (Figure 2.3). One end-cap contains holes for the introduction of electrons or ions into the trap, while the other has holes for ions ejected out of the trap towards the electron multiplier. In LC–MS systems, ions are generated in an external ion source. The ions are introduced to the trap in a pulsed mode and stored there. A helium bath gas (0.1 Pa) is present in the trap to stabilize the ion trajectories.

The acquisition of a mass spectrum by means of an ion trap with a external ion source requires a number of consecutive steps:

- The number of ions that can be stored in the ion trap without affecting the mass resolution and accuracy is limited due to space-charge effects (typically to 10^4 ions). In order to avoid problems, ion-current dependent ion injection times are applied via automatic-gain control (AGC, Thermo Finnigan) or ion-charge control (ICC, Bruker and Agilent Technology) procedures.
- An appropriate RF storage voltage is applied to the ring electrode to trap and accumulate ions with an *m/z* above a low-mass cut-off value. The ion-injection and trapping process may be influenced, *e.g.*, by filtering low-mass interferences in the RF-only ion guide prior to injection into the trap [40], or by the application of auxiliary RF waveforms to the end caps [41].
- At this stage, different MS experiments can be performed. In the full-scan mode, ions of different *m/z* are consecutively ejected from the trap towards the external detector by ramping the RF voltage at the ring electrode. Resonant ion ejection may be supported by additional waveforms applied to the end-cap electrodes. Ion ejection can be achieved with unit-mass resolution, or at enhanced resolution by slowing down the scan rate [42].

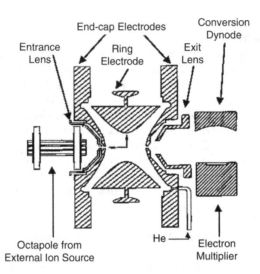

Figure 2.3: Ion-trap mass spectrometer with external ionization source.

- Alternatively, all ions but one selected m/z can be removed from the trap to perform SIM. The RF storage voltage at the ring electrode is adapted to store the ions of the selected m/z before these are ejected to the detector. In an ion trap, the use of SIM does not lead to a significant gain in S/N, because no gain in measurement of the selected ion is achieved.
- In both cases, the procedure ends by a RF pulse to remove all ions from the trap and to prepare the trap for the next sequence of events.

An important point of concern is the duty cycle, which must be as high as possible. Ions are injected into the ion trap during a limited period of time. During the time it takes to acquire the MS spectrum and to reset to initial conditions, the ions produced in the ion source are lost. For an API–ion trap MS system, the duty cycle can be defined from a practical point-of-view as the ratio of the number of ions injected into the ion trap and the total number of ions entering the RF-only multipole.

MS–MS and MS–MS^n in an ion trap

The m/z-selective instability mode for mass analysis in the ion trap allows additional experiments to be performed in the ion trap. When it is possible to selectively store ions of a particular m/z and eject other ions, the selection of a precursor ion for product-ion MS–MS is possible as well. The precursor ion is selected by applying two consecutive waveforms which eject all ions with m/z values on either side of the selected m/z. The isolated m/z is then excited by the application of a m/z-selective excitation waveform to the end-cap electrodes. The amplitude of the ion movement in the trap is increased, while the frequency of the movement remains the same. This results in more energetic collisions with the helium bath gas, which is constantly present in the trap. This effectively may lead to CID of the selected ions. The product ions may be ejected out and be detected due to an *m/z*-selective ion ejection process similar to the one described above. Alternatively, one of the product ions may be selected and subjected to a second stage of CID to obtain second-generation product ions. This process may be repeated a number of times to perform a multi-stage MS–MS experiment.

The CID process in the ion trap is fundamentally different from CID in the collision cell of a triple-quadrupole instrument:

- The collisions takes place with helium in the ion trap, and with nitrogen of argon in the triple quadrupole instrument. In a collision with a heavier neutral target effectively more translation energy can be converted in ion internal energy.
- A larger number of collisions (up to 100) can be performed in the ion trap, because of the potential longer residence time (up to 5 ms), than in the triple-quadrupole, where typically up to 10 collisions take place during the residence time of the ion in the collision cell (*ca.* 20 µs).
- The collision energy is due to an RF waveform (0–5 or 0–10 V) in the ion trap, and due to a 0–100 V acceleration or offset voltage in the triple quadrupole.
- The ion energetics allow dissociation reactions with low activation barriers only in the ion trap, while a large range of dissociation reactions is open in a triple quadrupole.
- The RF excitation waveform is *m/z*-selective, which implies that a product ion is not excited and will not further fragment in the ion trap, while the collision offset voltage is equally applied to the product ions in a triple quadrupole. This may lead to product ions from secondary fragmentation reactions.

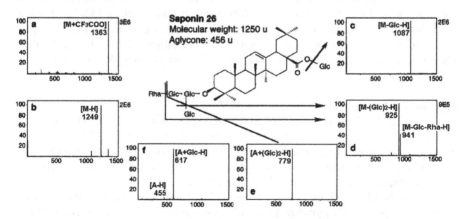

Figure 2.4: Demonstration of multi-stage MS–MS in an ion-trap instrument in the step-wise fragmentation of Saponin 25 (from [43], ©1998, Elsevier Science).

- The precursor ion is converted in the product ions with higher conversion efficiency in the ion trap than in the triple quadrupole.

The result of this is softer fragmentation in an ion trap, enabling step-wise fragmentation of complex structures in consecutive stages of MS–MS, as is illustrated with the stepwise losses of the sugar groups from a pentaglycosylated saponin in Figure 2.4 [43].

This step-wise fragmentation can be a disadvantage in the use of ion-trap for high-throughput screening, because a larger number of MS–MS steps take more time. The so-called wide-band excitation procedure was introduced by Thermo Finnigan to avoid such problems: instead of the highly-selective RF excitation waveform normally applied, an RF excitation waveform is applied which excites ions in a, for instance, twenty-m/z-units wide window. Ion-trap product-ion spectra acquired with wide-band excitation are often more similar to triple-quadrupole product-ion spectra obtained at moderate collision energies.

Data-dependent operation in MS–MS

Data-dependent acquisition (DDA) is a mode of operation, where the MS experiment performed in a particular scan is based on the data acquired in a previous scan. In a simple form, a DDA experiment switches the instrument from full-scan MS acquisition to full-scan product-ion MS–MS when the total-ion intensity or a selected-ion intensity exceeds a preset threshold. This avoids the need to perform two consecutive injections for the identification of unknowns in a mixture: first to obtain the m/z values for the intact protonated molecules of the unknowns, and second to acquire the product-ion MS–MS spectra of these unknowns in a time-scheduled procedure, switching between various preselected precursor ions as a function of the chromatographic retention time. The DDA was promoted by Thermo Finnigan upon the introduction of the API–ion trap combinations [44-46]. Similar procedures are available for other commercial ion-trap systems, as well as for triple-quadrupoles, *e.g.*, Information Dependent Acquisition (IDA) from Applied Biosystems MDS Sciex, Data-directed Analysis (DDA) from Waters, and Smart Select from Bruker.

Developments in ion-trap MS and MS–MS^n

Important developments of the relatively new technique of ion-trap MS are:
- The development of a MALDI source for an ion-trap instrument [47].
- The use of infrared multiphoton photodissociation (IRMPD) as an alternative to CID for the fragmentation of trapped ions [48-49]. IRMPD is more efficient in producing fragment ions over a large mass range.
- The study of gas-phase ion-molecule and/or ion-ion reactions, *e.g.*, reactions between protonated pyridine and multiple-charge oligonucleotide anions [50], and charge-state reduction reactions with multiple-charge proteins and oligonucleotides [51].

- The fragmentation of alkali adducts ions and other metal complexation products by ion-trap MS–MS^n [52-53].

Quadrupole–linear ion-trap hybrid instruments

A recent innovation is the commercial availability of linear two-dimensional ion traps [54]. The linear ion trap (LIT) is found to be less prone to space-charging effects, enabling a higher number of ions to be accumulated, which results in enhanced sensitivity. In the commercial instrument, the linear ion trap is the third quadrupole in a triple-quadrupole arrangement, *i.e.*, Q–q$_{coll}$–LIT. In that setup, it can be used to accumulate product ions generated by CID in a LINAC collision cell, providing enhanced sensitivity and lack of low-mass cut-off. Further stages of MS–MS can be performed in the linear ion-trap, which then has similar features as the three-dimensional ion-trap. Early reports described the application of the linear ion trap in metabolite identification and quantitative bioanalysis [55-56].

4.3   Time-of-flight instruments

While the principles of time-of-flight (TOF) mass spectrometry were well established for many years, significant breakthroughs in TOF technology and application were made in the 1990's [57-58]. This can be attributed to the emergence of MALDI as an ionization technique. TOF is considered as the ideal mass analyser for MALDI. This stimulated developments and research in TOF analysers, which in turn led to the rediscovery of the orthogonal-acceleration TOF (oaTOF).

In a TOF mass analyser, a package of ions is accelerated by a potential V into a field-free linear flight tube. The flight time t_{flight} needed for an ion with m/z to reach a detector placed at a distance d is measured. The flight time is related to the m/z:

$$t_{flight}^2 = \frac{m\,d^2}{2\,z\,e\,V} = \frac{m}{z}\left[\frac{d^2}{2\,e\,V}\right]$$

Pulsed ion introduction into the TOF analyser is required to avoid the simultaneous arrival of ions of different m/z. The pulse frequency is much higher with a TOF than with an ion trap. Some characteristics of the TOF mass analyser, especially relevant for on-line LC–MS application, are:
- High ion transmission.
- Fast spectrum-acquisition capabilities: Unlike quadrupole or ion-trap instruments, the acquisition of a mass spectrum is not based on scanning. Typical pulse rates of TOF instrument are 20–50 kHz. Spectra from different ion-introduction events are accumulated, resulting in improved S/N due to the averaging of random noise. The resulting mass spectra are stored at 4–10 spectra/s. The TOF is an integrating rather than a scanning instrument.
- An unlimited mass range. In practice, most systems commercially available for on-line LC–MS application are limited to m/z 10,000 or 20,000.

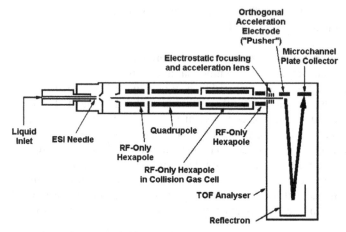

Figure 2.5: Quadrupole–time-of-flight hybrid instrument.

Due to the kinetic energy dispersion of the ions leaving the ion source, the resolution achievable in TOF-MS is limited. There are a number of ways to improve the resolution of a TOF mass analyser:

- The resolution is limited by the length of the flight tube. A longer tube will result in improved separation between ions with different m/z. An ion mirror or reflectron (see below) can be used to double the length of the flight tube without compromising the dimensions of the instrument.
- The initial kinetic energy spread can be diminished by delayed extraction or by orthogonal acceleration. In delayed extraction or time-lag focussing [59], ions are extracted from the source by a weak electrical field, electrostatically stored between two grids in order to even the kinetic energies, and then accelerated into the flight tube. Alternatively, the longitudinal kinetic energy spread can be reduced to accelerate the ions orthogonally to their direction of introduction [58]. Orthogonal-acceleration TOF-MS (oaTOF) is routinely applied in LC–MS.
- An electrostatic reflectron, consisting of a series of lens plates with different voltages, forms a retarding field. Ions with higher kinetic energies will penetrate the retarding field deeper, will spend more in time turning around, and will catch up with the ions with lower kinetic energy and reach the detector simultaneously.

These modifications result in a mass resolution in excess of 10,000 (FWHM) for modern TOF-MS systems, resulting in mass accuracies of better than 5 ppm in the combination of API–oaTOF-MS, *i.e.*, accurate-mass determination within ±5 mDa or better for a compound up to 1000 Da.

MS–MS in a time-of-flight instrument

In order to perform MS–MS with a TOF analyser, one has to combine it with another mass analyser to form a hybrid instrument. Combinations with quadrupole, ion-trap, and magnetic sector instruments have been described. The most successful among these hybrids is the quadrupole–time-of-flight (Q–TOF) instrument (Ch. 2.4.4). A MALDI–TOF–TOF tandem instrument has been described as well [60-61]. Although metastable fragmentation or post-source decay can be achieved in TOF-MS [62], the interpretation of the data is generally considered rather difficult. Therefore, alternative approaches for MS–MS in combination with MALDI have been investigated: the Q–TOF hybrid (Ch. 2.4.4), ion-traps (Ch. 2.4.2), or the ion-trap–time-of-flight hybrid (Ch. 2.4.5).

4.4   Quadrupole–time-of-flight hybrid instruments

The features of an oa-TOF mass analyser stimulated Waters [63-65] to build a quadrupole–time-of-flight hybrid (Q–TOF) instrument (Figure 2.5), especially for peptide sequencing analysis [65], but the instrument has found much wider application. In MS mode, the first quadrupole is operated in RF-only mode; ions are transmitted to the oa-TOF device for mass analysis. In product-ion MS–MS mode, the quadrupole provides precursor-ion selection at unit-mass resolution. The selected precursor ion is fragmented in a gas-filled hexapole collision cell, and the product ions are mass analysed at the orthogonal-acceleration TOF device. In this way, accurate-mass determination is available in both MS and MS–MS mode. The Q–TOF is commercially available from Waters. Subsequently, a quadrupole–time-of-flight instrument with similar features was developed by Applied Biosystems MDS Sciex [66] (QSTAR Pulsar). Both instruments have found wide application in LC–MS and peptide sequencing studies using nano-electrospray sample introduction. Principles and applications of quadrupole–time-of-flight hybrid instruments have been reviewed by Chernushevich *et al.* [67].

4.5   Ion-trap–time-of-flight hybrid instruments

A TOF mass analyser requires a pulsed ion introduction. In an electrospray–TOF combination, the duty cycle is an important issue. A significant improvement in the duty cycle can be achieved in an ion-trap–TOF hybrid instrument: the ions from a continuous ion source are accumulated in the ion trap between two ion introduction events. An ion-trap–TOF hybrid instrument was first described by the group of Lubman [68-69]. The system consists of an atmospheric-pressure ion source with a vacuum interface, a set of Einzel lenses, an ion-trap device, and a reflectron time-of-flight mass analyser. The system was applied for fast analysis in combination with a variety of separation techniques [70].

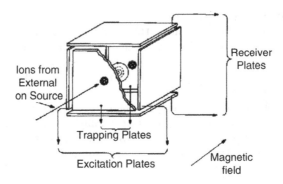

Figure 2.6: Cell of a Fourier-transform ion-cyclotron resonance mass spectrometer.

4.6   Fourier-transform ion-cyclotron resonance instruments

For many years, Fourier-transform ion-cyclotron resonance mass spectrometry (FT-ICR-MS) has been a tool used in fundamental studies of gas-phase ion-molecule reactions [71]. Due to its high-resolution and MS–MS capabilities, the application of FT-ICR-MS in combination with electrospray ionization for large biomacromolecules has been investigated more recently. At present, FT-ICR-MS can be considered as an important tool in the analysis of biomacromolecules and in applications where high resolution and/or high mass accuracy is important. The technique is gaining territory, also due to the introduction of user-friendly instrument by instrument manufacturers.

An FT-ICR-MS can be considered as an ion trap system, where the ions are trapped in the magnetic field rather than in a quadrupole field. In an FT-ICR-MS, the mass analysis is performed in a cubic cell placed in a strong magnetic field B [71]. The cell consists of two opposite trapping plates, two opposite excitation plates and two opposite receiver plates (Figure 2.6). An ion of mass m, velocity v, and with z elementary charges describes in such a cell a circle of radius r, perpendicular to the magnetic field. The cyclotron frequency ω_c, which is inversely proportional to the m/z value, is given by:

$$\omega_c = 2 \pi f = \frac{v}{r} = \frac{B e z}{m}$$

where f is the frequency in Hertz. When the ions, trapped in their cyclotron motion in the cell, are excited by means of a radio-frequency pulse, the radius of the circle is increased, and ions of one m/z value start moving in phase. The coherent movement of the ions generates an image current in the receiver plates. As the coherency is disturbed in time, the image current signal is decaying in time as well. The time-domain signal from the receiver plates contains all frequency information of the moving ions. By applying Fourier transformation, the time-domain signal can be transformed into a frequency-domain signal, that subsequently can be transformed in a regular mass spectrum by application of the equation above.

Important features of FT-ICR-MS are:

- An extremely (mass-dependent) high resolution, *i.e.*, in excess of 10^6 (FWHM).
- Extreme vacuum restraints, *e.g.*, 10^{-7} Pa in the cell.
- A very wide dynamic range of five orders-of-magnitude: *ca.* 10^2 elementary charges are required to produce a useful signal, while the charge capacity of an FT-ICR cell is *ca.* 10^6–10^7 elementary charges.
- The capabilities to select and excite ions via waveform pulses applied to the excitation plates and to study gas-phase ion-molecule and ion-ion reactions.

The development of FT-ICR-MS and its application in the analysis of biomacromolecules has been reviewed [72-73]. Some milestones and highlights in these development are:

- The selected accumulation of ions of noncovalent complexes [74] based on quadrupolar excitation of the ions in the trap.
- The mass spectrometry of very large biomolecules, such as the coliphage T4 DNA ion with a molecular weight of 10^8 Da [75].
- The use of high-field FT-ICR-MS instruments: FT-ICR-MS systems with 9.4-T [76] and 11.5-T [77] magnets have been reported. The commercially-available Bruker Apex-IV can be delivered with 4.7-, 7.0-, 9.4-, and 12-T actively shielded magnets.
- Application of FT-ICR-MS for on-line capillary electrophoresis–MS (CE–MS), *e.g.*, in the analysis of single cells [78]. A erythrocyte cell was lysed in a CE column. The α- and β-chains of hemoglobin, *ca.* 450 amol present in the cell, were detected with isotopic resolution.
- Excellent sensitivity, as indicated by the detection of *ca.* 7 amol of carbonic anhydrase in a crude extract from human blood [79]. Dissociation of 9 amol introduced via CE provided sequence specific ions, enabling identification from a protein database. The 42-Da mass difference observed was attributed to acetylation of the carbonic anhydrase.
- Unit-mass resolution and 3-Da mass accuracy for molecules with a molecular weight of 112 kDa [80].
- The use of in-trap hydrogen/deuterium (H/D) exchange to probe structure and conformation issues in relation to proteins [81].

MS–MS^n in an FT-ICR-MS instrument

Similarly to an ion-trap instrument, the FT-ICR-MS instrument has capabilities for MS–MS^n experiments. Targeted ions can be selectively trapped in the ion cell by the application of RF pulses to eliminate unwanted ions. A variety of ion excitation methods can be applied to induce fragmentation. Some of the ion-excitation methods applied are:

- Collision-induced dissociation (CID), similar to other MS–MS techniques. This requires the use of a pulsed-valve collision gas introduction. Alternative approaches, *i.e.*, very low energy and multiple excitation CID, have been developed [82].
- Laser photodissociation [83] or IRMPD [84], where 193-nm photons from a laser are applied for ion excitation.
- Surface-induced dissociation [85], which requires the introduction of an appropriate surface into the FT-ICR cell.
- Sustained off-resonance irradiation (SORI) [86-87], where ion excitation is achieved by application of an off-resonance electric pulse with a frequency of 500–2000 Hz away from the observed ion-cyclotron frequency. Due to the long duration of the pulse, the ions experience a series of acceleration–deceleration cycles, which in the presence of gas results in slow ion activation by sequential low-energy collisions. SORI promotes the applicability of CID to larger molecules. In a comparison of SORI and various CID techniques in the fragmentation of large multiple-charge ions [82], SORI was found to provide the highest efficiency, selectivity, and resolving power.
- Black-body infrared radiative dissociation (BIRD) [88], which takes advantage of the photon flux of the black-body field generated by the vacuum chamber walls. The efficiency of BIRD depends on the temperature of the vacuum chamber and the ion storage time (typically 10–1000 s).
- Electron-capture dissociation (ECD) [89], where the multiple-charge ion captures an electron to produce cationic products. The electron capture results in a radical site in the ion, leading to more and different backbone cleavages in the protein. It has been argued that the fragmentation actually takes place at the electron-capture site, *i.e.*, prior to internal energy redistribution.

Alternatively, the high resolving power of the FT-ICR-MS in the measurement of fragment ions can be exploited by performed fragmentation of the precursor ion, prior to introduction into the FT-ICR cell, *e.g.*, via in-source CID, or a hybrid instrument consisting of a quadrupole or LIT front-end and a FT-ICR-MS back end. Both types of instruments are commercially available.

4.7 Sector and related hybrid instruments

The single- and double-focussing sector instruments are the roots of mass spectrometry. Today, they still have distinct application areas, *e.g.*, in the analysis of dioxins and related compounds by GC–MS. Then, advantage is taken of the high-resolution SIM mode uniquely available on sector instruments. For other high-resolution applications, time-of-flight and FT-ICR-MS instruments are preferred. The role of sector instruments in LC–MS is rather limited. Specially-designed API sources have been developed for use in combination with the high acceleration voltage (see Ch. 5.8.4)

In a magnetic-sector mass analyser, ions with a particular kinetic energy are introduced into a magnetic field. When the magnetic force is counterbalanced by the centrifugal force, ions are transmitted to the detector. By variation of either the magnetic field or the acceleration voltage, ions with different m/z can be detected by a fixed detector behind a slit. Because ionization inevitably leads to ion kinetic energy dispersion, an electrostatic analyser (ESA) is required to improve the resolution of the magnetic-sector mass analyser. Such an instrument is usually called a double-focussing instrument.

In a double-focussing sector instrument, a collision cell can be placed either in the first or in the second field-free region. The detection of product ions in a BE geometry is possible by the B/E-linked scan mode. A sector instrument provides the ability to perform high-energy CID (Ch. 2.4.1). More advanced instrument configurations, e.g., three- or four-sector instruments with a collision cell in the third field-free region, enable better resolution in MS–MS. The power of high-energy CID is demonstrated by the occurrence of a variety of internal amino-acid cleavages, that are not observed in low-energy CID.

In addition, a number of hybrid instruments for MS–MS have been described:

- BE–quadrupole hybrid instruments, providing unit-mass resolution product-ion mass spectra [90].
- BE–time-of-flight hybrid instruments, enabling high-sensitivity accurate-mass determination of product ions [91-92].
- BE–quadrupole ion trap hybrid instruments, enabling multiple stages of MS–MS on a precursor ion selected at high resolution [93].

5. References

1. J.R. Chapman, *Practical Organic Mass Spectrometry*, 2nd Ed., 1993, Wiley, London.
2. E. de Hoffmann, J. Charette, V. Stroobant, *Mass Spectrometry, Principles and Applications*, 1996, Wiley, London.
3. F.W. McLafferty, D.B. Stauffer, *The Wiley/NBS Registry of Mass Spectral Data*, 1989, Wiley, London.
4. NIST/EPA/NIH Mass Spectral Library, National Institute of Standards and Technology, 100 Bureau Dr., Stop 2310, Gaithersburg, MD 20899-2310
5. W.M.A. Niessen (Ed.), *Current practice in gas chromatography–mass spectrometry*, 2001, Marcel Dekker Inc., New York, NY.
6. A.G. Harrison, *Chemical Ionization Mass Spectrometry*, 2nd Ed., 1992, CRC Press, Boca Raton, FL.
7. K.L. Busch, *Desorption ionization MS*, J. Mass Spectrom., 30 (1995) 233.
8. H. Jungclas, H. Danigel, L. Schmidt, J. Dellbrügge, *Combined LC to MS: An application of 252 Cf-PD-MS*, Org. Mass Spectrom., 17 (1982) 499.
9. J.G. Stroh. K.L. Rinehart, *LC–FAB-MS, recent developments*, LC-GC, 5 (1987) 562.
10. R.M. Caprioli (Ed.), *Continuous-Flow Fast Atom Bombardment Mass Spectrometry*, 1990, Wiley, New York.

11. K.K. Murray, *Coupling MALDI to liquid separations*, Mass Spectrom. Rev., 16 (1997) 283.
12. K. Tanaka, H. Waki, Y. Ido, S. Akita, Y. Yoshida, T. Yoshida, *Protein and polymer analyses up to m/z 100 000 by laser ionization TOF-MS*, Rapid Commun. Mass Spectrom., 2 (1988) 151.
13. M. Karas. F. Hillenkamp, *Laser desorption ionization of proteins with molecular masses exceeding 10 000 Daltons*, Anal. Chem., 60 (1988) 2299.
14. M. Karas, U. Bahr, A. Ingendoh, E. Nordhoff, B. Stahl, K. Strupat, F. Hillenkamp, *Principles and applications of UV-MALDI-MS*, Anal. Chim. Acta, 241 (1990) 175.
15. C. Fenselau, *MALDI-MS and strategies for protein analysis*, Anal. Chem., 69 (1997) 661A.
16. C.R. Blakley, J.J. Carmody, M.L. Vestal, *A new soft ionization technique for MS of complex molecules*, J. Am. Chem. Soc., 102 (1980) 5931.
17. J.B. Fenn, M. Mann, C.K. Meng, S.F. Wong, C.M. Whitehouse, *ESI – principles and practice*, Mass Spectrom. Rev., 9 (1990) 37.
18. J.V. Iribarne, B.A. Thomson, *On the evaporation of small ions from charged droplets*, J. Chem. Phys., 64 (1976) 2287.
19. J.F.J. Todd, *Recommendations for nomenclature and symbolism for mass spectroscopy*, Int. J. Mass Spectrom. Ion. Processes, 142 (1995) 209.
20. P. Price, *Standard definitions of terms relating to MS*, J. Am. Soc. Mass Spectrom., 2 (1991) 336.
21. O.D. Sparkman, *Mass Spec Desk Reference*, 2000, Global View Publishing, Pittsburgh, PA
22. C.K. Meng, M. Mann, J.B. Fenn, Presented at the 36th ASMS Conference on Mass Spectrometry and Allied Topics, June 5-10, 1988, San Francisco, CA, p. 771.
23. M. Mann, C.K. Meng, J.B. Fenn, *Interpreting mass spectra of multiply charged ions*, Anal. Chem., 61 (1989) 1702.
24. T.R. Covey, R.F. Bonner, B.I. Shushan, J.D. Henion, *The determination of protein, oligonucleotide and peptide molecular weights by ESI-MS*, Rapid Commun. Mass Spectrom., 2 (1988) 249.
25. S. Pleasance, P. Thibault, P.G. Sim, R.K. Boyd, *Cesium iodide clusters as mass calibrants in ESI-MS*, Rapid Commun. Mass Spectrom., 5 (1991) 307.
26. J.F. Anacleto, S. Pleasance, R.K. Boyd, *Calibration of ESI mass spectra using cluster ions*, Org. Mass Spectrom., 27 (1992) 660.
27. S. König, H.M. Fales, *Calibration of mass ranges up to m/z 10,000 in ESI-MS*, J. Am. Soc. Mass Spectrom., 10 (1999) 273.
28. M. Moini, *Ultramark 1621 as a calibrant/reference compound for MS. II. Positive- and negative-ion ESI*, Rapid Commun. Mass Spectrom., 8 (1994) 711.
29. J. Zaia, R.S. Annan, K. Biemann, *The correct molecular weight of myoglobin, a common calibrant for MS*, Rapid Commun. Mass Spectrom., 6 (1992) 32.
30. M. Moini, B.L. Jones, R.M. Rogers, L. Jiang, *Sodium trifluoroacetate as a tune/calibration compound for positive- and negative-ion ESI-MS in the mass range of 100–4000 Da*, J. Am. Soc. Mass Spectrom., 9 (1998) 977.
31. D.W. Ledman, R.O. Fox, *Water cluster calibration reduces mass error in ESI-MS of proteins*, J. Am. Soc. Mass Spectrom., 8 (1997) 1158.
32. C. Brunnee, *The ideal mass analyzer: fact or fiction?*, Int. J. Mass Spectrom. Ion Proc., 76 (1987) 125.

33. A.N. Tyler, E. Clayton, B.N. Green, *Exact mass measurement of polar organic molecules at low resolution using ESI and a quadrupole MS*, Anal. Chem., 68 (1996) 3561.
34. L. Yang, M. Amad, W.M. Winnik, A.E. Schoen, H. Schweingruber, I. Mylchreest, P.J. Rudewicz, *Investigation of an enhanced resolution triple quadrupole MS for high-throughput LC–MS–MS assays*, Rapid Commun. Mass Spectrom., 16 (2002) 2060.
35. F.W. McLafferty (Ed.), *Tandem Mass Spectrometry*, 1983, Wiley, New York, NY.
36. K.L. Busch, G.L. Glish, S.A. McLuckey, *Mass Spectrometry / Mass Spectrometry*, 1988, VCH Publishers, New York, NY.
37. B.A. Mansoori, E.W. Dyer, C.M. Lock, K. Bateman, R.K. Boyd, B.A. Thomson, *Analytical performance of a high-pressure RF-only quadrupole collision cell with a axial field applied by using conical rods*, J. Am. Soc. Mass Spectrom., 9 (1998) 775.
38. R.E. March, J.F.J. Todd (Ed), *Practical Aspects of Ion Trap Mass Spectrometry*, Vols 1, 2 and 3, 1995, CRC Press, Boca Raton, FL.
39. R.E. March, *An introduction to quadrupole ion trap MS*, J. Mass Spectrom., 32 (1997) 351.
40. R.D. Voyksner, H. Lee, *Investigating the use of an octapole ion guide for ion storage and high-pass mass filtering to improve quantitative performance of ESI ion-trap MS*, Rapid Commun. Mass Spectrom., 13 (1999) 1427.
41. H.G.J. Mol, R.C.J. van Dam, R.J. Vreeken, O.M. Steijger, *Determination of daminozide in apples and apple leaves by LC–MS*, J. Chromatogr. A, 833 (1999) 53.
42. J.D. Williams, K.A. Cox, R.G. Cooks, R.E. Kaiser, Jr., J.C. Schwartz, *High mass-resolution using a quadrupole ion-trap MS*, Rapid Commun. Mass Spectrom., 5 (1991) 327.
43. J.-L. Wolfender, S. Rodriguez, K. Hostettmann, *LC coupled to MS and NMR spectroscopy for the screening of plant constituents*, J. Chromatogr. A, 794 (1998) 299.
44. D.M. Drexler, P.R. Tiller, S.M. Wilbert, F.W. Bramble, J.C. Schwartz, *Automated identification of isotopically labeled pesticides and metabolites by intelligent 'real time' LC–MS–MS using a bench-top ion trap MS*, Rapid Commun. Mass Spectrom., 12 (1998) 1501.
45. P.R. Tiller, A.P. Land, I. Jardine, D.M. Murphy, R. Sozio, A. Ayrton, W.H. Schaefer, *Application of LC–MS^n analyses to characterize novel glyburide metabolites formed in vitro*, J. Chromatogr. A, 794 (1998) 15.
46. L.L. Lopez, X. Yu, D. Cui, M.R. Davis, *Identification of drug metabolites in biological matrices by intelligent automated LC–MS–MS*, Rapid Commun. Mass Spectrom., 12 (1998) 1756.
47. J. Qin, R.J.J.M. Steenvoorden, B.T. Chait, *A practical ion trap MS for the analysis of peptides by MALDI*, Anal. Chem., 68 (1996) 1784.
48. B.J. Goolsby, J.S. Brodbelt, *Characterization of β-lactams by photodissociation and CID techniques in a quadrupole ion trap*, J. Mass Spectrom., 33 (1998) 705.
49. A.H. Payne, G.L. Glish, *Thermally assisted IRMPD in a quadrupole ion trap*, Anal. Chem., 73 (2001) 3542.
50. W.J. Herron, D.E. Goeringer, S.A. McLuckey, *Ion-ion reactions in the gas phase: proton transfer reactions of protonated pyridine with multiply-charged oligonucleotide anions*, J. Am. Soc. Mass Spectrom., 6 (1995) 529.
51. J.L. Stephenson, Jr., G.J. van Berkel, S.A. McLuckey, *Ion-ion proton transfer*

reactions of bio-ions involving noncovalent interactions: holomyoglobin, J. Am Soc. Mass Spectrom., 8 (1997) 637.

52. M.R. Asam, G.L. Glish, *MS-MS of alkali cationized polysaccharides in a quadrupole ion trap*, J. Am. Soc. Mass Spectrom., 8 (1997) 987.

53. E.J. Alvarez, V.H. Vartanian, J.S. Brodbelt, *Metal complexation reactions of quinolone antibiotics in a quadrupole ion trap*, Anal. Chem., 69 (1997) 1147.

54. J.W. Hager, *A new LIT-MS*, Rapid Commun. Mass Spectrom., 16 (2002) 512.

55. G. Hopfgartner, C. Husser, M. Zell, *Rapid screening and characterization of drug metabolites using a new Q–LIT-MS*, J. Mass Spectrom., 38 (2003) 138.

56. Y.-Q. Xia, J.D. Miller, R. Bakhtiar, R.B. Franklin, D.Q. Liu, *Use of a Q–LIT-MS in metabolite identification and bioanalysis*, Rapid Commun. Mass Spectrom., 17 (2003) 1137.

57. E.W. Schlag (Ed.), *Time-of-Flight Mass Spectrometry and its Applications*, 1994, Elsevier, Amsterdam.

58. M. Guilhaus, D. Selby, V. Mlynski, *Oa-TOF-MS*, Mass Spectrom. Rev., 19 (2000) 65.

59. M.L. Vestal, P. Juhasz, S.A. Martin, *Delayed extraction MALDI-TOF-MS*, Rapid Commun. Mass Spectrom., 9 (1995) 1044.

60. C.E.C.A. Hop, *Design of an orthogonal TOF-TOF-MS for high-sensitivity MS–MS experiments*, J. Mass Spectrom., 33 (1998) 397.

61. K.F. Medzihradszky, J.M. Campbell, M.A. Baldwin, A.M. Falick, *The characteristics of peptide CID using a high-performance MALDI-TOF–TOF-MS–MS*, Anal. Chem., 72 (2000) 552.

62. B. Spengler, *Post-source decay analysis in MALDI-MS of biomolecules*, J. Mass Spectrom., 32 (1997) 1019.

63. H.R. Morris, T. Paxton, A. Dell, J. Langhorne, M. Berg, R.S. Bordoli, J. Hoyes, R.H. Bateman, *High-sensitivity CID-MS–MS on a novel Q–oaTOF-MS*, Rapid Commun. Mass Spectrom., 10 (1996) 889.

64. T. Keough, M.P. Lacey, M.M. Ketcha, R.H. Bateman, M.R. Green, *Oa single-pass TOF-MS for determination of the exact masses of product ions formed in MS–MS experiments*, Rapid Commun. Mass Spectrom., 11 (1997) 1702.

65. H.R. Morris, T. Paxton, M. Panico, R. McDowell, A. Dell, *A novel geometry MS, the Q–TOF, for low-femtomole/attomole-range biopolymer sequencing*, J. Protein Chem., 16 (1997) 469.

66. A. Shevchenko, I. Chernushevich, W. Ens, K.G. Standing, B. Thomson, M. Wilm, M. Mann, *Rapid 'de novo' peptide sequencing by a combination of nano-ESI, isotopic labeling and a Q–TOF-MS*, Rapid Commun. Mass Spectrom., 11 (1997) 1015.

67. I.V. Chernushevich, A.V. Loboda, B.A. Thomson, *An introduction to quadrupole–time-of-flight mass spectrometry*, J. Mass Spectrom., 36 (2001) 849.

68. S.M. Michael, B.M. Chien, D.H. Lubman, *Detection of ESI using an ion trap storage–reflectron TOF-MS*, Anal. Chem., 65 (1993) 2614.

69. B.M. Chien, S.M. Michael, D.M. Lubman, *The design and performance of an ion trap storage reflectron TOF-MS*, Int. J. Mass Spectrom. Ion Processes, 131 (1994) 149.

70. J.-T. Wu, M.G. Qian, M.X. Li, K. Zheng, P. Huang, D.M. Lubman, *On-line analysis of capillary separations interfaced to an ion-trap storage–reflectron TOF-MS*, J. Chromatogr. A, 794 (1998) 377.

71. N.M.M. Nibbering, *Basic ion chemistry studies using a FT-ICR-MS*, Trends Anal. Chem., 13 (1994) 223.

72. A.G. Marshall, C.L. Hendrickson, G.S. Jackson, *FT-ICR-MS: A primer*, Mass Spectrom., Rev., 17 (1998) 1.
73. R.D. Smith, *Evolution of ESI-MS and FT-ICR-MS for proteomics and other biological applications*, Int. J. Mass Spectrom., 200 (2000) 509.
74. J.E. Bruce, S.L. Van Orden, G.A. Anderson, S.A. Hofstadler, M.G. Sherman, A.L. Rockwood, R.D. Smith, *Selected ion accumulation of noncovalent complexes in a FT-ICR-MS*, J. Mass Spectrom., 30 (1995) 124.
75. R. Chen, X. Cheng, D.W. Mitchell, S.A. Hofstadler, Q. Wu, A.L. Rockwood, M.G. Sherman, R.D. Smith, *Trapping, detection and mass determination of coliphage T4 DNA ions of 10^8 Da by ESI-FT-ICR-MS*, Anal. Chem., 67 (1995) 1159.
76. M.W. Senko, C.L. Hendrickson, L. Paša-Tolić, J.A. Maro, F.M. White, S. Guan, A.G. Marshall, *ESI-FT-ICR-MS at 9.4 T*, Rapid Commun. Mass Spectrom., 10 (1996) 1824.
77. M.V. Gorshkov, L. Paša-Tolić, H.R. Udseth, G.A. Anderson, B.M. Huang, J.E. Bruce, D.C. Prior, S.A. Hofstadler, L. Tang, L.-Z. Chen, J.A. Willett, A.L. Rockwood, M.S. Sherman, R.D. Smith, *ESI-FT-ICR-MS at 11.5 T: Instrument design and initial results*, J. Am. Soc. Mass Spectrom., 9 (1998) 692.
78. S.A. Hofstadler, J.C. Severs, R.D. Smith, F.D. Swanek, A.G. Ewing, *Analysis of single cells with CE–ESI-FT-ICR-MS*, Rapid Commun. Mass Spectrom., 10 (1996) 919.
79. G.A. Valaskovic, N.L. Kelleher, F.W. McLafferty, *Attomole protein characterization by CE–MS*, Science, 273 (1996) 1199.
80. N.L. Kelleher, M.W. Senko, M.M. Siegel, F.W. McLafferty, *Unit resolution mass spectra of 112 kDa molecules with 3 Da accuracy*, J. Am. Soc. Mass Spectrom., 8 (1997) 380.
81. M.A. Freitas, C.L. Hendrickson, M.R. Emmett, A.G. Marshall, *High-field FT-ICR-MS for simultaneous trapping and gas-phase H/D exchange of peptide ions*, J. Am. Soc. Mass Spectrom., 9 (1998) 1012.
82. M.W. Senko, J.P. Speir, F.W. McLafferty, *CID of large multiply charged ions using FT-ICR-MS*, Anal. Chem., 66 (1994) 2801.
83. E.R. Williams, J.J.P. Furlong, F.W. McLafferty, *Efficiency of CID and 193 nm laser photoionization and photodissociation of peptide ions in FT-ICR-MS*, J. Am. Soc. Mass Spectrom., 1 (1990) 288.
84. D.P. Little, J.P. Spier, M.W. Senko, P.B. O'Connor, F.W. McLafferty, *IRMPD of large multiply charged ions for biomolecule sequencing*, Anal. Chem., 66 (1994) 2809.
85. E.R. Williams, K.D. Henry, F.W. McLafferty, J. Shabanowitz, D.F. Hunt, *SID of peptide ions in FT-MS*, J. Am. Soc. Mass Spectrom., 1 (1990) 413.
86. J.W. Gauthier, T.R. Trautman, D.B. Jacobson, *SORI for CID involving FT-ICR-MS. CID technique that emulates IRMPD*, Anal. Chim. Acta, 246 (1991) 211.
87. S.A. Hofstadler, J.H. Wahl, R. Bakhtiar, G.A. Anderson, J.E. Bruce, R.D. Smith, *CE–FT-ICR-MS with SORI for the characterization of protein and peptide mixtures*, J. Am. Soc. Mass Spectrom., 5 (1994) 894.
88. W.D. Price, P.D. Schnier, E.R. Williams, *MS–MS of large biomolecule ions by BIRD*, Anal. Chem., 68 (1996) 859.
89. R.A. Zubarev, D.M. Horn, E.K. Fridriksson, N.L. Kelleher, N.A. Kruger, M.A. Lewis, B.K. Carpenter, F.W. McLafferty, *ECD for structural characterization of multiply charged protein cations*, Anal. Chem., 72 (2000) 563.
90. A.E. Schoen, J.W. Amy, J.D. Ciupek, R.G. Cooks, P. Dobberstein, G. Jung, *A hybrid BEQQ-MS*, Int. J. Mass Spectrom. Ion Processes, 65 (1985) 125.

91. R.H. Bateman, M.R. Green, G. Scott, E. Clayton, *A combined magnetic sector–TOF-MS for structural determination studies by MS–MS*, Rapid Commun. Mass Spectrom., 9 (1995) 1227.

92. I. Lindh, W.J. Griffiths, T. Bergman, J. Sjövall, *ESI-CID of peptides: Studies on a magnetic sector–oa-TOF-MS*, Int. J. Mass Spectrom. Ion Processes, 164 (1997) 71.

93. J.A. Loo, H. Muenster, *Magnetic sector–ion trap MS with ESI for high sensitivity peptide sequencing*, Rapid Commun. Mass Spectrom., 13 (1999) 54.

TECHNOLOGY

3

STRATEGIES IN LC–MS INTERFACING

1. Introduction

In the past 10 years, the research efforts in the field of liquid chromatography–mass spectrometry (LC–MS) have changed considerably. Investigations into the coupling of LC and MS started some 30 years ago, in the early 1970's. While in the first 20 years most of the attention had to be given to solving interface problems, building new technology, and so on, most workers with LC–MS today are only concerned with the application of the available technique in their field of interest. Technological problems in interfacing appear to be solved, and from the wide variety of interfaces developed over the years basically only two remained, *i.e.*, electrospray (ESI) and atmospheric-pressure chemical ionization (APCI), which are both based on the principle of atmospheric-pressure ionization. Most of this book is devoted to the principles and applications of these two interfaces. However, for a good understanding of what LC–MS is today some knowledge of its history and its development is important, especially because from this knowledge one can answer the question why LC–MS interfaces are as they are today. Therefore, the first two chapters are devoted to the general history of LC–MS. This chapter deals with the history of LC–MS, not so much from a chronological perspective, but more from a strategic and developmental point of view, while in the next chapter a number of once successful, but now obsolete LC–MS interfaces are briefly discussed.

Table 3.1:
Initial objectives in developing on-line LC–MS
MS is a universal detector for LC Analysis of thermolabile analytes, not amenable to GC–MS Analysis of nonvolatile analytes Avoid analyte derivatization MS affords a low detection limit ($<10^{-12}$ g) Identification of the analytes Assessment of peak purity

Table 3.2:	
Requirements and specifications of LC–MS interfaces [3]	
LC operation	No restrictions to mobile phase composition Gradient elution Volatile and nonvolatile additives (buffers, ion-pair reagents) Free choice of LC column dimensions Flow-rate between 1 µl/min and 2 ml/min
Interface operation	Enrichment of sample to solvent High transfer efficiency No additional peak broadening, no loss in resolution No uncontrolled chemical modifications of the analytes
MS operation	High vacuum in mass analyser ($< 10^{-3}$ Pa) Free choice of ionization method (EI, CI or other) Free choice of CI reagent gas conditions Both positive and negative ion mode Low interference from solvents and solvent impurities
General	Low cost Operational simplicity Low detection limits Quantitation capabilities

Three objectives determined the developments throughout the past 30 years:
- the ability to couple conventional LC columns to a mass spectrometer.
- the ability to achieve analyte enrichment in order to enable electron ionization (EI).
- the ability to ionize analytes directly from the liquid phase.

The efforts with respect to the latter objective have broadened the scope and applicability range of MS: ionic and highly polar analytes, even biomacromolecules, are now amenable to MS analysis. The "*odd couple*" [1] is now moving borders and opening new application areas. LC–MS has become a technique that has found wide application, that can give solutions to analytical problems, and that will certainly continue to grow in the years to come.

2. History of LC–MS

An on-line combination of LC and MS offers the possibility to take advantage of both LC as a powerful and versatile separation technique and MS as a powerful and sensitive detection and identification technique. Fully exploiting the intrinsic properties of these two techniques results in an extremely powerful analytical tool, a true "hyphenated" technique.

The history of LC–MS starts in the early 1970's. In several laboratories, the possibilities of on-line LC and MS were investigated. Some breakthroughs in development reported in 1974 mark the actual beginning of the history of LC–MS. The technique has been extensively reviewed over the years [1-21].

Objectives in LC–MS research

Some of the older review papers [1-4] provide a view on the initial objectives in LC–MS research (see Table 3.1). In order to realize these objectives, a variety of LC–MS interfaces was developed. Requirements and specifications of LC–MS interfaces, as outlined in an early review paper by MacFadden [3], are summarized in Table 3.2.

Off-line or on-line LC–MS

During the first few years of the development of LC–MS, there was considerable debate on whether an on-line coupling should be pursued, or whether an off-line method, *i.e.*, fraction collection, evaporation of the solvent, and transfer of the analyte to the MS by means of a probe, would not be more appropriate, especially in real-life applications. The developments in the past 30 years have stopped and obsoleted this discussion, and not only because perhaps the most convenient way of fraction collection would be a fully-automated LC–MS-directed fractionation (*cf.* Ch. 9.5.4).

The versatility and high analytical power of on-line LC–MS have convincingly been demonstrated. On-line LC–MS cannot be eliminated anymore from a wide variety of routine analytical procedures in environmental, pharmaceutical, and biochemical laboratories.

Figure 3.1: Combined LC–MS, a difficult courtship (Reprinted from with permission [1], ©1982, Elsevier Science).

General problems in LC–MS coupling

Three major problems were met in developing an on-line combination of LC and MS:

- the apparent flow-rate incompatibility as expressed in the need to introduce 1 ml/min of a liquid effluent from a conventional 4.6-mm-ID LC column into the high vacuum of a mass spectrometer.
- the incompatibility with respect to mobile-phase composition as a result of the frequent use of nonvolatile mobile-phase additives in LC separation.
- the ionization of nonvolatile and/or thermally labile analytes.

The (almost) fundamental incompatibility between LC and MS was nicely pictured by Arpino [1] in Figure 3.1.

2.1 The start

The research on the possibilities of on-line LC and MS started in the early 1970's with several different approaches towards interfacing.

Obviously, the most straightforward way of coupling LC and MS would be inserting the column outlet directly into the ion source in the MS vacuum system. This approach, the capillary inlet interface, was theoretically and experimentally explored by Tal'roze *et al.* in 1972 [22]. It was subsequently investigated and applied by various other groups [23-28]. The capillary inlet interface is briefly discussed in Ch. 4.2.

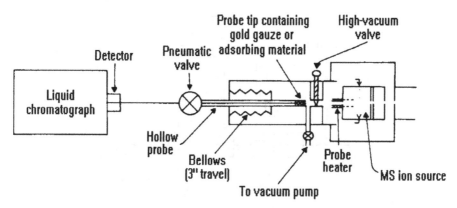

Figure 3.2: Stopped-flow LC–MS system (from ref 34. ©1973, American Chemical Society).

In order to avoid precipitation of non-volatile analytes at the inner wall of the tube, tapering of the outlet was investigated [29-31]. The difficulty with this approach is that tapering is art rather than science. The latter problem was solved by the introduction in 1980 of a small diaphragm as solvent restriction [32], leading to the direct liquid introduction interface (Ch. 4.5).

In 1972, Lovins *et al.* [33-34] described a stopped-flow LC–MS system (Figure 3.2). A fraction of the column effluent is collected in a tube, after which the flow through the column is stopped. The LC fraction is sucked by the MS vacuum out of the reservoir through the hollow probe towards the slightly heated (70–125 °C) probe-tip made of gold netting or an adsorbent. The solvent is evaporated while the analyte is concentrated on the probe tip. When sufficiently low pressure is reached, a high vacuum valve is opened and the probe is moved to the ion source. The analyte is evaporated from the tip and mass analysed. After completing the mass analysis the probe tip is removed from the high-vacuum ion source, the flow through the column is restarted, and the chromatographic elution is continued. The complete cycle time is 3–5 min. The system was applied to the identification of sulfa-drugs, pesticides, and polycyclic aromatic hydrocarbons [34].

A modified Pye Unicam moving-wire detector was described by Scott *et al.* [35] in 1974 to fit the vacuum requirements of a mass spectrometer (Figure 3.3). Part of the column effluent is deposited on to a wire, which transports the liquid along a heating element to evaporate the solvents, and through a series of vacuum locks to the ion source where the analyte is thermally desorbed from the wire prior to the ionization. Ionization is independent of the LC system. Therefore, conventional EI and CI spectra can be obtained [35]. This approach was subsequently adapted in 1976 by MacFadden [36] into the moving-belt interface (Ch.4.4).

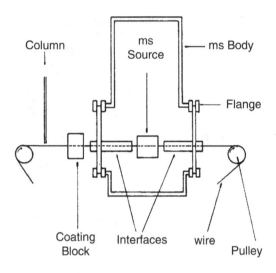

Figure 3.3: Moving-wire interface for LC–MS.

The fourth approach, based on APCI, was upon its introduction in 1974 far ahead of its time. APCI is a solvent-mediated chemical ionization method, initiated by electrons from a ^{63}Ni foil or a corona discharge electrode. For LC–MS, this approach was pioneered by the group of Horning [37-40] using a concentric pneumatic nebulizer. They demonstrated detection limits in the 500-pg range for anthracene using normal-phase LC coupled to APCI–MS. While this was the most promising of the early approaches, it lasted until the early 1990's before the actual breakthrough in LC–APCI–MS was achieved (see Ch. 5.3). This was primarily due to the lack of commercially available instrumentation for APCI.

A number of approaches directed at electrospraying solutions for MS sample introduction were investigated. These studies were not primarily directed at LC–MS interfacing, but considering the importance of ESI in current LC–MS developments these research efforts are quoted here as well. Between 1968 and 1971, Dole *et al.* [41-43] investigated the possibility of producing ions of intact high molecular-mass compounds by electrospraying protein solutions in an atmospheric-pressure region and analysing the ions using a ion-drift spectrometer. Between 1974 and 1978, Evans *et al.* [44-46] investigated the applicability of electrospraying solutions of saccharides, nucleosides, and small peptides in a vacuum chamber. Low vapour-pressure solvents like glycerol were used in this electrohydrodynamic ionization approach. Similar experiments were also reported by Zolotai *et al.* [47-48] in 1980 using both glycerol and aqueous salt solutions. The method is called 'field evaporation of ions from liquid solution'. These techniques hold closely with the theoretical concepts and subsequent experimental evidence for ion evaporation, introduced by Iribarne and Thomson [49-52] in 1976. These investigations find extensive application in both thermospray (TSP) and ESI interfacing (see Ch. 3.3.3).

2.2   Exploration of other strategies

While some of the early approaches were subsequently developed into commercially available LC–MS interfaces (Ch. 3.2.3), a wide variety of other approaches were investigated.

The use of semi-permeable silicone rubber membranes for on-line analyte-enrichment in LC–MS was described in 1974 by Westover et al. [53]. Polycyclic aromatic hydrocarbons were analysed by MS after evaporation of the reversed-phase LC column effluent and passage of the vapour through a membrane-based solvent-separator. This system, designed by Jones et al. [54], is limited to the analysis of non-polar analytes which are sufficiently volatile at 250°C, the maximum operating temperature of the membrane.

Interfacing micro-LC and MS via a capillary inlet interface connected to a GC–MS jet separator was described in 1978 by Takeuchi et al. [55]. In the period until 1982, this system was subsequently developed towards a vacuum nebulizer, in which the column effluent is pneumatically nebulized into a modified jet-separator type of device [56-57]. The instrumental developments of the vacuum nebulizer interfaces are discussed in Ch. 4.3. Pneumatic nebulization of column effluents directly into the CI source was described by a number of groups [58-61]. A so-called 'helium interface' for the introduction of organic solvents was described by Apffel et al. [59]. It was primarily applied to the analysis of pesticides in aqueous samples. Although the system was commercially available, it did not find wide application.

A series of highly complex interfaces were described between 1978 and 1980 by Blakley et al. [62-64] in an attempt to develop a system capable of the introduction of up to 1 ml/min aqueous mobile phase into an EI mass spectrometer. The systems contained extensive multistage vacuum systems and ingenious heating devices, e.g., laser evaporation or hydrogen-flame heaters, to achieve rapid solvent evaporation. These systems subsequently developed towards the TSP interface (Ch. 4.7).

A fractional sampling interface for combining LC with a plasma desorption time-of-flight (TOF) MS was described in 1982 by Jungclas et al. [65-67]. The column effluent is deposited in fractions by thermal nebulization on to one of the twelve 2-µm thick target foils which are placed on a sample carrousel. After collecting a particular fraction, the sample carrousel is rotated to collect the next fraction on the target. In this way, the 12 fractions of the column effluent can be collected and subsequently analysed by plasma desorption TOF-MS (Figure 3.4). The system was applied to the (quantitative) analysis of compounds of pharmaceutical interest, e.g., the antiarrhythmic drugs quinidine and verapamil, the anti-neoplastic agents etoposide and teniposide, and some Vinca rosea alkaloids [65-67].

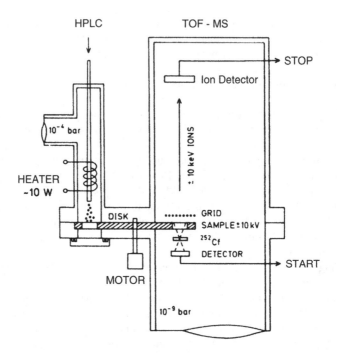

Figure 3.4: Schematic diagram of the disk interface for coupling LC and plasma-desorption mass spectrometry (from ref. 67, ©1983, Elsevier Science).

A mechanized off-line fraction collection from a microbore LC column on an indented sample holder, which is subsequently analysed by pulsed laser desorption MS, was described in 1983 by Huber *et al.* [68].

Between 1981 and 1983, Christensen *et al.* [69-70] developed a system in which preconcentration of the liquid stream is achieved by evaporation of part of the solvent as it flows down a resistively-heated wire. The concentrate is introduced into the MS by means of a special direct-liquid introduction (DLI) type interface. A schematic diagram of the continuous effluent preconcentrator is shown in Figure 3.5. The DLI interface probe consisted of a needle-valve type of restriction. The stem of the valve is a 0.10-mm tungsten wire grounded to a pencil-point, fitting in a 1-mm-OD and 0.12-mm-ID nickel tube, the end of which is peened to *ca.* 0.01 mm ID. After assembly a *ca.* 0.025-mm-ID circular orifice is formed in the seat. Precise positioning of the stem is important as it influences the spray quality. Any clogging of the orifice can be cleared by rapidly moving the stem in and out. The interface performs well with organic solvents, but spraying aqueous solvents is difficult. Conductive and radiative heating did not solve the problem of freezing of the solvent at the orifice, nor did the concentric introduction of a nebulizing gas. More successful is the appliance of ultrasonic vibrations, produced by using the nickel tube as a magnetostrictive oscillator. Additional heating of the interface tip from the ion source appeared to be necessary, especially with solvents with >80% water. The system, which was commercially available, was applied to the analysis of phenols [69], polycyclic aromatic hydrocarbons [69], aliphatic acids in shale oil process water [70], and the anti-convulsant valproic acid in human serum [70].

A so-called 'monodisperse aerosol generating interface' (MAGIC) was described in 1984 by Willoughby and Browner [71]. It consists of a cross-flow nebulizer (the monodisperse aerosol generator), a near atmospheric-pressure desolvation chamber, and a momentum separator. The desolvation chamber was kept at near atmospheric pressure because more efficient heat transfer to the droplets and thus more efficient droplet desolvation can be achieved at higher pressures. After desolvation of the droplets the resulting mixture of vapour and coagulated analyte molecules are expanded into a vacuum region. The resulting molecular beam of analyte particles is transported with a minimum of solvent vapours through a momentum separator, where the more volatile and lower mass compounds are pumped away, to the ion source. The particles are evaporated prior to the ionization by hitting a heated surface in the source. The interface is capable of an efficient analyte-enrichment. This system is the precursor of the particle-beam interface (PBI, Ch. 3.3.2 and 4.8).

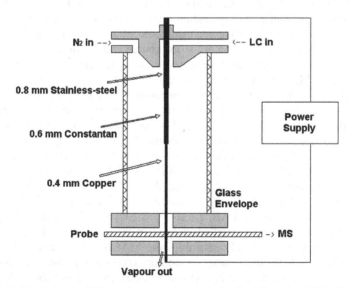

Figure 3.5: Continuous effluent preconcentrator wire interface for LC–MS

2.3   The breakthrough: commercial LC–MS interfaces

A number of the systems described above were adopted by instrument manufacturers and subsequently developed into commercial products, some of which found wide application.

The first commercial LC–MS interface, available in 1977, was the moving-belt interface, which was a modification by MacFadden *et al.* [36] of the moving-wire system described by Scott *et al.* [35]. The moving-belt interface, discussed in Ch. 4.4, was capable of introducing up to 1 ml/min of mobile phase and achieving solvent-independent analyte ionization by EI or CI. A similar system was described by Millington *et al.* [72].

The second commercial LC–MS interface, available in 1980, was based on a modification of restricted capillary inlet interfaces [29-31]. Melera [32] demonstrated that by using a diaphragm pinhole a stable and reliable system for DLI could be achieved. The design and performance of the resulting DLI interface was extensively studied and characterized by Arpino and coworkers [73-77]. The system is capable of introducing up to 50 μl/min of mobile phase. Analyte ionization is achieved in solvent-mediated CI. The DLI interface is discussed in Ch. 4.5.

Although the moving-belt interface and DLI found already some users outside the research laboratories, the actual breakthrough in the general acceptance of LC–MS as a powerful analytical technique was achieved by the introduction in 1983 of the TSP interface. The research efforts of the group of Vestal [62-64] finally led to a rather simple and easy to use interface device. This breakthrough resulted from the discovery in 1980 of a new ionization method [78] and the realization in 1983 that both the pumping and heating problem could be solved in much simpler ways, *i.e.*, by direct mechanical pumping at the ion source and by direct-electrical heating of the inlet capillary [79]. The developments, characteristics, and use of the TSP interface for LC–MS are discussed in more detail in Ch. 4.7.

Two approaches based on fast-atom bombardment (FAB) and introduced almost simultaneously were soon after their first description in 1985 and 1986 commercialized, *i.e.*, the frit-FAB [80] and the continuous-flow Cf-FAB [81]. Both systems are used to introduce part of the column effluent (typically 1-10 μl/min) into a FAB source. In the frit-FAB system, a capillary transfers the effluent to a stainless-steel or PFTE frit used as a FAB target, while in Cf-FAB system the effluent flows in a thin uniform film over the FAB target. A suitable FAB matrix, *e.g.*, glycerol, should be added to the mobile phase. Analyte molecules are directly desorbed and ionized from the liquid film by FAB. These approaches are discussed in Ch. 4.6.

From 1988, several commercial adaptations of the MAGIC [71] were described (Ch. 4.8). The PBI most closely resembles the MAGIC. It contains a more user-friendly and robust momentum separator and the cross-flow pneumatic nebulizer is replaced by a concentric pneumatic nebulizer [82-83]. In the thermabeam interface a TSP nebulizer is used [84]. The universal interface features the use of a TSP nebulizer and a countercurrent gas-diffusion membrane separator between the desolvation chamber and the momentum separator [85].

The research efforts in the late 1960's of the group of Dole (41-43) on electrospray sample introduction found continuation in the work of the group of Fenn (86-87) in 1984 and later. An LC–MS interface based on ESI introduction into an atmospheric-pressure ion (API) source was described by Whitehouse *et al.* [88] in 1985. The flow-rate limitations of the latter system were to some extent removed by the introduction of a pneumatically-assisted ESI interface (ionspray®) for LC–MS by Bruins *et al.* (89) in 1987. This system was developed for a Sciex API instrument which in those days was the only commercially available instrument equipped with an API source. A major breakthrough in ESI, and as a result of this also in the commercial availability of API instruments, was achieved in 1988 by the observation of multiple-charge ions from peptides and proteins [90-91]. This made the ESI interface to one of the most popular and powerful methods for LC–MS. The development of API interfacing for LC–MS is discussed in detail in Ch. 5.

Following the early research efforts in the mid 1970's of the group of Horning [37-40], the potential use of APCI in LC–MS continued to be investigated. The application of LC–APCI–MS using a DLI-type of device was described by Henion *et al.* [92] in 1982. Subsequently, promising results on LC–APCI–MS in qualitative and rapid quantitative analysis of phenylbutazone and its metabolites were described by Covey *et al.* [93] in 1986. The interface used in this study is known as the heated nebulizer. It consists of a concentric pneumatic nebulizer fitted in a large diameter heated quartz tube. The heated solvent and analyte vapour mixture from the tube is introduced in the API source, where APCI takes place, initiated by a corona discharge electrode. Most current APCI interfaces, discussed in Ch. 5.6, are based on this design, although a system in which TSP nebulization is used instead of the heated pneumatic nebulizer has also been described and used [94-95].

2.4 Further explorations

Currently, API based LC–MS interfaces, *i.e.*, ESI and APCI, are the most widely applied approaches, while other interfaces like TSP and Cf-FAB can be considered obsolete. Despite the successes of these commercially available interfaces, research towards newer and/or advanced interface strategies continues. These research efforts comprise among others the implementation of on-line LC–MS using matrix-assisted laser desorption/ionization (Ch. 5.9), the sonic spray (Ch. 5.7.1), and the laser spray (Ch. 5.7.2) interface.

3. Strategies in LC–MS interfacing

In developing on-line LC–MS, three fundamental compatibility problems had to be solved, *viz.*, the amount of solvent eluting from the LC column, the composition of the LC mobile phase, and the nature of the analytes. Interfaces have been developed to be placed between the LC column outlet and the ion source or the mass analyser in order to solve the incompatibility between the LC solvent flow and the MS high vacuum. The various successful approaches were briefly indicated in the previous section. In the research and development efforts described there, three development lines can be distinguished, concerning:

- efforts to introduce 1 ml/min of an aqueous mobile phase into the MS.
- efforts to achieve an enrichment of the analyte over the mobile-phase constituents in the interface.
- efforts to develop liquid-based ionization strategies.

In briefly discussing and exploring these lines of development, similarities between different LC–MS interfaces are recognized and the possibilities and limitations of the various LC–MS interfaces become clearer and can more readily be explained [12].

3.1   Introduction of 1 ml/min of an aqueous solvent

Research efforts to achieve the introduction of 1 ml/min of an aqueous mobile phase into the MS have to cope with two major problems:

- the need to modify the MS vacuum system in order to cope with the high gas loads associated with the introduction of 1 ml/min.
- the need to transfer of sufficient heat to achieve (complete) evaporation of the solvents.

The most obvious way of coupling LC and MS is by inserting the column outlet capillary directly into the MS vacuum system. With this capillary inlet interface (Ch. 4.2) ca. 10 µl/min of not-too-aqueous solvents can be introduced. The limitations in the applicability of this interface are due to the fact that under practical conditions the liquid-vapour interface is always located inside the transfer capillary. Upon evaporation of the relatively volatile mobile-phase constituents the less volatile analytes will precipitate at the inner wall of the tube. By introducing a restriction, e.g., tapering, at the outlet end of the transfer capillary, the position of the liquid-vapour interfaces can be shifted towards the outlet end. In practice, the most successful and versatile restriction in the transfer capillary outlet is a diaphragm with a small pinhole. This approach actually results in the nebulization of the column effluent into the vacuum of the mass spectrometer. Although this DLI interface (Ch. 4.5) can be used routinely, it suffers from two major drawbacks. The small pinholes, typically 3-5 µm ID, tend to clog easily. In obviating this problem by the use of larger pinholes or capillary restrictions, the second drawback became evident. The mobile phase is nebulized into a medium-pressure heated desolvation chamber. The heat transfer from the chamber walls to the solvent droplets is rather inefficient, which yield significant problems with larger amounts of liquid, i.e., when the amount of solvent introduced via the larger restriction increases. As a result, the amount of liquid that can be introduced in DLI is limited to *ca.* 50 µl/min. This limitation is not only valid for DLI, but for all other interface approaches in which nebulization of the column effluent into a medium-pressure desolvation region is applied like in vacuum nebulizers and the helium interface (*cf.* Ch. 4.3). There are two ways to solve these heat transfer problems:

* nebulization in a heated atmospheric-pressure region.
* preheating the liquid prior to nebulization.

Both solutions lead in turn to other problems, *i.e.*, with the transfer of ions or analytes from the atmospheric-pressure desolvation region into the MS high vacuum in the former approach, and with the need for a substantial increase of the pumping capacity at the ion source housing and the risk of thermal degradation of analytes in the preheated liquid in the latter approach.

Nebulization of column effluents as large as 1 ml/min into an atmospheric-pressure chamber connected by means of a restrictive pinhole or capillary to the MS vacuum system requires modification of the MS vacuum system. Aspects related to MS vacuum system for on-line LC–MS are treated in Ch. 5.2.

In a TSP interface (Ch. 4.7), heat is transferred directly from the electrically-heated vaporizer capillary to the liquid. The liquid starts to evaporate inside the vaporizer, resulting in a disruption of the liquid due to the formation of vapour bubbles that rapidly expand owing to the high temperature. In this way, the possibilities for analyte precipitation are greatly reduced because the liquid is efficiently kept away from the hot surface after the onset of the evaporation process. By appliance of a relatively high linear velocity of the liquid in the vaporizer capillary and a high heating power, the residence time of the analyte molecules in the heated vaporizer is relatively short, thereby minimizing the risk of thermal degradation. Although the amount of heat transferred to the liquid is sufficient for complete evaporation, part of the liquid emerges from the vaporizer as small droplets. The presence of microdroplets in the spray plume is considered vital in the transfer of labile compounds (Ch. 3.3.3).

While in the TSP interface the heated solvent is nebulized into a medium-pressure ion source region, other systems have been described in which nebulization into an atmospheric-pressure system is performed. In an atmospheric-pressure spray system, as described by Sakairi and Kambara [94-95], a TSP nebulizer is used for the efficient introduction and evaporation of a mobile phase into an APCI source. In other systems, a heated nebulizer is used to achieve sample introduction in an APCI source.

3.2 Analyte enrichment in interfacing

The LC–MS setup initially pursued closely resembled the successful GC–MS combination. Interfaces were developed to remove (most of) the mobile phase prior to the introduction of the analytes into the mass spectrometer. Early examples of this approach are the stopped-flow system, described by Lovins *et al.* [33-34], the moving-wire [35] and moving-belt interfaces [36], the membrane-based solvent separator [53-54], the effluent fractionation systems [65-68], and the continuous effluent preconcentrator-wire interface [69-70].

An evaluation of the applications of the moving-belt interface, which was the most widely used LC–MS interface based on analyte enrichment, demonstrates both the strength and the weakness of this approach. The advantage of complete analyte enrichment, *i.e.*, solvent removal prior to MS introduction, is a free choice of ionization method. Thus, EI mass spectra can be acquired. The disadvantage is that the analytes are transferred in solid state, *i.e.*, on a solid support like the belt, requiring vaporization prior to the ionization. Therefore, the analytes should have sufficient vapour pressure and thermal stability.

The drawback can be removed by performing gas-phase analyte enrichment instead of transfer via a solid support. Gas-phase analyte enrichment can be achieved by means of molecular-beam technologies, *i.e.*, in the PBI as well as in API interfaces. The enrichment is performed by nebulization of the column effluent in an atmospheric-pressure chamber, expansion of the vapour-droplet mixture into a low-pressure region via a nozzle, and subsequent sampling of the high-mass core portion of the expansion jet by means of a skimmer into a lower-pressure region. The nozzle-skimmer system essentially performs a momentum separation: preferentially sampling the high-mass analytes and removing the low-mass solvent and gas molecules. It is important to appreciate the similarities in PBI and API interface designs in this respect.

The important difference between PBI and API interfaces is that in the former ionization is performed at the low-pressure side of the interface, *i.e.*, by means of EI or CI after evaporative collisions of the analyte particles against the heated ion source walls. In API interfaces, the ionization is achieved in the high-pressure region and the ions generated are sampled into the MS.

3.3   Solvent-based ionization strategies

Perhaps the most fundamental progress in the LC–MS field is in the introduction of liquid-based ionization strategies. During the development of LC–MS interfaces, it was realized that the LC mobile phase was not necessarily disturbing the MS analysis, but could also be of help in the ionization or at least in preparing the analytes for the ionization. This was strongly advocated by Arpino and Guiochon in a paper with the subtitle "*Why the solvent should not be removed in LC–MS interfacing methods*" [75]. In DLI, the soft CI of highly labile analytes can be attributed to a desorption-CI effect: the analyte molecules, that are preferentially contained in the desolvating droplets, are transported to the ion plasma in the CI source [23-24, 96]. Contained in the desolvating droplet, the labile analyte molecule is gently and smoothly transferred from the liquid phase to the gas phase and subsequently ionized in solvent-mediated CI. This analyte transfer process is effective not only in DLI but also in the other nebulization interfaces, *i.e.*, pneumatic nebulization, TSP, ESI, heated nebulizer, and even particle-beam.

A second and even more important breakthrough in liquid-based ionization strategies was the observation of analyte ionization without a primary ionization source, *i.e.*, without filament, in the TSP interface [78]. The ionization mechanism is assumed to be the result of ion evaporation, as first described by Iribarne and Thomson [49-52]. The nebulization results in charged droplets. As a result of the evaporation of the neutral solvent molecules from the charged droplets, the charge-size ratio at the droplet surface increases. The increasing electrical field results in the emission of solvated analyte ions that are said to evaporate from the charged droplets. This mechanism is especially operative with ionic analytes; either ionic itself, such as quaternary ammonium compounds, or preformed ions in solution, such as acids and bases. These basic features of the ionization mechanism appear to be effective with other nebulization interface strategies such as TSP and ESI. The mechanisms are discussed in Ch. 6.3.

The observation of multiple-charge ions from proteins, first in ESI [90-91] and later also in TSP [97-98], can be considered as the third breakthrough in the development of liquid-based ionization strategies. The latter feature to a large extent explains the importance of ESI interfaces, which initially were not optimally appropriate to couple conventional LC columns to the MS. The possibility to achieve new soft ionization methods is considered to be more important than a splitless LC–MS coupling. The latter is certainly true for the Cf-FAB and frit-FAB interfaces, described by Ito *et al.* [80] and Caprioli *et al.* [81], respectively. It has been demonstrated that in continuous-flow systems the glycerol background and ion suppression effects are greatly reduced compared to static FAB (Ch. 4.6).

4. Conclusion

The role of the LC mobile phase has changed in time: from an active carrier in the LC process, which prior to MS should be removed as thoroughly as possible, via a transfer medium for nonvolatile and/or thermally labile analytes from the liquid to the gas phase, to a constituent essential in analyte ionization. Nevertheless, the LC mobile phase continues to put high demands and restrictions on the instrumentation in terms of vacuum (Ch. 5.2) and solvent compatibility (Ch. 6.6.3) problems. Gas-phase analyte-enrichment devices are nowadays routinely used to handle the vacuum problems. New ionization techniques like TSP, and ESI have solved the original problems related to analyte ionization. However, despite the introduction of some ingenious technological solutions, the fundamental incompatibility problems related to the composition of the LC mobile phase are not solved. The routine and prolonged use of nonvolatile mobile-phase additives like phosphate buffers and ion-pairing agents continues to be prohibited in LC–MS (see Ch. 6.6.3).

5. References

1. P.J. Arpino, *On-line LC–MS? An odd couple!*, Trends Anal. Chem., 1 (1982) 154.
2. P.J. Arpino, G. Guiochon, *LC–MS coupling*, Anal. Chem., 51 (1979) 682A.
3. W.H. McFadden, *LC–MS. Systems and applications*, J. Chromatogr. Sci., 18 (1980) 97.
4. D.E. Games, *Combined LC–MS*, Biomed. Mass Spectrom., 8 (1981) 454.
5. G. Guiochon, P.J. Arpino, *How to interface a LC column to a MS?*, J. Chromatogr., 271 (1983) 13.
6. J.D. Henion, *Micro LC–MS Coupling*, in: P. Kucera (Ed.), *Microcolumn High-Performance Liquid Chromatography*, 1984, Elsevier, Amsterdam, p. 260.
7. P.J. Arpino, *Ten years of LC–MS*, J. Chromatogr., 323 (1985) 3.
8. B.L. Karger, P. Vouros, *A chromatographic perspective of LC–MS*, J. Chromatogr., 323 (1985) 13.
9. A.P. Bruins, *Developments in interfacing microbore LC with MS*, J. Chromatogr., 323 (1985) 99.
10. T.R. Covey, E.D. Lee, A.P. Bruins, J.D. Henion, *LC–MS*, Anal. Chem., 58 (1986) 1451A.
11. K.B. Tomer, C.E. Parker, *Biochemical applications of LC–MS*, J. Chromatogr., 492 (1989) 189.
12. W.M.A. Niessen, U.R. Tjaden, J. van der Greef, *Strategies in developing interfaces for coupling LC–MS*, J. Chromatogr., 554 (1991) 3.
13. K.B. Tomer, M.A. Moseley, L.J. Deterding, C.E. Parker, *Capillary LC–MS*, Mass Spectrom. Rev., 13 (1994) 431.
14. W.M.A. Niessen, *Advances in instrumentation in LC–MS and related liquid-introduction techniques*, J. Chromatogr. A, 794 (1998) 407.
15. W.M.A. Niessen, *State-of-the-art in LC–MS*, J. Chromatogr. A, 856 (1999) 179.
16. B.A. Thomson, *API and LC–MS, together at last*, J. Am. Soc. Mass Spectrom., 9 (1998) 187.
17. A.L. Yergey, C.G. Edmonds, I.A.S. Lewis, M.L. Vestal, *Liquid Chromatography–Mass Spectrometry, Techniques and Applications*, 1989, Plenum Press, New York, NY.
18. M.A. Brown (Ed.), *Liquid Chromatography–Mass Spectrometry, Applications in Agricultural, Pharmaceutical, and Environmental Chemistry*, 1990, ACS Symposium Series, Vol 420, Washington, DC.
19. D. Barceló (Ed.), *Applications of LC–MS in Environmental Chemistry*, 1996, Elsevier Science, Amsterdam.
20. R.B. Cole (Ed.), *Electrospray Ionization Mass Spectrometry*, 1997, Wiley& Sons Ltd., Chichester.
21. B.N. Pramanik, A.K. Ganguly, M.L. Gross (Eds.), *Applied electrospray mass spectrometry*, 2002, Marcel Dekker Inc., New York.
22. V.L. Tal'roze, V.E. Skurat, I.G. Gorodetskii, N.B. Zolotai, Russ. J. Phys. Chem., 46 (1972) 456.
23. M.A. Baldwin, F.W. McLafferty, *LC–MS interface. I. Direct introduction of liquid solutions into a CI-MS*, Org. Mass Spectrom., 7 (1973) 1111.
24. M.A. Baldwin, F.W. McLafferty, *Direct CI of relatively involatile samples. Application to underivatized oligopeptides*, Org. Mass Spectrom., 7 (1973) 1353.
25. J.D. Henion, *Continuous monitoring of total micro LC eluant by DLI-LC–MS*, J. Chromatogr. Sci., 19 (1981) 57.

26. A.P. Bruins, B.F.H. Drenth, *Experiments with the combination of a micro-LC and a CI quadrupole MS, using a capillary interface for DLI. Some theoretical considerations concerning the evaporation of liquids from capillaries into vacuum*, J. Chromatogr., 271 (1983) 71.

27. H. Alborn, G. Stenhagen, *Direct coupling of packed fused-silica LC columns to a magnetic sector MS and application to polar thermolabile compounds*, J. Chromatogr., 323 (1985) 47.

28. W.M.A. Niessen, H. Poppe, *Open-tubular LC–MS with a capillary-inlet interface*, J. Chromatogr., 385 (1987) 1.

29. P.J. Arpino, M.A. Baldwin, F.W. McLafferty, *LC–MS systems providing continuous monitoring with nanogram sensitivity*, Biomed. Mass Spectrom., 1 (1974) 80.

30. R. Tijssen, J.P.A. Bleumer, A.L.C. Smit, M.E. van Kreveld, *Microcapillary LC in open-tubular columns with diameters of 10–50 μm. Potential application to CI-MS detection*, J. Chromatogr., 218 (1981) 137.

31. J.S.M. de Wit, C.E. Parker, K.B. Tomer, J.W. Jorgenson, *Direct coupling of open-tubular LC with MS*, Anal. Chem., 59 (1987) 2400.

32. A. Melera, *Design, operation and applications of a novel LC–MS CI interface*, Adv. Mass Spectrom., 8B (1980) 1597.

33. R.E. Lovins, J. Craig, F. Thomas, C.R. McKinney, *Quantitative protein sequencing using MS: A protein-sequenator–MS interface employing flash evaporative techniques*, Anal. Biochem., 47 (1972) 539.

34. R.E. Lovins, S.R. Ellis, G.D. Tolbert, C.R. McKinney, *LC–MS. Coupling of a LC to a MS*, Anal. Chem., 45 (1973) 1553.

35. R.P.W. Scott, C.G. Scott, M. Munroe, J. Hess, *Interface for on-line LC–MS analysis*, J. Chromatogr., 99 (1974) 395.

36. W.H. McFadden, H.L. Schwartz, S. Evans, *Direct analysis of LC effluents*, J. Chromatogr., 122 (1976) 389.

37. E.C. Horning, D.I. Carroll, I. Dzidic, K.D. Haegele, M.G. Horning, R.N. Stillwell, *LC–MS–computer analytical systems. A continuous-flow system based on API-MS*, J. Chromatogr., 99 (1974) 13.

38. E.C. Horning, D.I. Carroll, I. Dzidic, K.D. Haegele, M.G. Horning, R.N. Stillwell, *API-MS. Solvent-mediated ionization of samples introduced in solution and in a LC effluent stream*, J. Chromatogr. Sci., 12 (1974) 725.

39. D.I. Carroll, I. Dzidic, R.N. Stillwell, K.D. Haegele, E.C. Horning, *Atmospheric pressure ionization mass spectrometry: Corona-discharge ion source for use in liquid chromatograph–mass spectrometer–computer analytical system*, Anal. Chem., 47 (1975) 2369.

40. E.C. Horning, D.I. Carroll, I. Dzidic, S.N. Lin, R.N. Stillwell, J.P. Thenot, *API-MS. Studies of negative-ion formation for detection and quantification purposes*, J. Chromatogr., 142 (1977) 481.

41. M. Dole, R.L. Hines, L.L. Mack, R.C. Mobley, L.D. Ferguson, M.B. Alice, *Molecular beams of macroions*, J. Chem. Phys., 49 (1968) 2240.

42. L.L. Mack, P. Kralik, A. Rheude, M. Dole, *Molecular beams of macroions.II*, J. Chem. Phys., 52 (1970) 4977.

43. G.A. Clegg, M. Dole, *Molecular beams of macroions. III, Zein and polyvinylpyrrolidone*, Biopolymers, 10 (1971) 821.

44. D.S. Simons, B.N. Colby, C.A. Evans, Jr., *Electrohydrodynamic ionization MS – the*

ionization of liquid glycerol and non-volatile organic solutes, Int. J. Mass Spectrom. Ion Phys., 15 (1974) 291.

45. B.P. Stimpson, C.A. Evans, Jr., *Electrohydrodynamic ionization MS of biochemical materials*, Biomed. Mass Spectrom., 5 (1978) 52.

46. B.P. Stimpson, D.S. Simons, C.A. Evans, Jr., *Mass spectrometry of solvated ions generated directly from the liquid phase by electrohydrodynamic ionization*, J. Phys. Chem., 82 (1978) 660.

47. N.B. Zolotai, G.V. Karpov, V.L. Tal'roze, V.E. Skurat, G.I. Ramendik, Yu.V. Basyuta, *MS of the field evaporation of ions from liquid solutions in glycerol*, J. Anal. Chem. USSR, 35 (1980) 937.

48. N.B. Zolotai, G.V. Karpov, V.L. Tal'roze, V.E. Skurat, Yu.V. Basyuta, G.I. Ramendik, *MS of the field evaporation of ions from water and aqueous solutions. aqueous sodium iodide and saccharose solutions*, J. Anal. Chem. USSR, 35 (1980) 1161.

49. J.V. Iribarne, B.A. Thomson, *On the evaporation of small ions from charged droplets*, J. Chem. Phys., 64 (1976) 2287.

50. B.A. Thomson and J.V. Iribarne, *Field-induced ion evaporation from liquid surfaces at atmospheric pressure*, J. Chem. Phys., 71 (1979) 4451.

51. B.A. Thomson, J.V. Iribarne, P.J. Dziedzic, *Liquid ion evaporation–MS–MS for the detection of polar and labile molecules*, Anal. Chem., 54 (1982) 2219.

52. J.V. Iribarne, P.J. Dziedzic, B.A. Thomson, *Atmospheric-pressure ion evaporation–MS*, Int. J. Mass Spectrom. Ion Phys., 50 (1983) 331.

53. L.B. Westover, J.C. Tou, J.H. Mark, *Novel MS sampling device – Hollow fiber probe*, Anal. Chem., 46 (1974) 568.

54. P.R. Jones, S.K. Yang, *LC–MS interface*, Anal. Chem., 47 (1975) 1000.

55. T. Takeuchi, Y. Hirata, Y. Okumura, *On-line coupling of a micro LC and MS through a jet separator*, Anal. Chem., 50 (1978) 659.

56. S. Tsuge, Y. Hirata, T. Takeuchi, *Vacuum nebulizing interface for direct coupling of micro-LC and MS*, Anal. Chem., 51 (1979) 166.

57. S. Tsuge, *New approaches to interfacing LC and MS*, in: M.V. Novotny, D. Ishii (Ed.), *Microcolumn separations*, 1985, Elsevier, Amsterdam, p. 217.

58. K. Matsumoto, H. Kojima, K. Yasuda, S. Tsuge, *CI-MS using a glow-discharge ion source combined with a nebulizer sampling system*, Org. Mass Spectrom., 20 (1985) 243.

59. J.A. Apffel, U.A.Th Brinkman, R.W. Frei, E.I.A.M. Evers, *Gas-nebulized DLI interface for LC–MS*, Anal. Chem., 55 (1983) 2280.

60. A.L.C. Smit, R. Tijssen, J.F. Lambrechts, *A universal DLI LC–MS interface*, 13th Meeting of the British Mass Spectrometry Society, University of Warwick, September 19-22, 1983, Extended abstracts, p. 45.

61. F.S. Pullen, D.S. Ashton, M.A. Baldwin, *Corona-discharge ionization LC–MS interface for target compound analysis*, J. Chromatogr., 474 (1989) 335.

62. C.R. Blakley, M.J. McAdams, M.L. Vestal, *Crossed-beam LC–MS combination*, J. Chromatogr., 158 (1978) 261.

63. C.R. Blakley, M.J. McAdams, M.L. Vestal, *A new LC–MS interface using crossed-beam techniques*, Adv. Mass Spectrom., 7 (1978) 1616.

64. C.R. Blakley, J.J. Carmody, M.L. Vestal, *LC–MS for analysis of nonvolatile samples*, Anal. Chem., 52 (1980) 1636.

65. H. Jungclas, H. Danigel, L. Schmidt, *Quantitative 252Cf plasma desorption MS for*

pharmaceuticals. A new approach to coupling LC with MS, Org. Mass Spectrom., 17 (1982) 86.

66. H. Jungclas, H. Danigel, L. Schmidt, J. Dellbrügge, *Combined LC–MS: An application of 252 Cf-PD-MS*, Org. Mass Spectrom., 17 (1982) 499.

67. H. Jungclas, H. Danigel, L. Schmidt, *Fractional sampling interface for combined LC–MS with ^{252}Cf fission fragment-induced ionization*, J. Chromatogr., 271 (1983) 35.

68. J.F.K. Huber, T. Dzido, F. Heresch, *Mechanized off-line combination of microbore LC and laser MS*, J. Chromatogr., 271 (1983) 27.

69. R.G. Christensen, H.S. Hertz, S. Meiselman, E. White, V, *LC–MS interface with continuous sample preconcentration*, Anal. Chem., 53 (1981) 171.

70. R.G. Christensen, E. White, V, S. Meiselman, H.S. Hertz, *Quantitative trace analysis by reversed-phase LC–MS*, J. Chromatogr., 271 (1983) 61.

71. R.C. Willoughby, R.F. Browner, *Monodisperse aerosol generation interface for combining LC–MS*, Anal. Chem., 56 (1984) 2625.

72. D.S. Millington, D.A. Yorke, P. Burns, *A new LC–MS interface*, Adv. Mass Spectrom., 8 (1980) 1819.

73. P.J. Arpino, G. Guiochon, P. Krien, G. Devant, *Optimization of the instrumental parameters of a combined LC–MS, coupled by an interface for DLI. I. Performance of the vacuum equipment*, J. Chromatogr., 185 (1979) 529.

74. P.J. Arpino, P. Krien, S. Vajta, G. Devant, *Optimization of the instrumental parameters of a combined LC–MS, coupled by an interface for DLI. II. Nebulization of liquids by diaphragms*, J. Chromatogr., 203 (1981) 117.

75. P.J. Arpino, G. Guiochon, *Optimization of the instrumental parameters of a combined LC–MS, coupled by an interface for DLI. III. Why the solvent should not be removed in LC–MS interfacing methods*, J. Chromatogr., 251 (1982) 153.

76. P.J. Arpino, J.P. Bounine, M. Dedieu, G. Guiochon, *Optimization of the instrumental parameters of a combined LC–MS, coupled by an interface for DLI. IV. A new desolvation chamber for droplet focusing or townsend discharge ionization*, J. Chromatogr., 271 (1983) 43.

77. P.J. Arpino, C. Beaugrand, *Design and construction of LC–MS interfaces utilizing fused-silica capillary tubes as vacuum nebulizers*, Int. J. Mass Spectrom. Ion Processes, 64 (1985) 275.

78. C.R. Blakley, J.J. Carmody, M.L. Vestal, *A new soft ionization technique for MS of complex molecules*, J. Am. Chem. Soc., 102 (1980) 5931.

79. C.R. Blakley, M.L. Vestal, *TSP interface for LC–MS*, Anal. Chem., 55 (1983) 750.

80. Y. Ito, T. Takeuchi, D. Ishii, M. Goto, *Direct coupling of micro-LC with FAB-MS*, J. Chromatogr., 346 (1985) 161.

81. R.M. Caprioli, T. Fan, J.S. Cottrell, *Continuous-flow sample probe for FAB-MS*, Anal. Chem., 58 (1986) 2949.

82. R.F. Browner, P.C. Winkler, D.D. Perkins, L.E. Abbey, *Aerosols as microsample introduction media for MS*, Microchem. J., 34 (1986) 15.

83. P.C. Winkler, D.D. Perkins, W.K. Williams, R.F. Browner, *Performance of an improved monodisperse aerosol generation interface for LC–MS*, Anal. Chem., 60 (1988) 489.

84. R.C. Willoughby, F. Poeppel, Proceedings of the 36th ASMS Conference on Mass Spectrometry and Allied Topics, May 24-29, 1987, Denver, CO, p. 289.

85. M.L. Vestal, D. Winn, C.H. Vestal, J.G. Wilkes, in: M.A. Brown, *Liquid*

Chromatography–Mass Spectrometry. Applications in Agricultural, Pharmaceutical and Environmental Chemistry, ACS Symposium Series, Vol. 420, 1990, American Chemical Society, Washington, DC, p. 215.

86. M. Yamashita, J.B. Fenn, *ESI ion source. Another variation of the free-jet theme*, J. Phys. Chem., 88 (1984) 4451.

87. M. Yamashita, J.B. Fenn, *Negative ion production with the ESI ion source*, J. Phys. Chem., 88 (1984) 4671.

88. C.M. Whitehouse, R.N. Dreyer, M. Yamashita, J.B. Fenn, *ESI interface for LC–MS*, Anal. Chem., 57 (1985) 675.

89. A.P. Bruins, T.R. Covey, J.D. Henion, *Ion spray interface for combined LC–API-MS*, Anal. Chem., 59 (1987) 2642.

90. C.K. Meng, M. Mann, J.B. Fenn, Proceedings of the 36th ASMS Conference on Mass Spectrometry and Allied Topics, June 5-10, 1988, San Francisco, CA, p. 771.

91. T.R. Covey, R.F. Bonner, B.I. Shushan, J.D. Henion, *The determination of protein, oligonucleotide and peptide molecular weights by ESI-MS*, Rapid Commun. Mass Spectrom., 2 (1988) 249.

92. J.D. Henion, B.A. Thomson, P.H. Dawson, *Determination of sulfa drugs in biological fluids by LC–MS–MS*, Anal. Chem., 54 (1982) 451.

93. T.R. Covey, E.D. Lee, J.D. Henion, *High-speed LC–MS–MS for the determination of drugs in biological samples*, Anal. Chem., 58 (1986) 2453.

94. M. Sakairi, M. Kambara, *Characteristics of a LC–API-MS*, Anal. Chem., 60 (1988) 774.

95. M. Sakairi M. Kambara, *Atmospheric-pressure spray ionization for LC–MS*, Anal. Chem., 61 (1989) 1159.

96. M Dedieu, C. Juin, P.J. Arpino, G. Guiochon, *Soft negative ionization of nonvolatile molecules by DLI of liquid solutions into a CI-MS*, Anal. Chem., 54 (1982) 2372.

97. K. Chan, D. Wintergrass, K. Straub, *Determination of the charge state of ions in TSP mass spectra*, Rapid Commun. Mass Spectrom., 4 (1990) 139.

98. K. Straub, K. Chan, *Molecular weight determination of proteins from multiply-charged ions using TSP-MS*, Rapid Commun. Mass Spectrom., 4 (1990) 267.

4

HISTORY OF LC–MS INTERFACES

1. Introduction

Over 30 years of liquid chromatography–mass spectrometry (LC–MS) research has resulted in a considerable number of different interfaces (Ch. 3.2). A variety of LC–MS interfaces have been proposed and built in the various research laboratories, and some of them have been adapted by instrument manufacturers and became commercially available. With the advent in the early 1990's of interfaces based on atmospheric-pressure ionization (API), most of these interfaces have become obsolete. However, in order to appreciate LC–MS, one cannot simply ignore these earlier developments. This chapter is devoted to the older LC–MS interfaces, which is certainly important in understanding the history and development of LC–MS. Attention is paid to principles, instrumentation, and application of the capillary inlet, pneumatic vacuum nebulizers, the moving-belt interface, direct liquid introduction, continuous-flow fast-atom bombardment interfaces, thermospray, and the particle-beam interface. More elaborate discussions on these interfaces can be found in previous editions of this book.

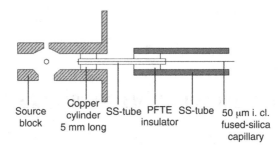

Figure 4.1: Schematic diagram of a capillary inlet interface. Redrawn from [7].

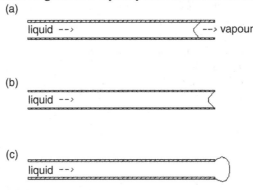

Figure 4.2: Competition between evaporation rate and liquid flow-rate.

2. Capillary inlet

The most obvious and therefore oldest way of coupling LC and MS is by means of a capillary between the LC column and the MS ion source. This approach, the capillary inlet interface, was pioneered and theoretically described by Tal'roze *et al.* [1-3] in the early seventies; a similar approach was used by the group of McLafferty [4-5]. A variety of laboratory-built capillary inlet interfaces have been described by several research groups [6-12]. A typical example is given in Figure 4.1 [7].

The theory of capillary inlet interfacing has been discussed by Tal'roze *et al.* [1-3] and others [7, 13-14]. The flow of a liquid through the capillary tube into the mass spectrometric vacuum system is the result of several counterbalancing effects: the capillary forces and the inlet pressure of the liquid which drives the liquid into the MS on one hand, and the vapour pressure of the liquid on the other. The flow-rate F_c (m³/s) of liquid entering the MS vacuum system can be calculated from:

$$F_c = \frac{\pi\, d_c^4}{128\, \eta_L\, L}\; [P_i - P_v + \frac{4\,\sigma}{d_c}]$$

where d_c is the internal diameter of the capillary, η_l the viscosity of the liquid, L the length of the capillary, P_i the inlet pressure, P_v the saturated vapour pressure of the liquid, and σ the surface tension of the liquid. Using this equation one can calculate that the flow-rate of water from a flask through a 1 m × 25 μm ID capillary at 50°C into a high-vacuum chamber will be ca. 1 nl/s. With such a small liquid flow-rate the pressure in the ion source is sufficiently low to obtain electron ionization (EI) spectra from the analytes, as is demonstrated in open-tubular LC–MS [11].

The practical value of this equation is limited because in most cases liquid mixtures are used, and because the temperature is not constant over the tube but is higher at the ion source side. Nevertheless, the equation is useful for gaining insight in the processes and the important experimental parameters.

The equation assumes that the evaporation of the liquid takes place at or near the end of the capillary. However, it can be calculated that the evaporation rate of water at 50°C from a 25-μm-ID tube is ca. 50 nl/s. Therefore, the evaporation does not take place at or near the end of the capillary, but somewhere inside the capillary. The competition between evaporation rate F_v and liquid flow-rate F_l is schematically depicted in Figure 4.2 [7]. The situation described above is marked (a) in Figure 4.2. By increasing the inlet pressure P_i it must be possible to go from the situation (a) to the ideal situation (b) or even to the situation (c). However, the resulting flow-rate will necessitate a larger pumping capacity of the vacuum system. The situation marked (c) does not result in stable ion source pressures, because the evaporation surface area is not constant.

In practical situations, the evaporation will always take place inside the capillary. This has consequences for the practical applicability. Nonvolatile impurities in the liquid stream precipitate at the position of the liquid-vapour interface and will ultimately block the inlet capillary. The most important group of nonvolatile components in the solvent stream is the analyte from the LC column. Therefore, the capillary inlet interface has a very narrow applicability range, limited to rather volatile analytes, i.e., rather nonpolar compounds with a molecular mass below ca. 400 Da [11], which are also readily amenable to GC–MS analysis.

3. Pneumatic nebulizer interfaces

In a pneumatic nebulizer, a high-speed gas flow is used to mechanically disrupt the liquid surface and to form small droplets which are subsequently dispersed by the gas to avoid droplet coagulation. Pneumatic nebulizers are widely used in various LC–MS interface strategies, especially coaxial nebulizers.

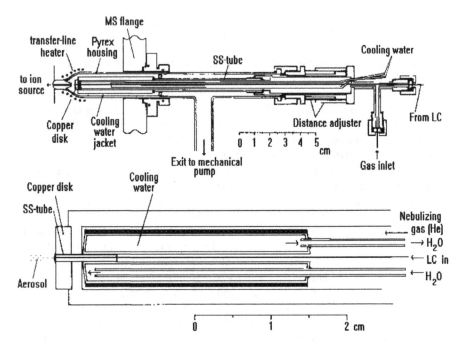

Figure 4.3: Vacuum nebulizer (Reprinted with permission from [18], ©1982, Springer-Verlag).

Pneumatic nebulizers can be used to nebulize the LC column effluent either in an atmospheric-pressure region or in a reduced-pressure region, *i.e.*, either directly into the ion source or into a reduced-pressure region separated from the ion source. The latter type is called a vacuum nebulizer.

The 'helium interface' described by Apffel *et al.* [15] was a commercially available pneumatic nebulizer, spraying directly into the reduced pressure ion source. No applications have been described other than those of the designers' group. It has been questioned by Bruins [16] whether, under the given experimental conditions with this interface, nebulization of the column effluent actually takes place. Considering the flow-rate (10-50 μl/min of preferentially organic solvents), the capillary diameter, and the temperature, Bruins [16] concludes that this interface most likely relies on complete evaporation of the column effluent while the sample vapour is subsequently swept into the source by the helium stream.

In a vacuum nebulizer the column effluent is nebulized into an evacuated chamber that is connected to the ion source by means of a heated tube. The design of these interfaces is based on jet separator interfaces used in packed column GC–MS. The development of these interfaces took place in the Japanese research group of

Tsuge and Yoshida [17-19]. In the first design, the heated transfer line between the jet and the ion source was rather long. Subsequently, the transfer line was shortened in later designs or replaced by a sampling orifice near the ion source. The later designs (Figure 4.3) were successfully applied to the LC–MS analysis of adenosine, amino acids, tripeptides, mono- and disaccharides [17-19]. Good results were obtained, although peaks due to thermal degradation are observed in some of the spectra reported. The vacuum nebulizers are designed for microbore LC–MS, thus applying flow-rates in the 10-50 µl/min range. Higher flow-rates cannot be introduced due to limitations in the heat transfer efficiency in the vacuum (cf. Ch. 3.3.1).

4. Moving-belt interface

In a moving-belt interface (MBI) [20], the column effluent is deposited onto an endless moving belt from which the solvent is evaporated by means of gentle heating and efficiently exhausting the solvent vapours. After removal of the solvents, the analyte molecules are (thermally) desorbed from the belt into the ion source and mass analysed. The MBI was reviewed in two papers [21-22].

The MBI was widely used in LC–MS applications between 1978 and 1990. The most important reasons for its success are the compatibility with a broad range of chromatographic conditions, while next to EI and chemical ionization (CI) other ionization methods, especially fast-atom bombardment (FAB) [23-24], can be employed as well.

A schematic diagram of a typical MBI for LC–MS is shown in Figure 4.4. It consists of an endless continuously moving belt which transports the analyte from the LC to the mass spectrometer while the mobile phase is removed via gentle heating and by evaporation under reduced pressure in two vacuum chambers. Desorption of the analyte is achieved by flash desorption at the tip of the belt which is positioned in the ion source. On the way back cleaning of the belt is performed by heating and washing.

MBI systems for LC–MS were commercially available from two instrument manufacturers, i.e., Finnigan MAT (currently Thermo Finnigan) and VG (currently Waters, [25]). The Finnigan system was used in ca. 65% of the application papers. The MBI is about equally used in combination with magnetic sector and with quadrupole instruments.

The MBI for LC–MS was used in a wide variety of applications, including the analysis of drugs and their metabolites, pesticides, steroids, alkaloids, polycyclic aromatic hydrocarbons, and others [26]. One example is briefly discussed here.

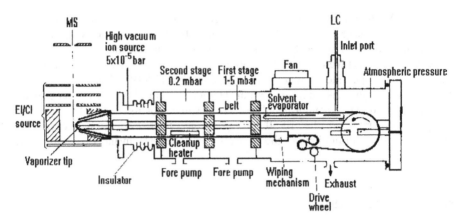

Figure 4.4: Moving-belt interface

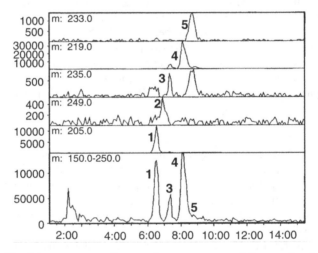

Figure 4.5: Total-ion current and mass chromatograms of the moving-belt ammonia CI LC-MS analysis of a post-mortem urine extract (Reprinted with permission from [54], ©1989, Preston Publications).

Identification of an intoxicating agent and its metabolites in post-mortem plasma and urine was described by Verheij *et al.* [27] by combining accurate mass determination with high resolution mass spectrometry in on-line LC–MS, EI, and CI spectra of the various peaks, library searching of the EI spectra, and information from the LC retention times.

In the LC-UV chromatogram of the post-mortem plasma and urine extracts, obtained in a routine benzodiazepine screening, three peaks were found that could not be identified by matching retention times and UV spectra. The compounds were not amenable to GC-MS. With MBI LC–MS both EI and CI spectra were obtained from the various peaks. The TIC and extracted mass chromatograms obtained with ammonia CI are given in Figure 4.5.

On-line accurate mass determination in LC–MS on the probable molecular ion in the spectrum of peak 1 led to an elemental composition of $C_7H_6Cl_2N_2O$ (M_r measured 203.989 and calculated 203.986). Computer library search identified this compound as N-dichlorophenyl urea. Interpretation of the spectrum confirmed this assignment. For peak 4 the molecular ion was found at m/z 218, indicating an additional methyl-group. From the shift of an intense peak at the low mass end from m/z 44 to 58, due to $[R_2N-C=O]^+$ where R is either H or CH_3, it was concluded that the additional methyl group most likely is at the N'-atom. Searching for the corresponding dimethyl-analogue was successful (peak 5 in Figure 4.5).

These three compounds elute in expected order from the reversed-phase LC system. An interesting aspect in the ammonia CI spectra of these three compounds is that an increasing intensity for the protonated molecule and a decreasing intensity for the ammoniated molecule is observed in the series urea – N'-methyl urea – N',N'-dimethyl urea, reflecting the increasing proton affinity in this series. The other two peaks in the chromatogram were identified as N-(dichlorophenyl)-N'-hydroxymethyl-N'-methyl urea (peak 2 in Figure 4.5) and N-(dichlorophenyl)-N-hydroxy-N'-methyl urea. From these results it was concluded that the intoxicating agent most likely was the common herbicide diuron, $i.e.$, N-(3,4-dichlorophenyl)-N',N'-dimethyl urea, although obviously the position of the chlorine atoms could not be determined with certainty.

The use of MBI for LC–MS stopped, because it is a complex mechanical device, requiring high operating skills. Renewal of the belt and belt memory are troublesome aspects as well. The MBI application field was taken over by the particle-beam interface (Ch. 4.8).

5. Direct liquid introduction

The Direct Liquid Introduction (DLI) interface (Figure 4.6) was developed [28] in order to solve the problems with in-capillary evaporation in the capillary inlet (Ch. 4.2). In a DLI interface, (part of) the column effluent is nebulized by the disintegration into small droplets of a liquid jet formed at a small diaphragm (a 2-5 μm ID pinhole or diaphragm, laser-drilled or electro-etched in 30-100 μm thick nickel or stainless-steel plate). After desolvation of the droplets in a desolvation chamber, the analytes can be analysed using solvent-mediated CI with the LC solvents as the reagent gas. The DLI interface was reviewed in a two-part paper [29-30]. DLI interfaces have been commercially available from Hewlett-Packard (currently Agilent Technologies) and from Nermag.

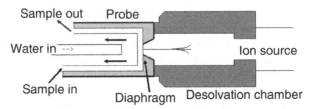

Figure 4.6: Direct liquid introduction interface.

The process of disintegrating liquid jets was theoretically described by Rayleigh in 1879 [31] and discussed in relation to LC–MS by Arpino *et al.* [13-14]. It is the result of surface instabilities on the jet. The minimum liquid flow-rate $F_{jet,min}$ required to form a liquid jet from a diaphragm with an internal diameter of d_{jet} can be estimated from:

$$F_{jet,min} = \frac{\pi \, d_{jet}^2}{4} \sqrt{\frac{8 \, \sigma}{\rho \, d_{jet}}} = 0.015 \sqrt{d_{jet}^3}$$

where σ is the surface tension and ρ is the density of the solvent. The right-hand term of this equation is valid for water-methanol and water-acetonitrile mixtures. It follows that the theoretical minimum liquid flow-rate for stable liquid-jet formation from a 3 μm pinhole is *ca.* 5 μl/min. Practical liquid flow-rates are 2–5 times higher because poor agreement exists between the theoretically calculated and the experimentally determined minimum jet flow-rates, especially with diaphragms smaller than 10 μm ID. Almost immediately after its formation, the liquid jet disintegrates into a mist of droplets with a narrow size distribution. Theory predicts that the droplets have a diameter of about twice the diaphragm diameter [32]. Liquid nebulization by means of a DLI interface is a delicate process. Small irregularities in the shape or a small burr at the edge of the pinhole can lead to problems. Clogging of the diaphragm is another frequently encountered problem. Regular renewal of the diaphragm is obligatory.

The next step in the DLI process is the desolvation. The droplets evaporate on their flight from the probe tip to the ion source through the desolvation chamber. Several designs of desolvation chambers were proposed, *e.g.*, with a convergent-divergent internal geometry [33]. Limitations in effective heat transfer in the desolvation chamber prevent the DLI interface from being used at flow-rates exceeding *ca.* 50 μl/min. Therefore, a large split ratio is needed, typically 1:100. The use of microbore columns appears to be favourable, since better mass detection limits can be obtained in those cases , as was demonstrated by Eckers *et al.* [34].

From the desolvation chamber, a mixture of solvent vapours, desolvated analytes, and residual tiny droplets enters the ion source. The vapours of the mobile-phase solvents act as reagent gas in solvent-mediated CI. High-energy electrons (100-400 eV) from a heated filament are used to generate the primary ions in the reagent gas, that upon ion-molecule reactions produce the protonated molecules or other ionic species that finally react with the analyte molecules. The performance of solvent-mediated CI is influenced by the experimental conditions, *e.g.*, the source pressure and temperature, and the reagent-gas composition.

Most applications of DLI LC–MS deal with qualitative analysis, where in most cases only molecular-mass information is obtained. DLI LC–MS found extensive application in the analysis of pesticides and related compounds [35], in the qualitative and quantitative determination of corticosteroids and metabolites in equine urine [36]. Highly labile compounds such as vitamin B12 (molecular weight 1354) and erythromycin A (molecular weight 733) were analysed by DLI negative-ion CI LC–MS [33]. As an example, the negative-ion CI spectrum of 92 ng vitamin B12 is shown in Figure 4.7.

The DLI interface was widely used in LC–MS applications between 1982 and 1985. The DLI interface did not survive the introduction of the thermospray interface, which removed some of the drawbacks of the DLI interface, *i.e.*, the flow-rate limitation of 50 µl/min and the problems with clogging of the diaphragms. Furthermore, thermospray added new ionization modes next to the solvent-mediated CI used in DLI.

6. Continuous-flow fast-atom bombardment

In a continuous-flow fast-atom bombardment (Cf-FAB) interface, typically a 5–15 µl/min liquid stream, mixed with 5% glycerol as FAB matrix, flows through a narrow-bore fused-silica capillary towards either a stainless-steel frit or a (gold-plated) FAB target. At the target or frit, a uniform liquid film is formed due to a subtle balance between solvent evaporation and solvent delivery. Ions are generated by bombardment of the liquid film by fast atoms or ions, common to FAB. The Cf-FAB interface for LC–MS have been reviewed [37-38].

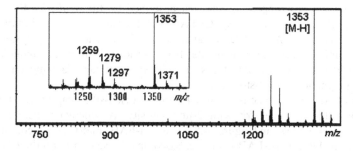

Figure 4.7: Negative-ion CI mass spectrum of vitamin B12 (M_r = 1354) by DLI LC–MS (Reprinted with permission from [64], ©1983, American Chemical Society).

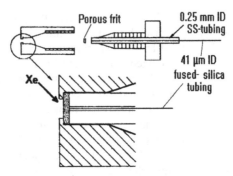

Figure 4.8: Frit-FAB interface for LC–MS (Reprinted with permission from [40], ©1988, Elsevier Science).

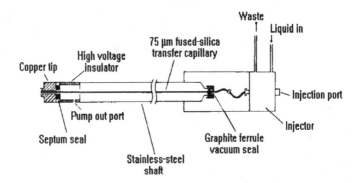

Figure 4.9: First prototype CF-FAB interface probe, built by Caprioli et al. [83] (Reprinted with permission from [41], ©1986, American Chemical Society).

A continuous-flow approach was first described by Ito *et al.* [39-40] in 1985, using a frit interface. Subsequently, Cf-FAB was described by Caprioli *et al.* [41] in 1986, and coaxial Cf-FAB, by de Wit *et al.* [42] in 1988. The first two systems were rapidly commercialized and available from various instrument manufacturers. Unlike most other LC–MS interfaces, the Cf-FAB interfaces are most frequently used with magnetic sector instruments.

The frit-FAB interface consists of a 40-μm-ID fused-silica capillary, which ends at a porous stainless-steel frit with 2-μm porosity. A schematic diagram of the interface tip is shown in Figure 4.8. The frit-FAB interface is commercially available from Jeol. A frit-FAB interface, based on the use of a 8 μm thick stainless-steel screen (2 μm mesh) as FAB target, was described Hogge *et al.* [43], and subsequently commercialized by Micromass (currently Waters) as the screen-wick interface with a thin screen with 2 μm pores at the tip of the interface probe [44].

In the Cf-FAB interface, first described by Caprioli *et al.* [41], an open capillary was used instead of a frit with the liquid flowing along a FAB target. The original design of the Cf-FAB probe is shown in Figure 4.9. The Cf-FAB system was commercialized by Kratos [45] (currently Shimadzu). Subsequently, Cf-FAB devices were developed by other manufacturers, *e.g.*, Finnigan MAT (currently Thermo Finnigan) and Micromass (formerly VG). In the Finnigan MAT design, a compressed filter paper pad, the 'wick', is implemented at the bottom side of the probe tip in order to collect excess liquid flowing from the probe [46].

A flat copper FAB target was used in these first experiments, while later a variety of FAB probe tips were described, differing in design, *e.g.*, flat, hemispherical, or conical, or flat with a drain channel, and/or material, *e.g.*, brass, copper, stainless-steel, gold, stainless-steel with a gold-plated channel.

The coaxial Cf-FAB interface was originally designed to couple open-tubular LC and MS [42]. It consists of two coaxial fused silica capillaries: a 10-μm-ID × 150-μm-OD open-tubular LC column surrounded by a 200-μm-ID × 350-μm-ID make-up or sheath tube. The matrix solvent is added close to the target. Either brass or stainless-steel targets were applied.

Cf-FAB in all its forms is a low flow-rate technique, *i.e.*, 1–15 μl/min. Therefore, one should use either a microbore or packed microcapillary column, or a conventional column in combination with a post-column splitting device [47-48].

Cf-FAB has some distinct advantages in terms of analyte ionization over static FAB, introduced by Barber *et al.* [49] in 1981, *e.g.*, improved detection limits and reduced ion suppression effects [50]. On the other hand, the mass range in Cf-FAB appears to be more limited compared to static FAB. In comparison to static FAB, the Cf-FAB technique is more convenient and easier to use. The system can be used in column-bypass mode, enabling a high sample throughput.

Cf-FAB was widely used to solve analytical problems concerning highly polar and/or ionic compounds, *e.g.*, carotenoids [51], acylcarnitine in the urine of a medium-chain acyl-CoA dehydrogenase deficient patient [52].

A quantitative bioassay for erythromycin 2'-ethylsuccinate (EM-ES, M_r 861 Da), a prodrug of the macrolide antibiotic erythromycin, using Cf-FAB LC–MS was described by Kokkonen *et al.* [53-54]. Reversed-phase LC of extracted plasma samples was performed at a flow-rate of 1 ml/min. In order to meet the flow-rate requirements of the Cf-FAB interface, *i.e.*, 15 μl/min, without splitting, the phase-system switching approach [53] was used. After post-column dilution of the column effluent with water, the eluent fraction of interest was enriched on a short precolumn, from which the compound of interest was desorbed and transferred to the Cf-FAB interface probe. A [$^2H_5$]-analogue was used as internal standard. Good linearity was observed in the range of 0.1 to 10 μg/ml EM-ES in plasma. The within-run precision was *ca.* 6%. The accuracy and inter-day precision, determined at 1.05 μg/ml in plasma, were 0.93±0.11 μg/ml and 12%, respectively (n=6). The determination limit was 0.1 μg/ml [54].

Cf-FAB was widely applied in the field of peptide characterization and the analysis of proteolytic digests. The Cf-FAB analysis of a tryptic digest of bovine ribonuclease B before and after treatment with N-glycanase was described by Mock *et al.* [55]. A fused-silica packed microcapillary column was used. A single injection of 100 pmol provided data covering *ca.* 70% of the sequence. Excellent data for small peptides were reported by Moseley, Deterding, and coworkers [56-57]. For example, the detection of 54 fmol Met-Leu-Phe (M_r 409 Da) and 850 fmol bradykinin (M_r 1060 Da) was demonstrated using a coaxial Cf-FAB system on a two- or four-sector mass spectrometer.

In all these application areas, Cf-FAB has almost completely lost territory to electrospray interfaces. The two most important disadvantages of Cf-FAB are the limitations in the maximum allowable flow-rate and the difficulty of achieving stable conditions by balancing the solvent flow-rate, viscosity and surface tension, and the temperature, wettability, and liquid-film properties of the target. Because of its easy implementation on sector instrument, some application of Cf-FAB is still reported, especially frit-FAB.

7. Thermospray interface

In a thermospray (TSP) interface [58], a jet of vapour and small droplets is formed out of a heated vaporizer tube into a low-pressure region. Nebulization takes place as a result of the disruption of the liquid by the expanding vapour that is formed upon evaporation of part of the liquid in the tube. Before the onset of the partial inside-tube evaporation a considerable amount of heat is transferred to the solvent, which assists in the desolvation of the droplets in the low pressure region. By applying efficient pumping directly at the ion source up to 2 ml/min of aqueous solvents can be introduced into the MS vacuum system. The ionization of the analytes takes place by means of ion-molecule reactions and ion evaporation processes. The CI reagent gas can be made either in a conventional way using energetic electrons from a filament or a discharge electrode, or in a process called TSP ionization, where the volatile buffer dissolved in the eluent is involved. Technique and applications of TSP LC–MS have been reviewed in two excellent review papers by Arpino [59-60] and in an extensive book chapter [61]. TSP has been the most widely applied LC–MS interface in the 1980's, which only in the early 1990's rapidly started to lose territory in favour of interfaces based on API.

The TSP interface was developed in the laboratories of Vestal at the University of Houston. It was the result of a long-term research project which started in the mid-70's, aiming at the development of an LC–MS interface which is compatible with 1 ml/min of aqueous mobile phase and capable to provide both EI and solvent-independent CI [62]. The initial interface was a highly complex system, which subsequently was greatly simplified with respect to vaporizer design and vacuum system [58, 62-65]. Developments in vaporizer design are summarized in Table 4.1. Finally, direct electrically-heated vaporizers were applied [64-65].

Table 4.1: Characteristics of thermospray vaporizers [64]				
heat supply	heated length (mm)	energy flux (W/cm²)	total power (W)	ref.
CO₂-laser	0.3	30000	25	[62]
Hydrogen flames	3	5000	50	[63]
Indirect electric	30	700	100	[58]
Direct electric	300	70	150	[64-65]

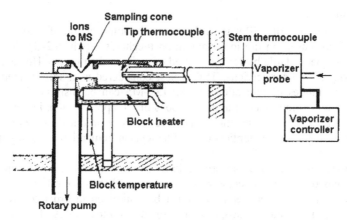

Figure 4.10: Schematic diagram of a thermospray interface (from C.R. Blakley *et al.*, Clin. Chem., 326 (1980) 1467, ©1980, Elsevier Science).

The vacuum system was significantly simplified as well. A 0.3-m³/s mechanical pump was connected directly to the outlet side of the TSP ion source, resulting in a considerable higher conductance of the pumping aperture at the source due to the highly directed flow of the liquid vapour jet [58].

Instrumentation
   A typical TSP system consists of a gas-tight cylindrical tube with the vaporizer probe at one end and the pump-out line at the other (see Figure 4.10). The source block is heated by cartridge heaters. A sampling cone acts as the ion entrance slit to the mass analyser. An electron entrance slit and a discharge electrode are positioned upstream of the sampling cone. A repeller or a retarding electrode is placed opposite or slightly downstream to the sampling cone. A temperature sensor is placed further downstream in order to monitor the vapour or jet temperature. A (liquid-nitrogen)

cold trap is placed between the ion source and the exhaust mechanical pump. Most of the solvent vapours are trapped by this cold trap in order to avoid contamination of the mechanical-pump oil.

Two types of commercial TSP vaporizers were used, *i.e.*, the capillary vaporizers developed by Vestal and coworkers [64-65] and commercially available from Vestec, and the flexible capillary vaporizers introduced by Finnigan MAT [66]. Problems with clogging of the Vestec type vaporizers were reported frequently. Replacement of the complete vaporizer is the final but expensive solution to clogging. With the Finnigan MAT type of vaporizer, the tip must be slightly squeezed to ensure the formation of a sufficiently stable liquid jet at a particular flow-rate, which inevitably leads to poor reproducibility between vaporizers.

Analyte ionization

The breakthrough of TSP was partly due to the introduction of a new ionization technique [58], based on the ion-evaporation mechanism (Ch. 6.2–3). Ammonium acetate or another volatile buffer is assumed to assist in the process. However, in the vast majority of the applications, TSP is best considered as a solvent-mediated CI method. Four modes of ionization in TSP LC–MS can be distinguished, *i.e.*, two liquid-based ionization modes (applied in 60% of the applications), ion-evaporation, and thermospray buffer ionization, and two electron-initiated ionization modes (applied in 40% of the applications): filament-on ionization, and discharge-on ionization.

With ionic analytes and preformed ions in solution, ion evaporation is most important; gas-phase ion-molecule reactions may lead to reneutralization reactions. Buffer composition and concentration must be optimized in order to promote ion evaporation and to reduce gas-phase reactions. In most cases, low ammonium acetate concentrations must be used.

With neutral compounds, TSP buffer ionization is predominant: ionization takes place either by gas-phase ion-molecule reactions or by a rapid proton-transfer reaction at the interface of the liquid droplet and the gas phase, *i.e.*, upon transition from the liquid to the gas phase due to the desolvation of the droplet. This behaviour can readily be described in terms of gas-phase chemical-ionization reactions. In TSP buffer ionization, the addition of a volatile buffer to the LC effluent is obligatory. In absence of a buffer, with non-aqueous mobile phases or with mobile phases that contain over 50% organic modifier either the filament-on or the discharge-on mode must be used. In the filament-on mode, high-energy electrons (0.4–1.0 keV) emitted from a heated filament are accelerated into the ion source. In the discharge-on mode, a continuous gas discharge is used to generate electrons. Solvent-mediated CI spectra are obtained in both filament-on and discharge-on mode.

A practical summary for the selection of the most appropriate ionization mode is as follows. For ionic compounds the concentration of the volatile electrolytes in the mobile phase should be carefully optimized. For neutral analytes, TSP buffer ionization or a electron-initiated ionization mode may be selected. In TSP buffer ionization, the presence of ammonium acetate or any other volatile buffer is required. Filament-on and discharge-on mode show a greater versatility in terms of applicability range when the buffer is left out, although with buffer the latter two modes may give enhanced performance. For most compounds, the positive-ion mode is more sensitive than the negative-ion mode.

Operation and optimization

For a proper operation of TSP, the careful optimization of a variety of mostly interrelated experimental parameters is required. The setup and systematic optimization of the TSP performance was described by various authors [65, 67-69]. The performance of the system can be checked under standard conditions, *e.g.*, a source block temperature of 250°C, a low repeller potential, and a flow-rate of 1.2 ml/min, using the mobile-phase composition to be applied in the analysis. With a Vestec type vaporizer, the stem and tip temperatures are set at 120 and 220°C, respectively, while in case of a Finnigan MAT type interface the vaporizer temperature is set at 100°C. Proper performance can be checked by the appearance of the reagent-gas spectrum, which depending on the solvent composition should contain several solvent cluster ions, and by column-bypass injections of some standard compounds, *e.g.*, adenosine and tertiary amines. Compared to previous experiences, no significant deviations in performance in terms of signal, signal-to-noise ratio, or signal stability should occur.

The performance of the TSP interface is determined by many interrelated experimental parameters, such as solvent composition, flow-rate, vaporizer temperature, repeller potential, and ion source temperature. These parameters have to be optimized with the solvent composition used in the analysis. This optimization procedure is often performed by column-bypass injections, in order to save valuable analysis time. However, for several compounds the spectral appearance may differ between column-bypass and on-column injection, owing to the influence of subtle differences in solvent composition or matrix effects.

In most TSP LC–MS applications, the mobile phase consists of an organic modifier, *e.g.*, methanol or acetonitrile, in water containing 0.05–0.1 mol/l of a volatile buffer. The latter is required as an ionizing agent in TSP buffer ionization. Ammonium acetate is applied in most cases. In filament-on and discharge-on modes the presence of ammonium acetate is not required, but it is still used in most cases.

Methanol and acetonitrile are most widely used as organic modifier (reversed-phase LC). For most compounds, a high water content is favourable in the TSP buffer ionization mode. At modifier contents exceeding 50% modifier, external ionization, *i.e.*, filament-on or discharge-on, must be applied. The typical flow-rates are in the range of 1.0 to 1.5 ml/min.

Careful adjustment of the vaporizer temperature for a particular solvent composition and flow-rate is necessary to avoid signal instabilities (too low vaporizer temperatures) and thermal degradation of the analyte (too high vaporizer temperatures). A sharper optimum for the vaporizer temperature (*ca.* 10°C wide) is generally found. Next to the vaporizer temperature, the source block temperature can be an important parameter, especially in the analysis of thermally labile compounds.

The TSP source is equipped with a repeller or retarder electrode, initially to improve the ion sampling efficiency. A high potential on the electrode can induce fragmentation, attributed to in-source collision-induced dissociation (CID) [70]. With an increase of the repeller potential, the reagent-gas spectrum changes significantly. The spectrum at low repeller potential, dominated by protonated methanol cluster ions at m/z 65 and 97, changes to a spectrum dominated by ions at m/z 19, 31 and 33 at high repeller potential. This indicates that the fragmentation in the analyte spectra might also be explained by changes in the type of ion-molecule reactions in the source [67, 71]. High repeller potentials for structural elucidation were applied, for instance, for indole alkaloids [72].

Selected applications

For some years (1987–1992), thermospray LC–MS was the most widely applied LC–MS interface. It has demonstrated its potential in qualitative as well as quantitative analysis in many application areas, such as drugs and metabolites, conjugates, nucleosides, peptides, natural products, pesticides. A few examples are given below.

Table 4.2: Full-scan detection limits (ng) of pesticides in thermospray LC–MS [73]		
Pesticides	**positive-ion mode**	**negative-ion mode**
organophosphorous (parathion group)	20 – 50	50 – 70
organophosphorous (paraoxon group)	1 – 2	50 – 70
carbamates	1 – 2	> 200
triazines	5 – 10	100
chlorinated phenoxy acids	> 200	1 – 10
phenylureas	2 – 5	10 – 20
quaternary ammonium	100	–

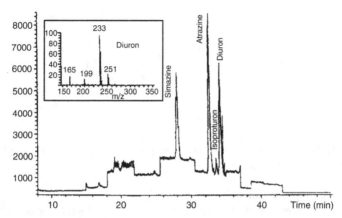

Figure 4.11: On-line SPE thermospray LC–MS for trace analysis of pesticides in river Meuse (Reprinted with permission from [74], ©1993, Elsevier Science).

Environmental applications

The majority of studies on TSP LC–MS analysis of pesticides, herbicides, and insecticides concerns the evaluation of interface performance, detection limits, and information content. Typical full-scan detection limits of various classes of pesticides are summarized in Table 4.2 [73].

In order to determine an individual pesticide at the regulatory level of 0.1 µg/l by means of a straightforward LC–MS method, an absolute detection limit of the method of 10 pg is needed (assuming a 100-µl injection and ignoring chromatographic dilution). From the data in Table 4.2, it can be concluded that the achievable absolute detection limits are far insufficient for the direct LC–MS analysis of pesticides. Off-line or on-line sample pretreatment methods must be used to achieve sufficient analyte preconcentration. An example of such a strategy in multi-residue pesticide analysis via TSP LC–MS is the on-line trace-enrichment by means of solid-phase extraction (SPE) on a 10×3.0-mm-ID C_{18}-packed precolumn, demonstrated by Bagheri et al. [74]. Method detection limits ranging from 2 to 90 ng/l were found for 39 carbamate, triazine, phenylurea, and organophosphorous pesticides, using only 50-ml water samples and one ion per compound for quantitation. Acceptable linearity was achieved over the concentration range tested (0.1–10 µg/l). For river Rhine water samples spiked at 1 µg/l, the RSD was 5–15% (n=6). Using this method, it was found that a river Meuse sample contained simazine at 1.2 µg/l, atrazine at 1.0 µg/l, isoproturon at 0.070 µg/l, and diuron at 2.0 µg/l. The corresponding chromatogram is shown in Figure 4.11 [74]. On-line SPE, using either precolumns or Empore disks, in multi-residue pesticide analysis has subsequently found extensive use in combination with a variety of LC–MS interfaces (Ch. 7.3.2).

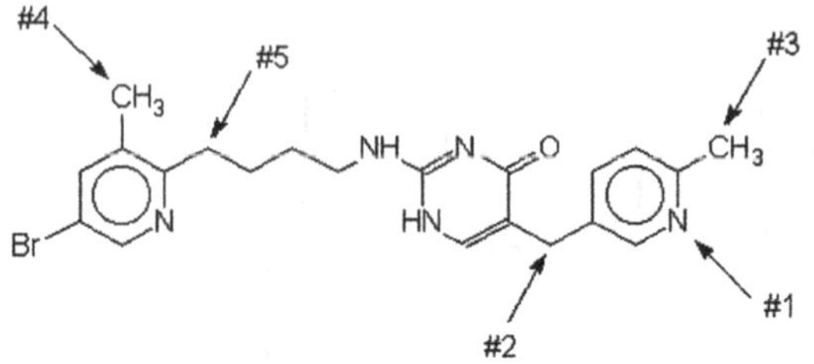

Figure 4.12: Structure of temelastine. The positions of metabolic oxidation are indicated (see text). Based on [75].

Pharmaceutical applications

In the pharmaceutical field, TSP LC–MS mainly found application in the identification of (Phase-I) metabolites and in quantitative bioanalysis. Two examples are discussed below.

TSP LC–MS–MS was applied in elucidating the Phase-I metabolism of the H_1-antagonist temelastine (441 Da). Temelastine is extensively metabolized, with phase I hydroxylation and phase II glucuronidation being two of the major routes [75]. Four hydroxylated, here referred to as hydroxy-#2 to #5 and one N-oxide species (all 457 Da) (Figure 4.12), were observed with TSP LC–MS, predominantly as protonated molecules. The N-oxide showed extensive loss of oxygen and fragmentation to a bromine containing fragment at m/z 242, while one of the hydroxylated metabolites showed the loss of both water and oxygen from the protonated molecule and a fragment ion at m/z 337/339. Differentiation of the other isomeric species was not possible from the LC–MS information.

In product-ion MS-MS using the ^{81}Br-containing [M+H]$^+$, the ions fragmented by cleavage between the CH_2 group and the exocyclic nitrogen of the pyrimidinone ring with transfer of a proton to the nitrogen, thus resulting in a peak at m/z 228 for hydroxy-#3 and at m/z 244 for hydroxy-#4 and hydroxy-#5. The latter two showed an additional loss of water resulting in a peak at m/z 226. However, the relative intensities of the peaks at m/z 226 and 244 are reversed for hydroxy-#4 and hydroxy-#5. Since this difference was reproducible, it could be used for differentiation between these two isomers. The N-oxide and the hydroxylated compound hydroxy-#2 also yielded an ion at m/z 228, but these compounds were distinguishable in LC–MS mode. The method was applied to the analysis of extracted faeces from humans dosed with temelastine [75].

The analysis of Phase-II glucuronide metabolites using TSP LC–MS was often not successful due to frequent solvolysis of the glycosidic bond.

TSP LC–MS in combination with coupled-column chromatography was used to separate and determine drug enantiomers of terbutaline (225 Da) in human plasma [76]. The (–)terbutaline enantiomer is pharmacologically active. The method was developed for single-dose pharmacokinetic studies to determine the plasma concentration of the drug enantiomers in the range of 0.4-40 nmol/l. [^{2}H$_6$]terbutaline, was used as internal standard. A schematic diagram of the instrumental set-up is shown in Figure 4.13. After off-line sample pretreatment by SPE, the sample was injected onto column 1, containing a phenyl stationary phase. The fraction of interest was heartcut to a loop and subsequently transferred to column 2, which was a β-cyclodextrin column used for the enantiomeric separation. Selected-ion monitoring (SIM) at the protonated molecules at m/z 226 and 232 was applied. Quantitation of both enantiomers at nmol/l level was possible [76]. No data on precision and accuracy were given. The same research group demonstrated an automated TSP LC–MS for the quantitative bioanalysis of some antiasthmatic drugs (terbutaline, bambuterol, and budesonide 21-acetate) [77]. This paper can be considered as the first demonstration of routine unattended quantitative analysis using LC–MS.

Biochemical applications
   The potential of TSP in analysing peptides was explored by Pilosof et al. [78] with the analysis of peptides like the α-melanocyte stimulating hormone (1665 Da, 4 nmol analysed) and glucagon (3483 Da, 2 nmol analysed). The abundance of the multiple-charge ions appeared to be correlated to the solution pH.
   The confirmation of the complete sequence of recombinant human interleukin-2 was elucidated from a tryptic digest of 7 nmol of reduced carboxymethylated interleukin-2 by Blackstock et al. [79]. The tryptic fragments were identified by molecular mass from either single-, double-, or triple-charge ions. The mobile phase contained 0.1% trifluoroacetic acid; a water-acetonitrile gradient was performed.

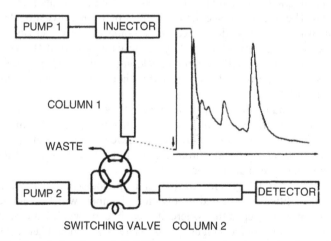

Figure 4.13: Coupled-column LC–MS system for the determination of terbutaline enantiomers in plasma (Reprinted with permission from [76], ©1988, Elsevier Science).

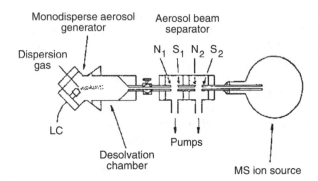

Figure 4.14: The prototype of the MAGIC (N nozzle, S skimmer). Reprinted from [80] with permission, ©1984, American Chemical Society.

8. Particle-beam interface

In a particle-beam interface (PBI), the column effluent is nebulized, either pneumatically or by TSP nebulization, into a near atmospheric-pressure desolvation chamber, which is connected to a momentum separator, where the high molecular-mass analytes are preferentially transferred to the MS ion source, while the low molecular-mass solvent molecules are efficiently pumped away. The analyte molecules are transferred in small particles to a conventional EI/CI ion source, where they disintegrate in evaporative collisions by hitting a heated target, *e.g.*, the ion source wall. The released molecules are ionized by EI or conventional CI.

The PBI was originally developed as a 'monodisperse aerosol generating interface for chromatography' (MAGIC) by the research group of Browner [80-81]. The design objective of MAGIC was the development of an LC–MS interface with EI capabilities, minimum peak distortion, and without a thermal desorption step, as is required in the MBI.

The system should be compatible with a wide range of mobile-phase compositions and with the typical flow-rates of conventional LC column. MAGIC is based on aerosol formation in order to readily achieve evaporation of the solvent and minimum band broadening of the chromatographic peaks and to avoid the need of thermal desorption of the analyte molecules.

The complete MAGIC system consists of a monodisperse aerosol generating (MAG) nebulizer, fitted in a heated desolvation chamber, which is connected to a two-stage aerosol-beam separator, where the analyte molecules are sampled and transported to the ion source (Figure 4.14). The MAG nebulizer consists of a diaphragm, at which a liquid jet is formed, and a gas stream at close distance of the diaphragm and perpendicular to the liquid jet. Small droplets with a narrow droplet-size distribution are generated by disruption of liquid jet due to Rayleigh instabilities and subsequent dispersion of the droplets by a gas stream perpendicular to the jet. During the desolvation of the droplets, the less volatile analyte molecules coagulate into small particles, typically 50–300 nm. Subsequently, solvent vapour and nebulization gas are separated from the particles by means of molecular-beam technology. The mixture is expanded at a nozzle into a vacuum chamber. The low-mass solvent molecules show a greater tendency to diffuse away from the centre of the expansion, while the heavier analyte particles are kept in the core of the vapour jet. The core of the jet is then sampled by a skimmer. By performing two subsequent expansion steps, the solvent vapour can be removed almost completely. The two-stage aerosol-beam separator developed is called a momentum separator. In the momentum separator, sufficient pressure reduction is achieved to generate EI and solvent-independent CI mass spectra.

The original MAGIC system was improved by Winkler et al. [82] by redesigning both the MAG nebulizer and the momentum separator. The 25-µm-ID glass diaphragm, prone to clogging, was replaced by a short piece of easily replaceable 25-µm-ID fused-silica capillary. An improved momentum separator was designed to reduce analyte losses due to particle sedimentation and poor nozzle-skimmer alignment. In the new design, the particles travel over a shorter distance, the aerodynamics of the nozzle and skimmers is improved, and their alignment is more readily performed [82].

In subsequent years (1988), the MAGIC system was commercialized, first by Hewlett-Packard (nowadays Agilent Technologies), and subsequently by other instrument manufacturers. Four commercial versions of the system have been available: (1) the particle-beam interface, featuring an adjustable concentric pneumatic nebulizer, (2) the thermabeam interface with a combined pneumatic-TSP nebulizer, (3) the universal interface, in which TSP nebulization and an additional gas diffusion membrane is applied, and (4) the capillary-EI interface, which resulted from systematic modifications to existing PBI systems by Cappiello [83]. The first system was most widely used, and is discussed in more detail below. For some years, PBI was widely used for environmental analysis, especially in the US.

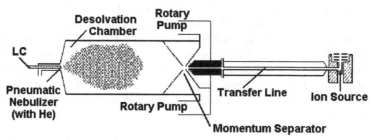

Figure 4.15: The particle-beam interface.

Instrumentation

The PBI comprises of (1) an adjustable concentric pneumatic nebulizer, (2) a heated desolvation chamber, (3) a two-stage momentum separator, and (4) a transfer line between the momentum separator and the EI ion source (Figure 4.15).

The concentric pneumatic nebulizer, used for solvent nebulization, consists of a 100-μm-ID fused-silica capillary for liquid introduction at flow-rates in the range of 0.1–0.5 ml/min, and a circumventing high helium flow (1–3 l/min). The relative positions of the nebulizer jacket and the liquid capillary outlet can be adjusted to optimize the spray performance. Micro-flow aerosol generators for introduction of 1–5 μl/min of liquid were described by Cappiello and Bruner [83-84].

After nebulization, the solvent is evaporated from the droplets in the externally-heated 200 mm × 60 mm ID desolvation chamber. The typical temperature of the chamber wall is 50–70°C. Due to the high pumping efficiency in the first stage of the momentum separator, the pressure in the desolvation chamber is generally between 20 and 30 kPa, *i.e.*, somewhat lower than atmospheric pressure.

The design of the momentum separator in commercial PBI systems is based on the improved design of Winkler *et al.* [82]. The system is evacuated by 300 l/min and 150 l/min mechanical pumps at the first and second pumping stage, respectively. Typical pressures in the various stages are: 25 kPa in the desolvation chamber, 10 kPa in the first pumping region, 30 Pa in the second pumping region, and 2 mPa in the ion-source housing.

The momentum separator can be considered as a two-stage differentially-pumped nozzle-skimmer system, thus featuring a nozzle, a skimmer, and a collimator. In this respect, the PBI resembles an atmospheric-pressure ionization system (Ch. 5.4). However, in the PBI, the analyte is sampled from a closed desolvation chamber, and the analyte is transferred to the high-vacuum region prior to ionization. Generally, little attention is paid to the design characteristics of the momentum separator with respect to optimized molecular-beam performance.

The submicron particles formed during desolvation strike the walls of the ion source, kept at *ca.* 250°C. Flash vaporization and/or disintegration of the particles takes place upon hitting the source wall; the released molecules are ionized by EI or CI. To support and/or enhance flash evaporation, in some systems, a stainless-steel [85] or teflon-coated [86] plug was installed in the ion source or a special independently heatable probe [84] was inserted into the GC-inlet side.

The major difference between the PBI and the thermabeam interface [87] is the use of a combined TSP and pneumatic nebulizer. The TSP nebulizer stimulates the evaporation of the aerosol resulting in smaller particles (*ca.* 50 nm). This is expected to enhance analyte volatilization and to minimize analyte degradation. Therefore, a somewhat smaller heatable stainless-steel desolvation chamber can be used in the thermabeam system. It was commercially available as part of the Waters Integrity® LC–MS system.

Analyte ionization

The most important feature of the PBI is the possibility to acquire on-line EI mass spectra, which may be library searched against commercially available EI libraries and/or readily interpreted along well-known rules. The PBI actually expands the applicability range of EI, leading to extension of the mass spectral libraries. Despite somewhat higher noise levels, good agreement with library spectra is generally achieved. An absolute amount of *ca.* 10–100 ng is generally required to obtain an interpretable spectrum. EI is used in over 90% of the applications of PBI, sometimes in combination with positive-ion CI. In most other applications, negative-ion CI or electron-capture negative ionization (ECNI) is used. ECNI is successfully applied in, for instance, the confirmation of ivermectin [88] in bovine milk, and the determination of chloramphenicol in muscle tissue [89].

Operation and optimization

The performance of the PBI is a function of a variety of experimental parameters, related to solvent nebulization, desolvation, particle-transfer through the momentum separator and the transfer tube, the evaporative collisions in the ion source, and the ionization process. The various parameters are highly interrelated. Systematic optimization of the various PBI parameters for (pharmaceutical) applications was reported by Voyksner *et al.* [90] and Tiller [91].

The PBI is preferentially operated at flow-rates between 0.1 and 0.5 ml/min, thus fitting the optimum flow-rates of a 2-mm-ID column. The PBI can be used in combination with both normal-phase and reversed-phase mobile phases. In reversed-phase LC, the optimum settings of most interface parameters are very much influenced by the water content of the mobile-phase. In general, the best response is obtained with a mobile phase with a high organic modifier content: a 70% loss in response between pure methanol and pure water was reported for methylene blue, furosemide, spectinomycin, and 2-chloro-4-nitrobenzamide [90].

The performance of the PBI can be enhanced by the use of (volatile) additives, such as ammonium acetate, formate, or oxalate, to the mobile phase [92]. They are assumed to act as carriers. Similarly, the use of additives with structures related to the target analyte structures, *e.g.*, phenoxyacetic acid in the analysis of chlorophenoxyacetic acids, was evaluated as well [93]. The carrier effects, exerted by either mobile-phase additives, coeluting compounds, and/or isotopically-labelled standards, is not really understood from a mechanistic point of view. It cannot be applied to consistently enhance the performance: for some compounds it works fine, while for others no effects are observed.

The optimum adjustment of the pneumatic nebulizer is to a large extent a matter of empirical trial-and-error. Relative gas and liquid flow-rates are readily optimized, although the optimization is indirect, *i.e.*, based on the signal obtained. The gas flow appears not to be very critical, as long as it is kept between 1 and 2 l/min.

Selected applications
The PBI has been widely applied for identification, confirmation of identity as well as quantitative analysis in a variety of application areas, especially in the analysis of pesticides. The interest of the US Environmental Protection Agency (US-EPA) in the use of the PBI for environmental monitoring of pesticides obviously contributed significantly to the proliferation of the system.

Environmental applications
Chlorophenoxy acetic acid (CPA) herbicides have been extensively studied by means of PBI LC–MS [85-86, 93-95]. In the EI mass spectra of CPA, generally a weak molecular ion is observed, while a fragment corresponding to the phenol generally is the base peak in the spectrum. It was shown that this major fragment is due to thermal decomposition of CPA in the ion source [85]. The resulting mass spectrum can be considered as a mixed spectrum of the intact molecule and its thermal decomposition products. ECNI of CPA in combination with a PBI was reported as well [93]. The base peak corresponds to the [M–HOCl]⁻ fragment. Phenoxyacetic acid (1.7 mg/l) was used as carrier in the mobile phase to enhance the response of the target compounds.

An interlaboratory comparison of the performance of thermospray and PBI LC–MS interfaces for the analysis of chlorinated phenoxyacid herbicides was reported by Jones *et al.* [94]. Except for Silvex, statistically significant differences were observed in the results from the two interfaces. PBI LC–MS exhibited a high positive bias, but a better %RSD at the highest concentration (500 µg/ml). A comparison of the official US-EPA method 515.1 for CPA analysis with on-line solid-phase extraction (SPE) in combination with GC with electron-capture detection (GC–ECD), LC–UV, and PBI LC–MS was reported by Bruner *et al.* [95]. In this method, liquid-liquid extraction (LLE), as prescribed in the US-EPA method, was replaced by SPE for sample preconcentration. In the LC methods, no derivatization was necessary. Detection limits were in the range of 0.07–0.8 ng/l for GC–ECD, 0.7–7 ng/l for PBI-LC–MS, and 6–80 ng/l for LC–UV. The most accurate methods were LC–UV and GC–ECD, although PBI LC–MS is still more accurate than the US-EPA 515.1 method.

The ability to obtain on-line (library-searchable) EI mass spectra of contaminants is an important advantage of the use of PBI LC–MS in environmental studies, as it enables identification of any unknown compounds detected. Given achievable absolute detection limits in the low nanogram range, analyte preconcentration is obligatory prior to PBI LC–MS analysis for environmental monitoring. Hogenboom *et al.* [96] reported on-line trace-enrichment of pesticides and related compounds from aqueous environmental samples by automated on-line SPE procedures in combination with PBI LC–MS. More than 100 compounds from different compound classes, such as (chlorinated) phenols, organophosphorous pesticides, CPA, phenylurea, and triazine herbicides, were determined. With sample volumes of only 10 ml, phenol and *m*-crosol were determined in surface water at 0.1 µg/l.

Food safety analysis

Delépine and Sanders [89] reported the determination of chloramphenicol (CAP) and three related compounds, *i.e.*, dehydro-CAP, nitroso-CAP, and nitrophenyl-aminopropanediol, in calf muscle tissue, using ECNI with methane as moderator gas. The quantification limit achieved is 2 μg/kg for CAP, requiring the extraction of 5 g of muscle. In a four-point calibration plot (2–20 μg/kg), the %RSD is around 6%, except for the 2 μg/kg level, where it is 12%.

A combination of liquid-liquid extraction and SPE was applied in the sample pretreatment of bovine milk and liver samples in the PBI LC–MS analysis of ivermectins [88]. ECNI with methane was performed to the intact molecular anion and some structure-informative fragments of two ivermectin components at *m/z* 874 and 860. The molecular anion and four fragment ions were used for regulatory confirmation. Signals were observed from on-column injections of 4 ng in extracts equivalent to 2 ml milk or 0.2 g liver.

Residues of oxytetracycline, tetracycline, and chlortetracyline in bovine milk were determined and confirmed after centrifugation of the milk, filtration over a 25 kDa cut-off filter, and SPE on a C_{18} cartridge. Methanol–oxalic acid–acetonitrile was used as mobile phase. Four ions for each tetracycline from the negative-ion mass spectra were used for confirmation at 100 ng/ml level [97]. Carson *et al.* [98] reported the determination of tetracycline residues in milk and oxytetracycline residues in shrimp. Off-line metal-chelate affinity chromatography on Cu^{2+}-loaded chelating Sepharose in combination with SPE on polymeric ENVI-ChromP material was used for sample pretreatment. LC is performed using a PLRP-s polymeric material and 5 mmol/l oxalic acid in the mobile phase. The method is validated with samples spiked at 30 ng/ml in milk and 100 ng/g in shrimps.

Vitamin analysis

The potential of PBI LC–MS in the analysis of various vitamins was explored by Careri *et al.* [99-100]. The fat-soluble vitamins A, D, and E were analysed in food and multivitamin preparations [99]. Absolute detection limits in SIM mode were 0.6–25 ng after fast reversed-phase separation using a 97% aqueous methanol as mobile phase. Mass spectra in EI, positive-ion and negative-ion CI were obtained and discussed. The mass-spectral and quantitative performance of PBI LC–MS in the analysis of eleven water-soluble vitamins was also explored [100]. Detection limits were determined in SIM mode under positive-ion CI, and were below 15 ng for ascorbic acid, nicotinamide, nicotinic acid, and pyridoxal, around 100 ng for dehydroascorbic acid, panthothenic acid, and thiamine, and above 200 ng for biotin, pyridoxamime, and pyridoxine. Riboflavine was not detected.

Perspectives

The PBI nowadays appears to be obsolete, because the most versatile approach to LC–MS appears to be via API interfacing, but PBI has some attractive features, especially the ability to obtain an EI mass spectrum for a compound eluting from an

LC–MS column. The EI mass spectrum may not be useful for all analyte molecules amenable to LC–MS, but it is certainly attractive for structure elucidation of a wide variety of small molecules. With the breakthrough of API interfaces, further development of LC–MS via a PBI was neglected. PBI in LC–MS was reviewed by Creaser and Stygall [101] and by Cappiello [102].

9. References

1. V.L. Tal'roze, V.E. Skurat, I.G. Gorodetskii. N.B. Zolotai, Russ. J. Phys. Chem., 46 (1972) 456.
2. V.L. Tal'roze, V.E. Skurat. G.V. Karpov, Russ. J. Phys. Chem., 43 (1969) 241.
3. V.L. Tal'roze, I.G. Gorodetsky, N.B. Zolotoy, G.V. Karpov, V.E. Skurat. V.Ya. Maslennikova, *Capillary system for continuous introducing of volatile liquids into analytical MS and its application*, Adv. Mass Spectrom., 7 (1978) 858.
4. M.A. Baldwin, F.W. McLafferty, *LC–MS interface. I. DLI of solutions into a CI-MS*, Org. Mass Spectrom., 7 (1973) 1111.
5. M.A. Baldwin, F.W. McLafferty, *Direct CI of relatively involatile samples. Application to underivatized oligopeptides*, Org. Mass Spectrom., 7 (1973) 1353.
6. J.D. Henion, *Continuous monitoring of total micro LC eluant by DLI LC–MS*, J. Chromatogr. Sci., 19 (1981) 57.
7. A.P. Bruins, B.F.H. Drenth, *Experiments with the combination of a micro-LC and a CI quadrupole MS, using a capillary interface for DLI. Some theoretical considerations concerning the evaporation of liquids from capillaries into vacuum*, J. Chromatogr., 271 (1983) 71.
8. H. Alborn, G. Stenhagen, *Direct coupling of packed fused-silica LC columns to a magnetic sector MS and application to polar thermolabile compounds*, J. Chromatogr., 323 (1985) 47.
9. P.J. Arpino, M.A. Baldwin, F.W. McLafferty, *LC–MS systems providing continuous monitoring with nanogram sensitivity*, Biomed. Mass Spectrom., 1 (1974) 80.
10. R. Tijssen, J.P.A. Bleumer, A.L.C. Smit, M.E. van Kreveld, *Microcapillary LC in open-tubular columns with diameters of 10–50 μm. Potential application to CI-MS detection*, J. Chromatogr., 218 (1981) 137.
11. W.M.A. Niessen and H. Poppe, *Open-tubular liquid chromatography–mass spectrometry with a capillary-inlet interface*, J. Chromatogr., 385 (1987) 1.
12. J.S.M. de Wit, C.E. Parker, K.B. Tomer, J.W. Jorgenson, *Direct coupling of open-tubular LC with MS*, Anal. Chem., 59 (1987) 2400.
13. P.J. Arpino, G. Guiochon, P. Krien, G. Devant, *Optimization of the instrumental parameters of a combined LC–MS, coupled by an interface for DLI. I. Performance of the vacuum equipment*, J. Chromatogr., 185 (1979) 529.
14. P.J. Arpino, C. Beaugrand, *Design and construction of LC–MS interfaces utilizing fused-silica capillary tubes as vacuum nebulizers*, Int. J. Mass Spectrom. Ion Processes, 64 (1985) 275.
15. J.A. Apffel, U.A.Th Brinkman, R.W. Frei, E.I.A.M. Evers, *Gas-nebulized DLI interface for LC–MS*, Anal. Chem., 55 (1983) 2280.
16. A.P. Bruins, *Developments in interfacing microbore LC with MS*, J. Chromatogr., 323

(1985) 99.

17. S. Tsuge, *New approaches to interfacing LC and MS*, in: M.V. Novotny, D. Ishii (Ed.), *Microcolumn separations*, 1985, Elsevier, Amsterdam, p. 217.

18. H. Yoshida, K. Matsumoto, K. Itoh, S. Tsuge, Y. Hirata, K. Mochizuki, N. Kokobun, Y. Yoshida, *Improvement of vacuum nebulizing interface for direct coupling micro-LC with MS and some applications to polar natural organic compounds*, Fres. Z. Anal. Chem., 311 (1982) 674.

19. T. Takeuchi, D. Ishii, A. Saito, T. Ohki, *Direct coupling of an ultra-micro-LC and a MS*, J. High Resolut. Chromatogr. Chromatogr. Commun., 5 (1982) 91.

20. W.H. McFadden, H.L. Schwartz, S. Evans, *Direct analysis of LC effluents*, J. Chromatogr., 122 (1976) 389.

21. N.J. Alcock, C. Eckers, D.E. Games, M.P.L. Games, M.S. Lant, M.A. McDowall, M. Rossiter, R.W. Smith, S.A. Wetswood, H. Wong., *LC–MS with transport interfaces*, J. Chromatogr., 251 (1982) 165.

22. P.J. Arpino, *Combined LC–MS. Part 1. Coupling by means of a MBI*, Mass Spectrom. Rev., 8 (1989) 35.

23. J.G. Stroh, J.C. Cook, R.M. Milberg, L. Brayton, T. Kihara, Z. Huang, K.L. Rinehart, Jr., I.A.S. Lewis, *Online LC–FAB-MS*, Anal. Chem., 57 (1985) 985.

24. J.G. Stroh, K.L. Rinehart, *LC–FAB-MS, recent developments*, LC-GC, 5 (1987) 562.

25. D.S. Millington, D.A. Yorke, P. Burns, *A new LC–MS interface*, Adv. Mass Spectrom., 8 (1980) 1819.

26. D.E. Games, *Combined LC–MS*, Biomed. Mass Spectrom., 8 (1981) 454.

27. E.R. Verheij, J. van der Greef, G.F. LaVos, W. van der Pol, W.M.A. Niessen, *Identification of diuron and four of its metabolites in human post-mortem plasma and urine by LC–MS with a MBI*, J. Anal. Toxicol., 13 (1989) 8.

28. A. Melera, *Design, operation and applications of a novel LC–MS CI interface*, Adv. Mass Spectrom., 8B (1980) 1597.

29. W.M.A. Niessen, *A review of DLI interfacing for LC–MS. Part 1: Instrumental aspects*, Chromatographia, 21 (1986) 277.

30. W.M.A. Niessen, *A review of DLI interfacing for LC–MS. Part 1: Mass spectrometry and applications*, Chromatographia, 21 (1986) 342.

31. Lord Rayleigh, Proc. Lond. Math. Soc., 10 (1879) 4-13.

32. N.R. Lindblad, J.M Schneider, *Production of uniform-sized liquid droplets*, J. Sci. Instrum., 42 (1965) 635.

33. M Dedieu, C. Juin, P.J. Arpino, G. Guiochon, *Soft negative ionization of nonvolatile molecules by introduction of liquid solutions into a CI-MS*, Anal. Chem., 54 (1982) 2372.

34. C. Eckers, D.S. Skrabalak, J.D. Henion, *On-line DLI interface for micro-LC–MS: Application to drug analysis*, Clin. Chem., 28 (1982) 1882.

35. R.D. Voyksner, J.T. Bursey, E.D. Pellizzari, *Analysis of selected pesticides by LC–MS*, J. Chromatogr., 312 (1984) 221.

36. D.S. Skralabak, K.K. Cuddy, J.D. Henion, *Quantitative determination of betamethasone and its major metabolite in equine urine by micro-LC–MS*, J. Chromatogr., 341 (1985) 261.

37. R.M. Caprioli (Ed.), *Continuous-flow fast-atom bombardment mass spectrometry*, 1990, Wiley, New York.

38. R.M. Caprioli, M.J.-F. Suter, *Cf_FAB: recent advances and applications*, Int. J. Mass

Spectrom. Ion Processes, 118/119 (1992) 449.

39. Y. Ito, T. Takeuchi, D. Ishii, M. Goto, *Direct coupling of micro-LC with FAB-MS*, J. Chromatogr., 346 (1985) 161.

40. T. Takeuchi, S. Watanabe, N. Kondo, D. Ishii, M. Goto, *Improvement of the interface for coupling of FAB-MS and micro-LC*, J. Chromatogr., 435 (1988) 482.

41. R.M. Caprioli, T. Fan, J.S. Cottrell, *Continuous-flow sample probe for FAB-MS*, Anal. Chem., 58 (1986) 2949.

42. J.S.M. de Wit, L.J. Deterding, M.A. Moseley, K.B. Tomer J.W. Jorgenson, *Design of a coaxial Cf-FAB probe*, Rapid Commun. Mass Spectrom., 2 (1988) 100.

43. L.R. Hogge, J.J. Balsevich, D.J.H. Olson, G.D. Abrams, S.L. Jacques, *Improved methodology for LC–Cf-FAB-MS: Quantitation of abscisic acid glucose ester using reaction monitoring*, Rapid Commun. Mass Spectrom., 7 (1993) 6.

44. A.R. Woolfitt, C.A.J. Harbach, *LC–MS analysis of an enkephalin mixture using an AutoSpec dynamic LSIMS interface*, Rapid Commun. Mass Spectrom., 7 (1993) 176.

45. A.E. Ashcroft, J.R. Chapman, J.S. Cottrell, *Cf-FAB-MS*, J. Chromatogr., 394 (1987) 15.

46. P.S. Kokkonen, E. Schröder, W.M.A. Niessen, J. van der Greef, *A new target for Cf-FAB LC–MS allowing higher flow-rates*, Org. Mass Spectrom., 25 (1990) 566.

47. J.E. Coutant, T.-M. Chen, B.L. Ackermann, *Interfacing microbore and capillary LC to Cf-FAB-MS for the analysis of glycopeptides*, J. Chromatogr., 529 (1990) 265.

48. T. Mizuno, K. Matsuura, T. Kobayashi, K. Otsuka, D. Ishii, *Pneumatic splitter for LC with FAB-MS*, Analyt. Sci., 4 (1988) 569.

49. H.R. Morris, M. Panico, M. Barber, R.S. Bordoli, R.D. Sedgwick, A. Tyler, *FAB: a new MS method for peptide sequence analysis*, Biochem. Biophys. Res. Commun., 101 (1981) 623.

50. R.M. Caprioli, W.T. Moore, T. Fan, *Improved detection of 'suppressed' peptides in enzymic digests analysed by FAB-MS*, Rapid Commun. Mass Spectrom., 1 (1987) 15.

51. R.B. van Breemen, H.H. Schmitz, S.J. Schwartz, *Cf_FAB LC–MS of carotenoids*, Anal. Chem., 65 (1993) 965.

52. D.L. Norwood, N. Kodo, D.S. Millington, *Application of Cf-FAB LC–MS to the analysis of diagnostic acylcarnitines in human urine*, Rapid Commun. Mass Spectrom., 2 (1988) 269.

53. P. Kokkonen, W.M.A. Niessen, U.R. Tjaden, J. van der Greef, *Phase-system switching in Cf-FAB LC–MS*, Rapid Commun. Mass Spectrom., 5 (1991) 19.

54. P.S. Kokkonen, W.M.A. Niessen, U.R. Tjaden, J. van der Greef, *Bioanalysis of erythromycin 2'-ethylsuccinate in plasma using phase-system switching Cf-FAB LC–MS*, J. Chromatogr., 565 (1991) 265.

55. K. Mock, J. Firth, J.S. Cottrell, *Application of on-line Cf-FAB HPLC–MS to the analysis of enzymatic digests of ribonuclease B*, Org. Mass Spectrom., 24 (1989) 591.

56. M.A. Moseley, L.J. Deterding, J.S.M. de Wit, K.B. Tomer, R.T. Kennedy, N. Bragg, J.W. Jorgenson, *Optimization of a coaxial Cf-FAB interface between capillary LC and magnetic sector MS for the analysis of biomolecules*, Anal. Chem., 61 (1989) 1577.

57. L.J. Deterding, M.A. Moseley, K.B. Tomer, J.W. Jorgenson, *Coaxial Cf-FAB in conjunction with MS–MS for the analysis of biomolecules*, Anal. Chem., 61 (1989) 2504.

58. C.R. Blakley, M.L. Vestal, *TSP interface for LC–MS*, Anal. Chem., 55 (1983) 750.

59. P.J. Arpino, *Combined LC–MS. Part II. Techniques and mechanisms of TSP*, Mass

Spectrom. Rev., 9 (1990) 631.
60. P.J. Arpino, *Combined LC–MS. Part III. Applications of TSP*, Mass Spectrom. Rev., 11 (1992) 3.
61. A.L. Yergey, C.G. Edmonds, I.A.S. Lewis, M.L. Vestal, *Liquid Chromatography–Mass Spectrometry, Techniques and applications*, (1990) Plenum Press, New York, NY, p. 31.
62. C.R. Blakley, M.J. McAdams, M.L. Vestal, *Crossed-beam LC–MS combination*, J. Chromatogr., 158 (1978) 261.
63. C.R. Blakley, J.J. Carmody, M.L. Vestal, *LC–MS for analysis of nonvolatile samples*, Anal. Chem., 52 (1980) 1636.
64. M.L. Vestal, G.J. Fergusson, *TSP LC–MS interface with direct electrical heating of the capillary*, Anal. Chem., 57 (1985) 2373.
65. D.A. Garteiz, M.L. Vestal, *TSP LC–MS interface: Principles and applications*, LC Mag., 3 (1985) 334.
66. W.H. McFadden, Spectra, 9 (1983) 23, and Anal. News, 10 (1984) 4.
67. C.E.M. Heeremans, R.A.M. van der Hoeven, W.M.A. Niessen, U.R. Tjaden, J. van der Greef, *Development of optimization strategies in TSP LC–MS*, J. Chromatogr., 474 (1989) 149.
68. C. Lindberg, J. Paulson, *Optimization of TSP conditions. effect of repeller potential and vaporizer temperature*, J. Chromatogr., 394 (1987) 117.
69. R.D. Voyksner, C.A. Haney, *Optimization and application of TSP LC–MS*, Anal. Chem., 57 (1985) 991.
70. W.H. McFadden, S.A. Lammert, *Techniques for increased use of TSP LC–MS*, J. Chromatogr., 385 (1987) 201.
71. C.E.M. Heeremans, R.A.M. van der Hoeven, W.M.A. Niessen, J. van der Greef, N.M.M. Nibbering, *Mechanisms of repeller-induced effects in TSP LC–MS*, Org. Mass Spectrom., 26 (1991) 519.
72. S. Auriola, T. Naaranlahti, R. Kostiainen, S.P. Lapinjoki, *Identification of indole alkaloids of Catharanthus roseus with LC–MS using CID with the TSP ion repeller*, Biomed. Environ. Mass Spectrom., 19 (1990) 400.
73. D. Barceló, G. Durand, R.J. Vreeken, G.J. de Jong, H. Lingeman, U.A.Th. Brinkman, *Evaluation of eluents in TSP LC–MS for identification and determination of pesticides in environmental samples*, J. Chromatogr., 553 (1991) 311.
74. H. Bagheri, E.R. Brouwer, R.T. Ghijsen, U.A.Th. Brinkman, *On-line low-level screening of polar pesticides in drinking and surface waters by TSP LC–MS*, J. Chromatogr., 647 (1993) 121.
75. I.G. Beattie, T.J.A. Blake, *The analysis and characterization of isomeric metabolites of temelastine by the combined use of TSP LC–MS and LC–MS–MS*, Biomed. Environ. Mass Spectrom., 18 (1989) 860.
76. L.-E. Edholm, C. Lindberg, J. Paulson, A. Walhagen, *Determination of drug enantiomers in biological samples by coupled column LC and LC–MS*, J. Chromatogr., 424 (1988) 61.
77. C. Lindberg, J. Paulson and A. Blomqvist, *Evaluation of an automated TSP LC–MS system for quantitative use in bioanalytical chemistry*, J. Chromatogr., 554 (1991) 215.
78. D. Pilosof, H.-Y. Kim, D.F. Dyckes, M.L. Vestal, *Determination of nonderivatized peptides by TSP LC–MS*, Anal. Chem., 56 (1984) 1236.
79. W.P. Blackstock, R.J. Dennis, S.J. Lane, J.I. Sparks, M.P. Weir, *The analysis of*

recombinant interleukin-2 by TSP LC–MS, Anal. Biochem., 175 (1988) 319.

80. R.C. Willoughby, R.F. Browner, *MAGIC for combining LC with MS*, Anal. Chem., 56 (1984) 2626.

81. R.C. Willoughby, Ph.D. Thesis: *Studies with an aerosol generating interface for LC–MS*, 1983, Georgia Institute of Technology, Atlanta, GA.

82. P.C. Winkler, D.D. Perkins, W.K. Williams, R.F. Browner, *Performance of an improved MAGIC interface for LC–MS*, Anal. Chem., 60 (1988) 489.

83. A. Cappiello, G. Famiglini, F. Mangani, P. Palma, *New trends in the application of EI to LC–MS interfacing*, Mass Spectrom. Rev., 20 (2001) 88.

84. A. Cappiello, F. Bruner, *Micro-flow-rate PBI for capillary LC–MS*, Anal. Chem., 65 (1993) 1281.

85. L.D. Betowski, C.M. Pace, M.R. Roby, *Evidence for thermal decomposition contributions to the mass spectra of CPA herbicides obtained by PBI LC–MS*, J. Am. Soc. Mass Spectrom., 3 (1992) 823.

86. A. Cappiello, G. Famiglini, *Analysis of thermally unstable compounds by a LC–MS PBI with a modified ion source*, Anal. Chem., 67 (1995) 412.

87. G.G. Jones, R.E. Pauls, R.C. Willoughby, *Analysis of styrene oligomers by PBI LC–MS*, Anal. Chem., 63 (1991) 460.

88. D.N. Heller, F.J. Schenck, *PBI LC–MS with ECNI for the confirmation of ivermectin residue in bovine milk and liver*, Biol. Mass Spectrom., 22 (1993) 184.

89. D. Delépine, P. Sanders, *Determination of CAP in muscle using a PBI for combining LC with ECNI-MS*, J. Chromatogr., 582 (1992) 113.

90. R.D. Voyksner, C.S. Smith, P.C. Knox, *Optimization and application of PBI LC–MS to compounds of pharmaceutical interest*, Biomed. Environ. Mass Spectrom., 19 (1990) 523.

91. P.R. Tiller, *Application of PBI MS to drugs. An examination of the parameters affecting sensitivity*, J. Chromatogr., 647 (1993) 101.

92. T.A. Bellar, T.D. Behymer, W.L. Budde, *Investigation of enhanced ion abundances from a carrier process in PBI LC–MS*, J. Am. Soc. Mass Spectrom., 1 (1990) 92.

93. M.J. Incorvia Mattina, *Determination of CPA using PBI LC–MS*, J. Chromatogr., 542 (1991) 385.

94. T.L. Jones, L.D. Betowski, B. Lesnik, T. Chlang, J.E. Teberg, *Interlaboratory comparison of TSP and PBI LC–MS interfaces: Evaluation of a CPA herbicide LC–MS analysis method*, Environ. Sci. Technol., 25 (1991) 1880.

95. F. Bruner, A. Berloni, P. Palma, *Determination of CPA in water. Comparison of official EPA method 515.1 and on-line SPE LC–UV and PBI-MS detection*, Chromatographia, 43 (1996) 279.

96. A.C. Hogenboom, I. Jagt, J.J. Vreuls, U.A.Th. Brinkman, *On-line trace level determination of polar organic microcontaminants in water using various precolumn–analytical column LC techniques with UV and MS detection*, Analyst, 122 (1997) 1371.

97. P.J. Kijak, M.G. Leadbetter, M.H. Thomas, E.A. Thomson, *Confirmation of oxytetracycline, tetracycline and chlortetracycline residues in milk by particle-beam liquid chromatography–mass spectrometry*, Biol. Mass Spectrom., 20 (1991) 789.

98. M.C. Carson, M.A. Ngoh, S.W. Hadley, *Confirmation of tetracycline residues in milk and oxytetracycline in shrimp by PBI LC–MS*, J. Chromatogr. B, 712 (1998) 113.

99. M. Careri, M.T. Lugari, A. Mangia, P. Manini, S. Spagnoli, *Identification of vitamins A, D and E by PBI LC–MS*, Fres. J. Anal. Chem., 351 (1995) 768.

100. M. Careri, R. Cilloni, M.T. Lugari, P. Manini, *Analysis of water-soluble vitamins by PBI LC–MS*, Anal. Commun., 33 (1996) 159.
101. C.S. Creaser and J.W. Stygall, *PBI LC–MS: Instrumentation and applications*, Analyst, 118 (1993) 1467.
102. A. Cappiello, *Is PBI an up-to-date LC–MS interface? State of the art and perspectives*, Mass Spectrom. Rev., 15 (1996) 283.

5

INTERFACES FOR
ATMOSPHERIC-PRESSURE IONIZATION

1. Introduction

Liquid chromatography–mass spectrometry (LC–MS) based on atmospheric-pressure ionization (API) was demonstrated as early as 1974 (Ch. 3.2.1). However, it took until the late 1980's before API was starting to be widely applied. Today, it can be considered by far the most important interfacing strategy in LC–MS. More than 99% of the LC–MS performed today is based on API interfacing. In this chapter, instrumentation for API interfacing is discussed. First, vacuum system for MS and LC–MS are briefly discussed. Subsequently, attention is paid to instrumental and practical aspects of electrospray ionization (ESI), atmospheric-pressure chemical ionization (APCI), and other interfacing approaches based on API. The emphasis in the discussion is on commercially available systems and modifications thereof. Ionization phenomena and mechanisms are dealt with in a separate chapter (Ch. 6). Laser-based ionization for LC–MS is briefly reviewed (Ch. 5.9).

2. Vacuum systems for mass spectrometry

The design of the vacuum system plays an important role in the development of API interfaces [1-3].

The vacuum system of a mass spectrometer, except that of some benchtop GC–MS systems, nowadays consists of two differentially pumped vacuum chambers, *i.e.*, the ion source housing and the analyser region, separated by means of a baffle containing a slit. Typical operating pressures are between 10^{-4} and 10^{-2} Pa in the ion source housing, and between 10^{-6} and 10^{-3} Pa in the analyser region. In quadrupole systems, somewhat higher pressures are permitted than in time-of-flight (TOF) and sector instruments, especially in the analyser region. Ion-trap systems are operated at *ca.* 0.1 Pa of helium as bath gas (Ch. 2.4.2). All systems consist of turbomolecular pumps, backed by mechanical fore-pumps.

In the initial development of LC–MS, the gas load to the vacuum system was a serious concern. A mobile-phase flow of 1 ml/min corresponds to a gas flow between 0.3 and 2.1 Pa m^3/s, depending primarily on the molecular mass of the solvent used. The effective pumping speed at the EI ion-source housing of a differentially pumped MS system is between 0.3 and 0.7 m^3/s, which allows the introduction of *ca.* 2 µl/min of water, which is only *ca.* 0.2% of the typical flow-rate of a conventional 4.6-mm-ID LC column. In order to introduce the complete effluent of a 4.6-mm-ID LC column into the ion-source housing, a substantial increase of the effective pumping efficiency at the ion-source housing is required. This can be done in various ways:

- Installation of larger pumps (not practical because of size restrictions [1]).
- Installation of a liquid-nitrogen trap (cryopump) inside the ion source housing [1]. Cryopumps can achieve very high pumping speeds, enabling the introduction of 30–120 µl/min of solvent into a system containing a liquid-nitrogen trap with a surface area of 300 cm^2.
- Installation of an additional mechanical pump at the outlet side of a highly gas-tight ion source, as is done in a thermospray (TSP) interface [4] (Ch. 4.7). Due to the highly directed flow of the vapour jet from the TSP vaporizer, a very high effective pumping speed is achieved for this exhaust pump. A liquid-nitrogen trap is enclosed between the pump and the ion source in order to avoid frequent cleaning of the pump oil. Flow-rates of 1–2 ml/min can be introduced.
- Application of an analyte-enrichment approach, as discussed below.

Analyte-enrichment interfaces for LC–MS show similarities with the interfaces used in packed column GC–MS, such as the jet separator, the Watson-Biemann fritted-glass, or the membrane interface. Various analyte-enrichment interfaces have been developed for LC–MS application, as discussed in Ch. 3.3.2. In API, gas-phase analyte-enrichment is performed by means of a molecular-beam system. The vaporized column effluent is sampled from an atmospheric-pressure spray chamber via a differentially pumped expansion chamber system [3].

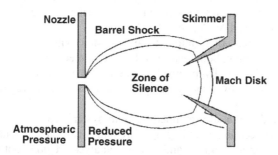

Figure 5.1: One-stage expansion chamber consisting of a nozzle and a skimmer, indicating the shock waves in supersonic expansion.

This is the most versatile approach to LC–MS, as it keeps most of the mobile phase out of the high-vacuum region. API interfaces are special gas-phase analyte enrichment interfaces, because analyte ionization takes place already in the spray chamber, and the ions generated are preferentially sampled into the vacuum system.

In most API systems, a two-stage expansion chamber setup is used, consisting of a nozzle and two skimmers. The nozzle acts as a restriction or leak between the API source and the first vacuum stage, pumped by a mechanical pump. At the nozzle, the gas expands into a low-pressure region (Figure 5.1). This results in a narrowed velocity distribution of the gas molecules and an increased gas velocity. This may lead to supersonic gas velocities and strong cooling of the vapour jet. When a gas mixture is expanded, high-mass particles show a lower momentum perpendicular to the axis of expansion than the low-mass particles. Therefore, the low-mass particles, *i.e.*, mobile-phase components, tend to diffuse away from the core of the expansion, and enrichment of high-mass species, *i.e.*, analyte molecules, occurs.

Initially, the highly directed flow of molecules in the core of the expanding beam (zone of silence) appears to be unaffected by the randomly moving molecules of the background gas in the low-pressure region. However, in the transition between directed and random motion, *i.e.*, in the surrounding barrel shock waves and the Mach disk (Figure 5.1), the gas molecules undergo multiple collisions, resulting in scattering of the beam.

The skimmer samples only part of the expanding supersonic beam, preferentially from within the zone of silence by means of the skimmer, thus penetrating the Mach disk (Figure 5.1). This optimum experimental setup is more readily achieved when the location of the Mach disk is further away from the nozzle, *i.e.*, when the pressure P_1 in the expansion chamber is lower, thus a Campargue-type supersonic beam system is used rather than the more efficient Fenn-type system. In the past few years, increasingly larger ion-sampling nozzles are applied in combination with larger pumps between the two skimmers.

Different vacuum systems have been applied in combination with API sources:
- An one-stage pumping system with a high-efficiency (cryogenic) pump was used in the original Sciex TAGA and API-III systems.
- A two-stage differentially pumped vacuum system with an ion optics region and a mass analyser region, both evacuated by turbomolecular pumps.
- A two-stage differentially pumped vacuum system consisting of a molecular-beam stage pumped by a rotary pump, and a mass analyser region pumped by a turbomolecular pump.
- A three-stage differentially-pumped vacuum system, consisting of a molecular-beam stage between ion-sampling aperture and a skimmer, evacuated by a rotary pump, an ion optics region, and a mass analyser region, both evacuated by turbomolecular pumps. This design is most frequently used today.
- A four-stage differentially-pumped vacuum system with two stages of pumping in the ion-optics region.

3. History of atmospheric-pressure ion sources

Although API sources for MS were already described in 1958, an important breakthrough resulting in a commercially-available system was due to the work of Horning et al. [5-6]. This research in 1974 led to an atmospheric-pressure corona discharge ion source for LC–MS [7-8]. The LC effluent is introduced via an inlet capillary by means of preheated gas into a heated glass evaporator, maintained at 250°C. A corona discharge needle with a sharpened point is positioned ca. 3 mm in front of the 25-µm-ID sampling aperture to the high vacuum of the mass spectrometer with a single-stage pumping system.

Since that time, several other API instruments were described. The most successful commercial instrument was the TAGA (trace atmospheric gas analysis) spectrometer, built by Sciex in Canada. Important features of this system are the 20-m^3/s cryopump, which allowed the use of a 100–200-µm-ID sampling aperture, and the air curtain to prevent clogging of the sampling aperture.

Despite the promising results of Horning et al. [7-8] and Henion et al. [9-10], it took until the early 1990s before the actual breakthrough of API took place. This is probably due to the very limited availability of API instruments from the major MS manufacturers. The actual breakthrough in the early 1990s was the direct result of the breakthrough achieved in ESI. A number of early API source designs from the late 1980s are reviewed below.

3.1   Fenn electrospray molecular-beam source

A schematic diagram of the ESI interface for LC–MS, developed by the group of Fenn [11-13], is shown in Figure 5.2. Sample solutions enter the spray chamber through a stainless-steel hypodermic needle at a flow-rate of 5–20 µl/min. The needle is kept at ground potential and the cylindrical electrode is set at –3.5 kV for positive-ion detection. The API source is sampled by a glass capillary of 120×0.5-mm-ID. The metallized inlet and outlet ends of the glass capillary are set at –4.5 kV and +40 V, respectively. For negative-ion detection the polarities of the various potential are reversed. The liquid is electrosprayed from the tip of the needle. The droplets formed are further dispersed by means of a countercurrent, heated nitrogen with a flow-rate of *ca.* 150 ml/min. The solvent vapour from the rapidly evaporating droplets is swept away by the gas, while the ions that come near the inlet of the glass capillary are entrained in dry bath gas and transported into the first vacuum chamber, forming a supersonic molecular beam. The low-pressure outlet of the sampling capillary can be considered as a nozzle.

The first vacuum chamber is evacuated down to *ca.* 0.05 Pa by means of a 1-m^3/s oil-diffusion pump. The core of the supersonic jet is sampled by a 2-mm-ID skimmer, kept at –20 V, and transported directly into the quadrupole analyser region. The countercurrent bath gas prevents the introduction of non-volatile contaminants into the high-vacuum region.

This source design was subsequently commercialized by Analytica of Branford. Initially, these ESI sources were produced as retrofits for existing instruments from various instrument manufacturers.

3.2   Bruins-Sciex ionspray source

The ionspray® interface, first described by Bruins *et al.* [14] in 1987, was introduced in order to combine the principles of ion evaporation (Ch. 3.2.3) and ESI. However, the prime ionization mechanisms of both approaches appeared to be similar. The main advantage of ionspray or pneumatically-assisted ESI over the conventional ESI is the higher flow-rates (up to 200 μl/min instead of 10 μl/min) that can be accommodated.

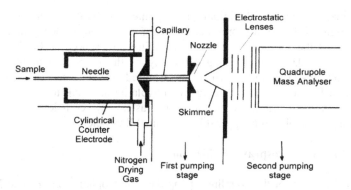

Figure 5.2: Schematic diagram of an early electrospray LC–MS interface and ion source. Reprinted from [13] with permission, ©1985, American Chemical Society.

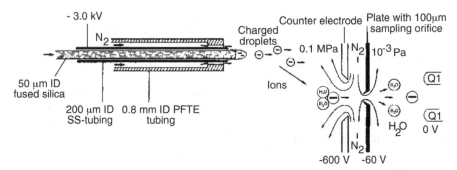

Figure 5.3: Schematic diagram of the ionspray interface. Reprinted from [14] with permission, ©1987, American Chemical Society.

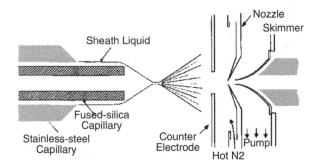

Figure 5.4: Schematic diagram of the Smith CE–MS source for electrospray (Reprinted from J.A. Loo *et al.*, Anal. Biochem., 179 (1989) 404 with permission, ©1989, Academic Press).

A schematic diagram of the initial ionspray interface, built for a Sciex TAGA, is shown in Figure 5.3. The ionspray needle (Ch. 5.5.2) is positioned generally 5–10 mm off-axis of the sampling orifice in the 4-litre API source (120×200-mm-ID). Because of the 20,000-l/s pumping efficiency of the cryopump in the analyser region, a 100-µm-ID sampling orifice can be used. A nitrogen curtain gas flows around the orifice to prevent clogging by nonvolatile material, to assist in droplet evaporation, and to decluster ion-solvent clusters by collisions. Ionspray is commercially available from Applied Biosystems MDS-Sciex.

3.3   Smith electrospray CE–MS source

An ESI interface for the coupling of capillary electrophoresis (CE) and MS was developed by the group of Smith [15-17]. A schematic diagram of the system is shown in Figure 5.4. A hot 2.5-l/min nitrogen curtain gas is used to clean the sampling aperture or nozzle. The instrument contains three differential pumping stages, *i.e.*, a mechanically pumped region between the 0.5-mm-ID nozzle and the 1.2-mm-ID skimmer, a high vacuum region containing an RF-only quadrupole (Ch. 5.4.5), pumped either by a 1500-l/s turbomolecular pump [15] or by a 30,000-l/s cryopump [16], and the quadrupole analyser region, pumped by a 500-l/s turbomolecular pump.

3.4   Chait electrospray source

A modified API source for a quadrupole MS was described by the group of Chait [18]. The system is based on the Fenn source design. The main difference is that the transport and desolvation of the ion-solvent clusters is affected by means of a heated 203×0.5-mm-ID stainless-steel transfer capillary. Further desolvation is achieved by means of collision activation in the low-pressure region (*ca.* 150 Pa) between the capillary exit and the skimmer. The vacuum system is a three-stage vacuum system. This system was commercialized by Finnigan MAT (nowadays Thermo Finnigan).

3.5   Hewlett-Packard orthogonal-sprayer source

In all API source designs described so far, the spray device is in axial position, or only slightly off-axis, relative to the sampling orifice or capillary. Hiraoka *et al.* [19] described an ESI source, where the sprayer is orthogonally positioned relative to the sampling orifice. This design allows higher flow-rates to be used.

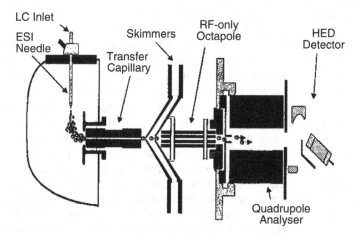

Figure 5.5: Schematic diagram of the orthogonal electrospray system. Reprinted with courtesy of Hewlett-Packard, nowadays Agilent Technologies.

An API system containing an orthogonally-positioned spray device was first introduced by Hewlett-Packard [20]. A schematic diagram of the system is shown in Figure 5.5. The orthogonal sprayer position significantly reduces the contamination of the sampling orifice. It can be used for high flow-rate ESI operation. Orthogonal positioning of the sprayer can be considered as an important developmental step in API LC–MS. Nowadays, most instrument manufacturers apply orthogonal sample introduction (Ch. 5.5.2).

4. Commercial atmospheric-pressure ion sources

A wide variety of API source designs have been available from the various instrument manufacturers. The various designs are briefly discussed below, in combination with some recent developments in API interfacing and laboratory-modified commercial systems. The discussion is illustrated with systematic diagrams of a variety of commercial API systems.

An API source can be considered to consist of five parts:
- The liquid introduction device (Ch. 5.5 and 5.6).
- The actual ion source region, where the ions are generated in an atmospheric-pressure region by means of electrospray ionization, APCI, or by other means.
- The ion-sampling aperture.
- The atmospheric-pressure to high-vacuum interface: the transition region.
- The ion-optical system, where the ions generated in the source are analyte-enriched and transported towards the mass analyser in the high-vacuum region.

The operational principle of most API systems is as follows. The LC column effluent is nebulized into an API source region. Nebulization is performed either pneumatically, *i.e.*, in heated nebulizer APCI (Ch. 5.6), by means of a strong electrical field, *i.e.*, in ESI, or by a combination of both, *i.e.*, in ionspray or pneumatically-assisted ESI (Ch. 5.5). Ions are produced from the evaporating droplets, either by gas-phase ion-molecule reactions initiated by electrons from a corona discharge, *i.e.*, in APCI (Ch. 6.4), or by the formation of microdroplets by solvent evaporation and repetitive electrohydrodynamic explosions and the desorption, evaporation or soft desolvation of ions from these droplets into the gas phase (Ch. 6.3). The ions generated, together with solvent vapour and the nitrogen bath gas, are sampled by a ion-sampling aperture into a first pumping stage. The mixture of gas, solvent vapour, and ions is supersonically expanding into this low-pressure region (Ch. 5.2). The core of the expansion is sampled by a skimmer into a second pumping stage, containing ion focussing and transfer devices to optimally transport the ions in a suitable manner to the mass analyser. From the vacuum point-of-view, it is not important whether a high flow-rate or a low flow-rate of liquid is nebulized, because the sampling orifice acts as the fixed restriction between the atmospheric-pressure region and the first pump stage. From the MS point-of-view, it is also not important whether the ions are generated by electrospray or APCI, although (slightly) different tuning of voltages in the ion optics might be needed due to some differences in the ion kinetic energies.

4.1   Sample introduction devices

The specific design of the various sample introduction devices or spray probes depends to a large extent on the technique applied, *i.e.*, ESI, APCI, or other. With respect to ESI, systems have been described for conventional pure ESI, pneumatically-assisted ESI or ionspray, ultrasonically-assisted ESI, thermally-assisted ESI, and micro- and nano-ESI (Ch. 5.5). The heated-nebulizer system (Ch. 5.6.2) is used in APCI and atmospheric-pressure photoionization (APPI).

Initially, the spray probes were positioned on-axis or only slightly off-axis with the ion-sampling orifice (Ch. 5.3.2). The major disadvantage of this setup is that any particulate or nonvolatile material in the spray may clog or start clogging the ion-sampling orifice. A number of measures were proposed to solve or avoid such contamination problems.

The curtain gas between the orifice plate and the curtain plate in the API source from Applied Biosystems MDS-Sciex [21] (Figure 5.6) has proved to be a reliable way of avoiding contamination of the ion-sampling orifice. The ESI needle is pointed a few millimetre next to the opening in the curtain plate. Ions are thus sampled from the outer regions of the spray plume.

A heated countercurrent gas flow is applied in a number of other sources, *e.g.*, in the initial Fenn ESI source (Figure 5.3) and in sources based on this design (Figure 5.5). The gas also assists in droplet evaporation. A cone gas is applied in a recent design of the Z-spray source from Waters (Figure 5.7).

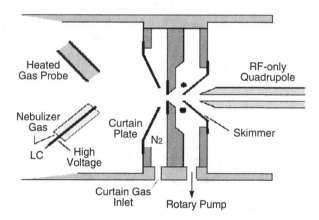

Figure 5.6: Turbo-ionspray source. Reprinted with courtesy from Applied Biosystems MDS-Sciex.

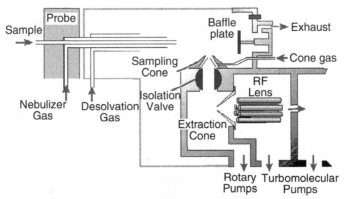

Figure 5.7: Z-spray electrospray source. Reprinted with courtesy from Waters.

A variety of devices to modify the flow direction of the aerosol has been described, e.g., a 'pepperpot' device, a cross-flow device, and orthogonal sample adapter. They all try to divert nonvolatile and particulate material away from the ion-sampling capillary, but often result in a significant signal reduction [22].

The most successful modification to reduce source contamination is the orthogonal positioning of the spray probe (Ch. 5.3.5, Figure 5.5, [20]). Orthogonal ESI was evaluated by Voyksner and Lee [23] in combination with an ion-trap MS.

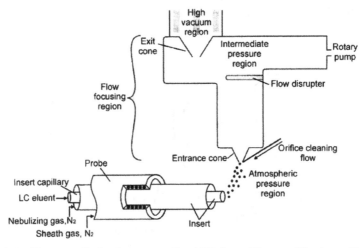

Figure 5.8: The so-called aQa-source for API from Thermo Finnigan. Reprinted from [24] with permission, ©2000, John Wiley & Sons, Ltd.

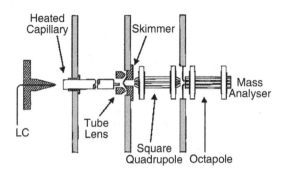

Figure 5.9: Source design of the LCD Deca with heated capillary and square-rod quadrupole. Reprinted with courtesy from Thermo Finnigan.

They tested the hypothesis that the sensitivity and dynamic range of ion trap is limited by charge residues in the form of charged particles and droplets entering the trap. At 20 µl/min, they found that an orthogonal ESI needle provided a six-fold reduction of the total ion current to the ion trap, which was accompanied by a six-fold increase of the analyte signal. At higher flow-rate the improvements are even more significant, *e.g.*, a 30-fold reduction in total ion current and a 20-fold increase in analyte signal at 200 µl/min. Most other manufacturers adopted orthogonal sample introduction.

A small solvent stream along the tip of the ion-sampling cone was applied in the 'aQa'-source (Figure 5.8), available on some single-quadrupole systems from Thermo Finnigan, to avoid cone contamination and clogging by nonvolatile sample constituents [24].

4.2   Application of heat in the API source

Over the years there has been some debate on the need to apply heat to the ESI ion source to assist in the evaporation of the droplets. Thermally-assisted solvent evaporation is especially important at higher flow-rates. A heated countercurrent gas is applied in the Fenn ESI source (Ch. 5.3.1, Figure 5.2) and in the orthogonal-sprayer API sources from Agilent Technologies and Bruker (Figure 5.5). A heated concurrent gas is applied as desolvation gas in the Z-spray source from Waters (Figure 5.7). The heater is typically set at 150°C. The ion-source block is also heated (typically 100°C).

In most API sources from Thermo Finnigan, a heated transfer capillary is used, similar to the device described by Chait *et al.* [18] (Figure 5.9, Ch. 5.3.4).

Initially, no heat was applied in the ionspray source. In the turboionspray source, heated gas is applied orthogonal to the sample introduction probe (Figure 5.6).

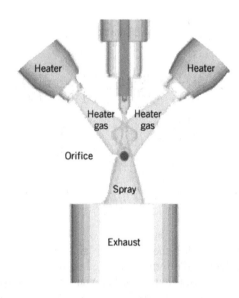

Figure 5.10: Turbo-V ionspray source. Reprinted with courtesy from Applied Biosystems MDS-Sciex.

In a more recent design, featuring an orthogonal ionspray probe, the heated gas is applied in a V-shape (Turbo-V ionspray), as shown in Figure 5.10. In the latter system, the plate containing the ion-sampling orifice contains a heater element as well.

4.3   Ion-sampling apertures

Four types of ion-sampling orifices are used in commercial API systems:
* A flat ion-sampling orifice (in the API systems from Sciex, Figure 5.6).
* An ion-sampling cone (in the API systems from Waters, Figure 5.7, and in the aQa-source from Finnigan, Figure 5.9).
* A 0.5-mm-ID glass capillary with metallized ends (in the API systems from Agilent Technologies and Bruker, Figure 5.5). The glass capillary electrically insulates the API source from the ion optical device. The gas flow through the capillary drags the ions towards the vacuum system, even against an electric field along the capillary.
* A 0.5-mm-ID heated stainless-steel capillary (in other API sources from Finnigan, Figure 5.9). The transport of ions through the (heated) capillary assists in the desolvation of ions. The transport of ions by means of a viscous gas flow through a capillary sampling device was studied by Lin and Sunner [25].

4.4   Transition-region fragmentation: In-source CID

Immediately after their production, the ions in the humid atmospheric-pressure ion source will attract solvent molecules by ion-dipole interactions. These solvated ions must be desolvated prior to entering the mass analyser. This is achieved by ion-molecule collisions in the transition region, especially between the ion-sampling aperture and the skimmer. A small potential difference between the nozzle and the skimmer is applied to enhance declustering.

Smith *et al.* [26-28] demonstrated that by a further increase of the nozzle-skimmer potential difference the internal energy of the ions can be increased and fragmentation of the multiple-charge protein ions can be induced as a result of collision-induced dissociation (CID). This is called in-source CID in this text.

Subsequently, Voyksner and Pack [29] demonstrated that in-source CID can also be achieved for small molecules. They found that mass spectra for compounds like aldicarb, propoxur, carbofuran, and cloxacillin obtained with a 30–50-V potential difference between nozzle and skimmer closely resembles those obtained in conventional CID at 30-eV collision energy. More recently, the internal-energy distribution of ions was compared for two different API sources using the dissociation of various substituted benzylpyridinium ions [30-31]. The fragmentation not only depends on the nozzle-skimmer potential difference, but also on the instrument configuration. This is an important observation with respect to the building in-source CID mass spectral libraries for general unknown screening in toxicology (Ch. 12.5). Tuning procedures to enable comparable in-source CID in instruments from different manufacturers have been proposed [32-33].

The influence of the pressure in the first pumping stage on analyte desolvation and fragmentation was systematically investigated for concanavalin A, a protein that forms multimers [34]. Under gentle conditions (pressure 1.7 mbar and 12 V nozzle potential), the tetramer was observed as broad peaks with poor S/N. An increase of the nozzle potential to 300 V stimulated the desolvation and improved the peak shape. However, fragmentation of the tetramer took place as well. An increase of the pressure in the first vacuum region to 4.1 mbar results in good desolvation without fragmentation, even at a high nozzle potential.

A nice application of in-source CID in the structural analysis of taxol-related compounds was described by Bitsch *et al.* [35]. Taxoid side chain fragments, generated by in-source CID, were further structurally characterized by means of MS–MS. The same approach of two-step fragmentation is applied in the elucidation of the DNA adduct of malondialdehyde (MDA) and the guanine base [36].

The parameter controlling the in-source CID depends on the source design, *e.g.*, the orifice or declustering potential in Sciex systems, the cone voltage in Waters systems, and the fragmentator voltage in systems from Agilent Technologies.

In some API source designs, an additional focussing device is applied, *e.g.*, a tube lens at the outlet side of the heated capillary (Figure 5.9), and a ring electrode between nozzle and skimmer (Figure 5.6).

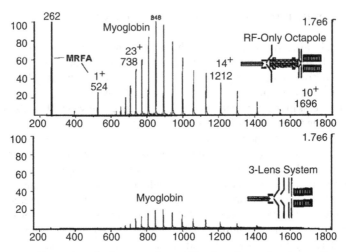

Figure 5.11: Effect of replacing a series of flat ion focussing lenses by a RF-only octapole. Reprinted from [38] with permission, ©1993, Finnigan Corporation.

4.5   Ion optical devices

The second pumping stage of the transition region contains an ion optical device to transfer as many ions as possible towards the mass analyser. Initially, a series of three flat lenses, as commonly used in EI/CI sources, was used (Figure 5.2). Subsequently, it was demonstrated that a better ion transmission in this region could be achieved by replacing the lens stack by an RF-only multipole (with either 4, 6, or 8 rods). The effects of the pressure on the ion transmission in an RF-only quadrupole device were discussed in detail by Douglas and French [37]. The higher transmission at higher pressure is attributed to collisional focussing, a mass rather than a *m/z* dependent process.

An important feature of such a RF-only device is the possibility to transport ions within the quadrupole field over a relatively long distance without large losses, enabling efficient pumping by a large turbomolecular pump in this region. The effects of an RF-only octapole focussing device, replacing a series of flat and conical lenses, on a myoglobin mass spectrum are shown in Figure 5.11 [38].

The RF-only multipole device applied depends on the instrument manufacturer:

- An RF-only quadrupole, the Q_0, is applied in systems from Applied Biosystems MDS Sciex (Figure 5.6).
- An RF-only hexapole is used in systems from Waters (Figure 5.7). In more recent Waters systems, it is replaced by an ion-tunnel device (see below).
- One or two RF-only octapoles are used in systems from Agilent Technologies and Bruker (Figure 5.5), and Thermo Finnigan (Figure 5.9).

- A combination of a Q-array and an RF-only octapole is used in the API source from Shimadzu. The Q-array is positioned prior to the skimmer.
- A square-rod RF-only quadrupole device next to an octapole has been used in some API source designs from Thermo Finnigan (Figure 5.9).

Generally, the RF-only multipole devices are used for ion transport and focussing. The use of API sources in combination with quadrupole ion traps has stimulated additional research in the potential of RF-only multipoles. Unit-mass resolution and mass accuracy can only be achieved in an quadrupole ion-trap when a limited number of ions (typically 10^4 ions) are stored (Ch. 2.4.2). The ion-current-dependent ion injection time designed to avoid problems with space-charge effects in practice translates to a competition between analyte ions and ions from solvent background and matrix interferences for storage in the ion trap. The RF-only multipole may be used as a high-pass mass filter to reduce low-mass interfering ions entering the ion trap. In addition, the RF-only multipole can actually be used to store ions, prior to their pulsed introduction into the ion trap [39]. In the quantitative analysis of the β-lactam antibiotic ceftiofur in milk, a ten-fold improvement in the detection limit was demonstrated by using such ion storage and high-pass mass filtering [39]. Good linearity between 2 and 200 ppb and an RSD within 8% for replicate analysis was achieved.

Obviously, an RF-only multipole device by itself has a charge-capacity limitation. This was investigated by Tolmachev et al. [40]. It was found that the charge-capacity limit is only determined by the number of poles and the RF voltage, but not by the mass and/or charge of the ions. The recently introduced linear ion traps take advantage of these features (Ch. 2.4.2).

An electrodynamic ion funnel interface, positioned between the outlet of a heated stainless-steel capillary and an RF-only octapole ion guide, thus replacing the skimmer, was described by Shaffer et al. [41-42]. The ion funnel interface consists of a series of ring electrodes with decreasing internal diameter. Both RF and DC voltages are applied to these electrodes. The device provides a more effective focussing and ion transmission. About ten-fold sensitivity improvement over a conventional nozzle-skimmer system was demonstrated.

Waters introduced a stacked-ring RF ion-transmission device (the MassTransit® ion tunnel), replacing the RF-only hexapoles [43]. The device consists of a series of constant-aperture and equally-spaced ring electrodes. An RF voltage is applied with 180° phase shift to adjacent plates, generating a field that constrains the ions to the centre region of the device. 60–80% improved ion transmission was observed.

5. Electrospray liquid introduction devices

A high electrical field applied to a solvent emerging from a capillary causes the solvent to break into fine threads which disintegrate into small droplets. This phenomenon, nowadays called electrospray, was first described by Zeleny [44] in 1917. Uniform droplets in the 1-μm-ID range are produced in a breakup process that results from autorepulsion of the electrostatically charged surface, which overcomes the cohesive forces of surface tension. ESI nebulization is widely used in painting, nuclear sciences, and spacecraft thrusters.

The interest in ESI nebulization in LC–MS results from the work of Fenn and coworkers [11-13]. In an ESI interface for LC–MS, the column effluent is nebulized into an atmospheric-pressure ion source. The nebulization is due to the application of a high electric field resulting from the 3-kV potential difference between the narrow-bore spray capillary, the 'needle', and a surrounding counter electrode. The solvent emerging from the needle breaks into fine threads which subsequently disintegrate in small droplets. Analyte ions are generated from these droplets by a variety of ionization process (Ch. 6.3).

ESI has been extensively reviewed. Two books have been published [45-46]. Fundamental principles of ESI were reviewed by Fenn et al. [47-48] and others [49-51]. Applications of ESI in the field of protein and peptide characterization and analytical biotechnology were discussed by several authors [52-55].

In this section, a variety of devices are discussed for sample introduction, i.e., ESI needle designs.

5.1   History

In the first ESI systems, 100–200-μm-ID stainless-steel hypodermic needles were used for sample introduction [11-13]. The system is restricted to liquid flow-rates in the range of 1-10 μl/min. The first needle modification, the so-called ionspray device [14], extends the applicability up to 200 μl/min (Figure 5.3). The column effluent flows through a 50-μm-ID fused-silica (or stainless-steel) capillary, which tightly fits inside a 200-μm-ID stainless-steel capillary, kept at ±3 kV, depending on the polarity of ionization. A 0.8-mm-ID PFTE tube with a narrower PFTE insert at the tip surrounds the stainless-steel capillary; nitrogen gas flows through the PFTE tube. The dimensions of the insert are chosen to provide linear gas velocities exceeding 200 m/s at the tip, which is needed for successful pneumatic nebulization. The relative position of the three concentric tubes must be adjusted to give a fine symmetric spray plume. In practice, the fused-silica capillary protrudes from the stainless-steel capillary, which again protrudes from the PFTE tube. Obviously, the commercial ionspray needles have a slightly different design.

For use in CE–MS, a coaxial ESI needle was developed by Smith et al. [17]. The inner tube is the 100-μm-ID fused-silica capillary where CE is performed, while the outer tube is a 0.25-mm-ID stainless-steel sheath-liquid capillary, which also serves as an electric connection, *i.e.*, an electrode required in the CE process. The fused-silica capillary protrudes 0.2–0.4 mm from the stainless-steel capillary (Figure 5.4). The CE separation can be performed in the optimum solvent, *e.g.*, in a pure aqueous buffer, while the sheath liquid ensures a suitable solvent composition for successful ESI. Coaxial ESI needles are widely used for CE–MS. The same device was extensively used in protein characterization studies [52, 55]. Generally, the aqueous protein solution was introduced at a flow-rate of *ca.* 0.5 µl/min through the inner fused-silica tube, while methanol or 2-propanol was introduced as a sheath liquid at a flow-rate of *ca.* 3 µl/min through the outer stainless-steel tube.

5.2   High flow-rate interfaces

Although liquid flow-rates in the range of 1–10 µl/min are readily compatible with 1-mm-ID microbore and especially 0.32-mm-ID packed microcapillary columns, various means to introduce higher flow-rates for more conventional LC were investigated. Ionspray (Figure 5.3) or pneumatically-assisted ESI was the first of these modifications, allowing flow-rates of 50–200 µl/min. A high-flow option for ionspray was described by Hopfgartner *et al.* [21] allowing flow-rates up to 1 ml/min. A conically shaped liquid shield (curtain plate) is used to catch the larger liquid droplets in the spray. Pneumatically-assisted ESI from a concentric needle is now routinely used in most LC–MS systems. Higher flow-rates demand for heat applied in the source to assist in the droplet evaporation (Ch. 5.4.2).

Other approaches to high-flow ESI, *e.g.*, thermally-assisted ESI allowing flow-rates up to 500 µl/min [56] and ultrasonically-assisted ESI [57], were not really successful in routine application.

5.3   Multichannel electrospray inlets

Multichannel ESI inlets have been developed for a number of reasons. They either divide the effluent from one LC system over several parallel ESI needles or introduce different solvent streams via separate needles into one source housing and one mass spectrometer.

A two- or four-sprayer device to spray one solution through a number of needles was described to study the dynamic range and the flow-rate limitations of an API system [58]. The device allowed flow-rates up to 1 ml/min and provided improved signal stability at higher flow-rates.

Multiple sprayers from several liquid streams were used to study gas-phase ion-ion and ion-molecule reactions in the ESI source. A seven-channel ESI device was applied to study gas-phase reactions of proteins [59], but also to facilitate the protonation of highly reactive pyrolytically-produced ketenes [60]. The analyte is fed through the centre capillary, while a reagent solution is introduced through the outer six channels. A dual-sprayer device was used to study the mechanism of matrix-related ion-suppression effects in quantitative analysis (Ch. 11.5.1, [61]).

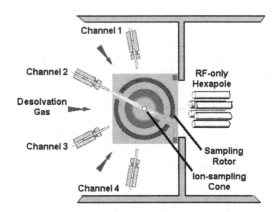

Figure 5.12: Four-channel multiplexed electrospray for four-channel parallel LC–MS, available from Waters. Reprinted with courtesy from Waters.

Various multiple sprayer devices were described to be used in combination with multiple separation systems to enhance sample throughput. A dual-sprayer ESI interface, enabling parallel LC–MS, was described [62]. Four- and eight-channel parallel introduction from four or eight LC systems into a multiplexed ESI source was introduced by Waters in 1999. The continuous ESI nebulization from all sprayers is sampled in succession using a rotating aperture, driven by a variable-speed step motor (Figure 5.12). Each sprayer is sampled for typically 0.1 s each 0.5–1 s. Initially, this device, the 'MUX', was implemented on TOF instruments, capable of fast data acquisition. A four-channel system was applied by de Biasi *et al.* [63] to perform high-throughput accurate molecular-mass determination of some drugs and their synthetic byproducts. More recently, this system was also made available for Waters (triple) quadrupole instruments. The data acquired from each sprayer are collected in separate data files; the multiple-sprayer device is 'indexed'. A nonindexed dual-sprayer device was developed and applied in high-throughput quantitative bioanalysis [64]. The ions generated from both sprayers within the same ion source are sampled through one orifice. Therefore, non-isobaric compounds must be introduced through the two inlets in order to obtain useful results.

A dual-inlet ESI source, the LockSpray®, was introduced by Waters, based on the rotating aperture of the MUX-source, for the co-introduction of a reference compound to act as a lock-mass for accurate-mass determination [65-66]. Dual-inlet devices to introduce a lock-mass compound have also been reported for sector [67] and Fourier-transform ion-cyclotron resonance MS (FT-ICR-MS) instruments [68].

5.4 Low flow-rate interfaces

Reduction of sample consumption during protein characterization by ESI-MS is the main objective in designing ESI needles for flow-rates lower than 1 µl/min. Such needles can also be applied in the on-line MS coupling of low flow-rate techniques like CE and nano-LC (Ch. 17.2 and Ch. 17.5.2).

A nonsheath low-flow ESI needle, enabling flow-rates lower than 0.25 µl/min from 60-mm × 5-, 10-, or 20-µm-ID etched fused-silica capillaries was described by Gale and Smith [69]. Due to the higher local electric field gradient at the smaller diameter tip, a lower needle potential can be used. An 8-fold more intense signal was observed for the 20+-myoglobin ion with a 2.5-fold less sample consumption.

A 5–20-µm-ID micro-ESI needle was described by Emmett and Caprioli [70]. The flow-rates used were 0.3–6.4 µl/min. Similar needles were successfully made by other as well. Robins and Guido [71] reported an integrated packed 150–250-µm-ID microcapillary LC column–micro-ESI device. A Teflon frit retains the column packing material. The last part of the fused-silica column tubing is drawn into a sharp tip to act as an ESI emitter.

5.5 Nano-electrospray needles

A nano-ESI needle device, produced by drawing heat softened 0.5-mm-ID glass capillaries into 1–3-µm-ID glass tips, was described by Wilm and Mann [72-73]. The tips can be used with flow-rates as low as 25–50 nl/min. In this setup, the needle is filled with ca. 1 µl of the protein solution to be investigated. The ESI generated at the needle may be stable for as long as 45 min, allowing the performance of a variety of MS and MS–MS experiments. The needles are positioned close to and in front of the ion-sampling orifice. The voltage required to achieve a stable nano-ESI is less than 1 kV, thus significantly lower than in high-flow ESI where typical voltages are applied between 3 and 5 kV.

Nano-ESI has become a very important technique in protein analysis (Ch. 17.2). Optimization of the nano-ESI needle designs has been the topic of extensive research. An important issue in these studies is related to establishing the electrical contact to the needle. A wide variety of tip coating procedures have been described, e.g., involving gold conductive epoxy [74], overcoated vacuum-sputtered gold [75], gold sputtering and electroplating [76], 'fairy-dust' coatings, i.e., covering a tip coated with polyimide with gold particles prior to drying [77], carbon coating [78], and coating with conductive polypropylene-graphite mixtures [79]. Long-term needle durability is a real problem, especially with needles used in CE–MS. The limited stability of metallized nano-ESI tips is related to the electrochemical processes that take place during ESI nebulization [80]. These involve oxidation of water at the metal-liquid interface in positive-ion mode and reduction of water in negative-ion mode. These electrolytic reactions cause the formation of oxygen or hydrogen at the surface, that may induce mechanical stress upon the coating. In addition, the coating may itself be oxidized and stripped off the surface. Such electrochemical reactions and their effect on various nano-ESI emitters were investigated using electrochemical techniques and scanning electron microscopy in combination with ESI durability studies [80]. Emitters with sputtered gold coatings lose their coating due to gas-formation related mechanical stress. Emitters where a gold coating is applied onto an adhesive layer of chromium or nickel allow [76] have excellent durability in positive-ion mode, while in negative-ion mode the adhesive layer is electrochemically dissolved. These problems do not occur with fairy-dust coatings [77]. These tips appear to be the most durable ones [80].

Vanhoutte *et al.* [81] compared various nano-ESI tips. With the uncoated non-tapered 20-μm-ID fused-silica tips, delivered with the Micromass NanoFlow™ probe, the sensitivity was dependent of the mobile-phase composition. Replacing these tips by uncoated tapered ones (20→9 μm ID) showed some improvement, while the best results, especially at low percentages of methanol in the mobile phase, were achieved with gold-coated tapered fused-silica tips.

Nano-ESI devices and needles are commercially available from Proxeon Biosystems (formerly Protana, Odense, Denmark), New Objective Inc. (Cambridge, MA), and NanoSeparations (Nieuwkoop, the Netherlands). Nano-ESI devices are also available from the instrument manufacturers.

5.6   Microfabricated microfluidic and chip-based electrospray devices

Microchip-based separation techniques are essential elements in the development of fully-integrated micro-total analysis systems. Given the importance of MS as an analytical tool in envisaged application of microchip technology, microchip–MS via nano-ESI interfacing is under investigation [82-83]. Potential advantages of on-line microchip–MS coupling comprise the reduction of sample consumption and sample losses due to handling, and the potential for multiplexing. Technology for on-line microchip–MS feature either spraying directly from an exposed channel at the side of the chip, or from an ESI emitter attached to the microchip (Ch. 17.5.5).

In the first experiments, sample solutions were sequentially sprayed at 100–200 nl/min from a series of parallel 60-μm wide, 25-μm deep, and 35–50 mm long channels on the glass microchip [84]. Each channel contains an individual ESI electrode. The detection limit for myoglobin was 60 nmol/l. Other devices were reported, *e.g.*, featuring electroosmotic sample delivery from a single 60×10-μm channels on a microchip with an attached nano-ESI needle [85] or via a liquid-junction coupling [86]. Further developments are comprised of the choice of new materials, *e.g.*, poly(dimethylsiloxane) (PDMS) [87-88], silicon chips with parylene polymer layers [89], and poly(methyl methacrylate) [90]. Sample handling like on-chip tryptic digestion [91], dialysis [92], and CE separation [93] was also reported.

An array of nano-ESI nozzles in monolithic silicon was used for the direct bioanalysis of drugs in plasma extracts [94] (Ch. 11.8 and Ch. 17.5.5). Significant progress is expected in this research area in the years to come.

6. APCI liquid introduction devices

In an APCI interface, the column effluent is nebulized into a heated vaporizer tube, where the solvent evaporation is almost completed. The gas-vapour mixture enters an API source, where APCI is initiated by electrons, generated at a corona discharge needle. The solvent vapour acts as reagent gas. Subsequently, the ions generated are sampled into the high vacuum of a mass spectrometer for mass analysis. APCI can be performed in exactly the same API sources as applied for ESI (Ch. 5.4). The most important modification is the need to implement the corona discharge needle into the source and to change the inlet probe.

6.1   History

The exploration of APCI for LC–MS started in the early 1970's with the research work of Horning et al. [5-8] on the use of a modified plasma chromatography–MS combination. In 1974, this research led to an APCI source, equipped with either a ^{63}Ni foil or a corona discharge needle as the primary source of electrons [7]. LC–APCI-MS was again demonstrated by Henion et al. [9] in 1982 in the analysis of sulfa drugs using a Direct Liquid Introduction interface (Ch. 4.5) and a Sciex TAGA instrument. Four years later, the same group demonstrated the applicability of LC–MS on a Sciex TAGA in combination with a prototype heated pneumatic nebulizer in the high-speed quantitative analysis of the drug phenylbutazone and three of its metabolites in plasma and urine [10]. After the introduction of ionspray by Bruins et al. [14], Sciex promoted their API-III instruments for LC–MS. Commercial API products from all major MS manufacturers were introduced in the late 1980's and early 1990's, when Fenn et al. [95] demonstrated multiple charging of protein by ESI. In most cases, these instruments are equipped with both ESI and APCI interfaces. However, APCI is not as widely used as ESI.

6.2   Nebulizers for APCI

The heated nebulizer applied in combination with APCI is a concentric pneumatic nebulizer attached to a heated quartz tube. Nitrogen is used as nebulizer gas. Heated nebulizers, comprising of three concentric tubes, i.e., a liquid tube in the centre, a nebulizer gas tube, and an auxiliary gas tube, and a heated vaporization zone, are available from all instrument manufacturers. A general schematic diagram of such a device is shown in Figure 5.13. With a typical liquid flow-rate of 1 ml/min, the required gas flow can be as high as 600 l/h. In some systems, up to 0.7 MPa gas pressure must be available. Covey et al. [96] described optimization of the temperature regimes in the heated nebulizer. They developed a newly-designed heated nebulizer, built in ceramics rather than quartz. The first part is heated to ca. 800°C, while the outlet part is kept at lower temperatures. This reduces memory effects in the heated nebulizer and results in an overall improvement in sensitivity.

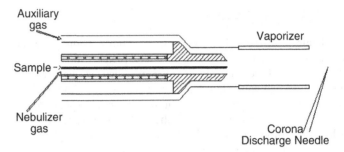

Figure 5.13: General schematic diagram of an APCI heated nebulizer probe.

7. Other atmospheric-pressure introduction devices

7.1   Sonic-spray interface

As part of the Hitachi family of API LC–MS interfaces [97], Hirabayashi *et al.* [98-99] described the sonic-spray ionization interface, which is based on the production of charged droplets in a pneumatic nebulizer from 30 μl/min of liquid using sonic gas velocities (3 l/min nitrogen, Mach 1). No heating is applied to the sprayer. In contrast to ESI or APCI, charged droplets are generated by the sonic-spray interface without heating or applying an electric field. It can be applied at flow-rates up to 1.5 ml/min. It is commercially available from Hitachi [100].

7.2   Laser spray interface

Explosive vaporization and mist formation occurs when a 100-μl/min aqueous effluent at the tip of a 0.1-mm-ID stainless-steel capillary is irradiated by a 10.6-μm infrared laser, as already demonstrated in TSP development (Ch. 4.7). Simultaneous application of a 3–4 kV voltage to the capillary results in strong signals of singly and multiple-charge ions with intensities more than one order of magnitude higher than from ESI. This new interfacing and ionization mode is named laser spray [101].

7.3   Atmospheric-pressure photoionization

A relatively widely-available alternative ionization technique is atmospheric-pressure photoionization (APPI) [102]. In APPI, the ionization process is initiated by photons from a discharge lamp rather than from a corona discharge electrode. APPI is promising in the analysis of relatively non-polar analytes. Commercial systems are available (Photospray® from Sciex, photoionization from Syagen Technology and other instrument manufacturers). Ionization under APPI is discussed in Ch. 6.5.

7.4   Combined electrospray–APCI source

A combined ESI–APCI source is available from several instrument manufacturers. It was developed for high-throughput characterization of combinatorial libraries [103]. The system uses the existing API source, and allows alternating ESI and APCI within one chromatographic run.

7.5   Surface-enhanced APCI

The nebulization in the heated nebulizer does not differ much from that in a TSP interface (Ch. 4.7). With TSP, a solvent-mediated analyte ionization can take place without a primary source of ionization. A similar process should be possible in APCI, *i.e.*, analyte ionization in the discharge-off mode. Over the years, several examples of this have been reported at conferences. Cristoni *et al.* [104] investigated the use of discharge-off APCI-MS in the analysis of peptides and proteins. Double-charge ions were favoured under these conditions. Subsequently, the same group [105-106] modified the APCI source by replacing the unused discharge needle by a gold surface to achieve surface-activated discharge-off APCI (SACI). Promising results were obtained for peptides and various small molecules. 10-fold improved signal-to-noise ratios (compared to ESI-MS) were reported for amphetamines in diluted urine samples. Various ionization phenomena were studied in detail [106].

8. API sources for other types of mass analysers

In Ch. 5.4, general API source designs, in most cases developed for use in combination with quadrupole mass analysers, are discussed. In most commercial systems, these ion sources are also fitted on ion-trap, TOF, and even magnetic sector instruments. However, significant initial research efforts were needed in the development of such systems. Some topics in the development of API for quadrupole ion-trap, TOF, FT-ICR-MS, and magnetic-sector instruments are highlighted in this section, paying attention to history and more recent developments.

8.1   Quadrupole ion-trap instruments

API on an ion-trap instrument requires a pulse-wise ion injection from an external ion source (Ch. 2.4.2). The first external API source for an ion-trap instrument was described by Van Berkel *et al.* [107-108] in 1990 (Figure 5.14). The system comprises of a two-stage differentially-pumped device. The pulsed ion introduction is achieved by the application of a suitable voltage to one of the half-plates of lens L2 (Figure 5.14). Similar systems were subsequently described by others.

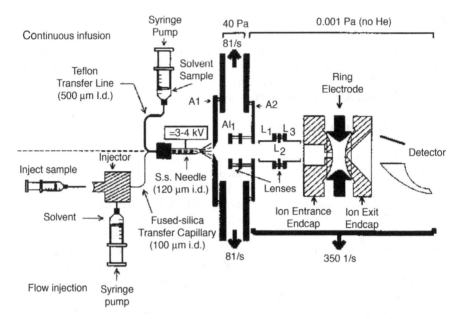

Figure 5.14: Schematic diagram of the ESI–ion-trap system. Reprinted from [108] with permission, ©1991, American Chemical Society.

A major breakthrough was the commercial introduction of dedicated quadrupole ion-trap instruments equipped with an API source for ESI and APCI, *i.e.*, the LCQ instruments from Thermo Finnigan, the Esquire from Bruker, and the LC–MSD-Trap from Agilent Technologies.

8.2   Time-of-flight instruments

The TOF mass analyser requires high-frequency (kHz) pulsed ion introduction (Ch. 2.4.3). For optimum mass resolution, orthogonal acceleration is preferred. Boyle *et al.* [109] first reported an API ion source for a linear TOF analyser via the on-axis ESI source from Analytica of Branford. Later, orthogonal acceleration and a reflectron TOF analyser was applied [110]. Similar systems were subsequently described by others.

Commercial ESI LC–TOF-MS systems are available from several instrument manufacturers. They are applied in two main areas: on-line accurate-mass determination for identification and confirmation of identity, and fast separations, which can be monitored due to the high spectrum-acquisition rates. The four- and eight-channel MUX source (Figure 5.12) takes full advantage of the high spectrum acquisition rates and provides accurate-mass determination of the analytes [63].

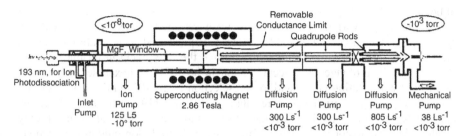

Figure 5.15: Schematic diagram of an external electrospray FT-ICR-MS source. Reprinted from [112] with permission, ©1990, Wiley & Sons Ltd.

8.3 Fourier-transform ion-cyclotron resonance instruments

The high-resolution characteristic of FT-ICR-MS and its potential in elucidating ion envelopes and isotope patterns of multiple-charge proteins in ESI mass spectra has stimulated the research into coupling of ESI and FT-ICR-MS. This in turn has stimulated further development of FT-ICR-MS as an MS tool (Ch. 2.4.6).

In developing API interfaces for FT-ICR-MS, one has to take account of the severe pressure restrictions: in the ion cell a pressure of 10^{-7}-Pa is required for high-resolution applications. ESI FT-ICR-MS can be performed in two ways, *i.e.*, with an external ion source in combination with a (quadrupole) ion guide, and with a probe mounted source which extends into the high field region. The first approach is more frequently applied.

FT-ICR-MS with an external electrospray ion source

The first ESI source for an FT-ICR-MS instrument was described by Henry *et al.* [111-113]. The ions from the external source are transmitted to the ICR-cell through a series of differentially pumped vacuum chambers by means of RF-only quadrupoles. A schematic diagram of such a system is shown in Figure 5.15. Due to the relatively high pressure in the FT-ICR cell, *i.e.*, 10^{-5} Pa instead of the required 10^{-7} Pa, the resolution is limited to *ca.* 5,000. Significant progress in resolving power, *e.g.*, a resolution in excess of 10^5 for carbonic anhydrase (29 kDa), is achieved by replacing the 2.8-T magnet with a 6.2-T magnet, and the installation of a 1500-l/s cryopump near the ICR cell [114].

Over the years, the group of Smith at the Pacific Northwest Laboratory has significantly contributed to the progress in instrument development in this field. Their first system consists of an external ion source containing a resistively-heated transfer capillary [18], six differential pumping stages, quadrupole RF-only ion guides in the fourth and sixth stages (the latter is 1 m long), two high-speed electromechanical shutters, and a 7-T magnet. The operating pressure in the ICR-cell is below 10^{-7} Pa. A mass resolution as high as 700,000 was achieved for the 4+ ion of bovine insulin [115].

Subsequently, various improvements were reported to existing FT-ICR-MS systems, *e.g.*, a dual-cell FT-ICR-MS system [116], external accumulation of ions in the first octapole ion guide in order to improve the duty cycle in on-line LC–FT-ICR-MS [117], prolonged ion accumulation in an RF-only hexapole, resulting in extensive fragmentation [118], and ion preselection in the RF multipole by RF-only resonant dipolar excitation (mass resolution 30–100) [119]. A recently-described ion-transfer system for FT-ICR-MS consists of an ion funnel (Ch. 5.4.5), an RF-only quadrupole for collisional focussing of the ions, an RF-only ion-guiding quadrupole, a selection quadrupole, where ions can be selected based on a linear RF/DC ramp or on resonant dipolar excitation, and a segmented accumulation quadrupole, acting as an axial potential well. This system contains four differential pumped stages [120]. Such techniques further expand the dynamic range, the duty cycle, and sensitivity of FT-ICR-MS. The detection limit of 10 zeptomol (~6000 molecules) of cytochrome *c* may serve as an example [120].

FT-ICR-MS systems equipped with external API sources are commercially available from Bruker Instruments, IonSpec, and Thermo Finnigan. In order to control the number of ions in the ICR cell, hybrid systems have been developed. Bruker offers a FT-ICR-MS hybrid with a quadrupole front-end (APEX-Qh) [121], whereas the LTQ-FT instrument from Thermo Finnigan features a linear-ion-trap (LIT, Ch. 2.4.2) front end [122]. In this way, MS–MS can be performed prior to ion introduction into the ICR-cell, avoiding problems with CID in the ICR-cell.

FT-ICR-MS with an in-field electrospray ion source

A probe-mounted ESI interface, which extended into the high-field region of the 1.5-T magnet of an FT-ICR-MS system, was described by Laude and coworkers [123-124]. By performing ESI in a radially homogeneous magnetic field, a more efficient ion injection into the FT-ICR cell is achieved. By redesigning the vacuum system, the pressure in the ICR-cell could be reduced to 2×10^{-6} Pa. Mass resolution in excess of 20,000 for the 4+-ion of mellitin is reported [124].

Because significant ion losses occur during transport of ions from the external API source via the RF-only ion guides, an in-field API source would be more attractive. However, the resolution with an in-field API source is often limited, due to limitations in the pumping efficiency in such a design.

8.4   Magnetic sector instruments

The major difficulties in coupling an API source to a sector instrument are related to the high acceleration voltage and the large pressure difference between ion source and analyser, which requires extensive pumping in order to avoid discharges, electrical breakdown in the instrument, and extensive collision activation of the ions. In general, an additional pumping stage is used, *i.e.*, a total number of four stages instead of three. Sometimes, precautions for electrical breakdown are incorporated.

Allen and Lewis [125] first demonstrated ESI on a sector instrument, using a laboratory-built source contained in a retractable probe. Larsen and coworkers [126-127] described adaptation of an ESI source from Analytica of Branford on a VG ZAB double-focussing sector instrument. Subsequently, the adaptation of the same source for double-focussing magnetic-sector instruments from other manufacturers was described [128-129]. Dobberstein and Schröder [130] reported the adaptation of an existing ESI source (Figure 5.9) for use on a magnetic-sector instruments. The Z-spray source (Figure 5.7) was adapted for magnetic-sector instruments as well.

Due to the ease of operation of TOF-MS in accurate-mass determination, the use of high-resolution sector instruments has diminished considerably in the past years.

9. Laser-induced ionization in LC–MS

In the recent past, a number of laser-based interface approaches were described. The role of the laser was different: from providing heat for the mobile-phase nebulization and subsequent solvent evaporation in the laser spray interface (Ch. 5.7.2), via laser-induced multiphoton ionization, to matrix-assisted laser desorption ionization (MALDI).

Pulsed sample introduction to time-of-flight MS

A pulsed sample introduction interface for LC–MS in a TOF instrument was described by Wang et al. [131-132]. Analyte ionization is performed by means of laser-induced multiphoton ionization. The interface is based TSP nebulization into a heated expansion chamber and a high-temperature pulsed nozzle. Experimental parameters in on-line LC–MS were evaluated [132].

Continuous-flow MALDI for LC–MS

Given the power of MALDI in peptide and protein analysis (Ch. 2.2.4), an on-line combination with LC would be highly desirable. One of the approaches to on-line LC–MALDI-MS is based on the frit-FAB interface (Ch. 4.6) and is investigated by Li et al. [133-135]. While initially the sample solution, consisting of a peptide or protein in 0.1% aqueous trifluoroacetic acid (TFA), methanol, ethylene glycol, and 3-nitrobenzyl alcohol (3-NBA) as the matrix (1:1:1:1), was continuously infused [133], post-column matrix addition was applied in on-line LC–MS experiments. Ethylene glycol in the solvent is required to reduce the evaporation rate of the liquid in order to achieve a uniform liquid film at the frit (Ch. 4.6). The interface can be used with flow-rates of 1–10 μl/min. On-line LC–MS results were demonstrated for a mixture of horse heart cytochrome c, chicken egg white lysozyme, and horse heart myoglobin, using either a 2.1-mm-ID column in combination with a post-column split or a 0.32-mm-ID packed microcapillary column. Poor resolution for larger peptides (>6 kDa) is due to adduct formation with 3-NBA as matrix.

Aerosol MALDI for LC–MS

Another approach to on-line LC–MALDI-MS is aerosol-MALDI, investigated by Murray and Russell [136-138]. The system is mainly used for constant infusion of samples and column-bypass sample introduction. The analyte is dissolved in methanol acidified by TFA, also containing the matrix compound. Various matrices were investigated, such as 4-nitroaniline, 3,5-dimethoxy-4-hydroxy cinnamic acid, α-cyano-4-hydroxy cinnamic acid, and 2-cyano-4-nitroaniline. The sample mixture is pneumatically nebulized at 0.5 ml/min into an evacuated expansion chamber (100 Pa). The aerosol beam is skimmed 20 mm downstream by means of a indirectly-heated 250×4-mm-ID copper tube. Further desolvation of the aerosol droplets as well as transfer to a second pumping region of the instrument is performed in this way. At the low-pressure exit (10^{-3} Pa) of the transfer tube, the beam crosses the pulsed laser beam and MALDI takes place. The third pumping region contains the linear TOF mass analyser. On-line LC–MS with the aerosol MALDI was demonstrated for the separation of bradykinin, gramicidin S, and myoglobin [138].

Off-line LC–MALDI–MS

The currently most frequently applied method for LC–MALDI-MS is automated post-column fractionation and on-plate collection in discrete spots of the LC column effluent. After the solvent is evaporated, the matrix solution can be added, and MALDI–MS analysis of the various spots can be performed. The procedure requires a liquid-handling robot, capable of disposition of effluent fractions at discrete spots on the MALDI target. A number of ways were proposed for deposition in discrete spots on the MALDI target, *e.g.*, blotting via direct contact between droplet and target [139-140], piezoelectric flow-through microdispensing [141], pulsed electrical-mediated droplet deposition [142], and a heated droplet interface [143]. Commercial LC–MALDI–MS devices were recently reviewed [144].

Atmospheric-pressure matrix-assisted laser desorption ionization

Next to conventional (vacuum) MALDI, atmospheric-pressure MALDI interfaces have been described, especially to enable MS–MS on MALDI-generated ions by ion-trap and Q–TOF instruments [145-146]. Atmospheric-pressure MALDI sources are commercially available from all major instrument manufacturers. First results on-line LC–atmospheric-pressure MALDI were reported as well [147].

Atmospheric-pressure laser ionization

The recently introduced atmospheric-pressure laser ionization system (APLI) can be considered as a modification of APPI (Ch. 5.7.3). In APLI, the one-step photoionization of APPI is replaced by a two-photon process in resonantly-enhanced multi-photon ionization [148]. Enhanced response for polycyclic aromatic hydrocarbons (relative to APCI) was demonstrated. Molecular ions rather than protonated molecules are generated in APLI (*cf.* Ch. 6.5).

10.   References

1. P.J. Arpino, G. Guiochon, P. Krien, G. Devant, *Optimization of the instrumental parameters of a combined LC–MS, coupled by an interface for DLI. I. Performance of the vacuum equipment*, J. Chromatogr., 185 (1979) 529.
2. A.P. Bruins, *MS with ion source operating at atmospheric pressure*, Mass Spectrom. Rev., 10 (1991) 53.
3. W.M.A. Niessen, A.P. Tinke, *LC–MS. General principles and instrumentation*, J. Chromatogr. A, 703 (1995) 37.
4. C.R. Blakley, M.L. Vestal, *TSP interface for LC–MS*, Anal. Chem., 55 (1983) 750.
5. E.C. Horning, M.G. Horning, D.I. Carroll, I. Dzidic, R.N. Stillwell, *New picogram detection system based on a MS with an external ionization source at atmospheric pressure*, Anal. Chem, 45 (1973) 936.
6. D.I. Carroll, I. Dzidic, R.N. Stillwell, M.G. Horning, E.C. Horning, *Subpicogram detection system for gas-phase analysis based upon API-MS*, Anal. Chem., 46 (1974) 706.
7. E.C. Horning, D.I. Carroll, I. Dzidic, K.D. Haegele, M.G. Horning, R.N. Stillwell, *LC–MS–computer analytical systems. A continuous-flow system based on API-MS*, J. Chromatogr., 99 (1974) 13.
8. D.I. Carroll, I. Dzidic, R.N. Stillwell, K.D. Haegele, E.C. Horning, *API-MS: Corona discharge ion source for use in LC–MS–computer analytical system*, Anal. Chem., 47 (1975) 2369.
9. J.D. Henion, B.A. Thomson, P.H. Dawson, *Determination of sulfa drugs in biological fluids by LC–MS–MS*, Anal. Chem., 54 (1982) 451.
10. T.R. Covey, E.D. Lee, J.D. Henion, *High-speed LC–MS–MS for the determination of drugs in biological samples*, Anal. Chem., 58 (1986) 2453.
11. M. Yamashita, J.B. Fenn, *ESI source. Another variation on the free-jet theme*, J. Phys. Chem., 88 (1984) 4451.
12. M. Yamashita, J.B. Fenn, *Negative ion production with the ESI source*, J. Phys. Chem., 88 (1984) 4671.
13. C.M. Whitehouse, R.N. Dreyer, M. Yamashita, J.B. Fenn, *ESI interface for LC–MS*, Anal. Chem., 57 (1985) 675.
14. A.P. Bruins, T.R. Covey, J.D. Henion, *Ionspray interface for combined LC–API-MS*, Anal. Chem., 59 (1987) 2642.
15. J.A. Olivares, N.T. Nguyen, C.R. Yonker, R.D. Smith, *On-line MS detection for capillary zone electrophoresis*, Anal. Chem., 59 (1987) 1230.
16. R.D. Smith, J.A. Olivares, N.T. Nguyen, H.R. Udseth, *Capillary zone electrophoresis–MS using an ESI interface*, Anal. Chem., 60 (1988) 436.
17. R.D. Smith, C.J. Barinaga, H.R. Udseth, *Improved ESI interface for capillary zone electrophoresis–MS*, Anal. Chem., 60 (1988) 1948.
18. S.K. Chowdhury, V. Katta, B.T. Chait, *An ESI-MS with new features*, Rapid Commun. Mass Spectrom., 4 (1990) 81.
19. K. Hiraoka, H. Fukasawa, F. Matsushita, K. Aizawa, *High-flow LC–MS interface using a parallel ionspray*, Rapid Commun. Mass Spectrom., 9 (1995) 1349.
20. K. Imatani, C. Smith, *An easy-to-use benchtop LC–MS detector*, Am. Lab., November 1996.
21. G. Hopfgartner, T. Wachs, K. Bean, J.D. Henion, *High-flow ESI LC–MS*, Anal. Chem.,

65 (1993) 439.
22. A.C. Hogenboom, M.P. Hofman, S.J. Kok, W.M.A. Niessen, U.A.Th. Brinkman, *Determination of pesticides in vegetables using large-volume injection column LC–ESI-MS–MS*, J. Chromatogr. A, 892 (2000) 379.
23. R.D. Voyksner, H. Lee, *Improvements in LC–ESI ion trap MS performance using an off-axis nebulizer*, Anal. Chem., 71 (1999) 1441.
24. S. Bajic, D.R. Doerge, L. Lu, E.B. Hansen, Jr., *Analysis of erythromycin by LC–MS using involatile mobile phases with a novel API source*, Rapid Commun. Mass Spectrom., 14 (2000) 156.
25. B. Lin, J. Sunner, *Ion transport by viscous gas flow through capillaries*, J. Am. Soc. Mass Spectrom., 5 (1994) 873.
26. J.A. Loo, H. R. Udseth, R.D. Smith, *Collisional effects on the charge distribution of ions from large molecules, formed by ESI-MS*, Rapid Commun. Mass Spectrom., 2 (1988) 207.
27. R.D. Smith, J.A. Loo, C.J. Barinaga, C.G. Edmonds, H.R. Udseth, *CID of large multiply charged polypeptides and proteins produced by ESI*, J. Am. Soc. Mass Spectrom., 1 (1990) 53.
28. R.D. Smith, C.J. Barinaga, *Internal energy effects in the CID of large biopolymer molecular ions produced by ESI-MS–MS of cytochrome c*, Rapid Commun. Mass Spectrom., 4 (1990) 54.
29. R.D. Voyksner. T. Pack, *Investigation of CID process and spectra in the transport region of an ESI single-quadrupole MS*, Rapid Commun. Mass Spectrom., 5 (1991) 263.
30. C. Collette, E. de Pauw, *Calibration of the internal energy distribution of ions produced by ESI*, Rapid Commun. Mass Spectrom., 12 (1998) 165.
31. C. Collette, L. Drahos, E. de Pauw, K. Vékey, *Comparison of the internal energy distributions of ions produced by different ESI sources*, Rapid Commun. Mass Spectrom., 12 (1998) 1673.
32. W. Weinmann, M. Stoertzel, S. Vogt, M. Svoboda, A. Schreiber, *Tuning compounds for ESI–in-source CID and mass spectra library searching*, J. Mass Spectrom., 36 (2001) 1013.
33. A.W.T. Bristow, W.F. Nichols, K.S. Webb, B. Conway, *Evaluation of protocols for reproducible ESI in-source CID on various LC–MS instruments and the development of spectral libraries*, Rapid Commun. Mass Spectrom., 16 (2002) 2374.
34. A. Schmidt, U. Bahr, M. Karas, *Influence of pressure in the first pumping stage on analyte desolvation and fragmentation in nano-ESI MS*, Anal. Chem., 73 (2001) 6040.
35. F. Bitsch, C.H.L. Shackleton, W. Ma, G. Park, M. Nieder, *Taxoid side-chain structure determination by ESI-MS–MS*, Rapid Commun. Mass Spectrom., 7 (1993) 891.
36. A.K. Chaudhary, M. Nokubo, T.D. Oglesby, L.J. Marnett, I.A. Blair, *Characterization of endogenous DNA adducts by LC–ESI-MS–MS*, J. Mass Spectrom., 30 (1995) 1157.
37. D.J. Douglas, J.B. French, *Collisional focusing effects in radio frequency quadrupoles*, J. Am. Soc. Mass Spectrom., 3 (1992) 398.
38. M. Hail, I. Mylchreest, Proceedings of the 41st ASMS Conference on Mass Spectrometry and Allied Topics, May 31 - June 4, 1993, San Francisco, CA, p. 745.
39. R.D. Voyksner, H. Lee, *Investigating the use of an octapole ion guide for ion storage and high-pass mass filtering to improve quantitative performance of ESI ion-trap MS*, Rapid Commun. Mass Spectrom., 13 (1999) 1427.

40. A.V. Tolmachev, H.R. Udseth, R.D. Smith, *Charge capacity limitations of RF ion guides in their use for improved ion accumulation and trapping in MS*, Anal. Chem., 72 (2000) 970.

41. S.A. Shaffer, D.C. Prior, G.A. Anderson, H.R. Udseth, R.D. Smith, *An ion funnel interface for improved ion focussing and sensitivity using ESI-MS*, Anal. Chem., 70 (1998) 4111.

42. S.A. Shaffer, A. Tolmachev, D.C. Prior, G.A. Anderson, H.R. Udseth, R.D. Smith, *Characterization of an improved electrodynamic ion funnel interface for ESI-MS*, Anal. Chem., 71 (1999) 2957.

43. K. Giles, B.H. Bateman, *Evaluation of a stacked-ring RF ion transmission device at intermediate pressures*, Proceedings of the 49th ASMS Conference in Mass Spectrometry and Allied Topics, May 27-31, 2001, Chicago, IL.

44. J. Zeleny, The electrical discharge from liquid points, and a hydrostatic method of measuring the electric intensity at their surfaces, Phys. Rev., 10 (1917) 1.

45. R.B. Cole (Ed.), *Electrospray Ionization Mass Spectrometry: Fundamentals, Instrumentation, and Applications*, 1997, Wiley Interscience.

46. B.N. Pramanik, A.K. Ganguly and M.L. Gross, *Applied electrospray mass spectrometry*, 2002, Marcel Dekker Inc., New York.

47. J.B. Fenn, M. Mann, C.K. Meng, S.F, Wong, C.M. Whitehouse, *ESI for MS of large biomolecules*, Science, 246 (1989) 64.

48. J.B. Fenn, M. Mann, C.K. Meng, S.F. Wong, C.M. Whitehouse, *ESI– principles and practice*, Mass Spectrom. Rev., 9 (1990) 37.

49. M. Hamdan. O. Curcuruto, *Development of the ESI technique*, Int. J. Mass Spectrom. Ion Processes, 108 (1991) 93.

50. S.J. Gaskell, *ESI: Principles and practice*, J. Mass Spectrom., 32 (1997) 677.

51. B.A. Thomson, *API and LC–MS - Together at last*, J. Am. Soc. Mass Spectrom., 9 (1998) 187.

52. R.D. Smith, J.A. Loo, C.G. Edmonds, C.J. Barinaga, H.R. Udseth, *New developments in biochemical MS: ESI*, Anal. Chem., 62 (1990) 882.

53. M. Mann, *ESI: Its potential and limitations as an ionization method for biomolecules*, Org. Mass Spectrom., 25 (1990) 575.

54. S.A. Carr, M.E. Hemling, M.F. Bean, G.D. Roberts, *Integration of MS in analytical biotechnology*, Anal. Chem., 63 (1991) 2802.

55. R.D. Smith, J.A. Loo, R.R. Ogorzalek Loo, M. Busman, H.R. Udseth, *Principles and practice of ESI-MS for large polypeptides and proteins*, Mass Spectrom. Rev., 10 (1991) 359.

56. E.D. Lee, J.D. Henion, *Thermally assisted ESI interface for LC–MS*, Rapid Commun. Mass Spectrom., 6 (1992) 727.

57. J.F. Banks, Jr., S. Shen, C.M. Whitehouse, J.B. Fenn, *Ultrasonically-assisted ESI for LC–MS determination of nucleosides from a transfer RNA digest*, Anal Chem., 66 (1994) 406.

58. R. Kostiainen, A.P. Bruins, *Effect of multiple sprayers on dynamic range and flow-rate limitations in ESI-MS*, Rapid Commun. Mass Spectrom., 8 (1994) 549.

59. J. Shiea, C.-H. Wang, *Applications of multiple channel ESI sources for biological sample analysis*, J. Mass Spectrom., 32 (1997) 247.

60. C.-M. Hong, F.-C. Tsai, J. Shiea, *A multiple channel ESI source used to detect highly reactive ketenes from a flow pyrolyzer*, Anal. Chem., 72 (2000) 1175.

61. R. King, R. Bonfiglio, C. Fernandez-Metzler, C. Miller-Stein, T. Olah, *Mechanistic investigation of ionization suppression in ESI*, J. Am. Soc. Mass Spectrom., 11 (2000) 942.

62. L. Zeng, D.B. Kassel, *Developments of a fully automated parallel LC–MS system for the analytical characterization and preparative purification of combinatorial libraries*, Anal. chem., 70 (1998) 4380.

63. V. de Biasi, N. Haskins, A. Organ, R. Bateman, K. Giles, S. Jarvis, *High-throughput LC–MSc analysis using a novel MUX ESI interface*, Rapid Commun. Mass Spectrom., 13 (1999) 1165.

64. D.L. Hiller, A.H. Brockman, L. Goulet, S. Ahmed, R.O. Cole T Covey, *Application of a non-indexed dual sprayer pneumatically assisted ESI source to the high throughput quantitation of target compounds in biological fluids*, Rapid Commun. Mass Spectrom., 14 (2000) 2034.

65. C. Eckers, J.-C. Wolff, N. J. Haskins, A.B. Sage, K. Giles, R. Bateman, *Accurate mass LC–MS on oa-TOF-MS using switching between separate sample and reference sprays. 1. Proof of concept*, Anal. Chem., 72 (2000) 3683.

66. J.-C. Wolff, C. Eckers, A.B. Sage, K. Giles, R. Bateman, *Accurate mass LC–MS on Q–oa-TOF-MS using switching between separate sample and reference sprays. 2. Applications using the dual ESIn source*, Anal. Chem., 73 (2001) 2605.

67. Y. Takahashi, S. Fujimaki, T. Kobayashi, T. Morita, T. Higuchi, *Accurate-mass determination by multiple sprayers nano-ESI-MS on a magnetic sector instrument*, Rapid Commun. Mass Spectrom., 14 (2000) 947.

68. J.C. Hannis, D.C. Muddiman, *A dual ESI source combined with hexapole accumulation to achieve high mass accuracy of biopolymers in FT-ICR-MS*, J. Am. Soc. Mass Spectrom., 11 (2000) 876.

69. D.C. Gale, R.D. Smith, *Small volume and low flow-rate ESI-MS of aqueous samples*, Rapid Commun. Mass Spectrom., 7 (1993) 1017.

70. M.R. Emmett, R.M. Caprioli, *Micro-ESI-MS: ultra-high-sensitivity analysis of peptides and proteins*, J. Am. Soc. Mass Spectrom., 5 (1994) 605.

71. R.H. Robins, J.E. Guido, *Design, construction and application of a simple packed capillary LC–ESI-MS system*, Rapid Commun. Mass Spectrom., 11 (1997) 1661.

72. M.S. Wilm, M. Mann, *ESI and Taylor-cone theory, Dole's beam of macromolecules at last?*, Int. J. Mass Spectrom. Ion Processes, 136 (1994) 167.

73. M.S. Wilm, M. Mann, *Analytical properties of the nano-ESI ion source*, Anal. Chem., 68 (1996) 1.

74. J.H. Wahl, D.R. Goodlett, H.R. Udseth, R.D. Smith, *Attomole level CE–MS protein analysis using 5-μm-ID capillaries*, Anal. Chem., 64 (1992) 3194.

75. G. Valaskovic, F.W. McLafferty, *Long-lived metallized tips for nanoliter ESI-MS*, J. Am. Soc. Mass Spectrom., 7 (1996) 1270.

76. J.F. Kelly, L. Ramaley, P. Thibault, *CE–ESI-MS at submicroliter flow-rates: practical considerations and analytical performance*, Anal. Chem., 69 (1997) 51.

77. D.R. Barnidge, S. Nilsson, K.E. Markides, *A design for low-flow sheathless ESI emitters*, Anal. Chem., 71 (1999) 4115.

78. Y.Z. Chang, Y.R. Chen, G.R. Her, *Sheathless CE–ESI-MS using a carbon-coated tapered fused-silica capillary with a beveled edge*, Anal. Chem., 73 (2001) 5083.

79. M. Wetterhall, S. Nilsson, K.E. Markides, J. Bergqvist, *A conductive polymeric material used for nano-ESI needle and low-flow sheathless ESI applications*, Anal.

Chem., 74 (2000) 239.
80. S. Nilsson, M. Svedberg, J. Pettersson, F. Björefors, K.E. Markides, L. Nyholm, *Evaluations of the stability of sheathless ESI-MS emitters using electrochemical techniques*, Anal. Chem., 73 (2001) 4607.
81. K. Vanhoutte, W. van Dongen, E.L. Esmans, *On-line nano-LC–ESI-MS: Effect of the mobile-phase composition and the ESI tip design on the performance of a NanoFlow^{TM} ESI*, Rapid Commun. Mass Spectrom., 12 (1998) 15.
82. N. Lion, T.C. Rohner, L. Dayon, I.L. Arnaud, E. Damoc, N. Youhnovski, Z.-Y. Wu, C. Roussel, J. Josserand, H. Jensen, J.S. Rossier, M. Przybylski, H.H. Girault, *Microfluidic systems in proteomics*, Electrophoresis, 24 (2003) 3533.
83. W.-C. Sung, H. Makamba, S.-H. Chen, *Chip-based microfluidic devices coupled with ESI-MS*, Electrophoresis, 26 (2005) 1783.
84. Q. Xue, F. Foret, Y.M. Dunayevskiy, P.M. Zavracky, N.E. McGruer, B.L. Karger, *Microchannel microchip ESI-MS*, Anal. Chem., 69 (1997) 426.
85. I.M. Lazar, R.S. Ramsey, S. Sundberg, J.M. Ramsey, *Sub-amol-sensitivity microchip nano-ESI source with TOF-MS detection*, Anal. Chem., 71 (1999) 3627.
86. D. Figeys, Y. Ning, R. Aebersold, *A microfabricated device for rapid protein identification by micro-ESI–ion-trap MS*, Anal. Chem., 69 (1997) 3153.
87. J.H. Chan, A.T. Timpermann, D. Qin, R. Aebersold, *Microfabricated polymer devices for automated sample delivery of peptides for analysis by ESI-MS–MS*, Anal. Chem., 71 (1999) 4437.
88. J.-S. Kim, D.R. Knapp, *Microfabricated PDMS multichannel emitter for ESI-MS*, J. Am. Soc. Mass Spectrom., 12 (2001) 463.
89. L. Licklider, X.-Q. Wang, A. Desai, Y.-C. Tai, T.D. Lee, *A micromachined chip-based ESI source for MS*, Anal. Chem., 72 (2000) 367.
90. C.H. Yuan, J. Shiea, *Sequential ESI analysis using sharp-tip channels fabricated on a plastic chip*, Anal. Chem., 73 (2001) 1080.
91. C. Wang, R. Oleschuk, F. Ouchen, J. Li, P. Thibault, D. J. Harrison, *Integration of immobilized trypsin bead beds for protein digestion within a microfluidic chip incorporating CE separations and an ESI-MS interface*, Rapid Commun. Mass Spectrom., 14 (2000) 1377.
92. N.X. Xu, Y.H. Lin, S.A. Hofstadler, D.Matson, C.J. Call, R.D. Smith, *A microfabricated dialysis device for sample cleanup in ESI-MS*, Anal. Chem., 70 (1998) 3553.
93. B. Zhang, H. Liu, B. L. Karger, F. Foret, *Microfabricated devices for CE–ESI-MS*, Anal. Chem., 71 (1999) 3258.
94. K.-M. Dethy, B.L. Ackermann, C. Delatour, J.D. Henion, G.A. Schultz, *Demonstration of direct bioanalysis of drugs in plasma using nano-ESI infusion from a silicon chip coupled with MS–MS*, Anal. Chem., 75 (2003) 805.
95. C.K. Meng, M. Mann, J.B. Fenn, Proceedings of the 36th ASMS Conference on Mass Spectrometry and Allied Topics, June 5-10, 1988, San Francisco, CA, p. 771.
96. T.R. Covey, R. Jong, H. Javahari, C. Liu, C. Thomson, Y. LeBlanc, *Design optimization of APCI instrumentation*, Proceedings of the 49th ASMS Conference on Mass Spectrometry and Allied Topics, May 27-32, 2001, Chicago, IL.
97. M. Sakairi, Y. Kato, *Multi-API interface for LC–MS*, J. Chromatogr. A, 794 (1998) 391.
98. A. Hirabayashi, M. Sakairi, H. Koizumi, *Sonic spray ionization method for API-MS*,

Anal. Chem., 66 (1994) 4557.
99. A. Hirabayashi, M. Sakairi, H. Koizumi, *Sonic spray MS*, Anal. Chem., 67 (1995) 2878.
100. D.A. Volmer, *Analysing thermally unstable compounds by LC–SSI–MS*, LC-GC Eur., 15 (2000) 838.
101. K. Hiraoka, S. Saito, J. Katsuragawa, I. Kudaka, *A new LC–MS interface: laser spray*, Rapid Commun. Mass Spectrom., 12 (1998) 1170.
102. D.B. Robb, T.R. Covey, A.P. Bruins, *APPI: an ionization method for LC–MS*, Anal. Chem., 72 (2000) 3653.
103. R.T. Gallagher, M.P. Balogh, P. Davey, M.R. Jackson, L.J. Southern, *Combined ESI–APCI source for use in high-throughput LC-MS applications*, Anal. Chem., 75 (2003) 973.
104. S. Cristoni, L.R. Bernardi, I. Biunno, F. Guidugli, *Analysis of peptides using partial (no discharge) APCI conditions with ion trap MS*, Rapid Commun. Mass Spectrom., 16 (2002) 1686.
105. S. Cristoni, L.R. Bernardi, I. Biunno, M. Tubaro, F. Guidugli, *Surface-activated no-discharge APCI*, Rapid Commun. Mass Spectrom., 17 (2003) 1973.
106. S. Cristoni, L.R. Bernardi, F. Guidugli, M. Tubaro, P. Traldi, *The role of different phenomena in SACI performance*, J. Mass Spectrom., 40 (2005) 1550.
107. G.J. van Berkel, G.L. Glish, S.A. McLuckey, *ESIn combined with ion trap MS*, Anal. Chem., 62 (1990) 1284.
108. S.A. McLuckey, G.J. van Berkel, G.L. Glish, E.C. Huang, J.D. Henion, *Ionspray LC–ion trap MS determination of biomolecules*, Anal. Chem., 63 (1991) 375.
109. J.G. Boyle, C.M. Whitehouse, J.B. Fenn, *An ion-storage TOF-MS for analysis of ESI ions*, Rapid Commun. Mass Spectrom., 5 (1991) 400.
110. J.G. Boyle, C.M. Whitehouse, *TOF-MS with a ESI ion beam*, Anal. Chem., 64 (1992) 2084.
111. K.D. Henry, E.R. Williams, B.H. Wang, F.W. McLafferty, J. Shabanowitz, D.F. Hunt, *FT-ICR-MS of large molecules by ESI*, PNAS, 86 (1989) 9075.
112. K.D. Henry, F.W. McLafferty, *ESI with FT-ICR-MS. Charge-state assignment from resolved isotopic peaks*, Org. Mass Spectrom., 25 (1990) 490.
113. K.D. Henry, J.P. Quinn, F.W. McLafferty, *High-resolution ESI mass spectra of large molecules*, J. Am. Chem. Soc., 113 (1991) 5447.
114. S.C. Beu, M.W. Senko, J.P. Quinn, F.W. McLafferty, *Improved FT-ICR-MS of large biomolecules*, J. Am Soc. Mass Spectrom., 4 (1993) 190.
115. B.E. Winger, S.A. Hofstadler, J.E. Bruce, H.R. Udseth, R.D. Smith, *High-resolution accurate mass measurements of biomolecules using a new ESI-FT-ICR-MS*, J. Am. Soc. Mass Spectrom., 4 (1993) 566.
116. M.V. Gorshkov, L. Paša-Tolić, J.E. Bruce, G.A. Anderson, R.D. Smith, *A dual-trap design and its applications in ESI-FT-ICR-MS*, Anal. Chem., 69 (1997) 1307.
117. M.W. Senko, C.L. Hendrickson, M.R. Emmett, S.D.-H. Shi, A.G. Marshall, *External accumulation of ions for enhanced ESI-FT-ICR-MS*, J. Am. Soc. Mass Spectrom., 8 (1997) 970.
118. K. Sannes-Lowery, R.H. Griffey, G.H. Kruppa, J.P. Spier, S.A. Hofstadler, *Multipole storage assisted dissociation, a novel in-source dissociation technique for ESI generated ions*, Rapid Commun. Mass Spectrom., 12 (1998) 1957.
119. M.E. Belov, E.N. Nikolaev, G.A. Anderson, K.J. Auberry, R. Harkewicz, R.D. Smith,

ESI-FT-ICR-MS using ion preselection and external accumulation for ultrahigh sensitivity, J. Am. Soc. Mass Spectrom., 12 (2001) 38.

120. M.E. Belov, E.N. Nikolaev, G.A. Anderson, H.R. Udseth, T.P. Conrads, T.D. Veenstra, C.D. Masselon, M.Y. Gorshkov, R.D. Smith, *Design and performance of an ESI interface for selective external ion accumulation coupled to a FT-ICR-MS*, Anal. Chem., 73 (2001) 253.

121. S.M. Patrie, J.P. Charlebois, D. Whipple, N.L. Kelleher, C.L. Hendrickson, J.P. Quinn, A.G. Marshall, B. Mukhopadhyay, *Construction of a hybrid Q–FT-ICR-MS for versatile MS–MS above 10 kDa*, J. Am. Soc. Mass Spectrom., 15 (2004) 1099.

122. J.E.P. Syka, J.A., Marto, D.L. Bai, S. Horning, M.W. Senko, J.C. Schwartz, B. Ueberheide, B. Garcia, S. Busby, T. Muratore, J. Shabanowitz, D.F. Hunt, *Novel linear Q–LIT–FT-ICR-MS: Performance characterization and use in the comparative analysis of histone H3 post-translational modifications*, J. Proteome Res., 3 (2004) 621.

123. S.A. Hofstadler, D.A. Laude, Jr., *ESI in the strong magnetic field of a FT-ICR-MS*, Anal. Chem., 64 (1992) 569.

124. S.A. Hofstadler, E. Schmidt, Z. Guan, D.A. Laude, Jr., *Concentric tube vacuum chamber for high magnetic field, high pressure ionization in a FT-ICR-MS*, J. Am. Soc. Mass Spectrom., 4 (1993) 168.

125. M.H. Allen, I.A.S. Lewis, *ESI on magnetic instruments*, Rapid Commun. Mass Spectrom., 3 (1989) 255.

126. C.-K. Meng, C.N. McEwen, B.S. Larsen, *ESI on a high-performance magnetic-sector MS*, Rapid Commun. Mass Spectrom., 4 (1990) 147.

127. B.S. Larsen, C.N. McEwen, *An ESI source for magnetic sector MS*, J. Am. Soc. Mass Spectrom., 2 (1991) 205.

128. R.T. Gallagher, J.R. Chapman, M.Mann, *Design and performance of an ESI source for a doubly-focusing magnetic sector MS*, Rapid Commun. Mass Spectrom., 4 (1990) 369.

129. R.B. Cody, J. Tamura, B.D. Musselman, *ESI–magnetic sector MS: Calibration, resolution and accurate mass measurements*, Anal. Chem., 64 (1992) 1561.

130. P. Dobberstein, E. Schröder, *Accurate mass determination of a high molecular weight protein using ESI with a magnetic sector instrument*, Rapid Commun. Mass Spectrom., 7 (1993) 861.

131. A.P.L. Wang, L. Li, *Pulsed sample introduction interface for combining flow injection analysis with multiphoton ionization TOF-MS*, Anal. Chem., 64 (1992) 769.

132. A.P.L. Wang, X. Guo, L. Li, *LC–TOF-MS with a pulsed sample introduction interface*, Anal. Chem., 66 (1994) 3664.

133. L. Li, A.P.L. Wang, L.D. Coulson, *Continuous-flow MALDI-MS*, Anal. Chem., 65 (1993) 493.

134. D.S. Nagra, L. Li, *LC–TOF-MS with continuous-flow MALDI*, J. Chromatogr. A, 711 (1995) 235.

135. R.M. Whittal, L.M. Russon, L. Li, *Development of LC–MS using continuous-flow MALDI-TOF-MS*, J. Chromatogr. A, 794 (1998) 367.

136. K.K. Murray, D.H. Russell, *Liquid sample introduction for MALDI*, Anal. Chem., 65 (1993) 2534.

137. K.K. Murray, T.M. Lewis, M.D. Beeson, D.H. Russell, *Aerosol MALDI for LC–TOF-MS*, Anal. Chem., 66 (1994) 1601.

138. X. Fei, K.K. Murray, *On-line coupling of gel permeation chromatography with*

MALDI-MS, Anal. Chem., 68 (1996) 3555.
139. T. Meyer, D. Waidelich, A.W. Frahm, *Separation and first structure elucidation of Cremophor® EL-components by hyphenated CE and delayed extraction-MALDI-TOF-MS*, Electrophoresis, 23 (2002) 1053.
140. T.J. Tegeler, Y. Mechref, K. Boraas, J.P. Reilly, M.V. Novotny, *Microdeposition device interfacing capillary electrochromatography and microcolumn LC with MALDI-MS*, Anal. Chem., 76 (2004) 6698.
141. T. Miliotis, S. Kjellström, J. Nilsson, T. Laurell, L.-E. Edholm, G. Marko-Varga, *Capillary LC interfaced to MALDI-TOF-MS using an on-line coupled piezoelectric flow-through microdispenser*, J. Mass Spectrom., 35 (2000) 369.
142. C. Ericson, Q.T. Phung, D.M. Horn, E.C. Peters, J.R. Fitchett, S.B. Ficarro, A.R. Salomon, L.M. Brill, A. Brock, *An automated noncontact deposition interface for LC–MALDI-MS*, Anal. Chem., 75 (2003) 2309.
143. B. Zhang, C. McDonald, L. Li, *Combining LC with MALDI-MS using a heated droplet interface*, Anal. Chem., 76 (2004) 992.
144. R. Mukhopadhyay, *The automated union of LC and MALDI-MS*, Anal. Chem., 77 (2005) 150A.
145. V.V. Laiko, M.A. Baldwin, A.L. Burlingame, *Atmospheric pressure MALDI-MS*, Anal. Chem., 72 (2000) 652.
146. V.V. Laiko, S.C. Moyer, R.J. Cotter, *Atmospheric pressure MALDI/ion trap MS*, Anal. Chem., 72 (2000) 5239.
147. J.M. Daniel, V.V. Laiko, V.M. Doroshenko, R. Zenobi, *Interfacing LC with atmospheric-pressure MALDI-MS*, Anal. Bioanal. Chem., 383 (2005) 895.
148. M. Constapel, M. Schellenträger, O.J. Schmitz, S. Gäb, K.J. Brockmann, R. Giese, Th. Benter, *APLI: a novel ionization method for LC–MS*, Rapid Commun. Mass Spectrom., 19 (2005) 326.

6

ATMOSPHERIC-PRESSURE IONIZATION

1. Introduction

In interface development for liquid chromatography–mass spectrometry (LC–MS), the mobile phase was initially considered as a disturbing and restricting factor: it should be evaporated prior to introduction into the mass spectrometer. Subsequently, strategies were considered that actually take advantage of the presence of the mobile phase (cf. Ch. 3.3.3). The ability of the direct liquid introduction (DLI) interface to transfer highly labile compounds, e.g., vitamin B12 (Figure 4.7), to the gas phase and make them available to chemical ionization (CI) is believed to be strongly supported by the presence of the mobile phase. The analytes included in the desolvating droplets are subjected to in-beam or direct chemical ionization [1-2].

The potential of using the inevitable presence of the mobile phase in interface development and in analyte ionization was recognized by Arpino and Guiochon [3]. The initial assumption in developing LC–MS interfaces was that ionization should follow the vaporization of the intact neutral compound. Although this is a viable approach for some compounds, it excludes the analysis of many others, especially highly polar, ionic, and high molecular-mass compounds. Arpino and Guiochon [3] reviewed a number of liquid-based ionization techniques, i.e., electrospray (ESI), electrohydrodynamic ionization (EHI), field desorption (FD), thermospray (TSP), and fast-atom bombardment (FAB), most of which at that time (1982) were still in an early stage of development. They concluded that all methods have some common features. In the positive-ion mode, they all generate cationized molecules for analytes that are readily soluble in the liquid matrix used and give little fragmentation. Desorption of preformed ions from the liquid phase appears to be the common mechanism. The energy needed in the desorption can be applied in a number of ways, as indicated by the variety of methods reviewed. In another paper, Vestal [4] reached similar conclusions. This concept opened new directions in LC–MS research.

In the current LC–MS interfaces, *i.e.*, ESI and atmospheric-pressure chemical ionization (APCI), interfacing and analyte ionization are closely interrelated. The column effluent is nebulized and ionization takes place in the aerosol generated, either with or without a primary external source of ionization. Ionization mechanisms of ESI, APCI, and atmospheric-pressure photoionization (APPI) are discussed in this chapter.

2.   History of electrospray ionization

2.1   First experiments of Dole

The first applications of ESI in MS date from 1968. Dole *et al.* [5-6] investigated the possibility to transfer macromolecules from the liquid phase to the gas phase by electrospraying dilute solutions in a nitrogen bath gas. The hypothesis of Dole and coworkers was that macro-ions can be produced by desolvating the charged droplets produced in electrospray. This ionization mechanism is called the *charge residue model*.

2.2   Electrohydrodynamic ionization

Whereas in the experiments of Dole and coworkers the sample solutions are sprayed in an atmospheric-pressure region, Evans *et al.* [7-8] investigated the applicability of electrospraying solutions in a vacuum. In EHI, charged droplets and/or (solvated) ions are emitted directly from the apex of a Taylor cone, as the result of the interaction of a strong electrostatic field with a liquid meniscus at the end of a capillary tube. Evans *et al.* [7-8] investigated EHI of polar and ionic organic molecules. Because the system is operated under high-vacuum conditions, the use of nonvolatile organic solvents, *e.g.*, glycerol, is required for analyte introduction. Sodium iodide is added to increase the conductivity. The compounds investigated are saccharides, nucleosides, and small peptides. Ions formed by attachment of a cation or an anion to the solvent or analyte molecules are emitted to the gas phase. Extensive series of cluster ions, *i.e.*, [M + (glycerol)$_n$ + Na]$^+$, are observed as well.

Zolotai *et al.* [9-10] performed similar experiments using glycerol or water as solvent and they introduced the term 'field evaporation of ions from solution'. Direct emission of preformed ions in solution is assumed to occur.

2.3 Ion evaporation experiments of Iribarne and Thomson

Iribarne and Thomson [11-14] investigated the direct emission of ions from liquid droplets. In their experimental setup, a liquid solution is pneumatically nebulized in an atmospheric-pressure chamber and the droplets produced are charged by random statistical charging [15] using an induction electrode positioned close to the nebulizer. Solvated singly-charged ions are formed in the evaporating spray. These ions are sampled from the chamber into a differentially pumped quadrupole mass spectrometer by means of a sampling orifice. Their theoretical description of the process is adapted later for TSP and ESI.

2.4 Thermospray ionization

With the development of the TSP interface for LC–MS (Ch. 4.7), Vestal *et al.* [4, 16-18] also introduced a new ionization technique. While the analyte ionization in their first experiments was initiated by electrons from a filament, they subsequently demonstrated that collision of the vapour-droplet beam from the TSP nebulizer with a nickel-plated copper plate leads to soft ionization of analytes. Next, the collision was found not to be a vital step in the process [18]. The presence of a volatile buffer or acid in the mobile phase appeared more important in TSP, *i.e.*, in charging the droplets generated by TSP, and in generation of preformed ions in solution. The ionization phenomena were explained in terms of the ion evaporation (IEV) model [4].

2.5 Soft desolvation or charge residue model

The explanation of TSP in terms of ion evaporation is criticized by the group of Röllgen [19-20]. According to Röllgen, the actual electrical fields needed for IEV are much higher than those calculated by Iribarne and Thomson, especially because they neglected the shielding of the field by the polarized water. Furthermore, the removal of a solvated ion from the charged surface is such an intervening event to a droplet surface under field stress that it would most likely generate a sequence of events, *e.g.*, the development of a jet in which a number of charges are removed from the droplet.

The alternative mechanism suggested is based on soft desolvation of the ions by solvent evaporation from small charged droplets produced by either electrohydrodynamic or mechanical instabilities (or both). This mechanism is similar to the charge residue hypothesis presupposed by Dole [5-6] in their ESI experiments. The desolvation is most effective when the droplets are small and the number of charges on a droplet is small as well.

3. Electrospray ionization

The original ESI experiments of Dole were continued by Fenn [21-22], implementing molecular beam technology. The liquid is electrosprayed into a bath gas. The dispersion of ions, solvent vapour, and bath gas is expanded into a vacuum chamber, forming a supersonic jet, the core of which is sampled to an MS system by means of a skimmer. A schematic diagram of the experimental setup is shown in Figure 5.2.

Yamashita and Fenn [21-22] tried in vain to reproduce the experiments of Dole [5-6], but nevertheless continued to investigate the sequence of electrospraying and droplet evaporation with low molecular-mass compounds. From their observation of single-charge solvated lithium ions, apparently emitted from a droplet containing *ca.* 1000 lithium chloride ion pairs (the concentration LiCl was 2.7×10^{-3} mol/l), they concluded that the hypothesis of Dole was not valid in their experiment. Fenn recognized the importance of the work of Iribarne and Thomson [11-14] in this field. Simultaneously, Aleksandrov [23] developed an ESI source fitted on a magnetic sector instrument. The breakthrough in ESI was in 1988, when the group of Fenn showed the generation of multiple-charge ions from proteins by ESI [24]. The generation of ions by ESI has been discussed in a number of review papers, *e.g.*, [25-30].

Below, an overview of the ionization processes in ESI is provided. A detailed in-depth discussion of the ESI mechanisms is beyond the purpose and scope of this chapter. While most mechanistic discussions in ESI are focussed on the ionization of proteins, we try to focus on the ionization of small molecules. ESI-MS of proteins is discussed in Ch. 16.2.3.

3.1   Overview

In an ESI interface for LC–MS, the column effluent from a reversed-phase (RP) LC, *i.e.,* a solvent mixture of methanol or acetonitrile and up to 10 mmol/l aqueous buffer or 0.1% aqueous acid, is nebulized into an API source. Pure ESI nebulization can only be achieved at flow-rates below 10 µl/min. Therefore, in most LC–MS applications, pneumatically-assisted ESI (Ch. 5.5.2) is performed:

the liquid flow is nebulized into small droplets by a combined action of a strong electric potential between needle and counter electrode, *e.g.*, 3 kV, and a high-speed concurrent N_2 flow (50–100 l/hr). Pneumatically-assisted ESI is often indicated with the Sciex trade name *ionspray*. The ESI nebulization process results in the formation of small droplets with an excess charge, *i.e.*, positive charges when the source is operated in positive-ion mode and negative charges in the negative-ion mode. These excess charges might be due to electrolyte ions or performed analyte ions. In their flight between the ESI needle and the ESI source block, neutral solvent molecules evaporate from the droplet surface. As a result, the droplet size decreases. This in turn reduces the distances between the excess charges at the droplet surface. After some time, the surface tension of the liquid can no longer accommodate the increasing Coulomb repulsion between the excess charges at the surface. At this point, a Coulomb explosion or field-induced electrohydrodynamic disintegration process leads to disintegration of the droplets. Surface disturbance of the droplet under field stress may grow out to Taylor cones, from which highly-charged microdroplets are emitted with a radius of less than 10% of that of the initial droplets. The processes of solvent evaporation and electrohydrodynamic droplet disintegration may be repeated a number of times, leading to smaller and smaller offspring droplets. At least three processes are responsible for the formation of gas-phase analyte ions from these microdroplets: (1) soft desolvation, (2) IEV, and (3) CI at the droplet surface or by gas-phase ion-molecule reactions. Finally, the gas phase ions generated by these processes can be mass analysed.

In the *charge-residue model* of Dole [5-6], analyte molecules are present in solution as preformed ions, *e.g.*, by choosing an appropriate pH below the pK_a of a basic molecule. The sequence of solvent evaporation and electrohydrodynamic droplet disintegration proceeds until microdroplets containing only one preformed analyte ion per droplet, *i.e.*, a solvated preformed analyte ion. After evaporation of the solvent, the preformed analyte ion is released to the gas phase.

In the *ion-evaporation model* of Iribarne and Thomson [11-14], the sequence of solvent evaporation and electrohydrodynamic droplet disintegration also leads to the production of microdroplets. Gas-phase ions can be generated from the highly-charged microdroplets, at which the local field strength is sufficiently high to allow preformed ions in solution to be emitted into the gas phase (IEV).

These two mechanism to some extent are complementary, because droplets that are not sufficiently charged to enable IEV may finally lead to gas-phase ion production by soft desolvation processes. It is in fact difficult to decide which of the two processes is most important in the actual ion production of a particular analyte under given experimental conditions. In this respect, the mechanistic discussions of Smith and Light-Wahl [31] on the conservation of noncovalent associates of proteins and drugs in ESI is of relevance. One of their questions is whether the preservation of noncovalent complexes is reasonable within the context of ionization mechanisms proposed for ESI. According to the IEV mechanism, highly desolvated multi-protonated molecules are generated. On both kinetic and thermodynamic grounds, it is difficult to accept that in such a process the necessary stripping of noncovalently associated solvent molecules can be achieved without influencing other noncovalent associations. The production of gas-phase highly-solvated multi-protonated molecules would be more reasonable. Therefore, they suggested a different ionization process, which in fact closely resembles the charge residue model of Dole. The initial highly-charged droplets in the 1-μm-ID range undergo (a series of) disintegrations, resulting in nano-droplets in the 10-nm-ID range. In these nano-droplets, the range of interactions and associations from the bulk solution are substantially maintained. These nano-droplets further shrink to yield the ions detected in MS. The shrinkage does not only involve charge-preserving solvent evaporation, but also the generation and emission of solvent cluster ions. An obvious prerequisite is that the noncovalently associated species remains associated while the residual solvent is removed.

Next to these two processes, there are at least two other gas-phase processes, which partly determine the analyte ionization and to a large extent determine the appearance of the mass spectrum of the analyte observed.

With respect to analyte ionization, ion-molecule reactions may take place between gas-phase buffer ions and neutral analyte molecules, either at the droplet surface, or in the gas phase, after soft desolvation of neutral molecules. These processes will be more relevant in ESI-MS of small molecules; most peptides and proteins are readily present as preformed ions in solution. Amad et al. [32] demonstrated that the analyte signal can be completely suppressed in a solvent with a proton affinity higher than that of the analyte. These types of gas-phase ion-molecule reaction are identical to the ionization processes in APCI (Ch. 6.4).

The appearance of the mass spectrum is not only determined by the generation of gas-phase analyte ions, but also by subsequent processes occurring in the time between ion production and mass analysis and detection. Gas-phase ions will be solvated again in the humid atmosphere of the API source. In their journey between generation and entrance into the mass analyser, the ions will experience several interactions. Voltages are applied to the ion-sampling orifice, which to some extent help in the transmission of ions through the atmospheric-pressure/vacuum transition region (Ch. 5.4.4), but also play a role in the declustering of the solvated analyte ions, and may induce collision-induced dissociation (CID) of analyte ions. Potential analytical use of such gas-phase processes was evaluated. Ogorzalek Loo et al. [33-35] studied gas-phase ion-molecule reactions between protein ions and organic bases [33-34] and gas-phase ion-ion reactions between protein ions and ions of opposite charge generated by means of a corona discharge or an ESI ion source [35].

ESI is best described as a mixed-mode ionization, where various processes contribute to the final result. The soft-desolvation and IEV models indicate the importance of generating preformed analyte ions in solution. For most analytes with basic functions, *e.g.*, amines and amides, or acidic functions, *e.g.*, carboxylic acid or aromatic phenols, preformed ions can be produced by the selection of an appropriate pH of the mobile phase. For basic compounds, acidic mobile-phase conditions are selected, and basic conditions for acidic compounds. Analyte derivatization has been described to introduce a basic site or a fixed charge from a quaternary ammonium group in analytes that lack such properties.

3.2   Electrospray nebulization

ESI nebulization is the result of charging a liquid at a needle tip by applying a high potential (*ca.* 3 kV), between the needle and a nearby counter electrode. The formation of the aerosol depends on the competition between coulomb repulsion and surface tension. Stable nebulization strongly depends on experimental parameters such as the potential difference applied, the inner and outer diameter as well as the shape of the needle, and the composition of the liquid sprayed.

The onset of ESI nebulization was carefully described [22]. When no voltage is applied to a 0.1-mm-ID horizontally positioned capillary, drops fall off under the influence of gravity. With increasing the potential, the droplet size reduces and the droplets begin to have a horizontal component in their movement as well as a higher speed. At higher potentials, the droplets are formed from a nearly horizontally liquid jet emerging from the liquid column that extends beyond the end of the capillary. At still higher potentials, this liquid column elongates and a fairly sharp point is formed at its tip, the 'Taylor cone'. This is the actual onset of ESI nebulization. The droplets are produced as a result of electrically affected Rayleigh instabilities at the surface of the liquid jet emerging from the Taylor cone. This is the *axial spray* mode. When the potential is further increased, successively two changes in the appearance of the spray are observed. First, a sudden transition takes place: the liquid cone vanishes and a fine mist of droplets is produced from a number of points at the sharp edge of the capillary tip (the *rim emission* mode). The second transition is the formation of a discharge in the source, which is a condition to be avoided.

The onset voltages V_{on} can be estimated [36]:

$$V_{on} \approx 2 \times 10^5 \sqrt{\sigma \, r_c} \, \ln \frac{4 \, d}{r_c}$$

where σ is the surface tension of the liquid, r_c is the inner diameter of needle, and d is the distance between capillary and counter electrode. For a typical system with r_c = 0.1 mm and d = 40 mm, this equation predicts onset voltages for methanol, acetonitrile, DMSO, and water to be 2.2 kV, 2.5 kV, 3.0 kV, and 4.0 kV, respectively. For stable ESI performance, the voltage should be set a few hundred volts higher than the onset voltage. Under these conditions, the current between the needle and the counter electrode is 0.1-1 μA. Higher currents indicate the formation of a discharge. The negative-ion mode is especially prone to discharge formation. To prevent this, the use of a scavenger gas with positive electron affinity, *e.g.*, oxygen [21-22] or SF_6 [37], or the addition of chlorinated solvents, such as chloroform, to the mobile phase [38] have been proposed.

The conductivity of the mobile phase is a major factor in the electrostatic disruption of the liquid surface during ESI nebulization. Stable ESI conditions can only be achieved with semiconducting liquids (conductivity $10^{-6}-10^{-8}$ $\Omega^{-1}m^{-1}$). Pure organic solvents like dichloromethane, benzene, and hexane are only suitable for ESI after mixing with >10% polar solvent (Ch. 6.3).

3.3   Electrochemical processes

ESI nebulization involves a variety of electrochemical processes at the needle and at the counter electrode [27, 30]. The ESI interface can be considered as a electrochemical cell, in which part of the ion transport takes place through the gas phase (Figure 6.1). In positive-ion mode, an enrichment of positive electrolyte ions occurs at the solution meniscus as the result of an electrophoretic charge separation. The liquid meniscus is pulled into a cone which emits a fine mist of droplets with an excess positive charge. Charge balance is attained by electrochemical oxidation at the capillary tip and reduction at the counter electrode. The topic arose significant discussion in 2000 and the discussion partners continued to disagree on the role of electrochemistry in ESI-MS [39].

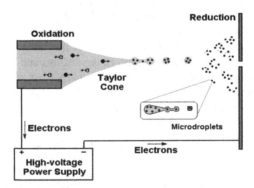

Figure 6.1: Schematic representation of the ESI source as electrochemical cell.

The group of Van Berkel [40-42] studied electrochemical processes in the ESI source, *e.g.*, by demonstrating that the radical cations observed in the ESI mass spectra of metalloporphyrins and polycyclic aromatic hydrocarbons have an electrochemical origin [40], and by proposing strategies to convert analytes into electrochemically-ionizable derivatives [42]. In a review on analytical applications of on-line electrochemistry–ESI-MS, Diehl and Karst [43] discussed topics related to the electrochemical processes in ESI as well.

3.4 Analyte concentration and properties

The total droplet or spray current in ESI nebulization depends on the conductivity of the liquid, which is proportional to the electrolyte concentration. At low analyte concentrations, the spray current is constant, because it is determined by the background electrolyte. The nature of the electrolyte has little influence on the current, because of its limited influence on the conductivity. Between 10^{-4} and 10^{-2} mol/l, the spray current is increased linearly with the electrolyte concentration. However, at an analyte-dependent concentration of *ca.* 10^{-4} mol/l, a fairly abrupt departure from linearity is observed [36, 44]. First, the ion current of the analyte levels off, and then even starts to decrease. This is attributed to various processes. As in the ESI nebulization process a more-or-less fixed amount of excess charge is generated on the droplets, a high analyte concentration may result in charge depletion: not all the analyte molecules can be charged given the limited number of charges available [27, 45]. However, Zook and Bruins [46] questioned this charge depletion model, because the saturation effect occurs at the same analyte concentration, irrespective of the electrolyte concentration and the applied voltage. They proposed saturation of the droplet surface with preformed analyte ions; not all analyte ions can find a favourable position at the droplet surface. Support for this comes from the observation of proton-bound dimers at higher analyte concentrations. In addition, upon the analysis of an equimolar mixture of a solvophilic analytes and a surface-active analyte, the former shows a lower response than the latter [47-48]. This indicates that analytes at the droplet surface are preferentially converted into gas-phase ions. It may also be concluded that surface-active sample constituents may suppress the response of less-surface-active analyte molecules [49] (Ch. 11.5). Theoretical models have been developed to describe the dependence of the analyte ion current on the analyte and electrolyte concentration [47].

The properties of the analyte affect the gas-phase ion production in ESI in a number of ways, *e.g.*, in its solvophobicity, and in the ability to accommodate one of more charges. In the analysis of quaternary ammonium compounds, the selected ion current increases with increasing chain length of the alkyl groups.

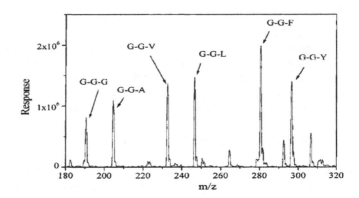

Figure 6.2: ESI mass spectrum of an equimolar mixture of six tripeptides (Gly-Gly-X, where X is Gly (G), Ala (A), Val (V), Leu (L), Phe (F), or Tyr (Y)). The response increases as the non-polar character of the side-chain of the C-terminal amino acid increases. Reprinted from [49] with permission. ©2000, American Chemical Society.

This effect can be explained from an increase in solvophobicity, *i.e.*, less work is required to remove the analyte from the droplet. At a further increase of the chain length, the signal will diminish, primarily because of the lower solubility of the analytes [44]. Similar results were obtained for a series of tripeptides (Gly-Gly-X, where X is Gly, Ala, Val, Leu, Phe, or Tyr). The peptides with the more polar side chain, *e.g.*, Gly, Ala, and Tyr, show lower response in ESI than the ones with the more nonpolar side chain, *e.g.*, Val, Leu, and Phe [49] (see Figure 6.2).

The charge state of an analyte ion depends on the number of sites that can accommodate a charge, *e.g.*, a proton or alkali cation in positive-ion mode. Protonation normally occurs at basic sites, *i.e.*, nitrogen atoms in the molecule. In a protein, these sites are the *N*-terminal and the three basic amino acid residues (Lys, Arg, His). At first approximation, the maximum number of positive charges a protein can carry is equal to the number of basic amino acids plus the amino terminus [50]. When the observed maximum is less than the number of basic sites, steric hindrance due to disulfide bridges and/or protein folding is assumed to reduce the actual number of available protonation sites. An additional constraint is that the bonding energy of one charge at any site on the molecule should be equal to or exceed the electrostatic repulsion energy due to the Coulomb repulsion by all other charges on the molecule [44, 51]. However, experimental evidence indicates that the charge-state distribution of proteins in the ESI mass spectra is determined by a number of parameters (Ch. 16.2.3).

3.5 Wrong-way-around electrospray

Kelly *et al.* [52] observed both positive-ion and negative-ion mass spectra of myoglobin at pH 3.5, whereas this would not have been expected from the general concept of ESI involving preformed ions in solution. The term *wrong-way-around* ESI was introduced for this phenomenon [53]. It refers to the observation of abundant protonated analyte molecules in a mass spectrum acquired under strong basic conditions and/or abundant deprotonated molecules acquired under strong acidic conditions. Numerous other examples can be found in literature. In search for explanations, Mansoori *et al.* [53] measured the pH of collected sprayed solutions. In contrast to other observations [54], only small changes in the solution pH were observed (less than 1 pH unit). Earlier, Gatlin and Tureček [54] investigated the pH changes in ESI with a pH-dependent equilibrium system, *i.e.*, the dissociation of Fe^{2+}(bpy)$_3$ and Ni^{2+}(bpy)$_3$ complexes. They concluded that a 10^3–10^4-fold increase in the $[H_3O^+]$ concentration takes place upon solvent evaporation. The results of Mansoori *et al.* [53] are confirmed by Zhou *et al.* [55], who monitored pH changes in the ESI plume by means of laser-induced fluorescence and a pH-sensitive fluorescent dye (Figure 6.3). Spraying in positive-ion mode a solution with a initial pH of 6.9 results in a pH of 5.7 at a position 8 mm downstream. The pH change is most pronounced in the first mm of the spray plume.

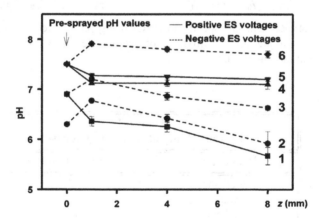

Figure 6.3: Plots of the pH in the spray plume, monitored by means of a pH-sensitive fluorescent dye, as a function of the axial distance from the emitter tip. ESI needle voltages applied are (1) + 4.0 kV, (2) – 3.0 kV, (3) – 3.3 kV, (4) + 4.0 kV, (5) + 3.5 kV, and (6) – 3.0 kV. Reprinted from [55] with permission, ©2002, American Chemical Society.

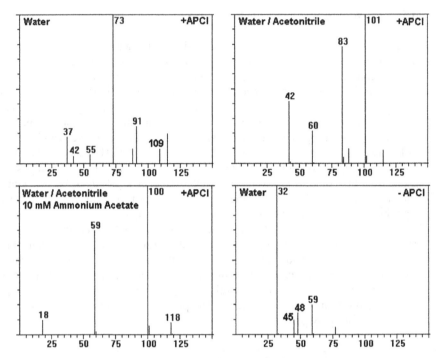

Figure 6.4: Reagent gas spectra in APCI using pure water in positive-ion and negative-ion mode and 1:1 mixtures of acetonitrile–water with and without 10 mmol/l ammonium acetate.

Table 6.1 Proton affinity of some reagent gases	
Reagent gas	**PA (kJ/mol)**
methane	536
water	697
methanol	773
acetonitrile	787
ammonia	854
pyridine	921

Further studies on caffeine showed that at near-neutral pH in a solution of low ionic strength, the protonated caffeine results from enrichment of electrolytically produced protons in the surface layer of the droplets [56]. At high pH, protonation of caffeine is due to gas-phase proton transfer from ammonium ions. At neutral and high pH at high ionic strength, protonated caffeine is generated by discharge-induced ionization. It could not be decided whether processes in the gas phase or at the gas-liquid interface were involved.

The observation of wrong-way-around ESI indicates that the pH influence of the mobile phase is more complicated than initially thought (Ch. 6.3.1). In terms of method development for the LC–MS, this is actually good news (Ch. 6.6.3).

4. Atmospheric-pressure chemical ionization

A major impetus to the use of APCI in LC–MS was given by the research group of Horning [57] (Ch. 5.6.1). APCI is based on solvent-mediated CI by ion-molecule reactions in an API source, initiated by electrons produced in the corona discharge. Instrumentation for APCI is discussed in Ch. 5.6.

4.1 Ionization by a corona discharge

Conventional low- and medium-pressure CI is based on a chemical reaction between a reagent-gas ion and an analyte molecule (Ch. 2.2.2). The reagent-gas ion is first produced by interaction of the reagent gas and energetic electrons, followed by a series of ion-molecule reactions. In APCI, initial ionization results from electrons produced by a corona discharge. The corona discharge needle is kept at 1–5 kV, generating a discharge current of 1–5 µA. The energetic electrons from the discharge are assumed to start a sequence of reactions:

$$N_2 + e^- \rightarrow N_2^{+\bullet} + 2\,e^-$$

In the presence of only traces of water, the nitrogen molecular ions enter a series of ion-molecule reactions [57], resulting in protonated water clusters:

$$N_2^{+\bullet} + H_2O \rightarrow H_2O^{+\bullet} + 2\,N_2$$

$$H_2O^{+\bullet} + H_2O \rightarrow H_3O^+ + HO\bullet$$

$$H_3O^+ + nH_2O \rightarrow H_3O^+.(H_2O)_n$$

The charge transfer in the first reaction is likely to occur because the ionization potential of water (12.6 eV) is lower than that of nitrogen (15.6 eV). When an APCI-MS system is run with pure water as mobile phase, a series of protonated water clusters $[(H_2O)_n + H]^+$ can be observed in the low m/z region with the cluster ion with n=4 being especially abundant due to the magic numbers determining the stability of such clusters (Figure 6.4). If APCI-MS is done in combination with LC, the solvent introduced in the mobile phase from the LC and ion-molecule reactions, especially proton transfer reactions, can take place between various mobile-phase constituents.

In RPLC–APCI-MS, where the mobile phase consists of a mixture of water and methanol or acetonitrile, and eventually a buffer, the formation of protonated water clusters can be considered as a starting point in a series of even-electron ion-molecule reactions. The protonated water clusters transfer their proton to any species in the gas mixture with a higher proton affinity (Table 6.1). The mass spectrum of acetonitrile (MeCN)–water mixture shows protonated MeCN–water clusters, $[(MeCN)_m (H_2O)_n + H]^+$, with m-values of 1–3, and n-values of 0–1. The addition of ammonium acetate to MeCN–water results in the observation of mixed solvent clusters, $e.g.$, $[(MeCN)_m + NH_4]^+$ and $[(MeCN)_m (H_2O) + NH_4]^+$ (Figure 6.4). Similar reactions can take place with other solvents.

4.2   Solvent-mediated (atmospheric-pressure) chemical ionization

In solvent-mediated CI, the composition of the reagent gas is derived from the mobile-phase constituents.

Positive-ion mode
In the positive-ion mode, the ion-molecule reactions are determined by the proton affinities of the reactants. The proton affinity (PA) or gas-phase basicity of molecule M is defined as the exothermicity of its protonation reaction:

$$PA_{(M)} \equiv \Delta H_{f(M)} + \Delta H_{f(H^+)} - \Delta H_{f(MH^+)}$$

Typical values of proton affinity for some reagent gases and mobile-phase constituents are given in Table 6.1. The proton-transfer reaction between the protonated solvent cluster SH^+, generated by the sequence of events indicated in Ch. 6.4.1, and the analyte molecule M is:

$$SH^+ + M \rightarrow S + MH^+$$

This reaction only proceeds, if the proton affinity of the analyte M exceeds that of the reagent gas S. Based on the PA values in Table 6.1, the reagent gas of MeCN–water is primarily determined by MeCN-related solvent clusters, while in an ammonium-acetate containing mobile phase the reagent gas is primarily determined by NH_4^+-related ions. This means, that in MeCN–water, any analyte with a proton affinity exceeding that of acetonitrile (797 kJ/mol) can be protonated. The addition of ammonium acetate results in a more selective ionization: only analytes with a proton affinity exceeding 854 kJ/mol can be protonated. Analytes with proton affinities below that of ammonia generally are not observed in APCI-MS with an ammonium-acetate containing mobile phase.

Next to protonated, adduct formation may take place:

$$M + SH^+ \rightarrow S + MH^+$$

leading for instance to an ammoniated molecule $[M+NH_4]^+$. Adduct formation with for instance NH_4^+ is observed when the proton affinity of the analyte is ± 40 kJ/mol from that of NH_3. Adduct ions to some extent broaden the applicability range of APCI-MS with a particular mobile-phase composition.

In summary, protonated solvent clusters are generated by ion-molecule reactions initiated by the corona discharge. These cluster ions act as reagent gas ions for the solvent-mediated APCI. The composition of the reagent gas is determined by the mobile-phase constituent with the highest proton affinity.

Some general considerations

This representation of APCI-MS ionization is very useful in practice, but some additions should be made to the description.

In the above discussion, it was assumed that the PA of the solvent clusters of the type $[(MeCN)_m\,(H_2O)_n + H]^+$ are identical. This of course is not really true: the various solvent cluster ions observed each have their own proton affinity, but in predicting analyte ionization in APCI-MS, these differences may be ignored.

Like with ESI (Ch. 6.3.1), the appearance of the mass spectrum is not only determined by the actual ionization event, but also by gas-phase processes occurring between ionization and entrance into the mass analyser. The theory outlined above predicts the formation of protonated or ammoniated analytes according to thermodynamic parameters. Decomposition and/or reformulation of the analyte ion composition may take place during the transport of ions, *i.e.*, the ions in the mass spectrum do not necessarily exactly reflect the ions initially generated in the ion-molecule reaction.

Apparent fragmentation of ions observed in APCI-MS spectra in most cases should not be considered as actual fragmentation, because the fragment ions are thermal degradation products, generated in the heated nebulizer, and subsequently ionized in the APCI source. An interesting example of this is the thermally-induced reduction of the NO_2-group to an NH_2-group, observed in the positive-ion APCI-MS analysis of aromatic nitro compounds [58]. A fragment due to the loss of 30 Da was observed. H/D exchange shows that this loss is not due to the loss of an NO•, but rather due to the indicated reduction.

Negative-ion mode

In negative-ion mode, a similar treatment holds, with superoxide O_2^- as an important initial ionic species, which is involved in ion-molecule reactions with the mobile-phase constituents. The proton-transfer reaction leading to the deprotonated analyte molecule $[M–H]^-$:

$$M + [S–H]^- \rightarrow [M–H]^- + S$$

Table 6.2 Gas-phase acidities (kJ/mol) of some compounds	
Compound	ΔH_{acid} (kJ/mol)
water	1607
methanol	1557
acetonitrile	1528
superoxide O_2^-	1449
acetic acid	1429
formic acid	1415
trifluoroacetic acid	1323

The reaction is determined by the relative gas-phase acidities ΔH_{acid} of the analyte and the reagent gas molecules, defined as:

$$\Delta H_{acid} \equiv \Delta H_{[S-H]^-} + \Delta H_{H^+} - \Delta H_S$$

The proton-transfer or abstraction reaction proceeds if the gas-phase acidity of the solvent-related anion exceeds that of the analyte molecule. Gas-phase acidities for some compounds are given in Table 6.2. Again, adduct formation, *i.e.*, attachment of anion A^- via the formation of a hydrogen bond, can take place:

$$M + A^- \rightarrow MA^-$$

For most polar molecules, the latter reaction is thermodynamically favourable, leading to the generation of $[M + HCOO]^-$ or $[M + CH_3COO]^-$ in mobile phases containing formic and acetic acid, respectively.

4.3   Electron-capture negative ionization APCI

While electron-capture negative ionization (ECNI) is an important ionization technique in GC–MS, for long it has not been very popular in LC–MS. Singh *et al.* [59] demonstrated that it is possible to perform ECNI in an APCI source. Low-energy, thermal electrons, generated in the initial step of the ionization process, can be captured by compounds with favourable electron affinities. After their conversion to pentafluorobenzyl derivatives, steroids and prostaglandins could be detected with 25–100 times improved detection limits compared to conventional negative-ion APCI. The strategy of analyte derivatization to introduce a high electron-affinity group has been applied by various other groups, *e.g.*, to achieve a 20-fold improved sensitivity for neurosteroids derivatized with 2-nitro-4-trifluoromethylphenylhydrazine [60].

5. Atmospheric-pressure photoionization

APPI was introduced as a new ionization technique for LC–MS in 2000 by two groups simultaneously [61-63]. APPI was reviewed [64-65]. Two types of commercial APPI sources are available: the Photospray® system at Sciex instruments, based on the work of the group of Bruins [61], and APPI and Dual APCI–APPI or ESI–APPI sources from Syagen Technology, which can be fitted on instruments from a variety of manufacturers and are based on the work of the group of Syage [62-63, 66-67].

The basic principle of APPI is the formation of a analyte molecular ion by absorption of a photon and ejection of an electron.

$$M + h\nu \rightarrow M^{+\cdot}$$

This is the direct-APPI approach, promoted by the group of Syage [66]. In the observations and experimental setup of the group of Bruins [61], the direct-APPI process is not sufficiently efficient. Therefore, an easily ionizable compound, the dopant D, is added to the mobile phase or to the nebulizing gas to enhance the response. Toluene [61] or anisole [68] are frequently used as dopant. With a dopant, the APPI takes place via a charge-exchange reaction between the dopant molecular ion and the analyte molecule:

$$D + h\nu \rightarrow D^{+\cdot}$$

$$D^{+\cdot} + M \rightarrow D + M^{+\cdot}$$

The latter reaction only proceeds when the electron affinity of the analyte is higher and/or the ionization potential of the analyte is lower than that of the dopant.

Whereas in both direct-APPI and dopant-APPI, an analyte molecular ion $M^{+\cdot}$ would be expected, rather than a protonated molecule $[M+H]^+$ is observed, for quite many analytes. This is assumed to be due to interaction with the mobile-phase constituents. In the direct-APPI approach, the protonated molecule is formed due to a reaction of the analyte molecular ion with the solvent S:

$$M^{+\cdot} + S \rightarrow [M+H]^+ + [S–H]\cdot$$

In dopant-APPI, the formation of $[M+H]^+$ is attributed to internal proton rearrangement in the solvated dopant ion clusters [69]:

$$D^{+\cdot} + S \rightarrow [S+H]^+ + [D–H]\cdot$$

$$[S+H]^+ + M \rightarrow [M+H]^+ + S$$

The formation of $[M+H]^+$ is especially important in protic solvents and in an APPI source with a long reaction length [70]. Similar processes are applicable in negative-ion APPI [71].

The group of Kostiainen [72-74] reported a number of comparative studies into the performance of APPI relative to APCI and ESI, *e.g.*, in the LC–MS analysis of flavonoids [72], anabolic steroids [73], and naphthalenes [74]. A variety of solvents were compared for the toluene-doped APPI of naphthalenes, *i.e.*, some pure solvents like hexane, chloroform, water, methanol, and acetonitrile, and 1:1 water–methanol and water–acetonitrile mixtures without or with 0.1% acetic acid, ammonium acetate or ammonium hydroxide. When the proton affinity of the solvent exceeds that of $C_7H_7^+$, protonated analytes were observed, while radical analyte cations were observed when the proton affinity of the solvent was lower than that of $C_7H_7^+$.

APPI has now found its place in LC–MS, next to ESI and APCI. Commercial APPI sources are available and applied in various application areas [64-65]. Dual APCI/APPI and ESI/APPI provide extended applicability range of LC–MS, which is especially useful in screening of combinatorial libraries and other studies in early drug discovery [63, 75]. Another attractive feature of APPI, which so far has only been explored in the coupling of capillary electromigration techniques with MS [76-77], is that it apparently is less disturbed by the presence of phosphate buffers and surfactants.

6. LC–MS by means of ESI and APCI

The performance of any LC–MS system is determined by many, often highly interrelated parameters. For most analytes, ESI is primarily determined by liquid-phase chemistry, whereas APCI is determined by gas-phase chemistry.

In general, ESI is considered more difficult to operate than APCI. ESI is more sensitive to solvent composition and additives. APCI does not appear to need significant optimization of interface parameters. Nevertheless, at least 90% of the LC–MS applications are performed using ESI.

6.1   Hardware issues

Most API interfaces consume huge amounts of N_2, *i.e.*, as countercurrent or curtain gas, and for nebulization in pneumatically-assisted ESI. Common sources of N_2 are generators, boil-off of liquid N_2, or gas cylinders.

In most ESI interfaces, the position of the spray needle is relatively fixed, while in some systems, the needle position can be optimized. Researchers do not agree on the importance of optimization of the spray needle position. Some instrument designs are more sensitive to the needle position than others. The needle position becomes more critical at lower flow-rates.

In modern interface designs, orthogonal liquid introduction is performed (Ch. 5.4.1). This means that the sampling of ions is done from the outer regions of the spray. According to a study by Hiraoka [78-79], this is not favourable. The abundance of multiple-charge ions maximizes in the central region of the spray, while single-charge ions are preferentially observed in the peripheral regions.

The ESI needle potential is an important parameter in optimizing the performance. A typical value is ±3 kV. When the voltage is applied to the needle, its polarity should be *plus* in the positive-ion mode and *minus* in the negative-ion mode. The needle potential in a Sciex ionspray system (4–5 kV) is generally higher than with the other systems (2.5–4 kV). In negative-ion mode, the needle potential must be a little lower, in order to prevent discharge formation (Ch. 6.3.2).

6.2   Flow-rate

The mobile-phase flow-rate is an important parameter, especially in ESI-MS. In early ESI interfaces, the flow-rate was restricted to 10 µl/min, sufficient or even too high in protein characterization, but rather low in LC–MS. Such a flow-rate is ideally suited for use in combination with packed microcapillary columns (320-µm-ID columns). Current pneumatically-assisted ESI devices can be operated with flow-rates up to 1 ml/min or even higher. However, the optimum flow-rate giving the best response per injected amount is in the range of 100–400 µl/min for most commercial systems. Such a flow-rate is nicely compatible with 1–2-mm-ID LC columns.

The influence of the flow-rate in ESI on the analyte response is a complicated topic, affecting both the ionization efficiency and behaviour of ESI-MS as a detector in LC. One of the factors determining the ionization efficiency in ESI is the initial droplet size. Larger droplets are generated at higher flow-rates. These larger droplets need longer evaporation time and a larger number of evaporation–disintegration events (Ch. 6.3.1). Preformed ions in solution are more readily transferred to the gas-phase from smaller droplets. Smaller droplets provided an improved surface-to-volume ratio. As a result, better ionization efficiency is achieved at lower flow-rates. In addition, an ESI needle operated at a lower flow-rate can be positioned closer to the ion-sampling orifice, whereby a larger fraction of the ions produced in the ESI process can be sampled into the vacuum of the MS. Using nano-ESI from gold-coated pulled glass capillaries with an emitter tip with a 1–5-µm internal diameter (Ch. 5.5.5), Mann *et al.* [80-81] found a 500-fold improved overall transfer efficiency of ions compared to conventional ESI. Typical flow-rates in static nano-ESI are 20–100 nl/min, whereas capillaries with emitter tip diameters (5–20 µm) are applied with packed microcapillary or nano-LC columns (typical flow-rate 0.1–2 µl/min). The improved surface-to-volume ratio also results in greater tolerance towards salts and buffers and enhanced performance in the ionization of compounds like oligosaccharides and noncovalent protein complexes [82].

It is generally believed that under ESI conditions, the mass spectrometer acts as a concentration-sensitive detector, *i.e.*, the response of the detector is independent of the flow-rate [47, 83]. This is difficult to prove experimentally. A proper experimental design for such an experiment is difficult, because it is difficult to achieve a constant mass-flow of analyte while changing the solvent flow-rate. The behaviour of the detection system as a concentration-sensitive device can be explained from the flow-rate dependence of the ionization efficiency and the splitting at the sampling orifice. Only a fixed part of the source volume is sampled. In addition, the concentration of ions in the spray plume generated in ESI nebulization is limited by space charging [84]. An excellent discussion on this topic is provided by Abian *et al.* [85].

It is frequently argued that a reduction of the column ID is favourable for the sensitivity. However, this is only true when the same amount of sample (in mass) would be injected onto the column. Obviously, the loading capacity of the column decreases with the column diameter squared (Table 1.1). Evaluating this leads to the conclusion that a reduction of the column ID is only favourable when the available amount of sample is limited. A good example of this is provided by Abian *et al.* [86] as well.

From a practical point of view, the discussion on flow-rate can be summarized as follows. In LC–APCI-MS, the typical flow-rate is 0.5–1.0 ml/min. For routine applications of LC–ESI-MS in many fields, extreme column miniaturization comes with great difficulties in sample handling and instrument operation. In these applications, LC–MS is best performed with a 2-mm-ID column, providing an optimum flow-rate of 200 µl/min, or alternatively with conventional 3–4.6-mm-ID columns in combination with a moderate split. In sample limited cases, further reduction of the column inner diameter must be considered. Packed microcapillary and nano-LC columns with micro-ESI and nano-ESI are routinely applied in proteomics studies (Ch. 17.5.2).

6.3   Mobile-phase composition

Initially, considerable attention was paid to the selection of the best solvent composition. Nowadays, it is readily understood that the operation of LC–MS implies compromises in the performance of both LC and MS. A particular mobile-phase composition might be ideal in terms of analyte ionization, but if this mobile phase yields infinite retention or no retention at all, it cannot be applied. A free selection of the solvent composition is not possible. Always, a compromise must be found between LC separation and MS ionization.

Solvent selection
In most applications, RPLC–MS is performed (Ch. 1.3.2). The polar mobile phase consists of a mixture of methanol or acetonitrile and water, eventually containing a buffer. The stationary phase in most cases is a nonpolar C_8- or C_{18}-

bonded silica-based material. In RPLC, nonpolar compounds are more retained than polar compounds. Elution of compounds with decreasing polarity can be achieved by means of a solvent gradient with an increasing amount of organic modifier. For both drugs [86], pesticides [87], and other compounds, a better response in both ESI and APCI can be achieved with methanol instead of acetonitrile as the organic modifier. The effect appears to be more pronounced in the presence of ammonium acetate.

The analyte response is also influenced by the organic modifier content, although the latter is primarily determined by the separation. In both ESI and APCI, highly aqueous mobile phases are generally not favourable. The organic modifier content needed is determined by the polarity of the analyte and the type of stationary phase applied. The selection of the RPLC stationary phase, *e.g.*, the use of polymeric instead of silica-based material, or C_{18} instead of C_8 material, can help in optimization of the modifier content. In this respect, the use of hydrophilic interaction chromatography (HILIC, Ch. 1.4.5) can be advantageous in the analysis of highly polar compounds [88]. In HILIC, a polar stationary-phase material, *e.g.*, aminopropyl-modified silica, is used in combination with a mobile phase consisting of a water–organic mixture. Polar compounds are more retained than nonpolar compounds. Elution of compounds with increasing polarity can be achieved by means of a gradient with a decreasing organic modifier content. Thus, the elution order between RPLC and HILIC is reversed.

Pure organic mobile phases, as applied in normal-phase LC (NPLC) and with some chiral stationary phases (Chiralpak AD and AS, Ch. 1.4.3), are generally not applicable in LC–ESI-MS. Post-column addition of a mixture of 5 mmol/l aqueous ammonium acetate and methanol or 2-propanol is required. The alcohol assures miscibility with the organic solvent [89]. Examples of chiral NPLC–MS for quantitative bioanalysis are discussed in Ch. 11.7.4. Post-column addition of the aqueous phase via a sheath liquid interface has been proposed [90].

Pure organic mobile phases, *e.g.*, mixtures of hexane and methanol, dioxane, or isopropanol, are readily compatible with LC–APCI-MS. Because some users dislike the use of hexane, ethoxynonafluorobutane has been suggested as an inflammable alternative [91]. Nonaqueous RPLC with a propionitrile–hexane solvent gradient in combination with positive-ion APCI was performed in the LC–MS analysis of triacylglycerols (TAGs) [92] (Ch. 21.3.1). Another example is the LC–MS analysis of polychlorinated *n*-alkanes on bare silica and with chloroform as the mobile phase [93]. Under these conditions, chloride-enhanced APCI can be applied to generate [M+Cl]⁻ adduct ions, which suppresses the loss of Cl• from the polychlorinated *n*-alkanes.

A fundamental discussion on solvent selection for APCI is provided by Kolakowski *et al.* [94-95].

Buffers

For compounds with groups that can be protonated or deprotonated, *i.e.*, compounds that show liquid-phase acid-base behaviour, a buffer must be added to the RPLC mobile phase in order to avoid problems with poor retention, poor resolution, and/or poor repeatability in retention time. Phosphate buffers are applied for this purpose in RPLC with UV or fluorescence detectors, because of the low UV cut-off (<200 nm). In LC–MS, the use of the nonvolatile phosphate buffers is not recommended. Although most of the modern ESI source will no longer show clogging due to phosphate buffers or other nonvolatile additives, their use may lead to signal suppression, the formation of adduct ions, and background noise. Volatile mobile-phase additives are preferred in LC–MS.

Below, the discussion is focussed at LC–ESI-MS. In general, ESI-MS is far more prone to problems with mobile-phase compatibility than APCI. One should remember that the reagent gas in APCI is determined by the mobile-phase composition, and can therefore have distinct influence on the performance of LC–APCI-MS (Ch. 6.4.2). In positive-ion APCI-MS, the use of ammonium formate or acetate limits the applicability range of LC–MS to compounds that have relatively high proton affinities. A buffer is needed in RPLC for compounds that shows liquid-phase acid-base behaviour. From the perspective of extending the applicability range of LC–MS by the use of APCI-MS rather than ESI-MS, these compounds are not relevant, because they generate preformed ions in solution and may thus well be detected by ESI-MS. The compounds without nitrogen atoms or lacking acidic functions, that do not show liquid-phase acid-base behaviour, are more important in this respect. For these less polar compounds, the use of ammonium acetate should be avoided. Therefore, as long as ammonium-containing buffers are used, a major extension in applicability range by the use of combined ESI/APCI sources should not be expected. Similarly, the addition of acetate or formate to the mobile phase limits the applicability range in negative-ion APCI, as has been demonstrated by Schaefer and Dixon [96]. Acetate and formate have a low gas-phase acidity (Table 6.2). In the presence of trifluoroacetate (TFA), negative-ion APCI is virtually impossible.

In LC–ESI-MS, the role of the mobile phase pH is complicated. In practice, often a compromise must be struck between analyte retention and ionization. From the perspective of generating preformed ions in solution, the optimum conditions for the ESI analysis of basic compounds, *e.g.*, amines, would be an acidic mobile phase with a pH at 2 units below the dissociation constant pK_a of the analytes, while for acidic compounds, *e.g.*, carboxylic acid or aromatic phenols, a basic mobile phase with a pH two units above the pK_a of the analytes is preferred [97]. These conditions are unfavourable for an analyte retention in RPLC. The analytes elute virtually unretained. In RPLC, it is important to reduce protolysis of basic and acidic analytes, *i.e.*, to assure that the compounds are present as neutrals in the mobile phase. Fortunately, ionization processes in ESI are more complicated. Processes like wrong-way-around electrospray [53] (Ch. 6.3.5) and gas-phase ionization (Ch. 6.3.1) are of great help in this respect. In some cases, for instance in the analysis of small acidic analytes, the separation must be performed at low pH, *i.e.*, ion-suppressed RPLC. As an example, good ESI response was demonstrated for phenol and naphthol glucuronides from ion-suppressed RPLC [98].

Despite these considerations, the first approach in method development for ESI-MS is the formation of preformed ions in solution, *i.e.*, protonation of basic analytes or deprotonation of acidic analytes. Thus, for basic analytes, mixtures of ammonium salts and volatile acids like formic and acetic acid are applied. Alternatively, formic or acetic acid may be added to the mobile phase, just to set a low pH for the generation of preformed ions in solution. The latter approach is successful if sufficient hydrophobic interaction between preformed analyte ions and the reversed-phase material remains. The concentration of buffer is kept as low as possible, *i.e.*, at or below 10 mmol/l in ESI-MS. The buffer concentration is obviously determined by the buffer capacity needed to achieve stable pH conditions upon repetitive injection of the samples. Constantopoulos *et al.* [99] derived an equilibrium partitioning model to predict the effect of the salt concentration on the analyte response in ESI. If the salt concentration is below 10^{-3} mol/l, the analyte response is proportional to its concentration. The response is found to decrease with increasing salt concentration.

Systematic studies on the influence of mobile-phase additives and the pH of the mobile phase have been reported frequently, *e.g.*, for tetracyclines [100], nucleoside antiviral agents [101], and a variety of basic and acidic drugs [102].

Mallet *et al.* [102] studied the influence of a number of mobile-phase additives on the response of eight acidic drugs in negative-ion mode and of eight basic drugs in positive-ion mode. The influence of the type and concentration of the additive was investigated for formic, acetic, and trifluoroacetic acid, and ammonium hydroxide, formate, biphosphonate, and bicarbonate. Some of their results for compounds analysed in the negative-ion mode are pictured in Figure 6.5. Perhaps the most striking is that the acidic mobile-phase additives only have a minor influence on the response of the basic compounds, and conversely, a basic additive on the response of acidic compounds. This is in contradiction to results reported by others, *e.g.*, [97, 100-101], for other compound classes.

Most compounds in the study of Mallet *et al.* [102] respond similarly to the addition of acetic formic acid, although the effect is compound dependent. In negative-ion mode (Figure 6.5), most compounds are ion suppressed by formic and acetic acid additives. Ammonium hydroxide results in (significant) response enhancement for some compounds, *e.g.*, raffinose, while the response of other compounds, *e.g.*, malic acid and etidronic acid, is strongly suppressed.

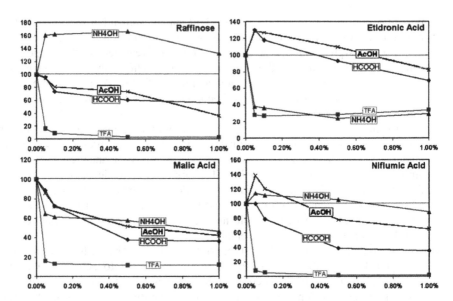

Figure 6.5: Effect of additives, formic acid (HCOOH), trifluoroacetic acid (TFA), ammonium hydroxide (NH4OH), and acetic acid (AcOH) on the relative response of raffinose, etidronic acid, malic acid, and niflumic acid in ESI-MS (compared to 50% aqueous methanol, which is 100%). Data from [102].

The ion suppression might be due to gas-phase reneutralization of the ion-evaporated anions. The response of raffinose is also enhanced by ammonium formate and bicarbonate. Ammonium biphosphate suppresses the response of all acidic compounds. In the positive-ion mode, some compounds, *e.g.*, propranolol, terfenadine, and reserpine, show enhanced response at acid concentrations below 0.5%, while others show response suppression, *e.g.*, risperidone. Surprisingly, the response of some bases, *e.g.*, risperidone and terfenadine, is enhanced by addition of ammonium hydroxide. The response of pipenzolate, a quaternary ammonium compound, is not significantly influenced by additives, except TFA. Most basic compounds are suppressed by the addition of ammonium formate, bicarbonate, and biphosphonate [102].

Ion-pairing agents

Finding the optimum mobile-phase composition for RPLC of analytes that readily form protonated or deprotonated molecules in solution is not always straightforward. Next to RPLC, other approaches available for the separation of ionic compounds are ion-suppressed RPLC, ion-exchange LC, and ion-pair LC.

In ion-pair LC, relatively stable liquid-phase ion pairs are formed between the analyte ions and a mobile-phase additive, indicated as an ion-pairing agent. In LC–UV, large tri- or tetra-alkylammonium compounds and dodecylsulfate are applied as ion-pairing agents for negative-charge and positive-charge analyte ions, respectively. These additives are not sufficiently volatile and may yield significant signal suppression. They can be replaced by various alternatives, *e.g.*, short-chain perfluorinated alkylcarboxylic acids, such as pentafluoropropionic or heptafluorobutyric acid (HFBA). TFA can be used as an ion-pairing agent as well, but may lead to significant signal suppression (see below). Short-chain tri- and tetra-alkylammonium compounds can be applied as ion-pairing agents in the analysis of acidic analytes, *e.g.*, triethylammonium bicarbonate for the analysis of nucleic acids (Ch. 22.2.3). Because any ion-pairing agent leads to response suppression and more background noise, their use should be avoided as much as possible.

TFA is frequently applied as an additive in LC–MS, for instance in the RPLC separation of peptides (Ch. 16.3.2). TFA is applied as an ion-pairing agent and to mask secondary retention effects of RPLC stationary phases. Without TFA in the mobile phase, the peptides would be almost irreversibly adsorbed. TFA results in significant signal suppression due to both ion-pairing and surface-tension effects. The TFA anion more-or-less masks the positive charge on an analyte molecule at the droplet surface and thereby prohibits IEV of that ion. In the "TFA-fix", a post-column addition of propionic acid in 2-propanol (75:25, v/v) is used to counteract the suppression [103-104]. This approach has been found to be successful in some cases, but not in all.

Other additives

Several compounds show significant affinity to sodium, which is present at a concentration of 10^{-5}–10^{-4} mol/l in most solvent applied in LC–MS. A sodiated molecule $[M+Na]^+$ is observed next to or instead of $[M+H]^+$. The relative abundance of these two ions may vary in time and with the experimental conditions. For some compounds, the response of the sodiated molecule is better than that of the protonated species. Post-column addition of 1–5% formic acid in some cases greatly reduces the abundance of the sodiated molecule.

Addition of ammonium acetate is sometimes performed to reduce the formation of sodiated and potassiated molecules in favour of ammoniated and/or protonated molecules. In this way, one may direct the adduct formation of analytes towards just one adduct, which is favourable in quantitative analysis. $[M+Na]^+$ and $[M+K]^+$ of many analytes are difficult to fragment in triple-quadrupole MS–MS, thereby inhibiting the use of selected-reaction monitoring (SRM) for quantitative analysis.

Mobile-phase additives can also influence the relative abundance of the various adduct ions. Karlsson [105] performed post-column addition of alkali cations to enhance ESI-MS of carbohydrates and other compounds without nitrogen atoms. For most analytes, the adduct formation increased with the size of the cation. Optimum concentration of the cation in the solution was *ca.* 5×10^{-5} mol/l. Alkali-metal affinities and alkali-metal influence on fragmentation in MS–MS have been studied by others as well [106-107].

The group of Brodbelt [108-110] studied metal complexation as an alternative to protonation. Initially, complexes of a deprotonated analyte with Cu^{2+}, Co^{2+}, or Zn^{2+} and a 2,2'-bipyridine auxiliary ligand $[(Analyte–H)\,Metal^{2+}\,(Ligand)]^+$ were studied for a variety of compounds [108]. Signal enhancement and structure characterization of analyte–metal-ion complexes were also studied for various compound classes, *e.g.*, tetracyclines [109] and flavonoid glycosides [110].

Zhao *et al.* [111] evaluated various small alkylammonium additives with respect to their ability to reduce or direct the adduct formation of the cholesterol-lowering agent simvastatin, and to reduce its fragmentation in MS–MS. With the additives, the appearance of the MS and MS–MS spectrum can be greatly influenced. Without an additive, $[M+Na]^+$ and $[M+K]^+$ are the most abundant ions, whereas addition of ammonium acetate or an alkylammonium additive reduces their abundance in favour of $[M+H]^+$ and $[M+Alkylammonium+H]^+$. Methylamine gave the best performance and was used as an additive in the quantitative bioanalysis of simvastatin [111] (Ch. 11.4.3).

Some molecules tend to show multimer formation $[M_n+Adduct]^+$ in ESI, where the adduct can be H, Na, NH_4 or otherwise. Some examples are given by Kamel *et al.* [100-101]. In most cases, only adduct-bound dimers are observed, perhaps of the limited scan range chosen. These multimers can give problems with the unambiguous determination of the molecular mass of an unknown and with the linearity of calibration curve in quantitative analysis. Stefansson *et al.* [112] reported that the multimer formation of artemisinin, lasalocid, and deoxynivalenol could be reduced by the addition of primary amines.

Solving mobile-phase incompatibilities

The most appropriate way to solve mobile-phase incompatibility problems in LC–MS is to change the mobile-phase composition to LC–MS compatible conditions. This is the approach taken in most cases. Unfortunately, it is not always possible to do so. In some cases, specialized LC column materials demand a particular mobile-phase composition. This is the case with for instance some chiral columns, which will only provide adequate enantiomeric separation in a predefined mobile phase. The retention behaviour in high-performance anion-exchange chromatography (HPAEC) is significantly influenced by the cation (sodium or ammonium) in the mobile phase. In these cases, mobile-phase incompatibility problems can be solved by means of additional liquid-phase technology. Two examples are given here.

The phase-system switching approach is a column-switching technique that can be used in target-compound analysis to switch from an LC-favourable mobile-phase composition to an MS-compatible mobile-phase composition. After elution from the LC column, the target analyte is heartcut from the chromatogram and adsorbed onto a short trapping column. After washing away the phosphate buffers and other hydrophilic mobile-phase constituents, the analytes are eluted from the trapping column to ESI-MS for detection. An example of this approach is the bioanalytical quantification of metoprolol enantiomers in plasma after chiral separation in a phosphate buffer on an α_1-acid glycoprotein chiral stationary phase [113].

HPAEC is a highly efficient separation technique for oligosaccharides (Ch. 20.4.1). The separation is based on anion exchange of sugar anions at pH 13 on Dionex CarboPac columns. The solvent system consists of gradients of up to 1 mol/l sodium acetate in 0.1 mol/l aqueous sodium hydroxide. The increasing sodium-acetate concentration acts as a displacer, inducing the subsequent elution of oligosaccharides with higher degree of polymerisation (DP). On-line HPAEC–MS can be achieved by the use of an anion micromembrane or an anion self-regeneration suppressor to achieve a post-column exchange of sodium ions (up to ~0.6 mol/l) for hydronium ions. As an example, Torto et al. [114] reported the on-line monitoring of the enzymatic degradation of polysaccharides and the on-line monitoring of the hydrolysis of wheat starch using on-line microdialysis introduction into a HPAEC–MS system.

7. Matrix effects in LC–MS

An important issue in the method development for quantitative analysis using LC–MS is the possible occurrence of matrix effects. A matrix effect is an (unexpected) suppression or enhancement of the analyte response due to coeluting matrix constituents. It can be easily detected by comparing responses between a standard solution and a spiked pre-treated sample (post-extraction spike). Detailed studies on matrix effects revealed that the ion suppression or enhancement is frequently accompanied by significant deterioration of the precision of the analytical method [115-116]. Therefore, it can be useful to discriminate the two type of matrix effects. The "absolute matrix effect" indicates the difference in response between the solvent standard and the post-extraction spike, while the "relative matrix effect" indicates the difference in response between various lots of post-extraction spiked samples [116]. Unless counteraction is taken, an absolute matrix effect will primarily affect the accuracy of the method, whereas a relative matrix effect will primarily affect the precision of the method. Although matrix effects are discussed in significantly more detail in Ch. 11.5, some general issues are discussed here. Matrix effects in veterinary residue analysis and quantitative pesticide analysis by LC–MS were reviewed by Antignac *et al.* [117] and Niessen *et al.* [118], respectively.

As discussed in Ch. 6.6.3, some mobile-phase additives are also known to suppress or enhance analyte response. It appears useful to discriminate between effects due to the mobile-phase composition and effects due to the actual analyte matrix. Some additives were found to reduce the matrix suppression [119].

King *et al.* [120] showed that the matrix effect primarily is a liquid-phase and not a gas-phase process. Nonvolatiles in sample or mobile phase prevent preformed analyte ion to escape from the droplet to the gas phase. Compounds with high proton affinities in positive-ion mode or with low gas-phase acidity in negative-ion mode may also suppress the analyte response in ESI-MS [27]. Other mechanistic studies indicate that components with higher surface activity may suppress the ionization of the analyte [48-49]. In addition, compounds that form strong ion-pairs with the preformed analyte ions, *e.g.*, TFA, are known to suppress the response of these analytes.

Matrix effects are compound and matrix dependent. They are due to co-eluting matrix constituents. Components responsible for the matrix effects are often not ionised by ESI or APCI, and therefore cannot be detected by MS.

There are two approaches to deal with matrix effects in quantitative analysis. One may eliminate the sample constituents responsible for the matrix effects by improving sample pretreatment and/or chromatography. Alternatively, one may reduce or eliminate the influence that matrix effects have on the accuracy and/or precision of the method.

7.1   Remove matrix constituents

There are two ways to remove the sample constituents causing the matrix effect: improvement of the sample pretreatment or of the chromatography. A more selective analyte extraction procedure or a more extensive sample cleanup prior to injection can reduce the amount of matrix components that are introduced into the analytical system. With large numbers of samples, improvement of the sample pre-treatment generally is the most effective method in reducing or eliminating matrix effects.

In the bioanalytical practice, where one or only a few target analytes must be determined, 20–50-mm long columns are used offering limited chromatographic resolution. Often, only separation from the solvent front is pursued. In multiresidue analysis for environmental or food-safety applications, longer columns (100–250 mm) are applied, providing far better separation. Therefore, improving the chromatographic resolution provides only limited gain in reducing matrix suppression in multiresidue pesticide analysis. The most important issue is to achieve sufficient analyte retention and to move the more polar compounds sufficiently away from the solvent front. In this respect, the use of HILIC may be of interest (Ch. 1.4.5 and Ch. 11.7.3).

7.2   Eliminate effects on accuracy and/or precision

There are a number of ways to eliminate the effects of matrix interferences on accuracy and/or precision of the method. The change from ESI-MS to APCI-MS has been shown to be successful in some cases [115-116]. Similarly, the use of the negative-ion instead of the positive-ion mode, or vice versa, may provide reduction of matrix effects.

Different mobile-phase additives may influence the extent of matrix suppression in a particular method [119, 121].

The most adequate method to eliminate matrix effects is the use of isotopically-labelled internal standards in combination with matrix-matched calibration samples. An isotopically-labelled internal standard shows (almost) identical behaviour to the target analyte in both sample pretreatment, chromatography, and analyte ionization. Unfortunately, isotopically-labelled internal standards are available for only a limited number of target analytes, they are expensive, and often difficult to obtain for other target compounds, especially with sufficient high isotopic purity (D_0 contamination). When an analogue internal standard is used instead, the ionization of the standard and the analyte may be differently affected by the matrix. Therefore, an almost co-eluting analogue internal standard is preferred [122].

While the selection of an isotopically-labelled or analogue internal standard is relatively easy in quantitative bioanalysis, the situation is more complicated in multiresidue analysis. It is difficult to select appropriate analogue standards for a wide variety of target compounds, while isotopically-labelled standards are often not available for all target compounds. In addition, if one would introduce one standard per target compound, this would seriously limit the sensitivity of the method as it doubles the number of SRM transitions that have to be monitored. Another problem in multiresidue analysis is the selection of appropriate blanks for the production of the matrix-matched standards and the number of matrices that might have to be studied. When no adequate blank matrix is available, the standard addition method is the only way to achieve sufficiently accurate and precise results [123]. This method is time-consuming and labourious.

8. References

1. M.A. Baldwin, F.W. McLafferty, *Direct CI of relatively involatile samples. Application to underivatized oligopeptides*, Org. Mass Spectrom., 7 (1973) 1353.
2. M. Dedieu, C. Juin, P.J. Arpino, G. Guiochon, *Soft negative ionization of nonvolatile molecules by introduction of liquid solutions into a CI-MS*, Anal. Chem., 54 (1982) 2372.
3. P.J. Arpino, G. Guiochon, *Optimization of the instrumental parameters of a combined LC–MS, coupled by an interface for DLI. III. Why the solvent should not be removed in LC–MS interfacing methods*, J. Chromatogr., 251 (1982) 153.
4. M.L. Vestal, *Ionization techniques for nonvolatile molecules*, Mass Spectrom. Rev., 2 (1983) 447.
5. M. Dole, R.L. Hines, L.L. Mack, R.C. Mobley, L.D. Ferguson, M.B. Alice, *Molecular beams of macroions*, J. Chem. Phys., 49 (1968) 2240.
6. J. Gieniec, L.L. Mack, K. Nakamae, C. Gupta, V. Kumar, M. Dole, *ESI-MS of macromolecules: Application of an ion-drift spectrometer*, Biomed. Mass Spectrom., 11 (1984) 259.
7. D.S. Simons, B.N. Colby, C.A. Evans, Jr., *EHI-MS – the ionization of liquid glycerol and non-volatile organic solutes*, Int. J. Mass Spectrom. Ion Phys., 15 (1974) 291.
8. B.P. Stimpson, C.A. Evans, Jr., *EHI-MS of biochemical materials*, Biomed. Mass Spectrom., 5 (1978) 52.
9. N.B. Zolotai, G.V. Karpov, V.L. Tal'roze, V.E. Skurat, G.I. Ramendik Yu.V. Basyuta, *MS of the field evaporation of ions from liquid solutions in glycerol*, J. Anal. Chem. USSR, 35 (1980) 937.
10. N.B. Zolotai, G.V. Karpov, V.L. Tal'roze, V.E. Skurat, Yu.V. Basyuta, G.I. Ramendik, *MS of the field evaporation of ions from water and aqueous solutions, aqeous sodium iodide and saccharose solutions*, J. Anal. Chem. USSR, 35 (1980) 1161.
11. J.V. Iribarne, B.A. Thomson, *On the evaporation of small ions from charged droplets*, J. Chem. Phys., 64 (1976) 2287.
12. B.A. Thomson, J.V. Iribarne, *Field-induced IEV from liquid surfaces at atmospheric pressure*, J. Chem. Phys., 71 (1979) 4451.
13. B.A. Thomson, J.V. Iribarne, P.J. Dziedzic, *Liquid IEV-MS–MS for the detection of polar and labile molecules*, Anal. Chem., 54 (1982) 2219.
14. J.V. Iribarne, P.J. Dziedzic, B.A. Thomson, *AP-IEV-MS*, Int. J. Mass Spectrom. Ion Phys., 50 (1983) 331.
15. E.E. Dodd, *The statistics of liquid spray and dust electrification by the Hopper and Laby method*, J. Appl. Phys., 24 (1953) 73.
16. C.R. Blakley, J.J. Carmody, M.L. Vestal, *A new soft ionization technique for MS of complex molecules*, J. Am. Chem. Soc., 102 (1980) 5931.
17. C.R. Blakley, J.J. Carmody, M.L. Vestal, *LC–MS for analysis of nonvolatile samples*, Anal. Chem., 52 (1980) 1636.
18. C.R. Blakley, M.L. Vestal, *TSP interface for LC–MS*, Anal. Chem., 55 (1983) 750.
19. F.W. Röllgen, E. Bramer-Weger, L. Bütfering, *Field ion emission from liquid solutions: IEV against electrohydrodynamic disintegration*, J. Physique, 48 (1987)

C6-253.
20. G. Schmelzeisen-Redeker, L. Bütfering, F.W. Röllgen, *Desolvation of ions and molecules in TSP-MS*, Int. J. Mass Spectrom. Ion Proc., 90 (1989) 139.
21. M. Yamashita, J.B. Fenn, *ESI ion source. Another variation of the free-jet theme*, J. Phys. Chem., 88 (1984) 4451.
22. M. Yamashita, J.B. Fenn, *Negative ion production with the ESI ion source*, J. Phys. Chem., 88 (1984) 4671.
23. M.L. Aleksandrov, L.N. Gall, V.N. Krasnov, V.I. Nikolaev, V.A. Pavlenko, V.A. Shkurov, G.I. Baram, M.A. Gracher, V.D. Knorre, Y.S. Kusner, Bioorg Khim., 10 (1984) 710.
24. C.K. Meng, M. Mann, J.B. Fenn, Proceedings of the 36th ASMS Conference on Mass Spectrometry and Allied Topics, June 5-10, 1988, San Francisco, CA, p. 771.
25. J.B. Fenn, M. Mann, C.K. Meng, S.F. Wong, C.M. Whitehouse, *ESI – principles and practice*, Mass Spectrom. Rev., 9 (1990) 37.
26. M. Mann, *ESI: Its potential and limitations as an ionization method for biomolecules*, Org. Mass Spectrom., 25 (199) 575.
27. P. Kebarle, L. Tang, *From ions in solution to ions in the gas phase. The mechanism of ESI-MS*, Anal. Chem., 65 (1993) 972A.
28. R.B. Cole, *Some tenets pertaining to ESI-MS*, J. Mass Spectrom., 35 (2000) 763.
29. P. Kebarle, *A brief overview of the present status of the mechanisms involved in ESI-MS*, J. Mass Spectrom., 35 (2000) 804.
30. N.B. Cech, C.G. Enke, *Practical implications of some recent studies in ESI fundamentals*, Mass Spectrom. Rev., 20 (2001) 362.
31. R.D. Smith, K.J. Light-Wahl, *The observation of non-covalent interactions in solution by ESI-MS: promise, pitfalls and prognosis*, Biol. Mass Spectrom., 22 (1993) 493.
32. M.H. Amad, N.B. Cech, G.S. Jackson, C.G. Enke, *Importance of gas-phase proton affinities in determining the ESI response for analytes and solvents*, J. Mass Spectrom., 35 (2000) 784.
33. R.R. Ogorzalek Loo, J.A. Loo, H.R. Udseth, J.L. Fulton, R.D. Smith, *Protein structural effects in gas phase ion-molecule reactions with diethylamine*, Rapid Commun. Mass Spectrom., 6 (1992) 159.
34. R.R. Ogorzalek Loo and R.D. Smith, *Investigation of the gas-phase structure of electrosprayed proteins using ion-molecule reactions*, J. Am. Soc. Mass Spectrom., 5 (1994) 207.
35. R.R. Ogorzalek Loo, H.R. Udseth, R.D. Smith, *A new approach for the study of gas-phase ion-ion reactions using ESI*, J. Am. Soc. Mass Spectrom., 3 (1992) 695.
36. M.G. Ikonomou, A.T. Blades, P. Kebarle, *ESI–ionspray: A comparison of mechanisms and performance*, Anal. Chem., 63 (1991) 1989.
37. M.G. Ikonomou, A.T. Blades, P. Kebarle, *ESI-MS of methanol and water solutions. Suppression of electric discharge with SF_6 gas.*, J. Am. Soc. Mass Spectrom., 2 (1991) 497.
38. R.B. Cole, A.K. Harrata, *Solvent effect on analyte charge state, signal intensity, and stability in negative ion ESI-MS; Implications for the mechanism of negative-ion formation*, J. Am. Soc. Mass Spectrom., 4 (1993) 546.
39. J. Fernandez de la Mora, G.J. van Berkel, C.G. Enkie, R.B. Cole, M. Martinez-

Sanchez, J.B. Fenn, *Electrochemical processes in ESI-MS*, J. Mass Spectrom., 35 (2000) 939.

40. G.J. Van Berkel, S.A McLuckey, G.L. Glish, *Electrochemical origin of radical cations observed in ESI mass spectra*, Anal. Chem., 64 (1992), 1586.

41. F. Zhou, G.J. Van Berkel, *Characterization of an ESI ion source as a controlled-current electrolytic cell*, Anal. Chem., 67 (1995) 2916.

42. G.J. Van Berkel, J.M.E. Quirke, R.A. Tigani, A.S. Dilley, T.R. Covey, *Derivatization for ESI-MS. 3. Electrochemically ionizable derivatives*, Anal. Chem., 70 (1998) 1544.

43. G. Diehl, U. Karst, *On-line electrochemistry–MS and related techniques*, Anal. Bioanal. Chem., 373 (2002) 390.

44. J.B. Fenn, *Ion formation from charged droplets: roles of geometry, energy, and time*, J. Am. Soc. Mass Spectrom., 4 (1993) 524.

45. C.G. Enke, *A predictive model for matrix and analyte effects in ESI of singly-charged ionic analytes*, Anal. Chem., 69 (1997) 4885.

46. D.R. Zook, A.P. Bruins, *On cluster ions, ion transmission, and linear dynamic range limitations in ESI-MS*, Int. J. Mass Spectrom. Ion Processes, 162 (1997) 129.

47. L. Tang, P. Kebarle, *Dependence of ion intensity in ESI-MS on the concentration of the analytes in the electrosprayed solution*, Anal. Chem., 65 (1993) 3654.

48. S. Zhou, K.D. Cook, *A mechanistic study of ESI-MS: charge gradients within ESI droplets and their influence on ion response*, J. Am. Soc. Mass Spectrom., 12 (2001) 206.

49. N.B. Cech, C.G. Enke, *Relating ESI response to non-polar character of small peptides*, Anal. Chem., 72 (2000) 2717.

50. R.D. Smith, J.A. Loo, C.G. Edmonds, C.J. Barinaga, H.R. Udseth, *New developments in biochemical MS: ESI*, Anal. Chem., 62 (1990) 882.

51. S.F. Wong, C.K. Meng J.B. Fenn, *Multiple charging in ESI of poly(ethylene glycols)*, J. Phys. Chem., 92 (1988) 546.

52. M.A. Kelly, M.M. Vestling, C.C. Fenselau, P.B. Smith, *ESI analysis of proteins: A comparison of positive-ion and negative-ion mass spectra at high and low pH*, Org. Mass Spectrom., 27 (1992) 1143.

53. B.A. Mansoori, D.A. Volmer, R.K. Boyd, *'Wrong-way-around' ESI of amino acids*, Rapid Commun. Mass Spectrom., 11 (1997) 1120.

54. C.L. Gatlin, F. Tureček, *Acidity determination in droplets formed by electrospraying methanol–water solutions*, Anal. Chem., 66 (1994) 712.

55. S. Zhou, B.S. Prebyl, K.D. Cook, *Profiling pH changes in the ESI plume*, Anal. Chem., 74 (2002) 4885.

56. A. Zhou, K.D. Cook, *Protonation in ESI-MS: Wrong-way-round or right-way-round?*, J. Am. Soc. Mass Spectrom., 11 (2000) 961.

57. D.I. Carroll, I. Dzidic, E.C. Horning, R.N. Stillwell, *API-MS*, Appl. Spectrosc. Rev., 17 (1981) 337.

58. T. Karancsi, P. Slégel, *Reliable molecular mass determination of aromatic nitro compounds: elimination of gas-phase reduction occurring during APCI*, J. Mass Spectrom., 34 (1999) 975.

59. G. Singh, A. Gutierrez, K. Xu, A.I. Blair, *LC–ECNI-APCI-MS: Analysis of pentafluorobenzyl derivatives of biomolecules and drugs in the attomole range*,

Anal. Chem., 72 (2000) 3007.
60. T. Higashi, N. Takido, K. Shimada, *Studies on neurosteroids XVII. Analysis of stress-induced changes in neurosteroid levels in rat brains using LC–ECNI-APCI-MS*, Steroids, 70 (2005) 1.
61. D.B. Robb, T.R. Covey, A.P. Bruins, *APPI: an ionization method for LC–MS*, Anal. Chem., 72 (2000) 3653.
62. J.A. Syage, M.D. Evans, K.A. Hanold, *Photoionization MS*, Am. Lab., 32 (2000) 24.
63. J.A. Syage, M.D. Evans, *Photoionization MS – a powerful new tool for drug discovery*, Spectrosc., 16 (2001) 14.
64. A. Raffaelli, A. Saba, *APPI-MS*, Mass Spectrom. Rev., 22 (2003) 318.
65. S.J. Bos, S.M. van Leeuwen, U. Karst, *From fundamentals to applications: recent developments in APPI-MS*, Anal. Bioanal. Chem., 384 (2006) 85.
66. K.A. Hanold, S.M. Fischer, P. Cormia, C.E. Miller, J.A. Syage, *APPI: I. General properties for LC–MS*, Anal Chem. 76, 2842 (2004).
67. J.A. Syage, K.A. Hanold, T.C. Lynn, J.A. Horner, R.A. Thakur, *APPI: II. Dual source ionization*, J. Chromatogr. A, 1050 (2004) 137.
68. T.J. Kauppila, R. Kostiainen, A.P. Bruins, *Anisole, a new dopant for APPI-MS of low proton affinity, low ionization energy compounds*, Rapid Commun. Mass Spectrom., 18 (2004) 808.
69. D.B. Robb, M.W. Blades, *Effects of solvent flow, dopant flow, and lamp current on dopant-assisted APPI for LC–MS. Ionization via proton transfer*, J. Am. Soc. Mass Spectrom., 16 (2005) 1275.
70. J.A. Syage, *Mechanism of [M+H]⁺ formation in photoionization MS*, J. Am. Soc. Mass Spectrom., 15 (2004) 1521.
71. T.J Kauppila, T. Kotiahoa, R. Kostiainen, A.P Bruins, *Negative-ion APPI-MS*, J. Am. Soc. Mass Spectrom., 15 (2004) 203.
72. J.-P. Rauha, H. Vuorela, R. Kostiainen, *Effect of eluent on the ionization efficiency of flavonoids by ESI, APCI and APPI-MS*, J. Mass Spectrom., 36 (2001) 1269.
73. A. Leinonen, T. Kuuranne, R. Kostiainen, *LC–MS in anabolic steroid analysis - optimization and comparison of three ionization techniques: ESI, APCI and APPI-MS*, J. Mass Spectrom., 37 (2002) 693.
74. T.J. Kauppila, T. Kuuranne, E.C. Meurer, M.N. Eberlin, T. Kotiaho, R. Kostiainen, *Ionization and solvent effects in APPI of naphthalenes*, Anal. Chem., 74 (2002) 5470.
75. Y. Cai, D. Kingery, O. McConnell, A.C. Bach, II, *Advantages of APPI-MS in support of drug discovery*, Rapid Commun. Mass Spectrom., 19 (2005) 1717.
76. S.L. Nilsson, C. Andersson, P.J.R Sjöberg, D. Bylund, P. Petersson, M. Jörntén-Karlsson, K.E. Markides, *Phosphate buffers in CE–MS using APPI and ESI*, Rapid Commun. Mass Spectrom., 17 (2003) 2267.
77. R. Mol, G.J. de Jong, G.W. Somsen, *APPI for enhanced compatibility in on-line micellar electrokinetic chromatography-MS*, Anal. Chem., 77 (2005) 5277.
78. K. Hiraoka, *How are ions formed from electrosprayed charged liquid droplets?*, Rapid Commun. Mass Spectrom., 6 (1992) 463.
79. K. Hiraoka, I. Kudaka, *Species-selectivity effects in the production of ESI ions*, Rapid Commun. Mass Spectrom., 7 (1993) 363.

80. M.S. Wilm, M. Mann, *ESI and Taylor-cone theory, Dole's beam of macromolecules at last?*, Int. J. Mass Spectrom. Ion Processes, 136 (1994) 167.
81. M.S. Wilm, M. Mann, *Analytical properties of the nano-ESI ion source*, Anal. Chem., 68 (1996) 1.
82. M. Karas, U. Bahr, T. Dülcks, *Nano-ESI-MS: addressing analytical problems beyond routine*, Fres. J. Anal. Chem., 366 (2000) 669.
83. G. Hopfgartner, K. Bean, J.D. Henion, R.A. Henry, *Ionspray MS detection for LC: a concentration- or a mass-flow-sensitive device?*, J. Chromatogr., 647 (1993)51.
84. M. Busman, J. Sunner, C.R. Vogel, *Space-charge-dominated MS ion sources: Modelling and sensitivity*, J. Am. Soc. Mass Spectrom., 2 (1991) 1.
85. J. Abian, A. J. Oosterkamp, E. Gelpí, *Comparison of conventional, narrow-bore and capillary LC–MS for ESI-MS: practical considerations*, J. Mass Spectrom., 34 (1999) 244.
86. D. Temesi, B. Law, *The effect of LC eluent composition on MS responses using ESI*, LC-GC Intern., 12 (1999) 175.
87. R.B. Geerdink, A. Kooistra-Sijpersma, J. Tiesnitsch, P.G.M. Kienhuis, U.A.Th. Brinkman, *Determination of polar pesticides with APCI-MS using methanol and/or acetonitrile for SPE and gradient LC*, J. Chromatogr. A, 863 (1999) 147.
88. W. Naidong, *Bioanalytical LC–MS–MS methods on underivatized silica columns with aqueous/organic mobile phases*, J. Chromatogr. B, 796 (2003) 209.
89. A.P. Zavitsanos, T. Alebic-Kolbah, *Enantioselective determination of terazosin in human plasma by NPLC–ESI-MS*, J. Chromatogr. A, 794 (1998) 45.
90. L. Charles, F. Laure, P. Raharivelomanana, J.-P. Bianchini, *Sheath liquid interface for the coupling of NPLC with ESI-MS and its application to the analysis of neoflavonoids*, J. Mass Spectrom., 40 (2005) 75.
91. M.Z. Kagan, M. Chlenov, C.M. Kraml, *NPLC separations using ethoxynonafluorobutane as hexane alternative. II. LC–APCI-MS applications with methanol gradients*, J. Chromatogr. A, 1033 (2004) 321.
92. W.C. Byrdwell, E.A. Emken, *Analysis of triglycerides using APCI-MS*, Lipids, 30 (1995) 173.
93. Z. Zencak, M. Oehme, *Chloride-enhanced APCI-MS of polychlorinated n-alkanes*, Rapid Commun. Mass Spectrom., 18 (2004) 2235.
94. B.M. Kolakowski, J.S. Grossert, L. Ramaley, *The importance of both charge exchange and proton transfer in the analysis of polycyclic aromatic compounds using APCI-MS*, J. Am. Soc. Mass Spectrom., 15 (2004) 301.
95. B.M. Kolakowski, J.S. Grossert, L. Ramaley, *Studies on the positive-ion mass spectra from APCI of gases and solvents used in LC and direct liquid injection*, J. Am. Soc. Mass Spectrom., 15 (2004) 311.
96. W.H. Schaefer, F. Dixon, Jr., *Effect of LC mobile phase components on sensitivity in negative LC–APCI-MS*, J. Am. Soc. Mass Spectrom., 7 (1996) 1059.
97. S. Gao, Z.-P. Zhang, H.T. Karnes, *Sensitivity enhancement in LC–APCI-MS using derivatization and mobile phase additives*, J. Chromatogr. B., 825 (2005) 98.
98. R. Andreoli, P. Manini, E. Bergamaschi, A. Mutti, I. Franchini, W.M.A. Niessen, *Determination of naphthalene metabolites in human urine by LC–MS with ESI*, J. Chromatogr. A, 847 (1999) 9.
99. T.L. Constantopoulos, G.S. Jackson, C.G. Enke, *Effects of salt concentration on*

analyte response using ESI-MS, J. Am. Soc. Mass Spectrom., 10 (1999) 625.

100. A.M. Kamel, P.R. Brown, B. Munson, *ESI-MS of tetracycline, oxytetracycline, chlorotetracycline, minocycline, and methacycline*, Anal. Chem., 71 (1999) 968.

101. A.M. Kamel, P.R. Brown, B. Munson, *Effects of mobile-phase additives, solution pH, ionization constant, and analyte concentration on the sensitivities and ESI mass spectra of nucleoside antiviral agents*, Anal. Chem., 71 (1999) 5481.

102. C.R. Mallet, Z. Lu, J.R. Mazzeo, *A study of ion suppression effects in ESI from mobile phase additives and SPE*, Rapid Commun. Mass Spectrom., 18 (2004) 49.

103. A. Apffel, S. Fischer, G. Goldberg, P.C. Goodley, F.E. Kuhlmann, *Enhanced sensitivity for peptide mapping with LC–ESI-MS in the presence of signal suppression due to TFA-containing mobile phases*, J. Chromatogr. A, 712 (1995) 177.

104. F.E. Kuhlmann, A. Apffel, S.M. Fisher, G. Goldberg, P.C. Goodley, *Signal enhancement for gradient RPLC–ESI-MS analysis with TFA and other strong acid modifiers by postcolumn addition of propionic acid and isopropanol*, J. Am. Soc. Mass Spectrom., 6 (1995) 1221.

105. K.E. Karlsson, *Cationization in micro-LC–ESI-MS*, J. Chromatogr. A, 794 (1998) 359.

106. S.M. Blair, J.S. Brodbelt, A.P. Marchand, H.-S. Chong, S. Alihodzic, *Evaluation of alkali metal binding selectivities of caged aza-crown ether ligands by micro-ESI-quadrupole ion trap MS*, J. Am. Soc. Mass Spectrom., 11 (2000) 884.

107. D.A. Volmer, C.M. Lock, *ESI and CID of antibiotic polyether ionophores*, Rapid Commun. Mass Spectrom., 12 (1998) 157.

108. E.J. Alvarez, J.S. Brodbelt, *Metal complexation as an alternative to protonation in ESI of pharmaceutical compounds*, J. Am. Soc. Mass Spectrom., 9 (1998) 463.

109. V.H. Vartanian, B. Goolsby, J.S. Brodbelt, *Identification of tetracycline antibiotics by ESI in a quadrupole ion trap*, J. Am. Soc. Mass Spectrom., 9 (1998) 1089.

110. M. Satterfield, J.S. Brodbelt, *Structural characterization of flavonoid glycosides by CID of metal complexes*, J. Am. Soc. Mass Spectrom., 12 (2001) 537.

111. J.J. Zhao, A.Y. Yang, J.D. Rogers, *Effects of LC mobile phase buffer contents on the ionization and fragmentation of analytes in LC–ESI-MS–MS determination*, J. Mass Spectrom., 37 (2002) 421.

112. M. Stefansson, P.J.R. Sjöberg, K.E. Markides, *Regulation of multimer formation in ESI-MS*, Anal. Chem., 68 (1996) 1792.

113. A. Walhagen, L.-E. Edholm, C.E.M. Heeremans, R.A.M. van der Hoeven, W.M.A. Niessen, U.R. Tjaden, J. van der Greef, *Coupled column LC–MS: TSP LC–MS and LC–MS–MS analysis of metoprolol enantiomers in plasma using phase-system switching*, J. Chromatogr., 474 (1989) 257.

114. N. Torto, A. Hofte, R.A.M. van der Hoeven, U.R. Tjaden, L. Gorton, G. Marko-Varga C. Bruggink, J. Van der Greef, *Microdialysis introduction HPAEC–ESI-MS for monitoring of on-line desalted carbohydrate hydrolysates*, J. Mass Spectrom., 33 (1998) 334.

115. B.K. Matuszewski, M.L. Constanzer, C.M. Chavez-Eng, *Matrix effects in quantitative LC–MS–MS analysis of biological fluids: a method for determination of finasteride in human plasma at pg/ml concentrations*, Anal. Chem., 70 (1998) 882.

116. B.K. Matuszewski, M.L. Constanzer, C.M. Chavez-Eng, *Strategies for the assessment of matrix effect in quantitative bioanalytical methods based on HPLC–MS–MS*, Anal. Chem., 75 (2003) 3019.

117. J.-P. Antignac, K. de Wasch, F. Monteau, H. De Brabander, F. Andre, B. Le Bizec, *The ion suppression phenomenon in LC–MS and its consequences in the field of residue analysis*, Anal Chim Acta, 529 (2005) 129.

118. W.M.A. Niessen, P. Manini, R. Andreoli, *Matrix effects in quantitative pesticide analysis using LC–MS*, Mass Spectrom. Rev., in press.

119. B.K. Choi, D.M. Hercules, A.I. Gusev, *LC–MS–MS signal suppression effects in the analysis of pesticides in complex environmental matrices*, Fres. J. Anal. Chem., 369 (2001) 370.

120. R.C. King, R. Bonfiglio, C. Fernandez-Metzler, C. Miller-Stein, T.V. Olah, *Mechanistic investigation of ionization suppression in ESI*, J. Am. Soc. Mass Spectrom., 11 (2000) 942.

121. T. Benijts, R. Dams, W. Lambert, A. De Leenheer, *Countering matrix effects in environmental LC–ESI-MS–MS water analysis for endocrine disrupting chemicals*, J. Chromatogr. A, 1029 (2004) 153.

122. R. Kitamura, K. Matsuoka, E. Matsushima, Y. Kawaguchi, *Improvement in precision of the LC–ESI-MS–MS analysis of 3'-C-ethynylcytidine in rat plasma*, J. Chromatogr. B, 754 (2001) 113.

123. S. Ito, K. Tsukada, *Matrix effect and correction by standard addition in quantitative LC–MS analysis of diarrhetic shellfish poisoning toxins*, J. Chromatogr. A, 943 (2001) 39.

APPLICATIONS: SMALL MOLECULES

7

LC–MS ANALYSIS OF PESTICIDES

1. Introduction

LC–MS plays an important role in the analysis of pesticides and related compounds, *e.g.*, herbicides, insecticides, acaricides, as well as their degradation products and metabolites. The analysis of pesticides is relevant for environmental studies, food safety, toxicology, and occupational health. Pesticides have to be analysed in environmental samples, such as different water compartments, soil, sediments, sludge, and animal tissue like fish, in food, especially fruit and vegetables, and in (human) body fluids and tissues. Many modern pesticides and related compounds are not amenable to GC–MS analysis, or only after derivatization. Therefore, LC–MS has been evaluated as an alternative. Various review papers on the analysis of pesticides and related compounds in various sample matrices were published [1-8].

This chapter is devoted to the analysis of pesticides and related compounds. LC–MS characteristics of various classes of pesticides are described, *i.e.*, the mass spectral information obtained using electrospray ionization (ESI) and atmospheric-pressure chemical ionization (APCI). Next, typical strategies with the analysis of pesticides in environmental samples and in fruit and vegetables are discussed.

179

Table 7.1: Suitability of LC–MS ionization mode for various classes of pesticides [9]				
Compound class	APCI+	APCI–	ESI+	ESI–
acetanilide herbicides	+	+	+	–
carbamate insecticides	++	–	++	–
chlorophenoxy acid herbicides	–	+	–	++
organochlorine insecticides	–	–	–	–
organophosphate insecticides	++	+	++	–
phenylurea herbicides	++	+/–	+	+/–
quaternary ammonium herbicides	–	–	++	–
sulfonylurea herbicides	++	+	++	+
triazine herbicides	++	–	+	–

2. Mass spectrometry of pesticides and herbicides

Many papers on the LC–MS analysis of pesticides and related compounds deal with the characterization of interface and ionization performance, the improvement of detection limits by variation of experimental conditions, and the information content of the mass spectra. As far as ESI and APCI are concerned, this type of information is reviewed for various pesticide classes in this section (see Ch. 4.7.4 for results with thermospray and Ch. 5.6.1 with particle-beam interfacing).

Thurman *et al.* [9] evaluated the performance of APCI and ESI in both positive-ion and negative-ion mode in the analysis of 75 pesticides from various compound classes. Part of their results is summarized in Table 7.1.

2.1   Carbamates

Carbamate pesticides are used as insecticide, acaricide, and herbicide. Various subclasses can be distinguished. Aryl *N*-methyl carbamates (1), such as carbaryl, carbofuran, propoxur, and oxime *N*-methyl carbamates(2), such as aldicarb, methomyl, oxamyl, have most widely been studied. LC–MS is the method-of-choice for carbamates, since their thermal lability prohibits GC analysis.

(1)   (2)

Table 7.2: Comparison of SIM detection limits in ng (quantitation ion) of some carbamates [11]			
Interface	Carbofuran (M$_r$ 221 Da)	Carbaryl (M$_r$ 201 Da)	Aldicarb (M$_r$ 190 Da)
Moving-belt with EI	25 (165)	25 (145)	6 (89)
Thermospray	0.8 (239)	0.8 (219)	0.9 (208)
Particle-beam with EI	55 (164)	10 (144)	> 500
ESI	0.3 (222)	1.0 (202)	1.5 (191)
APCI	0.05 (222)	0.05 (145)	0.07 (116)

A comparison of various LC–MS interfaces in the analysis of carbofuran was reported by Honing *et al.* [10]. The progress in LC–MS interface performance can be read from a comparison of absolute detection limits in selected-ion monitoring (SIM) of three representative carbamates, as collected by Pleasance *et al.* [11], using data from various other authors as well (see Table 7.2).

Carbamates are analysed in the positive-ion mode. In ESI, carbamates generally show protonated and/or ammoniated or sodiated molecules [11-19]. Fragmentation can easily be induced by in-source CID, as was systematically investigated by Voyksner and Pack [12]. *N*-methyl carbamates showed a characteristic loss of methyl isocyanate (CH$_3$–N=C=O, neutral loss of 57) [12, 18], while in *N*-oxime carbamates such as aldicarb (M 190 Da), the fragmentation is directed by the N-oxime group rather than by the carbamate group, resulting in a loss of carbamic acid (CH$_3$NHCOOH, neutral loss of 75) and a fragment at *m/z* 89 due to [CH$_3$SC(CH$_3$)$_2$]$^+$.

In APCI mass spectra of carbamates, fragment ions are observed, which are most likely due to thermal decomposition in the heated nebulizer interface and subsequent ionization of the thermal decomposition products [11, 14, 20-23]. For example, base peaks were observed at *m/z* 163 for oxamyl, due to the loss of methyl isocyanate, at *m/z* 168 for propoxur, due to the loss of propylene, and at *m/z* 157 for aldicarb, due to the loss of H$_2$S. The APCI mass spectra of aldicarb and two of its metabolites, aldicarb sulfoxide and aldicarb sulfone, showed significant fragmentation. Major fragments for aldicarb were due to the loss of carbamic acid (to *m/z* 116) and due to charge retention at [CH$_3$–S–C(CH$_3$)$_2$]. For aldicarb sulfoxide and aldicarb sulfone, the loss of carbamic acid resulted in the base peaks of the spectra (at *m/z* 132 and 148, respectively).

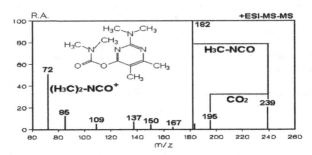

Figure 7.1: MS–MS product-ion mass spectrum of pirimicarb (based on ref. 26).

Product-ion MS–MS spectra of carbamates were investigated by Chiu *et al.* [24], using thermospray ionization and either protonated or ammoniated molecules as precursor ions. The fragmentation is similar to that in in-source CID. For *N*-methyl carbamates, the loss of 57 due to methyl isocyanate is a characteristic feature. In *N*-oxime carbamates, the fragmentation is directed from the oxime rather than from the carbamate group.

The ion-trap multistage MS–MS mass spectra of protonated carbofuran (*m/z* 222) showed the loss of methyl isocyanate in the first stage to *m/z* 165, which in turn showed *m/z* 123 due to the loss of propylene, and a weak *m/z* 137, supposedly due to the loss of ethylene after double hydrogen rearrangement [25]. The *m/z* 123 could be further fragmented to protonated phenol due to the loss of CO [25].

Extensive rearrangements occurs in the fragmentation of the protonated pirimicarb (see Figure 7.1): primary fragment ions are due to losses of CO_2 or $CH_3–N=C=O$. The latter requires a methyl rearrangement to the ring. Both rearrangements were confirmed using accurate mass determination in a quadrupole–time-of-flight hybrid (Q–TOF) instrument [26].

2.2   Organophosphorous pesticides

Organophosphorous pesticides (OPP) have extensively been studied with LC–MS. The compound class can be subdivided in various sub-classes, such as phosphates (**3**), phosphonates (**4**), phosphorothionates (**5**), phosphorothioates (**6**), and phosphorodithioates (**7**) (R_1 is an alkyl, R_2 is an alkyl or aryl substituent).

ESI mass spectra of 7 OPP, *i.e.*, oxydemeton methyl, trichlorfon, dimethoate, dichlorvos, demeton-*s*-methyl, fenitrooxon and fenamiphos, and two degradation products, fenamiphos sulfoxide and sulfone, were tabulated by Molina *et al.* [27]. Without in-source CID, the sodiated molecule was the base peak for all OPP studied, while with in-source CID a number of fragment ions appeared, *e.g.*, $[(CH_3O)_2PO]^+$ at m/z 109 for trichlorfon, dichlorvos, demeton-*s*-methyl, and fenitrooxon, and $[(CH_3O)_2POSC_2H_4]^+$ at m/z 169 for oxydemeton-methyl and demeton-*s*-methyl [27]. Extensive fragmentation of dimethoate under ESI conditions was reported by Slobodník *et al.* [14], with a base peak at m/z 199, while only low-abundance fragments were observed next to the protonated molecule for fenamiphos. Chlorpyrifos could be analysed as a protonated molecule at m/z 350 at low cone voltages, while at higher cone voltages or in MS–MS, fragmentation occurred due to cleavage between 3,5,6-trichloropyridinol and the diethylphosphorothionate with charge retention on either side [28].

APCI is frequently applied in the analysis of OPP, despite the fact that (potentially thermally-induced) fragmentation occurs for many compounds. Some OPP can be analysed in positive-ion mode, others in negative-ion mode. The APCI mass spectra of twelve OPP in positive-ion mode and nine OPP in negative-ion mode were tabulated by Kawasaki *et al.* [29]. In positive-ion mode, the protonated molecule and a number of fragments were observed. In negative-ion mode, no deprotonated molecule was observed, but a fragment due to the loss of an alkyl group, *e.g.*, methyl for fenitrothion and for parathion. In a subsequent paper [21], the fragmentation observed for a number of OPP was studied in more detail.

Protonated molecules for dimethoate, fenamiphos, fenthion, coumaphos, and chlorpyrifos, next to some fragments, were reported by Slobodník *et al.* [14, 22]. For dimethoate (229 Da), the base peak was either a fragment at m/z 199 [14] or the protonated molecule at m/z 230 [22].

In the positive-ion and negative-ion mass spectra of twelve OPP, tabulated by Lacorte and Barceló [30], significant fragmentation was observed, even at low cone voltages (20 V). Further fragmentation occurred at a higher cone voltage (40 V). A base-peak protonated molecule was only observed for dichlorvos, fenthion, and diazinon. For all other compounds, a fragment peak was most abundant, *e.g.*, $[M+H–CH_3OH]^+$ for mevinphos, $[(CH_3O)_2P(OH)]^+$ for parathion-methyl and parathion-ethyl, $[(CH_3CH_2O)_2P(OH)_2]^+$ for chlorfenvinphos, and the 3-methyl benzotriazine heterocyclic ring for azinphos-methyl and azinphos-ethyl. Surprisingly, sodium adducts were observed in APCI for mevinphos, azinphos-methyl, malathion, azinphos-ethyl, and chlorfenvinphos. In negative-ion mode, the $[(RO)_2PX_2]^-$-anion (with R is methyl or ethyl and X is O or S) was observed as base peak for all compounds, except parathion-ethyl, fenthion, and diazinon. In most cases, a second intense fragment was observed as well [30].

Group-specific fragments like $[(CH_3O)_2PO_2]^+$, $[(CH_3O)_2PO]^+$, and $[(CH_3O)_2PS]^+$ at m/z 125, 109, and 125, respectively, as well as the loss of such groups from the protonated molecules were observed for fenthion and temephos as well as their degradation products [31].

Positive-ion and negative-ion APCI mass spectra were compared for fenitrothion, malathion, parathion-ethyl, and vamidothion [32]. Significant fragmentation was observed, *e.g.*, due to losses of NO• or CH_4 for fenitrothion, loss of ethanol for malathion, loss of NO• for parathion-ethyl, and loss of $(HS)P(=O)(OCH_3)_2$ for vamidothion in positive-ion APCI. In negative-ion APCI, losses of $O=P(OCH_3)_2$ or $S=P(OCH_3)_2$ were observed for fenitrothion and parathion-ethyl, and charge retention at $[(CH_3O)_2PS_2]^-$ or $[(CH_3O)_2POS]^-$ resulted in the base peak for malathion and vamidothion, respectively. Vamidothion and malathion showed best detection limits in positive-ion APCI, while fenitrothion and parathion-ethyl were best analysed in negative-ion APCI [32].

The effects of the nebulizer temperature (between 100 and 500°C) and the cone-voltage (between 10 and 60 V) on the response and the fragmentation of twelve OPP were studied by Lacorte *et al.* [33]. Higher temperature and cone voltage induced excessive fragmentation. Therefore, cone voltages below 40 V must be applied. Optimization of the nebulizer temperature was somewhat more complex, because temperatures between 400 and 500°C provided better response than lower temperatures, but also more fragmentation. The positive-ion APCI mass spectra, acquired with 20-V and 40-V cone voltages, of the twelve compounds were tabulated [33].

2.3   Triazines

Triazine herbicides, *i.e.*, 1,3,5-triazines, are another important compound class, frequently studied by LC–MS. While triazines are readily amenable to GC–MS, this is not true for the hydroxy- and des-alkyl degradation products.

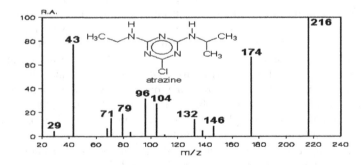

Figure 7.2: Product-ion mass spectrum of protonated atrazine.

Table 7.3: Fragmentation of atrazine ([M+H]$^+$ at m/z 216) in MS-MS		
m/z	**fragment**	**interpretation**
174/176	MH–42	loss of C_3H_6
146/148	MH–42–28	loss of C_3H_6 and C_2H_4
138	MH–42–36	loss of C_3H_6 and HCl
132/134	MH–42–42	loss of C_3H_6 and HN=C=NH (ring opening)
110	MH–42–36–28	loss of C_3H_6 and HCl and C_2H_4
104/106	MH–42–42–28	loss of C_3H_6 and HN=C=NH and C_2H_4
96	MH–42–42–36	loss of C_3H_6 and HN=C=NH and HCl
79/81		HN=CCl–NH$_3$$^+$
71		NC–NH$_2$–C$_2$H$_5$$^+$
43		HN=C=NH$_2$$^+$, C$_3H_7$$^+$

Triazines are analysed in positive-ion mode only. In ESI, protonated molecules are observed. Under in-source CID conditions, limited fragmentation due to the loss of an alkyl side chain is observed [9, 12, 14, 22, 34]. In APCI, triazines show an abundant protonated molecule and some fragmentation due to the loss of the alkyl side chain, e.g., for terbutylazine and terbutryn [9, 14, 22, 32, 35-36].

The fragmentation of protonated triazines was investigated in detail by Nélieu et al. [37], using a triple-quadrupole instrument and deuterated ammonia. The MS-MS product-ion mass spectrum of protonated atrazine is shown in Figure 7.2. The fragmentation is explained in Table 7.3. Stepwise fragmentation of triazines can be observed in multistage MS–MS in an ion-trap, as demonstrated for propazine [38], and for atrazine, simazine, and terbutylazine [25].

2.4   Phenylureas

Due to the thermal lability of the urea group, phenylureas are not amenable to GC–MS. They are frequently analysed by LC–MS. The general structure is shown below. The phenyl ring is substituted with halogen(s), methoxy, methyl, trifluoromethyl, or 2-propyl substitution. The R_1 side chains are methyl groups for most phenylureas, while the R_2 side chain can be methyl like in diuron, methoxy like in linuron, butyl like in neburon, or a proton like in monomethylmetoxuron.

The appearance of ESI mass spectra depends on the solvent conditions, especially on the extent of sodium contamination. Sodiated molecules were observed as the base peak for chlortoluron, isoproturon, diuron, linuron, and diflubenzuron [34], while protonated molecules and only weak sodiated molecules were observed for monuron, diuron, and neburon [14]. Some phenylureas can also be analysed in negative-ion ESI, where deprotonated molecules as well as acetate or formate adducts can be observed, depending on the mobile-phase composition [9].

Phenylureas can be analysed by APCI as protonated molecules without fragmentation [14, 23, 32, 35-36]. Isoproturon was also detected as a deprotonated molecule in negative-ion mode, but the positive-ion mode was more sensitive [32]. Diuron could be analysed in both positive-ion and negative-ion mode, resulting in protonated or deprotonated molecules at low cone voltages and additional fragmentation, i.e., the loss of dichloroaniline to $[(CH_3)_2N=C=O]^+$ at m/z 72 in positive-ion mode, and the loss of dimethylamine to m/z 186 in negative-ion mode [36]. Diflubenzuron provided a protonated molecule next to an intense fragment due to the loss of chlorophenyl–N=C=O [35], while in negative-ion mode, next to the deprotonated molecule, the loss of water and HF was observed [39]. In in-source CID, an intense fragment was observed at m/z 156 due to deprotonated difluorobenzamide as well as various other fragments.

In MS-MS, only a few intense fragment ions are observed, i.e., $[(CH_3)_2N–C=O]^+$ at m/z 72 and $[(CH_3)_2NH_2]^+$ at m/z 46 for N-dimethyl phenylureas, and $[(CH_3O)(CH_3)N=C=O]^+$ at m/z 88 for N-methyl-N-methoxy phenylureas [40].

2.5   Halogenated phenoxy acids

Chlorinated phenoxy acid (CPA) herbicides can analysed by GC–MS only after derivatization. For LC–MS, negative-ion ESI is the method of choice. Deprotonated molecules are detected as the most abundant ions under these conditions, often next to a phenolate fragment ion due to the loss of the acid side chain. Next to the deprotonated molecule, weak formic acid adducts $[M+HCOO]^-$ were observed for MCPA, 2,4-D, MCPP, and MCPB [41].

The fragmentation of CPA by in-source CID and its effect on the signal-to-noise ratio was investigated by Crescenzi et al. [42]. Inducing in-source CID resulted in an increasing abundance of the fragment ion as well as an improvement in signal-to-noise ratio for most CPA investigated. Negative-ion MS and MS–MS mass spectra of various CPA, including 2,4-D, dichlorprop, fluazifop, MCPA, and mecoprop, were tabulated by Køppen and Spliid [43]. In the negative-ion multistage MS–MS of MCPA, 2,4-D, mecoprop, and dichlorprop, the phenolate anion was observed in the first stage of MS–MS for all compounds. In the second stage of MS–MS, the loss of HCl was observed for dichlorprop [26]. Clofibric acid was analysed at deprotonated molecule in negative-ion APCI. A fragment due to the phenolate anion was observed as well [44].

Halogenated aryloxyphenoxypropionic acids are a new class of herbicides used for the selective removal of grass species. In commercial preparations, they are present as alkyl esters. In negative-ion ESI, haloxifop, fluazifop, and diclofop all show similar behaviour. The deprotonated molecule is the base peak in the spectrum. Weak formate and acetate adducts occur, and a fragment due to the loss of the propionate part [45]. The analysis of fluazifop and its butyl ester, fenoxaprop, quizalofop and haloxyfop and their ethyl esters, and diclofop and its methyl ester was reported. The free acids were analysed in negative-ion mode, and the esters in positive-ion mode. The esters showed sodium and potassium adducts next to the protonated molecules. The adduct formation was suppressed by the addition of 25 mmol/l formic acid to the mobile phase. The influence of the orifice potential on the appearance of their mass spectra was studied [46-47]

Deprotonated molecules for 2,4-D, MCPA, and mecoprop were observed next to significant fragmentation to the phenolate fragment in negative-ion APCI [32, 44]. Chlorine/hydrogen exchange was observed for MCPA. The influence of fragmentor voltage, the vaporizer temperature, the corona current, and the capillary voltage was systematically investigated [44].

2.6   Sulfonylureas

Sulfonylureas form a group of selective herbicides. The general structure is given below in Table 7.4. R_1 and R_2 generally are substituted heterocyclic rings, e.g., 4,6-dimethylpyrimidin-2-yl and 2-(benzoic acid methyl ester) for sulfometuron methyl and 4-methoxy-6-methyl-1,3,5-triazin-2-yl and 1-(2-chlorophenyl) for chlorsulfuron, respectively. The compounds are thermally labile and cannot readily be derivatized and are therefore not amenable to GC–MS.

Early ESI spectra of sulfonylureas were reported by Reiser and Fogiel [48]. Abundant protonated and sodiated molecules were observed, the ratio of which appears to be concentration dependent. Subsequently, positive-ion ESI data were reported [49-51]. The positive-ion ESI mass spectra of 8 sulfonylureas were tabulated by Volmer et al. [49]. The data for sulfometuron methyl and chlorsulfuron are summarized in Table 7.4. The mass spectra for seven sulfonylureas were studied as a function of the cone voltage by Di Corcia et al. [50]. Protonated molecules without significant fragmentation were only achieved at very low cone voltages. When a voltage of 25 V was applied, at least three fragment ions per compounds are formed, which can be used for confirmatory purposes. Protonated molecules and little fragmentation was reported for twelve sulfonylurea herbicides [51].

Table 7.4: Positive-ion ESI mass spectra of 2 sulfonylureas [49]		
R_1—N—C—N—S—R$_2$ (with O, O, O on the structure)	Sulfometuron methyl (M_r 364 Da)	Chlorsulfuron (M_r 357 Da)
Identification	**m/z (%RA)**	**m/z (%RA)**
$[M+K]^+$	403 (80)	–
$[M+Na]^+$	387 (90)	380 (65)
$[M+H]^+$	365 (100)	358 (100)
$[R_2SO_2]^+$	199 (75)	–
$[R_1N=C=O+NH_4]^+$	–	184 (55)
$[R_1N=C=O+H]^+$	150 (95)	167 (40)
$[R_1NH_2+H]^+$	–	141 (50)

The $[R_1N=C=O+H]^+$ ion at m/z 167 was observed as a common fragment ion for sulfonylureas containing the 4-methoxy-6-methyl-1,3,5-triazine substituent [49-51].

Sulfonylurea herbicides can also be analysed in negative-ion ESI mode. Deprotonated molecules were observed for chlorsulfuron, metsulfuron-methyl, thifensulfuron-methyl, and tribenuron-methyl [43, 52]. A major fragment for all four compounds was found at m/z 139, due to the 4-methoxy-6-methyl-1,3,5-triazine group. Eight sulfonylureas were analysed in negative-ion mode in a multiresidue study [47].

Product-ion MS–MS mass spectra of sulfonylureas, based on a protonated molecule generated by ESI, were reported by Li et al. [53]. For most compounds, only two or three product ions were detected, e.g., m/z 141 and 167 for chlorsulfuron, and m/z 150 and 199 for sulfometuron methyl. The $[R_1N=C=O+H]^+$ fragment is a common fragment in MS–MS. Negative-ion MS–MS mass spectra of chlorsulfuron, metsulfuron-methyl, thifensulfuron-methyl, and tribenuron-methyl were tabulated [43, 52]

2.7   Quaternary ammonium herbicides

Positive-ion ESI is the obvious mode-of-choice in the analysis of quaternary ammonium herbicides and plant growth regulators, although APCI data were reported as well. ESI mass spectra of paraquat and diquat were reported by Song and Budde [54], obtained under CE–MS conditions. The appearance of the mass spectra depended on the solvent composition. In acetic acid or sodium acetate, the base peak was due to the doubly-charged ions at m/z 92 $[M]^{2+}$ for diquat and at m/z 93 for paraquat. In addition, peaks were detected at m/z 183 and 184 for diquat and m/z 185 and 186 for paraquat. The ions at m/z 184 and 186 must be formed by a one-electron reduction of the M^{2+} ion. The ions at m/z 183 and 185, i.e., the base peaks when ammonium is present, were assumed to result from a proton transfer reaction to NH_3 in solution or during desolvation, i.e., [Cation–H]$^+$.

Under the conditions applied by Marr and King [55], the doubly-charged cation was the most abundant ion for paraquat. Furthermore, peaks were observed at m/z 185 and 171 due to [Cation–H]$^+$ and [Cation–CH$_3$]$^+$, respectively. For diquat, the ion at m/z 183 due to [Cation–H]$^+$ was the base peak, while the doubly-charged cation was observed as well. In MS–MS of [Cation–H]$^+$, the loss of C_2H_2 was observed for diquat• and the losses of either CH$_3$• or HCN for paraquat [55].

The influence of mobile-phase constituents on the mass spectra of paraquat and diquat was investigated in more detail by Taguchi et al. [56]. In the end, diquat was analysed as the [Cation–H]$^+$-ion at m/z 183, and paraquat as the [Cation/TFA]$^+$-ion pair. The mobile phase was 7% methanol in water with 25 mmol/l TFA and a post-column addition of 75% propionic acid in methanol.

Mass spectra for paraquat, diquat, mepiquat, chlormequat, and difenzoquat, obtained at two different cone voltages in both ESI and APCI, were tabulated by Castro et al. [57]. An acetonitrile in 15 mmol/l aqueous heptafluorobutyric acid (HFBA) gradient was applied, with post-column addition of acetonitrile. Under these conditions, no doubly-charged ions were observed for paraquat, difenzoquat, and diquat. In ESI, the [Cation–H]$^+$-ion was most abundant for paraquat and diquat, while the [Cation]$^+$ was most abundant for mepiquat, chlormequat, and difenzoquat. In APCI, the [Cation–CH$_3$]$^+$-ion was most abundant for paraquat, the [Cation–H]$^+$-ion for diquat, the [Cation]$^+$-ion for mepiquat, chlormequat and difenzoquat [57]. In a subsequent study, SIM and positive-ion ESI was applied using the [Cation]$^+$-ion for mepiquat, chlormequat, and difenzoquat, and the [Cation–H]$^+$-ion for diquat and paraquat [58].

The MS–MS spectra of the chlormequat and [D$_9$]-chlormequat were reported by Hau et al. [59] (see Figure 7.3). The major fragments are due to the loss of the chloroethyl moiety, leading to two ions at m/z 58 and 59 due to [(CH$_3$)$_2$N=CH$_2$]$^+$ and [(CH$_3$)$_3$N]$^{+•}$, respectively. The chloroethyl ions at m/z 63 were observed as well. Interestingly, the relative abundance of the fragments due to [(CH$_3$)$_2$N=CH$_2$]$^+$ and [(CH$_3$)$_3$N]$^{+•}$ were reversed in the [D$_9$]-analogues due to a heavy-atom effect [59]. This effect is illustrated by the spectra in Figure 7.3.

Ion-trap MS–MS spectra for paraquat, diquat, difenzoquat, mepiquat, and chlormequat were reported [60-61]. The fragmentation pathways were discussed in considerable detail [60]. The interpretation of the product-ion MS–MS spectra was checked and studied in more detail using accurate product-ion determination via MS–MS on a Q–TOF instrument. The elucidation of the fragments observed for paraquat, diquat, mepiquat, chlormequat, and difenzoquat was tabulated [62].

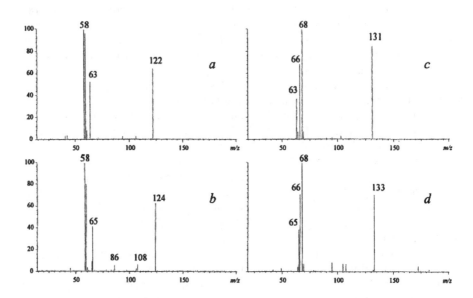

Figure 7.3: ESI–MS–MS product-ion spectra of chlormequat, showing (a) the ^{35}Cl isotope peak and (b) the ^{37}Cl isotope peak of unlabelled chlormequat; (c) the ^{35}Cl isotope and (d) the ^{37}Cl isotope peak of [D$_9$]-labelled chlormequat. Reprinted from [59] with permission. ©2000, Elsevier Science B.V.

2.8   Miscellaneous pesticide classes

Chloracetanilide herbicides like alachlor, metolachlor and metazachlor can be analysed in positive-ion mode. In ESI, protonated and sodiated molecules are observed as well as the loss of methanol [16, 34, 47, 63]. The MS–MS spectrum of alachlor was studied, using MS–MS and in-source CID on a orthogonal-acceleration time-of-flight mass spectrometer [63].

Diphenyl-ether herbicides such as aclonifen, bifenox and lactofen are relatively new herbicides, used for weed control in the growth of seeded legumes, such as soybeans. MS and MS–MS of five neutral diphenyl-ethers herbicides and three acid metabolites was reported [64]. Negative-ion ESI is preferred for the acidic compounds [47, 64] and either negative-ion APCI [64] or positive-ion ESI [47] is used for the neutral ones. Structure informative fragmentation was observed in negative-ion MS–MS [64].

The positive-ion ESI mass spectra of six *imidazolinone herbicides, e,g,,* imazapyr and imazaquin, showed protonated molecules at low cone voltages (40 V), while fragmentation was observed at higher cone voltages [51, 65-67]. However, in a multiresidue study, imazamethabenz methyl was analysed in positive-ion ESI, while for imazapyr, imazethapyr, imazamethabenz, and imazaquin negative-ion ESI was preferred [47]. Interpretation of the MS–MS spectra of imazapyr [65], imazamethabenzmethyl [66], and imazethapyr [67] was reported as well.

In negative-ion ESI, *bentazone* could be analysed as a deprotonated molecule at m/z 239 [68]. In negative-ion APCI, bentazone and its 6-hydroxy- and 8-hydroxy-degradation products could be analysed as deprotonated molecules [44]. Bentazone showed a weak fragment due to the loss of SO_2, while the base peak in the spectrum of 6-hydroxybentazone at low fragmentor voltages was due to $[M–H_2O+HCOO]^-$ at m/z 283. The influence of fragmentor voltage, the vaporizer temperature, the corona current, and the capillary voltage was systematically investigated [44]. In MS–MS, losses of a propyl radical (to m/z 196), of SO_2 (to m/z 175), and of both groups (to m/z 132) were observed [43, 68].

Dinoseb and dinoterb have been detected as deprotonated molecules in negative-ion APCI [32]. The fragments in negative-ion MS–MS of dinoseb at m/z 222, 207, 193, and 163 were not readily explained. The ions at m/z 193 and 163 could be due to subsequent losses of $NO_2\bullet$ and $NO\bullet$ [43].

Negative-ion ESI was applied for the detection of *glyphosate and AMPA* after ion chromatography. Deprotonated molecules were observed, while glyphosate showed additional fragments due to the loss of water or CO_2 [69]. The negative-ion ESI multistage ion-trap MS–MS spectra of glyphosate, glufosinate, AMPA and methylphosphinicopropionic acid was studied in detail by Goodwin *et al.* [70].

The *conazole fungicides* triadimenol, tebuconazole, flusilazole, penconazole and propiconazole can be analysed as protonated molecules in positive-ion ESI or APCI. Flusilazole and propiconazole could be fragmented in ion-trap MS–MS spectra, the other three components cannot [71]. Imazalil can be analysed as a protonated molecule over a wide range of fragmentor voltages in both ESI and APCI [19]. Flutriafol was analysed as a deprotonated molecule in negative-ion APCI. A fragment due to the loss of fluorobenzene was observed [72]. Prochloraz showed a weak deprotonated molecule at m/z 374 and a fragment apparently due to (thermally-induced) exchange of one Cl atom by an O atom at m/z 356, while the base peak was due to 2,4,6-trichlorophenolate at m/z 195 [72].

Carbendazim can be analysed as a protonated molecule at m/z 192 in both positive-ion ESI and APCI [19, 23]. At higher fragmentor voltages, the loss of methanol to m/z 160 or of $C_2H_2O_2$ to m/z 134 is observed.

Clofentazine can be analysed in positive-ion APCI, but a more intense response can be achieved in negative-ion electron-capture conditions, resulting in $M^{-\bullet}$ [82].

Thiabendazole showed a protonated molecule at m/z 202 in positive-ion APCI and ESI [19, 23]. At higher fragmentor voltages, a fragment due to the loss of HCN was observed [19].

3. Strategies in environmental analysis

Initially, LC–MS strategies in environmental analysis were focussed at target compound analysis of a limited number of pesticides from within one compound class in different environmental water compartments, mainly ground and surface water. Later on, this focus broadened in a number of ways, *e.g.*, including other matrices like soil and sediments, including pesticide degradation products, and aiming at multiresidue screening, involving compounds from a variety of classes. This change in focus is somewhat reflected in the various sections of this chapter.

At the same time, there was a growing interest in other potential hazardous compounds in the environment, including pesticide degradation products, pharmaceuticals, surfactants, aromatic sulfonates, and endocrine disruptors (Ch. 8). The analysis of waste water, prior and after sewage treatment plants, is an example of this.

3.1   General considerations

Three aspects are especially of concern in the development of analytical strategies for the analysis of pesticides in environmental matrices:
- The achievable concentration detection limit: in order to determine an individual pesticide at the regulatory level of 0.1 µg/l (in the EU) by means of a straightforward LC–MS method, an absolute detection limit of the method of *ca.* 10 pg is needed.
- The obligation to not only detect an unknown compound at this level, but also perform unambiguous identification of the unknown.
- The continuous influx of samples, demanding automated and unattended operation of the methods for continuous monitoring purposes.

Given the generally poorly defined ways detection limits are quoted in the literature, the highly compound-dependent response in LC–MS methods, and the wide variety of analytes of interest, statements on actual detection limits are difficult to make. Reviewing the current state-of-the-art allows a number of conclusions to be drawn:
- There can be significant differences in response between compounds from different compound classes. Thirty-fold differences in the response factor for compounds readily amenable to LC–MS were reported [9].
- Within a particular compound class, quite significant differences between the responses of favourable and unfavourable analytes can be found. Thurman *et al.* [9] reported sevenfold and fourfold differences in the response factors between various phenylurea and triazine herbicides, respectively.
- Some compound classes are preferentially analysed in positive-ion mode, while for others the negative-ion mode has to be preferred. Actually, within some compound classes, *e.g.*, the organophosphorous pesticides (Ch. 7.2.2), the

positive-ion mode is preferred for some compounds, while the negative-ion mode provides better response for others.

- The achievable absolute detection limits are often insufficient for the direct LC–MS detection of pesticides at regulatory levels, e.g., 0.1 µg/l of individual pesticides within the EU. Therefore, additional technology, especially with respect to preconcentrating sample pretreatment, is required.

3.2 Sample pretreatment strategies

Liquid-liquid extraction

Within the field of environmental analysis, liquid-liquid extraction (LLE, Ch. 1.5.2) is not very popular anymore, given the need to use (large amounts of) organic extraction solvents which may be harmful to the environment. In addition, LLE methods are rather labourious and time consuming.

As an example: 2.5 l of estuarine water were extracted with two times 100 ml dichloromethane for the analysis of chloracetanilide, triazine, and phenylurea herbicides. The extract was evaporated to dryness and the residue was dissolved in 400 µl of methanol, from which 10 µl was injected in LC–MS. Typical absolute detection limits in SIM are between 10 and 500 pg [74].

Off-line solid-phase extraction

The selection of packing materials for SPE (Ch. 1.5.3) is especially important in multi-residue analysis. In general, nonspecific trapping on hydrophobic surfaces like C_{18}-bonded silica or styrene–divinylbenzene copolymers, e.g., PLRP-S, is preferred.

Graphitized carbon black (GCB) materials are frequently used for off-line SPE as well. GCB behaves as a nonspecific sorbent, but because positively-charged adsorption sites are present, it also acts as an anion exchanger, which is useful for the enrichment of acidic compounds [75]. Examples involve the enrichment of chlorinated phenoxy acids [42], aryloxyphenoxypropionic herbicides [45], sulfonylurea herbicides [50], and imidazolinone herbicides [65].

The use of dual-SPE approaches has also been described, e.g., extraction of N-methylcarbamates insecticides and their metabolites from urine by means of GCB and further clean-up using NH_2-modified silica [18], or the use of RP-102 extraction cartridges with further clean-up of extracts on a strong anion-exchange column in the determination of selected sulfonylurea and imidazolinone herbicides [51].

The stability of desethylatrazine, fenamiphos, fenitrothion, and fonofos adsorbed on disposable SPE cartridges was studied under different storage conditions [76-77]. Complete recovery of all compounds was possible after one-month storage at –20°C, while degradation of fenamiphos and fentrothion occurred after one-month at 4°C.

Particle-loaded membrane extraction disks, the so-called Empore disks, form an alternative to SPE cartridges [27, 78-79]. With Empore disks, higher sampling flow-rates are possible. They show excellent sorbent capacity. Recoveries at 0.1-µg/l level of a series of organophosphorous pesticides on C_{18} or styrene-divinylbenzene Empore disks were found to be comparable to those obtained with conventional C_{18} cartridges [27]. The Empore disks can be used off-line as well as on-line.

On-line solid-phase extraction

On-line SPE via valve-switching techniques, pioneered by the group of Brinkman [80], is frequently used in the environmental analysis of aqueous samples, as it enables the rapid treatment of large samples. The precolumn enables selective sorption of the analytes, removal of hydrophilic interferences, and significant analyte preconcentration. In most cases, either 1–10 mm × 2–4.6 mm ID SPE cartridges packed with 30–50 μm ID particles are used. Typical applications of on-line SPE–LC–MS are the determination of acidic herbicides [41], sulfonylureas [49], and organophosphorous pesticides [30]. Instead of SPE cartridges, Empore disks can be used on-line as well [79].

Alternative SPE approaches involve the use of:
- Restricted-access materials (RAM, Ch. 1.5.6), as demonstrated for the analysis of triazines in water [81] and of azole pesticides in urine [82],
- Molecularly-imprinted polymers, *e.g.*, in the analysis of triazines [83].
- Large-volume injection with analyte preconcentration on top of the analytical column [84]. With the current detection limits, achievable in target-compound LC–MS–MS, large-volume injection is often successfully applied in order to avoid time-consuming sample pretreatment.
- Solid-phase microextraction (SPME), *e.g.*, for the determination of carbamate and triazine pesticides from water, soil leachates, and slurries [85-87].
- Combined SPE sample pretreatment and LC separation for target-compound analysis of up to eight components on one single short column (typically 10–20 mm × 2–4.6 mm ID high-pressure packed with 8–15 μm ID particles) [22, 38, 40]. 4–15 ml of an aqueous environmental sample is preconcentrated before the column is eluted with a fast gradient program. The limited chromatographic resolution is compensated by the selectivity of the selected reaction monitoring (SRM) mode. The potential of this approach was demonstrated in the analysis of six triazine or eight phenylurea herbicides in positive-ion APCI on a triple-quadrupole [40] or an ion-trap [38] instrument and in pesticide degradation studies [63, 88].
- The coupled-column approach (*cf.* Ch. 1.4.6), where a particular peak of interest is heart-cut from the chromatogram developed on the first column and sent to a second column for a second stage of analysis, either via a sample loop or via a short trapping column. Coupled-column LC–MS was successfully applied as a tool in reducing matrix effects [84]. Sancho *et al.* [28, 89] reported the use of coupled-column LC for biological monitoring of occupational pesticide exposure (see Ch. 7.8).
- Turbulent-flow chromatography (Ch. 1.5.5). Eleven priority pesticides were

enriched on a carbon column, separated on a monolithic column (Ch. 1.4.4) and analysed by LC–APCI–MS–MS [90].

Immunoaffinity-based pretreatment

Immunoaffinity-based sample pretreatment (Ch. 1.4.2) was for instance applied in the analysis of carbofuran at 40-pg/ml level from water or at 2.5 ng/g from a potato extract [91], the analysis of carbendazim at 100-ppb level in soil extracts, and at 25-ppt levels in lake water samples [92], and for the ppt-level determination of various pesticides in sediments and natural waters [35].

4. Environmental target compound analysis

Numerous examples are available of the use of LC–MS in the target compound analysis, directed at the determination of a small number of compounds, in most cases from just one compound class, in environmental samples. Initially, the detection was performed in SIM, using one ion per compound. In this way, absolute detection limits were for instance achieved in the range of 0.15–0.5 ng for carbamates [93], of 2–50 ng for OPP [21], or of 10–100 pg for OPP after off-line SPE on an Empore disk [27], of 0.2–0.5 ng for triazines [93], and of 0.1–0.2 ng for phenylureas [93]. Later on, MS–MS strategies were implemented (see Ch. 7.4.2).

4.1   Quaternary ammonium herbicides as target compounds

As an example of the type of studies performed in environmental target compound analysis, the optimization and method development for the quantitative LC–MS analysis of quaternary ammonium compounds, especially paraquat and diquat, is briefly reviewed. The type of ions observed for paraquat and diquat strongly depends on the experimental conditions (Ch. 7.2.7).

Initially, methods for the analysis of diquat and paraquat were developed, based on CE–MS [54, 94], providing detection limits of $ca.$ 50 ng/l. The first LC–MS method was reported by Marr and King [55]. The analytes were separated using ion-pair LC with 10 mmol/l heptafluorobutyric acid (HFBA) as ion-pair agent. SRM was applied with the transitions m/z 185→158 for paraquat and m/z 186→157 for diquat. Detection limits in direct analysis were 5 µg/l for paraquat and 1 µg/l for diquat. Caffeine was used as internal standard.

Taguchi $et\ al.$ [56] reported a method based on SPE using ENVI-8 disk material, LC separation on a C_1-column using a mobile phase of 7% methanol in water, containing 25 mmol/l TFA. A post-column addition of propionic acid in methanol was applied (TFA-fix, Ch. 6.6.3). Detection was based on the [Cation–H]$^+$-ion at m/z 183 for diquat, and the [Cation/TFA]$^+$-ion pair for paraquat. Detection limits were 0.1 and 0.2 µg/l.

Castro *et al.* [57] studied the influence of various ion-pair agents on the response of diquat and paraquat. The ion-pair LC separation is based on a 0.5–40% acetonitrile gradient in 15 mmol/l aqueous HFBA. The compounds were analysed in tap water after a Sep-Pak sample pretreatment. The detection limits were 0.9 and 4.7 µg/l in ESI and 0.1 and 1.8 µg/l in APCI for diquat and paraquat, respectively. In a subsequent study [58], on-line SPE is performed on ENVI-8 disks, after addition of 15 mmol/l HFBA to the filtered drinking water sample. Detection limits were 50 and 60 ng/l for diquat and paraquat, respectively. Further improvement of the detection limits to 30 ng/l for both compounds was achieved by the use of on-line SPE–LC–MS–MS. The intra-day and inter-day precision for diquat were 9.4% and 12.8%, respectively [60]. In another study from the same group [95], an oa-TOF-MS, operated in full-spectrum acquisition mode, and a triple-quadrupole instrument, operated in SRM mode, were compared. The detection limits with the triple-quadrupole instrument were at least tenfold better than those obtained with the oa-TOF instrument, *i.e.*, 60 and 3 ng/l in tap water for paraquat and diquat, respectively.

4.2   Confirmation of identity

While initially target compound analysis based on the detection of one target ion in SIM was performed, later on, also stimulated by discussions with regulatory bodies, one or two additional ions per compound were included for confirmation of identity. Discussion on confirmation criteria were for instance reported by Li *et al.* [53] in the determination of sulfonylureas in soil, and by Geerdink *et al.* [96] in the determination of triazines in surface water. The current practice generally involves the use of one ion for quantitation, preferentially the protonated or deprotonated molecule, and two fragment ions for confirmation of identity, or the monitoring of two SRM transitions. This is also in agreement with the principle of identification points, outlined in the EU regulations for residues of veterinary drugs in food [97], which is adapted by many researchers in pesticide analysis. Hernández *et al.* [98] evaluated the potential of various MS approaches, *i.e.*, triple-quadrupole MS–MS in multi-channel SRM mode, oa-TOF and Q–TOF instruments, in full-spectrum MS and MS–MS mode, respectively, in gaining sufficient identification points for the confirmation of identity in the multiresidue target analysis of pesticides in environmental water. With the triple-quadrupole MS–MS, four or five identification points can be achieved in most cases. With the Q–TOF instrument, up to 20 identification points can be reached. The oa-TOF instrument only provides a sufficient number of identification points in favourable cases, *i.e.*, for compounds showing sufficient sensitivity, isotopic patterns, and/or easy fragmentation by in-source CID.

Examples of the application of the three-ion-criteria in SIM mode can be found in reports on the analysis of acidic pesticides [42], sulfonylureas [49-51], and imidazolinone herbicides [51]. Rodriguez and Orescan [51] actually apply confirmation criteria based on (a) the LC–MS retention time of analyte, which must be within 1% of that of the standards, (b) the monitoring of the protonated molecule and two fragment ions, and (c) the ion abundance ratio of the fragments which should be within 20% of ion ratio in a standard. A typical chromatogram obtained for a 20 µg/l standard solution, containing all sixteen target herbicides, is shown in Figure 7.4.

With the advent of triple-quadrupole and ion-trap MS–MS systems, confirmation of identity is based on single- or multiple-transition SRM procedures, as for instance demonstrated for sulfonylureas [53], triketone, and various other herbicides [99].

Instead of fragmentation in MS–MS, the determination of accurate mass of the target analyte using an orthogonal-acceleration time-of-flight (oa-TOF) instrument can also be used for confirmation of identity. This was first demonstrated in a multiresidue analysis based on on-line SPE–LC–oa-TOF-MS [100]. Mass accuracies better than 5 mDa or 20 ppm were routinely obtained for protonated molecules, sodium adducts, and fragment ions of the target compounds investigated.

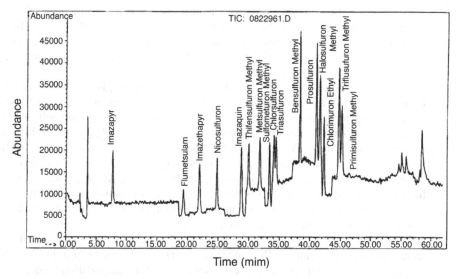

Figure 7.4: Reconstructed total-ion-current chromatogram of the time-scheduled SIM analysis of 16 target herbicides from the sulfonylurea and imidazolinone class. Injection of a 20 µg/l standard mixture. Reprinted from [51] with permission. ©1998, American Chemical Society.

5. Environmental multiresidue screening

In more recent years, environmental analysis has been more directed to multiresidue screening, either in screening for pesticides and their degradation products from a variety of component classes, or in an even broader screening, searching for any contaminant. The former case can be described as a multiresidue target analysis. In the latter case, not only pesticides are searched for, but also other classes of endocrine disruptors, surfactants, pharmaceuticals, and pesticide degradation products. As such, it can be described as a general contaminant screening.

5.1   Multiresidue target analysis

Multiresidue target analysis is directed at the detection of multiple target compounds from various compound classes. In most cases, quantitative analysis as well as confirmation of identity is required. An important issue in multiresidue target analysis is the sample pretreatment, which should allow the isolation of all target compounds, which may significantly differ in polarity, from the environmental matrix. SPE is used most often in such cases. Various MS acquisition strategies can be applied to perform this task. Some selected examples are briefly discussed below.

Obviously, the full-spectrum acquisition mode can be applied. This is demonstrated in the determination of sixteen carbamate, urea and thiourea pesticides and herbicides, using a single quadrupole instrument [101] and by the use of accurate-mass determination using an oa-TOF instrument, providing somewhat better confirmation of identity [100].

SIM with three compound specific ions can be applied, as for instance demonstrated by Wang and Budde [101] for sixteen carbamate, urea and thiourea pesticides and herbicides, by Rodriguez and Orescan [51] for selected sulfonylurea, imidazolinone, and sulfonamide herbicides, and by Yu *et al.* [17] for 52 carbamates, thiocarbamates, and phenylureas. In the latter case, computer-controlled optimization of the MS measurement conditions is performed. The most prominent ion, either a protonated or an ammoniated molecule, is used for quantitation. The dwell time for each ion in SIM was 0.02 s with an inter-channel delay of 0.02 s, *i.e.*, 2 s are needed for each data point in the chromatogram. Detection limits ranged from 0.09 to 19 μg/l with 50-μl injections.

Instead of multi-channel SIM, multi-channel SRM can be applied as well. This is demonstrated by the group of Hernández [102-103] in the on-line SPE–LC–MS–MS analysis of 35 target pesticides and some of their degradation products in environmental water samples at 25-ng/l level, using SPE of only 1.33 ml of water onto a 10 × 2 mm PRP-1 SPE cartridge column.

5.2   General contaminant screening

General contaminant screening using LC–MS is somewhat more complex than using GC–MS. This is mainly due to the fact that ESI is a soft ionization technique, which can provide both positive-ion and negative-ion modes. In order to get a screening as comprehensive as possible, both positive-ion and negative-ion mode should be used. In addition, other ionization strategies, like APCI and atmospheric-pressure photoionization (APPI), should be applied to detect the less polar compounds that are not prone to ESI [9]. In addition, both full-spectrum MS and MS–MS data should be acquired to enable identification of unknowns detected. Procedures that comprehensive have not yet been described.

A possible strategy for general contaminant screening may be illustrated by the situation in the Netherlands, where the drinking water of a significant part of the population originates from the rivers Rhine and Meuse. Continuous monitoring systems are placed in the rivers where they enter the Netherlands. A number of continuous analytical and biological tests are performed on water samples at these stations. An important tool at these monitoring stations is an integrated analytical system called SAMOS (System for the Automated Monitoring of Organic pollutants in Surface water [104]), developed within the framework of the Rhine Basin Program [105]. SAMOS enables the automated unattended analysis of filtered 100-ml surface-water samples by means of SPE coupled on-line with an LC system equipped with the UV photo-diode-array (DAD) detector. The data are automatically evaluated, allowing detection and quantitation of any compound found at concentrations above the regulatory level of 0.1 µg/l. Any compound detected by SAMOS at or above the regulatory level is subsequently identified and/or confirmed by MS. For confirmation of identity, multiresidue target analysis approaches have been used in most cases. Various analytical strategies involving LC–MS have been developed and applied for general contaminant screening. The use of LC–MS is now far more implemented in many environmental laboratories than it was in the early 1990s, when the SAMOS system was developed.

An important issue is the need to acquire both MS and MS–MS data for the unknowns. In general, this requires two injections with data-processing in between. The first run is done in full-spectrum LC–MS mode. Precursor m/z-values for relevant peaks in this chromatogram have to be determined in (manual) data processing. The m/z-values found are then used in a time-scheduled product-ion MS–MS procedure using multiple precursor-ions. Alternatively, data-dependent acquisition (DDA, Ch. 2.4.2) can be used, as demonstrated by Drexler et al. [106]. An alternative to DDA in a triple-quadrupole instrument is the RF product-ion analysis mode (RFD), proposed by Kienhuis and Geerdink [107].

Hogenboom et al. [100] described a procedure for identification of unknown contaminants based on the use of on-line SPE–LC–MS in an oa-TOF instrument. The accurate mass of the peak of interest and the possible elemental compositions ed from it were searched against a pesticide database containing 800 entries. er identification was based on retention time, UV spectra in relation to the le-bond equivalent, and the expected isotopic pattern. This procedure was fied by Bobeldijk et al. [26] and Ibáñez et al. [108], where LC–MS data on tion time, accurate mass, isotopic pattern, and possible elemental composition supplemented with product-ion MS–MS mass spectra acquired on a Q–TOF ment. Injection of a standard of the identified compound was used for final rmation. A similar approach was reported by Thurman et al. [109], based on the from an oa-TOF instrument in combination with multistage MS–MS in a ion-instrument. These studies demonstrated the successful identification of own compounds in environmental samples, although the data do not always to a structure.

esticide degradation and metabolism

bviously, potential hazardous effects are not only related to pesticides, but bly also to their metabolites in living systems and/or degradation products in environment. Considerable attention has been paid in recent years to the acterization of degradation products of pesticides and to the monitoring and titative analysis of such compounds in environmental samples. Selected results strategies are reviewed in this section.

Chlorophenols and nitrophenols

henols of environmental interest are derived from a wide variety of industrial ces, or present as biodegradation products of humic substances, tannins, and ns, and as degradation products of many chlorinated phenoxyacid herbicides organophosphorous pesticides. Phenols, especially chlorophenols, are persistent, toxic at a few μg/l. Therefore, phenols are listed at the US-EPA list of priority tants and the EU Directive 76/464/EEC as dangerous substances. The samples analysed can be surface waters or industrial effluents.

uig et al. [110] compared the performance of LC–MS using thermospray, ESI, APCI. The best results were obtained with APCI, which was subsequently used velop an automated method for trace level determination of 19 priority phenols. ine SPE–LC–MS in negative-ion APCI mode was applied for 16 compounds. od detection limits in SIM on [M–H]⁻, using 50-ml water samples, were in the e of 20–40 ng/l for the monochlorophenols, and below 5 ng/l for the other pounds. Three other compounds (phenol, 4-methylphenol, and 2,4-thylphenol) were separated on a graphitized nonporous carbon column in pure anol and analysed in negative-ion ESI mode (detection limit 50-75 ng/l with a l water sample) [111]. The separation by means of a ternary gradient and subsequent APCI–MS detection of seventeen chlorophenols was reported by Jáuregui et al. [112]. [M–H]⁻ was the main ion for mono- and dichlorophenols, while for higher chlorophenols [M–H–HCl]⁻ was most abundant. Detection limits in 0.2 g soil samples after soxhlet extraction ranged between 0.01 µg/g for pentachlorophenol and 0.3–0.7 µg/g for monochlorophenols. By processing 10 g of soil in microwave-assisted extraction followed by an SPE clean-up, lower concentration detection limits, ranging between 0.007 ng/g and 0.3 ng/g, were demonstrated in a similar method by Alonso et al. [113]. Negative-ion APCI is obviously the method of choice for chlorophenols.

6.2 Pesticide degradation products

LC–MS plays only a minor role in the identification of pesticide degradation products. Both GC–MS and LC–MS were applied in some studies on the degradation of OPP [31, 78, 114] and of alachlor [115].

A photochemical reactor coupled on-line with ESI MS–MS was used to introduce samples into the reactor either by column-bypass injection or after LC separation [116]. The system, schematically drawn in Figure 7.5, enables the on-line monitoring of *in situ* generated photodegradation products. In order to study the photodegradation of alachlor in surface water, Hogenboom et al. [63] described a system for the regular sampling from a large-volume photochemical reaction vessel. The samples were preconcentrated using SPE and subsequently analysed by LC–MS analysis using either a triple-quadrupole MS–MS or an oa-TOF-MS instrument. Other studies in this area concern the identification of degradation products of pirimiphos methyl in water under ozone treatment [117], and the photodegradation products of the triazine herbicides terbutylazine, simazine, terbutryn, and terbumeton in water using a Q–TOF instrument [118].

6.3 Ionic chloracetanilide metabolites

A variety of neutral degradation products of the chloracetanilide herbicide alachlor was identified [63, 115]. However, the ionic metabolites such as the oxoethanesulfonic acid derivative appear to be of more significance, as they are readily leached to groundwater. While alachlor itself is amenable to GC–MS, its ionic metabolites are not. Initially, GC–MS, LC-UV-DAD, and fast-atom bombardment MS–MS were applied in the analysis and identification of such metabolites [119]. Subsequently, the potential of LC–ESI–MS in this area was recognized [120]. Both oxanilic acid and oxoethanesulfonic acid metabolites of alachlor, acetochlor and metolachlor were identified in surface water and ground water, and subsequently determined with detection limits at the 0.01-µg/l level using off-line SPE in combination with LC–MS [120].

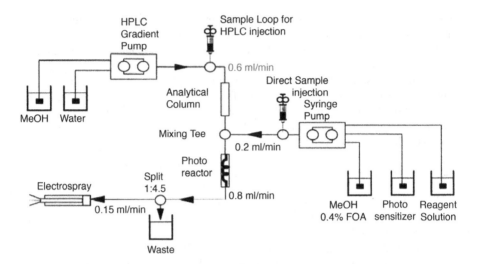

Figure 7.5: Experimental setup for the on-line LC–MS monitoring of *in situ* generated pesticide photodegradation products. Reprinted from [116] with permission. © 1998, Elsevier Science B.V.

The quantitative analysis of the metabolites in groundwater is an important application of LC–MS. Vargo *et al.* [121-122] described a method based on off-line SPE and LC–MS–MS, enabling the detection at 0.1-ppb level in groundwater.

7. Pesticide residues in fruit and vegetables

While in the 1990s most LC–MS applications in pesticide analysis concerned environmental analysis, in the first decade of the 2000s the analysis of pesticide residues in (citrus) fruit and vegetables is more prominent. The determination of pesticide residues in fruit and vegetables was reviewed by Picó *et al.* [4, 8]. Selected examples are reviewed here.

Barnes and coworkers pioneered in this area. They reported the analysis of diflubenzuron in mushrooms [39], and of diflubenzuron and clofentezine in various fruit drinks [73], and the development of a multiresidue study for ten pesticides in fruit, involving ionization polarity switching in LC–APCI–MS [123]. In these studies, significant attention is paid to matrix-dependent ion suppression or enhancement effects (Ch. 6.7), which is observed even in APCI. Matrix effects in food analysis must be studied in detail for each fruit or vegetable. Obviously, optimization of the sample pretreatment procedures plays an important role in method development for pesticide residue analysis in food.

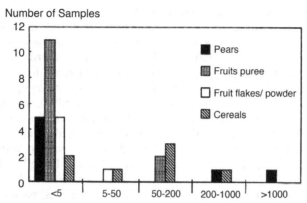

Figure 7.6: Results of a limited survey expressed as concentration ranges of chlormequat residues (mg/kg) found in various food commodities. Reprinted from [66] with permission. ©2000, Elsevier Science B.V.

In general, liquid extraction with or without a successive clean-up step using SPE is used for sample pretreatment. While Startin *et al.* [124] applied a methanol extraction of chlormequat residues from pears without further clean-up, Hau *et al.* [59] applied a methanol-water (1:1, *v/v*) extraction followed by a clean-up step using a LiChrolut SCX SPE cartridge for the same compound in the same matrix. Using a combination of MS and MS–MS data, Startin *et al.* [124] achieved a detection limit of 40 µg/kg, while Hau *et al.* [59] reported a detection limit of 1 µg/kg, demonstrating the advantage of additional clean-up. Hau *et al.* [59] applied their method to pears, pear juice concentrates, fruit purees, and cereal products from various sources. Results of their survey are summarized in Figure 7.6. Similar detection limits for chlormequat and mepiquat in pear, tomato, and wheat flour were reported by Riediker *et al.* [125] using injection of extracts in an on-line SPE–LC–MS–MS system, using an SCX SPE cartridge. Detection limits of 1 µg/kg chlormequat in tomatoes were reported by Careri *et al.* [126] using liquid extraction, no clean-up, and LC–MS on a SCX-column. The recommended method for chlormequat, issued by the EU [127], implies the use of an isotopically-labelled internal standard in order to minimize matrix effects.

The group of Picó [19, 72, 87, 128-130] reported various methods for the determination of pesticides and post-harvest fungicides in (citrus) fruit. LC–APCI–MS was applied after a two-step liquid extraction for the detection of the fungicides benomyl, carbendazim, imazalil, thiabendazole, and thiophanate methyl in oranges down to 20-µg/kg levels [19]. LC–APCI–MS methods were described for the determination of carbendazim, the insecticides imidacloprid and methiocarb, and the acaricide hexythiazox in peaches and nectarines [128] and for the fungicides dichloran, flutriafol, *o*-phenylphenol, prochloraz, and tolclofos methyl in oranges, lemons, bananas, peppers, chards, and onions [72].

Stir-bar sorptive extraction (SBSE) and matrix solid-phase dispersion (MSPD) were compared to ethyl acetate liquid extraction in the method development for bitertanol, carbendazim, fenthion, flusilazole, hexythiazox, imidacloprid, methidathion, methiocarb, pyriproxyfen, and trichlorfon in oranges, using LC–APCI–MS. MSPD was found to be an excellent extraction and preconcentration method, applicable to a wide range of pesticides [129]. SPME was evaluated for sample pretreatment in the determination of the post-harvest fungicides dichloran, flutriafol, *o*-phenylphenol, prochloraz, and tolclofos methyl in cherry, lemon, orange, and peach samples using LC–MS–MS on an ion-trap instrument. The method provides detection limits ranging between 0.5 and 10 µg/kg, depending on the target analyte [87]. Pressurized liquid extraction was evaluated for the analysis of benzimidazoles and azoles, organophosphorous, carbamates, neonicotinoids, and acaricides in oranges and peaches [130] and enabled analysis down to 0.025 and 0.25 mg/kg, *i.e.*, well below established maximum residue levels.

Some examples of multiresidue target analysis involving a wide variety of pesticides and their degradation products in fruits and vegetables based on time-scheduled multichannel SRM are the detection and confirmation of 38 pesticides in crude extracts from grape, kiwi fruit, lemon, spinach, and strawberry [131], the multiresidue target analysis of 74 pesticides in thirteen different fruits and vegetables [132], and the detection and confirmation of 57 pesticides and degradation products in a wide variety of matrices involving eleven out of the fourteen matrix-type groups, similar to the ones by the EU when establishing MRLs [133]. The time-scheduled multichannel SRM for the multiresidue target analysis of 74 pesticides in fruits ands vegetables, reported by Ortelli *et al.* [132], may serve as an example. Optimum experimental conditions (cone voltage and collision energy) were determined for each target compound. Two SRM channels were optimised for each compound, enabling quantitation as well as confirmation of identity. Time-scheduled acquisition was performed in eleven acquisition groups using the most abundant transition with a dwell time of 50 ms per transition. This enables quantification by comparison to two level standard solutions. Samples exceeding the MRL were confirmed in a second analysis using both transitions for the relevant compound(s) with a dwell time of 500 ms each. Quantitation at 0.01 mg/kg was possible for most target compounds. Significant matrix effects were observed, indicating the importance of the use of matrix-matched standards for each individual matrix. In order to reduce the number of matrix-matched standards, 57 pesticides and metabolites were determined in fourteen different commodity groups of fruits and vegetables by others [133]. Commodity groups are fruits or vegetables which share common features, *e.g.*, high water content, high fat content, or high acid content (SANCO/825/00). Matrix effects in the analysis of twenty pesticides in eight different vegetables were evaluated in order to select one representative matrix for calibration purposes [134]. A cucumber blank extract was selected for this purpose.

Thurman and Ferrer [109, 135-137] recently promoted the use of oa-TOF-MS in the multiresidue analysis and identification of pesticides and their metabolites and degradation products in fruits and vegetables.

8. Biological monitoring of pesticide exposure

Surprisingly, LC–MS currently only plays a minor role in the analysis of pesticides in biological samples, such as urine, plasma, serum, and whole blood. From recent reviews dealing with the biological monitoring of pesticide exposure [138-140], it can be concluded that GC–MS is still of major importance in this area. LC–MS appears to be an useful tool in biological monitoring for clinical and forensic toxicology (Ch. 12.5).

Some examples of the use of LC–MS in the analysis of pesticides and metabolites in human urine for exposure monitoring are:

- The analysis of some *N*-methylcarbamates and their metabolites in human urine by off-line SPE, LC–UV-DAD and confirmation by ESI LC–MS [141].
- Coupled-column LC–MS for the determination of chlorpyrifos and its main metabolite 3,5,6-trichloro-2-pyridinol in human serum and urine [28].
- Mixed-mode reversed-phase–weak anion exchange LC for the determination of various metabolites of chlorpyrifos in human urine [142].
- Quantitative analysis of the organophosphorous pesticides acephate, azinphos, chlorpyrifos, coumaphos, diazinon, isazofos, malathion, methamidophos, parathion, pirimiphos, and their *O,O*-dimethyl analogues in human urine using LC–MS–MS [143].
- High-throughput analysis of nineteen exposure markers from various classes of pesticides, *i.e.*, OPP, pyrethroids, herbicides, and DEET, in human urine by LC–MS, in order to estimate low-dose human exposure values [144].
- Determination of 4-nitrophenol and 3-methyl-4-nitrophenol, urine exposure markers of OPP like parathion, parathion methyl, and fenitrothion, using coupled-column LC–LC–MS–MS. A second method was developed for the determination of other parathion and parathion methyl metabolites, such as dimethyl thiophosphate, dimethyl phosphate, 4-nitrophenyl sulfate, and 4-nitrophenol glucuronide [145].

9. References

1. D. Barceló (Ed.), *Applications of LC–MS in Environmental Chemistry*, 1996, Elsevier Science, Amsterdam.
2. M. Careri, P. Manini, M. Maspero, *LC–MS in environmental analysis*, Ann. Chim. (Rome), 84 (1994) 475.
3. J. Slobodník, B.L.M. van Baar, U.A.Th. Brinkman, *Column LC–MS: Selected*

techniques in environmental applications for polar pesticides and related compounds, J. Chromatogr. A, 703 (1995) 81.

4.   Y. Picó, G. Font, J.C. Moltó, J. Mañes, *Pesticide residue determination in fruit and vegetables by LC–MS*, J. Chromatogr. A, 882 (2000) 153.

5.   A.C. Hogenboom, W.M.A. Niessen, U.A.Th. Brinkman, *The role of column LC–MS in environmental trace-level analysis*, J. Sep. Sci., 24 (2001) 331.

6.   R.B. Geerdink, W.M.A. Niessen, U.A.Th. Brinkman, *Trace-level determination of pesticides in water by means of LC and GC*, J. Chromatogr. A, 970 (2002) 65.

7.   W.L. Budde, *Analytical MS of herbicides*, Mass Spectrom. Rev., 23 (2004) 1.

8.   Y. Picó, C. Blasco, G. Font, *Environmental and food applications of LC–MS–MS in pesticide-residue analysis: an overview*, Mass Spectrom. Rev, 23 (2004) 45.

9.   E.M. Thurman, I. Ferrer, D. Barceló, *Choosing between APCI and ESI interfaces for the LC–MS analysis of pesticides*, Anal. Chem., 73 (2001) 5441.

10. M. Honing, D. Barceló, B.L.M. van Baar, U.A.Th. Brinkman, *Limitations and perspectives in the quantification of carbofuran using various LC–MS interface systems*, Trends Anal. Chem., 14 (1995) 496.

11. S. Pleasance, J.F. Anacleto, M.R. Bailey, D.H. North, *An evaluation of API techniques for the analysis of N-methyl carbamate pesticides by LC–MS*, J. Am. Soc. Mass Spectrom., 3 (1992) 378.

12. R.D. Voyksner, T. Pack, *Investigation of CID process and spectra in the transport region of an ESI single-quadrupole MS*, Rapid Commun. Mass Spectrom., 5 (1991) 263.

13. H.-Y. Lin, R.D. Voyksner, *Determination of environmental contaminants using an ESI interface combined with a ion trap MS*, Anal. Chem., 65 (1993) 451.

14. J. Slobodník, A.C. Hogenboom, J.J. Vreuls, J.A. Rontree, B.L.M. van Baar, W.M.A. Niessen, U.A.Th. Brinkman, *Trace-level determination of pesticide residues using on-line SPE–column LC with API-MS–MS*, J. Chromatogr. A, 741 (1996) 59.

15. M. Honing, J. Riu, D. Barceló, B.L.M. van Baar, U.A.Th. Brinkman, *Determination of ten carbamate pesticides in aquatic and sediment samples using LC-ESI and TSP-MS*, J. Chromatogr. A, 733 (1996) 283.

16. A. Di Corcia, M. Nazzaru, R. Rao, R. Samperi, E. Sebastiani, *Simultaneous determination of acidic and non-acidic pesticides in natural water by LC–MS*, J. Chromatogr. A, 878 (2000) 87.

17. K. Yu, J. Krol, M. Balogh, I. Monks, A fully automated LC–MS method development and quantification protocol targeting 52 carbamates, thiocarbamates and phenylureas, Anal. Chem., 75 (2003) 4103.

18. J.M. Fernandez, P.P. Vazquez, J.L.M. Vidal, *Analysis of N-methylcarbamates insecticides and some of their main metabolites in urine with LC using DAD and ESI-MS*, Anal. Chim. Acta, 412 (2000) 131.

19. M. Fernández, R. Rodríguez, Y. Picó, J. Mañes, *LC–MS determination of post-harvest fungicides in citrus fruits*, J. Chromatogr. A, 912 (2001) 301.

20. S. Kawasaki, F. Nagumo, H. Ueda, Y. Tajima, M. Sano, J. Tadano, *Simple, rapid and simultaneous measurement of eight different types of carbamate pesticides in serum using LC–APCI-MS*, J. Chromatogr., 620 (1993) 61.

21. H. Itoh, S. Kawasaki, J. Tadano, *Application of LC–APCI-MS to pesticide analysis*, J. Chromatogr. A, 754 (1996) 61.

22. A.C. Hogenboom, J. Slobodník, J.J. Vreuls, J.A. Rontree, B.L.M. van Baar, W.M.A.

Niessen, U.A.Th. Brinkman, *Single short-column LC with APCI-MS–MS detection for trace environmental analysis*, Chromatographia, 42 (1996) 506.

23. K.A. Barnes, R.J. Fussell, J.R. Startin, M.K. Pegg, S.A. Thorpe, S.L. Reynolds, *LC–APCI-MS with ionization polarity switching for the determination of selected pesticides*, Rapid Commun. Mass Spectrom., 11 (1997) 117.

24. K.S. Chiu, A. Van Langenhove, C. Tanaka, *LC–MS–MS analysis of carbamate pesticides*, Biomed. Environ. Mass Spectrom., 18 (1989) 200.

25. D. Baglio, D. Kotzias, B.R. Larsen, *API multiple MS analysis of pesticides*, J. Chromatogr. A, 854 (1999) 207.

26. I. Bobeldijk, J.P.C. Vissers, G. Kearney, H. Major, J.A. van Leerdam, *Screening and identification of unknown contaminants in water with LC and Q–TOF-MS*, J. Chromatogr. A, 929 (2001) 63.

27. C. Molina, M. Honing, D. Barceló, *Determination of OPP in water by SPE followed by LC–ESI-MS*, Anal. Chem., 66 (1994) 4444.

28. J.V. Sancho, O.J. Pozo, F. Hernández, *Direct determination of chlorpyrifos and its main metabolite 3,5,6-trichloro-2-pyridinol in human serum and urine by coupled-column LC–ESI-MS*, Rapid Commun. Mass Spectrom., 14 (2000) 1485.

29. S. Kawasaki, H. Ueda, H. Itoh, J. Tadano, *Screening of OPP using LC–APCI-MS*, J. Chromatogr., 595 (1992) 193.

30. S. Lacorte, D. Barceló, *Determination of ppt levels of OPP in groundwater by automated on-line SPE followed by LC–ESI-MS using positive and negative ion modes of operation*, Anal. Chem., 68 (1996) 2464.

31. S. Lacorte, G. Jeanty, J.-L. Marty, D. Barceló, *Identification of fenthion and temephos and their transformation products in water by LC with DAD and APCI-MS detection*, J. Chromatogr. A, 777 (1997) 99.

32. C. Aguilar, I. Ferrer, F. Borrull, R.M. Marcé, D. Barceló, *Comparison of automated on-line SPE followed by LC–MS with APCI and PBI MS for the determination of a priority pesticides in environmental waters*, J. Chromatogr. A, 794 (1998) 147.

33. S. Lacorte, C. Molina, D. Barceló, *Temperature and extraction voltage effect on fragmentation of OPP in LC–APCI-MS*, J. Chromatogr. A, 795 (1998) 13.

34. C. Molina, G. Durand, D. Barceló, *Trace determination of herbicides in estuarine waters by LC–ESI-MS*, J. Chromatogr. A, 712 (1995) 113.

35. I. Ferrer, M.-C. Hennion, D. Barceló, *Immunosorbents coupled on-line with LC–ESI-MS for the ppt level determination of pesticides in sediments and natural waters using low preconcentration volumes*, Anal. Chem., 69 (1997) 4508.

36. I. Ferrer, D. Barceló, *Simultaneous determination of antifouling herbicides in marina water samples by on-line SPE–LC–MS*, J. Chromatogr. A, 854 (1999) 197.

37. S. Nélieu, M. Stobiecki, L. Kerhoas, J. Einhorn, *Screening and characterization of atrazine metabolites by LC–MS–MS*, Rapid Commun. Mass Spectrom., 8 (1994) 945.

38. A.C. Hogenboom, W.M.A. Niessen, U.A.Th. Brinkman, *Rapid target analysis of microcontaminants in water by on-line single-short-column LC–APCI-ion-trap-MS*, J. Chromatogr. A, 794 (1998) 201.

39. K.A. Barnes, J.R. Startin, S.A. Thorpe, S.L. Reynolds, R.J. Fussell, *Determination of the pesticide diflubenzuron in mushrooms by LC–APCI-MS*, J. Chromatogr. A, 712 (1995) 85.

40. A.C. Hogenboom, P. Speksnijder, R.J. Vreeken, W.M.A. Niessen, U.A.Th. Brinkman, *Rapid target analysis of microcontaminants in water by on-line single-*

short-column LC combined with APCI-MS–MS, J. Chromatogr. A, 777 (1997) 81.

41. S. Chiron, S. Papilloud, W. Haerdi, D. Barceló, *Automated on-line SPE followed by LC–ESI-MS for the determination of acidic herbicides in environmental waters*, Anal. Chem., 67 (1995) 1637.

42. C. Crescenzi, A. Di Corcia, S. Marchese, R. Samperi, *Determination of acidic pesticides in water by a benchtop LC–ESI-MS*, Anal. Chem., 67 (1995) 1968.

43. B. Køppen, N.H. Spliid, *Determination of acidic herbicides using LC–ESI-MS–MS detection*, J. Chromatogr. A, 803 (1998) 157.

44. T.C.R. Santos, J.C. Rocha, D. Barceló, *Determination of rice herbicides, their transformation products and clofibric acid using on-line SPE–LC with DAD and APCI-MS detection*, J. Chromatogr. A, 879 (2000) 3.

45. A. Laganà, G. Fago A. Marino, M. Mosso, *Soil column extraction followed by LC–ESI-MS for the efficient determination of aryloxyphenoxypropionic herbicides in soil samples at ng/g levels*, Anal. Chim. Acta, 375 (1998) 107.

46. G. D'Ascenzo, A. Gentili, S. Marchese D. Perret, *Determination of arylphenoxypropionic herbicides in water by LC–ESI-MS*, J. Chromatogr. A, 813 (1998) 285.

47. R. Curini, A. Gentili, S. Marchese, A. Marino, D. Perret, *SPE–LC–ESI-MS for monitoring of herbicides in environmental water*, J. Chromatogr. A, 874 (2000) 187.

48. R.W. Reiser, A.J. Fogiel, *LC–MS analysis of small molecules using ESI and FAB ionization*, Rapid Commun. Mass Spectrom., 8 (1994) 252.

49. D. Volmer, J.G. Wilkes, K. Levsen, *LC–MS multiresidue determination of sulfonylureas after on-line trace enrichment*, Rapid Commun. Mass Spectrom., 9 (1995) 767.

50. A. Di Corcia, C. Crescenzi, R. Samperi, L. Scappaticcio, *Trace analysis of sulfonylurea herbicides in water: Extraction and purification by a Carbograph 4 cartridge, followed by LC with UV detection, and confirmatory analysis by an ESI-MS*, Anal. Chem., 69 (1997) 2819.

51. M. Rodriguez, D.B. Orescan, *Confirmation and quantitation of selected sulfonylurea, imidazolinone, and sulfonamide herbicides in surface water using LC–ESI-MS*, Anal. Chem., 70 (1998) 2710.

52. R. Bossi, B. Køppen, N.H. Spliid, J.C. Streibig, *Analysis of sulfonylurea herbicides in soil water at sub-part-per-billion levels by negative-ion ESI-MS followed by confirmatory MS–MS*, J. AOAC, 81 (1998) 775.

53. L.Y.T. Li, D.A. Campbell, P.K. Bennett, J.D. Henion, *Acceptance criteria for ultratrace LC–MS–MS: Quantitative and qualitative determination of sulfonylurea herbicides in soil*, Anal. Chem., 68 (1996) 3397.

54. X. Song, W.L. Budde, *CE–ESI-MS of the herbicides paraquat and diquat*, J. Am. Soc. Mass Spectrom., 7 (1996) 981.

55. J.C. Marr, J.B. King, *A simple LC–ESI-MS method for the direct determination of paraquat and diquat in water*, Rapid Commun. Mass Spectrom., 11 (1997) 479.

56. V.Y. Taguchi, S.W.D. Jenkins, P.W. Crozier, D.T. Wang, *Determination of diquat and paraquat in water by LC–ESI-MS*, J. Am. Soc. Mass Spectrom., 9 (1998) 830.

57. R. Castro, E. Moyano, M.T. Galceran, *Ion-pair LC–API-MS for the determination of quaternary ammonium herbicides*, J. Chromatogr. A, 830 (1999) 145.

58. R. Castro, E. Moyano, M.T. Galceran, *On-line ion-pair SPE–LC–API-MS for the analysis of quaternary ammonium herbicides*, J. Chromatogr. A, 869 (2000) 441.

59. J. Hau, S. Riediker, N. Varga, R.H. Stadler, *Determination of the plant growth regulator chlormequat in food by LC–ESI-MS–MS*, J. Chromatogr. A., 878 (2000) 77.
60. C.S Evans, J.R Startin, D.M Goodall, B.J Keely, *MS–MS analysis of quaternary ammonium pesticides*, Rapid Commun. Mass Spectrom., 15 (2001) 699.
61. R. Castro, E. Moyano, M.T. Galceran, *Determination of quaternary ammonium pesticides by LC–ESI-MS–MS*, J. Chromatogr. A, 914 (2001) 111.
62. O. Núñez, E. Moyano, M.T. Galceran, *High mass accuracy in-source CID–MS–MS and multi-step MS as complementary tools for fragmentation studies of quaternary ammonium herbicides*, J. Mass Spectrom., 39 (2004) 873.
63. A. C. Hogenboom, W. M. A. Niessen, U. A. Th. Brinkman, *Characterization of photodegradation products of alachlor in water by on-line SPE–LC combined with MS–MS and oaTOF-MS*, Rapid Commun. Mass Spectrom., 14 (2000) 1914.
64. A. Laganà, G. Fago, L. Fasciani, A. Marino, M. Mosso, *Determination of diphenyl-ether herbicides and metabolites in natural waters using LC with DAD and MS–MS detection*, Anal. Chim. Acta, 414 (2000) 79.
65. A. Laganà, G. Fago, A. Marino, *Simultaneous determination of imidazolinone herbicides from soil and natural waters using soil column extraction and off-line SPE followed by LC with UV detection or LC–ESI-MS*, Anal. Chem., 70 (1998) 121.
66. G. D'Ascenzo, A. Gentili, S. Marchese, A. Marino, D. Perret, *Optimization of LC–MS apparatus for determination of imidazolinone herbicides in soil at levels of a few ppb*, Rapid Commun. Mass Spectrom., 12 (1998) 1359.
67. G. D'Ascenzo, A. Gentili, S. Marchese, D. Perret, *Development of a method based on LC–ESI-MS for analyzing imidazolinone herbicides in environmental water at ppt levels*, J. Chromatogr. A, 800 (1998) 109.
68. R.J.C.A. Steen, A.C. Hogenboom, P.E.G. Leonards, R.A.L. Peerboom, W.P. Cofino, U.A.Th. Brinkman, *Ultratrace-level determination of polar pesticides and their transformation products in surface and estuarine water samples using column LC–ESI-MS*, J. Chromatogr. A, 857 (1999) 157.
69. K.-H. Bauer, T.P. Knepper, A. Meas, V. Schatz, M. Voihsel, *Analysis of polar organic micropollutants in water with ion chromatography–ESI-MS*, J. Chromatogr. A, 837 (1999) 117.
70. L. Goodwin, J.R. Startin, D.M. Goodall, B.J. Keely, *MS–MS analysis of glyphosate, glufosinate, aminomethylphosphonic acid and methylphosphinicopropionic acid*, Rapid Commun. Mass Spectrom., 17 (2003) 963.
71. R. Jeannot, H. Sabik, E. Sauvard, E. Genin, *Application of LC–MS with DAD and MS–MS for monitoring pesticides in surface waters*, J. Chromatogr. A, 879 (2000) 51.
72. C. Blasco, Y. Picó, J. Mañes, G. Font, *Determination of fungicide residues in fruits and vegetables by LC–APCI-MS*, J. Chromatogr. A, 947 (2002) 227.
73. K.A. Barnes, R.J. Fussell, J.R. Startin, S.A. Thorpe, S.L. Reynolds, *Determination of the pesticides diflubenzuron and clofentezine in plums, strawberries and blackcurrent-based fruit drinks by LC–APCI-MS*, Rapid Commun. Mass Spectrom., 9 (1995) 1441.
74. M.A. Aramendía, V. Boráu, I. García, C. Jiménez, F. Lafont, J.M. Marinas, F.J. Urbano, *Qualitative and Quantitative analyses of phenolic compounds by LC and detection with APCI-MS*, Rapid Commun. Mass Spectrom., 10 (1996) 1585.
75. A. DiCorcia, M. Marchetti, *Method development for monitoring pesticides in environmental waters: SPE followed by LC*, Environ. Sci. Technol., 26 (1992) 66.

76. I. Ferrer, D. Barceló, *Stability of pesticides stored on polymeric SPE cartridges*, J. Chromatogr. A, 778 (1997) 161.
77. C. Aguilar, I. Ferrer, F. Borrull, R. M. Marcé, D. Barceló, *Monitoring of pesticides in river water based on samples previously stored in polymeric cartridges followed by on-line SPE–LC–DAD and confirmation by APCI-MS*, Anal. Chim Acta, 386 (1999) 237.
78. S. Lacorte, D. Barceló, *Determination of OPP and their transformation products in river waters by automated on-line SPE followed by TSP LC–MS*, J. Chromatogr. A, 712 (1995) 103.
79. D. Barceló, S. Chiron, S. Lacorte, E. Martinez, J.S. Salau, M.C. Hennion, *Solid-phase sample preparation and stability of pesticides in water using Empore disks*, Trends Anal. Chem., 13 (1994) 352.
80. H. Bagheri, E.R. Brouwer, R.T. Ghijsen, U.A.Th. Brinkman, *On-line low-level screening of polar pesticides in drinking and surface waters by TSP LC–MS*, J. Chromatogr., 647 (1993) 121.
81. P. Önnerfjord, D. Barceló, J. Emnéus, L. Gorton, G. Marko-Varga, *On-line SPE in LC using restricted access pre-columns for the analysis of s-triazines in humic-containing waters*, J. Chromatogr. A, 737 (1996) 35.
82. J. Martinez Fernandez, J.L. Martinez Vidal, P. Parrilla Vazquez, A. Garrido Frenich, *Application of RAM column in coupled-column RPLC with UV detection and ESI-MS for determination of azole pesticides in urine*, Chromatographia, 53 (2001) 503.
83. R. Koeber, C. Fleischer, F. Lanza, K.-S. Boos, B. Sellergren, D. Barceló, *Evaluation of a multidimensional SPE platform for highly selective on-line cleanup and high-throughput LC–MS analysis of triazines in river water samples using molecularly imprinted polymers*, Anal. Chem., 73 (2001) 2437.
84. E. Dijkman, D. Mooibroek, R. Hoogerbrugge, E. Hogendoorn, J.-V. Sancho, O. Pozo, F. Hernández, *Study of matrix effects on the direct trace analysis of acidic pesticides in water using various LC modes coupled to MS–MS detection*, J. Chromatogr. A, 926 (2001) 113.
85. D.A. Volmer, J.P.M. Hui, *Rapid SPME LC–MS–MS analysis of N-methylcarbamate pesticides in water*, Arch. Environ. Contam. Toxicol., 35 (1998) 1.
86. M. Moder, P. Popp, R. Eisert, J. Pawliszyn, *Determination of polar pesticides in soil by SPME coupled to LC–ESI-MS*, Fres. J. Anal. Chem., 363 (1999) 680.
87. C. Blasco, G. Font, J. Mañes, Y. Picó, *SPME LC–MS–MS to determine post-harvest fungicides in fruits*, Anal. Chem., 75 (2003) 3606.
88. A.C. Hogenboom, W.M.A. Niessen, U.A.Th. Brinkman, *On-line SPE–short-column LC combined with various MS–MS scanning strategies for the rapid study of transformation of pesticides in surface water*, J. Chromatogr. A, 841 (1999) 33.
89. J.V. Sancho, O.J. Pozo, F.J. López, F. Hernández, *Different quantitation approaches for xenobiotics in human urine samples by LC–ESI-MS–MS*, Rapid Commun. Mass Spectrom., 16 (2002) 639.
90. A. Asperger, J. Efer, T. Koal, W. Engewald, *Trace determination of priority pesticides in water by means of high-speed on-line SPE–LC–MS–MS using TFC columns for enrichment and a short monolithic column for fast LC separation*, J. Chromatogr. A, 960 (2002) 109.
91. G.S. Rule, A.V. Mordehai, J.D. Henion, *Determination of carbofuran by on-line IAC with coupled-column LC–MS*, Anal. Chem., 66 (1994) 230.

92. K.A. Bean, J.D. Henion, *Determination of carbendazim in soil and lake water by IAC and coupled-column LC–MS–MS*, J. Chromatogr. A, 791 (1997) 119.

93. D.R. Doerge, S. Bajic, *Analysis of pesticides using LC–APCI-MS*, Rapid Commun. Mass Spectrom., 6 (1992) 663.

94. E. Moyano, D.E. Games, M.T. Galceran, *Determination of quaternary ammonium herbicides by CE–MS*, Rapid Commun. Mass Spectrom., 10 (1996) 1379.

95. O. Núñez, E. Moyano, M.T. Galceran, *TOF high resolution versus triple quadrupole MS–MS for the analysis of quaternary ammonium herbicides in drinking water*, Anal. Chim. Acta, 525 (2004) 183.

96. R.B. Geerdink, W.M.A. Niessen, U.A.T. Brinkman, *MS confirmation criterion for product-ion spectra generated in flow-injection analysis - Environmental application*, J. Chromatogr. A, 910 (2001) 291.

97. SANCO 2002/657/EC (Commission Decision of 12 August 2002 implementing Council Directive 96/23/EC concerning the performance of analytical methods and the interpretation of results).

98. F. Hernández, M. Ibáñez, J.V. Sancho, O.L. Pozo, *Comparison of different MS techniques combined with LC for confirmation of pesticides in environmental water based on the use of identification points*, Anal. Chem., 76 (2004) 4349.

99. L.G. Freitas, C.W. Gotz, M. Ruff, H.P. Singer, S.R. Muller, *Quantification of the new triketone herbicides, sulcotrione and mesotrione, and other important herbicides and metabolites, at the ng/l level in surface waters using LC–MS–MS*, J. Chromatogr. A, 1028 (2004) 277.

100. A.C. Hogenboom, W.M.A. Niessen, D. Little, U.A.Th. Brinkman, *Accurate mass determinations for the confirmation and identification of organic microcontaminants in surface water using on-line SPE–LC–ESI-TOF-MS*, Rapid Commun. Mass Spectrom., 13 (1999) 125.

101. N. Wang, W.L. Budde, *Determination of carbamate, urea, and thiourea pesticides and herbicides in water*, Anal. Chem., 73 (2001) 997.

102. F. Hernández, J.V. Sancho, O.J. Pozo, A. Lara, E. Pitarch, *Rapid direct determination of pesticides and metabolites in environmental water samples at sub-µg/l level by on-line SPE–LC–ESI-MS–MS*, J. Chromatogr. A, 939 (2001) 1.

103. J.V. Sancho, O.J. Pozo, F. Hernández, *LC and MS–MS: a powerful approach for the sensitive and rapid multiclass determination of pesticides and transformation products in water*, Analyst, 129 (2004) 38.

104. U.A.Th. Brinkman, J. Slobodník and J.J. Vreuls, *Trace-level detection and identification of polar pesticides in surface water: The SAMOS approach*, Trends Anal. Chem., 13 (1994) 373.

105. P. J. M. van Hout, U. A. Th. Brinkman, *The Rhine Basin Program*, Trends Anal. Chem., 13 (1994) 382.

106. D.M. Drexler, P.R. Tiller, S.M. Wilbert, F.Q. Bramble, J.C. Schwartz, *Automated identification of isotopically labeled pesticides and metabolites by intelligent 'real time' LC–MS–MS using a bench-top ion trap mass spectrometer*, Rapid Commun. Mass Spectrom., 12, 1501–1507

107. P.G.M. Kienhuis, R.B. Geerdink, *LC–MS–MS analysis of surface and waste water with APCI*, Trends Anal Chem, 19 (2000) 249 and 460.

108. M. Ibáñez, J.V. Snacho, O.J. Pozo, W.M.A. Niessen, F. Hernández, *Use of Q–TOF-MS in the elucidation of unknown compounds present in environmental water*, Rapid

Commun. Mass Spectrom., 19 (2005) 169.

109. E.M. Thurman, I. Ferrer, A.R. Fernández-Alba, *Matching unknown emperical formulas to chemical structure using LC–MS TOF accurate mass and database searching*, J. Chromatogr. A, 1067 (2005) 127.

110. D. Puig, D. Barceló, I. Silgoner, M. Grasserbauer, *Comparison of three different LC–MS interfacing techniques for the determination of priority phenolic compounds in water*, J. Mass Spectrom., 31 (1996) 1297.

111. D. Puig, I. Solgoner, M. Grasserbauer, D. Barceló, *Ppt level determination of priority methyl-, nitro- and chlorophenols in river water samples by automated on-line SPE–LC–MS using APCI and ESI interfaces*, Anal. Chem., 69 (1997) 2756.

112. O. Jáuregui, E. Moyano, M.T. Galceran, *LC–APCI-MS for chlorinated phenolic compounds. Application to the analysis of polluted soil*, J. Chromatogr. A, 823 (1998) 241.

113. M.C. Alonso, D. Puig, I. Silgoner, M. Grasserbauer, D. Barceló, *Determination of priority phenolic compounds in soil samples by various extraction methods followed by LC–APCI-MS*, J. Chromatogr. A, 823 (1998) 231.

114. S. Lacorte, S.B. Lartiges, P. Garrigues, D. Barceló, *Degradation of OPP and their transformation products in estuarine waters*, Environ. Sci. Technol., 29 (1995) 431.

115. S. Mansiapan, E. Benfenati, P. Grasso, M. Terreni, M. Pregnolato, G. Pagani, D. Barceló, *Metabolites of alachlor in water: Identification by MS and chemical synthesis*, Environ. Sci. Technol., 31 (1997) 3637.

116. D.A. Volmer, *Investigation of photochemical behaviour of pesticides in a photolysis reactor coupled on-line with a LC–ESI-MS–MS system. Application to trace and confirmatory analyses in food samples*, J. Chromatogr. A, 794 (1998) 129.

117. S. Chiron, A. Rodriguez, A. Fernández-Alba, *Application of GC and LC–MS to the evaluation of pirimiphos methyl degradation products in industrial water under ozone treatment*, J. Chromatogr. A, 823 (1998) 97.

118. M. Ibáñez, J.V. Sancho, O.J. Pozo, F. Hernández, *Use of Q–TOF-MS in environmental analysis: Elucidation of transformation products of triazine herbicides in water after UV exposure*, Anal. Chem., 76 (2004) 1328.

119. D.A. Aga, E.M. Thurman, M.E. Yockel, L.R. Zimmerman, T.D. Williams, *Identification of a new sulfonic acid metabolite of metalochlor in soil*, Environ. Sci. Technol., 30 (1996) 592.

120. I. Ferrer, E.M. Thurman, D. Barceló, *Identification of ionic chloroacetanilide-herbicide metabolites in surface water and groundwater by HPLC–MS using negative ionspray*, Anal. Chem., 69 (1997) 4547.

121. J.D. Vargo, *Determination of sulfonic acid degradates of chloroacetanilide and chloroacetamide herbicides in groundwater by LC–MS–MS*, Anal. Chem., 70 (1998) 2699.

122. R.A. Yokley, L.C. Mayer, S.-B. Huang, J.D. Vargo, *Analytical method for the determination of metolachlor, acetochlor, alachlor, dimethenamid, and their corresponding ethanesulfonic and oxanillic acid degradates in water using SPE and LC–ESI-MS–MS*, Anal. Chem., 74 (2002) 3754.

123. K.A. Barnes, R.J. Fussell, J.R. Startin, M.K. Pegg, S.A. Thorpe, S.L. Reynolds, *LC–APCI-MS with ionization polarity switching for the determination of selected pesticides*, Rapid Commun. Mass Spectrom., 11 (1997) 117.

124. J.R. Startin, S.J. Hird, M.D. Sykes, J.C. Taylor, A.R.C. Hill, *Determination of*

residues of the plant growth regulator chlormequat in pears by ion-exchange LC–ESI-MS, Analyst, 124 (1999) 1011.

125. S. Riediker, H. Obrist, N. Varga, R.H. Stadler, *Determination of chlormequat and mepiquat in pear, tomato, and wheat flour using on-line SPE (Prospekt) coupled with LC–ESI-MS–MS*, J. Chromatogr. A, 966 (2002) 15.

126. M. Careri, L. Elviri, A. Mangia, I. Zagnoni, *Rapid method for determination of chlormequat residues in tomato products by ion-exchange LC–ESI-MS–MS*, Rapid Commun. Mass Spectrom., 16 (2002) 1821.

127. CEN, *LC–MS–MS method of analysis of chlormequat and mepiquat cation*, CEN/TC275/WG 4 N Brussels 146 (2002) 1.

128. C. Blasco, M. Fernández, Y. Picó, G. Font, J. Mañes, *Simultaneous determination of imidacloprid, carbendazim, methiocarb and hexythiazox in peaches and nectarines by LC–MS*, Anal. Chim. Acta, 461 (2002) 109.

129. C. Blasco, G. Font, Y. Picó, *Comparison of microextraction procedures to determine pesticides in oranges by LC–MS*, J. Chromatogr. A, 970 (2002) 201.

130. C. Blasco, G. Font, Y. Picó, *Analysis of pesticides in fruits by pressurized liquid extraction and LC–ion trap–triple stage MS*, J. Chromatogr. A, 1098 (2005) 37.

131. M.J. Taylor, K. Hunter, K.B. Hunter, D. Lindsay, S. Le Bouhellec, *Multiresidue method for rapid screening and confirmation of pesticides in crude extracts of fruits and vegetables using isocratic LC–ESI-MS–MS*, J. Chromatogr. A, 982 (2002) 225.

132. D. Ortelli, P. Edder, C. Corvi, *Multiresidue analysis of 74 pesticides in fruits and vegetables by LC–ESI-MS–MS*, Anal. Chim. Acta, 520 (2004) 33.

133. C. Jansson, T. Pihlström, B.-G. Österdahl, K.E. Markides, *A new multi-residue method for analysis of pesticide residues in fruit and vegetables using LC–MS–MS detection*, J. Chromatogr. A, 1023 (2004) 93.

134. J.L Martínez Vidal, A. Garrido Frenich, T. López López, I. Martínez Salvador, L. Hajjaj el Hassani, M. Hassan Benajiba, *Selection of a representative matrix for calibration in multianalyte determination of pesticides in vegetables by LC–ESI-MS–MS*, Chromatographia, 61 (2005) 127.

135. I. Ferrer, E.M. Thurman A.R. Fernández-Alba, *Quantitation of accurate mass analysis of pesticides in vegetables by LC–TOF-MS*, Anal. Chem., 77 (2005) 2818.

136. E.M. Thurman, I. Ferrer, J.A. Zweigenbaum, J.F. García-Reyes, M. Woodman, A.R. Fernández-Alba, *Discovering metabolites of post-harvest fungicides in citrus with LC–TOF-MS and ion trap MS–MS*, J. Chromatogr. A, 1082 (2005) 71.

137. I. Ferrer, J.F. García-Reyes, M. Mezcua, E.M. Thurman, A.R. Fernández-Alba, *Multiresidue pesticide analysis in fruits and vegetables by LC–TOF-MS*, J. Chromatogr. A, 1082 (2005) 81.

138. D.B. Barr, L.L. Needham, *Analytical methods for biological monitoring of exposure to pesticides: a review*, J. Chromatogr. B, 778 (2002) 5.

139. P. Manini, R. Andreoli, W.M.A. Niessen, *LC–MS in occupational toxicology: A novel approach to the study of biotransformation of industrial chemicals*, J. Chromatogr., 1058 (2004) 21.

140. F. Hernández, J.V. Sancho, O.J. Pozo, *Critical review of the application of LC–MS to the determination of pesticide residues in biological samples*, Anal. Bioanal. Chem., 382 (2005) 934.

141. E. Lacassie, P. Marquet, J.-M. Gaulier, M.-F. Dreyfuss, G. Lachâtre, *Sensitive and specific multiresidue methods for the determination of pesticides of various classes in*

clinical and forensic toxicology, For. Sci. Intl., 121 (2001) 116.

142.   W. Bicker, M. Lämmerhofer, W. Lindner, *Determination of chlorpyrifos metabolites in human urine by RP/IEC LC–ESI-MS–MS*, J. Chromatogr. B, 822 (2005) 160.

143.   J. Martínez Fernández, P. Parrilla Vázquez, J.L. Martínez Vidal, *Analysis of N-methylcarbamates insecticides and some of their main metabolites in urine with LC using DAD and ESI-MS*, Anal. Chim. Acta, 412 (2000) 131.

144.   A.O. Olsson, J.V. Nguyen, M.A. Sadowski, D.B. Barr, *An LC–ESI-MS–MS method for quantification of specific OPP biomarkers in human urine*, Anal. Bioanal. Chem., 376 (2003) 808.

145.   A.O. Olsson, S.E. Baker, J.V. Nguyen, L.C. Romanoff, S.O. Udunka, R.D. Walker, K.L. Flemmen, D.B. Barr, *A LC–MS–MS multiresidue method for quantification of specific metabolites of OPP, synthetic pyrethroids, selected herbicides, and DEET in human urine*, Anal. Chem., 76 (2004) 2453.

146.   F. Hernández, J.V. Sancho, O.J. Pozo, *An estimation of the exposure to OPP through the simultaneous determination of their main metabolites in urine by LC–MS–MS*, J. Chromatogr. B, 808 (2004) 229.

8

ENVIRONMENTAL APPLICATIONS OF LC–MS

1. Introduction

LC–MS plays an important role in the development of new analytical strategies for environmental analysis. In the past few years, the perspective of environmental analysis has changed. For many years, most attention appeared to be given to the analysis of pesticides in the environment. In recent reviews, Richardson [1-2] focuses attention to emerging contaminants, including endocrine disrupting chemicals, pharmaceuticals, algal toxins, drinking water disinfection byproducts, organotin, and natural organic matter. Environmental analysis is definitively broadening its perspective. This is reflected in this chapter, reviewing the use of LC–MS in relation to a number of other environmental contaminants, such as surfactants and pharmaceuticals. The overview cannot be and is not meant to be comprehensive.

2. Natural organic matter

In the LC–UV analysis of environmental water samples (ground or surface water), a huge interfering background is observed, which elutes with the solvent gradient. The background is caused by humic and fulvic acids. Humic and fulvic acids are part of the natural organic matter, also containing amino acids, carbohydrates, lipids, lignins, and waxes. Electrospray (ESI) and, to a lesser extent atmospheric-pressure chemical ionization (APCI), MS can be used to characterize these complex macromolecular structures. Fulvic acids are assumed to have a molecular-weight distribution in the range of 200–2000 Da, humic acid of 1000–100,000 Da. Due to the complex character of these compounds, high-resolution MS has to be used for a successful characterization. Solvent conditions significantly influence the data obtained, leading to different molecular-weight distributions as a result of self-association or cleavage thereof.

McIntyre *et al.* [3] reported ESI mass spectra of fulvic acid from a double-focussing sector instrument. Each nominal mass peak was shown to consist of several different ions. Subsequently, Brown and Rice [4] reported the ESI-FT-ICR-MS analysis of fulvic acid reference standards. Complex mass spectra were obtained in the positive-ion mode, with ion distributions on the order of m/z 500–3000, indicating an average molecular weight in the range of 1700–1900 Da. In the negative-ion mode, complex multiple-charge ion patterns were observed [4]. A sinusoidal spectral distribution centred around 450 Da was observed for fulvic acids using a Q–TOF hybrid instrument [5]. This study was criticized by McIntyre *et al.* [6] with respect to poor referencing to previous literature, poor MS resolution, and insufficient calibration against representative standards. Leenheer *et al.* [7] discussed the complexity of the data interpretation due to the generation of multiple-charge ions. The fulvic acids show losses of water and CO_2 in MS–MS [5, 7]. Pfeifer *et al.* [8] studied humic substances by means of APCI and ESI and correlated their data with size-exclusion chromatography. They showed a comparison of spectra of an aquatic fulvic acid fraction (HO 10FA), obtained under different conditions (Figure 8.1). The mass spectrum obtained in the positive-ion ESI mode exhibited a molecular-mass distribution from m/z 500–3000, centred at 1440 Da. In negative-ion mode, the effect of multiply-charging is evident from the distribution from m/z 100–1500, centred at 740 Da. Doubly- and triply-charged ions can be recognized from high-resolution MS data. In positive-ion and negative-ion APCI, distributions were observed that centred at 310 and 235 Da, respectively. APCI probably causes fragmentation. Kujawinski *et al.* [9] optimized solvent and instrument conditions for the positive-ion ESI-FT-ICR-MS characterization of two fulvic acid samples. Due to a mass resolution of ~80,000 at m/z 300, they could resolve individual compounds. Peaks were observed at every nominal mass between m/z 400 and 1200, with four to eight peaks per nominal mass. Primarily, singly-charged ions were observed.

These data indicate that lower than expected molecular-mass distributions are observed in ESI. Stenson *et al.* [10] addressed this issue, using high resolution measurements on a FT-ICR-MS at 9.4 T. Their studies confirm that primarily singly-charged ions are observed, and no significant fragmentation appears to occur for the humic and fulvic acid mixtures. The mixtures contain molecular families of compounds differing from each other in the degree of saturation, functional group substitution, and number of CH_2 groups.

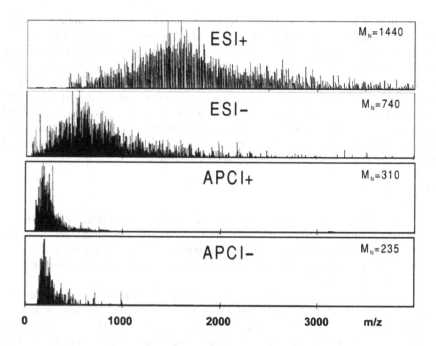

Figure 8.1: Mass spectra of a (HO 10FA) fulvic acid fraction using positive-ion and negative-ion ESI and APCI (Reprinted from [8] with permission, ©2001, Elsevier Science).

The apparent low-MW bias in the ESI spectra is not due to inadequate accounting for high charge states. Obviously, the high-MW components may not readily ionize under electrospray conditions. In a further study [11], molecular formulas were assigned for 4626 individual fulvic acids in the Suwannee River reference sample. These individual components could be assigned to 1 of 266 distinct homologous series, differing in oxygen content and double-bond equivalence. MS–MS was used to propose plausible structures consistent with degraded lignins.

The analysis of freshwater humic substances was recently reviewed [12].

3. Endocrine disrupting compounds

In the past few years, considerable attention has been given to the presence of endocrine disrupting compounds (EDC) in the aquatic environment. EDC are not defined by chemical nature but rather by their biological effect. The endocrine system is an intricate hormone system that regulates development, growth, reproduction, and behaviour [1-2]. The two relevant classes of compounds that can disrupt or interfere with the normal function comprise natural compounds, e.g., hormones found in humans and animals and phytoestrogens found in some plants, and man-made substances, *e.g.*, synthetic hormones for birth control, but also many other compounds like alkylphenols, phthalate esters, and polychlorinated and polybrominated compounds, which can exert endocrine disrupting activity. Recently, Petrovic *et al.* [13] provided an overview of the current state-of-the-art in the MS analysis of EDC in aquatic environmental samples. LC–MS is primarily involved in the analysis of alkylphenolic compounds, bisphenol A, phthalate esters, and synthetic and natural steroids. Here, the discussion is focussed at the environmental analysis of steroids. Alkylphenolic compounds are discussed in Ch. 8.4.2.

3.1   Steroids

The LC–MS analysis of steroids is discussed from a more general perspective in Ch. 13. In environmental analysis, SPE on C_{18}- or carbon-materials are generally applied for analyte extraction and preconcentration. Gradient elution using 20–100% acetonitrile in water on a C_{18}-column is used in combination with either negative-ion ESI or positive-ion APCI.

López de Alda and Barceló [14] reported the determination of natural and synthetic estrogens and progestogens in influents and effluents from a sewage treatment plant (STP), surface water, and drinking water. The estrogens were determined as $[M–H]^-$ in negative-ion ESI, the progestogens as $[M+Na]^+$ in positive-ion mode ESI or APCI. Detection limits of <1 ng/l water have been achieved.

Laganà *et al.* [15] reported the LC–MS analysis of 17-estradiol, estriol, and 17-ethinyl estradiol in STP effluents, using SPE on ENVI-CARB material, positive-ion APCI, and SRM. Recoveries ranging from 84 to 93% were reported. The precision was better than 11 to 8%. Quantitation limits were 0.5 ng/l for 17-estradiol and 17-ethinyl estradiol, and 1 ng/L for estrone and estriol.

Ferguson *et al.* [16] developed a method, based on immunoaffinity extraction coupled to LC–ESI-MS, for the determination of β-estradiol, estrone, and α-ethinyl estradiol in wastewater. The immunoaffinity extraction not only removed interfering sample matrix components, but also lowered the isobaric noise in SIM traces. Detection limits in the range of 0.07–0.18 ng/l were achieved with recoveries better than 90% and a precision better than 5%.

Petrovic *et al.* [17] reported the use of pressurized liquid extraction, on-line clean-up of the extracts on LiChrospher ADS C_4 restricted-access material (RAM), and LC–ESI-MS for the determination of steroids in sediment. Detection limits were achieved of 0.5 ng/g for progestogens and 1–5 ng/g for estrogens, with an overall precision better than 19%. In another study [18], ultrasonic extraction of lyophilized sediment, clean-up on C_{18} cartridges, and LC–MS analysis were applied. Detection limits were in the 0.04–1 ng/g range. Estrogens and progestogens were found in the low ng/g range in river sediments from Catalonia. Maximum concentrations as high as 22.8 ng/g ethinyl estradiol and 11.9 ng/g estrone were found.

Benijts *et al.* [19] applied negative-ion ESI on an ion-trap system for the determination of the estrogens estrone, estradiol, estriol, ethinyl estradiol, and diethylstilbestrol in environmental water samples. With manual off-line SPE on 50-ml samples, the detection limits ranged from 3.2 to 10.6 ng/l.

Isobe *et al.* [20] extended the scope by analysing both estrogens like estradiol, estrone, and estriol, and their sulfate and glucuronide conjugates in river water, lake water, and STP samples. SPE in combination with negative-ion LC–ESI-MS–MS in SRM mode was used. Method detection limits between 0.1 and 3.1 ng/l were reported. While the free steroids and some of the sulfates were detected in environmental samples, most of the conjugates were below the detection limit.

Beck *et al.* [21] reported the LC–MS analysis of estrogenic compounds in coastal surface water of the Baltic Sea. SPE on Oasis HLB cartridges was applied to isolate the analytes from 5 l of water. Detection limits were from 0.02–1 ng/l. Estrone and ethinyl estradiol were found at 0.1 and 17 ng/l, respectively.

4. Surfactants

Surfactants are widely used for industrial purposes and in households. Although legislation prescribes the biodegradability of the surfactants used, complete biodegradation is often not achieved within the relevant time scale. As a result, they are found in surface water. Surfactants or their degradation products may have toxic effects, detrimental ecological effects, *e.g.*, endocrine disrupting properties (Ch. 8.3). The difficulties in determining surfactants are partly inherent to their properties, *i.e.*, their high polarity and high water-solubility. Their low volatility excludes the use of GC–MS. The LC–UV analysis of many surfactants is difficult because of the lack of a chromophore. Various LC–MS interfaces have been applied in surfactant analysis over the years. At present, ESI or APCI are applied.

4.1 Anionic surfactants

The anionic surfactants of major importance are the linear alkylbenzene sulfonates (LAS). Other important classes are alkyl sulfates, sulfonates, and ethoxy-sulfates (AES). Few LC–MS studies have been published.

Popenoe *et al.* [22] reported the negative-ion LC–ESI-MS analysis of alkyl sulfates and AES in environmental matrices. The mixture could be completely resolved by generating mass chromatograms of the individual components, both by alkyl and by ethoxylate chain length. Detection limits below 10 ng/l were achieved. Scullion *et al.* [23] applied positive-ion APCI to generate $[M+Na]^+$ ions of LAS, but in the analysis of environmental water samples, the LAS were removed by an anion-exchange material prior to LC–MS analysis.

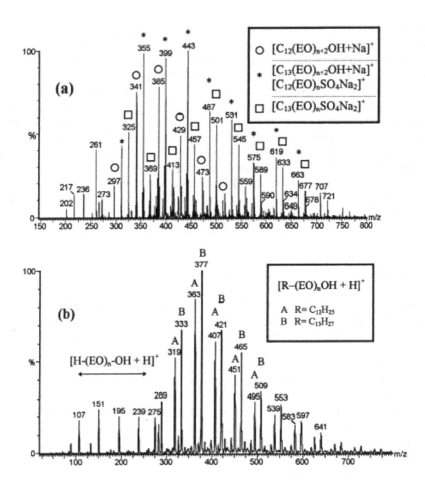

Figure 8.2: Mass spectra of Dobanol 23PESO4 (1000 ppm in 50% methanol with 1% acetic acid) obtained by means of (a) electrospray ionization, and (b) APCI (Reprinted with permission from [24], ©1999, American Society for Mass Spectrometry).

Jewett *et al.* [24] compared the use of positive-ion and negative-ion ESI and positive-ion APCI in the analysis of AES, *i.e.*, $CH_3(CH_2)_x(OCH_2CH_2)_nOSO_3{}^-Na^+$. In positive-ion ESI, the sodiated sodium salt $[M+Na]^+$ and desulfated molecule $[M-NaSO_3+2H]^+$ were observed for each AES component, while in positive-ion APCI only the latter is observed. The positive-ion ESI and APCI spectra of the industrial AES mixture Dobanol 23PESO4 are shown in Figure 8.2. In negative-ion

ESI, the deprotonated molecule $[M-Na]^+$ is observed. The ions generated in APCI could be used as precursor ions in MS–MS to obtain structural information on alkyl chain length and ethoxylate (EO) number, while no significant fragmentation was observed from the sodiated molecules generated in ESI. The signals of individual components can be used to estimate their relative concentrations in the mixture.

Cuzzola *et al.* [25-26] identified and characterized Fenton oxidation products of lauryl sulfate and AES. The Fenton reaction is a frequently applied oxidative treatment in STPs. The degradation products of anionic surfactants have been objects of study as well, because their biodegradation products might involve the loss of the sulfate group, resulting in essentially nonionic surfactants (see below). According to Schröder [27], the biodegradation of nonylphenol ethoxysulfates does not involve a loss of the sulfate.

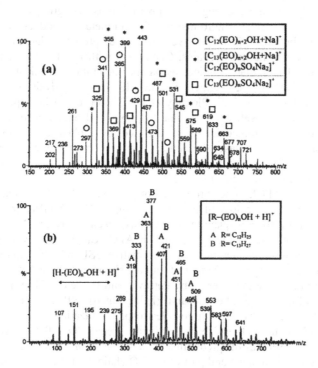

Figure 8.3: Mass spectra of an NPEO mixture with an average EO number of four acquired by (a) ESI, and (b) APCI. In the electrospray spectrum, sodiated molecules are observed. In the APCI, various ions are observed: protonated molecules (□), ammoniated molecules (●), and fragments due to the loss of nonene from the protonated molecule (▲) (Reprinted from [28] with permission, ©2001, Wiley & Sons Ltd.).

4.2   Nonionic surfactants

Important classes of nonionic surfactants are aliphatic poly-ethoxylate alcohols (AEO), and octyl or nonyl phenol polyethoxylates (OPEO and NPEO). The alkylphenol ethoxylates (APEO) attracted special attention due to their supposedly endocrine disrupting properties (Ch. 8.3). LC–MS analysis may also involve nonylphenolethoxycarboxylates (NPEC), biodegradation products of NPEO, and halogenated analogues, generated in chlorine disinfection treatments in drinking water production plants.

Because of the lack of a chromophore, LC–MS is extremely useful in the analysis of AEO. The LC–MS analysis of NPEO and their metabolites using ESI and APCI was reviewed by Petrović and Barceló [28]. In positive-ion ESI-MS, sodiated molecules are observed of AEO and NPEO in most cases. In general, the response increases with increasing the number of ethoxylate groups (for n is 1 to 6). Moriwaki et al. [29] studied the formation of alkali metal ion adducts of NPEO. Not only $[M+Na]^+$, but also $[2M+Na]^+$ were observed, probably due to hydrophobic interactions between the nonyl chains. Complexes with Li^+, K^+, Rb^+, and Cs^+ were studied as well. The sodiated molecules hamper quantitative analysis of NPEO in SRM, because they do not fragment in MS–MS. In the analysis of the technical NPEO mixture Igepal CO-720, Plomley et al. [30] observed both protonated and sodiated molecules as well as peaks due to polyglycol ethers, also present in the sample. Ammoniated molecules were detected by Cohen et al. [31] using a methanol–water mobile phase containing 5 mmol/l ammonium acetate and 0.5 mmol/l trifluoroacetic acid. Adduct formation of NPEO in ESI-MS was investigated in significant detail by Jonkers et al. [32].

In positive-ion APCI, either protonated or ammoniated molecules are observed, and fragments due to the loss of nonene. Low-mass NPEO suffer from enhanced background noise [30]. For quantitative analysis, electrospray LC–MS is preferred, while the protonated molecules generated in APCI can be readily fragmented in MS–MS for structure elucidation [28]. The positive ESI and APCI mass spectra of an NPEO with an average EO number of four are shown in Figure 8.3.

Positive-ion APCI was preferred by Petrović and Barceló [33] in the analysis of AEO and NPEO, while phenols and (halogenated) NPEC were analysed as deprotonated molecules by negative-ion electrospray ionization.

In MS–MS of the technical NPEO mixture Igepal CO-720 [30], two types of fragments were observed: (a) predominant fragments at m/z 121 and 165, after loss of the nonyl chain as nonene and part of the EO chain, and (b) a series of fragments at m/z 89, 133 and 177, due to losses of nonylphenol and part of the EO chain.

The mass spectrometric characterization of an AEO surfactant (dodecyl alcohol with 9 EO) by means of multi-stage ion-trap MS–MS was reported by Levine et al. [34]. The fragmentation observed include (a) consecutive losses of ethylene oxide units, (b) loss of dodecene followed by consecutive losses of ethylene oxide units, (c) loss of dodecanol followed by consecutive losses of ethylene oxide units, and (d) loss of HO–EO$_6$–H followed by consecutive losses of ethylene oxide units.

Product-ion MS–MS spectra of NPEC showed an intense fragment at m/z 218 due to the loss of the ethoxycarboxylate side chain (resulting in the nonylphenolate anion). Similar fragments were observed for the monochlorinated and monobrominated NPEO at m/z 253 and 297 [35].

Considerable attention has been given to the LC separation of AEO and APEO. Both reversed-phase and normal-phase LC have been applied. In reversed-phase LC, the use of methanol as modifier is preferred over the use of acetonitrile. A separation based on different alkyl chains irrespective of the EO-number can be achieved [36]. In normal-phase LC, the EO-number is more important than the alkyl chain length. Aminosilica columns are more successful than bare silica, especially with a acetonitrile–water–dichloromethane mobile phase [36].

Alternative LC methods in the separation of NPEO resulting in a separation based on EO number include the use of an alumina column using an ethylene oxide–n-hexane mixture as mobile phase [37], of a cyanosilica column and a mobile phase gradient of toluene and a 10:88:2 mixture of 0.5 mmol/l sodium acetate in toluene, methanol, and water [38], and of a poly(vinyl alcohol) column and 10–55% acetonitrile in 30 mmol/l aqueous ammonium acetate as mobile phase [39]. Ion-pair LC–MS, using 5 mmol/l triethylamine in the mobile phase, was applied in the analysis of phenols and NPEC [33].

The LC–MS analysis of AEO and especially APEO and its related compounds in environmental waters, sediments, and STP sludge has been reported. Some examples are discussed below.

Plomley et al. [30] applied precursor-ion scanning on m/z 121 and 133 as well as SRM on multiple transitions in the analysis of the technical NPEO mixture Igepal CO-720 in STP samples. Determination limits in the range of 50 ng/l were reported.

Ferguson et al. [40] reported the analysis of APEO metabolites in estuarine water and sediments. Compounds like nonylphenols, octylphenols, their mono-, di-, and triethoxylates, halogenated nonylphenols, and NPEC were found in water samples after SPE. Individual APEO metabolite concentrations of 1–320 ng/l in water and 5–2000 ng/g in sediment were found.

Petrović and Barceló [33] reported the analysis of AEO, APEO, and their degradation products in sewage sludge. Ultrasonic solvent extraction with 70% methanol in dichloromethane was applied to achieve recoveries better than 84%. SPE was performed for clean-up and fractionation into two fractions. In the sewage sludge samples analysed, 25–600 µg/g nonylphenol, 10–190 µg/g AEO, and 2–135 µg/g NPEO were detected.

Cohen et al. [31] reported routine analysis of AEO and NPEO in wastewater and sludge (40 samples per day). Detection limits for individual components were achieved in the range of 1–10 µg/l in wastewater and around 100 µg/kg in sludge.

Petrović et al. [41] reported the determination of halogenated APEO and their degradation products formed during chlorine disinfection in the presence of bromide. These compounds were analysed in sludge, river sediments, surface, drinking, and wastewater.

Complete separation between each individual NPEO from river water was achieved by Shao *et al.* [42] via the on-line combination of a C_{18} precolumn, a silica analytical column and elution with an acetonitrile gradient. Using positive-ion electrospray in SIM mode, the detection limits varied between 0.5 and 2 ng/l for individual NPEO with EO>2 and between 5 and 0.5 µg/l for EO<3.

The simultaneous determination of NPEO and NPEC in surface and drinking water was reported by Houde *et al.* [43]. SPE of a 100-ml sample is performed on GCB prior to LC on a C_8-column in isocratic elution. Detection limits range from 0.01–0.05 µg/l for NP(1–17)EO and are 0.01 µg/l for NP(1–2)EC in SRM.

Petrović *et al.* [35] reported the low ng/l determination of nonylphenol, NPEC, and their halogenated derivatives in water and sludge from a drinking water treatment plant near Barcelona (Spain). Detection limits of 1–2 ng/l in water and of 0.5–1.5 µg/kg in sludge were reported.

Following previous studies on lauryl sulfate and AES, Cuzzola *et al.* [44] also studied the Fenton oxidation products of AEO and NPEO. The aerobic biodegradation of AEO and NPEO and anaerobic biodegradation of NPEO were studied by Schröder [27] by means of FIA–MS and LC–MS and MS–MS in positive-ion and/or negative-ion APCI. Methyl ethers of AEO are persistent in aerobic conditions. NPEO degradation results in NPEC. Anaerobic biodegradation of NPEO results in nonylphenols.

5. Pharmaceuticals

In recent years, considerable attention has also been paid to the occurrence of pharmaceuticals in environmental compartments, especially wastewater, STP effluents, surface, and ground water. In 1999, over 50 individual pharmaceuticals and personal care products or their metabolites had been identified in environmental samples. Considering the polar nature of most compounds involved, LC–MS plays an important role. Obviously, a wide variety of compound classes are involved, including steroidal hormones and other EDC (Ch. 8.3) [13, 45], antibiotics, antineoplastic drugs like methotrexate [46], cholesterol-reducing drugs [47-48], non-steroidal antiinflammatory drugs [49-51], and various other pharmaceutical compounds [50-52]. Environmental analysis of pharmaceuticals [53] and antimicrobials [54] using LC–MS were recently reviewed.

Antimicrobial compounds attracted special attention in this respect. In an early study, Hirsch *et al.* [55] determined macrolide antibiotics, sulfonamides, tetracyclines, betalactam antibiotics, trimethoprim, and chloramphenicol in different water compartments using LC–ESI–MS–MS. Three different gradient LC methods were developed for the 18 target compounds. Quantitation limits of 50 ng/l were achieved for tetracyclines and 20 ng/l for all other compounds after SPE.

Subsequently, the LC–MS analysis of various classes of antibiotics (Ch. 14.2–4) in environmental samples was reported by others. Hartig *et al.* [56] reported the detection of sulfamethoxazole and sulfadiazine in municipal waste waters at 30–2000 ng/l and 10–100 ng/l, respectively, after SPE and LC–MS in SRM mode. Lindsey *et al.* [57] reported the determination of trace levels of sulfonamide and tetracycline antibiotics in ground and surface water at 0.05-μg/l concentrations. Bruno *et al.* [58] reported a method for the determination of trace levels of the eight widely used penicillins in aqueous environmental samples, using a Carbograph 4 cartridge and *in situ* conversion into methyl esters prior to their LC–ESI-MS analysis. The derivatization results in a significant improvement in the response. Limits of quantification range from 2 to 24 ng/l in river water. Zhu *et al.* [59] performed SPE on both polymeric and C_{18} cartridges prior to positive-ion LC–ESI-MS–MS on an ion-trap instrument to determine oxytetracycline, tetracycline, and chlortetracycline in lagoon water at 3–4 μg/l. Pfeifer *et al.* [60] determined sulfonamides and trimethoprim in manure at 5 μg/kg. Hamscher *et al.* [61] determined persistent tetracycline residues in soil fertilized with liquid manure. Loke *et al.* [62] reported the determination of oxytetracycline and its degradation products in manure-containing anaerobic test systems. Castiglioni *et al.* [63] reported the multiresidue analysis of various classes of pharmaceuticals including sulfonamides, penicillins, and quinolones in urban wastewater. Sulfonamide antibiotics and various pesticides were routinely monitored in surface waters by Stoob *et al* [64].

6. Haloacetic acids

As a result of the disinfection of drinking water by means of ozone, chlorine dioxide, chloramine, and chlorine, a variety of disinfection byproducts may occur in drinking water, including oxyhalides, haloacetic acids, and halogenated AEO and APEO metabolites (Ch. 8.4.2). The LC–MS analysis of disinfection byproducts in drinking water was recently reviewed by Zwiener and Richardson [65].

The haloacetic acids (HAA) are comprised of mono-, di- and trichloroacetic acid, mono-, di-, and tribromoacetic acid, and bromo-chloroacetic acid, bromo-dichloroacetic acid, and dibromo-chloroacetic acid. Toxicological studies showed that these compounds have carcinogenic properties and may have adverse reproductive consequences. HAA have no strong chromophore for sensitive UV detection; electrochemical detection has been described. Analysis by GC–MS requires derivatization. Due to their relatively low molecular mass, the LC–MS analysis can be hindered by low-mass background interferences.

Hashimoto and Otsuki [66] reported the negative-ion LC–ESI-MS analysis of all nine acids after separation on a crosslinked polystyrene resin with a mobile phase of 20% acetonitrile in water containing 3% acetic acid. Detection limits after extraction from 200 ml water was between 3 and 70 mg/l.

In order to avoid problems with low-mass background interferences, Ells *et al.* [67-68] demonstrated the use of high-field asymmetric waveform ion mobility spectrometry (FAIMS) in combination with negative-ion ESI-MS. FAIMS is a tuneable ion filter, which continuously transmits selected ions from a complex mixture. The power of FAIMS is demonstrated in Figure 8.4, where the mass spectra for a 500-fold diluted EPA 552.1 standard solution without and with FAIMS are compared. In this way, detection limits achieved initially were in the range of 0.5–1 µg/l [67], but in a subsequent paper between 5 and 36 ng/l [68].

Magnuson and Kelty [69] determined HAA in several water matrices at 0.13–0.64 µg/l after microscale liquid-liquid extraction and the formation of stable 1:1 associates with perfluoroheptanoic acid. Takino *et al.* [70] compared di-*n*-butylamine, *N,N*-dimethyl-*n*-butylamine, and tri-*n*-butylamine as ion-pairing agents. The first compound was selected. Detection limits for the HAA ranged from 24 to 118 ng/l after sample filtration. Debré *et al.* [71] evaluated the possibility of omitting LC separation of HAA via the use of the high mass accuracy of a time-of-flight MS. The reported detection limits of 24–86 µg/l are within a factor of 10 of the levels required in regulatory monitoring. Loos and Barceló [72] determined HAA after SPE of 50-ml water samples at pH 1.8. In negative-ion LC–ESI-MS deprotonated HAA and fragment ions due to the loss of CO_2 were observed. Triethylamine was used as volatile ion-pairing agent. Quantitation limits between 0.1 and 2.4 µg/l were reported. In tap water in Barcelona (Spain), a total concentration of HAA of 70 µg/l was measured. Significant higher levels were found in various swimming pools.

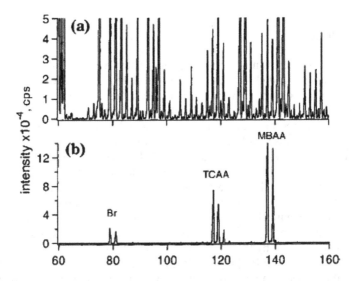

Figure 8.4: Comparison of mass spectra of haloacetic acids in a 500-fold diluted EPA 552.1 standard in electrospray MS acquired (a) without, and (b) with the use of high-field asymmetric waveform ion mobility spectrometry (FAIMS) (Reproduced from [67] with permission, ©1999, American Chemical Society).

7. Aromatic sulfonates

Aromatic sulfonates are widely applied in a variety of industrial processes including concrete furnishing, textile processing, tannery industry, and as precursors in the production of dyes. High concentrations of chlorinated aromatic mono- and disulfonic acid waste products from DDT manufacturing were observed in aqueous samples from the Stringfellow hazardous waste site (47 miles east of Los Angeles, CA). Aromatic sulfonates are water soluble and resistant to microbial degradation. Therefore, they are not easily eliminated in water treatment plants.

Suter *et al.* [73] developed an LC–MS method based on the in-source CID of the aromatic sulfonates: a loss of SO_2 and the formation of the $SO_3^{-\bullet}$ radical ion are observed. Absolute detection limits were ~1 ng, similar to UV detection. Unlike UV detection, LC–MS provides a similar response for benzene and naphthalene sulfonates. The method was applied to landfill leachates, and allowed the identification of an unknown aromatic sulfonic acid.

Storm *et al.* [74] evaluated various trialkylamines as ion-pairing agents for the LC separation of aromatic sulfonates. Tributylamine was preferred. In time-scheduled SRM, 19 aromatic sulfonic acids could be determined with detection limits of 3–74 µg/l. Socher *et al.* [75] demonstrated the applicability of ion-exchange chromatography with an ammonium acetate salt gradient up to 500 mmol/l in combination with negative-ion ESI as well as APCI LC–MS.

Pocurull *et al.* [76] applied on-line ion-pair SPE of only 15 ml water on a PLRP-S cartridge and ion-pair LC using triethylamine as ion-pairing agent in the LC–MS analysis of naphthalene monosulfonates in tap water, seawater, and river water, providing detection limits of 0.05–1 µg/l in SIM on ions corresponding to the loss of SO_2 and the formation of the $SO_3^{-\bullet}$ radical ion.

8. References

1. S.D. Richardson, *Water analysis: Emerging contaminants and current issues*, Anal. Chem., 75 (2003) 2831.
2. S.D. Richardson, T.A. Ternes, *Water analysis: Emerging contaminants and current issues*, Anal. Chem., 77 (2005) 3807.
3. C. McIntyre, B.D. Batts, D.R. Jardine, *ESI-MS of groundwater organic acids*, J. Mass Spectrom., 32 (1997) 328.
4. T.L. Brown, J.A. Rice, *Effect of experimental parameters on the ESI FT-ICR-MS of fulvic acid*, Anal. Chem., 72 (2000) 384.
5. G. Plancque, B. Amekraz, V. Moulin, P. Toulhoat, C. Moulin, *Molecular structure of fulvic acids by ESI-Q–TOD-MS*, Rapid Commun. Mass Spectrom., 15 (2001) 827.
6. C. McIntyre, D. Jardine, C. McRae, *ESI-MS of aquatic fulvic acids*, Rapid Commun. Mass Spectrom., 15 (2001) 1974.
7. J.A. Leenheer, C.E. Rostad, P.M. Gates, E.T. Furlong, I. Ferrer, *Molecular resolution and fragmentation of fulvic acid by ESI-MS–MSⁿ*, Anal. Chem., 73 (2001) 1461.

8. T. Pfeifer, U. Klaus, R. Hoffmann, M. Spiteller, *Characterization of humic substances using APCI and ESI-MS combined with SEC*, J. Chromatogr. A, 926 (2001) 151.

9. E.B. Kujawinski, P.G. Hatcher, M.A. Freitas, *High-resolution FT-ICR-MS of humic and fulvic acids: Improvements and comparisons*, Anal. Chem., 74 (2002) 413.

10. A.C. Stenson, W.M. Landing, A.G. Marshall, W.T. Cooper, *Ionization and fragmentation of humic substances in ESI-FT-ICR-MS*, Anal. Chem., 74 (2002) 4397.

11. A.C. Stenson, A.G. Marshall, W.T. Cooper, *Exact masses and chemical formulas of individual Suwannee river fulvic acids from ultrahigh resolution ESI-FT-ICR-MS*, Anal. Chem., 74 (2002) 4397.

12. S. McDonald, A.G. Bishop, P.D. Prenzler, K. Robards, *Analytical chemistry of freshwater humic substances*, Anal. Chim. Acta, 527 (2004) 105.

13. M. Petrović, E. Eljarrat, M.J. López de Alda, D. Barceló, *Recent advances in the MS analysis of EDC in aquatic environmental samples*, J. Chromatogr., 974 (2002) 23.

14. M.J. Lopéz de Alda, D. Barceló, *Determination of steroid sex hormones and related synthetic compounds considered as EDC in water by LC–DAD–MS*, J. Chromatogr. A, 892 (2000) 391.

15. A. Laganà, A. Bacaloni, G. Fago, A. Marino, *Trace analysis of estrogenic chemicals in sewage effluent using LC–MS–MS*, Rapid Commun. Mass Spectrom., 14 (2000) 401.

16. P.L. Ferguson, C.R. Iden, A.E. McElroy, B.J. Brownawell, *Determination of steroid estrogens in wastewater by IAC coupled with LC–ESI-MS*, Anal. Chem., 73 (2001) 3890.

17. M. Petrović, S. Tavazzi, D. Barceló, *Column-switching system with restricted access pre-column packing for an integrated sample cleanup and LC–MS analysis of alkylphenolic compounds and steroid sex hormones in sediment*, J. Chromatogr. A, 971 (2002) 37.

18. M.J. López de Alda, A. Gil, E. Paz, D. Barceló, *Occurrence and analysis of estrogens and progestogens in river sediments by LC–ESI-MS*, Analyst, 127 (2002) 1299.

19. T. Benijts, R. Dams, W. Günther, W. Lambert, A. De Leenheer, *Analysis of estrogenic contaminants in river water using LC coupled to ion trap MS*, Rapid Commun. Mass Spectrom., 16 (2002) 1358.

20. T. Isobe, H. Shiraishi, M. Yasuda, A. Shinoda, H. Suzuki, M. Morita, *Determination of estrogens and their conjugates in water using SPE–LC–MS–MS*, J. Chromatogr. A, 984 (2003) 195.

21. I.-C. Beck, R. Bruhn, J. Gandrass, W. Ruck, *LC–MS–MS analysis of estrogenic compounds in coastal surface water of the Baltic Sea*, J. Chromatogr. A, 1090 (2005) 98.

22. D.D. Popenoe, S.J. Morris, III, P.S. Horn, K.T. Norwood, *Determination of alkyl sulfates and alkyl ethoxysulfates in waste-water treatment plant influents and effluents and in river water using LC–ESI-MS*, Anal. Chem., 66 (1994) 1620.

23. S.D. Scullion, M.R. Clench, M. Cooke, A.E. Ashcroft, *Determination of surfactants in surface water by SPE, LC and LC–MS*, J. Chromatogr. A, 733 (1996) 207.

24. B.N. Jewett, L. Ramaley, J.C.T. Kwak, *API-MS techniques for the analysis of alkyl ethoxysulfate mixtures*, J. Am. Soc. Mass Spectrom., 10 (1999) 529.

25. A. Cuzzola, A. Raffaelli, A. Saba, S. Pucci, P. Salvadori, *Identification and characterization of Fenton oxidation products of surfactants by ESI-MS and SPME GC–MS. 1. Lauryl sulphate*, Rapid Commun. Mass Spectrom., 13 (1999) 2140.

26. A. Cuzzola, A. Raffaelli, A. Saba, P. Salvadori, *Identification and characterization of*

Fenton oxidation products of surfactants by ESI-MS and SPME GC–MS. 2. Fatty alcohol polyethoxy sulphates, Rapid Commun. Mass Spectrom., 14 (2000) 834.

27. H. Fr. Schroder, *Tracing of surfactants in the biological wastewater treatment process and the identification of their metabolites by FIA-MS and LC–MS–MS*, J. Chromatogr. A, 926 (2001) 127.

28. M. Petrović, D. Barceló, *Analysis of ethoxylated nonionic surfactants and their metabolites by LC–API-MS*, J. Mass Spectrom., 36 (2001) 1173.

29. H. Moriwaki, T. Nakano, S. Tsunoi, M. Tanaka, *Detection of 1:1 and 2:1 complexes of NPEO with alkali metal cations by ESI-MS*, Rapid Commun. Mass Spectrom., 15 (2001) 2208.

30. J.B. Plomley, P.W. Crozier, V.Y. Taguchi, *Characterization of NPEO in sewage treatment plants by combined precursor-ion scanning and SRM*, J. Chromatogr. A, 854 (1999) 245.

31. A. Cohen, K. Klint, S. Bøwadt, P. Persson, J.Å. Jönsson, *Routine analysis of alcohol and NPEO in wastewater and sludge using LC–ESI-MS*, J. Chromatogr. A, 927 (2001) 103.

32. N. Jonkers, H. Govers, P. De Voogt, *Adduct formation in LC–ESI–MS of NPEO: MS, theoretical and quantitative analytical aspects*, Anal. Chim. Acta, 531 (2005) 217.

33. M. Petrović, D. Barceló, *Determination of anionic and nonionic surfactants, their degradation products, and EDC in sewage sludge by LC–MS*, Anal. Chem., 72 (2000) 4560.

34. L.H. Levine, J.L. Garland, J.V. Johnson, *LC–ESI-quadrupole ion trap MS for characterization and direct quantification of amphoteric and nonionic surfactants in aqueous samples*, Anal. Chem., 74 (2002) 2064.

35. M. Petrović, D. Barceló, A. Diaz, F. Ventura, *Low nanogram per liter determination of halogenated NPEO, NPEC, and their non-halogenated precursors in water and sludge by LC–ESI-MS*, J. Am. Soc. Mass Spectrom., 14 (2003) 516.

36. P. Jandera, M. Holčapek, G. Theodoridis, *Investigation of chromatographic behaviour of ethoxylated alcohol surfactants in NP and RP systems using LC–MS*, J. Chromatogr. A, 813 (1998) 299.

37. Á. Kósa, A. Dobó, K. Vékey, E. Forgács, *Separation and identification of NPEO oligomers by LC with UV and MS detection*, J. Chromatogr. A, 819 (1998) 297.

38. D.Y. Shang, M.G. Ikonomou, R.W. MacDonald, *Quantitative determination of NPEO surfactants in marine sediment using NPLC–ESI-MS*, J. Chromatogr. A, 849 (1999) 467.

39. M. Takino, S. Daishima K. Yamaguchi, *Determination of NPEO oligomers by LC–ESI-MS in river water and nonionic surfactants*, J. Chromatogr. A, 904 (2000) 65.

40. P.L. Ferguson, C.R. Iden, B.J. Brownawell, *Analysis of AEO metabolites in the aquatic environment using LC–ESI-MS*, Anal. Chem., 72 (2000) 4322.

41. M. Petrović, A. Diaz, F. Ventura, D. Barceló, *Simultaneous determination of halogenated derivatives of APEO and their metabolites in sludges, river sediments, and surface, drinking, and waste-waters by LC–MS*, Anal. Chem., 73 (2001) 5886.

42. B. Shao, J.-y. Hu, M. Yang, *Determination of NPEO in the aquatic environment by NPLC–ESI-MS*, J. Chromatogr. A, 950 (2002) 167.

43. F. Houde, C. DeBlois, D. Berryman, *LC–MS–MS determination of NPEO and NPEC in surface water*, J. Chromatogr. A, 961 (2002) 245.

44. A. Cuzzola, A. Raffaelli, A. Saba, S. Pucci, P. Salvadori, *Identification and*

characterization of Fenton oxidation products of surfactants by ESI-MS and SPME GC–MS. 3. AEO and NPEO, Rapid Commun. Mass Spectrom., 15 (2001) 1198.

45. T.R. Croley, R.J. Hughes, B.G. Koenig, C.D. Metcalfe, R.E. March, *MS applied to the analysis of estrogens in the environment*, Rapid Commun. Mass Spectrom., 14 (2000) 1087.

46. R. Turci, G. Micoli, C. Minoia, *Determination of methotrexate in environmental samples by SPE and LC: UV or MS–MS detection?*, Rapid Commun. Mass Spectrom., 14 (2000) 685.

47. X.-S. Miao, C.D. Metcalfe, *Determination of pharmaceuticals in aqueous samples using positive and negative voltage switching microbore LC–ESI-MS–MS*, J. Mass Spectrom., 38 (2003) 27.

48. X.-S. Miao, C.D. Metcalfe, *Determination of cholesterol-lowering statin drugs in aqueous samples using LC–ESI-MS–MS*, J. Chromatogr. A, 998 (2003) 133.

49. S. Marchese, A. Gentili, D. Perret, G. D'Ascenzo, F. Pastori, *Q–TOF versus triple-quadrupole MS for the determination of non-steroidal antiinflammatory drugs in surface water by LC–MS–MS*, Rapid Commun. Mass Spectrom., 17 (2003) 879.

50. M. Farré, I. Ferrer, A. Ginebreda, M. Figueras, L. Olivella, L. Tirapu, M. Vilanova, D. Barceló, *Determination of drugs in surface water and wastewater samples by LC–MS: methods and preliminary results including toxicity studies with Vibrio fischeri*, J. Chromatogr. A, 938 (2001) 187.

51. X.-S. Miao, B.G. Koenig, C.D. Metcalfe, *Analysis of acidic drugs in the effluents of sewage treatment plants using LC–ESI-MS–MS*, J. Chromatogr. A, 952 (2002) 139.

52. T. Ternes, M. Bonerz, T. Schmidt, *Determination of neutral pharmaceuticals in wastewater and rivers by LC–ESI-MS–MS*, J. Chromatogr. A, 938 (2001) 175.

53. S,.-C. Kim, K. Carlson, *LC–MS^2 for quantifying trace amounts of pharmaceutical compounds in soil and sediment matrices*, Trends Anal. Chem., 24 (2005) 635.

54. M.S. Díaz-Cruz, D. Barceló, *LC–MS^2 trace analysis of antimicrobials in water, sediment and soil*, Trends Anal. Chem., 24 (2005) 645.

55. R. Hirsch, T.A. Ternes, K. Haberer, A. Mehlich, F. Ballwanz, K.-L. Kratz, *Determination of antibiotics in different water compartments by LC–ESI-MS–MS*, J. Chromatogr. A, 815 (1998) 213.

56. C. Hartig, T. Storm, M. Jekel, *Detection and identification of sulphonamide drugs in municipal waste water by LC–ESI-MS–MS*, J. Chromatogr. A, 854 (1999) 163.

57. M.E. Lindsey, M. Meyer, E.M. Thurman, *Analysis of trace levels of sulfonamide and tetracycline antimicrobials in groundwater and surface water using SPE and LC–MS*, Anal. Chem., 73 (2001) 4640.

58. F. Bruno, R. Curini, A. Di Corcia, M. Nazzari, R. Samperi, *Method development for measuring trace levels of penicillins in aqueous environmental samples*, Rapid Commun. Mass Spectrom., 15 (2001) 1391.

59. J. Zhu, D.D. Snow, D.A. Cassada, S.J. Monson, R.F. Spalding, *Analysis of oxytetracycline, tetracycline, and chlortetracycline in water using SPE and LC–MS–MS*, J. Chromatogr. A, 928 (2001) 177.

60. T. Pfeifer, J. Tuerk, K. Bester, M. Spiteller, *Determination of selected sulfonamide antibiotics and trimethoprim in manure by ESI and APCI-MS–MS*, Rapid Commun. Mass Spectrom., 16 (2002) 663.

61. G. Hamscher, S. Sczesny, H. Höper, H. Nau, *Determination of persistent tetracycline residues in soil fertilized with liquid manure by LC–ESI-MS–MS*, Anal. Chem., 74

(2002) 1509.
62. M.-L. Loke, S. Jespersen, R. Vreeken, B. Halling-Sørensen, J. Tjørnelund, *Determination of oxytetracycline and its degradation products by LC–MS–MS in manure-containing anaerobic test systems*, J. Chromatogr. B, 783 (2003) 11.
63. S. Castiglioni, R. Bagnati, D. Calamari, R. Fanelli, E. Zuccato, *A multiresidue analytical method using SPE and LC–MS–MS to measure pharmaceuticals of different therapeutic classes in urban wastewaters*, J. Chromatogr. A, 1092 (2005) 206.
64. K. Stoob, H.P. Singer, C.W. Goetz, M. Ruff, S.R. Mueller, *Fully automated online SPE–LC–MS–MS. Quantification of sulfonamide antibiotics, neutral and acidic pesticides at low concentrations in surface waters*, J. Chromatogr. A, 1097 (2005) 138.
65. C. Zwiener, S.D. Richardson, *Analysis of disinfection by-products in drinking water by LC–MS and related MS techniques*, Trends Anal. Chem., 24 (2005) 613.
66. S. Hashimoto, A. Otsuki, *Simultaneous determination of haloacetic acids in environmental waters using LC–ESI-MS*, J. High Resolut. Chromatogr., 21 (1998) 55.
67. B. Ells, D.A. Barnett, K. Froese, R.W. Purves, S. Hrudey, R. Guevremont, *Detection of chlorinated and brominated byproducts of drinking water disinfection using ESI–FAIMS–MS*, Anal. Chem., 71 (1999) 4747.
68. B. Ells, D.A. Barnett, R. W. Purves, R. Guevremont, *Detection of nine chlorinated and brominated haloacetic acids at ppt levels using ESI–FAIMS–MS*, Anal. Chem., 72 (2000) 4555.
69. M.L. Magnuson, C.A. Kelty, *Microextraction of nine haloacetic acids in drinking water at microgram per liter levels with ESI-MS of stable association complexes*, Anal. Chem., 72 (2000) 2308.
70. M. Takino, S. Daishima, K. Yamaguchi, *Determination of haloacetic acids in water by LC–ESI-MS using volatile ion-pairing reagents*, Analyst, 125 (2000) 1097.
71. O. Debré, W.L. Budde, X. Song, *Negative-ion ESI of bromo- and chloroacetic acids and an evaluation of exact mass measurements with a bench-top TOF-MS*, J. Am. Soc. Mass Spectrom., 11 (2000) 809.
72. R. Loos, D. Barceló, *Determination of haloacetic acids in aqueous environments by SPE followed by ion-pair LC–MS detection*, J. Chromatogr. A, 938 (2001) 45.
73. M.J.-F. Suter, S. Riediker, W. Giger, *Selective determination of aromatic sulfonates in landfill leachates and groundwater using micro-LC–MS*, Anal. Chem., 71 (1999) 897.
74. T. Storm, T. Reemstra, M. Jekel, *Use of volatile amines as ion-pairing agents for LC–MS–MS determination of aromatic sulfonates in industrial wastewater*, J. Chromatogr. A, 854 (1999) 175.
75. G. Socher, R. Nussbaum, K. Rissler, E. Lankmayr, *Analysis of sulfonated compounds by ion-exchange LC–MS*, J. Chromatogr. A, 912 (2001) 53.
76. E. Pocurull, C. Aguilar, M.C. Alonso, D. Barceló, F. Borrull, R.M. Marcé, *On-line SPE–ion-pair LC–ESI-MS for the trace determination of naphthalene monosulphonates in water*, J. Chromatogr. A, 854 (1999) 187.

9

LC–MS IN DRUG DISCOVERY
AND DEVELOPMENT

1. Introduction

The most important application area of LC–MS is in drug development within the pharmaceutical industry. LC–MS is involved in almost every step. Drug development involves a series of specialized research efforts, which should in the end lead to the introduction of a novel drug onto the market. Traditionally, the drug development process consists of four steps: drug discovery, preclinical development, clinical development, and manufacturing [1].

The drug discovery phase is directed at generating a novel lead candidate with suitable pharmaceutical properties in terms of efficacy, bioavailability, and toxicity for preclinical evaluation. It may start with research on a new biological target, often a protein-like substance (receptor or enzyme), which needs to be characterized. Natural product sources and/or synthetic compound libraries are screened for activity towards the target. Potential lead compounds emerging from this screening are optimized, based on exploratory metabolism studies and drug safety evaluations. The introduction of combinatorial synthesis has resulted in a significant increase of the number of compounds to be screened for biological activity as well as to be analysed for identity, purity, and in toxicology studies.

In the preclinical development stage, data are collected in order to enable the filing of an investigational new drug (in the US) or a clinical trial application (in Europe). These data are comprised of details on the composition of the drug and the

233

synthetic processes involved in its production. In addition, animal toxicity data and protocols for early phase clinical trials must be obtained. Analytical research in this phase is directed at the synthetic process, formulation, metabolism, and toxicity. The structure, physical and chemical characteristics as well as the stereochemical identity of the drug candidate are fully characterized. Bioanalytical methods are developed in order to evaluate adsorption, distribution, metabolism, and excretion (ADME) of the drug candidate.

The clinical development stage consists of three distinct phases (I–III) and should lead to filing a new drug application (in the US) or a marketing authorization application (in Europe). Each phase involves process scale-up and activities directed at investigating pharmacokinetics, drug delivery, and safety. During Phase-I studies, the safety and pharmacokinetic profile of the compound is defined. Efficacious doses are estimated. Phase-II studies are directed at establishing efficacy, determining the effective dose range, and obtaining safety and tolerability data. Phase-III studies should complete the human safety and efficacy research and secure approval.

The manufacturing phase often starts during the Phase-III clinical studies. In this phase, issues related to large-scale production, formulation, and packaging are addressed. Standardized analytical procedures for routine monitoring and release by quality control should be established. Long-term stability studies are now performed.

One of the major concerns in this complete process is the factor time. The costs associated with the first three stages of drug development can amount up to $500,000,000 for a single new chemical entity. The patent of the drug has a limited lifetime. Speeding up the drug development process allows a longer period for sales of the patented compound, and thereby better perspectives on return of investment and profits. Significant efforts have been put in accelerating each stage of the drug development process in order to reduce the time between the discovery of a new chemical entity and the introduction of the drug onto the market. This leads to a severe demand for high-throughput analytical methods, applicable especially in early stages of drug development. In addition, reduction of operational costs during the drug development stages is another incentive [1].

LC–MS in drug development

At present, LC–MS is involved in almost every stage in the drug development process. LC–MS is applied as a tool in checking proper progress in the synthesis of new chemical entities, *e.g.*, via open-access LC–MS, in performing rapid determination of molecular mass and assessing sample purity in relation to bioactivity screening of combinatorial libraries, in assisting in the identification of reaction byproducts and degradation products as well as drug metabolites. Because of its specificity and sensitivity, LC–MS–MS has rapidly become the technique-of-choice in quantitative bioanalysis. Numerous companies have series of dedicated LC–MS instruments available to perform quantitative bioanalysis of drugs and their metabolites in biological fluids and tissue extracts. Similar work is performed by contract research laboratories sponsored by the pharmaceutical industries.

An important aspect of the drug development process is that in early developmental phases the analysis is directed at a relatively small number of samples of a wide variety of compounds, while at a later stage the number of analytes is significantly reduced but the number of samples to be analysed per compound increases exponentially. Routinely applicable, robust, and sensitive analytical methods are required in the various phases of the process.

The discussion on the application of LC–MS in drug development is divided over three chapters in this text. This chapter deals with the drug development process in general, with a special focus on LC–MS in drug discovery, impurity profiling, and drug stability studies. Two subsequent chapters highlight two other important topics in the drug development process, *i.e.*, identification of metabolites (Ch. 10), and quantitative bioanalysis of drugs and metabolites (Ch. 11).

It is virtually impossible to review the current applications of LC–MS in the pharmaceutical research area, mainly because for confidentiality reasons most of the work is not published in the public domain. The discussion is not meant as a complete overview of the pharmaceutical applications of LC–MS, but only shows a number of typical examples and highlights developments with an emphasis on technology rather than results.

2.   Open-access LC–MS for synthetic chemists

The traditional approach to drug discovery is based on rational drug design. Lead compounds, often structurally derived from natural products, are synthesized one at a time and are further optimized by a series of synthesis and screening steps. An important aspect is the structural characterization of reaction intermediates and end-products. MS and, given the nature of many of the compounds involved, LC–MS can play an important role in this. Often, molecular-mass determination with a soft ionization method leads to sufficient answers to the synthetic chemists.

An important opening towards the application of LC–MS in checking the proper progress of the chemical synthesis was the introduction of open-access LC–MS methods, which converts an LC–MS instrument into a walk-up "black box" for synthetic chemists in need for rapid confirmation of the proper progress of their synthesis by molecular mass determination of their products. A remote computer serves as a log-in to the system. After entering the sample identification code and selecting from a menu the type of LC–MS experiments to be performed, the computer indicates the position(s) in the autosampler rack to be used for the sample. The sample is run automatically, *e.g.*, by a fast wide-range gradient LC–MS run in both positive-ion and negative-ion mode and at both a high and a low in-source collision-induced dissociation (CID) potential. The resulting spectra are placed onto the LIMS network or sent to the chemist by electronic mail.

This approach was pioneered by Hayward *et al.* [2] for the automated analysis of potential agricultural chemicals by means of column-bypass thermospray (TSP) MS–MS. The method was found to provide the required information in approximately 70% of the MS structural confirmations performed at the Agricultural Research Division of American Cyanamid. Subsequently, an automated routine MS characterization of potential drug compounds was reported by Tiller and Lane [3] at Glaxo. Both particle-beam (PBI) in positive-ion chemical ionization (CI) mode and TSP interfacing were investigated. The TSP system was found to be more robust, to require less maintenance, and to be easier to use.

A substantial impetus to these automated approaches was given by Pullen and coworkers [4-6] at Pfizer Central Research. Based on automated column-bypass TSP-MS, an 'Open-Access' service to synthetic chemist was developed [4]: the chemist logs a sample into a queue on the mass spectrometer, the sample is run, and the mass spectrum is printed for later collection by the synthetic chemist. MS data are made available without intervention of a specialized mass spectrometrist. Subsequently, the system was extended to allow unattended and automated LC separation on-line with the TSP-MS analysis [5]. The success of the approach is indicated by the fact that in 1995 at Pfizer a staff of two took care of seven open-access instruments, providing structural confirmation on over 120,000 newly synthesized compounds. Seven more specialized mass spectrometrists used five other instruments to generate about 10,000 mass spectra for samples requiring more elaborate attention [6]. While previously described systems were based on TSP, the use of atmospheric-pressure chemical ionization (APCI) [7] and electrospray (ESI) [8] in open-access LC–MS was also described.

Dedicated open-access software modules are currently offered by most instrument manufacturers to be used in combination with mainly single-quadrupole and orthogonal-acceleration time-of-flight instruments. Open-access LC–MS systems are widely applied within pharmaceutical industry by synthetic organic chemists. Open-access MS was reviewed by Spreen and Schaffter [9] and Mallis *et al.* [10]. The most recent progress in this area is the implementation of on-line LC–NMR within the complete strategy, as described by Pullen *et al.* [8, 11].

3.   Characterization of combinatorial libraries

The approach of rational drug design based on the synthesis of one compound at a time is a time- and labour-intensive process. Many novel compounds are identified as possible lead compounds in receptor-based assays. Therefore, automated procedures can be applied for screening of a wide variety of compounds. In recent years, the developments in combinatorial chemistry have led to changes in drug discovery approaches [1]. The rationale in the application of combinatorial synthesis is to accelerate the discovery of molecules showing affinity against a target such as an enzyme or receptor through the simultaneous synthesis of a large number of structurally-related analogues. In addition to discovery of relevant new chemical entities as lead compounds, combinatorial synthesis can also play an important role in lead optimization, *i.e.*, the synthesis of related compounds to evaluate and optimize structural features responsible for the required biological activity as well as to evaluate and reduce the side effects.

Combinatorial libraries may contain hundreds or thousands of compounds, which must be screened for biological activity using a high-throughput screening method and for which occurrence, structure, and purity must be assessed. MS and especially LC–MS has been found to be a very powerful tool in the analytical support of such activities. LC–MS is often involved in fast characterization of products by molecular mass during synthesis, in characterization of the combinatorial libraries prior to or after biological screening, and in guiding sample purification for the isolation of possible new lead compounds [12-14]. In addition, LC–MS can play a role in the actual bioactivity screening (Ch. 9.4).

The commercial softwares, initially developed by instrument manufacturers for open-access operation, were adapted to enable unattended data acquisition and automated data processing for large series of samples from an autosampler supporting the 96-well microtitre plate format, which is the sample format of choice in combinatorial synthesis. Initially, mainly Gilson 215 or 233 XL autosamplers were used, but other systems have become available from other instrument manufacturers. The complete system is under control of the MS data system. It consists of a 96-well-plate autosampler, an LC pumping system, eventually a UV-photodiode-array detector (DAD) in series and/or evaporative light scattering (ELSD) detector in parallel, and the mass spectrometer equipped with ESI, APCI, and/or atmospheric-pressure photoionization (APPI).

The commercial softwares allow rapid and unattended analysis of large series of samples (up to 60 samples per hour in column-bypass or flow-injection mode, and up to 15 samples per hour in fast gradient LC–MS mode). Sample lists can be imported from spreadsheets. For reporting of processed data, a databrowser allows rapid decisions on whether the expected products are present in the various wells of the 96-well plates. An example of the typical screen layout of such a databrowser is shown in Figure 9.1. Based on the expected molecular mass, the instrument decides whether the compound is present in a well or not. The software evaluates data acquired in both positive-ion and negative-ion mode. This procedure results in a reduction of data: the sample positions are colour-coded in the screen representation of the microtitre plate (green if the expected compound was detected, and red if not). The chromatogram and mass spectra for each sample position can be viewed by clicking the well.

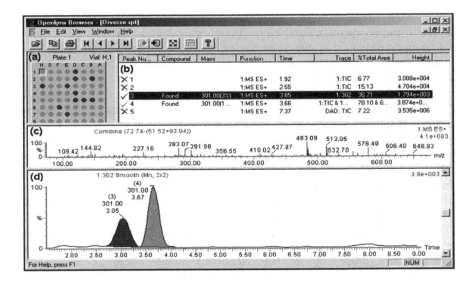

Figure 9.1: Typical screen layout of a databrowser (Micromass OpenLynx®) enabling overview and easy access to the data acquired in the analysis of samples from a 96-well plate. The screen shows (a) a representation of the 96-well plate with colour-coded samples (green is found, red is not found), (b) the peak list for the selected well on the plate (position H1 in this particular case), (c) a back-ground subtracted spectrum of the component selected in the peak list, and (d) the smoothed total-ion chromatogram acquired for the selected sample. Reprinted with permission from J. Chromatogr. A, 1000 (2003) 413. ©2003, Elsevier Science B.V.

Next to the softwares developed and distributed by the MS instrument manufacturers, related approaches were also developed within pharmaceutical industries. Görlach, Richmond and coworkers [15-16] from Novartis (Switzerland) reported the application of in-house developed software (RackViewer, written in Visual Basic) enabling automated high-throughput flow-injection analysis of combinatorial chemistry samples. An important issue in this work is purity assessment, without performing LC separation. The purity of each sample is calculated from the ratio of the summed ion currents of ions related to the expected compound and the total ion current. The purity of each sample is visualized by different colours in the databrowser layout of RackViewer program. Data input concerning the compounds to be expected as well as reports of results are networked to chemists at the company's research facilities in various countries. Performing the analysis in a central laboratory demands for efficient logistic operations in order to achieve secure sending of many samples from the remote laboratories to the central facility, where analysis is performed. In this way, over 70,000 samples were analysed in a two-year period [16].

In such a high-throughput flow-injection approach, directed at both confirmation of identity and purity assessment, sample carry-over is a serious concern [16-17]. Estimation of sample carry-over was implemented into the RackViewer data processing software. Initially, the median carry-over was estimated to be 0.88% [16]. With the growing number of samples presented for analysis, a further reduction of the sample measurement duty cycle was required, while maintaining the carry-over below 1%. Using recently introduced high-speed autosamplers, the duty cycle could actually be reduced from 168 s [16] to 44 s [17]. Further optimization of syringe and loop wash steps enabled a further reduction of the median inter-sample carry-over to at best 0.01%.

Purity assessment of combinatorial libraries is an important issue. It has led to the use of alternative detectors next to MS, such as UV-DAD and ELSD. Structural characterization and purity assessment of compound libraries obtained by combinatorial parallel synthesis using LC–APCI–MS and MS–MS, UV-VIS DAD, and NMR has been reported by Duléry et al. [18].

Automated data processing, interpretation, and reporting via electronic mail for high-throughput open-access LC–MS was also described by the group of Siegel [19-20] for both quadrupole MS and high-resolution Fourier-transform ion-cyclotron resonance MS (FT-ICR-MS). One of the important issues in fully-automated data processing is the prediction of the ions related to the compound of interest which will occur in the mass spectra. Both ESI and APCI are prone to adduct formation, leading to the possible occurrence of a variety of other ionic species next to the protonated or deprotonated molecule. Useful listings of frequently occurring adducts ions as well as artefact peaks have been reported by Tong et al. [19].

Natural product extracts have been found to provide a valuable source of molecular diversity, complementary to that provided by traditional or combinatorial organic synthesis. Natural product extracts may include amino acids, proteins and antibiotics from microbes, alkaloids from plants and some bacteria, steroids from marine animals, plants, and fungal sources, pigments from microbes and plants, pyrimidines and purines from microbes, and carbohydrates, fats, and terpenes from all sources, including terrestrial animals. High-throughput screening methods applied to natural products have been reviewed by Strege [21]. The inherent diversity of natural product extracts has not only stimulated the evaluation of their biological activity, but also presented significant challenges for separation and detection to enable rapid characterization.

Initially, high-throughput characterization of combinatorial libraries was developed for and performed with single quadrupole instruments, providing unit-mass resolution and nominal molecular-mass determination. The application of high-resolution mass spectrometers, such as time-of-flight (TOF) and FT-ICR-MS, further enhances the power of the approach [22].

A TOF instrument does not only enable more accurate mass determination, but also enables a further enhancement of the sample throughput resulting from the faster data acquisition (typically 10 spectra/s). The potential of TOF-MS in this respect was fully exploited in a multiplexed LC–MS interface featuring four or eight multichannel parallel ESI inlets ([23], MUX, Ch. 5.5.3 and Figure 5.12).

The potential of the extreme high resolution achievable with FT-ICR-MS in characterization of complex mixtures, such as combinatorial libraries, was readily recognized [24-27]. Good examples are the characterization of a 19-component octapeptide library to differentiate between Lys and Gln [24], or of a peptide library containing over 10,000 compounds at a resolution of 130,000 [25].

An interesting approach to the identification of compounds in combinatorial libraries using multistage accurate-mass determination using FT-ICR-MS in multistage MS–MS mode was proposed by Wu [26]. Accurate-mass determinations of precursor and product ions in multistage MS–MS were performed to obtain a unique elemental composition of a compound with a molecular mass of 517 Da containing *C*, *H*, *N*, *O*, *S*, and *F*. When a unique elemental composition for this compound should be obtained in a direct single-stage accurate-mass measurement, a mass accuracy better than 0.02 ppm would be required.

4.   LC–MS in high-throughput bioactivity screening

In most cases, the implementation of LC–MS in high-throughput screening is directed at confirming the identity and/or assessing the purity of ligands, but not in the actual *in vitro* bioactivity screening. In the *in vitro* bioactivity screening, the specific noncovalent interactions between a biopolymer and the drug candidates are studied in order to identify compounds that selectively bind to the active site of the biopolymer. Actual biological activity of the ligands selected in this way must be certified by means of an *in vivo* biological assay.

ESI-MS can be applied for *in vitro* bioactivity screening. This may involve the study of noncovalent complexes between drug candidates and various biopolymers, such as peptides, proteins, RNA, and DNA. Alternatively, ESI-MS may be applied as a detector in combination with various liquid-phase separation techniques applied in the study of drug–biopolymer interactions [29].

The role of ESI-MS in the detection and characterization of noncovalent complexes is discussed in Ch. 16.5. The bioaffinity characterization–MS approach, as proposed by the group of Smith [30], may serve as an illustrative example. The various steps in the process are combined in the measurement cell of an FT-ICR-MS instrument. The affinity target and ligand library are electrosprayed directly from solution. The resulting ion population is trapped in the ICR cell and subjected to a number of consecutive measurement steps. The noncovalent receptor-ligand complexes are first identified in the mass spectrum and isolated by selected-ion accumulation. The accumulated ions are dissociated to release and trap the ligands that show significant affinity to the receptor. These ligands can then be further characterized using MS–MS related approaches. This concept avoids the need for time-consuming steps related to the use of solid supports. Competitive binding of various inhibitors to carbonic anhydrase II was investigated along this line [31-32].

While these experiments involve the study of noncovalent complexes in the gas phase, liquid-phase chemistry is applied more often, both in conventional screening assays involving fluorescence or radioactivity detection and in MS-based approaches. For liquid-phase screening assays, there are two general strategies, one involving the direct detection of the complex or the ligand after dissociation of the complex, the other involving indirect detection of a reporter ligand, which is released as a result of the interaction between the candidate drug and a reporter ligand–receptor complex.

The use of on-line affinity CE–MS is an early example of a direct detection approach [33]. The receptor is present in the electrophoresis buffer and the library of ligands is injected as the sample. Ligands that show relatively strong binding to the receptor are retained and can thus be separated from compounds that do not interact. The on-line MS detection allows direct characterization of the interacting ligands.

The use of on-line immunoaffinity extraction in combination with coupled-column LC–MS–MS was demonstrated for the characterization of benzodiazepine libraries [34]. The benzodiazepine library was injected onto a Protein G column loaded with benzodiazepine antibodies. The benzodiazepine-antibody complexes are stripped of the column at low pH and eluted to a restricted-access material in order to separate the benzodiazepines from the antibodies. In the last step, the compounds interacting with the benzodiazepine antibodies are separated on a C_8-reversed-phase column and identified by MS–MS.

In frontal affinity chromatography (FAC), ligands are continuously infused into a column, in which the receptor of interest is immobilized onto the solid support. Compounds pass through the column: nonbinding ligands at the void volume, and ligands with increasing affinity at larger volumes. ESI-MS can be applied to detect the ligands and obtain a response-time curve from the extracted-ion chromatograms of the various ligands [35-36]. A recent application of FAC–MS involves the study of competitive ligands for the ATP and substrate sites of protein kinase C and the measurement of ligand binding to both active and inactive kinases [37]. Combined affinity methods and MS was reviewed by Kelly *et al.* [38].

On-line ultrafiltration and ESI-MS can assist in the identification of lead compounds in the rapid screening of combinatorial libraries [39-40]. The procedure consists of three steps: (1) the ligands from a library mixture are bound to a macromolecular receptor, (2) the ligand–receptor complexes are purified by ultrafiltration, *i.e.*, unbound ligands are washed away, and (3) the complexes are dissociated with methanol to characterize the ligands by ESI-MS. In this way, the selective binding of warfarin, salicylate, furosemide, and thyroxine to human serum albumin was investigated and high-affinity ligands to adenosine deaminases were searched for. As an example, the selective binding of *erythro*-9-(2-hydroxy-3-

nonyl)adenine to calf intestine adenosine deaminase is demonstrated in Figure 9.2 [39]. The versatility of this approach was also demonstrated by Wieboldt *et al.* [41], who applied ultrafiltration to benzodiazepine-antibody-complexes and the subsequent characterization of the most active benzodiazepines by LC–MS–MS. On-line pulsed ultrafiltration–MS has recently been reviewed [42].

The separation between ligand–receptor complexes and free ligands may also be achieved by means of size-exclusion chromatography (SEC). After incubation of the drug candidates with the receptor, the mixture is injected onto SEC to isolate the complexes. The SEC column effluent, containing the receptor and its complexes with the ligands, is denatured and subsequently analysed by ESI-MS or LC–MS. In evaluation of the affinity of some combinatorial peptide libraries to various receptors, Kaur *et al.* [43] performed denaturation and concentration of the SEC effluent in a reversed-phase cartridge prior to ESI-MS–MS analysis, enabling confirmation of the ligand by the molecular mass and of its structure from the MS–MS spectrum.

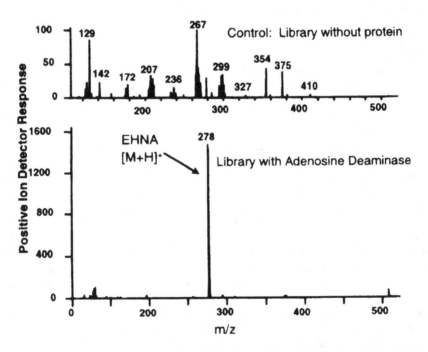

Figure 9.2: Identification of *erythro*-9-(2-hydroxy-3-nonyl)adenine as the highest affinity ligand for adenosine deaminase in a combinatorial library of 20 compounds using pulsed ultrafiltration electrospray MS. Reprinted from [39] with permission. ©1997, American Chemical Society.

The use of SEC-spin columns for this purpose has been reported as well [44]. SEC-spin columns are short columns that are centrifuged to speed up the chromatographic process. With these SEC-spin columns, the power of a multidimensional screening assay, involving both ESI-MS and nuclear-magnetic resonance (NMR) spectroscopy, was demonstrated as well, *e.g.*, in the screening of a library of 32,000 compounds for the identification of molecules exhibiting specific binding to the RGS4 protein [45].

High-throughput screening of potential antagonists of *E. coli* dihydrofolate reductase was reported by Annis *et al.* [46]. Rapid SEC is integrated with reversed-phase LC–ESI-MS on a TOF instrument and novel data processing algorithms enabled screening of up to 250,000 compounds per day.

Rather than measuring protein-ligand complexes or dissociated ligands directly, a reported ligand can be applied to indirectly determine the interaction between ligands and the receptor. On-line monitoring of biospecific interactions in a homogeneous biochemical assay using ESI-MS has been reported by Hogenboom *et al.* [47]. A scheme of the continuous-flow system is shown in Figure 9.3. First, the ligand to be investigated, *i.e.*, biotinylated compounds or digoxin, is injected into a continuous-flow system and allowed to react with the receptor protein, *i.e.*, streptavidin or anti-digoxigenin. Next, a reporter ligand is added to saturate the remaining free binding sites of the affinity protein. The concentration of the free reporter ligand is determined by ESI-MS. Along these lines, continuous-flow biochemical assays can be tailored for a wide variety of reactions, not only involving ligand–receptor complexes, but also in the study of enzyme inhibition. On-line MS detection enables the determination of binding constants as well as identification of unknown ligands.

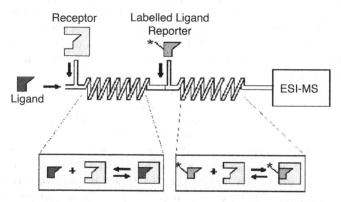

Figure 9.3: Scheme of the continuous-flow system applied for on-line monitoring of ligand-receptor interactions in a homogeneous biochemical assay. Reprinted with permission from J. Chromatogr. A, 1000 (2003) 413. ©2003, Elsevier Science B.V.

The versatility of this homogenous continuous-flow bioassay format is best illustrated with some examples. Van Elswijk *et al.* [48] developed a system to identify ligands for soluble orphan receptors. Ligands and receptor are mixed in a first reaction coil. A restricted-access column is used to selectively trap unbound ligands, while the ligand–receptor complexes are dissociated in a second reaction coil. In a second restricted-access column, the released ligands are trapped and separated from the receptors. Finally, the ligands are eluted to the MS and detected and identified. In another study by the same group [49], an on-line β-estrogen receptor bioassay was developed to screen for estrogenic compounds in *Punica granatum*.

Next to ligand–receptor bioassays, enzyme inhibition assays have also been reported. The on-line monitoring of the glutathione-S-transferase catalysed reaction between 1-chloro-2,4-dinitrobenzene and a H-γ-Glu-Cys-Xxx-OH library using ESI on FT-ICR-MS, as described by Wigger *et al.* [50], can be considered as an example. Enzyme inhibition assays are also suitable to be performed in a continuous-flow system, as demonstrated by de Boer [51] in ESI-MS monitoring of reaction products of a substrate of cathepsin B in order to screen plant extracts for potential inhibitors.

5.   Screening and identification of drug impurities

An important step in the drug development process is impurity profiling, *i.e.*, screening for, toxicology assessment, and identification of any compounds present in the drug substances above the limits established by the International Conference on Harmonization (ICH), which is active in the coordination of technical requirements for the registration of pharmaceuticals in the United States, Japan, and the European Union [52-54]. Impurities in drugs result from many sources, *e.g.*, raw materials and reagents, reaction byproducts, and due to degradation during manufacture and storage. The regulatory bodies demand qualification of the impurities present in drugs and their formulations down to the 0.1% w/w level or 1 mg/day, whichever is lower (for drug administered at up to 2 g per day). The qualification involves acquisition and evaluation of data which establish the biological safety of the individual impurity or a given impurity profile at the level(s) specified. The 0.1% level implies that a limit of quantification of *ca.* 0.05% is required for the impurities.

In principal, two types of related substances or impurities must be considered and are discussed here separately: reaction byproducts present as a result of the synthesis and the production process (Ch. 9.5.2), and degradation products of the active drug substance in a formulation (Ch. 9.5.3). These sections on selected examples are preceded by a discussion on general topics related to the search for impurities by LC–MS (Ch. 9.5.1). Finally, LC–MS controlled fractionation in preparative LC is discussed (Ch. 9.5.4).

5.1   General issues in impurity profiling

Impurity profiling is concerned with the detection of minor components in the presence of a major component. This may lead to experimental difficulties related to, *e.g.*, column overloading, limited linearity and/or dynamic range of the detection system, and the different detection characteristics of the impurities. Some of these problems are especially important in the application of LC–MS in the impurity profiling studies [55]. ESI-MS is known for its limited dynamic range. Therefore, special precautions must be taken, *e.g.*, by diverting away the major component from the ion source by valve-switching techniques, in order to avoid source contamination and ion suppression. Wolff *et al.* [56] reported a valve-switching system, which was actuated by the signal of the in-line UV detector, to divert away major components in order to enhance the sensitivity for minor impurities. Rudaz *et al.* [57] reported a heart-cut method to selectively transfer the minor process impurities to the MS. In addition, structural analogues can show significantly different response factors in ESI-MS [58]. Therefore, the major role of LC–MS is in structure elucidation and identification rather than in quantitative assessment of impurities.

Efficient peak purity assessment from the complex LC–MS data is another issue of concern. In LC–MS, the analyte peaks have to be searched for against a relatively high background of solvent-related ion current. Therefore, the search for trace impurities is often difficult. The use of base-peak chromatograms can be helpful in this respect, but the implementation of the base-peak chromatograms in most commercial MS software packages is rather poor: in most cases the m/z range to be searched for base peaks cannot be specified. Approaches developed for LC–UV-DAD, *e.g.*, univariate approaches such as spectral overlays, iso-absorbance plots, and derivative transformations, and numerical approaches such as absorbance ratioing, purity parameters, and multiple absorbance ratio correlation, cannot be used unaltered in LC–MS. Powerful approaches based on multivariate analysis, *e.g.*, principal component analysis and factor analysis, have been applied to LC–MS data [59-60]. Barbarin *et al.* [61] applied some of these alternative methods like the Contour chromatogram and the component detection (CODA) algorithms in the characterization of related substances in trimethoprim tablets.

Bryant *et al.* [62] investigated the potential of LC–MS in purity assessment of (deliberately) coeluting peaks of a major component and a number of related impurities. As they essentially pursued the detection of the impurities based on differences in m/z, the voltage setting of the sampling cone had to be carefully optimized. At low voltages, the positive-ion mass spectrum contains intact cationized analyte species, together with a number of solvent cluster ions. At higher voltages, the latter disappear, but fragment ions may appear in addition to intact cationized species. At still higher voltages, the fragment ions become increasingly more intense. Therefore, intermediate voltages were applied. In this way, coeluting impurities of famciclovir and ropinirole with very similar UV spectra spiked at 0.1%

level could be detected. Due to the differences in response factors between even closely related components, quantitative data are unreliable: a ropinirole-related impurity spiked at the 0.1% level shows a 0.7% relative abundance in the mass spectrum of ropinirole. Antonovich and Keller [63] reported more elaborate studies to evaluate the ability to detect and identify coeluting impurities of azimilide, 5-aminosalicylic acid, digoxin, and digitoxin by LC–MS. At the 1% level, 75% of the coeluting impurities were detected; only 35% at 0.1%.

Detection and identification of related substances cannot be performed by LC–MS alone. In practice, additional experiments are required using various LC gradients and/or orthogonal phase systems, alternative detection techniques like evaporative light scattering detection or optical rotation detection, and other spectroscopic techniques, like UV-DAD and NMR. Obviously, this discussion focusses primarily on the role of LC–MS in the impurity profiling and identification. Developments in LC–MS and especially mass analysers are important in strengthening the power of LC–MS in structure elucidation.

The importance of high-resolution MS in providing accurate-mass determination has been recognised in an early stage. Haskins et al. [64] confirmed the identity of four poorly-separated reaction byproducts, which were sometimes observed in cimetidine batches, via accurate-mass measurement (at 8-9 ppm) by means of LC–MS on a FT-ICR-MS instrument. In a subsequent study, the potential of an quadrupole–time-of-flight (Q–TOF) hybrid instrument, providing mass accuracy better than 5 ppm, was evaluated [65]. Similar accuracy was achieved for product ions in MS–MS operation.

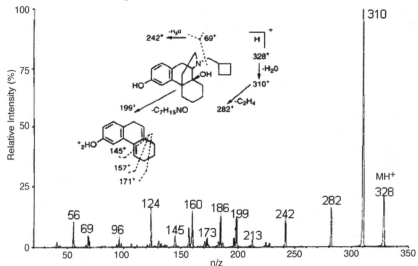

Figure 9.4: Product-ion mass spectrum of butorphanol. Reprinted from [68] with permission, ©1996, Elsevier Science BV.

Paul *et al.* [66] reported the use of an enhanced resolution triple-quadrupole instrument for the identification of a degradation product of cabergoline. The mass accuracy was within 3 ppm in this case using a peak-matching procedure on peaks of ammoniated pol(ethylene)-glycol introduced simultaneously.

The use of high-resolution inductively-coupled plasma MS was used for the sulfur-specific detection of cimetidine impurities. The structure of most of the impurities was subsequently confirmed by LC–ESI-MS [67].

Obviously, MS–MS plays an important role in the identification of related substances by LC–MS. The product-ion MS–MS spectrum of the related substances will provide the information required for the identification. In addition, a triple-quadrupole instrument provides additional scan modes that can be performed for structure-specific screening. In the impurity profiling of the intranasal analgesic butorphanol tartrate, Volk *et al.* [68] demonstrated the application of precursor-ion and neutral loss analysis for screening. The product-ion MS–MS spectrum of the parent compound is shown in Figure 9.4 showing a series of structure-informative fragments. The screening strategy is based on the premise that targeted degradation products will retain part of the original butorphanol structure, and will therefore undergo similar fragmentation. Precursor-ion scans, *e.g.*, using m/z 199 due to butorphanol three-membered ring core and subsequent fragmentation to m/z 171, 157, and 145, as well as neutral-loss scans, *e.g.*, loss of ethene (28 Da) or cyclobutane (58 Da), can be performed to screen for the degradation products. In this way, five degradation products in long-term stability studies could be identified (see Table 9.1).

Nicolas and Scholz [69] described another use of MS–MS in impurity profiling. In the routine monitoring of drug impurities by means of LC, variation in retention time can lead to uncertainty with respect to the identity of a particular component. The use of the precursor m/z and at least three product ion m/z are used either as MS–MS fingerprint or as diagnostic ions to trace and confirm related substances.

Next to MS–MS, NMR plays an important role in identification of related substances. In the last fifteen years, the on-line coupling of LC and NMR has rapidly become popular [70]. Therefore, on-line LC–NMR, often in conjunction to LC–MS, has been applied in the identification of related substances, *e.g.*, of a protease inhibitor in dosage formulations [71]. LC–ESI-MS was used with alternating scans with low and high cone voltages, in order to induce fragmentation by means for in-source CID. The LC–MS data led to tentative structure assignments. However, unambiguous identification was not possible for three of the six degradation products. LC–NMR was applied to confirm the proposed structures and to select the correct structure from the isomeric or isobaric structures proposed. LC–NMR was performed under conditions identical to the LC–MS studies.

Table 9.1:
Degradation products of butorphanol identified
by electrospray LC–MS-MS strategies [68]

Proposed structure	Relative retention	Molecular mass	R_1	R_2	R_3
Norbutorphanol	0.58	259	H	H	OH
Hydroxy-butorphanol	0.78	343	$CH_2-(C_4H_7)$	OH	OH
Ring-contracted butorphanol	0.9	313	$CH_2-(C_4H_7)$	H	OH
Butorphanol	1	327	$CH_2-(C_4H_7)$	H	OH
Keto-butorphanol	1.3	341	$CH_2-(C_4H_7)$	=O	OH
$\Delta 1,10\alpha$-butorphanol	1.5	309	$CH_2-(C_4H_7)$	H	H

5.2 Identification of reaction byproducts

In this section, selected examples of the use of LC–MS in the detection and identification of reaction byproducts are discussed. ESI-MS is used in most cases, although the use of APCI-MS was described as well, e.g., [54, 61, 63].

Raffaelli et al. [72] applied EI, FAB, and ESI, in the identification of a reaction by-product of pyrazolotriazolopyrimidines. The impurity was due to an ethyl substitution at one of the rings. The position was established using ¹H-NMR.

Lehr et al. [54] described the identification of two process impurities in trimethoprim from different manufacturers. The impurities were isolated by preparative LC and identified using both MS and NMR. In both impurities, one of the three methoxy substituents at the phenyl ring was replaced, either by an ethoxy group or by a bromine atom. NMR was required to determine which group was replaced. Related substances in trimethoprim tablets were also investigated by Barbarin et al. [61].

5.3   Degradation products in drug substances

A low-level degradation product was found during a package screening study in film-coated tablets of the H_2-receptor antagonist famotidine [73]. Using LC–MS in positive-ion APCI, the product was found to have a molecular mass of 349 Da, corresponding to the addition of a carbon atom. Using LC–MS–MS, it was found that the carbon was added to the side of the N-(aminosulfonyl)-propanimid-amide moiety of famotidine. The tentative structure was confirmed by means of synthesis.

Bartlett *et al.* [74] applied LC–ESI-MS(–MS) to identify a number of degradation products in the bulk drug form of isradipine. In this case, MS–MS was of little use, because the major fragment ions correspond to losses of the two ester moieties (methyl and *i*-propyl) without much further fragmentation of the skeleton.

Zhao *et al.* [75] observed three unknown products in the LC analysis of severely stressed losartan tablets and applied LC–APCI-MS–MS to identify these products as an aldehyde and two dimeric derivatives of losartan, present at sub-0.1% level. The structural assignment was further confirmed by comparing the MS–MS spectra of the degradation products with those of the synthesized products.

Many other examples are available in the literature, but in general no additional technologies are introduced to profile and identify the impurities.

The identification of photodegradation products of various drugs by means of LC–MS has been reported as well, *e.g.*, photodegradation products of irinotecan in infusion solutions [76]. Brum and Dell'Orco [77] described a system for real-time on-line monitoring of photolysis and applied to idoxifene. Drug stability studies by means of LC–MS have been reported as well, *e.g.*, with respect to the polymyxins B_1, E_1, and E_2 in aqueous solution [78].

5.4   MS-directed fractionation in preparative LC

In the search for related substances, isolation of the impurities by preparative LC can be one of the steps to be performed. The characterization of combinatorial libraries by means of LC–MS reveals that the compounds generated in this way are often not sufficiently pure for successful biological screening. Therefore, there is a need for high-throughput preparative purification procedures. This led to the development of automated LC–MS controlled fractionation systems to be used in preparative LC.

Zeng *et al.* [79] proposed an automated analytical/preparative LC–MS system for this purpose. Samples from parallel synthesis are analysed by rapid analytical LC–MS using a 5–10 min gradient on a C_{18}-column. Purity assessment is performed as part of the automated post-acquisition data processing. Any sample falling below the set purity threshold, *e.g.*, 90%, is subjected to automated on-line preparative LC–MS, where the triggering of the fraction collection is based on the real-time MS signal. In this way, unattended purification at milligram level was achieved for several compounds from libraries. A further development of this approach is the use of parallel columns, two for analytical and two for preparative LC, in combination with a dual-inlet ESI ion source [80]. This system enabled a sample throughput of 200 samples in analytical mode during daytime and 200 samples in preparative mode overnight.

Preparative purification based on MS-directed fraction collection has subsequently been described by others [81-83]. The same approach can obviously also be applied for the purification of drug metabolites [84] or process impurities in drug substances [85]. Integrated automated systems for LC–MS-controlled fractionation in preparative chromatography have become commercially available from various instrument manufacturers. A schematic diagram of one of these commercially available setups, capable of rapidly switching between analytical and preparative LC, is shown in Figure 9.5. Such a system requires good software to control the valve switching involved in the fraction collection, but adequate hardware as well.

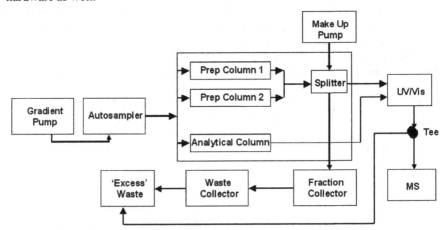

Figure 9.5: Schematic diagram of the Waters system for MS-directed fraction collection in preparative-LC. Reprinted with permission from J. Chromatogr. A, 1000 (2003) 413. ©2003, Elsevier Science B.V.

6.   **References**

1.   M.S. Lee, E.H. Kerns, *LC–MS applications in drug development*, Mass Spectrom. Rev., 18 (1999) 187.
2.   M.J. Hayward, J.T. Snodgrass, M.L. Thomson, *Flow-injection TSP-MS in the automatic analysis of agricultural chemicals*, Rapid Commun. Mass Spectrom., 7 (1993) 85.
3.   P.R. Tiller, S.J. Lane, *Automated routine TSP-MS in the analysis of potential drug compounds*, Rapid Commun. Mass Spectrom., 7 (1993) 1055.

4. D.V. Brown, M. Dalton, F.S. Pullen, G.L. Perkins, D. Richards, *An automated, open-access service to synthetic chemists: TSP-MS*, Rapid Commun. Mass Spectrom., 8 (1994) 632.
5. F.S. Pullen, D.S. Richards, *Automated LC–MS for chromatographers*, Rapid Commun. Mass Spectrom., 9 (1995) 188.
6. F.S. Pullen, G.L. Perkins, K.I. Burton, R.S. Ware, M.S. Taegue, J.P. Kiplinger, *Putting MS in the hands of the end users*, J. Am. Soc. Mass Spectrom., 6 (1995) 394.
7. L.C.E. Taylor, R.L. Johnson, R. Raso, *Open access APCI-MS for routine sample analysis*, J. Am. Soc. Mass Spectrom., 6 (1995) 387.
8. F.S. Pullen, A.G. Swanson, M.J. Newman, D.S. Richards, *LC–MS via PBI en ESI interface, solvent split to on-line NMR spectroscopy*, Rapid Commun. Mass Spectrom., 9 (1995) 1003.
9. R.C. Spreen, L.M. Schaffter, *Open access MS: A walk-up MS service*, Anal. Chem., 68 (1996) 414A.
10. L.M. Mallis, A.B. Sarkahian, J.M. Kulishoff, Jr., W.L. Watts, Jr., *Open-access LC–MS in drug discovery environment*, J. Mass Spectrom., 37 (2002) 889.
11. R.M. Holt, M.J. Newman, F.S. Pullen, D.S. Richards, A.G. Swanson, *LC–NMR–MS: Further advances in hyphenated technology*, J. Mass Spectrom., 32 (1997) 64.
12. C. Enjalbal, J. Martinez, J.-L. Aubagnac, *MS in combinatorial chemistry*, Mass Spectrom. Rev., 19 (2000) 139.
13. A. Triolo, M. Altamura, F. Cardinale, A. Sisto, C.A. Maggi, *MS and combinatorial chemistry: a short outline*, J. Mass Spectrom., 36 (2001) 1249.
14. X. Cheng, J. Hochlowski, *Current application of MS to combinatorial chemistry*, Anal. Chem., 74 (2002) 2679.
15. G. Hegy, E. Görlach, R. Richmond, F. Bitsch, *High-throughput ESI-MS of combinatorial chemistry racks with automated contamination surveillance and result reporting*, Rapid Commun. Mass Spectrom., 10 (1996) 1894.
16. R. Richmond, E. Görlach, *The automatic visualisation of carry-over in high-throughput flow-injection analysis MS*, Anal. Chim Acta, 390 (1999) 175.
17. R. Richmond, *The analytical characterization of sub-minute measurement duty cycles in flow injection analysis MS, by their carry-over*, Anal. Chim Acta, 403 (2000) 287.
18. B.D. Duléry, J. Verne-Mismer, E. Wolf, C. Kugel, L. Van Hijfte, *Analysis of compound libraries obtained by high-throughput parallel synthesis: strategy of quality control by LC, MS and NMR techniques*, J. Chromatogr. B, 725 (1999) 39.
19. H. Tong, D. Bell, K. Tabei, M.M. Siegel, *Automated data massaging, interpretation and E-mailing modules for high-throughput open-access MS*, J. Am. Soc. Mass Spectrom., 10 (1999) 1174.
20. N. Huang, M.M. Siegel, G.H. Kruppa, F.H. Laukien, *Automation of a FT-ICR-MS for acquisition, analysis and e-mailing of high resolution exact mass ESI mass spectral data*, J. Am. Soc. Mass Spectrom., 10 (1999) 1166.
21. M.A. Strege, *LC–ESI-MS analysis for the integration of natural products with modern high-throughput screening*, J. Chromatogr. B, 725 (1999) 67.
22. K.F. Blom, *Strategies and data precision requirements for the MS determination of structures from combinational mixtures*, Anal. Chem., 69 (1997) 4354.
23. V. de Biasi, N. Haskins, A. Organ, R. Bateman, K. Giles, S. Jarvis, *High-throughput LC–MS analysis using a novel MUX ESI interface*, Rapid Commun. Mass Spectrom., 13 (1999) 1165.

24. B.E. Winger and J.E. Campana, *Characterization of combinational peptide libraries by ESI-FT-ICR-MS*, Rapid Commun. Mass Spectrom., 10 (1996) 1811.

25. J.P. Nawrocki, M. Wigger, C.H. Watson, T.W. Hayes, M.W. Senko, S.A. Benner, J.R. Eyler, *Analysis of combinatorial libraries using ESI-FT-ICR-MS*, Rapid Commun. Mass Spectrom., 10 (1996) 1860.

26. A.S. Fang, P. Vouros, C.C. Stacey, G.H. Kruppa, F.H. Laukien, E.A. Wintner, T. Carell, J. Rebek, Jr., *Rapid characterization of combinatorial libraries using ESI-FT-ICR-MS*, Comb. Chem. High Throughput Screen., 1 (1998) 23.

27. S.-A. Poulsen, P.J. Gates, G.R. L. Cousins, J.K.M. Sanders, *ESI-FT-ICR-MS of dynamic combinatorial libraries*, Rapid Commun. Mass Spectrom., 14 (2000) 44.

28. Q. Wu, *Multistage accurate MS: A "basket in a basket" approach for structure elucidation and its application to a compound from combinatorial synthesis*, Anal. Chem., 70 (1998) 865.

29. M.M. Siegel, *Early discovery drug screening using MS*, Curr. Top. Med. Chem., 2 (2002) 13.

30. J.E. Bruce, G.A. Anderson, R. Chen, X. Cheng, D.C. Gale, S.A. Hofstadler, B.L. Schwartz, R.D. Smith, *Bio-affinity characterization MS*, Rapid Commun. Mass Spectrom., 9 (1995) 644.

31. X.H. Cheng, R.D. Chen, J.E. Bruce, B.L. Schwartz, G.A. Anderson, S.A. Hofstadler, D.C. Gale, R.D. Smith, J.M. Gao, G.B. Sigal, M. Mammen, G.M. Whitesides, *Using ESI-FT-ICR-MS to study competitive binding of inhibitors to carbonic anhydrase*, J. Am. Chem. Soc., 117 (1995) 8859.

32. J.M. Gao, X.H. Cheng, R.D. Chen, G.B. Sigal, J.E. Bruce, B.L. Schwartz, S.A. Hofstadler, G.A. Anderson, R..D Smith, G.M. Whitesides, *Screening derivatized peptide libraries for tight binding inhibitors to carbonic anhydrase II by ESI-MS*, J. Med. Chem., 39 (1996) 1949.

33. Y.-H. Chu, D.P. Kirby, B.L. Karger, *Free solution identification of candidate peptides from combinatorial libraries by affinity CE–MS*, J. Am. Soc. Soc, 117 (1995) 5419.

34. M.L. Nedved, S. Habibi-Goudarzi, B. Ganem, J.D. Henion, *Characterization of benzodiazepine "combinatorial" chemical libraries by on-line IAC, coupled column LC–ESI-MS–MS*, Anal. Chem., 68 (1996) 4228.

35. D.C. Schriemer, D.R. Bundle, L. Li, O. Hindsgaul, *Micro-scale FAC with MS detection: A new method for the screening of compound libraries*, Angew. Chem., Int. Ed., 37 (1998) 3383.

36. B. Zhang, M.M. Palcic, D.C. Schriemer, G. Alvarez-Mailla, M. Pierce, O. Hindsgaul, *FAC coupled to MS for screening mixtures of enzyme inhibitors*, Anal. Biochem., 2999 (2001) 173.

37. J.J. Slon-Usakiewicz, J.-R. Dai, W. Ng, J.E. Foster, E. Deretey, L. Toledo-Sherman, P.R. Redden, A. Pasternak, N. Reid, *Global kinase screening. Applications of FAC coupled to MS in drug discovery*, Anal. Chem., 77 (2005) 1268.

38. M.A. Kelly, T.J. McLellan, P.J. Rosner, *Strategic use of affinity-based MS techniques in the drug discovery process*, Anal. Chem., 74 (2002) 1.

39. R.B. van Breemen, C.-R. Huang, D. Nikolic, C.P. Woodbury, Y.-Z. Zhao, D.L. Venton, *Pulsed ultrafiltration MS: a new method for screening combinatorial libraries*, Anal. Chem., 69 (1997) 2159.

40. Y.Z. Zhao, R.B. Van Breemen, D. Nikolic, C.R. Huang, C.P. Woodbury, A. Schilling, D.L. Venton, *Screening solution-phase combinatorial libraries using pulsed*

ultrafiltration ESI-MS, J. Med. Chem., 40 (1997) 4006.

41. R. Wieboldt, J. Zweigenbaum, J.D. Henion, *Immunoaffinity ultrafiltration with LC–ESI-MS for screening small-molecule libraries*, Anal. Chem., 69 (1997) 1683.

42. B.M. Johnson, D. Nikolic, R.B. van Breemen, *Applications of pulsed ultrafiltration–MS*, Mass Spectrom. Rev., 21 (2002) 76.

43. S. Kaur, L. McGuire, D. Tang, G. Dollinger, V. Huebner, *Affinity selection and MS-based strategies to identify lead compounds in combinatorial libraries*, J. Prot. Chem., 16 (1997) 505.

44. Y.M. Dunayevskiy, J.-J. Lai, C. Quinn, R. Talley, P. Vouros, *MS identification of ligands selected from combinatorial libraries using gel filtration*, Rapid Commun. Mass Spectrom., 11 (1997) 1178.

45. F.J. Moy, K. Haraki, D. Mobilio, G. Walker, R. Powers, K. Tabei, H. Tong, M.M. Siegel, *MS/NMR: A structure-based approach for discovering protein ligands and for drug design by coupling SEC, MS, and NMR*, Anal. Chem., 73 (2001) 571.

46. D.A. Annis J. Athanasopoulos, P.J. Curran, J.S. Felsch, K. Kalghatgi, W.H. Lee, H.M. Nash, J.-P.A. Orminati, K.E. Rosner, G.W. Shipps, Jr., G.R.A. Thaddupathy, A.N. Tyler, L. Vilenchik, C.R. Wagner, E.A. Wintner, *An affinity selection–MS method for the identification of small molecule ligands from self-encoded combinatorial libraries. Discovery of a novel antagonist of E. coli dihydrofolate reductase*, Int. J. Mass Spectrom., 238 (2004) 77.

47. A.C. Hogenboom, A.R. de Boer, R.J.E. Derks, H. Irth, *Continuous-flow, on-line monitoring of biospecific interactions using ESI-MS*, Anal. Chem., 73 (2001) 3816.

48. D.A. van Elswijk, U.R. Tjaden, J. van der Greef, H. Irth, *MS-based bioassay for the screening of soluble orphan receptors*, Int. J. Mass Spectrom., 210/211 (2001) 625.

49. D.A. van Elswijk, U.P. Schobel, E.P. Lansky, H. Irth, J. van der Greef, *Rapid dereplication of estrogenic compounds in pomegranate (Punica granatum) using on-line biochemical detection coupled to MS*, Phytochem., 65 (2004) 233.

50. M. Wigger, J.P. Nawrocki, C.H. Watson, J.R. Eyler, S.A. Benner, *Assessing enzyme substrate specificity using combinatorial libraries and ESI-FT-ICR-MS*, Rapid Commun. Mass Spectrom., 11 (1997) 1749.

51. A.R. de Boer, T. Letzel, D.A. van Elswijk, H. Lingeman, W.M.A. Niessen, H. Irth, *On-line coupling of LC to continuous-flow enzyme assay based on ESI-MS*, Anal. Chem., 76 (2004) 3155.

52. International Conference on Harmonization of Technical Requirements for Registration of Pharmaceuticals for Human Use, 1995, ICH Steering Committee, March 30, 1995.

53. J.C. Berridge, *Impurities in drug substances and drug products: new approaches to quantification and qualification*, J. Pharm. Biomed. Anal., 14 (1995) 7.

54. G.J. Lehr, T.L. Barry, G. Petzinger, G.M. Hanna, S.W. Zito, *Isolation and identification of process impurities in trimethoprim drug substance by LC, LC–APCI-MS and NMR*, J. Pharm. Biomed. Anal., 19 (1999) 373.

55. W.M.A. Niessen, *LC–MS and CE–MS strategies in impurity profiling*, Chimia, 53 (1999) 478.

56. J.-C. Wolff, L. Barr, P. Moss, *New ultraviolet signal actuated switching valve for the measurement of low level impurities by LC–MS*, Rapid Commun. Mass Spectrom., 13 (1999) 2376.

57. S. Rudaz, S. Souverain, C. Schelling, M. Deleers, A. Klomp, A. Norris, T.L. Vu, B. Ariano, J.-L. Veuthey, *Development and validation of a heart-cutting LC–MS method*

for the determination of process-related substances in cetirizine tablets, Anal. Chim. Acta, 492 (2003) 271.

58. K. Lazou, T. De Geyter, L. De Reu, Y. Zhao, P. Sandra, *Applicability of a benchtop LC–DAD–MSD for quality control in the pharmaceutical industry*, LC–GC Eur., 13 (2000) 340.

59. D. Lincoln, A.F. Fell, N.H. Anderson, D. England, *Assessment of chromatographic peak purity of drugs by multivariate analysis of DAD and MS data*, J. Pharm. Biomed. Anal., 10 (1992) 837.

60. W. Windig, J. Phalp, A.W. Payne, *A noise and background reduction method for component detection in LC–MS*, Anal. Chem., 68 (1996) 3602.

61. N. Barbarin, J.D. Henion, Y. Wu, *Comparison between LC–UV detection and LC–MS for the characterization of impurities and/or degradants present in trimethoprim tablets*, J. Chromatogr. A, 970 (2002) 141.

62. D.K. Bryant, M.D. Kingswood, A. Belenguer, *Determination of LC peak purity by ESI-MS*, J. Chromatogr. A, 721 (1996) 41.

63. R.S. Antonovich, P.R. Keller, *Applicability of MS to detect coeluting impurities in LC*, J. Chromatogr. A, 971 (2002) 159.

64. N.J. Haskins, C. Eckers, A.J. Organ, M.F. Dunk, B.E. Winder, *The use of ESI-FT-ICR-MS in the analysis of trace impurities*, Rapid Commun. Mass Spectrom., 9 (1995) 1027.

65. C. Eckers, N.J. Haskins, J. Langridge, *The use of LC combined with a Q–TOF-MS for the identification of trace impurities in drug substance*, Rapid Commun. Mass Spectrom., 11 (1997) 1916.

66. G. Paul, W. Winnik, N. Hughes, H. Schweingruber, R. Heller, A. Schoen, *Accurate mass measurement at enhanced mass-resolution on a triple quadrupole MS for the identification of a reaction impurity and collisionally-induced fragment ions of cabergoline*, Rapid Commun. Mass Spectrom., 17 (2003) 561.

67. E.H. Evans, J.-C. Wolff, C. Eckers, *Sulfur-specific detection of impurities in cimetidine drug substance using LC coupled to high resolution inductively coupled plasma MS and ESI-MS*, Anal. Chem., 73 (2001) 4722.

68. K.J. Volk, S.E. Klohr, R.A. Rourick, E.H. Kerns, M.S. Lee, *Profiling impurities and degradants of butorphanol tartrate using LC–MS–MS substructural techniques*, J. Pharm. Biomed. Anal., 14 (1996) 1663.

69. E.C. Nicolas, T.H. Scholz, *Active drug substance impurity profiling: Part II. LC–MS–MS fingerprinting*, J. Pharm. Biomed. Anal., 16 (1998) 825.

70. V. Exarchou, M. Krucker, T.A. van Beek, J. Vervoort, I.P. Gerothanassis, K. Albert, *LC–NMR coupling technology: recent advancements and applications in natural products analysis*, Magn. Reson. Chem., 43 (2005) 681.

71. S.X. Peng, B. Borah, R.L.M. Dobson, Y.D. Liu, S. Pikul, *Application of LC–NMR and LC–MS to the identification of degradation products of a protease inhibitor in dosage formulations*, J. Pharm. Biomed. Anal., 20 (1999) 75.

72. A. Raffaelli, S. Pucci, G. Uccello-Barretta, F. Russo, S. Guccione, *Identification of an impurity in the synthesis of pharmacologically active pyrazolotriazolopyrimidines by a combined spectrometric approach*, Rapid Commun. Mass Spectrom., 10 (1996) 1939.

73. X.-Z. Qin, D.P. Ip, K. H,-C. Chang, P.M. Dradransky, M.A. Brooks, T. Sakuma, *Pharmaceutical applications of LC–MS. 1. Characterization of a famotidine degradate in a package screening study by LC–APCI-MS*, J. Pharm. Biomed. Anal., 12 (1994)

221.

74. M.G. Bartlett, J.C. Spell, P.S. Mathis, M.F.A. Elgany, B.E. El Zeany, M.A. Elkawy, J.T. Stewart, *Determination of degradation products from the calcium-channel blocker isradipine*, J. Pharm. Biomed. Anal., 18 (1998) 335.

75. Z. Zhao, Q. Wang, E.W. Tsai, X.-Z. Qin, D. Ip, *Identification of losartan degradates in stressed tablets by LC–MS–MS*, J. Pharm. Biomed. Anal., 20 (1999) 129.

76. H.M. Dodds, J. Robert, L.P. Rivory, *The detection of photodegradation products of irinotecan (CPT-11, Campto(R), Camptosar(R)), in clinical studies, using LC–APCI-MS*, J. Pharm. Biomed. Anal., 17 (1998) 785.

77. J. Brum, P. Dell'Orco, *On-line MS: real-time monitoring and kinetics analysis for the photolysis of idoxifene*, Rapid Commun. Mass Spectrom., 12 (1998) 741.

78. J.A. Orwa, C. Govaerts, K. Gevers, E. Roets, A. Van Schepdael, J. Hoogmartens, *Study of the stability of polymyxins B_1, E_1 and E_2 in aqueous solution using LC–MS*. J. Pharm. Biomed. Anal., 29 (2002) 203.

79. L. Zeng, L. Burton, K. Yung, B. Shushan, D.B. Kassel, *Automated analytical/preparative LC–MS system for the rapid characterization and purification of compound libraries*, J. Chromatogr. A, 794 (1998) 3.

80. L. Zeng, D.B. Kassel, *Developments of a fully automated parallel LC–MS system for the analytical characterization and preparative purification of combinatorial libraries*, Anal. Chem., 70 (1998) 4380.

81. D.M. Drexler, P.R. Tiller, *'Intelligent' fraction collection to improve structural characterization by MS*, Rapid Commun. Mass Spectrom., 12 (1998) 895.

82. G. Siuzdak, T. Hellenbeck, B. Bothner, *Preparative MS with ESI*, J. Mass Spectrom., 34 (1999) 1087.

83. J.P. Kiplinger, R.O. Cole, S. Robinson, E.J. Roskamp, R.S. Ware, H.J. O'Connell, A. Brailsford, J. Batt, *Structure-controlled automated purification of parallel synthesis products in drug discovery*, Rapid Commun. Mass Spectrom., 12 (1998) 658.

84. R.S. Plumb, J. Ayrton, G.J. Dear, B.C. Sweatman, I.M. Ismael, *The use of preparative LC–MS–MS directed fraction collection for the isolation and characterisation of drug metabolites in urine by NMR and LC–MS*, Rapid Commun. Mass Spectrom., 13 (1999) 845.

85. W.M.A. Niessen, J. Lin, G.C. Bondoux, *Developing strategies for isolation of minor impurities with MS-directed fractionation*, J. Chromatogr. A, 970 (2002) 131.

LC–MS IN DRUG METABOLISM STUDIES

1. Introduction

Metabolism studies play an important role in drug discovery and development. The advent of combinatorial drug synthesis has increased the need for a fast assessment of drug metabolism [1]. LC–MS is an important tool in the identification of drug metabolites [2-4]. The soft ionization conditions in electrospray ionization (ESI) or atmospheric-pressure chemical ionization (APCI) facilitate the detection of metabolism in biological samples. Many common biotransformation, *e.g.*, oxidation, hydroxylation, hydrolysis, and reduction, can be detected from just the knowledge of the molecular mass of the metabolites. Further confirmation and/or structure elucidation of metabolites is possible using the variety of MS–MS platforms.

This chapter is devoted to analytical strategies for the identification of drug metabolites using LC–MS. The focus in this chapter is on technology and method development. After a general introduction to metabolite identification, attention is paid to strategies in data interpretation, and automated MS and MS–MS data acquisition. Changes in the drug discovery process have resulted in changes in drug metabolism studies and in the position of such studies in the drug development process. Nowadays, metabolism is evaluated at an earlier stage in drug development than it used to. Early evaluation of pharmacokinetic properties including absorption, distribution, metabolism, and excretion (ADME) is done to terminate a development process of a potential lead as early as possible. This resulted in more integrated and faster approaches in drug metabolism studies.

2. General considerations

Drugs are xenobiotic compounds that are biotransformed in living systems into less toxic, less active, and more hydrophilic analogues, that can be excreted via the urine. Biotransformation takes place in two steps: Phase-I metabolites are generated due to oxidation, dehydrogenation, and dealkylation reactions, and Phase-II metabolism involves conjugation of the Phase-I metabolites with glucuronic acid, sulfate or glutathione. For a drug to be effective and safe, the biotransformation should not lead to a too rapid clearance from the body, to the formation of unwanted active and/or toxic metabolites, or to unwanted drug-drug interactions.

Table 10.1: Mass shifts in common Phase-I metabolism			
Process	**Δ (Da)**	**Group involved**	**Biotransformation reaction**
Hydroxylation	+16	Aromatic	Ar–H → Ar–OH
		Aliphatic	R–CH$_2$ → R–CHOH
		N-Hydroxylation	R–NH–CO–R → R–NOH–CO–R
		Epoxidation	R–CH=CH–R → R–CHOCH–R
		Sulfoxidation	R–S–R → R–SO–R
Dealkylation	−14	N-alkyl	R–NH–CH$_3$ → R–NH$_2$
		O-alkyl	R–O–CH$_3$ → R–OH
		S-alkyl	R–S–CH$_3$ → R–SH
Amine Oxidase	−30	Nitro	R–NO$_2$ → R–NH$_2$
Dehydrogenase	− 2	Alcohol	R–CH$_2$–OH → R–COH
	+16	Aldehyde	R–COH → R–COOH
	+14	Alcohol	R–CH$_2$–OH → R–COOH

Table 10.2: Mass shifts in common Phase II metabolism		
Conjugate	Δ (in Da)	
Glucuronic Acid	+176	Alcohols: R–OH → R–O–Glu
		Carboxylic Acids: R–COOH → R–COO–Glu
		Thiols: R–SH → R–S–Glu
		Amines: R–NH$_2$ → R–NH–Glu
Sulfate	+ 80	Alcohol: R–OH → R–OSO$_3$H
		Aromatic Amine: Ar–NH$_2$ → Ar–NH–SO$_3$H
Glutathione	+ 309	R–CH=CH$_2$ → R–CH$_2$–CH$_2$–S–G
	+ 162	R–CH=CH$_2$ → R–CH$_2$–CH$_2$–S–Mercapturic Acid
	+ 309	Epoxide: Ar–O → Ar–(OH)–S–G
	+ 162	Epoxide: Ar–O → Ar–(OH)–S-Mercapturic Acid
	+ 262	R–CH$_2$–NO$_2$ → R–CH$_2$–S–G
Amino Acid	+ 57 Gly	
	+ 71 Ala	
	+ 107 Tau	R–COOH → R–CO–AA
	+ 114 Orn	
	+ 145 Gln	

2.1   Phase-I metabolism

The generation of Phase-I metabolites can be considered as a preparation for the Phase-II metabolism. The detection and identification of Phase-I metabolites is important, because possibly toxic metabolites are formed. Some metabolites are even more active than the drug itself, either in the action the drug was administered for or in toxic side effects. Administration of a prodrug with more favourable properties is sometimes performed. The prodrug is rapidly transformed in the actual active substance. Moreover, in quite a number of cases the analytical strategy is directed at identification of the Phase-I metabolites, even after chemical or enzymatic deconjugation of the Phase-II metabolites.

The Phase-I biotransformation reactions often lead to a mass shift relative to the parent compound (Table 10.1).

2.2   Phase-II metabolism

Phase-II metabolism is the actual detoxification step. A bulky polar group is attached to the parent drug or one of its Phase-I metabolites. In most cases, Phase-II metabolites are no longer biologically active. Glucuronidation to a glucuronic acid conjugate is the most important reaction, but other conjugates can be formed as well. The mass shifts due to the conjugate formation are summarized in Table 10.2.

2.3   General approach in metabolite identification

Both *in vitro* and *in vivo* metabolism has to be studied. Most of the results reviewed are from *in vitro* studies. The metabolites are generated by the use of recombinant enzymes, *e.g.*, related to the cytochrome P450 (CYP) complex, by the use of subcellular fractions, *e.g.*, microsomes, cytosols, or S9 fractions, or by the use of hepatocytes or liver slices. In the case of *in vivo* metabolism studies, the metabolites are generated in living animals or humans, and analysis is performed in urine, plasma, bile, and/or faeces.

After generation, the metabolites must be extracted and isolated. A variety of sample pretreatment methods can be applied, such as protein precipitation, liquid extraction, liquid-liquid extraction (LLE), solid-phase extraction (SPE), and microdialysis. Cleanup prior to LC–MS analysis is essential to reduce endogenous interferences and small peaks at every *m/z* in the mass spectra. The cleanup step also facilitates the data processing and peak detection in the chromatogram.

A variety of tools are available for the detection and characterization of the metabolites. Unless advanced data-acquisition strategies (Ch. 10.4.4) are applied, at least two LC–MS analyses are required. The first injection is performed in full-spectrum LC–MS mode, preferable using an MS system with improved full-spectrum sensitivity, such as (linear) ion trap and time-of-flight (TOF) instruments. The analysis of a blank extract next to real samples greatly facilitates peak detection and avoids further work on endogenous peaks. The data are processed to find the relevant *m/z*-values of potential metabolites. These *m/z*-values are used as precursor ion *m/z* in a subsequent (time-scheduled) product-ion LC–MS–MS analysis. Again, instruments with improved full-spectrum sensitivity, such as ion-traps, quadrupole–linear ion trap hybrids (Q–LIT), or quadrupole–TOF hybrids, are favourable.

The data from the first LC–MS run provide insight in the type of Phase-I metabolites generated. By interpretation of the product-ion MS–MS spectra from the second run, the metabolic changes can be pinpointed to certain structural elements in the molecule. Further studies using NMR must solve issues related to positional isomers and for confirmation. Often, fraction collection after (semi-)preparative LC is performed to isolate sufficient compound for NMR. LC–MS–MS directed fractionation (Ch. 9.5.4) was developed for the isolation of drug metabolites and subsequent characterization by NMR [5-6]. Synthesis of the identified metabolites and subsequent confirmation of LC retention time, MS, and product-ion MS–MS spectra is an essential step in metabolite identification.

Because many drugs are quite polar, and their metabolites are even more polar, ESI is the method of choice for analyte ionization. As the biotransformation may lead to significant changes in the ionization characteristics of the analytes, data acquisition in both positive-ion and negative-ion mode is recommended.

3. Identification of Phase-I metabolites

3.1 Prediction of metabolites

Based on the expertise from toxicology and/or prediction of biotransformation products by computer-aided molecular modelling, an educated guess can be made on possible metabolites of a particular parent drug. This information can be applied for the synthesis and a target-oriented detection strategy.

An example of this approach was described by Carini *et al.* [7] in investigating the *in vitro* metabolism of an NO-releasing nonsteroidal anti-inflammatory drug (NCX 4016). Possible metabolites were postulated (Figure 10.1). An LC–MS system was developed for the separation of these components in extracts of various rat liver subcellular fractions (S9, microsomes, cytosol). LC–UV-DAD analysis of extracts after 90-min incubation revealed that only HBA, SA, and HBN were detected. In addition, an unknown metabolite was detected, which was identified as a glutathione conjugate at the benzyl carbon of HBA or HBN. Quantitative analysis was performed to study the kinetics of the biotransformation [7].

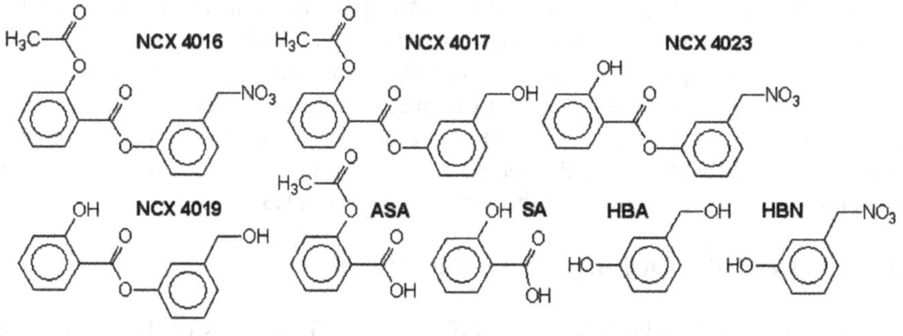

Figure 10.1: Structures of parent compound NCX4016 and of its postulated metabolites. Redrawn from [7].

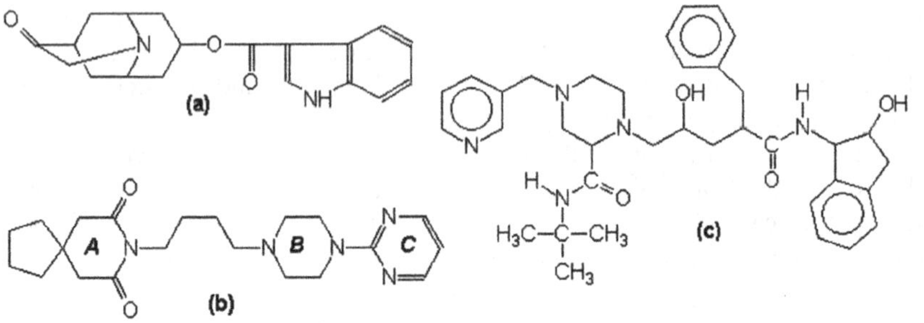

Figure 10.2: Structures of some of the parent drugs discussed in Ch. 10.3. (a) dolasetron, (b) buspirone, (c) indinavir.

In principle, the prediction of metabolites is one of the tools underlying the metabolite identification software tools like Metabolite ID, available from various instrument manufacturers. From a menu, one selects the expected biotransformation reactions. After data-acquisition, the software will interrogate the LC–MS data for the predicted metabolites and prepare LC–MS–MS experiments based on this information. Such a Metabolite ID software was evaluated by Ramanathan *et al.* [8].

3.2   Radioactive labelling and detection

Radioactive labelling is frequently applied in drug metabolism studies. It plays an important role in studies related to ADME and bioavailability. Ackermann *et al.* [9] studied the metabolism of dolasetron (Figure 10.2a), an antagonist of the serotonin 5-HT$_3$ receptor. They compared the results from packed microcapillary LC–ESI–MS and [^{14}C] radioprofiles. In LC–MS, five metabolites were detected. The identity of these metabolites was confirmed in LC–MS-MS analysis. In the LC-scintillation counting [^{14}C] radioprofile, four metabolites were detected. A comparison of peak area data from the two methods revealed that the ESI response did not provide an accurate assessment of the quantity of each metabolite, because the response in ESI-MS is highly compound-dependent. A more recent example is the identification of metabolites of the anti-cancer drug YH3945 [10].

3.3   Metabolite identification from mass shifts

Phase-I metabolism leads to mass shifts relative to the mass of the parent drug (Table 10.1). In principle, the type of biotransformation can be derived from this mass shift, while more information on the actual site of the modification can be derived from mass shifts of the fragment ions in the MS–MS spectrum. Identification approaches are based on the fact that metabolites generally retain a significant part of the parent core structure. Data interpretation based on profile groups was proposed by Kerns *et al.* [11]. Profile groups correlate specific product ions and neutral losses with the presence, absence, substitution, and molecular connectivity of specific substructures of the drug and their modifications. Depending on structure and fragmentation characteristics of the parent compound, various strategies can be followed in the interpretation of the data [1].

The concept can be illustrated by the identification of the metabolites of the anxiolytic drug buspirone [1, 12]. Buspirone (Figure 10.2b) is considered as a combination of three subgroups: the azaspirone decane dione (A), the butyl piperazine (B), and pyrimidine (P). Whereas buspirone itself is annotated at A-B-P, a metabolite with a substitution in the pyrimidine ring can be indicated as A-B-P$_s$. It was found that in buspirone the azaspirone decane dione and the pyrimidine part of the molecule are prone to (multiple) substitution, such as hydroxylation, hydroxylation and methoxylation, and glucuronidation. The loss of the complete substructure is also observed. In this way, a total number of 26 buspirone metabolites were identified and categorized.

In vitro metabolism by rat liver microsomes of the anti-inflammatory drug tiaramide results in three metabolites, corresponding to mass shifts of +14, +16, and –44 Da [4]. The biotransformation sites can to some extent be derived from mass shifts of the fragments in the MS–MS spectrum, given and interpreted in Figure 10.3. Data on the MS and MS–MS data of the parent compound and its three metabolites are summarized in Table 10.3. A proper understanding of the fragmentation of the parent compound is of utmost importance in this approach.

For metabolite M1, a mass shift of +14 Da is observed. According to Table 10.1, this may correspond to dealkylation or transformation of an –CH$_2$OH group into an –COOH group. From the MS–MS data, it can be concluded the transformation can only take place in the hydroxyethyl group of the structure, because the fragments related to the benzothiazol-2-one part of the molecule are not shifted. The +14-shift is not observed in the fragment m/z 312, because in M1 the loss of formic acid is observed rather than the loss of ethanal. Metabolite M3 is formed by the loss of the hydroxyethyl part of the molecule. This explains why the benzothiazol-2-one-related fragments are unaltered, and no loss of water or ethanal is observed anymore. Metabolite M2 results from oxidation (mass shift +16 Da), which occurs at the piperazine ring. In principle, this could correspond to either an hydroxy-group at one of the *C*-atoms or the formation of an *N*-oxide. The latter is the case [4]. Experiments enabling the discrimination of *N*-oxide formation from hydroxylation are discussed in Ch. 10. 4.8.

The same data-interpretation strategy is applied by Gangl *et al.* [13] in a study on the metabolism of the HIV-protease inhibitors ritonavir and indinavir (see Figure 10.2c). After the *in vitro* generation by incubation with human liver microsomes, the metabolites were extracted by means of LLE. Gradient elution LC–MS on an ion-trap instrument was applied to search for possible metabolites, which were subsequently fragmented in time-scheduled triple-quadrupole MS–MS.

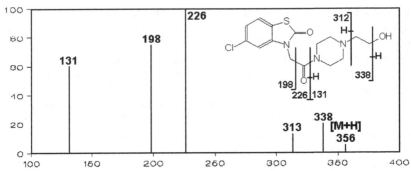

Figure 10.3: Product-ion MS–MS spectrum of tiaramide and interpretation of the fragments. Reprinted from [4] with permission. ©2003, John Wiley & Sons, Ltd.

Table 10.3: Mass shifts in the MS and MS–MS spectra of tiaramide and its three metabolites. Based on [4]						
m/z	**MS**			**MS–MS**		
Parent	356	338	312	226	198	131
M1	+14		+12	=	=	+14
M2	+16	+16	+16	=	=	+16
M3	–44			=	=	–44

In total nine ritonavir metabolites were found, among which four were monohydroxylated metabolites (mass shift of +16 Da). The data interpretation based on mass shifts in MS–MS fragment ions allowed the unambiguous identification of some of the metabolites, while for others further study was required, *e.g.*, with respect to the exact position of the modification. Pseudo-MS^3 via combined in-source collision-induced dissociation (CID) and MS–MS on the triple quadrupole was also applied to further elucidate the identity of some of the metabolites.

For parent drugs that lack backbone fragmentation, this approach cannot be applied. In that case, the identification of important fragments related to certain substructures and/or metabolic sites can help in data interpretation. This can be illustrated by data from a study on the *in vitro* metabolism of the H$^+$/K$^+$ ATPase inhibitor KR-60436 by incubation with rat or human liver microsomes [14]. The product-ion MS–MS spectrum of the parent compound is shown in Figure 10.4a. Unfortunately, the interpretation indicated by the authors in the spectrum is not correct, because the OCF$_3$-group may be lost either as a radical, resulting in an odd-electron ion at *m/z* 349, or as neutral trifluoromethanol, resulting in an even-electron fragment ion at *m/z* 348. The fragment at *m/z* 350 most likely results from a cleavage of the pyridine ring. A more likely interpretation is given in Figure 10.4b. The erroneous interpretation did not affect the identification of the metabolites.

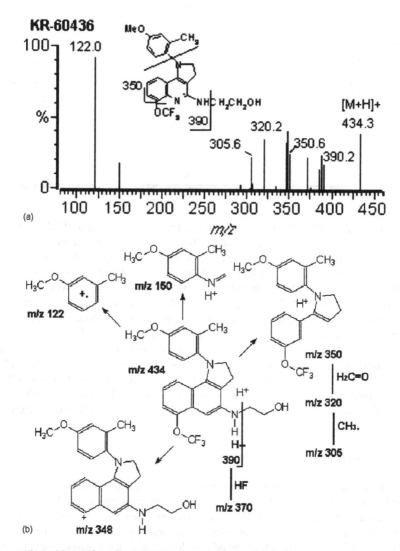

Figure 10.4: (a) Product-ion MS–MS spectrum, and (b) alternative interpretation of the fragmentation for the parent compound KR-60436. Spectrum reprinted from [14] with permission. ©2002, Elsevier Science B.V.

Table 10.4: Mass shifts in some of the MS–MS fragment ions of the seven metabolites of KR-60436. Based on [14]						
Parent	434	390	350	150	122	
M1	+16	+16		=	=	
M2	+14	–2			=	
M3	–14	–14	–14		–14	
M4	–16		–16		–14	
M5	–44		=	=	=	
M6	–2	–2	–2		=	
M7	–46		–2	=	=	

The mass shifts of some fragment peaks in the MS–MS spectra of KR-60436 and its metabolites are summarized in Table 10.4 [14]. From the interpretation of the data, four biotransformation processes can be recognized, each related to a specific site in the structure (see the structure in Table 10.4): (A) hydroxylation (mass shift of +16 Da), (B) double-bond formation (–2 Da), (C) loss of the hydroxyethyl side chain (–44 Da), and (D) demethylation (–14 Da). Combination of these processes takes place as well, *e.g.*, in M2, where both double-bond formation and hydroxylation takes place. From the MS–MS data, it can be concluded that hydroxylation at (A) blocks the fragmentation of the pyridine ring, *i.e.*, no loss of 84 Da is observed, and that double-bond formation at (B) blocks the formation of the fragment at *m/z* 150. The identification of the metabolites was as follows: M1 and M2 contained a hydroxy-group in (A). M2, M4, M6, and M7 contained a double bond in (B). M3 and M4 were demethylated at D. In M5 and M7, the hydroxyethyl group was lost [14].

3.4   Importance of LC separation

An important issue in metabolite identification is adequate separation by LC. In general, the metabolites are more polar than the parent compound and will therefore elute earlier. The formation of four monohydroxylated ritonavir metabolites [13] emphasizes the importance of separation: without separation these four isomers cannot be properly identified. A good resolution also diminishes problems with analyte suppression due to coeluting interferences (Ch. 11.5). Furthermore, some labile metabolites, *e.g.*, N-oxides, glucuronides, sulfates, are prone to fragmentation due to in-source CID or thermal dissociation. This may in fact result in the original drug. A nice example is discussed by Liu and Pereira [15]: a carbamoyl glucuronide metabolite of a drug was found to interfere in the quantitation of the parent drug, unless adequate separation of the two is achieved. Similar behaviour may be observed with lactone rings and their open-ring acids, the E- and Z-isomers of methoximes, thiols and their disulfides, phenolic drugs and their prodrugs, and amine drugs and their N-oxides.

In order to achieve sufficient resolution, relatively slow solvent gradients on typically 10–15 cm long columns are used, resulting in analysis times up to one hour. A number of trends can be recognized with respect to separation.

For a number of reasons, e.g., reduced sample consumption and enhanced sensitivity due to a lower solvent load to the ESI interface, the use of packed-capillary columns [9, 16-18] has been promoted for metabolism studies. These columns are applied in combination with a short trapping precolumn to enable larger injection volumes without compromising the performance of the analytical column.

In order to take full advantage of the enhanced sensitivity of nano-ESI (Ch. 5.5.5), and the ability to perform spectrum averaging and multiple MS–MS experiments on Q–LIT and Q–TOF instruments, Staack et al. [19] proposed RPLC fractionation into 20 fractions, which are subsequently analysed by a commercial microchip-nano-ESI system at 200 nl/min.

While good separation efficiency is needed in metabolite identification studies, modern drug discovery demands higher sample throughput in early metabolite identification and ADME studies. These demands can be met by means of:

- Fast solvent gradients (< 2 min) on short (20 mm) LC columns, packed with 3-μm-ID particles and operated at flow-rates of 1.0–2.5 ml/min [20]. With this system, the analysis of liver microsomal incubation was performed. Separation of isomeric metabolites and the parent compound within 2 min was demonstrated.
- The use of 50×4.6 mm-ID monolithic columns, operated at 4 ml/min and with a 5 min solvent gradient [21]. With this approach, six monohydroxylated metabolites of debrisoquine were separated within 1 min.
- Column switching featuring a 50×1-mm-ID turbulent-flow column (TFC, 50-μm porous particles) for the on-line SPE of up to 200 μl microsomal incubates at a flow-rate of 4 ml/min and a 20×4-mm-ID C_{18} column (3 μm particles) for the analytical gradient separation at a flow-rate of 2 ml/min. This system enables the separation of isomeric metabolites of venlafaxine, haloperidol, and adatanserin with a total cycle time of 8 min [22]. On-column isocratic focussing after 20 multiple injections onto TFC prior to LC–MS was also reported [23].
- Ultra-performance LC on columns packed with 1.7-μm particles [24]. The smaller particles enable a shorter analysis time without compromising the chromatographic resolution. Improved resolution reduces ion suppression, and thereby enhances the overall sensitivity in metabolite analysis. A mass spectrometer with fast acquisition potential is a prerequisite.

4. Tools in metabolite identification

4.1   MS–MS instrumentation: Triple quadrupole and ion trap

A variety of MS–MS instruments are available for metabolite identification. So far, results from triple-quadrupole and ion-trap instrument were discussed. The performance of these instruments was evaluated by Gangl *et al.* [13] in the study on ritonavir metabolism (Ch. 10.3.3). The ion trap shows better performance in full-spectrum MS and MS–MS, which is often necessary in metabolite identification studies. However, a fundamental limitation of the ion trap in full-spectrum MS–MS operation may be its inability to trap fragment ions with *m/z* less than one-third of the precursor *m/z*. This means that highly informative fragment peaks at the low-*m/z* end of spectrum are missed. For this reason, pseudo-MS^3 on a triple quadrupole was preferred over MS^3 on an ion trap for further structure elucidation.

Pseudo-MS^3 by a combination of in-source CID and MS–MS on one of the fragments generated in the first fragmentation step is actually applied frequently in drug metabolite studies, *e.g.*, in the study on the metabolism of bosentan [25].

In other cases, multistage MS–MS in an ion-trap can be extremely powerful in structure elucidation of drug metabolites, as demonstrated in the identification of a dextromethorphan metabolite using up to five stages of MS–MS in an ion trap [26]. Multistage MS–MS was required to elucidate the triple-quadrupole MS–MS spectrum and to correlate various fragments into different fragmentation routes. These fragmentation routes were represented as genealogical maps.

In general, LC–MS–MS does not allow the determination of positional isomerism, *e.g.*, in hydroxylation at an aromatic ring. NMR spectroscopy is the method-of-choice to elucidate such questions. Yoshitsugu *et al.* [27] reported the LC–MS–MS elucidation of positional isomerism in glucuronide conjugates of vintoperol. Ring fragmentation in the vintoperol structure was found to lead to key fragments enabling the discrimination between the various isomers.

The enhanced resolution and accurate-mass capability of a recently introduced triple-quadrupole instrument can also be helpful in metabolite identification [28].

4.2   Precursor-ion and neutral-loss analysis modes

Because metabolites generally retain a significant part of the core structure of the parent drug, structure-related MS–MS analysis modes can be applied to screen biological samples for metabolites with certain features in common with the parent drug. The precursor-ion analysis mode can be applied to search for metabolites that share common fragments, while the neutral-loss analysis mode can be used to screen for metabolites that share common neutral losses. In this way, common key fragments or losses are applied as characteristic markers for drug-related compounds.

In the product-ion MS–MS spectra of dolasetron and metabolites (Ch. 10.3.2), an ion at m/z 166, representing the loss of the indole-3-carboxylic acid, was found to be a common fragment ion. The precursor-ion analysis mode, with m/z 166 as common product ion, was applied to search for possible additional metabolites. All five previously identified metabolites were found, but no other compounds [9].

In a study on the *in vivo* metabolism of epothilone B in tumour-bearing nude mice [29], the precursor-ion analysis mode was applied as well. In the product-ion MS–MS spectrum of this 16-membered macrolide, a fragment at m/z 166 was recognized as a key fragment. It is related to the thiazole ring which is not significantly affected by biotransformation. The resulting chromatogram obtained in the analysis of a mouse liver homogenate obtained 1.5 h after administration of epothilone B is shown in Figure 10.5. Three metabolites are detected and subsequently identified from product-ion MS–MS spectra.

The product-ion MS–MS spectrum of S9788 revealed a fragment ion at m/z 341, due to loss of the metabolically stable di(4-fluorophenyl)ethylamino part of the molecule [30]. This can be applied in the neutral-loss analysis mode to screen biological samples for metabolites of S9788, as demonstrated in the analysis of a rat bile sample in Figure 10.6. Next to the already known metabolites (i) through (v), three new minor metabolites were found (A, B, and C).

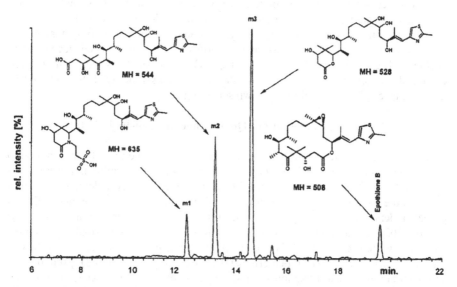

Figure 10.5: Precursor-ion MS–MS chromatogram (common product-ion m/z 166) of a mouse liver homogenate obtained 1.5 h after administration of epothilone B. Reprinted from [29] with permission. ©2001, John Wiley & Sons, Ltd.

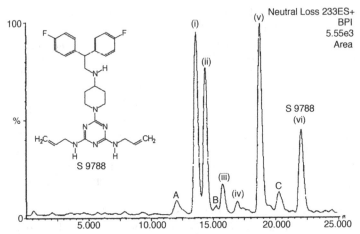

Figure 10.6: Neutral-loss LC–MS–MS chromatogram (neutral loss 233 Da) of a 6-8 h bile sample from a biliary cannulated female Sprague-Dawley rat dosed intravenously with [^{14}C]-S9788 at 3 mg/kg. Reprinted from [30] with permission. ©1995, John Wiley & Sons, Ltd.

The precursor-ion analysis mode and especially the neutral-loss analysis mode are frequently applied in screening for Phase-II metabolites (Ch. 10.5.2).

Conventional precursor-ion and neutral-loss analysis modes are only available in triple-quadrupole and Q–LIT instruments. Software tools have been developed to enable such experiments on ion-trap and Q–TOF instruments as well. Due to their enhanced full-spectrum sensitivity, neutral-loss analysis on Q–TOF and Q–LIT instruments can be powerful in unknown screening.

4.3   MS–MS instrumentation: Quadrupole–time-of-flight hybrid

Another invaluable tool in metabolite identification is the Q–TOF, which, due to the orthogonal-acceleration TOF, provides accurate-mass determination in both MS and MS–MS mode. The power of a Q–TOF instrument is nicely demonstrated in a study on the metabolism of bosentan [31]. For the identification of the metabolites, a proper understanding of the MS–MS spectrum of the parent drug is important. For the m/z-202 fragment in the bosentan MS–MS spectrum, two possible structures could be written, consisting of different parts of the parent structure, as shown in Figure 10.7. Using hydrogen/deuterium (H/D) exchange (Ch. 10.4.8), pseudo-MS^3, and accurate-mass determination on a sector instrument, the correct identity of the m/z-202 fragment could be postulated in a previous study [25]. By using the Q–TOF, the same conclusion was reached more readily.

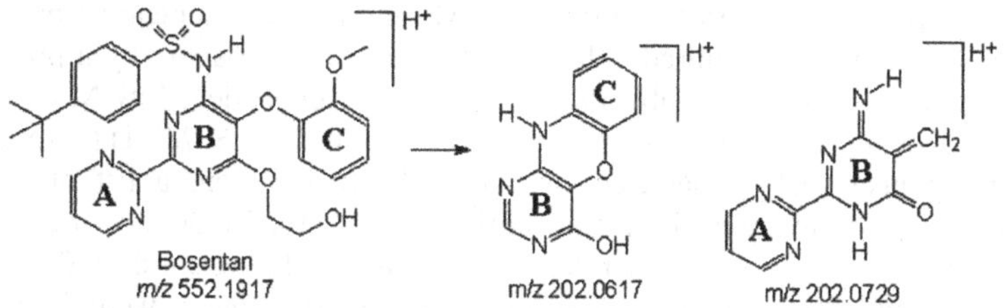

Figure 10.7: Structures of bosentan and of its two possible m/z-202 fragments. The measured m/z of the fragment was 202.0725, indicating that m/z 202.0729 is the correct structure (Δ 0.4 mDa). Based on [31].

TOF and Q–TOF instruments are frequently applied in metabolism studies. The identity of the epothilone B metabolites found by precursor-ion analysis (Figure 10.5, Ch. 10.4.2) was confirmed using accurate-mass determination on a LC–TOF-MS instrument [29]. Some other examples are the characterization of metabolites of moclobemide and remikiren [32], the identification of ketobemidone Phase-I and Phase-II metabolites [33], and the identification of *in vitro* metabolites of ethoxidine [34]. The use of a five-channel multiplexed ESI interface (four channels for parallel LC–MS and one channel for lock-mass compound infusion) on a Q–TOF instrument was recently described to speed up metabolite identification and to enhance the efficient use of the costly instrument [35].

4.4   Data-dependent acquisition

The search for and identification of unknown metabolites using LC–MS requires two injections with data-processing in between (Ch. 10.2.3): The first run in full-spectrum LC–MS mode, and the second in a time-scheduled product-ion MS–MS procedure. This procedure was applied in the identification of ritonavir metabolites [13] (Ch. 10.3.3). The full-spectrum LC–MS run was performed using an ion-trap instrument, while the subsequent LC–MS–MS run was performed using a triple-quadrupole instrument (Ch. 10.4.1).

Data-dependent acquisition (DDA) approaches (Ch. 2.4.2) can be used to simplify this procedure. In DDA, real-time data-controlled switching between a full-spectrum survey LC–MS mode and product-ion LC–MS–MS mode is performed, based on preset rules and a intensity threshold.

DDA was applied in a number of studies to indinavir metabolism [36-38]. To study the *in vitro* metabolism of indinavir by incubation with rat liver S9 fractions, a DDA procedure of six scan function was applied [36]. Upon detection of an ion with an abundance above the preset threshold in the survey scan, a full-spectrum product-ion MS–MS spectrum of the most abundant ion in the MS spectrum was acquired, followed by a product-ion MS³ experiment on the most abundant ion in the MS–MS spectrum. After returning to the survey LC–MS mode, a list-dependent product-ion MS–MS spectrum was acquired, when applicable, followed again by a product-ion MS^3 experiment on the most abundant ion in the list-dependent MS–MS spectrum. In the list-dependent MS–MS experiment, the switching to MS–MS is performed if a sufficient intense peak is detected for ion with an m/z from a list of predicted metabolites, *i.e.*, showing positive mass shifts of 2, 14, 16, 18, 30, 32, or 36 Da, or a negative mass shift of 2 or 14 Da relative to the m/z of the parent drug (Table 10.1) [36]. Similar and more advanced experiments related to indinavir were described by others [37-38].

Six known and two new metabolites of glyburide, a sulfonylurea drug applied in the treatment of insulin-independent *Diabetes mellitus*, were identified using DDA in LC–MS^n on an ion-trap instrument [39]. DDA was also applied as an alternative to precursor-ion analysis in the screening and identification of moclobemide and remikiren metabolites [32]. More recently, ion-trap MS–MS in DDA mode was applied to perform the simultaneous metabolite identification and assessment of metabolic stability [40-41]

Increasing speed in data-acquisition on newer instruments enables more advanced DDA experiments to be performed. The use of automated DDA-MS–MS of multiple precursor ions on an ion-trap instrument is combined with a post-acquisition search through the complete data set for specific neutral losses or common fragment ions. The method is applied to search for metabolites of MEN 15916 [42] and revealed both mono-, di-, and trihydroxy-metabolites, as well as some unexpected metabolites (a carboxylic acid, a *N*-dealkylated metabolite, and its hydroxy-analog).

As an alternative to DDA, the use of high-resolution LC in combination with Q–TOF-MS has been proposed [43]. Survey-MS data and product-ion MS–MS are acquired simultaneously by collision-energy switching (low/high). In this setup, no precursor-ion selection is performed. Post-acquisition processing allows detection of expected metabolites by generating accurate-mass extracted-ion chromatograms, and via both precursor-ion and neutral-loss type searching through the data.

4.5   MS–MS instrumentation: Quadrupole–linear-ion-trap hybrid

A relatively new and powerful tool in metabolite identification is the Q–LIT instrument. In this triple-quadrupole instrument, the third quadrupole can be applied as a scanning quadrupole, but also as a linear ion trap (Ch. 2.4.2). Potential and additional features of a Q–LIT in metabolite identification have been discussed by various groups [44-45]. The advantages of triple-quadrupole MS–MS spectra at ion-trap sensitivity as well as the enhanced sensitivity and speed in DDA experiments with the Q–LIT was demonstrated for the collagenase inhibitor trocade [44]. Various enhanced scan functions of the Q–LIT were applied in the identification of 6-aminobutylphthalide metabolites in rat brains, using microdialysis and LC–MS–MS [45]. Simultaneous quantification of a parent compound and screening for its predicted metabolites by means of SRM-triggered DDA for MS–MS has been reported [46-47].

4.6 Multi-instrument strategies in metabolite identification

Summing up the various MS tools discussed leads to a general protocol for metabolite identification. It can be argued that such a protocol should imply the use of various types of mass spectrometers. In a nice paper, Clarke *et al.* [48] outlined such a multi-instrument strategy, which after slight modification comprises of:

• Acquisition and interpretation of the product-ion MS–MS spectrum of the parent drug using a triple-quadrupole and/or a Q–TOF instrument.

• The interpretation of the parent drug fragmentation in combination with predicted metabolism from toxicology expertise enables the proposal of various precursor-ion and neutral-loss analysis experiments to screen for metabolites in biological samples. A triple-quadrupole or Q–LIT is used in these experiments.

• Based on the results of the previous step, product-ion MS–MS spectra of the observed metabolites are obtained using triple-quadrupole or Q–LIT instruments. At this stage, elaborate data interpretation is required to identify the observed metabolites as much as possible.

• The data interpretation will result in a number of questions concerning the identity of some compounds and/or fragments. Such questions can be addressed by means of multistage MS–MS on an ion-trap instrument. In addition, accurate-mass determination of the intact metabolites as well as of some of the product ions observed using a Q–TOF will help in solving the identification problems.

• Finally, the identity of the identified metabolites must be confirmed by synthesis, re-injection in LC–MS and LC–MS–MS, and NMR spectroscopy.

This protocol clearly indicates that identification of unknowns and related substances often demands data from a variety of MS instruments. The power of Q–LIT, providing triple-quadrupole MS–MS and MS^3 in the LIT, for fragmentation studies was promoted [49]. The complementary character of ion-trap and Q–TOF has been evaluated in a study on the *in vivo* metabolism of an cathepsin K inhibitor (NVP-AAV490) in the rat [50]. Some of the remaining questions may be solved by additional tools outlined below.

4.7 Combined LC–MS and LC–NMR

Given the role of NMR in drug metabolism studies, on-line LC–NMR or LC–NMR–MS is highly attractive [51]. This is demonstrated in a number of applications, *e.g.*, in identification of urinary metabolites of acetaminophen [52], and paracetamol [53], and the novel non-nucleoside reverse transcriptase inhibitor GW 420867 [54]. Nowadays, this expensive technique is frequently applied. It is commercially available.

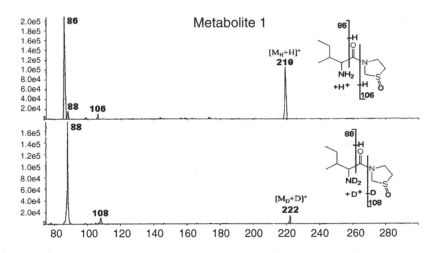

Figure 10.8: *H/D* exchange to facilitate metabolite identification. Product-ion MS–MS spectra of an oxidized metabolite (structure from [15]) obtained with (bottom spectrum) and without (top) D_2O in the LC mobile phase. Reprinted from [59] with permission (©2001, John Wiley & Sons, Ltd.) and modified.

Various tools have been developed to enhance the potential of the technique, such as mass-directed peak selection [55], on-line SPE for on-line preconcentration [56] or post-column peak trapping and stop-flow NMR [57].

4.8   Hydrogen/deuterium exchange

H/D exchange is widely applied in the elucidation of fragmentation mechanisms. *H/D* exchange can be achieved in LC–MS in a number of ways, *i.e.*, by means of a sandwich-plug technique in column-bypass injection, by the use of deuterated mobile phases in micro-LC–MS, or by addition of ND_3 to the nebulizer gas in the ESI source [58-59]. Determination of the number of exchangeable hydrogens on a metabolite facilitates its structure elucidation. It enables discrimination between various types of oxidative metabolites: whereas hydroxylated metabolites do show *H/D* exchange, *N*-oxide and *S*-oxide metabolites do not. In this way, five metabolites of a particular drug have been identified as sulfoxide, sulfone, carbamoyl glucuronide, *N*-glucuronide, and the *N*-glucoside [15, 59]. Product-ion MS–MS spectra of the sulfoxide metabolite with and without *H/D* exchange are shown in Figure 10.8. The metabolite exhibits a +16-Da shift relative to the parent drug. From the shift of the small fragment peak from *m/z* 90 / 92 in the parent drug to *m/z* 106 / 108 in the metabolite, it may be concluded the oxidation took place at the thiazolidine side of the molecule. Therefore, a sulfoxide was proposed, also because a double oxidized metabolite without *H/D* exchange was detected as well [59]. *H/D* exchange in metabolite identification by LC–MS was reviewed [60]. In the same paper, they discussed various chemical derivatization strategies that can be used to facilitate metabolite identification, *e.g.*, reactions to determine hydroxylation and glucuronidation sites [60].

4.9 Stable-isotope labelling and isotope cluster technique

While *H/D* exchange can be considered as a dynamic labelling technique, labelling of drugs by either stable or radioactive labels can be applied to facilitate detection in the complex biological matrix and/or structure elucidation. Radioactive labelling is performed for radioactivity detection (Ch. 10.3.2).

This isotope cluster technique [61] was used to facilitate the identification of omeprazole metabolites in rat urine [62]. Omeprazole is a selective inhibitor of gastric acid secretion and is used in the treatment of acid-related disorders. A 1:1 mixture of natural [^{32}S]-omeprazole and labelled [^{34}S]-omeprazole was administered to a rat. More than 40 related substances were found in the LC–MS chromatogram of the rat urine sample. All these compounds were related to omeprazole, because they showed the expected isotopic cluster (two peaks of similar height two *m/z* units apart). The summed background-subtracted mass spectrum of the entire sample is shown in Figure 10.9. This spectrum represents a good overview of the metabolic pattern and the metabolic routes omeprazole is involved in.

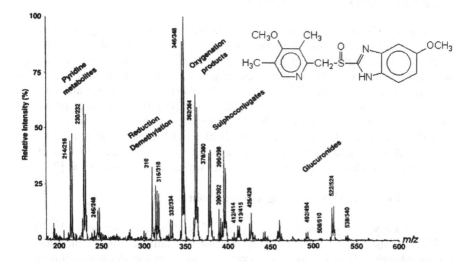

Figure 10.9: Summed background-subtracted spectrum based on the LC–ESI–MS analysis of a rat urine extract containing a total amount of *ca.* 8 μg omeprazole metabolites. Reprinted from [62] with permission. ©1992, John Wiley & Sons Ltd.

The isotope-cluster technique was also applied in a high-throughput screening of reactive metabolites generated in microsomal incubations [63-64]. Mixtures of glutathione and $[^{13}C_2\text{-}^{15}N]$-labelled glutathione were used to trap these reactive metabolites. Due to the specific isotope pattern, these compounds are readily recognized in the LC–MS data. After incubation and isolation of the glutathione conjugates by SPE, reactive metabolites were detected for the 12 model drug compounds by neutral-loss analysis (Ch. 10.4.2 and Ch. 10.5.2), taking advantage of the isotopic doublet differing 3 Da in m/z. The highly specific detection allowed direct injection of the samples after cleanup without LC separation. The method was applied by others as well [65].

4.10   Element-specific metabolite detection by LC–ICP-MS

ICP-MS and on-line LC–ICP-MS are primarily used for element speciation in environmental, food-related, and clinical applications. Its use in metabolism studies is more recent [66-67]. Bromine-specific detection by LC–ICP-MS in conjunction with LC–MS has been applied to selectively find the bromine-containing metabolites of 4-bromoaniline [66] in rat urine, while chlorine and sulfur-specific detection was applied in a metabolite study on diclofenac [67]. Pharmaceutical applications of LC–ICP-MS were reviewed [68].

5. Identification of Phase-II metabolites

Important Phase-II metabolites are summarized in Table 10.2 together with the mass shift related to their formation. Conjugation of (metabolized) drugs into glucuronic acid, sulfate, and glutathione conjugates is the most important route for the excretion of these compounds, especially via urine. Often, a chemical or enzymatic hydrolysis step of conjugates is performed prior to the analysis and structure elucidation of the Phase-I metabolites, e.g., [69]. However, this approach is not generally applicable and may lead to ambiguous results, e.g., due to artefact formation. Therefore, the direct MS analysis of intact and underivatized conjugates is preferred. LC–ESI-MS can play significantly role in this. An excellent review on the structure elucidation of Phase-II metabolites was given by Levsen et al. [70].

5.1   LC–MS(–MS) of conjugated metabolites

Direct analysis of glucuronide conjugates of gemfibrozil was reported by Xia et al. [71]. In this study, the fragments m/z 85 and 113, attributed to fragmentation of the gemfibrozil side chain, are most likely due to secondary fragmentation of the glucuronate anion. This is confirmed by the absence of these fragments in the MS^3 spectrum, where the conjugate was fragmented into the parent drug, which was subsequently fragmented further. The detection and subsequent identification of a glutathione conjugate of one of the metabolites of NCX4016 was also reported [7] (Ch. 10.3.1). The easy in-source fragmentation of an carbamoyl glucuronide metabolite and its potential influence of quantitation of conjugated metabolites was discussed as well [15] (Ch. 10.3.4). Glutathione conjugates were also studied in relation to stable-isotope trapping of minor reactive metabolites [63-65] (Ch. 10.4.9).

An interesting, but complicating aspect of conjugation by glutathione (γ-L-glutamyl-L-cysteinylglycine) is the secondary metabolism of these conjugates, resulting in losses of the Glu and/or Gly parts of the structure as well as the subsequent formation of an *N*-acetyl analogue and various oxidized forms [70].

5.2   Neutral-loss and precursor-ion analysis

The glucuronide, sulfate, and glutathione conjugates are readily fragmented in MS–MS by the formation of either conjugate-specific fragment ions, e.g., m/z 175 and its secondary fragments at m/z 85 and 113 for glucuronides in the negative-ion mode, or conjugate-specific neutral losses, e.g., 80 and 176 Da for sulfates and glucuronides, respectively. Conjugate-specific fragmentation by MS–MS has been summarized by Kostiainen *et al.* [3]; an abstract is provided in Table 10.5.

An example of the use of LC–MS in the neutral-loss analysis mode in the analysis of Phase-II metabolites of S12813 was reported by Brownshill *et al.* [72]. Three glucuronides, three sulfates, and a disulfate were generated by incubation of S12813 with Wistar rat liver slices. In LC–MS analysis, five of these conjugates were detected in positive-ion mode with poor signal-to-noise (S/N) ratio and poor peak shape. However, in LC–MS-MS in neutral-loss analysis mode (176 or 80 Da), the glucuronide and sulfate conjugates, respectively, were detected with good peak shape and excellent S/N ratio, as demonstrated in Figure 10.10.

Table 10.5: Conjugate-specific fragmentations of drug conjugates [3]			
Conjugate	**Mode**	**Neutral Loss**	**Precursor-Ion**
Glucuronides	+/–	176 ($-C_6H_8O_6$)	
Phenolic Sulfates	+	80 ($-SO_3$)	
Aliphatic Sulfates	–		97 (HSO_4^-)
Aryl–Glutathione	+	275 ($C_{10}H_{17}N_3O_6$) or 129 ($C_5H_7NO_3$)	
Aliphatic–Glutathione	+	129 ($C_5H_7NO_3$)	
N-Acetylcysteines	+	129 ($C_5H_7NO_3$)	

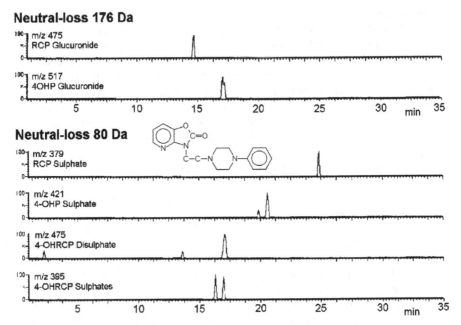

Figure 10.10: Chromatograms demonstrating the detection of sulfate and glucuronide conjugated metabolites of S12813, obtained by LC–MS-MS in neutral-loss analysis (80 an 176 Da) mode. Reprinted from [72] with permission. ©1994, John Wiley & Sons, Ltd.

Neutral-loss and precursor-ion analysis modes are regularly applied in combination with LC–MS for the detection and/or characterization of Phase-II metabolites [44, 71, 73-74]. Accurate-mass neutral-loss analysis on a Q–TOF instrument was applied to screen for glutathione conjugates [74].

6. Additional topics in metabolism studies

Next to identification, there are a number of other issues related to drug metabolism. First of all, quantitative metabolic profiling is of utmost importance in ADME studies and in assessing the efficacy of new chemical entities as drugs. In practice, quantitative bioanalysis (Ch. 11) of the parent compounds and its major metabolites must be performed in order to evaluate pharmacokinetics of the drug involved. The increasing demands for high sample throughput in early drug discovery more and more leads to integration of the qualitative and quantitative bioanalysis, as demonstrated in several studies [21, 75-76].

Electrochemical oxidation reactions have been found to mimic cytochrome P450 catalysed oxidation reaction to some extent, and can therefore be used as a tool in performing preliminary drug metabolism studies. Combined electrochemistry–mass spectrometry was explored for drug metabolite studies [77-78].

6.1 Metabolic stability screening

High-throughput metabolic stability tests using liver microsomal preparations is one of the tools to speed up processes of lead optimization in early drug discovery. The extent of CYP metabolism of new chemical entities is investigated by fast, preferentially automated quantitative analysis of microsomal incubation samples.

The development of on-line combination of pulsed ultrafiltration (PUF) and MS was reported by Van Breemen et al. [79-80]. Relevant CYP substrates, such as imipramine and chlorpromazine, were flow-injected into an ultrafiltration chamber, where rat liver microsomes were trapped in. Negative-ion ESI–MS was applied for the on-line detection of the metabolites.

A fully automated LC–MS system for the analysis of liver incubation samples, consisting of a 96-well plate autosampler, an generic LC system, and single-quadrupole ESI–MS, was reported [81]. Via automated data acquisition and processing, the system provides a sample throughput of 75 compounds per week.

Further increase of the sample throughput can be achieved by the implementation of fast LC strategies. TFC sample pretreatment prior to LC–MS–MS resulted in a cycle time of 8 min per sample [21]. Fast solvent gradients (< 2 min) on short LC columns provided the analysis of liver microsomal incubation within 2 min [82].

Gu and Lim [76] described the use of DDA for microsomal stability studies and metabolite profiling. First, the relevant *m/z* of an unknown lead compounds is determined. This information is used in SIM experiments for quantitation. Once the peak area of the compound tested falls below a certain threshold, *e.g.*, 60% of the initial value, data-dependent product-ion MS–MS analysis is performed for metabolite identification. Some results for adatanserin are shown in Figure 10.11.

More recently, the development of high-speed SPE systems used in combination with capillary LC columns was described for testing the microsomal stability of tracers for positron emission tomography [17]. Automated method development for generic fast gradient LC–MS–MS using the Sciex Autometon® software was applied in metabolic stability tests performed with a robotic system. A sample throughput of 96 compounds per day on one dedicated LC–MS system was achieved [83].

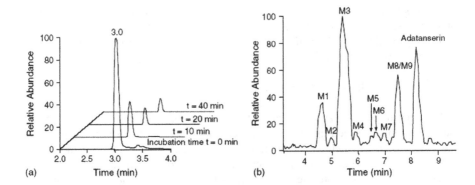

Figure 10.11: Results for adatanserin of (a) a microsomal metabolite stability study, and (b) the metabolite profiling study performed after 10 min incubation. Metabolites M1 and M2 are dihydroxy-analogues, M3, M4, M5, and M9 are hydroxy-analogues, and M6, M7, and M8 still have to be identified. Reprinted from [76] with permission. © 2001, John Wiley & Sons, Ltd.

6.2   Membrane permeability in Caco-2 cell monolayers

An important parameter in efficacy tests of oral drugs is their transport across the intestinal epithelial cell barrier. An *in vitro* model mimicking this process is Caco-2 cell monolayers. This provides a tool for rapid evaluation of the transport of drugs across a cell barrier. LC–MS has been applied in this area. As the transport experiments are performed in Hanks' balanced salt solution, direct injection of the Caco-2 cell solution into MS is not practical. Excessive source contamination can be avoided by the use of automatic valve-switching to direct the first part of the column effluent to waste [84]. LC–MS was applied to quantify drugs from the *in vitro* permeability tests in Caco-2 monolayers, *e.g.*, in a test whether drugs are substrates for P-glycoprotein, a membrane efflux pump located on the epithelium of intestinal cells [84]. In order to improve the sample throughput in such a test, the use of sample pooling and cassette dosing [85] and of four-channel multiplexed ESI systems [86] has been investigated. Fully automated nano-ESI sample introduction via a microchip-ESI device for the analysis of Caco-2 samples was described as well [87].

6.3   Drug-drug interaction via cytochrome P450 inhibition screening

Another important aspect of drug development is the evaluation of potential drug-drug interactions. An early indication of drug-drug interactions can be obtained by studying the *in vitro* inhibition or induction of the metabolism of model compounds (probe substrates) by CYP isoenzymes. In general, the isoenzymes most commonly responsible for drug metabolism are studied. LC–MS is applied to quantitatively monitor the effect of the test compound on the metabolism of probe substrates of the CYP isoenzymes investigated. Ayrton *et al.* [86] described the development of an 'open-access' facility for this type of studies. The metabolism of seven probe substrates by seven CYP isoenzymes was used to monitor induction or inhibition by test compounds, *i.e.*, phenacetin for CYP1A2, coumarin for CYP2A6, s-mephenytoin for CYP2C19, bufuralol for CYP2D6, chlorzoxazone for CUP2E1, midazolam for CYP3A4, and tolbutamide for CYP2C8, 9, and 10. In a series of papers, Bu *et al.* [87-90] reported the development and validation of high-throughput CYP inhibition screening via a cassette-probe-dosing strategy using the SPE–MS–MS strategy initially developed for Caco-2 cell monolayer permeability studies [85]. More recent development in this area concerns the use of ultrafast gradient LC on monolithic columns, enabling the analysis of six CYP substrates and an internal standard in only 24 s [91]. A high-throughput combined metabolic stability testing and CYP inhibition screening was realized by the use of a SAGIAN™ core robotics system. Microsomal incubation studies are performed using LC–MS, while CYP inhibition assays are performed using a fluorescence plate reader [92].

7. References

1. M.S. Lee, E.H. Kerns, *LC–MS applications in drug development*, Mass Spectrom. Rev., 18 (1999) 187.
2. E.J. Oliveira, D.G. Watson, *LC–MS in the study of the metabolism of drugs and other xenobiotics*, Biomed. Chromatogr., 14 (2000) 351.
3. R. Kostiainen, T. Kotiaho, T. Kuuranne, S. Auriola, *LC–API-MS in drug metabolism studies*, J. Mass Spectrom., 38 (2003) 357.
4. Z. Tozuka, H. Kaneko, T. Shiraga, Y. Mitani, M. Beppu, S. Terashita, A. Kawamura, A. Kagayama, *Strategies for structural elucidation of drugs and drug metabolites using (MS)^n fragmentation in an ESI ion trap*, J. Mass Spectrom., 38 (2003) 793.
5. R.S. Plumb, J. Ayrton, G.J. Dear, B.C. Sweatman, I.M. Ismail, *The use of preparative LC–MS–MS directed fraction collection for the isolation and characterization of drug metabolites in urine by NMR and LC–MS^n*, Rapid Commun. Mass Spectrom., 13 (1999) 845.
6. G. J. Dear, R. S. Plumb, B. C. Sweatman, I. M. Ismail, J. Ayrton, *MS–MS linked fraction collection for the isolation of drug metabolites from biological matrices*, Rapid Commun. Mass Spectrom., 13 (1999) 886.
7. M. Carini, G. Aldini, M. Orioli, R. Maffei Facino, *In vitro metabolism of NCX 4016 by rat liver: LC and LC–MS studies*, J. Pharm. Biomed. Anal., 29 (2002) 1061.
8. R. Ramanathan, D.L. McKenzie, M. Tugnait, K. Siebenaler, *Application of semi-automated metabolite identification software in the drug discovery process for rapid identification of metabolites and the cytochrome P450 enzymes responsible for their formation*, J. Pharm. Biomed. Anal., 28 (2002) 945.

9. B.L. Ackermann, T.A. Gillespie, B.T. Regg, K.F. Austin, J.E. Coutant, *Application of packed-capillary LC–MS with ESI to the study of the human biotransformation of the anti-emetic drug dolasetron*, J. Mass Spectrom., 31 (1996) 681.

10. J. Lee, S. La, B.R. Ahn, T.C. Jeong, D.-H. Kim, *Metabolism of YH3945, a novel anti-cancer drug, in rats using the ^{14}C-labeled compound*, Rapid Commun. Mass Spectrom., 18 (2004) 1901.

11. E.H. Kerns, K.J. Volk, S.E. Hill, M.S. Lee, *Profiling new taxanes using LC–MS–MS substructural analysis techniques*, Rapid Commun. Mass Spectrom., 9 (1995) 1539.

12. E.H. Kerns, R.A. Rourick, K.J. Volk, M.S. Lee, *Buspirone metabolite structure profile using a standard LC–MS protocol*, J. Chromatogr. B, 698 (1997) 133.

13. E. Gangl, I. Utkin, N. Gerber, P. Vouros, *Structural elucidation of metabolites of ritonavir and indinavir by LC–MS*, J. Chromatogr. A, 974 (2002) 91.

14. S.J. Choi, H.Y. Ji, S.K. Baek, H.-Y. Lee, J.-K. Choi, H.G. Cheon, D.H. Lee, H. Lim, H.S. Lee, *LC–MS–MS identification of in vitro metabolites of a new H^+/K^+ ATPase inhibitor, KR-60436 produced by rat and human liver microsomes*, J. Pharm. Biomed. Anal., 28 (2002) 897.

15. D.Q. Liu, T. Pereira, *Interference of a carbamoyl glucuronide metabolite in quantitative LC–MS–MS*, Rapid Commun. Mass Spectrom., 16 (2002) 142.

16. G.J. Dear, J. Ayrton, R. Plumb, I.J. Fraser, *Rapid identification of drug metabolites using capillary LC–ion-trap MS*, Rapid Commun. Mass Spectrom., 13 (1999) 456.

17. M. Lavén, K. Markides, B. Långström, *Analysis of microsomal metabolic stability using high-flow-rate extraction coupled to capillary LC–MS*, J. Chromatogr. B, 806 (2004) 119.

18. D.J. Foltz, J. Castro-Perez, P. Riley, J.R. Entwisle, T.R. Baker, *Narrow-bore sample trapping and LC combined with Q–TOF-MS for ultra-sensitive identification of in vivo and in vitro metabolites*, J. Chromatogr. B, 825 (2005) 144.

19. R.F. Staack, E. Varesio, G. Hopfgartner, *The combination of LC–MS–MS and chip-based infusion for improved screening and characterization of drug metabolites*, Rapid Commun. Mass Spectrom., 19 (2005) 618.

20. C.E.C.A. Hop, P.R. Tiller, L. Romanyshyn, *In-vitro metabolite identification using fast gradient LC–MS–MS*, Rapid Commun. Mass Spectrom., 16 (2002) 212.

21. G.J. Dear, R. Plumb, D. Mallett, *Use of monolithic silica columns to increase analytical throughput for metabolite identification by LC–MS–MS*, Rapid Commun. Mass Spectrom, 15 (2001) 152.

22. H.K. Lim, K.W. Chan, S. Sisenwine, J.A. Scatina, *Simultaneous screen for microsomal stability and metabolite profile by direct injection TFC–LC and automated MS–MS*, Anal. Chem., 73 (2001) 2140.

23. J.L. Herman, *The use of TFC and the isocratic focusing effect to achieve on-line cleanup and concentration of neat biological samples for low-level metabolite analysis*, Rapid Commun. Mass Spectrom., 19 (2005) 696.

24. J. Castro-Perez, R. Plumb, J.H. Granger, I. Beattie, K. Joncour, A. Wright, *Increasing throughput and information content for in-vitro drug metabolism experiments using ultra-performance LC–Q–TOF-MS*, Rapid Commun. Mass Spectrom., 19 (2005) 843.

25. G. Hopfgartner, W. Vetter, W. Meister, H. Ramuz, *Fragmentation of bosentan (Ro 47-0203) in ESI-MS after CID at low energy: a case of radical fragmentation of an even-electron ion*, J. Mass Spectrom., 31 (1996) 69.

26. R.J. Strife, L.C. Robosky, G. Garrett, M.M. Ketcha, J.D. Shaffer, N. Zhang, *Ion-trap*

MS^n genealogical mapping – approaches for structure elucidation of novel products of consecutive fragmentations of morphinans, Rapid Commun. Mass Spectrom., 14 (2000) 250.

27. H. Yoshitsugu, T. Fukuhara, M. Ishibashi, T. Nanbo, N. Kagi, *Key fragments for the identification of positional isomer pairs in glucuronides from the hydroxylated metabolites of RT-3003 (Vintoperol) by LC–MS*, J. Mass Spectrom., 34 (1999) 1063.

28. M. Jemal, Z. Ouyang, W. Zhao, M. Zhu, W.W. Wu, *A strategy for metabolite identification using triple-quadrupole MS with enhanced resolution and accurate mass capability*, Rapid Commun. Mass Spectrom., 17 (2003) 2732.

29. W. Blum, R. Aichholz, P. Ramstein, J. Kühnöl, J. Brüggen, T. O'Reilly, A. Flörsheimer, *In-vivo metabolism of epothilone B in tumor-bearing nude mice: identification of three new epothilone B metabolites by capillary LC–MS–MS*, Rapid Commun. Mass Spectrom., 15 (2001) 41.

30. P.J. Jackson, R.D. Brownsill, A.R. Taylor, B. Walther, *Use of ESI and neutral-loss LC–MS–MS in drug metabolism studies*, J. Mass Spectrom., 30 (1995) 446.

31. G. Hopfgartner, I.V. Chernushevich, T.R. Covey, J.B. Plomley, R. Bonner, *The exact mass measurement of product ions for the structural elucidation of drug metabolites with a Q–TOF-MS*, J. Am. Soc. Mass Spectrom., 10 (1999) 1305.

32. G. Hopfgartner and F. Vilbois, *The impact of accurate mass measurements using Q–TOF-MS on the characterization and screening of drug metabolites*, Analusis, 28 (2000) 906.

33. I. Sundström, M. Hedeland, U. Bondesson, P.E. Andrén, *Identification of glucuronide conjugates of ketobemidone and its phase I metabolites in human urine utilizing accurate mass and Q–TOF-MS*, J. Mass Spectrom., 37 (2002) 414.

34. A. Deroussent, M. Ré, H. Hoellinger, E. Vanquelef, O. Duval, M. Sonnier, T. Cresteil, *In vitro metabolism of ethoxidine by human CYP1A1 and rat microsomes: identification of metabolites by LC–ESI-MS–MS and accurate mass measurements by TOF-MS*, Rapid Commun. Mass Spectrom., 18 (2004) 474.

35. L. Leclercq, C. Delatour, I. Hoes, F. Brunelle, X. Labrique, J. Castro-Perez, *Use of a five-channel multiplexed ESI-Q–TOF-MS for metabolite identification*, Rapid Commun. Mass Spectrom., 19 (2005) 1611.

36. L.L. Lopez, X. Yu, D. Cui, M.R. Davis, *Identification of drug metabolites in biological matrices by intelligent automated LC–MS–MS*, Rapid Commun. Mass Spectrom., 12 (1998) 1756.

37. X. Yu, D. Cui, M.R. Davis, *Identification of in vitro metabolites of Indinavir by "intelligent automated LC–MS–MS" (INTAMS) utilizing triple-quadrupole MS–MS*, J. Am. Soc. Mass Spectrom., 10 (1999) 175.

38. M.R. Anari, R.I. Sanchez, R. Bakhtiar, R.B. Franklin, T.A. Baillie, *Integration of knowledge-based metabolic predictions with DDA LC–MS–MS for drug metabolism studies: application to studies on the biotransformation of indinavir*, Anal. Chem., 76 (2004) 823.

39. P.R. Tiller, A.P. Land, I. Jardine, D.M. Murphy, R. Sozio, A. Ayrton, W.H. Schaefer, *Application of LC–MS^n analyses to characterize novel glyburide metabolites formed in vitro*, J. Chromatogr. A, 794 (1998) 15.

40. E. Kantharaj, A. Tuytelaars, P.E.A. Proost, Z. Ongel, H.P. van Assouw, R.A.H.J. Gilissen, *Simultaneous measurement of drug metabolic stability and identification of metabolites using ion-trap MS*, Rapid Commun. Mass Spectrom., 17 (2003) 2661.

41. E. Kantharaj, P.B. Ehmer, A. Tuytelaars, A. Van Vlaslaer, C. Mackie, R.A.H.J. Gilissen, *Simultaneous measurement of metabolic stability and metabolite identification of 7-methoxymethylthiazolo[3,2-a]pyrimidin-5-one derivatives in human liver microsomes using LC–ion-trap-MS*, Rapid Commun. Mass Spectrom., 19 (2005) 1069.

42. A. Triolo, M. Altamura, T. Dimoulas, A. Guidi, A. Lecci, M. Tramontana, *In-vivo metabolite detection and identification in drug discovery via LC–MS–MS with DDA scanning and post-acquisition data mining*, J. Mass Spectrom., 40 (2005) 1572.

43. M. Wrona, T. Mauriala, K.P. Bateman, R.J. Mortishire-Smith, D. O'Connor, *'All-in-One' analysis for metabolite identification using LC–Q–TOF with collision energy switching*, Rapid Commun. Mass Spectrom., 19 (2005) 2597.

44. G. Hopfgartner, C. Husser, M. Zell, *Rapid screening and characterization of drug metabolites using a new Q–LIT-MS*, J. Mass Spectrom., 38 (2003) 138.

45. J.-p. Qiao, Z. Abliz, F.-m. Chu, P.-l. Hou, F. Liang, Y. Chang, Z.-r. Guo, *Application of a novel Q–LIT-MS to study the metabolism of 6-aminobutylphthalide in rat brains*, Rapid Commun. Mass Spectrom., 18 (2004) 3142.

46. A.C. Li, D. Alton, M.S. Bryant, W.Z. Shou, *Simultaneously quantifying parent drugs and screening for metabolites in plasma pharmacokinetic samples using SRM-DDA on a QTrap instrument*, Rapid Commun. Mass Spectrom., 19 (2005) 1943.

47. W.Z. Shou, L. Magis, A.C. Li, W. Naidong, M.S. Bryant, *A novel approach to perform metabolite screening during the quantitative LC–MS–MS analyses of in vitro metabolic stability samples using a Q–LIT-MS*, J. Mass Spectrom., 40 (2005) 1347.

48. N.J. Clarke, D. Rindgen, W.A. Korfmacher, K.A. Cox, *Systematic LC–MS metabolite identification in drug discovery. A four-step strategy to characterize metabolites by LC–MS techniques early in the pharmaceutical discovery process*, Anal. Chem., 73 (2001) 430A.

49. M.-Y. Zhang, N. Pace, E.H. Kerns, T. Kleintop, N. Kagan, T. Sakuma, *Q–LIT in fragmentation mechanism studies: application to structure elucidation of buspirone and one of its metabolites*, J. Mass Spectrom., 40 (2005) 1017.

50. W. Blum, T. Buhl, E. Altmann, J. Kuhnol, P. Ramstein, R. Aichholz, *Complementary use of ion trap and Q–TOF-MS in combination with capillary LC: early characterization of in vivo metabolites of the cathepsin K inhibitor NVP-AAV490 in rat*, J. Chromatogr. B, 787 (2003) 255.

51. D.A. Jayawickrama, J.V. Sweedler, *Hyphenation of capillary separations with NMR*, J. Chromatogr. A, 1000 (2003) 819.

52. J.P. Shockcor, S.E. Unger, I.D. Wilson, P.J.D. Foxall, J.K. Nicholson, J.C. Lindon, *Combined HPLC, NMR, and ion-trap MS with application to the detection and characterization of xenobiotic and endogenous metabolites in human urine*, Anal. Chem., 68 (1996) 4431.

53. K.I. Burton, J.R. Everett, M.J. Newman, F.S. Pullen, D.S. Richards, A.G. Swanson, *On-line LC–NMR–MS: a new technique for drug metabolism structure elucidation*, J. Pharm. Biomed. Anal., 15 (1997) 1903.

54. G.J. Dear, J. Ayrton, R. Plumb, B.C. Sweatman, I.M. Ismail, I.J. Fraser, P.J. Mutch, *A rapid and efficient approach to metabolite identification using NMR, LC–MS and LC–NMR–MSⁿ*, Rapid Commun. Mass Spectrom., 12 (1998) 2023.

55. G.J. Dear, R.S. Plumb, B.C. Sweatman, J. Ayrton, J.C. Lindon, J.K. Nicholson, I.M. Ismail, *Mass directed peak selection, an efficient method of drug metabolite*

identification using LC–NMR–MS, J. Chromatogr. B, 748 (2000) 281.

56. M. Godejohann, L.-H. Tseng, U. Braumann, J. Fuchser, M. Spraul, *Characterization of a paracetamol metabolite using on-line LC–SPE–NMR–MS and a cryogenic NMR probe*, J. Chromatogr. A, 1058 (2004) 191.

57. D. Bao, V. Thanabal, W.F. Pool, *Determination of tacrine metabolites in microsomal incubate by LC–NMR–MS with a column trapping system*, J. Pharm. Biomed. Anal., 28 (2002) 23.

58. N. Ohashi, S. Furuuchi, M. Yoshikawa, *Usefulness of H/D exchange in the study of drug metabolism using LC–MS–MS*, J. Pharm. Biomed. Anal., 18 (1998) 325.

59. D.Q. Liu, C.E.C.A. Hop, M.G. Beconi, A. Mao, S.-H. Lee Chiu, *Use of on-line H/D exchange to facilitate metabolite identification*, Rapid Commun. Mass Spectrom., 15 (2001) 1823.

60. D.Q. Liu, C.E.C.A. Hop, *Strategies for characterization of drug metabolites using LC–MS–MS in conjunction with chemical derivatization and on-line H/D exchange approaches*, J. Pharm. Biomed. Anal., 37 (2005) 1.

61. G.N. Thompson, P.J. Pacy, G.C. Ford, D. Halliday, *Practical considerations in the use of stable isotope labelled compounds as tracers in clinical studies*, Biomed. Environ. Mass Spectrom., 18 (1989) 321.

62. L. Weidolf, T.R. Covey, *Studies on the metabolism of omeprazole in the rat using LC–ESI-MS and the isotope cluster technique with [^{34}S]omeprazole*, Rapid Commun. Mass Spectrom., 6 (1992) 192.

63. Z. Yan, G.W. Caldwell, *Stable-isotope trapping and high-throughput screenings of reactive metabolites using the isotope MS signature*, Anal. Chem., 76 (2004) 6835.

64. Z. Yan, N. Maher, R. Torres, G.W. Caldwell, N. Huebert, *Rapid detection and characterization of minor reactive metabolites using stable-isotope trapping in combination with MS–MS*, Rapid Commun. Mass Spectrom., 19 (2005) 3322.

65. A. Mutlib, W. Lam, J. Atherton, H. Chen, P. Galatsis, W. Stolle, *Application of stable isotope labeled glutathione and rapid scanning MS in detecting and characterizing reactive metabolites*, Rapid Commun. Mass Spectrom., 19 (2005) 3482.

66. J.K. Nicholson, J.C. Lindon, G. Scarfe, I.D. Wilson, F. Abou-Shakra, J. Castro-Perez, A. Eaton, S. Preece, *HPLC-ICP-MS for the analysis of xenobiotic metabolites in rat urine: application to the metabolites of 4-bromoaniline*, Analyst, 125 (2000) 235.

67. O. Corcoran, J.K. Nicholson, E.M. Lenz, F. Abou-Shakra, J. Castro-Perez, A.B. Sage, I.D. Wilson, *Directly coupled LC–ICP-MS and TOF-MS for the identification of drug metabolites in urine: application to diclofenac using chlorine and sulfur detection*, Rapid Commun. Mass Spectrom., 14 (2000) 2377

68. P.S. Marshall, B. Leavens, O. Heudi, C. Ramirez-Molina, *LC–ICP-MS in the pharmaceutical industry: selected examples*, J. Chromatogr. A, 1056 (2004) 3.

69. S. M. R. Stanley, *Equine metabolism of buspirone studied by LC–MS*, J. Mass Spectrom., 35 (2000) 402.

70. K. Levsen, H.-M. Schiebel, B. Behnke, R. Dötzer, W. Dreher, M. Elend, H. Thiele, *Structure elucidation of Phase II metabolites by MS–MS: An overview*, J. Chromatogr. A, 1067 (2005) 55.

71. Y.-Q. Xia, J.D. Miller, R. Bakhtiar, R.B. Franklin, D.Q. Liu, *Use of a Q–LIT-MS in metabolite identification and bioanalysis*, Rapid Commun. Mass Spectrom., 17 (2003) 1137.

72. R. Brownshill, J.-P. Combal, A. Taylor, M. Bertrand, W. Luijten, B. Walther, *The*

application of ESI and neutral-loss MS to the identification of metabolites of S12813 in rat liver slices, Rapid Commun. Mass Spectrom., 8 (1994) 361.

73. D.Q. Liu, Y.-Q. Xia, R. Bakhtiar, *Use of a LC–ion trap MS/triple quadrupole MS system for metabolite identification*, Rapid Commun. Mass Spectrom., 16 (2002) 1330.

74. J. Castro-Perez, R. Plumb, L. Liang, E. Yang, *A high-throughput LC–MS–MS method for screening glutathione conjugates using exact mass neutral loss acquisition*, Rapid Commun. Mass Spectrom., 19 (2005) 798.

75. G. K. Poon, G. Kwei, R. Wang, K. Lyons, Q. Chen, V. Didolkar, C. E. C. A. Hop, *Integrating qualitative and quantitative LC–MS analysis to support drug discovery*, Rapid Commun. Mass Spectrom., 13 (1999) 1943.

76. M. Gu, H.-K. Lim, *An DDA system for simultaneous screening of microsomal stability and metabolite profiling by LC–MS*, J. Mass Spectrom., 36 (2001) 1053.

77. U. Jurva, A.P. Bruins, H.V. Wikström, *In vitro mimicry of metabolic oxidation reactions by electrochemistry–MS*, Rapid Commun. Mass Spectrom., 14 (2000) 529.

78. U. Jurva, H.V. Wikström, L. Weidolf, A.P. Bruins, *Comparison between electrochemistry–MS and cytochrome P450 catalyzed oxidation reactions*, Rapid Commun. Mass Spectrom., 17 (2003) 800.

79. R.B. van Breemen, D. Nikolic, J.L.Bolton, *Metabolic screening using on-line ultrafiltration MS*, Drug Metab. Disp., 26 (1998) 85.

80. B.M. Johnson, D. Nikolic and R.B. van Breemen, *Applications of PUF-MS*, Mass Spectrom. Rev., 21 (2002) 76.

81. W.A. Korfmacher, C.A. Palmer, C. Nardo, K. Dunn-Meynell, D. Grotz, K. Cox, C.-C. Lin, C. Elicone, C. Liu, E. Duchoslav, *Development of an automated MS system for the quantitative analysis of liver microsomal incubation samples: a tool for rapid screening of new compounds for metabolic stability*, Rapid Commun. Mass Spectrom., 13 (1999) 901.

82. P.R. Tiller, L.A. Romanyshyn, *LC–MS–MS quantification with metabolite screening as a strategy to enhance the early drug discovery process*, Rapid Commun. Mass Spectrom., 16 (2002) 1225.

83. L.E. Chovan, C. Black-Schaefer, P.J. Dandliker, Y.Y. Lau, *Automatic MS method development for drug discovery: application in metabolic stability assays*, Rapid Commun. Mass Spectrom., 18 (2004) 3105.

84. G.W. Caldwell, S.M. Easlick, J. Gunnet, J.A. Masucci, K. Demarest, *In-vitro permeability of eight β-blockers through Caco-2 monolayers utilizing LC–ESI-MS*, J. Mass Spectrom., 33 (1998) 607.

85. H.-Z. Bu, M. Poglod, R.G. Micetich, J.K. Khan, *High-throughput Caco-2 cell permeability screening by cassette dosing and sample pooling approaches using direct injection–on-line guard cartridge extraction–MS–MS*, Rapid Commun. Mass Spectrom., 14 (2000) 523.

86. E.N. Fung, I. Chu, C. Li, T. Liu, A. Soares, R. Morrison, A.A. Nomeir, *Higher-throughput screening for Caco-2 permeability utilizing a MUX LC–MS–MS*, Rapid Commun. Mass Spectrom., 17 (2003) 2147.

87. C.K. Van Pelt, S. Zhang, E. Fung, I. Chu, T. Liu, C. Li, W.A. Korfmacher, J. Henion, *A fully automated nano-ESI-MS–MS method for analysis of Caco-2 samples*, Rapid Commun. Mass Spectrom., 17 (2003) 1573.

88. J. Ayrton, R. Plumb, W. J. Leavens, D. Mallett, M. Dickins, G. J. Dear, *Application*

of a generic fast gradient LC–MS–MS method for the analysis of cytochrome P450 probe substrates, Rapid Commun. Mass Spectrom., 12 (1998) 217.

89.   H.-Z. Bu, L. Magis, K. Knuth, P. Teitelbaum, *High-throughput CYP inhibition screening via cassette probe-dosing strategy. I. Development of direct injection–on-line guard cartridge extraction–MS–MS for the simultaneous detection of CYP probe substrates and their metabolites*, Rapid Commun. Mass Spectrom., 14 (2000) 1619.

90.   H.-Z. Bu, L. Magis, K. Knuth, P. Teitelbaum, *High-throughput CYP inhibition screening via cassette probe-dosing strategy - II. Validation of a direct injection–on-line guard cartridge extraction–MS–MS method for CYP2D6 inhibition assessment*, J. Chromatogr. B, 753 (2001) 321.

91.   H.-Z. Bu, K. Knuth, L. Magis, P. Teitelbaum, *High-throughput CYP inhibition screening via cassette probe-dosing strategy: III. Validation of a direct injection–on-line guard cartridge extraction–MS–MS method for CYP2C19 inhibition evaluation*, J. Pharm. Biomed. Anal., 25 (2001) 437.

92.   H.-Z. Bu, K. Knuth, L. Magis, P. Teitelbaum, *High-throughput CYP inhibition screening via cassette probe-dosing strategy. IV. Validation of a direct injection–on-line guard cartridge extraction–MS–MS method for simultaneous CYP3A4, 2D6 and 2E1 inhibition assessment*, Rapid Commun. Mass Spectrom., 14 (2000) 1943.

93.   S.X. Peng, A.G. Barbone, D.M. Ritchie, *High-throughput cytochrome P450 inhibition assays by ultrafast gradient LC–MS–MS using monolithic columns*, Rapid Commun. Mass Spectrom., 17 (2003) 509.

94.   K.M. Jenkins, R. Angeles, M.T. Quintos, R. Xu, D.B. Kassel, R.A. Rourick, *Automated high throughput ADME assays for metabolic stability and cytochrome P450 inhibition profiling of combinatorial libraries*, J. Pharm. Biomed. Anal., 34 (2004) 989.

11

QUANTITATIVE BIOANALYSIS USING LC–MS

1. Introduction

Quantitation of drugs and their metabolites in biological matrices currently is one of the most important applications of LC–MS. This can be attributed to the greatly enhanced selectivity and reliability of the analysis, as compared to LC–UV. Selective reaction monitoring (SRM) in tandem mass spectrometry (MS–MS) has become the method-of-choice in quantitative bioanalysis. Ample examples are described in literature. Even a larger number of successful examples are hidden in the archives of pharmaceutical companies and contract research organizations.

The use of high-speed LC–MS–MS with a heated-nebulizer atmospheric-pressure chemical ionization (APCI) interface in a bioassay for phenylbutazone and its metabolites in equine urine and plasma, described by Covey *et al.* [1] in 1986, can be considered as a breakthrough in LC–MS. Phenylbutazone, hydroxy-phenylbutazone, and oxy-phenylbutazone were separated on a 33×4.6-mm-ID column packed with 3-μm particles in only 1 min. SRM of four product ions per analyte was applied. The results for phenylbutazone in a 60-sample, 48-h phenylbutazone pharmacokinetic study are shown in Figure 11.1. A set of 5 standards, 11 plasma samples, and 11 urine samples were analysed *in duplo* within 60 min [1].

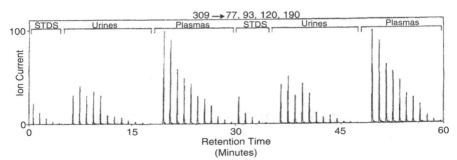

Figure 11.1: Ion current of the phenylbutazone daughter ions for a 60-sample, 48-h pharmacokinetic study in the horse using high-speed LC–MS–MS. Reprinted from [1] with permission, ©1986, American Chemical Society.

In this application, the potential of the reliable and robust performance of atmospheric-pressure ionization (API) in LC–MS and the selectivity gain by the use of SRM became evident. The high selectivity of SRM enables significant reduction of the chromatographic analysis time. Some review papers were published [2-4].

2. General considerations

Quantitative bioanalysis plays an important role in the various stages of the drug development process ([5], Ch. 9.1).

In the first stage, *i.e.*, drug discovery and lead optimization, quantitative bioanalysis is involved in early-ADME studies, where absorption, distribution, metabolism, and elimination of potential lead compounds is investigated. A number of experiments are involved. Experiments to evaluate drug absorption include membrane permeability tests using Caco-2 cell monolayers (Ch. 10.6.2) and *in vivo* pharmacokinetic profiling, to assess bioavailability of the drug. Experiments to study drug distribution include evaluation of *in vitro* protein binding and *in vivo* tissue distribution. Metabolic stability screening and cytochrome P450 inhibition screening (Ch. 10.6.1) are performed to obtain a preliminary evaluation of drug metabolism. Elimination of drugs and their metabolites is evaluated by means of quantitative bioanalysis of drugs in biological fluids.

In the second stage, *i.e.*, preclinical studies, further ADME studies are performed. In addition, attention is paid to metabolite identification (Ch. 10.3-5) and metabolite profiling.

In the third stage, *i.e.*, the various phases of clinical trials, pharmacokinetics is studied in detail in order to obtain the therapeutic index, to study drug-drug interactions, and to design dosage regimes.

In the preclinical and clinical stages, quantitative bioanalysis is done in a way different from the drug discovery stage. The discovery stage is characterized by a high throughput of different compounds. For each target, a limited number of samples have to be analysed. Generic methods are preferred, because proper method development would take too long and would be too expensive. Preclinical and clinical bioanalysis concentrates on a limited number of compounds. For each target, a large number of samples have to be analysed. Proper method development is required to optimize the sample throughput and the quality of the results. The analysis is performed according to Good Laboratory Practice (GLP) rules.

In addition, quantitative bioanalysis is involved in therapeutic drug monitoring in cases, where the drug administered has a low therapeutic index and/or due side effects and/or toxic metabolites, a narrow safe plasma concentration, or when patient compliance with a certain drug must be evaluated (Ch. 12.2).

In most applications, the bioanalysis involves the analysis of a number of drugs, or one drug and (some of) its metabolites in biological fluids, especially whole blood, plasma, serum, or urine. However, other matrices are studied as well: various tissues (skin, liver, brain, thyroid gland), faeces, hair, tear fluid, cerebrospinal fluid, semen. In most studies, the analysis of samples from human origin or from rats is performed, although the analysis of samples from rabbits, mice, minipigs, dogs, and monkeys is also performed.

For an analytical method to deliver valid analytical results, proper characterization and validation of the method is necessary. In drug development, and especially with respect to preclinical and clinical studies, the generated data will be part of the submission files for approval and registration of the new drug. This means that GLP and FDA guidelines must be fulfilled [6]. Obviously, validation issues are dealt with differently in drug discovery stages, as compared to preclinical and clinical trials.

3. Method development

In the past years, method development for quantitative bioanalysis using LC–MS has changed significantly. While in the past most attention was given to finding the optimum mobile-phase composition for the ionization technique selected, nowadays it is realized that parameters related to sample pretreatment, chromatography, analyte ionization, and mass spectrometric analysis are all strongly interrelated. One cannot change one parameter without influencing many others. As is discussed in Ch. 11.5.3, the choice between electrospray ionization (ESI), APCI, or another ionization technique should be made on the basis of analytical results with spiked pretreated samples. Only in this way the matrix effects can be properly evaluated. In addition, optimum solvent conditions in analyte ionization are generally less important that achieving an appropriate chromatographic separation. Some important issues in method development are discussed below.

3.1   Sample pretreatment

A wide variety of sample pretreatment methods are applied in quantitative bioanalysis using LC–MS. The importance of sample pretreatment is evident from the discussion on matrix effects in Ch. 11.5. In general, the three major goals of sample pretreatment are [7]:

- Removal of unwanted matrix components, that may interfere in the LC–MS analysis, *e.g.*, compounds with high surface activity and nonvolatile compounds like proteins and salts.
- Preconcentrate or enrich the analyte to improve the limits of quantitation.
- Exchange the analyte into a (more) favourable solvent composition.

Obviously, the sample pretreatment should avoid hydrolytic, enzymatic, or any other degradation of the target analytes. The most important sample pretreatment methods used in quantitative bioanalysis are:

- Dilution, filtration (Ch. 11.6.1), or protein removal by ultrafiltration.
- Protein precipitation (Ch. 11.6.2).
- Liquid-liquid extraction (LLE, Ch. 11.6.3).
- Solid-phase extraction (SPE), which can be done off-line or on-line by means of cartridges, cartridge columns, or extraction disks (Ch. 11.6.4), and eventually by using special packing materials, such as restricted access materials (RAM) or turbulent-flow chromatography (TFC) (Ch. 11.6.5).

Examples of bioanalytical LC–MS involving these sample-pretreatment strategies are given below. Sometimes, various sample pretreatment strategies must be combined to achieve the desired results, *e.g.*, a combination of protein precipitation and SPE.

In some cases, precolumn analyte derivatization is applied in bioanalytical LC–MS. Derivatization of L-368,899 with trifluoroacetic anhydride was performed to eliminate carryover and adsorption effects [8]. Derivatization of 5-fluorouracil with 4-bromomethyl-7-methoxycoumarin [9] and conversion of sampatrilat to its dimethyl ester [10] was performed in order to enhance the ionization efficiency.

3.2   Chromatography

The aim of the LC step is the separation of the target compounds from each other and from matrix interferences. In bioanalytical LC–MS, this means that at least isobaric and/or isomeric target compounds must be separated from each other, and that the target compounds must be separated from metabolites and other related substances that may interfere.

The separation must be performed using a mobile phase that is compatible with the selected ionization method, which at least means that preferably no nonvolatile mobile-phase additives should be applied (Ch. 6.6.3). In some cases, additives must be used to mask secondary stationary-phase effects. Some mobile-phase issues for generic LC–MS were addressed by Law and Temesi [11].

Reversed-phase LC (RPLC) is applied most often in LC–MS, although attention is also paid to the use of hydrophilic interaction chromatography (HILIC, Ch. 11.7.3) and the need for chiral separations (see Ch. 11.7.4).

The column inner diameter is determined by the amount of sample available and the LC–MS interface selected. In general, flow-rates between 200 and 400 µl/min are considered optimum for (pneumatically-assisted) ESI. This explains the frequent use of 2-mm-ID columns. In sample-limited analysis, e.g., in the analysis of mouse plasma samples, microbore (1 mm ID) or packed-microcapillary columns (320 µm ID) are applied at relatively low flow-rates [12-13]. For APCI, 4.6-mm-ID columns are preferred, operated at typically 1 ml/min. The LC system should provide symmetric peaks with a width that enables the acquisition of 10–20 data points for each compound in order to enable an accurate determination of the peak area.

Short columns (20–50 mm length) are generally applied, which provide only limited chromatographic resolution. This may lead to erroneous data, e.g., when the existence of various hydroxylated metabolites is ignored, e.g., in the bioanalysis of risperidone and its 9-hydroxy metabolite [14] (Ch. 11.4.2), or with co-eluting labile metabolites. In the fast gradient LC–MS analysis of a particular drug, a shoulder was observed, which was first attributed to column deterioration [15]. When a more shallow gradient was applied, the shoulder was found to be a separate compound, i.e., a carbamoyl glucuronide metabolite of the parent drug. This was found to readily fragment in APCI or ESI. One of the fragments was the parent drug itself. Unless the metabolite is separated from the parent drug, this will lead to an overestimation of the parent drug concentration. Other compounds that are prone to this type of behaviour are lactone rings and their open-ring acids, the E- and Z-isomers of methoximes, thiols and their disulfides, phenolic drugs and their prodrugs, and amine drugs and their N-oxides [16].

A bioanalytical method is generally developed using control plasma or urine. However, in the analysis of post-dose samples coeluting metabolites may significantly disturb the accuracy of the method [8, 17]. Matuszewski et al. [8] reported 4–7-times higher peak areas of the analogue internal standard in post-dose urine samples than in control urine samples. By changing the chromatographic conditions, at least seven early eluting isobaric metabolites were detected.

3.3   Choice between ESI and APCI

Unless the analyte under investigation is thermally labile, most drugs can be analysed with either ESI or APCI. Depending on basic and/or acidic properties, either the positive-ion or the negative-ion mode is applied. ESI is applied in the majority of the studies. However, various studies indicate that APCI is less prone to matrix effects than ESI. For this reason, the use of APCI instead of ESI should be considered, whenever the quantitation limits and the linear dynamic range achievable by LC–APCI–MS are adequate (see Ch. 11.5).

Recently, atmospheric-pressure photoionization (APPI) was introduced and applied in a few bioanalytical studies, e.g., in the determination of idoxifene and its metabolites [18]. The influence of mobile-phase composition and flow-rate as well as matrix effects in the APPI analysis of clozapine and lonafarnib was investigated by Hsieh et al. [19]. Subsequently, LC–APPI–MS was applied a pharmacokinetic studies of 42 drug candidates in rats.

3.4   Internal standard

In order to obtain sufficient accurate results with good precision, an internal standard (IS) must be used in quantitative bioanalysis using LC–MS. Either an isotopically-labelled internal standard (ILIS) or an analogue internal standard (ANIS) can be applied.

The ILIS is the ideal IS, because its physicochemical properties are (almost) identical to those of the analyte. In selecting an ILIS, adequate separation between the isotopic patterns is required. In addition, the ILIS selected should have an isotopic purity with respect to the nonlabelled compound (D_0) of better than 99.9%. The labels must be resistant towards exchange in the chemical environments they have to exist in. Problems with partial D-loss by H/D-exchange in aqueous solvents were met in the use of $[^{13}CD_3]$-labelled analogues of rofecoxib [20]. A $[^{13}C_7]$-labelled analogue was applied instead.

With respect to ANIS, LC–MS does not pose additional selection rules compared to other techniques. The ANIS should be as much alike as possible. Therefore, a small difference in aliphatic substitution is generally preferred over changes in conjugation, or changes in the number or nature of the polar groups. With an ANIS, proper attention should be paid to the possibility of common fragment or adduct ions. Almost co-elution of ANIS and analyte provides the best results.

It has been frequently demonstrated that an ILIS provides (somewhat) better precision than an ANIS, e.g., [14, 21-22]. It is also generally believed that ion-suppression by matrix effects can always be corrected for by an ILIS. This is generally true. However, in the bioanalysis of mevalonic acid in human plasma and urine, Jemal et al. [23] observed that when large sample volumes of certain batches of urine were analysed, the response ratio of the analyte and its $[D_7]$-ILIS changed from the expected value. This indicates that the analyte and its ILIS can be differently affected by the matrix.

Suppression of the response of the internal standard by increasing concentrations of a co-eluting analyte in a calibration series was studied by Sojo et al. [24] using fexofenadine and its $[D_6]$-analogue, dapsone and its $[D_4]$-analogue, and pseudoephidrine and its $[D_3]$-analogue. The effects were observed at flow-rates higher than 100 µl/min in ESI-MS for all analyte pairs in solvent-based calibration samples, and for fexofenadine also in extracted plasma samples.

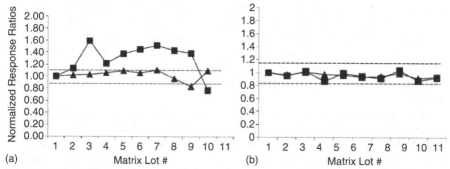

Figure 11.2: Differential suppression test for two potential internal standards (■ and ▲), plasma samples from ten different sources, and sample pretreatment by means of (a) LLE, and (b) SPE. Reprinted from [26] with permission. ©2002, John Wiley & Sons, Ltd.

Ion enhancement and suppression effects between target compounds and their co-eluting ILIS was investigated for both APCI-MS and ESI-MS [25]. In APCI-MS, seven out of nine pairs were found to enhance the response, while in ESI-MS all nine pairs were found to suppress each other's response. Therefore, it should be validated whether the analyte response factor is constant, *e.g.*, within 5%RSD, over the calibration range at the ILIS concentration chosen.

Differential ion suppression tests were proposed by Avery [26] as a tool in selecting an appropriate ANIS. After the pretreatment of blank plasma samples from various sources with the method of choice (protein precipitation, SPE, and LLE were tested), a 1:1 mixture of the analyte and a potential ANIS is added. The samples are analysed by LC–MS. Differential suppression can be monitored by plotting the analyte–ANIS ratio, normalized to a commercial control sample. Results of the differential suppression tests for two potential ANIS after LLE and SPE of the ten different plasma samples are shown in Figure 11.2.

3.5   Mass spectrometry

In quantitative bioanalysis, the application of SRM in a triple-quadrupole instrument is the method-of-choice. SRM provides ultimate selectivity, because only compounds are detected, which after sample pretreatment of the matrix are detected within a selected retention time window in the chromatogram as the structure-specific fragment at a particular m/z in MS_2, generated by collision-induced dissociation (CID) from a precursor ion with a particular m/z of the compound of interest, selected in MS_1. The ultimate selectivity results in good sensitivity and low detection limits. Software procedures for the automatic optimization of instrument parameters for SRM have been reported, *e.g.*, [27-28]. Such procedures are now commercially available in the software of the various instrument manufacturers.

In developing an SRM method, a number of issues must be considered.

The precursor ion is preferentially selected at unit-mass resolution or better (≤ 0.7 *m/z*-unit full width at half maximum (FWHM)). When the detection limit achieved is not adequate, for selectivity reasons it is better to degrade the resolution in MS_2 than in MS_1.

A triple-quadrupole instrument with enhanced-resolution possibilities was introduced by Thermo Electron. It has been demonstrated [29-32] that the (only) twofold loss in response at higher resolution (FWHM 0.1–0.2 *m/z*-unit instead of 0.7) is compensated by the reduced chemical noise. In a comparison between unit-mass and enhanced resolution, a twofold improvement in sensitivity was found [31] (Ch.11.4.5). As a result, similar or better detection limits can be achieved. In an evaluation of the performance of such an instrument in quantitative bioanalysis, the importance of an accurate precursor-ion selection in MS_1 was emphasized: if the precursor ion selection is done with a 0.2 *m/z*-unit wide window, the *m/z* set must be at the accurate *m/z* ($\pm$ 0.1 unit) of the analyte [30].

In principle, a longer dwell time leads to an improved signal-to-noise ratio (S/N). In practice, a compromise must be struck. The dwell time at the relevant *m/z* must be optimized to achieve 10–20 data-points over the chromatographic peak.

Crosstalk between SRM transitions may occur if short dwells time are applied in the analysis of compounds with the same product ion *m/z*. Special collision cells have been designed to reduce crosstalk and allow the use of very short dwell times without reduction in response. This is important in high-throughput bioanalysis. A linear accelerating high-pressure collision cell (LINAC), featuring a reduction of the rod distance from begin to end as well as an axial DC field to accelerate the ions through the collision cell, was introduced for this purpose by Sciex. Waters introduced a travelling-wave ion tunnel device, where the ions are pushed through by switching-on voltages at consecutive plates in the ion tunnel device. Both devices greatly diminish crosstalk and allow dwell times as low as 10 ms without compromising the analyte response.

The SRM transitions for the analyte and its IS are preferentially selected as a common neutral loss. In many automatic optimization procedures for SRM, the selection of the SRM transition is based on the maximum response. However, from selectivity point of view, additional criteria based on structure specificity, selectivity, and enhanced S/N in the analysis of real samples are important. The selection of an SRM transition by evaluating the background noise and the absence of interferences was reported by Woolf *et al.* [33] for the bioanalysis of an indinavir metabolite.

The use of two SRM transitions, one with an intense fragment ion for the low end of the calibration and one with a less intense fragment ion for the high end, has been proposed to expand the linear dynamic range of a bioanalytical method [34]. Proper attention has to be paid to validation issues related to this approach.

Although SRM in a triple-quadrupole instrument is generally preferred, some applications have been reported based on the use of selected-ion monitoring (SIM). An obvious reason for the use of SIM instead of SRM seems to be the lack of fragmentation of the selected precursor ion, *e.g.*, when this is a sodiated molecule $[M+Na]^+$ [35-36]. However, even in such cases SRM is often preferred over SIM, as an improvement in S/N due to the fragmentation of isobaric and/or isomeric interference is achieved, although the analyte itself is not fragmented [37]. Other examples of the use of SIM in bioanalytical LC–MS are the determination of nefazodone and its metabolites [38] and celecoxib [39] in human plasma. Constanzer *et al.* [40] compared SIM and SRM in the analysis of the substance C inhibitor aprepitant. In terms of absolute response for a low calibration standard, the SIM mode was two-times more sensitive, while the S/N in SRM was two-times better. Good cross validation between SIM and SRM was observed. Both instruments could be applied to achieve similar precision and accuracy in the concentration range between 10 and 5000 ng/ml. However, in SRM, a lower limit of quantification (LOQ) of 1 ng/ml could be achieved.

As an alternative to a triple-quadrupole, ion-trap instruments have been used in some cases. Due to the limited number of quantitative applications [41], the potential of quadrupole–linear-ion-trap hybrid (Q–LIT) cannot yet be evaluated.

4. Selected applications

Quantitative bioanalysis has been reported for a wide variety of drugs and their metabolites for various stages of drug development. In this section, procedures and results for five compounds (classes) are reviewed. The five compounds were selected, because a number of studies were reported for each compound. This allows to illustrate the diversity in strategies in quantitative bioanalysis.

4.1 Reserpine

Reserpine is an alkaloid used in the treatment of hypertension and psychosis in humans, but also as a tranquillizer in horses.

In order to study the pharmacokinetics of reserpine, reliable and fast analytical methodology is required. Reserpine can be analysed by LC with fluorescence detection down to 100 pg/ml. As a basic drug, reserpine is ideally suited for analysis by ESI or APCI. The quantitative bioanalysis of reserpine in equine plasma was reported using rescinnamine as ANIS [42]. In the method development, two sample pretreatment methods were compared: LLE and SPE. The three-step LLE procedure provided an acceptable recovery of 60% at 10 pg/ml, but is tedious and time consuming (sample throughput: 60 samples in ~2 days). SPE utilizing membrane disks was automated using a Gilson ASPEC (60 samples overnight) with recoveries between 52 and 58% in the range of 100–500 pg/ml. With LC–ESI-MS–MS in SRM mode, monitoring m/z 609→195 for reserpine and m/z 635→221 for the IS, the LOQ was 50 pg/ml with LLE pretreatment and 200 pg/ml with SPE.

A similar method was developed for the analysis of reserpine in FVB/N mouse plasma [43]. The use of a semi-automated 96-well LLE procedure enabled a reduction of the pretreatment time down to ~3 h for 96 samples! In order to solve protein-binding issues, disodium EDTA was employed as a protein-bound release agent. The LOQ was 20 pg/ml, utilizing 0.1 ml plasma.

4.2   Risperidone

Risperidone is an antipsychotic agent. The compound is metabolized into two major metabolites, of which the 9-hydroxy analogue is biologically active and the 7-hydroxy analogue is not. Bioanalytical methods for risperidone should have a sub-ng/ml LOQ, discriminate between the 7- and 9-hydroxy metabolites, and should be selective towards other psychotropic drugs, which may be used simultaneously by patients.

The simultaneous determination of risperidone and its 9-hydroxy metabolite by LC–MS was reported [44]. The analytes were extracted from plasma (adjusted to pH 10.5) by LLE with 15% dichloromethane in pentane. LC separation was achieved on a 50×4.6-mm-ID phenyl-hexyl column (5 µm) using an isocratic mobile phase consisting of 50% acetonitrile, 45% methanol, and 5% 0.15 mmol/l aqueous ammonium acetate (AmOAc). Positive-ion ESI-MS was applied in SRM mode. Good linearity was obtained between 0.1 and 100 ng/ml. The LOQ was 0.1 ng/ml. The method was applied to study pharmacokinetic parameters and for therapeutic drug monitoring in patients.

In an alternative method, special attention was paid to the separation of the 9-hydroxy from the isomeric 7-hydroxy metabolite [14]. Because the LLE described before [44] resulted in the co-extraction of basic endogenous compounds, SPE at pH 6 on a mixed-mode hydrophobic–cation exchange material (Oasis HLB) was applied. An acetonitrile–0.01 mmol/l aqueous ammonium formate (pH 4) gradient was applied on a 100×4.6-mm-ID C_{18}-column (3 μm). Both ILIS and ANIS were tested. Chromatograms of plasma spiked at 0.1 ng/ml with risperidone and the 9-hydroxy metabolite, as well as plasma taken from a volunteer 3 h after oral administration of 2 mg risperidone, are shown in Figure 11.3. Neglecting the non-active 7-hydroxy metabolite results in a ~6% overestimation of the biologically-active 9-hydroxy metabolite. The method was validated according to FDA protocols.

In another study [45], the quantitation of risperidone and its 9-hydroxy metabolite in EDTA-anticoagulated plasma was reported. After protein precipitation with acetonitrile, the supernatant was injected onto an on-line SPE–LC–MS system. The sample was loaded in a weak solvent (15% acetonitrile in 10 mmol/l aqueous AmOAc) to a 12.5×4.6-mm-ID C_{18} SPE-column (5 μm) for 1 min at 0.7 ml/min. After valve-switching, the SPE column was backflushed with a stronger solvent (80% acetonitrile in 10 mmol/l aqueous AmOAc) to the 30×2.1-mm-ID C_{18} analytical column (3.5 μm) at 0.35 ml/min. Positive-ion ESI-MS was applied in SRM mode, monitoring transitions for risperidone, 9-hydroxy risperidone, and an ANIS. No separation of the 7-hydroxy and 9-hydroxy metabolites was achieved under these conditions. The method was validated.

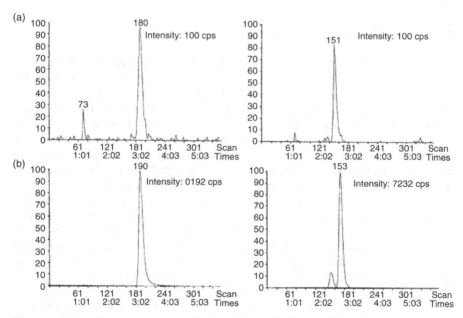

Figure 11.3: Chromatograms of risperidone (left panels) and monohydroxy-risperidone (right panels). (a) plasma spiked at 0.1 ng/ml with risperidone and 9-hydroxy risperidone, and (b) plasma from a volunteer 3 h after oral administration of 2 mg risperidone, showing the separation between the 7-hydroxy and 9-hydroxy metabolite. Reprinted from [14] with permission. ©2003, Elsevier Science BV.

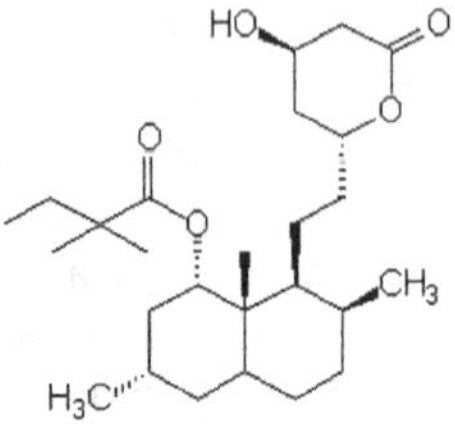

Simvastatin (Lactone)

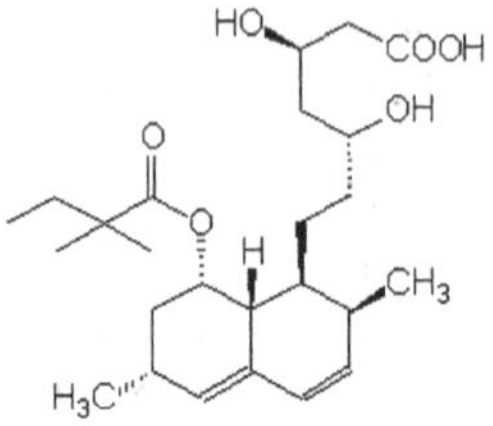

Simvastatin (β-hydroxy acid)

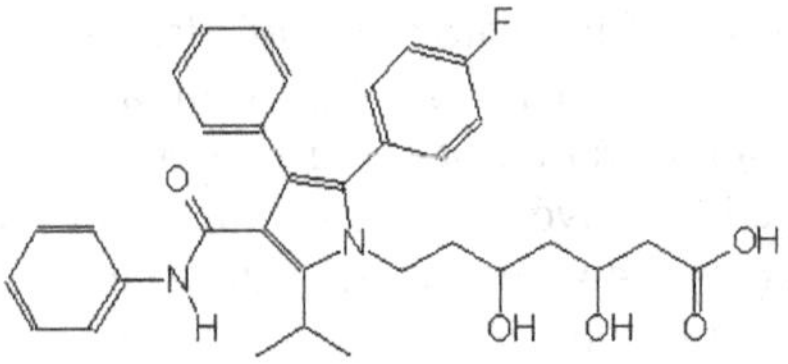

Atorvastatin

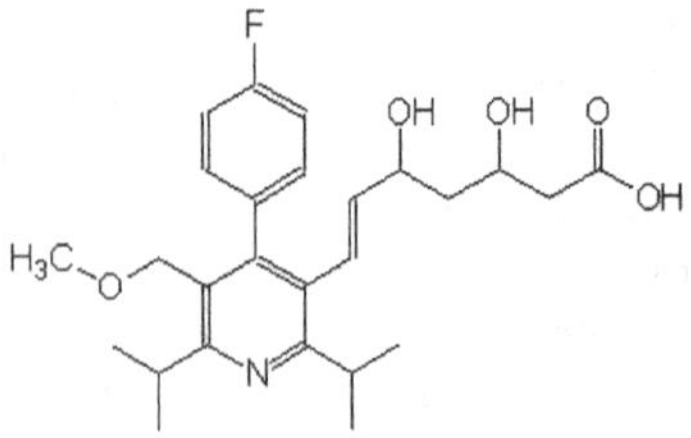

Cerivastatin

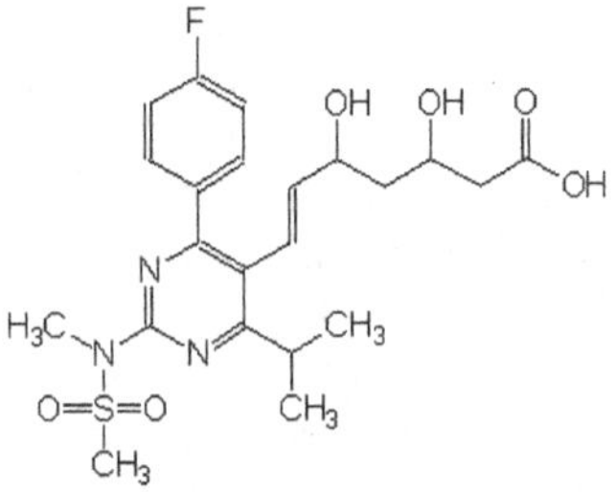

Rosuvastatin

4.3   Cholesterol-reducing drugs

Simvastatin, lovastatin, atorvastatin, cerivastatin, and rosuvastatin are members of a class of drugs that act as cholesterol-reducing agents. In addition, these compounds could also play a role in cancer therapy. The drugs are administered either as a lactone prodrug (simvastatin, lovastatin), which in the liver is rapidly converted into the active β-hydroxy acid, or as the β-hydroxy acid itself (atorvastatin, cerivastatin, and rosuvastatin), which may be converted into the lactone by biotransformation. In general, both the β-hydroxy acid and its lactone need to be determined. The lactone is readily converted into the free acid, unless special care is taken (pH in the range of 5, working temperature 4°C).

With respect to positive-ion ESI-MS, it should be noted that simvastatin and lovastatin do not contain N-atoms. In the negative-ion mode, the deprotonated molecule is observed for the simvastatin lactone and its β-hydroxy acid [46-47]. In positive-ion ESI-MS, strong sodium adducts are observed [46]. Control of the sodium content and thereby of the relative abundance of $[M+Na]^+$ relative to other adducts was not completely possible during sample pretreatment and analysis [47]. In the presence of AmOAc, the formation of $[M+Na]^+$ and $[M+K]^+$ is somewhat suppressed in favour of $[M+H]^+$ and $[M+NH_4]^+$ [47]. Various mobile-phase additives were investigated to improve the sensitivity, to direct the adduct formation, and/or to direct the fragmentation of simvastatin in ESI-MS [48]. The best performance was achieved by the addition of 2 mmol/l of methylamine to the aqueous phase (adjusted to pH 4.5 by means of acetic acid). This resulted in a fourfold enhancement in the response of simvastatin in the positive-ion mode without detrimental effect on the response of the β-hydroxy acid in the negative-ion mode [48].

Fragmentation of [M+Na]$^+$ of simvastatin in MS–MS results in the loss of the ester chain, e.g., the loss of 116 Da for simvastatin [46]. MS–MS on [M+H]$^+$ or [M+NH$_4$]$^+$ results in more extensive fragmentation, where the fragment due to the loss of the ester chain is the base peak [47]. In the negative-ion mode, two complement fragments related to the loss of the ester chain are observed, i.e., the charge is retained on either side of the cleavage, e.g., fragments for simvastatin ([M–H]$^-$ at m/z 435) are observed at m/z 115 and 319 [46-47].

One of the challenges in the determination of lovastatin and its β-hydroxy acid in mouse and rat plasma was the small amount (0.1 ml) of plasma available [46]. The analytes were isolated from plasma by SPE. The reconstituted extracts were separated using a 50×2-mm-ID C$_{18}$ column (5 μm) and 1 mmol/l aqueous AmOAc (pH 4.0)–acetonitrile gradient. They were analysed by ESI-MS in SRM mode with polarity switching, i.e., in positive-ion mode for lovastatin and in negative-ion mode for its β-hydroxy acid. Simvastatin and its β-hydroxy acid were used as ANIS. An LOQ of 0.5 ng/ml was achieved with intra-day and inter-day precision better than 7%. The method was validated.

The bioanalysis of simvastatin and its β-hydroxy acid in human plasma was reported using an [$^{13}CD_3$]-ILIS [47]. The analytes were extracted from plasma using solid-supported liquid-liquid extraction (SS-LLE) in methyl-t-butyl ether (MTBE). The interconversion between the lactone and its β-hydroxy acid was significantly reduced due to the use of SS-LLE. The compounds were separated using a 50×2-mm-ID C$_{18}$ column (5 μm) and a 25% 1 mmol/l AmOAc (pH adjusted to 4.5 by formic acid) in acetonitrile. Detection by ESI-MS was performed by polarity switching. An LOQ of 0.5 ng/ml was reported with a linearity over three orders. The method was validated and applied in clinical studies. Automated LLE in 96-well format, via a Multiprobe II Workstation and a Tomtec Quadra 96-channel liquid handler (Ch. 11.6.3), was subsequently reported by the same group [49]. This automatic sample pretreatment procedure reduced the total analysis time by two-third, and made the procedure significantly less labour intensive. Conversion between the lactone and the β-hydroxy acid was found to be negligible. The LOQ was 0.05 ng/ml with intra-day and inter-day precision better than 7.5%.

A simple method was described for the determination of simvastatin in human plasma in the range of 0.1–20 ng/ml, based in LC–MS detection of [M+Na]$^+$ in SIM on a single quadrupole system [36]. No attention was paid to the possible conversion of simvastatin into its β-hydroxy acid.

The bioanalysis of atorvastatin and its active 2-hydroxy and 4-hydroxy metabolites in plasma was described [50]. A two-step LLE with diethyl ether was applied to isolate the analytes from plasma. The reconstituted analytes were separated using a 150×2-mm-ID C$_{18}$ column (4 μm) and a mobile phase consisting of 30% 0.1% aqueous acetic acid in acetonitrile. Positive-ion ESI-MS in SRM mode was applied, using [D_5]-ILIS. The LOQ was 0.24 ng/ml. The method was validated.

The development and validation of the bioanalysis of atorvastatin, its two hydroxy metabolites, and their lactone forms in human serum was reported, using a $[D_5]$-ILIS [51]. Samples were extracted at pH 5.0 with MTBE. The analytes were separated on a 50×2-mm-ID YMC Basic column (5 μm) and an aqueous formic acid–methanol gradient, and analysed in positive-ion ESI-MS in SRM mode. The LOQ was 0.5 ng/ml.

Cerivastatin is administered in its acid form. It forms seven acid and lactone biotransformation products. The simultaneous bioanalysis of all eight components in human serum was reported using $[D_3]$-ILIS for cerivastatin and its lactone [52]. The method was similar to the method developed for atorvastatin [51], described above. The LOQ was 0.01 ng/ml for cerivastatin and its lactone, and between 0.05 and 0.5 ng/ml for the other biotransformation products. The method was validated.

Another method was developed and validated for only cerivastatin in human plasma [53]. It was based on LLE with 30% dichloromethane in diethyl ether, separation using a 100×3-mm-ID Xterra C_{18} column (3.5 μm) and a 30% water in acetonitrile mobile phase (containing 0.03% formic acid) at a flow-rate of 0.4 ml/min, and ESI-MS in SRM mode. The LOQ was 0.01 ng/ml. With an analysis time of only 2 min, a sample throughput of more than 400 plasma samples per day could be achieved.

The validated bioanalysis of *rosuvastatin* in human plasma by automated SPE in 96-well format with Oasis HLB material and positive-ion LC–ESI–MS–MS was reported using a $[D_6]$-ILIS [54]. The stability of rosuvastatin and its potential conversion into the lactone due to sample pretreatment was thoroughly investigated. The method was applied in pharmacokinetic studies during clinical trials. A similar method was applied by the same group in the bioanalysis of the *N*-desmethyl metabolite of rosuvastatin [55].

4.4   Methylphenidate

Methylphenidate (MPH, Ritalin®) is a central nervous system stimulant that is used for the treatment of attention deficit disorders, with and without hyperactivity, and narcolepsy. MPH has two chiral centres and is marketed as a racemic mixture. It is known that *d-threo*-MPH is pharmacologically more active than *l-threo*-MPH. The drug is rapidly metabolized in humans to the inactive ritalinic acid. High-throughput analysis with chiral selectivity is demanded for the bioanalysis of MPH and its major metabolite.

d-threo-MPH   Ritalinic acid

Comparison between LC–MS, GC–MS, and immunoassays for the urinary screening of MPH abuse demonstrated LC–MS was superior in both cost and performance [56].

The group of Bakhtiar [57-59] described the chiral bioanalysis of MPH in various matrices, utilizing a number of sample pretreatment strategies, separation on a Chirobiotic V columns, and the use of positive-ion APCI-MS in SRM mode. The chiral selectivity of the Chirobiotic V column is based on the use of the macrocyclic glycopeptide antibiotic vancomycin. The column can be used in both aqueous and organic mobile phase.

Initially, MPH was extracted from plasma by LLE with cyclohexane. The reconstituted analytes were analysed using LC–MS [57]. A $[^{13}CD_8]$-analogue was used as ILIS. LC was performed using a 150×4.6-mm-ID Chirobiotic V column (5 μm) and methanol with 0.05% ammonium trifluoroacetate as mobile phase at a flow-rate of 1.0 ml/min. The first 3 min of the run were diverted to waste. The total run time was only 7.5 min. The plasma-concentration–time profile for a child with ADHD after an oral administration of 17.5 mg of racemic MPH showed considerably higher plasma levels of *d-threo*-MPH, as expected. The LOQ of this method was 87 pg/ml. This method was also applied in toxicokinetic studies in rat, rabbit, and dog [59]. In a further study [58], a semi-automated LLE in 96-well plate format was developed and validated. No chiral column was used in this case.

In order to study the effects of chronic administration on the cognitive function of children by means of nonhuman primates, a method was developed for the determination of MPH plasma levels under chronic and acute dosing conditions [60]. MPH and ritalinic acid were isolated from plasma in SPE on Oasis HLB. LC was performed using a 150×2-mm-ID C_{18} column (3 μm) and 20% acetonitrile in 0.1% aqueous formic acid at a flow-rate of 0.2 ml/min. Positive-ion ESI-MS was performed in SIM mode on a single-quadrupole system. $[D_3]$-MPH was used as ILIS. The LOQ was 0.25 ng/ml in monkey plasma.

A high-throughput bioanalytical method for the analysis of MPH in rat plasma was reported [61]. Samples were protein precipitated with acetonitrile in a 96-well plate format. $[D_3]$-MPH and $[D_5]$-ritalinic acid were used as ILIS. High-speed LC separation was performed using a 25×4.6-mm-ID Chromalith Flash RP-18e monolithic column and a mobile phase of 25% acetonitrile in 0.1% aqueous formic acid at a flow-rate of 3.5 ml/min. The analysis time was 15 s with base-line resolution between the analytes. The column was connected to ESI-MS via a post-column split, delivering 2.1 ml/min for detection. The autosampler enabled sample injection every 17 s. In this way, eight 96-well plates were analysed in 3 h and 45 min. Representative chromatograms obtained during such a study are shown in Figure 11.4. The LOQ was 0.1 ng/ml for MPH and 0.5 ng/ml for ritalinic acid.

The analysis of MPH from human urine was used as a model to demonstrate the direct-infusion quantitative bioanalysis from a polymer-based microfluidic chip electrospray emitter [62]. A calibration curve in the range 0.4–800 ng/ml was acquired.

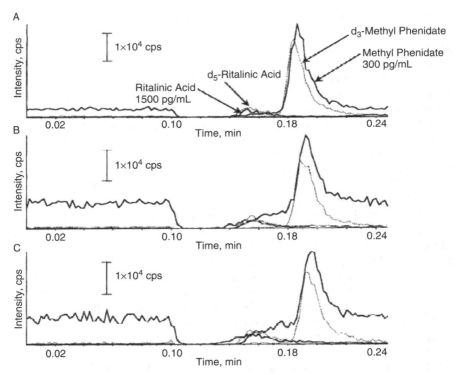

Figure 11.4: Representative chromatograms for the MPH analysis of low QC samples (0.3 ng/ml) at (a) 7.75 min (sample 32), (b) 93 min (sample 368), and (c) 188 min (sample 752) in the rapid sequential analysis of 768 samples in 3h and 45 min. Reprinted from [61] with permission. ©2003, Elsevier Science BV.

4.5   Loratadine

Loratadine (LOR, trade name Claritin®) is a selective peripheral histamine H_1-receptor antagonist. It is a long-acting and nonsedating tricyclic antihistamine, used for the relief of nasal and nonnasal symptoms of seasonal allergies and skin rash. The compound is extensively metabolised in the liver to an active metabolite, descarboethoxyloratadine or desloratadine (DCL).

The bioanalysis of LOR and DCL was described after LLE from human plasma with toluene, and back-extraction of DCL into 20% aqueous formic acid [63]. The LOR-containing toluene fraction was evaporated to dryness and combined with the DCL fraction. The separation was performed using a 150×2.1-mm-ID C_{18} column (5 μm) and a 10–90% acetonitrile gradient in 0.1% formic acid in 2 min at a flow-rate of 0.3 ml/min. ESI-MS in SRM mode was applied. The LOQ was 0.1 ng/ml for LOR and DCL. The method was applied in pharmacokinetic studies.

At the same time, the bioanalysis of LOR and DCL in rat, rabbit, mouse, and dog plasma was reported by others [64]. In order to get more reliable toxicology data, the bioanalysis in these four preclinical species is done simultaneously instead of on separate days. The sample pretreatment was SPE in a 96-well plate format, using a Tomtec Quadra liquid handling system and an Empore C_{18} 96-well extraction disk plate. Four-channel parallel LC was done with four 100×2-mm-ID C_8 columns (5 μm) and a mobile phase of 85% methanol in 25 mmol/l aqueous AmOAc (adjusted to pH 3.5). The mobile phase was delivered at a flow-rate of 800 μl/min and split into 200 μl/min over each of the four columns. A multi-injector system was applied with four injection needles. A post-column split was applied to deliver 60 μl/min per column to a four-channel multiplexed ESI source (Ch. 5.5.3). The interspray step time was 50 ms. Positive-ion ESI-MS was performed in SRM mode with a dwell time of 50 ms for each of the four transitions, *i.e.*, LOR, DCL, and their $[D_4]$-ILIS, with 20 ms interchannel delay. The total cycle time was thus 1.24 s. The LOQ was 1 ng/ml for both analytes. QC samples showed precision ranging from 1 to 16% and accuracy from –8.44 to 10.5%. The interspray crosstalk was less than 0.08% at concentrations as high as 1000 ng/ml.

The same group reported the bioanalysis of LOR and DCL on a triple quadrupole with enhanced resolution [29], using the same sample pretreatment and chromatographic phase system, running on one column instead of four. SRM at unit-mass resolution with 0.7 *m/z*-unit FWHM in both MS_1 and MS_2 was compared to SRM at enhanced resolution with 0.1 *m/z*-unit FWHM in MS_1 and 0.5 *m/z*-unit in MS_2. While the absolute signal in enhanced resolution was about one third of that at unit-mass resolution, an LOQ of ~0.06 ng/ml was achieved in both modes. Operating at enhanced resolution helps in removing isobaric coeluting interferences.

Subsequently, the validation of the LC–MS–MS method for LOR, DCL, and 3-hydroxy-desloratadine, running on a standard triple quadrupole system and applying an automated 96-well SPE method, was reported by this group [65]. An LOQ of 25 pg/ml was achieved.

The bioanalysis of LOR and DCL was also reported with hydrophilic-interaction LC (HILIC, Ch. 1.4.5), *i.e.*, utilizing a 50×3-mm-ID silica column (5 μm) and an aqueous acetonitrile mobile phase (90% acetonitrile containing 0.1% TFA) [66]. In HILIC, the retention order is reversed relative to RPLC, *i.e.*, highly polar analytes are more strongly retained. The analytes were extracted from alkalized human plasma by LLE with hexane. The extract was evaporated to dryness and reconstituted in 0.1% aqueous TFA. The flow-rate was 0.5 ml/min; the run time 2 min. Positive-ion ESI-MS in SRM mode was used, with a $[D_3]$-ILIS. The LOQ was 10 pg/ml for LOR and 25 pg for DCL, with precision in the range of 3.5–9.4%. More recently, an automated 96-well plate LLE procedure was adapted [67]. LOR and DCL were extracted from plasma using MTBE. The organic MTBE layer was directly injected into the HILIC system.

The bioanalysis of LOR in human plasma by means of LC–MS–MS on an ion-trap instrument was also described [68]. Sample pretreatment was LLE using 5% *i*-pentanol in *i*-octane. LC was performed using a 150×2.1-mm-ID phenyl column (5 μm) with a mobile phase consisting of 69% acetonitrile in 0.2% aqueous formic acid at a flow-rate of 0.2 ml/min. Itraconazole was used as ANIS. The validated method provided an LOQ of 0.1 ng/ml.

5. Matrix effects

An important issue in the quantitative bioanalysis using LC–MS via ESI or APCI is the possible occurrence of matrix effects. A matrix effect is an (unexpected) suppression of enhancement of the analyte response, detected by comparing a standard solution and a spiked pretreated sample (absolute matrix effect). Detailed studies on matrix effects by Matuszewski *et al.* [69-70] revealed that ion suppression or enhancement is frequently accompanied by significant deterioration of the precision. In principle, a repeatable and reproducible ion suppression in a bioanalytical method should not be a significant problem. When a sample matrix, *e.g.*, "plasma", after pretreatment and LC yields different and/or (strongly) varying responses (relative matrix effect), the matrix effect has a much greater influence on the reliability of the analytical data. This issue is nicely pictured in Figure 11.5, where precision data (%RSD) are plotted as a function of the analyte concentration for a single lot and for five lots of plasma samples [70].

From reviewing the LC–MS literature, it first appears that matrix effects are only observed with ESI-MS. However, already in 1988, Gelpi *et al.* [71] discussed sample effects observed in the thermospray LC–MS analysis of labelled serotonin. More recently, Hajšlová and Zrostlíková [72] reviewed matrix effect in the ultratrace GC–MS and LC–MS analysis of pesticide residues in food and biotic matrices. The discussion in this section is focussed on matrix effects in LC–MS, especially with respect to quantitative bioanalysis. A minireview was written by Annesley [73].

5.1   Understanding matrix effects

Response suppression or enhancement effects may be exerted by any coeluting component entering the API source via LC. Some mobile-phase additives are known to suppress or enhance analyte response. It is useful to discriminate between effects due to the analytical system, *e.g.*, mobile-phase composition, source parameters, and effects due the actual analyte matrix.

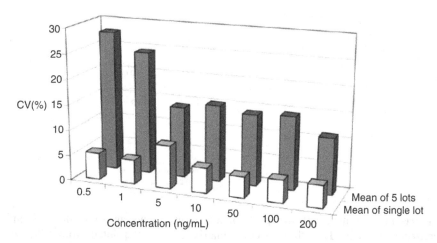

Figure 11.5: Precision (%CV) of a bioanalytical method at various analyte concentrations, determined in either a single plasma lot or in five different plasma lots. While for the single plasma lot the precision is acceptable, it is not when five different plasma lots are taken into account. Reprinted from [70] with permission. ©2003, American Chemical Society.

Matrix-related ion suppression was first reported by Buhrmann *et al.* [74]. They compared three different sample pretreatment methods in the analysis of the platelet-activating factor receptor antagonist SR27417 in human plasma and evaluated extraction efficiency, ion suppression, and overall process efficiency. Buhrmann *et al.* [74] emphasized the importance of the sample pretreatment as a tool to avoid or minimize matrix effects.

In two separate studies, Matuszewski *et al.* [21, 69] discussed matrix effects in more detail.

In the bioanalysis of finasteride in human plasma [69], the method development was initially directed at the use of ESI-MS. An ANIS was used. While a precision in the finasteride peak area better than 5%RSD was obtained upon injection of standard solutions, precision values ranging between 17 and 43%RSD were obtained with extracts of five different sources of control human plasma. The matrix effects were further evaluated in two ways. The response for the ANIS in post-extraction spiked plasma samples was evaluated for the five different plasma extracts. In addition, the effect of enhancing the selectivity of the extraction and the efficiency of the LC separation was evaluated. Although these measures showed some improvement, the precision and accuracy of the ESI-MS method remained inadequate to support clinical studies. Therefore, an LC–MS method based on the use of APCI was successfully developed and validated.

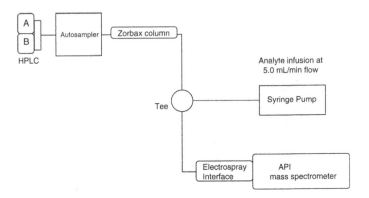

Figure 11.6: Scheme of the post-column infusion system to evaluate sample pretreatment methods with respect to matrix effects. Reprinted from [76] with permission. ©2000, American Society for Mass Spectrometry.

In the study on indinavir in human plasma [21], two sample pretreatment methods, *i.e.*, dilution of urine and LLE with MTBE, two chromatographic systems, *i.e.*, with low and high capacity factor, and the use of ILIS and ANIS were compared. It was concluded that the potential effect of co-eluting 'unseen' endogenous species should be evaluated during method development and validation of bioanalytical LC–MS methods.

The group of King [75-76] studied the matrix effects in more detail. They proposed a post-column infusion system (Figure 11.6), which enables the evaluation of the matrix effects of different sample pretreatment procedures. By continuous post-column infusion of the analyte of interest and injection onto an LC column of a blank matrix extract with the sample pretreatment to be evaluated, the matrix effects on the analyte can be observed as a function of the chromatographic retention time. An example of the results is shown in Figure 11.7. An interesting aspect of this infusion method is that matrix effects can be monitored beyond the analytical run time. By doing so, one can actually observe long-term ion suppression effects, which may in high-throughput repetitive injection act upon subsequent samples.

From these results, accumulation of matrix effects in a long sample series can be anticipated [75]. The post-column infusion system is an elegant approach to monitor matrix effects in a pragmatic and phenomenological manner, *i.e*, without addressing questions related to the molecular or mechanistic causes of the matrix effects.

In a subsequent study, King *et al.* [76] studied mechanistic aspects of ion suppression by a number of experiments. By the use of a dual ESI–APCI source and a dual-sprayer ESI source, it was demonstrated that the matrix effect primarily is a liquid-phase and not a gas-phase process.

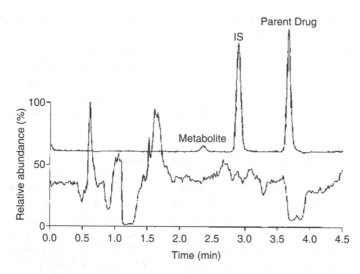

Figure 11.7: Typical results of a post-column infusion experiment. In the top chromatogram the separation of the parent drug, its metabolite, and an ANIS is shown. The bottom chromatogram shows the matrix effect on the response of the parent drug. The ion suppression and ion enhancement effects prevent the reliable determination of the target compounds. Reprinted from W.M.A. Niessen, J. Chromatogr. A, 1000 (2003) 413 with permission. ©2003, Elsevier Science BV.

By collecting residues from the curtain plate in a Sciex API source (Figure 5.6) and re-injection, it was demonstrated that nonvolatiles present in the liquid flow, originating from nonvolatile mobile-phase additives or sample constituents actually increase the amount of analyte found at the curtain plate. The nonvolatiles prevent the preformed analyte ions to desorb from the droplet into the gas phase. This study emphasizes the importance of removing nonvolatile sample constituents during sample pretreatment [76]. It was concluded that APCI is less prone to matrix effects than ESI, which is also confirmed by others, *e.g.*, [77-79], but one should certainly not exclude the occurrence of matrix effects in APCI-MS.

Although the presence of nonvolatiles is certainly important with respect to matrix effects, it does not tell the complete story. In early mechanistic studies, it was shown that compounds with high proton affinities may also suppress the analyte response, potentially via gas-phase processes [80]. Other mechanistic studies indicate the importance of surface activity in the ESI process [81-82]. Matrix components with higher surface activity may suppress the ionization of the analyte. This again stresses the importance of a pragmatic approach to monitoring matrix effects via the post-column infusion setup (Figure 11.6). The extent of matrix effects in the bioanalysis of a particular compound also depends on the source design of the

LC–MS system utilized [83]: using an instrument from another manufacturer may solve problems with matrix effects in some cases.

The post-column infusion setup has been widely applied in matrix effects studies, *e.g.*, in developing methods for the quantitative screening of drug discovery compounds by fast-gradient (1 min run time) LC–MS [83-85], in the optimization of the sample pretreatment in the analysis of methadone, comparing four off-line and three on-line sample pretreatment methods [77], and in evaluating various protein precipitating additives [86].

5.2   Quantitative description of matrix effects

Matuszewski *et al.* [70] discussed methods to adequately evaluate matrix effects. A fairly complete description can be achieved by acquiring calibration plots with three sample sets for plasma samples from five different sources. The first sample set consists of the analyte and the IS in mobile phase, the second sample set consists of post-extraction spiked plasma samples, and the third set of pre-extraction spiked plasma samples. From the peak areas acquired for these calibration plots one can calculate the percentage matrix effect %ME from:

$$\%ME = \frac{Area\ of\ Post-extraction\ Spike}{Area\ of\ Standard} * 100$$

and the recovery of the sample pretreatment method %RE from:

$$\%RE = \frac{Area\ of\ Pre-extraction\ Spike}{Area\ of\ Post-extraction\ Spike} * 100$$

and finally the overall process efficiency %PE from:

$$\%PE = \frac{Area\ of\ Pre-extraction\ Spike}{Area\ of\ Standard} * 100 = \frac{(\%ME \times \%RE)}{100}$$

These terms differ somewhat from the terms introduced earlier by Buhrmann *et al.* [74]. By the definition give here, ion enhancement can be accounted for as well.

In practice, the absolute %ME has limited relevance. In bioanalytical practice, it is more important to demonstrate the absence of relative matrix effects, *i.e.*, between different sources of the biofluid [70]. Alternatively, relative matrix effects may be evaluated by a comparison of precision expressed as %RSD in repetitive injection of standards and post-extraction spiked samples. If significant differences in the %RSD exist, a relative matrix effect is present between the different sample batches.

The discussion leads to the recommendation to actually evaluate plasma samples from five different sources during method validation. The consequences in terms of the number of analyses to be performed to evaluate the matrix effect are discussed as well [70].

5.3 Reduction or elimination of matrix effects

From the previous discussion, it can be concluded that matrix effects are due to co-eluting matrix constituents. It is important to state that such components often are not ionized by ESI, which means that they cannot be detected in ESI-MS. There are two approaches to deal with matrix effects: (1) remove interfering matrix constituents, and (2) eliminate the effects on accuracy and/or precision (Ch. 6.7).

Improvement of the sample pretreatment and the chromatographic resolution and/or analysis time is a way to remove interfering matrix interferences.

As the matrix effect is (strongly) compound dependent, an ANIS may be differently affected by the matrix than the analyte. The precision of a method can be significantly improved by modifying the mobile-phase conditions in such a way that analyte and ANIS co-elute [87]. An improvement in the intra-day precision from 5.2–16.2%RSD with separation between 2-C-ethynylcytidine as analyte and 3'-C-ethylcitidine as ANIS to 2.7–4.2%RSD without separation was achieved in this way.

The most adequate method to eliminate matrix effects is the use of an ILIS. An ILIS shows (almost) identical behaviour to the target analyte in both sample pretreatment, chromatography, and analyte ionization. The data reported in the bioanalysis of mevalonic acid [23] indicate that this issue needs proper attention during method development and validation (Ch. 11.3.4).

If no ILIS or proper ANIS are available, or no adequate blank matrices are available, the standard addition method should be applied to obtain accurate and precise results [88].

5.4 Other issues related to matrix effects

The effect of dosing vehicle excipients such as PEG400, propylene glycol, Tween 80, and hydroxypropyl-β-cyclodextrin on the accuracy of pharmacokinetic analysis by LC–MS was investigated as well [89-90]. Due to matrix effects, the analysis may indicate a 2–5-fold increase in the calculated plasma clearance. This would result in a false rejection of the compound in a drug discovery screen. Similar effects of dosing vehicle excipients were also investigated via post-column infusion studies by others [79, 85, 91].

Ion suppression effects can also be caused by polymers from plastic tubes or Li-heparin, a commonly used anticoagulant [83].

Matrix effects experienced in the analysis of microsomal incubation products (Ch.10.6.1) were evaluated by Zheng et al. [92]. The individual effects of the Tris buffer, NADPH, and the microsomes on the ESI-MS response of 27 different drugs were investigated in direct injection MS–MS experiments. The more polar analytes showed up to 5-fold ion suppression. Therefore, an automated Oasis-HLB SPE procedure was developed in 96-well plate format. Direct injection of protein-precipitated incubations yielded similar results. Additional use of fast LC separation prior to MS–MS analysis gave no further improvement.

5.5   Matrix effects in other application areas

Matrix effects are not only relevant in quantitative bioanalysis, but in any other application area of LC–MS as well. As long as the method concerns the analysis of a limited number of target compounds, strategies outlined above can be applied to monitor and reduce matrix effects. The situation is more complicated when multiresidue analysis must be performed, targeted either on a large number of known analytes or even unknown analytes. Two examples are given here.

Müller *et al.* [93] compared different sample pretreatment methods for serum for systematic toxicological analysis or general unknown screening (Ch. 12.5). In this case, a relatively long gradient-elution LC is applied. The post-column infusion method with two target analytes was applied to evaluate matrix effects after LLE, mixed-mode SPE, protein precipitation, and a combination of protein precipitation and mixed-mode SPE. It was found that severe matrix effects are observed in the first part of the chromatogram, but not in the more relevant parts of the gradient run. These experiments also nicely stress the importance of performing LC instead of direct injection of pretreated samples into the MS–MS instrument (Ch. 11.7.1).

Becker *et al.* [94] evaluated matrix effects in the multiresidue analysis of 15 penicillins and cephalosporins in bovine muscle and kidney tissue and in milk. Comparison of the responses between standard solution and post-extraction spiked samples for all compounds in all three matrices showed different behaviour of different analytes and in different matrices. For one target compound, cefquinone, the standard addition method had to be applied to get sufficiently accurate and precise results.

6. Sample pretreatment

Proper sample pretreatment methods are important to reduce or eliminate matrix effects. In this section, advances in sample pretreatment methods in relation to LC–MS quantitative bioanalysis are briefly reviewed (see also Ch. 1.5). Contrary to environmental analysis, where on-line sample pretreatment is often used (Ch. 7.3.2), off-line sample pretreatment appears to be preferred in quantitative bioanalysis. This is partly due to the composition of biological samples, where the presence of especially proteins may cause clogging of the SPE columns, cartridges or disks used, and partly due to the fact that decoupling sample pretreatment and LC–MS analysis generally allows for a higher sample throughput.

Sample pretreatment in 96-well plate format has almost become state-of-the-art. This approach was pioneered for SPE in bioanalytical LC–MS by Kaye *et al.* [95] and Allanson *et al.* [96]. The sample pretreatment procedure for the 5-HT receptor agonist 311C90 and its desmethyl metabolite in human plasma, initially performed on the Gilson ASPEC, could be reduced from 5 hr for 60 samples to about 1 hr for 96 samples, utilizing the Porvair Microlute™ SPE in combination with a Multiprobe robotic sample processor [96]. Subsequently, other pretreatment procedures were adapted to 96-well plate format.

6.1 Sample filtration

A potential problem with the analysis of biological fluids, especially plasma samples from *in vivo* studies, is the risk of clogging of SPE cartridges and/or analytical columns. Therefore, filtration of such samples prior to LC or SPE is recommended. Combined filtering and protein precipitation in 96-well plate format was described [97-98]. The samples are collected and stored frozen in sealed 96-well polypropylene filter plates. Prior to SPE and LC–MS analysis, the seals are removed and the plate is placed on top of a 96-well SPE manifold. As the plasma thaws, it passes through the filter and into the SPE device.

In order to study plasma protein binding of drug candidates, a 96-well plate ultrafiltration device was used in combination with fast LC–MS [99]. All liquid handling was done using a robotic liquid handler. After preparation, the plates were centrifuged at 3000*g* for 45 min.

6.2 Protein precipitation

Protein precipitation is a very popular sample pretreatment method, because it is a very fast and almost generic approach (Ch. 1.5.1).

The effectiveness of various protein precipitation additives in terms of protein removal and matric effects was investigated [86]. Acetonitrile, trichloroacetic acid (TCA), and zinc sulfate were found most effective in removing proteins (applied in a 2:1 additive-to-plasma ratio). By a post-column infusion setup (Figure 11.6), these three methods were further evaluated with respect to matrix effect for five different mobile-phase compositions. As both buffered, acidified, and pure methanol–water mobile phases were compared, actual conclusions are difficult. In the pure methanol–water mobile phases, the use of TCA enhances the response, probably by generating acidic conditions more favourable in ESI-MS. With buffered or acidified mobile phases, the difference in matrix effects between acetonitrile or TCA as protein precipitation additive was less pronounced. A similar comparison between acetonitrile, perchloric acid, and TCA as protein precipitation additives was reported by others [100]. With acid precipitation, lower analyte recovery and higher %RSD was observed. Therefore, precipitation by acetonitrile was preferred.

Protein precipitation was automated into a 96-well plate format by means a robotic liquid handler by Watt *et al.* [101]. The procedure is described in more detail in Ch. 1.5.1. It enabled a 4-fold improvement in sample throughput on the LC–MS instrument, compared to previous manual protein precipitation procedures. The method was applied as a generic sample pretreatment method in combination with a generic LC–MS method to a variety of drug candidates [102].

Protein precipitation in a 96-well format has also been reported by others, *e.g.*, in the determination of MPH in rat plasma [61] (Ch. 11.4.4).

6.3   Liquid-liquid extraction

LLE is a simple and fast sample pretreatment (Ch. 1.5.2), which is efficient in the removal on nonvolatiles, and successful in reducing matrix effects [69-70, 75].

From the selected applications in Ch. 11.4, it can be concluded that LLE is frequently applied, *e.g.*, in the extraction of risperidone and its 9-hydroxy metabolite [44], of cholesterol-reducing drugs [51-53], MPH [58], and of LOR and DCL [63, 66, 68]. Next to an one-step LLE followed by evaporation to dryness of the organic layer, two-step LLE procedures with back-extraction to an aqueous phase have been reported as well, *e.g.*, for reserpine [42], atorvastatin [50], and LOR [63].

LLE in 96-well plate format was pioneered by the group of Henion [103-104]. A typical procedure is discussed in Ch. 1.5.2. With this approach, four 96-well plates could be prepared by one person in 90 min. Subsequently, it took ~11 hr to analyse these four plate with LC–MS, providing an analysis time of 1.2 min per sample.

Nowadays, LLE in 96-well plate format is frequently applied, for instance for reserpine in FVB/N mouse plasma [43], for simvastatin and its β-hydroxy acid [49], for MPH [58], and for LOR and DCL [67] (Ch. 11.4).

Solid-supported LLE (SS-LLE, Ch. 1.5.2) in 96-well plate format was applied in the extraction of simvastatin from human plasma [47] and in several other studies.

6.4   Solid-phase extraction

SPE is an important sample pretreatment method in quantitative bioanalysis. As discussed in Ch. 1.5.3, SPE can be performed in single cartridges, in 96-well plate format, and in on-line mode, either on top of the analytical column, or preferably on a precolumn in a column-switching system.

An example of the use of cartridge-SPE is the determination of lovastatin and its β-hydroxy acid in mouse and rat plasma [46] (Ch. 11.4.3). The 0.1-ml plasma was spiked with 50 μl of the analytes in 90% acetonitrile, and then diluted with 0.4 ml of water. The SPE cartridges (1 ml, 100 mg) were conditioned with two 1-ml portions of methanol and water. The diluted sample was loaded onto the cartridge. After applying low vacuum for 1 min, the cartridge was washed with 1 ml of water, 1 ml of 5% formic acid and again 1 ml of water. After drying for 1 min, the cartridge was first eluted with 1 ml 70% aqueous methanol, followed by low vacuum for 1 min and drying for 15s. Finally, the cartridge was eluted with 1 ml of acetonitrile. The combined eluates were evaporated to dryness and reconstituted in mobile phase. The extracts were analysed by LC–MS [46]. Numerous other examples are available. This type of SPE procedure can be automated by means of a Gilson ASPEC, as demonstrated for reserpine [42], or a Zymark RapidTrace automation system.

Unfortunately, ASPEC procedures are rather slow. Therefore, SPE procedures in 96-well plate format were developed [95-96]. Again, both cartridge and disk SPE systems have been used. An example of a 96-well plate SPE procedure is discussed in detail in Ch. 1.5.3. In this application, the use of the 96-well plate SPE procedure reduced sample work-up time for the determination of fentanyl in plasma from ~3.5 hr to ~2 hr [105].

The 96-well plate SPE procedures have become very popular. In Ch. 11.4, applications of SPE in 96-well plate format were described for LOR and DCL [29, 64-65]. Many other examples are available in literature. The use of SPE in a 384-well plate format was reported for the determination of methotrexate and its 7-hydroxy metabolite in human plasma and urine [106]. The work-up time for the 384-well plate is similar as for the 96-well plate,

Mixed-mode SPE procedures using Oasis-HLB materials (Ch. 1.5.3) were described above for the determination of risperidone and its hydroxylated metabolites [14], and of MPH and its metabolite ritalinic acid [60]. The mixed-mode materials are also available in 96-well plate format, as described above for the analysis of rosuvastatin in human plasma [54] and in the method development for metabolic incubation studies with 27 candidate drugs [92].

A comparison between manual LLE, automated 96-well plate LLE, and automated 96-well plate SPE procedures for the analysis of a carboxylic acid containing analyte was reported by Jemal et al. [107]. Their time analysis of the three extraction procedures is shown in Table 11.1. The same LOQ of 50 pg/ml could be achieved by either method.

Table 11.1: Time analysis (in min) of manual LLE, automated 96-well plate LLE, and automated 96-well plate SPE procedures
Reprinted from [107] with permission. ©1999, Elsevier Science BV

	Manual LLE	96-well LLE	96-well SPE
Standard curve preparation	25	8	8
Labelling of tubes	15	5	0
Sample transfer and extraction	180	65	43
Drying	40	20	45
Reconstitution and transfer	30	5	5
Total time	290	103	101
Analyst time	250	< 10	< 5

Until recently, on-line SPE via valve-switching techniques was not very popular in bioanalytical LC–MS. With the introduction of the Spark Holland Prospekt II, on-line SPE–LC–MS is promoted for quantitative bioanalysis. The typical procedure and column-switching setup for on-line SPE–LC–MS is discussed Ch. 1.5.4.

An example of SPE–LC–MS is the determination of Ro 48-6791, a short-acting agent for conscious sedation, and its major metabolite in human plasma [108]. In this particular case, the rationale for the use of the precolumn was in enlarging the permissible injection volume, because a 300-μm-ID microcapillary packed column was used in LC. The determination of risperidone and 9-hydroxyrisperidone in plasma and saliva from adult and pediatric patients using SPE–LC–MS was discussed in Ch. 11.4.2 [45].

An elegant example of SPE–LC–MS was reported by Siethoff et al. [109] for the determination of capecitabine and its active metabolite 5-fluorouracil in human plasma. The mixture was injected onto a 10 × 2.1 mm ID Xterra C_{18} column (5 μm). 5-fluorouracil and its [$^{15}N_2$]-ILIS passed through this column. They were analysed on a 150×2.1-mm Hypercarb column (5 μm). After 0.5 min, valve-switching was performed to transfer capecitabine and its ANIS fluvastatin to a second analytical column (50×2.1-mm-ID XTerra C_{18}, 3.5 μm). Both columns were connected to the ESI-MS by means of another switching valve.

6.5   Alternative SPE-related procedures

A number of related sample pretreatment procedures have been described, such as solid-phase microextraction (SPME), RAM stationary-phases, molecularly imprinted polymers (MIP), and TFC.

SPME is a powerful tool in GC–MS. Adaptation to LC–MS involves two operational modes that are in use, i.e., fiber SPME and (automated) in-tube SPME [110].

RAM columns allow the direct injection of plasma samples without protein precipitation (Ch. 1.5.6). Often, on-line RAM–LC–MS is described, following a procedure identical to on-line SPE–LC–MS (Ch. 1.5.4) [111].

In TFC, SPE is performed at very high flow-rates on either columns packed with 50 μm Cyclone HTLC particles or monolithic columns (Ch. 1.4.4). The high linear flow results in similar chromatographic efficiency in much shorter analysis time. There is no need for protein precipitation prior to the analysis: plasma samples are just centrifuged and then injected. One-step TFC–MS was found to perform similar as a manual LLE and a 96-well plate SPE method [112].

A typical two-column setup featuring two six-port switching valves (Ch. 1.5.5 and Figure 1.4) was evaluated. To support the claim that no sample pretreatment prior to analysis is required with the TFC system, Herman [113] analysed 1000 compounds in plasma, urine, intestinal perfusates, brain homogenates, and cerebrospinal fluid. Similar results in terms of peak area were obtained for the five different matrices.

Sample throughput can be increased by applying two- or four-channel staggered parallel TFC, as demonstrated by Herman [114]. The application of staggered four-channel parallel TFC–MS in quantitative bioanalysis under GLP was successfully validated by King *et al.* [115] using four 50×2.1-mm-ID C_{18} columns (5 μm) operated at 1 ml/min with a 2-min gradient.

6.6   Do we need sample pretreatment?

The answer to the question "*Do we need sample pretreatment in the quantitative bioanalysis?*" is a clear YES! Over the years, a number of papers claim direct plasma injection in their title, suggesting that they do not perform sample pretreatment prior to their LC–MS analysis. However, a closer reading reveals that in all these cases either TFC–MS is performed, or another type of high-flow sample pretreatment, in most cases using Oasis-HLB precolumns [116-118].

7.   Liquid chromatography

7.1   Do we need LC separation?

One can ask whether we need an LC separation prior to MS and especially MS–MS detection. Most researchers in the field answer this question with a clear YES! again. It has clearly been demonstrated that reliable quantitation of compound elution in the solvent front is not possible. An important function of the short analytical columns applied in high-throughput quantitative bioanalysis is to assure at least a minimum separation between the solvent front and the analyte. Additional issues related to the need to separate the parent drug from its metabolites are addressed in Ch. 11.3.2.

Nevertheless, a number of examples of flow-injection or column-bypass injection into the ESI-MS system are available. Nifedipine was determined in this way with an LOQ of 0.5 ng/ml [119]. Good precision and accuracy was achieved in the validated quantitative bioanalysis of topiramate in human plasma, using LLE and flow-injection–MS–MS [120].

The direct coupling of SPE to MS without LC separation was also described. However, as gradient elution from the SPE column was performed, some separation is actually achieved on the SPE column (Ch. 8.3.2). Different base/organic followed by acid/organic washing steps on Oasis-HLB mixed-mode materials were evaluated to reduce matrix interferences in SPE–MS [121].

An interesting approach to quantitative bioanalysis, proposed by Henion *et al.* [122-123] and commercialized by Advion Biosciences, is the direct infusion of plasma extracts by means of nano-ESI, positioned on a multi-sprayer microchip. A new sprayer tip is applied for each individual sample. This approach has been applied in the bioanalysis of verapamil and its active metabolite [122], in the analysis of midazolam in human plasma [123], and in the analysis of a parent drug and its metabolite using SRM, full-spectrum MS–MS, and MS³ on a Q–LIT instrument [124]. In the latter case, LLE was needed to minimize matrix effects, and MS–MS and MS³ was applied to improve the selectivity of the assay.

7.2   Monolithic columns

In bioanalytical LC–MS, the use of monolithic columns (Ch. 1.4.4) is promoted for high-throughput applications. After initial results by others [125-126], Hsieh *et al.* [127-128] nicely demonstrated their potential in bioanalytical LC–MS: shorter analysis time without sacrificing separation (Figure 1.2). Adequate separation of a parent drug, its major metabolite, and an ANIS was achieved within 45 s. Similar plasma-concentration–time profiles were obtained as under conventional conditions. The application of monolithic LC in the high-throughput analysis of MPH and its metabolite ritalinic acid was discussed in Ch. 11.4.4 [61]. Parallel extraction and LC–MS analysis of monolithic TFC columns has been applied in GLP-regulated quantitative bioanalysis, featuring injection-to-injection cycle times as low as 24 s [129]. The use of a monolithic packed-capillary column was proposed to avoid the disadvantages of the high flow-rate and solvent consumption applied in monolithic columns [130].

7.3   Hydrophilic interaction chromatography

HILIC (Ch. 1.4.5) was promoted for application in bioanalytical LC–MS by Naidong *et al.* [66-67, 131-132]. In initial studies [131], 5–8-fold and 20-fold gain in sensitivity was achieved in HILIC for basic and acidic compounds, respectively, compared to reversed-phase LC. The use of HILIC in the analysis of LOR and DCL was discussed in Ch. 11.4.5 [66]. Different from RPLC, an organic solvent is a weak solvent in HILIC. This opens the possibility to direct injection of organic extracts from LLE, as demonstrated with MTBE extracts after 96-well plate LLE of LOR and DCL [67]. The potential of HILIC in bioanalysis with LC–MS was reviewed by Naidong [132].

7.4   Chiral separations

An important issue in drug development is the assessment of chirality of drugs and the separate evaluation of the toxicology of both enantiomers. The use of chiral separation in combination with LC–MS is important. Phase systems for chiral separations are discussed in Ch. 1.4.3.

Columns like Chiralpak are used with normal-phase mobile phases. Post-column addition of an aqueous–alcohol solvent is required in ESI-MS. This column type was used with ESI-MS in, for instance, the enantioselective determination of felodipine [133], and with APCI-MS in the analysis of Org-4428 in human plasma [134].

Chirobiotic columns are used with polar-organic mobile phase. Chirobiotic T was for instance used for enantiomers of salbutamol and its sulfate metabolites [135], for the high-throughput analysis of albuterol in dog plasma [136], while Chirobiotic V was used for MPH (Ch. 11.4.4) [57].

Chiral columns like Chiral AGP and Cyclobond, used with highly-aqueous mobile phases, were for instance applied for the enantioselective determination of ketoprofen [137], of chlorpheniramine and its major metabolites [138].

Chiral LC–MS method is generally combined with racemic sample pretreatment methods, such as LLE [57, 133, 138], 96-well plate SPE [135, 137], or TFC [136].

Desai and Armstrong [139] discussed the optimization of mobile-phase conditions for enantioselective separations in combination with LC–MS. Their conclusions can be summarized as: (a) polar organic mobile phases as used with Chirobiotic T and V columns are best compatible with ESI-MS, (b) normal-phase mobile phases require post-column addition, (c) high-aqueous mobile phases compromise ESI-MS sensitivity, and (d) ammonium trifluoroacetate provides response enhancement for analytes with amine or amide groups.

8. Increasing sample throughput

Significant progress has been made in quantitative bioanalysis using LC–MS. Improvements in sample pretreatment, LC column technology, and MS instrumentation resulted in a significant higher sample throughput. Some approaches to further improvements in sample throughput are briefly discussed in this section.

In terms of sample pretreatment, the implementation of 96-well based methods for protein precipitation [97, 101-102], LLE [103-104], and SPE [95-96] greatly contributed to improving the sample throughput. For SPE, 384-well plate methodology has even been demonstrated [106]. Staggered parallel high-flow TFC–MS is another way to increase the sample throughput [113-115].

In terms of LC technology, the reduction of column dimensions, the introduction of monolithic columns [125-128], and the use of generic LC conditions [140] contributed to the increase in sample throughput. Romanyshyn et al. [141-142] discussed various aspects of high-throughput bioanalytical LC–MS on ultra-short columns, using either ballistic gradients of fast isocratic elution. The recently introduced ultra-performance LC on columns packed with 1.7-μm particles will result in further progress in this area [143].

With respect to MS, progress in collision cells, e.g., LINAC cell and travelling-wave ion tunnels (Ch. 11.3.5), has enabled shorter dwell times in SRM without response losses. Enhanced-resolution triple quadrupole and Q–LIT instruments show promising results, but have not yet been as widely applied in routine bioanalytical LC–MS.

In some reports, the limits of high sample throughput have been explored and challenged. Some examples are briefly described here.

Zweigenbaum *et al.* [104] developed a method to analyse 1152 urine samples for 5 benzodiazepines and an ANIS within 12 hr. LLE in 96-well plate format using a Tomtec Quadra robotic liquid handler was performed prior to LC on a 15×2.1-mm-ID C_{18} column (3 μm) with 33% acetonitrile in water as mobile phase at 1 ml/min, which is about threefold the optimum flow-rate. ESI-MS was performed in SRM mode. Due to speed limitations of the autosampler used, four autosamplers had to be used in parallel to perform analyte injection. The overall cycle time between injection was ~37 s.

Onorato *et al.* [144] reported the use of 30×1-mm-ID C_{18} column (3 μm), operated at 0.7 ml/min and 70°C, to achieve a separation of idoxifene and its major metabolite within 10 s/sample. Sequential injections were performed using a Gilson multiprobe liquid handler, capable of aspirating eight samples simultaneously. Sample pretreatment was performed by LLE in 96-well plate format.

Hsieh *et al.* [127-128] demonstrated the separation of a parent drug, its major metabolite, and an ANIS within 30 s using a monolithic column, operated at a flow-rate of 4 ml/min, coupled to APCI-MS.

There are various other ways to further increase the effective sample throughput, *i.e.*, pooling of samples prior to analysis, performing cassette-dosing experiment, and/or parallel analysis.

In cassette-dosing experiments, *in vivo* pharmacokinetics of multiple drug candidates are investigated simultaneously in one animal (n-in-one-study). The rationale is not only high-throughput analysis, but also reduction of the number of test animals required. Olah *et al.* [145-146] reported the simultaneous quantitation of up to twelve drug candidates in a concentration range between 1 and 1000 ng/ml in two different animals. On-line SPE using a Prospekt valve-switching system was applied in these studies. Potential risks of cassette-dosing experiments are drug-drug interactions and saturation of receptors. To avoid these risks and still increase the sample throughput, pooling of plasma samples can be performed: each animal is subjected to only one drug candidate, while after pooling different plasma samples several drug candidates are determined simultaneously [147].

Three general approaches are available for parallel sample analysis.

The staggered parallel TFC–LC–MS approach was already discussed Ch. 11.6.5.

Four staggered parallel columns were coupled to one ESI-MS system via a switching valve, enabling the subsequent detection of the chromatogram from each column [148]. The overall run time was reduced from 4.5 to 1.65 min per sample.

The third parallel approach involves the use of a four-channel multiplexed ESI source (Ch. 5.5.3). In this system, four parallel LC columns are connected to four individual sprayers placed in one ESI source. The sprayers are indexed, *i.e.*, the data from each sprayer are acquired in a separate datafile. Initial results were reported by Bayliss *et al.* [149] and Yang *et al.* [64]. In the latter study, LOR and DCL were studied in four different sample matrices (Ch. 11.4.5).

Reviewing all these systems, one starts thinking about combining different approaches into one system. Deng *et al.* [150] reported a sample throughput of 120 samples per hour using a combination of the simultaneous SPE on four 96-well plates using a Zymark robot system and a Tecan Genesis liquid handler. The extracts were injected by means of a four-injector autosampler onto four parallel 100×4.6-mm-ID monolithic columns, operated at 4 ml/min. After a 1:4 split, the column effluents were fed into a four-channel multiplexed ESI source for LC–MS analysis in SRM mode. In this way, twelve 96-well plates containing 1152 samples could be analysed within 10 hr, resulting in a throughput of ~30 s/sample. The cycle time of the SPE matches the LC–MS analysis time.

The future will tell whether systems like this will be used as routine tools in quantitative bioanalysis, or whether perhaps an even faster approach will become state of the art.

9. References

1. T.R. Covey, E.D. Lee, J.D. Henion, *High-speed LC–MS–MS for the determination of drugs in biological samples*, Anal. Chem., 58 (1986) 2453.
2. M. Jemal, *High-throughput quantitative bioanalysis by LC–MS–MS*, Biomed. Chromatogr., 14 (2000) 422.
3. G. Hopfgartner, E. Bourgogne, *Quantitative high-throughput analysis of drugs in biological matrices by MS*, Mass Spectrom. Rev., 22 (2003) 195.
4. M.J. Berna, B.L. Ackermann, A.T. Murphy, *High-throughput chromatographic approaches to LC–MS–MS bioanalysis to support drug discovery and development*, Anal. Chim. Acta, 509 (2004) 1.
5. M.S. Lee, *LC–MS applications in drug development*, 2002, John Wiley & Sons, New York, USA.
6. FDA Guidance for Industry. Bioanalytical method validation (http://www.fda.gov/cder/guidance/4252fnl.htm).
7. D..A. Wells, *High throughput bioanalytical sample preparation. Methods and automation strategies*, 2003, Elsevier Science, Amsterdam, the Netherlands.
8. B.K. Matuszewski, C.M. Chavez-Eng, M.L. Constanzer, *Development of LC–MS–MS methods for the determination of a new oxytocin receptor antagonist (L-368,899) extracted from human plasma and urine: a case of lack of specificity due to the presence of metabolites*, J. Chromatogr. B, 716 (1998) 195.
9. K. Wang, M. Nano, T. Mulligan, E.D. Bush, R.W. Edom, *Derivatization of 5-fluorouracil with 4-bromomethyl-7-methoxycoumarin for determination by LC–MS*, J. Am. Soc. Mass Spectrom., 9 (1998) 970.
10. R.F. Venn, B. Kaye, P.V. Macrae, K.C. Saunders, *Clinical analysis of sampatrilat, a combined renal endopeptidase and angiotensin-converting enzyme inhibitor - I: Assay in plasma of human volunteers by API-MS following derivatization with BF3-methanol*, J. Pharm. Biomed. Anal., 16 (1998) 875.
11. B. Law, D. Temesi, *Factors to consider in the development of generic bioanalytical LC–MS methods to support drug discovery*, J. Chromatogr. B, 748 (2000) 21.
12. I.J. Fraser, G.J. Dear, R. Plumb, M. L'Affineur, D. Fraser, A.J. Skippen, *The use of capillary LC–ESI-MS for the analysis of small volume blood samples from serially bled mice to determine the pharmacokinetics of early discovery compounds*, Rapid

Commun. Mass Spectrom., 13 (1999) 2366.

13. R. S. Plumb, H. Warwick, D. Higton, G. J. Dear, D. N. Mallett, *Determination of 4-hydroxytamoxifen in mouse plasma in the pg/mL range by gradient capillary LC–MS–MS*, Rapid Commun. Mass Spectrom., 15 (2001) 297.

14. B.M.M. Remmerie, L.L.A. Sips, R. de Vries, J. de Jong, A.M. Schothuis, E.W.J. Hooijschuur, N.C. van de Merbel, *Validated method for the determination of risperidone and 9-hydroxyrisperidone in human plasma by LC–MS–MS*, J. Chromatogr. B, 783 (2003) 461.

15. D.Q. Liu, T. Pereira, *Interference of a carbamoyl glucuronide metabolite in quantitative LC–MS–MS*, Rapid Commun. Mass Spectrom., 16 (2002) 142.

16. M. Jemal, Y.-Q. Xia, *The need for adequate chromatographic separation in the quantitative determination of drugs in biological samples by LC–MS–MS*, Rapid Commun. Mass Spectrom., 13 (1999) 97.

17. T.K. Majumdar, S. Vedananda, F.L.S. Tse, *Rapid analysis of MMI270B, an inhibitor of matrix metalloproteases in human plasma by LC–MS–MS: matrix interference in patient samples*, Biomed. Chromatogr., 18 (2004) 77

18. C. Yang, J.D. Henion, *APPI LC–MS determination of idoxifene and its metabolites in human plasma*, J. Chromatogr. A, 970 (2002) 155.

19. Y. Hsieh, K. Merkle, G. Wang, J.-M. Brisson, W.A. Korfmacher, *LC–APPI-MS–MS analysis for small molecules in plasma*, Anal. Chem., 75 (2003) 3122.

20. C.M. Chavez-Eng, M.L. Constanzer, B.K. Matuszewski, *LC–MS–MS evaluation and determination of stable isotope labeled analogs of rofecoxib in human plasma samples from oral bioavailability studies*, J. Chromatogr. B, 767 (2002) 117.

21. I. Fu, E.J. Woolf, B.K. Matuszewski, *Effect of the sample matrix on the determination of indinavir in human urine by LC with turbo ion spray MS–MS detection*, J. Pharm. Biomed. Anal., 18 (1998) 347.

22. E. Stokvis, H. Rosing, L. López-Lázaro, J.H.M. Schellens, J. H. Beijnen, *Switching from an analogous to a stable ILIS for the LC–MS–MS quantitation of the novel anticancer drug Kahalalide F significantly improves assay performance*, Biomed. Chromatogr., 18 (2004) 400.

23. M. Jemal, A. Schuster, D.B. Whigan, *LC–MS–MS methods for quantitation of mevalonic acid in human plasma and urine: method validation, demonstration of using a surrogate analyte, and demonstration of unacceptable matrix effect in spite of use of a stable isotope analog internal standard*, Rapid Commun. Mass Spectrom., 17 (2003) 1723.

24. L.E. Sojo, G. Lum, P. Chee, *Internal standard signal suppression by co-eluting analyte in isotope dilution LC–ESI-MS*, Analyst, 128 (2003) 51.

25. H. R. Liang, R. L. Foltz, M. Meng, P. Bennett, *Ionization enhancement in APCI and suppression in ESI between target drugs and ILIS in quantitative LC–MS–MS*, Rapid Commun. Mass Spectrom., 17 (2003) 2815.

26. M.J. Avery, *Quantitative characterization of differential ion suppression on LC–API-MS bioanalytical methods*, Rapid Commun. Mass Spectrom., 17 (2003) 197.

27. K.M. Whalen, K.J. Rogers, M.J. Cole, J.S. Janiszewski, *AutoScan: an automated workstation for rapid determination of MS–MS conditions for quantitative bioanalytical MS*, Rapid Commun. Mass Spectrom., 14 (2000) 2074.

28. D.M. Higton, *A rapid, automated approach to optimisation of SRM conditions for quantitative bioanalytical MS*, Rapid Commun. Mass Spectrom., 15 (2001) 1922.

29. L. Yang, M. Amad, W.M. Winnik, A.E. Schoen, H. Schweingruber, I. Mylchreest, P. J. Rudewicz, *Investigation of an enhanced resolution triple quadrupole MS for high-throughput LC–MS–MS assays*, Rapid Commun. Mass Spectrom., 16 (2002) 2060.

30. M. Jemal, Z. Ouyang, *Enhanced resolution triple-quadrupole MS for fast quantitative bioanalysis using LC–MS–MS: investigations of parameters that affect ruggedness*, Rapid Commun. Mass Spectrom., 17 (2003) 24.

31. X. Xu, J. Veals, W.A. Korfmacher, *Comparison of conventional and enhanced mass resolution triple-quadrupole MS for discovery bioanalytical applications*, Rapid Commun. Mass Spectrom., 17 (2003) 832.

32. N. Hughes, W.M. Winnik, J.-J. Dunyach, M. Amad, M. Splendore, G. Paul, *High-sensitivity quantitation of cabergoline and pergolide using a triple-quadrupole MS with enhanced mass-resolution capabilities*, J. Mass Spectrom., 38 (2003) 743.

33. E. Woolf, H.M. Haddix, B. Matuszewski, *Determination of in-vivo metabolite of a HIV protease inhibitor in human plasma by LC–MS–MS*, J. Chromatogr. A, 762 (1997) 311.

34. M.A. Curtis, L.C. Matassa, R. Demers, K. Fegan, *Expanding the linear dynamic range in quantitative LC–MS–MS by the use of multiple product ions*, Rapid Commun. Mass Spectrom., 15 (2001) 963.

35. M. Jemal, R.B. Almond, D.S. Teitz, *Quantitative bioanalysis utilizing LC–ESI-MS via SIM of the $[M+Na]^+$*, Rapid Commun. Mass Spectrom., 11 (1997) 1083.

36. H. Yang, Y. Feng, Y. Luan, *Determination of simvastatin in human plasma by LC–MS*, J. Chromatogr. B, 785 (2003) 369.

37. R. Andreoli, P. Manini, E. Bergamaschi, A. Mutti, I. Franchini, W.M.A. Niessen, *Determination of naphthalene metabolites in human urine by LC–ESI-MS*, J. Chromatogr. A, 847 (1999) 9.

38. M. Yao, V.R. Shah, W.C. Shyu, N.R. Srinivas, *Sensitive LC–MS assay for the simultaneous quantitation of nefazodone and its metabolites in human plasma using SIM*, J. Chromatogr. B, 718 (1998) 77.

39. M. Abdel-Hamid, L. Novotny, H. Hamza, *LC–MS determination of celecoxib in plasma using SIM and its use in clinical pharmacokinetics*, J. Chromatogr. B, 753 (2001) 401.

40. M.L. Constanzer, C.M. Chavez-Eng, J. Dru, W.F. Kline, B.K. Matuszewski, *Determination of a novel substance P inhibitor in human plasma by LC–APCI-MS detection using single and triple quadrupole detectors*, J. Chromatogr. B, 807 (2004) 807.

41. Y.-Q. Xia, J.D. Miller, R. Bakhtiar, R.B. Franklin, D.Q. Liu, *Use of a Q–LIT-MS in metabolite identification and bioanalysis*, Rapid Commun. Mass Spectrom., 17 (2003) 1137.

42. M.A. Anderson, T. Wachs, J.D. Henion, *Quantitative ionspray LC–MS–MS determination of reserpine in equine plasma,* J. Mass Spectrom., 32 (1997) 152.

43. J. Ke, M. Yancey, S. Zhang, S. Lowes, J. Henion, *Quantitative LC-MS-MS determination of reserpine in FVB/N mouse plasma using a "chelating" agent (disodium EDTA) for releasing protein-bound analytes during 96-well LLE*, J. Chromatogr. B, 742 (2000) 369.

44. M. Aravagiri, S.R. Marder, *Simultaneous determination of risperidone and 9-hydroxyrisperidone in plasma by LC–ESI-MS*, J. Mass Spectrom., 35 (2000) 718.

45. J. Flarakos, W. Luo, M. Aman, D. Svinarov, N. Gerber, P. Vouros, *Quantification of*

risperidone and 9-hydroxyrisperidone in plasma and saliva from adult and pediatric patients by LC–MS, J. Chromatogr. A, 1026 (2004) 175.

46. Y. Wu, J. Zhao, J.D. Henion, W.A. Korfmacher, A.P. Lapiguera, C.-C. Lin, *Microsample determination of lovastatin and its hydroxy acid metabolite in mouse and rat plasma by LC–ESI-MS–MS*, J. Mass Spectrom., 32 (1997) 379.

47. J.J. Zhao, I.H. Xie, A.Y. Yang, B.A. Roadcap, J. D. Rogers, *Quantitation of simvastatin and its β-hydroxy acid in human plasma by LLE and LC–MS–MS*, J. Mass Spectrom., 35 (2000) 1133.

48. J.J. Zhao, A.Y. Yang, J.D. Rogers, *Effects of LC mobile phase buffer contents on the ionization and fragmentation of analytes in LC–ESI-MS–MS determination*, J. Mass Spectrom., 37 (2002) 421.

49. N. Zhang, A. Yang, J.D. Rogers, J.J. Zhao, *Quantitative analysis of simvastatin and its β-hydroxy acid in human plasma using automated LLE based on 96-well plate format and LC–MS–MS*, J. Pharm. Biomed. Anal., 34 (2004) 175.

50. W.W. Bullen, R.A. Miller, R.N. Hayes, *Development and validation of a LC–MS–MS assay for atorvastatin, ortho-hydroxy atorvastatin and para-hydroxy atorvastatin in human, dog and rat plasma*, J. Am. Soc. Mass Spectrom., 10 (1999) 55.

51. M. Jemal, Z. Ouyang, B.-C. Chen, D. Teitz, *Quantitation of the acid and lactone forms of atorvastatin and its biotransformation products in human serum by LC–ESI-MS–MS*, Rapid Commun. Mass Spectrom., 13 (1999) 1003.

52. M Jemal, S. Rao, I. Salahudeen, B.-C. Chen, R. Kates, *Quantitation of cerivastatin and its seven acid and lactone biotransformation products in human serum by LC–ESI-MS–MS*, J. Chromatogr. B, 736 (1999) 19.

53. N.V.S. Ramakrishna, M. Koteshwara, K.N. Vishwottam, S. Puran, S. Manoj, M. Santosh, *Simple, sensitive and rapid LC–MS–MS method for the quantitation of cerivastatin in human plasma – application to pharmacokinetic studies*, J. Pharm. Biomed. Anal., 36 (2004) 505.

54. C.K. Hull, A.D. Penman, C.K. Smith, P.D. Martin, *Quantification of rosuvastatin in human plasma by automated SPE using MS–MS detection*, J. Chromatogr. B, 772 (2002) 219.

55. C.K. Hull, P.D. Martin, M.J. Warwick, E. Thomas, *Quantification of the N-desmethyl metabolite of rosuvastatin in human plasma by automated SPE followed by LC–MS–MS detection*, J. Pharm. Biomed. Anal., 35 (2004) 609.

56. J. Eichhorst, M. Etter, J. Lepage, D.C. Lehotay, *Urinary screening of methylphenidate (Ritalin) abuse: a comparison of LC–MS–MS, GC–MS, and immunoassay methods*, Clin. Biochem., 37 (2004) 175.

57. L. Ramos, R. Bakhtiar, T. Majumdar, M. Hayes, F.L.S. Tse, *LC–APCI-MS–MS enantiomeric separation of dl-threo-methylphenidate, (Ritalin®) using a macrocyclic antibiotic as the chiral selector*, Rapid Commun. Mass Spectrom., 13 (1999) 2054.

58. L. Ramos, R. Bakhtiar, F.L.S. Tse, *LLE using 96-well plate format in conjunction with LC–MS–MS for quantitative determination of methylphenidate (Ritalin«) in human plasma*, Rapid Commun. Mass Spectrom., 14 (2000) 740.

59. R. Bakhtiar, L. Ramos, F.L.S. Tse, *Quantification of methylphenidate in rat, rabbit and dog plasma using a chiral LC–MS–MS method. Application to toxicokinetic studies*, Anal. Chim. Acta, 469 (2002) 261.

60. D.R. Doerge, C.M. Fogle, M.G. Paule, M. McCullagh, S. Bajic, *Analysis of methylphenidate and its metabolite ritalinic acid in monkey plasma by LC–ESI-MS,*

Rapid Commun. Mass Spectrom., 14 (2000) 619.

61.   N. Barbarin, D.B. Mawhinney, R. Black, J.D. Henion, *High-throughput SRM LC–MS determination of methylphenidate and its major metabolite, ritalinic acid, in rat plasma employing monolithic columns*, J. Chromatogr. B, 783 (2003) 73.

62.   Y. Yang, J. Kameoka, T. Wachs, J.D. Henion, H.G. Craighead, *Quantitative MS determination of methylphenidate concentration in urine using an ESI source integrated with a polymer microchip*, Anal. Chem., 76 (2004) 2568.

63.   F.C.W. Sutherland, A.D. de Jager, D. Badenhorst, T. Scanes, H.K.L. Hundt, K.J. Swart, A.F. Hundt, *Sensitive LC–MS–MS method for the determination of loratadine and its major active metabolite descarboethoxyloratadine in human plasma*, J. Chromatogr. A, 914 (2001) 37.

64.   L. Yang, T.D. Mann, D. Little, N. Wu, R.P. Clement, P.J. Rudewicz, *Evaluation of a four-channel MUX ESI MS–MS for the simultaneous validation of LC–MS–MS methods in four different preclinical matrixes*, Anal. Chem., 73 (2001) 1740.

65.   L. Yang, R.P. Clement, B. Kantesaria, L. Reyderman, F. Beaudry, C. Grandmaison, L. Di Donato, R. Masse, P.J. Rudewicz, *Validation of a sensitive and automated 96-well SPE LC–MS–MS method for the determination of desloratadine and 3-hydroxydesloratadine in human plasma*, J. Chromatogr. B, 792 (2003) 229.

66.   W. Naidong, T. Addison, T. Schneider, X. Jiang, T.D.J. Halls, *A sensitive LC–MS–MS method using silica column and aqueous-organic mobile phase for the analysis of loratadine and descarboethoxy-loratadine in human plasma*, J. Pharm. Biomed. Anal., 32 (2003) 609.

67.   W. Naidong, W. Zhou, Q. Song, S. Zhou, *Direct injection of 96-well organic extracts onto a HILIC–MS–MS system using a silica stationary phase and an aqueous/organic mobile phase*, Rapid Commun. Mass Spectrom., 18 (2004) 2963.

68.   I.I. Salem, J. Idrees, J.I. Al Tamimi, *Determination of loratadine in human plasma by LC–ESI-ion-trap-MS–MS*, J. Pharm. Biomed. Anal., 34 (2004) 141.

69.   B.K. Matuszewski, M.L. Constanzer, C.M. Chavez-Eng, *Matrix effects in quantitative LC–MS–MS analysis of biological fluids: a method for determination of finasteride in human plasma at pg/ml concentrations*, Anal. Chem., 70 (1998) 882.

70.   B.K. Matuszewski, M.L. Constanzer, C.M. Chavez-Eng, *Strategies for the assessment of matrix effect in quantitative bioanalytical methods based on LC–MS–MS*, Anal. Chem., 75 (2003) 3019.

71.   E. Gelpí, J. Abián, F. Artigas, *Effects of sample matrix and LC eluent composition on the thermospray response for polar compounds*, Rapid Commun. Mass Spectrom., 2 (1988) 232.

72.   J. Hajšlová, J. Zrostlíková, *Matrix effect in the (ultra)trace analysis of pesticide residues in food and biotic matrices*, J. Chromatogr. A, 1000 (2003) 181.

73.   T.M. Annesley, *Ion suppression in MS*, Clin. Chem., 49 (2003) 1041.

74.   D.L. Buhrman, P.I. Price, P.J. Rudewicz, *Quantitation of SR 27417 in human plasma using LC–ESI-MS–MS: A study of ion suppression*, J. Am. Soc. Mass Spectrom., 7 (1996) 1099.

75.   R. Bonfiglio, R.C. King, T.V. Olah, K. Merckle, *The effects of sample preparation methods on the variability of ESI response for model drug compounds*, Rapid Commun. Mass Spectrom., 13 (1999) 1175.

76.   R.C. King, R. Bonfiglio, C. Fernandez-Metzler, C. Miller-Stein, T.V. Olah, *Mechanistic investigation of ionization suppression in ESI*, J. Am. Soc. Mass

Spectrom., 11 (2000) 942.

77.   S. Souverain, S. Rudaz, J.-L. Veuthey, *Matrix effect in LC–ESI–MS and LC–APCI–MS with off-line and on-line extraction procedures*, J. Chromatogr. A, 1058 (2004) 61.

78.   K.A. Barnes, R.J. Fussell, J.R. Startin, M.K. Pegg, S.A. Thorpe, S.L. Reynolds, *LC–APCI–MS–MS with ionization polarity switching for the determination of selected pesticides*, Rapid Commun. Mass Spectrom., 11 (1997) 117.

79.   C. Chin, Z.P. Zhang, H.T. Karnes, *A study of matrix effects on an LC–MS–MS assay for olanzapine and desmethyl olanzapine*, J. Pharm. Biomed. Anal., 35 (2004) 1149.

80.   P. Kebarle, L. Tang, *From ions in solution to ions in the gas phase. The mechanism of ESI–MS*, Anal. Chem., 65 (1993) 972A.

81.   N.B. Cech, C.G. Enke, *Relating ESI response to nonpolar character of small peptides*, Anal. Chem., 72 (2000) 2717.

82.   S. Zhou, K.D. Cook, *A mechanistic study of ESI–MS: Charge gradients within electrospray droplets and their influence on ion response*, J. Am. Soc. Mass Spectrom., 12 (2001) 206.

83.   H. Mei, Y. Hsieh, C. Nardo, X. Xu, S. Wang, K. Ng, W.A. Korfmacher, *Investigation of matrix effects in bioanalytical LC–MS–MS assays: application to drug discovery*, Rapid Commun. Mass Spectrom., 17 (2003) 97.

84.   Y. Hsieh, M. Chintala, H. Mei, J. Agans, J.-M. Brisson, K. Ng, W.A. Korfmacher, *Quantitative screening and matrix effect studies of drug discovery compounds in monkey plasma using fast-gradient LC–MS–MS*, Rapid Commun. Mass Spectrom., 15 (2001) 2481.

85.   P.R. Tiller, L.A. Romanyshyn, *Implications of matrix effects in ultra-fast gradient or fast isocratic LC–MS in drug discovery*, Rapid Commun. Mass Spectrom., 16 (2002) 92.

86.   C. Polson, P. Sarkar, B. Incledon, V. Raguvaran, R. Grant, *Optimization of protein precipitation based upon effectiveness of protein removal and ionization effect in LC–MS–MS*, J. Chromatogr. B, 785 (2003) 263.

87.   R. Kitamura, K. Matsuoka, E. Matsushima, Y. Kawaguchi, *Improvement in precision of the LC–ESI–MS–MS analysis of 3'-C-ethynylcytidine in rat plasma*, J. Chromatogr. B, 754 (2001) 113.

88.   S. Ito, K. Tsukada, *Matrix effect and correction by standard addition in quantitative LC–MS analysis of diarrhetic shellfish poisoning toxins*, J. Chromatogr. A, 943 (2001) 39.

89.   X.S. Tong, J. Wang, S. Zheng, J.V. Pivnichny, X. Shen, M. Donnelly, K. Vakerich, C. Nunes, J. Fenyk-Melody, *Effect of signal interference from dosing excipients on pharmacokinetic screening of drug candidates by LC–MS*, Anal. Chem., 74 (2002) 6305.

90.   J. Schuhmacher, D. Zimmer, F. Tesche, V. Pickard, *Matrix effects during analysis of plasma samples by ESI and APCI–MS: practical approaches to their elimination*, Rapid Commun. Mass Spectrom., 17 (2003) 1950.

91.   W.Z. Shou, W. Naidong, *Post-column infusion study of the 'dosing vehicle effect' in the LC–MS–MS analysis of discovery pharmacokinetic samples*, Rapid Commun. Mass Spectrom., 17 (2003) 589.

92.   J.J. Zheng, E.D. Lynch, S.E. Unger, *Comparison of SPE and fast LC to eliminate MS matrix effects from microsomal incubation products*, J. Pharm. Biomed. Anal., 28

(2002) 279.

93. C. Müller, P. Schäfer, M. Störtzel, S. Vogt, W. Weinmann, *Ion suppression effects in LC–ESI-transport-region-CID-MS with different serum extraction methods for systematic toxicological analysis with mass spectra libraries*, J. Chromatogr. B, 773 (2002) 47.

94. M. Becker, E. Zittlau, M. Petz, *Residue analysis of 15 penicillins and cephalosporins in bovine muscle, kidney and milk by LC–MS–MS*, Anal. Chim. Acta, 520 (2004) 19.

95. B. Kaye, W.J. Heron, P.V. Mcrae, S. Robinson, D.A. Stopher, R.F. Venn, W. Wild, *Rapid SPE technique for the high-throughput assay of darifenacin in human plasma*, Anal. Chem., 68 (1996) 1658.

96. J.P. Allanson, R.A. Biddlecombe, A.E. Jones, S. Pleasance, *The use of automated SPE in the '96 well' format for high throughput bioanalysis using LC–MS–MS*, Rapid Commun. Mass Spectrom., 10 (1996) 811.

97. R.A. Biddlecombe, S. Pleasance, *Automated protein precipitation by filtration in the 96-well format*, J. Chromatogr. B, 734 (1999) 257.

98. M. Berna, A. T. Murphy, B. Wilken, B. Ackermann, *Collection, storage, and filtration of in vivo study samples using 96-well filter plates to facilitate automated sample preparation and LC–MS–MS analysis*, Anal. Chem., 74 (2002) 1197.

99. E.N. Fung, Y.-H. Chen, Y.Y. Lau, *Semi-automatic high-throughput determination of plasma protein binding using a 96-well plate filtrate assembly and fast LC–MS–MS*, J. Chromatogr. B, 795 (2003) 187.

100. S. Souverain, S. Rudaz, J.-L. Veuthey, *Protein precipitation for the analysis of a drug cocktail in plasma by LC–ESI-MS*, J. Pharm. Biomed. Anal., 35 (2004) 913.

101. A.P. Watt, D.Morrison, K.L. Locker, D.C. Evans, *Higher throughput bioanalysis by automation of a protein precipitation assay using a 96-well format with detection by LC–MS–MS*, Anal. Chem., 72 (2000) 979.

102. D. O'Connor, D.E. Clarke, D. Morrison, A.P. Watt, *Determination of drug concentrations in plasma by a highly automated, generic and flexible protein precipitation and LC–MS–MS method applicable to the drug discovery environment*, Rapid Commun. Mass Spectrom., 16 (2002) 1065.

103. S. Steinborner, J. Henion, *LLE in the 96-well plate format with SRM LC–MS quantitative determination of methotrexate and its major metabolite in human plasma*, Anal. Chem., 71 (1999) 2340.

104. J. Zweigenbaum, K. Heinig, S. Steinborner, T. Wachs, J. Henion, *High-throughput bioanalytical LC–MS–MS determination of benzodiazepines in human urine: 1000 samples per 12 hours*, Anal. Chem., 71 (1999) 2294.

105. W.Z. Shou, X. Jiang, B.D. Beato, W. Naidong, *A highly automated 96-well SPE and LC–MS–MS method for the determination of fentanyl in human plasma*, Rapid Commun. Mass Spectrom., 15 (2001) 466.

106. G. Rule, M. Chapple, J.D. Henion, *A 384-well SPE for LC–MS–MS determination of methotrexate and its 7-hydroxy metabolite in human urine and plasma*, Anal. Chem., 73 (2001) 439.

107. M Jemal, D. Teitz, Z. Ouyang, S. Khan, *Comparison of plasma sample purification by manual LLE, automated 96-well LLE, and automated 96-well SPE for analysis by LC–MS–MS*, J. Chromatogr. B, 732 (1999) 501.

108. M. Zell, C. Husser, G. Hopfgartner, *Low picogram determination of Ro 48-6791 and its major metabolite, Ro 48-6792, in plasma with column-switching microbore*

LC–ESI-MS–MS, Rapid Commun. Mass Spectrom., 11 (1997) 1107.

109. C. Siethoff, M. Orth, A. Ortling, E. Brendel, W. Wagner-Redeker, *Simultaneous determination of capecitabine and its metabolite 5-fluorouracil by column switching and LC–MS–MS*, J. Mass Spectrom., 39 (2004) 884.

110. T. Kumazawa, X.-P. Lee, K. Sato, O. Suzuki, *SPME and LC–MS in drug analysis*, Anal. Chim. Acta, 492 (2003) 49.

111. S. Souverain, S. Rudaz, J.-L. Veuthey, *RAM and large particle supports for on-line sample preparation: an attractive approach for biological fluids analysis*, J. Chromatogr. B, 801 (2004) 141.

112. D. Zimmer, V. Pickard, W. Czembor, C. Müller, *Comparison of TFC with automated SPE in 96-well plates and LLE used as plasma sample preparation techniques for LC–MS–MS*, J. Chromatogr. A, 854 (1999) 23.

113. J. L. Herman, *Generic method for on-line extraction of drug substances in the presence of biological matrices using TFC*, Rapid Commun. Mass Spectrom., 16 (2002) 421.

114. J.L. Herman, *Generic approach to high throughput ADME screening for lead candidate optimization*, Int. J. Mass Spectrom., 238 (2004) 107.

115. R.C. King, C. Miller-Stein, D.J. Magiera, J. Brann, *Description and validation of a staggered parallel LC system for GLP-level quantitative analysis by LC–MS–MS*, Rapid Commun. Mass Spectrom., 16 (2002) 43.

116. M. Jemal, M. Huang, X. Jiang, Y. Mao, M.L. Powell, *Direct injection versus LLE for plasma sample analysis by LC–MS–MS*, Rapid Commun. Mass Spectrom., 13 (1999) 2125.

117. J.-T. Wu, H. Zeng, M. Qian, B.L. Brogdon, S.E. Unger, *Direct plasma sample injection in multiple-component LC–MS–MS assays for high-throughput pharmacokinetic screening*, Anal. Chem., 72 (2000) 61.

118. M. Kollroser, C. Schober, *Simultaneous analysis of flunitrazepam and its major metabolites in human plasma by LC–MS–MS*, J. Pharm. Biomed. Anal., 28 (2002) 1173.

119. J. Dankers, J. van den Elshout, G. Ahr, E. Brendel, C. van der Heiden, *Determination of nifedipine in human plasma by flow-injection–tandem mass spectrometry*, J. Chromatogr. B, 710 (1998) 115.

120. S. Chen, P.M. Carvey, *Validation of LLE followed by flow-injection negative-ion ESI-MS assay to Topiramate in human plasma*, Rapid Commun. Mass Spectrom., 15 (2001) 159.

121. J. Ding, U.D. Neue, *A new approach to the effective preparation of plasma samples for rapid drug quantitation using on-line SPE–MS*, Rapid Commun. Mass Spectrom., 13 (1999) 2151.

122. J.-M. Dethy, B.L. Ackermann, C. Delatour, J.D. Henion, G.A. Schultz, *Demonstration of direct bioanalysis of drugs in plasma using nano-ESI infusion from a silicon chip coupled with MS–MS*, Anal. Chem., 75 (2003) 805.

123. J.T. Kapron, E. Pace, C.K. Van Pelt, J.D. Henion, *Quantitation of midazolam in human plasma by automated chip-based infusion nano-ESI-MS–MS*, Rapid Commun. Mass Spectrom., 17 (2003) 2019.

124. L.A. Leuthold, C. Grivet, M. Allen, M. Baumert, G. Hopfgartner, *Simultaneous SRM, MS–MS and MS³ quantitation for the analysis of pharmaceutical compounds in human plasma using chip-based infusion*, Rapid Commun. Mass Spectrom., 18

(2004) 1995.

125. R. Plumb, G. Dear, D. Mallett, J. Ayrton, *Direct analysis of pharmaceutical compounds in human plasma with chromatographic resolution using an alkyl-bonded silica rod column*, Rapid Commun. Mass Spectrom., 15 (2001) 986.

126. J.-T. Wu, H. Zeng, Y. Deng, S.E. Unger, *High-speed LC–MS–MS using a monolithic column for high-throughput bioanalysis*, Rapid Commun. Mass Spectrom., 15 (2001) 1113.

127. Y. Hsieh, G. Wang, Y. Wang, S. Chackalamannil, J.-M. Brisson, K. Ng, W.A. Korfmacher, *Simultaneous determination of a drug candidate and its metabolite in rat plasma samples using ultrafast monolithic column LC–MS–MS*, Rapid Commun. Mass Spectrom., 16 (2002) 944.

128. Y. Hsieh, G. Wang, Y. Wang, W.A. Korfmacher, *Direct plasma analysis of drug compounds using monolithic column LC–MS–MS*, Anal. Chem., 75 (2003) 1812.

129. S. Hsieh, T. Tobien, K. Koch, J. Dunn, *Increasing throughput of parallel on-line extraction LC–ESI-MS–MS system for GLP quantitative bioanalysis in drug development*, Rapid Commun. Mass Spectrom., 18 (2004) 285.

130. J. Ayrton, R.A. Clare, G.J. Dear, D.N. Mallett, R.S. Plumb, *Ultra-high flow-rate capillary LC with MS detection for the direct analysis of pharmaceuticals in plasma at sub-ng/ml concentrations*, Rapid Commun. Mass Spectrom., 13 (1999) 1657.

131. W. Naidong, W. Shou, Y.-L. Chen, X. Jiang, *Novel LC–MS–MS methods using silica columns and aqueous-organic mobile phases for quantitative analysis of polar ionic analytes in biological fluids*, J. Chromatogr. B, 754 (2001) 387.

132. W. Naidong, *Bioanalytical LC–MS–MS methods on underivatized silica columns with aqueous/organic mobile phases*, J. Chromatogr. B, 796 (2003) 209.

133. B. Lindmark, M. Ahnoff, B.-A. Persson, *Enantioselective determination of felodipine in human plasma by chiral normal-phase LC–ESI-MS*, J. Pharm. Biomed. Anal., 27 (2002) 489.

134. J.E. Paanakker, J. de Jong, J.M.S.L. Thio, H.J.M. van Hal, *Validation of an LC–MS assay for the quantification of the enantiomers of Org 4428 in human plasma*, J. Pharm. Biomed. Anal., 16 (1998) 981.

135. K.B. Joyce, A.E. Jones, R.J. Scott, R.A. Biddlecombe, S. Pleasance, *Determination of the enantiomers of salbutamol and its 4-O-sulphate metabolites in biological matrices by chiral LC–MS–MS*, Rapid Commun. Mass Spectrom., 12 (1998) 1899.

136. S.T. Wu, J. Xing, A. Apedo, D.B. Wang-Iverson, T.V. Olah, A.A. Tymiak, N. Zhao, *High-throughput chiral analysis of albuterol enantiomers in dog plasma using on-line sample extraction/polar organic mode chiral LC–MS–MS detection*, Rapid Commun. Mass Spectrom., 18 (2004) 2531.

137. T.H. Eichhold, R.E. Bailey, S.L. Tanguay, S.H. Hoke, II, *Determination of (R)- and (S)-ketoprofen in human plasma by LC–MS–MS following automated SPE in the 96-well format*, J. Mass Spectrom., 35 (2000) 504.

138. K.M. Fried, A.E. Young, S. Usdin Yasuda, I.W. Wainer, *The enantioselective determination of chlorpheniramine and its major metabolites in human plasma using chiral LC on a β-cyclodextrin chiral stationary phase and MS detection*, J. Pharm. Biomed. Anal., 27 (2002) 479.

139. M.J. Desai, D.W. Armstrong, *Transforming chiral LC methodologies into more sensitive LC–ESI-MS without losing enantioselectivity*, J. Chromatogr. A, 1035 (2004) 203.

140.   J. Ayrton, G.J. Dear, W.J. Leavens, D.N. Mallett, R.S. Plumb, *Use of generic fast gradient LC–MS–MS in quantitative bioanalysis*, J. Chromatogr. B, 709 (1998) 243.

141.   L.A. Romanyshyn, P.R. Tiller, *Ultra-short columns and ballistic gradients: considerations for ultra-fast LC–MS–MS analysis*, J. Chromatogr. A, 928 (2001) 41.

142.   L.A. Romanyshyn, P.R. Tiller, R. Alvaro, A. Pereira, C.E.C.A. Hop, *Ultra-fast gradient vs. fast isocratic LC in bioanalytical quantification by LC–MS–MS*, Rapid Commun. Mass Spectrom., 15 (2001) 313

143.   R. Plumb, J. Castro-Perez, J. Granger, I. Beattie, K. Jonceur, A. Wright, *UPLC coupled to Q–TOF-MS*, Rapid Commun. Mass Spectrom., 18 (2004) 2331.

144.   J.M. Onorato, J.D. Henion, P.M. Lefebvre, J.P. Kiplinger, *SRM LC–MS determination of idoxifene and its pyrrolidinone metabolite in human plasma using robotic high-throughput, sequential sample injection*, Anal. Chem., 73 (2001) 119.

145.   T.V. Olah, D.A. McLaoughlin, J.D. Gilbert, *The simultaneous determination of mixtures of drug candidates by LC–APCI-MS–MS as an in-vivo drug screening procedure*, Rapid Commun. Mass Spectrom., 11 (1997) 17.

146.   D.A. McLoughlin, T.V. Olah, J.D. Gilbert, *A direct technique for the simultaneous determination of 10 drug candidates in plasma by LC–APCI-MS interfaced to a Prospekt SPE system*, J. Pharm. Biomed. Anal., 15 (1997) 1893.

147.   Y. Hsieh, M.S. Bryant, J.-M. Brisson, K. Ng, W.A. Korfmacher, *Direct cocktail analysis of drug discovery compounds in pooled plasma samples using LC–MS–MS*, J. Chromatogr. B, 767 (2002) 353.

148.   C.K. Van Pelt, T.N. Corso, G.A. Schultz, S. Lowes, J.D. Henion, *A four-column parallel LC system for isocratic or gradient LC–MS analyses*, Anal. Chem., 73 (2001) 582.

149.   M.K. Bayliss, D. Little, D.N. Mallett, R.S. Plumb, *Parallel ultra-high flow rate LC with MS detection using a MUX ESI source for direct, sensitive determination of pharmaceuticals in plasma at extremely high throughput*, Rapid Commun. Mass Spectrom., 14 (2000) 2039.

150.   Y. Deng, J.-T. Wu, T.L. Lloyd, C.L. Chi, T.V. Olah, S.E. Unger, *High-speed gradient parallel LC–MS–MS with fully automated sample preparation for bioanalysis: 30 seconds per sample from plasma*, Rapid Commun. Mass Spectrom., 16 (2002) 1116.

12

CLINICAL APPLICATIONS OF LC–MS

1. Introduction

In an editorial in *Clinical Chemistry*, Kinter [1] questioned why MS is not used more commonly in routine analysis in the clinical laboratory. He indicated that the answer to this question is based on several misconceptions, including the following: (a) MS is not amenable to most analytes of interest in clinical assays, (b) MS is too slow and difficult to automate for the sample throughput needed in the clinical laboratory, (c) MS is too expensive, and (d) MS is difficult to operate. Kinter [1] then takes the edge off these arguments. With the material collected in this book, we could do the same.

In the past few years, LC–MS has rapidly conquered the clinical laboratory, and further developments in this area are expected for the near future. Recent reviews [2-3] highlighted the importance of such developments. Application areas for LC–MS in the clinical laboratory are therapeutic drug monitoring, neonatal screening, reference methods, and toxicology. Today, one should add the rapidly developing area of clinical biomarker discovery to this list [4] (Ch. 18.6).

In this chapter, we give a taste of the current state-of-the-art of LC–MS in clinical applications. Certainly not all possibilities and applications are discussed in this chapter. Topics discussed are therapeutic drug monitoring (Ch. 12.2) and neonatal screening (Ch. 13.2). In addition, the LC–MS analysis of various classes of drugs of abuse, and systematic toxicological analysis are discussed.

2. Therapeutic drug monitoring

Therapeutic drug monitoring (TDM) is important in the prolonged medical treatment with drugs that have a narrow (patient-dependent) therapeutic range, show a great inter- and/or intra-individual variability in the absorption and clearance, and/or may interact with other medication. It allows to achieve optimum therapeutic efficacy, while minimizing the risk of toxicity. In general, TDM demands for a fast, simple, automated method, that is robust and sufficiently sensitive. For some drugs, the response time required from the TDM system is short. As a large number of monitoring assays are available within the clinical laboratory, and it is often difficult to predict when a particular method is due to be applied, these methods must above all be simple and robust, *i.e.*, do not require extensive start-up time and optimization of parameters. Immunoassays are often applied for these purposes. However, as many of these methods show cross-reactivity to other components, *e.g.*, metabolites of the drug tested, alternative approaches are continuously under investigation. LC–MS is one of the alternatives investigated.

2.1   Immunosuppressive drugs

LC–MS has been tested as a tool for TDM for immunosuppressive drugs for already several years. These drugs are applied to patients that received transplants of solid organs (liver, kidney) or bone marrow in order to reduce the reactions of their immune systems against these transplants. Cyclosporin A (CsA), a cyclic peptoid drug, is frequently used for this purpose. However, because CsA can give unwanted reactions like nephrotoxicity and hypertension, alternatives have been searched for. The macrolide compounds sirolimus (rapamycin), tacrolimus (FK506), and, more recently, everolimus (SDZ-RAD) are important alternative immunosuppressive agents. LC–MS has successfully been applied to the TDM of these compounds in patients' whole blood samples.

CsA, sirolimus, tacrolimus, and/or everolimus have to be determined in whole blood (EDTA-anticoagulation is often applied). The method is directed at either a single target or developed as a multiresidue approach. In practice, the methods described for either purpose show little difference. Initial results using electrospray ionization (ESI) were mainly reported by two groups: a group at the Medizinische Hochschule in Hannover (Germany) [5-8] and a group at the University of Queensland (Australia) [9-14]. The methods proposed consist of: (1) protein precipitation, (2) on-line solid-phase extraction (SPE) on a cartridge or guard column, (3) reversed-phase LC (RPLC) separation, and (4) positive-ion ESI-MS detection in either selected-ion monitoring (SIM) or selected-reaction monitoring (SRM) mode.

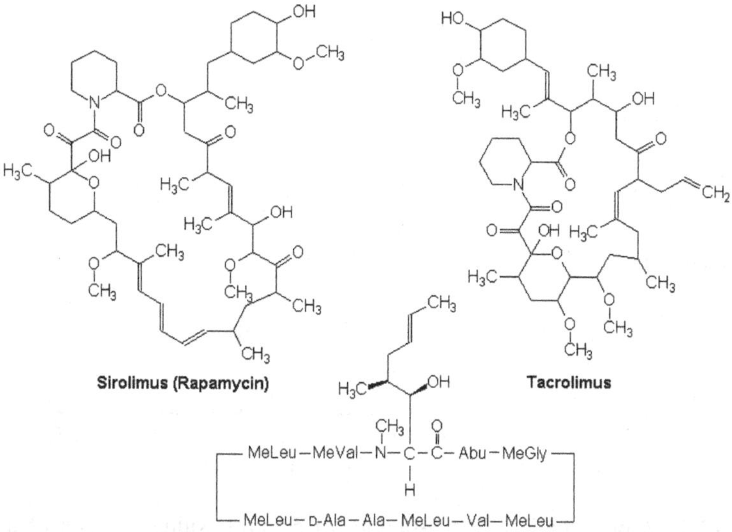

Sirolimus (Rapamycin)   Tacrolimus

MeLeu—MeVal—N—C—C—Abu—MeGly
                    | | ‖
                   CH₃ H   O

MeLeu—D-Ala—Ala—MeLeu—Val—MeLeu

Cyclosporin

In the protein precipitation step, 250–1000 µl EDTA-anticoagulated whole blood is treated with methanol or acetonitrile, eventually containing 15–100 mmol/l zinc sulfate. The internal standard (IS) is added to the solvent applied for protein precipitation. Compounds proposed as IS are: 28-*O*-acetylsirolimus [5], cyclosporin D [11], and 32-demethoxysirolimus [12], ritonavir [15], nor-rapamycin [16], and ascomycin or cyclosporin D [17]. The mixture is vortexed and centrifuged. The supernatant is injected onto an on-line SPE system with a C_8 or C_{18} cartridge. The SPE column is eluted onto a C_8 or C_{18} analytical column for isocratic LC separation using 80–90% of methanol in water, 1% aqueous formic acid, or 2 mmol/l aqueous ammonium acetate. While initially 150–250-mm long columns were applied, much shorter columns are used in more recent applications [15-17]. In most cases, this method is used as a single-residue method, but Volosov *et al.* [15] showed that the same method could be applied for three components as well. Due to the very short column used in this case, CsA, sirolimus, tacrolimus, and the internal standard were hardly separated (Figure 12.1).

Two other groups [16-17] demonstrated that similar good results could be obtained without the on-line SPE step. Holt *et al.* [16] performed an off-line liquid-liquid extraction (LLE) of the basified supernatant with 1-chlorobutane, evaporation to dryness and reconstitution of the sample in LC mobile phase, while Keevil *et al.* [17] directly injected the supernatant of the protein precipitation step onto the 33×3.0-mm-ID cyano column.

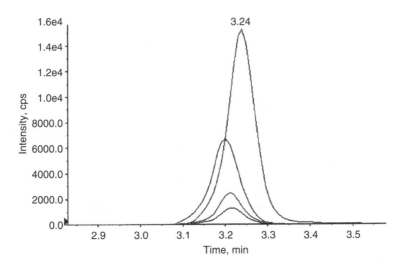

Figure 12.1: Representative chromatogram of a multiresidue analysis of a spiked blood sample containing sirolimus, tacrolimus, CsA, and ritonavir (IS). Reprinted from [15] with permission. ©2001, The Canadian Society of Clinical Chemists.

Table 12.1:			
Accuracy (% deviation), intra-assay, and inter-assay precision (%CV) obtained in the validation of an assay for sirolimus in whole blood. Data from [16]			
	low: 3 µg/l	medium: 15 µg/l	high: 30 µg/l
Accuracy	3.7%	–0.3%	–0.6%
Intra-assay precision	7.3%	1.9%	2.6%
Inter-assay precision	8.1%	3.8%	3.0%

All target compounds are generally analysed in positive-ion ESI-MS, although the use of positive-ion atmospheric-pressure chemical ionization (APCI) has been demonstrated as well [15]. Either SIM on the ammoniated $[M+NH_4]^+$ or sodiated molecules $[M+Na]^+$ is applied, or SRM using the $[M+NH_4]^+$ as precursor ions. For CsA, the $[M+NH_4]^+$ at m/z 1219.8 is fragmented to the protonated molecule $[M+H]^+$ at m/z 1202.8. For sirolimus and tacrolimus, the losses of 67 Da (NH_3, CH_3OH, and H_2O) or 53 Da (NH_3 and $2 \times H_2O$) are monitored, respectively.

Considerable attention is paid to validation of the analytical results. Typical linear dynamic ranges are 0.2–100 µg/l for sirolimus and tacrolimus, and 1–1000 μg/l for CsA. An extended calibration range of 1–5000 μg/l was recently demonstrated for CsA [17]. The most elaborate validation of their method was reported by Holt *et al.* [16]. Some of their results are summarized in Table 12.1. Keevil *et al.* [17] reported a 14-h method stability test.

Several authors pay attention to the evaluation of matrix effects (Ch. 12.5). This runs from a statement that no matrix effects were observed in the multiresidue method, where four components are almost coeluting [15] (Figure 12.1), to an elaborate evaluation [16]. Holt *et al.* [16] extracts 10 blood samples from patients receiving either CsA or tacrolimus, but no sirolimus. The extract is evaporated to dryness and reconstituted in a solvent containing both sirolimus and the IS at a concentration equivalent to a 10-μg/l calibrator. The peak areas of these extracts and a 10-μg/l standard sirolimus solution after LC–MS analysis were found to differ less than 1%. From this result, they deduced the absence of matrix effects.

Finally, as part of the validation, the LC–MS results are compared to the results obtained by various immunoassay methods [13, 15-17]. Overestimation of the concentration of CsA, tacrolimus, and sirolimus in the immunoassay is reported, due to cross reactivity with other components, *e.g.*, metabolites of the analytes.

One can conclude that TDM of immunosuppressive drugs by LC–MS is feasible. A throughput of 100 samples/day was reported by Holt *et al.* [16]. With the use of a shorter column (33 mm instead of 150 mm), a sample throughput as high as 30 samples in 1.5 h was reported by Keevil *et al.* [17]. Validation of the method shows both accuracy and precision are adequate.

Separate method development was reported for the more recently introduced immunosuppressive drug everolimus. Segarra *et al.* [18] reported a method based on SPE of protein precipitated 1-ml whole blood samples, LC analysis on a 250×4-mm-ID C_8 column (3 μm) using a mobile phase consisting of methanol–0.1% aqueous formic acid, and 1 μmol/l sodium formate at a flow-rate of 0.4 ml/min. Under these conditions, everolimus and the analogue IS were detected as $[M+Na]^+$ in a single-quadrupole ESI-MS system in SIM mode. Linearity was tested between 0.1 and 100 μg/l. The mean day-to-day precision was 8.0%. McMahon *et al.* [19] reported the analysis of both everolimus and CsA using SPE in a 96-well plate format (Ch. 1.5.3). Everolimus was detected as M^- in negative-ion APCI, while CsA was detected as $[M+H]^+$ in the positive-ion mode. Linearity was tested between 0.4 and 250 μg/l for everolimus and 7 and 1500 μg/l for CsA. Subsequently, an alternative method was reported by the same group [20]. LLE with methyl *t*-butyl ether (MTBE) was performed in 96-well plate format. The loss of NH_3 from $[M+NH_4]^+$ was monitored in SRM mode for CsA, and the loss of 67 Da (NH_3, H_2O and CH_3OH) from $[M+NH_4]^+$ of everolimus. Four 96-well plates could be extracted and analysed in less than 28 h.

More recent developments in the TDM of immunosuppressive drugs comprise of optimization of LC conditions [21-22]. Hatsis and Volmer [22] evaluated the use of a cyano column (see also [17]) instead of the commonly applied C_{18} column, as the cyano column will yield smaller retention time differences between sirolimus and tacrolimus on one hand and CsA on the other.

The simultaneous determination of various immunosuppressive drugs has become quite common now. Deters *et al.* [23] reported TDM of CsA, sirolimus, tacrolimus, and everolimus in whole blood. Oellerich *et al.* [24-25] reported the TDM of all four compounds in whole blood [24] as well as of CsA and sirolimus in pediatric patients with kidney transplants [25]. Koal *et al.* [26] reported the simultaneous determination of all four compounds in whole blood using on-line SPE–LC–MS (Ch. 1.5.4) after protein precipitation. A perfusion column (Ch. 1.4.1) was applied as SPE column to improve the sample throughput. Lower limit of quantification (LOQ) were 10 µg/l for CsA and 1 µg/l for the other three components. Ceglarek *et al.* [27] reported the utilization of turbulent-flow LC (Ch. 1.5.5) in combination with LC–MS for the rapid TDM of CsA, sirolimus, and tacrolimus. The total analysis time was less than 3 min per sample for both methods. Poquette *et al.* [28] reported simultaneous clinical analysis of sirolimus, tacrolimus, and cyclosporin using an isocratic LC method in combination with a single-quadrupole MS method.

TDM of immunosuppressants by LC–MS was recently reviewed [29-30]. Despite the high investment needed for LC–MS, TDM of immunosuppressive drugs by LC–MS is more cost effective than the conventional immunoassays. Savings of ~40% per test have been indicated [24, 31]. In addition, due to the absence of cross reactivity, LC–MS is generally more accurate. An interlaboratory study for the LC–MS analysis of sirolimus was performed [32].

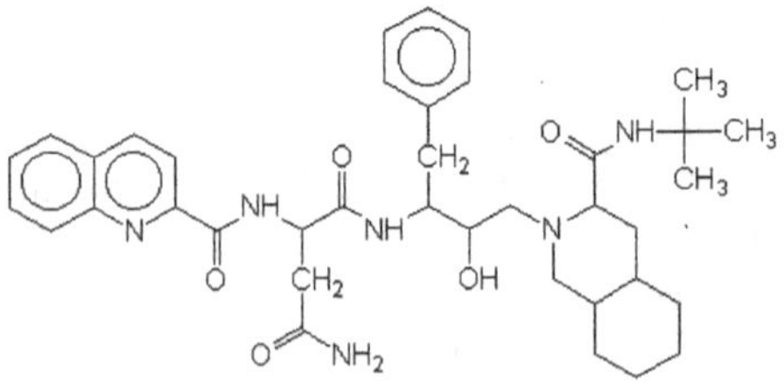

Saquinavir

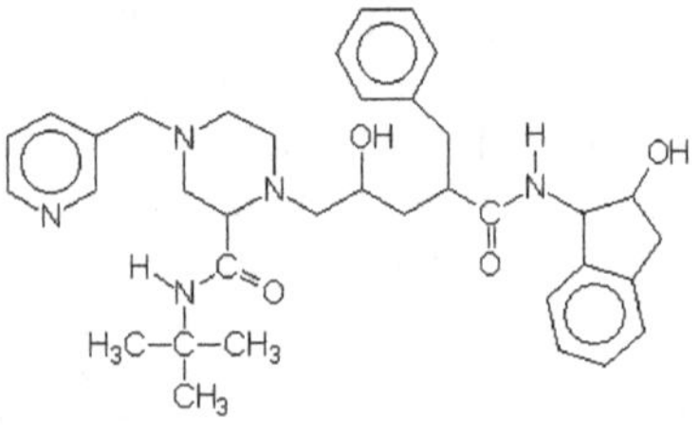

Indinavir

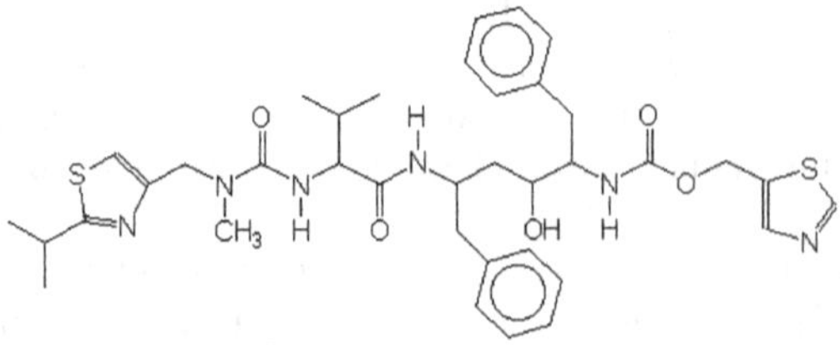

Ritonavir

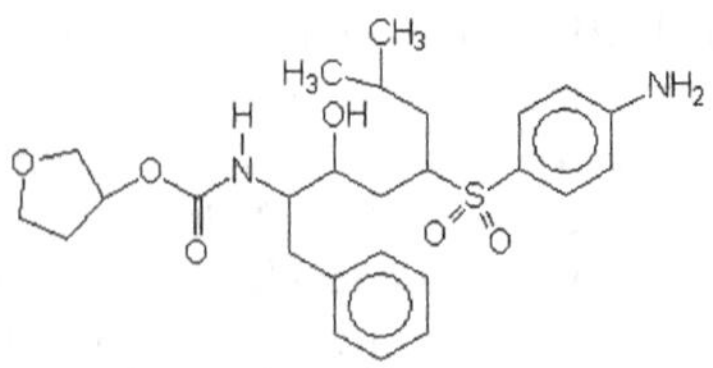

Amprenavir

2.2 TDM of HIV protease inhibitors

In recent years, various drugs have been developed for the treatment of the human immunodeficiency virus (HIV) infection. The two most important compound classes are the HIV selective protease inhibitors, nucleoside reverse transcriptase inhibitors (NRTI), and the non-nucleoside reverse transcriptase inhibitors (nNRTI). Methods developed for TDM of these compounds are discussed here. TDM is required because of a high inter- and intra-individual variability.

Crommentuyn *et al.* [33] reported a TDM method enabling the simultaneous monitoring of six protease inhibitors, *i.e.*, amprenavir (AMP), indinavir (IND), lopinavir (LOP), nelfinavir (NEL), ritonavir (RIT) and saquinavir (SAQ), and the pharmacologically-active M8-metabolite of nelfinavir (NM8) in human plasma. Only 100 µl of plasma is protein precipitated with a mixture of methanol and acetonitrile, containing the IS. The diluted supernatants were analysed by positive-ion LC–ESI–MS in time-scheduled SRM mode with a dwell time of 50 ms per transition (nine transitions in total). LC was performed on a 50×2.0-mm-ID C_{18} column (5 µm) protected with a 10×2.0-mm-ID precolumn and a step gradient starting with 35% methanol in aqueous ammonium acetate (pH 5.0) as solvent A and stepping to 85% methanol in solvent A after 0.2 min. This composition was maintained for 1.6 min, after which the initial conditions were applied again. Reconditioning of the column was allowed for 3.5 min. $[D_6]$-IND was applied as internal standard for AMP and IND, and $[D_5]$-SAQ for the other analytes. The validated concentration ranges are summarized in Table 12.2 and compared with results reported by others. These validated concentration ranges were adequate in daily practice: the method is applied in TDM and for pharmacokinetic studies. Intra-day and inter-day precision was always better than 10%. LC–MS analysis time is 5 min. A very similar method was subsequently developed for the new HIV protease inhibitors atazanavir (ATA) and tipranavir (TIP) [34]. This method could be easily combined with the previous method.

Similar methods were reported by others [35-37]. Frerichs *et al.* [35] performed a LLE and a linear gradient between 25 and 80% acetonitrile in a 5 mmol/l acetate buffer. A RIT-analogue was applied as internal standard. Calibrated concentration ranges are summarized in Table 12.2. LC–MS analysis time is 3.5 min. The chromatogram obtained from a patient sample, after administration of AMP, RIT, and LOP is shown in Figure 12.2.

Rentsch [36] reported the determination of AMP, IND, LOP, NEL, RIT, and SAQ together with the nNRTI efavirenz and nevirapine by means of two similar LC–MS methods. Two slightly different SPE procedures were applied for sample pretreatment. Two LC–MS methods were applied, both based on the use of APCI on an ion-trap instrument: LOP, NEL, RIT, SAQ, and the IS in positive-ion and nevirapine in the negative-ion mode in method A, and AMP, efavirenz, IND, and the IS in negative-ion mode in method B. The LC-MS analysis time was 21 min.

Table 12.2: Validated concentration ranges (µg/ml) of some of the reported TDM methods for the simultaneous determination of HIV protease inhibitors			
reference:	[33]	[35]	[36]
AMP	0.1–10	0.016–10	0.01–12
IND	0.01–10	0.016–4	0.01–12
LOP	0.1–20	0.016–4	0.9–15.8
NEL	0.05–10	0.016–4	0.09–4
NM8	0.01–5	0.008–5	n.d.
RIT	0.05–10	0.05–5	0.5–11.8
SAQ	0.01–10	0.016–4	0.09–4

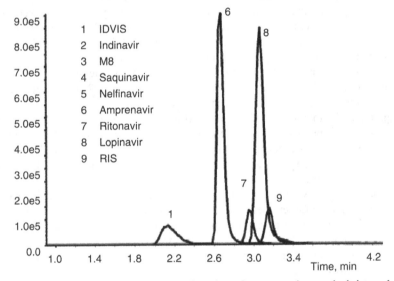

Figure 12.2: Chromatogram of a sample taken from a patient, administered with AMP, RIT, and LOP. The plasma concentrations determined were 7.20, 10.1, and 1.28 µg/ml, respectively. Reprinted from [35] with permission. ©2002, Elsevier Science BV.

Dickinson *et al.* [37] reported the rapid simultaneous determination of AMP, IND, LOP, NEL, RIT, SAQ, and atazanavir (ATA) in human plasma. The analysis time is 10 min per sample.

As lymphocytes, monocytes, and macrophages are the primary targets for viral infection, the penetration of antiviral agents is important. Lymphocytes and monocytes are indicated as peripheral blood mononuclear cells (PBMC). The preparation of control PBMC from blood was reported in considerable detail by Jemal *et al.* [38]. In addition, a validated assay for the determination of ATA in PBMC was developed. The determination of protease inhibitors in human PBMC was reported by several groups [39-40]. After LLE, the analytes were analysed by LC–MS. Both methods enable the intra-cellular determination of the analytes and can be applied for TDM and pharmacokinetic studies.

2.3   Nucleoside reverse transcriptase inhibitors

The NRTI and nNRTI form two additional classes of antiviral compounds applied against HIV. Examples of NRTI are zidovudine (AZT), lamivudine (3TC), didanosine (ddI), and stavudine (d4T). Examples of nNRTI are efavirenz and nevirapine.

Lamivudine (3TC)   Stavudine (d4T)   Zidovudine (AZT)

Didanosine (ddI)   Abacavir

The simultaneous determination of 3TC and AZT in human serum and seminal plasma was reported [41-42]. The sample pretreatment consisted of an automated ultrafiltration step. For isocratic LC, a 150×2-mm-ID C_{18} column was used and 15% acetonitrile in water at 0.3 ml/min. The eluate of the first 2 min after injection were directed to waste. ESI-MS in SRM mode was applied in the positive-ion mode for 3TC and negative-ion mode for AZT. Polarity switching was done in between the two peaks. The LOQ was 2.5 ng/ml.

The determination of d4T in human plasma was reported by Wiesner *et al.* [43]. After sample pretreatment by means of SPE, LC–MS was performed on a 150×2-mm-ID C_{18} column (5 μm) with a mobile phase of 10% acetonitrile and 10%

methanol in 10 mmol/l aqueous ammonium acetate at a flow-rate of 0.3 ml/min. Negative-ion APCI-MS was applied. The LOQ was 4 ng/ml.

The determination of ddI [44] and d4T [45] in human serum was reported by on-line SPE–LC–MS with positive-ion ESI-MS. 3TC and ddI were applied as IS. The analysis time was less than 5 min for both methods. The LOQ was 10 ng/ml for both compounds. Another group described the determination of ddI and d4T in human plasma, bronchoalveolar lavage fluid (BALF), alveolar cells, PBMC, seminal plasma, cerebrospinal fluid (CSF), and tonsil tissue [57]. Depending on the matrix, either isocratic or gradient LC was applied after SPE sample pretreatment. Positive-ion ESI-MS in SRM mode was applied. The LOQ for both compounds were 2.0 ng/ml in plasma, 0.5 ng/ml in CSF, 0.4 ng/ml in alveolar cells, BALF, and PBMC, 1 ng/ml in seminal plasma, and 0.01 ng/mg in tonsil tissue.

Pharmacologically more relevant than the analysis in plasma is the determination of the intracellular phosphorylated anabolites. The NRTI undergo intracellular conversion to the active triphosphate, which inhibits viral reverse transcriptase and acts as a chain terminator of the proviral DNA. Obviously, the LC methods for the phosphorylated anabolites differ from those applied to the drugs.

The simultaneous quantitation of the triphosphates of AZT, 3TC, and d4T in PBMC was reported [47]. The compounds were extracted, purified by SPE on an ion-exchange material, and treated with alkaline phosphatase to convert them into nucleosides. The LC–MS method was similar to the methods reported above. A similar procedure involving enzymatic dephosphorylation was applied by others in the sub-ppb analysis of intracellular ddI-triphosphate in CEM-T4 cells [48]. The LOQ was 0.02 ng/ml.

The determination of abacavir and its mono-, di-, and tri-phosphorylated anabolites in PBMC was reported by Fung et al. [49] using dimethyl-hexyl-amine (DMHA) ion-pairing agent. As the DMHA forms adducts with the analytes, they can be analysed in positive-ion ESI mode.

The mono-, di-, and tri-phosphorylates of d4T were analysed by ion-pair LC–MS after lysis of the PBMC cell in Tris/methanol and centrifugation [50]. The supernatant was injected into the LC system with a 150×2.1-mm-ID C_{18} column (5 µm) and a mobile-phase gradient of 70 to 35% solvent A (10 mmol/l DMHA and 3 mmol/l ammonium formate adjusted at pH 11.5) in solvent B (50% acetonitrile in 20 mmol/l DMHA and 6 mmol/l ammonium formate). Negative-ion ESI-MS was performed in SRM mode. The method enabled the direct measurement of the chain terminator ratio (d4T-triphosphate/deoxythymidine-triphosphate). Subsequently, the same group [51] reported modifications of this method, including simplifications of the sample pretreatment, replacement of the LC column for another type, and reduction of the column inner diameter from 2 mm ID to 0.32 mm ID. This improved method was applied to the determination of the phosphorylates of d4T, 3TC, and ddI. The sample throughput is 200 samples per week. The determination of intracellular AZT-triphosphate in PBMC [52], and the validation of the method for the determination of the ddI and d4T triphosphates was reported separately [53].

2.4 Non-nucleoside reverse transcriptase inhibitors

Next to NRTI, efavirenz and nevirapine are applied as nNRTI. These compounds are often administered in combination therapy. Reports on the LC–MS of nNRTI concern the simultaneous determination with protease inhibitors [34, 36, 54-55] or NRTI [56-57].

Efavirenz   Nevirapine

The determination of 3TC, d4T, and efavirenz in human serum was reported by Fan et al. [56]. The compounds were extracted from serum using SPE. A gradient of 7–93% acetonitrile in 20 mmol/l aqueous ammonium acetate (pH 4.5) was applied and a 150×2.0-mm-ID C_6 column (3 μm) at a flow-rate of 0.2 ml/min. 3TC and d4T were analysed in positive-ion, and efavirenz and aprobarbital (IS) in the negative-ion ESI-MS mode with SRM. The LOQ were 1.1 ng/ml for 3TC, 12.3 ng/ml for d4T, and 1.0 ng/ml for efavirenz. The analysis of nevirapine in human plasma was reported by Chi et al. [57]. Positive-ion LC–ESI-MS of the supernatant was performed after protein precipitation with perchloric acid. The LOQ was 25 ng/ml.

An interesting development for TDM applications is the determination of various antiretroviral drugs (protease inhibitors and nNRTI) in dried blood spots of HIV/AIDS patient whole blood samples [55].

3. Neonatal screening for inherited metabolic disorders

Organic acidemias are a group of inherited metabolic disorders, caused by defects of mitochondrial fatty acid oxidation and mitochondrial catabolism of branched-chain amino acids. More than 200 of these single-gene disorders are known. They may carry significant clinical consequences to the affected neonate or young infant. Early detection and diagnosis of these metabolic diseases is important. Organic acidemias lead to an accumulation of potentially toxic acyl-Coenzyme A (acylCoA) esters in the mitochondria. Through the formation of acylcarnitines, L-carnitine plays an important role in removing these acylCoA esters. This leads to an increased concentration of circulating acylcarnitines as well as an increased excretion of acylcarnitines in urine. Profiling of acylcarnitines in urine, plasma, or blood spots is a powerful screening tool for the diagnosis of these disorders.

Similar MS–MS technology as developed for acylcarnitine profiling can be applied for the selective screening for a number of fatty acid oxidation defects, organic acidemias, and aminoacidopathies [58]. Routine determination of free carnitine and acylcarnitines, amino acids, acylglycines, purines and pyrimidines, bile acids, and very-long-chain fatty acids can now be applied for neonatal diagnosis.

3.1   Acylcarnitines

Acylcarnitine profiling by means of MS was first introduced by Millington and coworkers, using fast-atom bombardment (FAB) [59], thermospray [60], and/or continuous-flow FAB [61]. In MS–MS, the acylcarnitine or butyl-derivatized acylcarnitine is fragmented to a common fragment ion at m/z 85, while the methyl-derivative is fragmented to an ion with m/z 99.

Rashed *et al.* [62] were the first to demonstrate the power of ESI-MS in acylcarnitine profiling. Circles of 5 mm were punched out from blood spots collected on Guthrie or PKU cards. The samples were extracted with 400 µl methanol in an ultrasonic bath for 30 min. The supernatant is transferred to a new vial, evaporated to dryness, and redissolved in 200 µl of *n*-butanol with 2 drops of acetyl chloride. After capping, the vials were heated at 65°C for 30 min. After again evaporation to dryness, the *n*-butyl derivatized acylcarnitines were reconstituted in 200 µl of methanol and transferred to the autosampler. Every six minutes, a 20 µl sample was injected in a liquid stream of 50% methanol in water with a flow-rate of 50 µl/min. The positive-ion ESI-MS–MS was operating in precursor-ion analysis mode with m/z 85 as the common fragment ion and monitoring between m/z 250 and 550. The method was fully automated. In comparison to the FAB method, the ESI-MS–MS allowed for a much higher sample-throughput and required less and simpler maintenance. Recently, Rashed [58] reviewed ten years of diagnosis and screening for inherited metabolic diseases.

The introduction of 96-well plates and improvements in LC–MS instrumentation allowed a further reduction of the sample throughput to an analysis time of 1.3 min per sample [63]. Dedicated software tools assisted in organizing the workflow and in performing automatic data processing. Diagnostic parameters for acylcarnitines in a number of diseases were established [63]. Using this approach, Chace *et al.* [64] reported the rapid diagnosis of the medium-chain acyl-CoA dehydrogenase (MCAD) deficiency from dried blood spots (Figure 12.3).

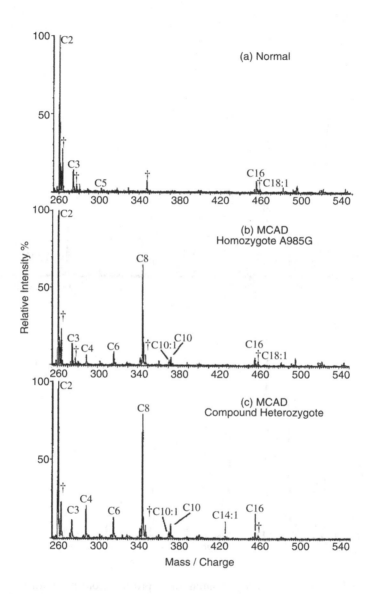

Figure 12.3: Precursor-ion MS–MS acylcarnitine profiles (common fragment ion *m/z* 85) with profiles from (a) a healthy newborn, (b) a newborn with MCAD deficiency (homozygous for A985G mutation), and (c) a newborn with MCAD deficiency (heterozygous for A985G mutation). Reprinted from [64] with permission. ©1997, American Association for Clinical Chemistry.

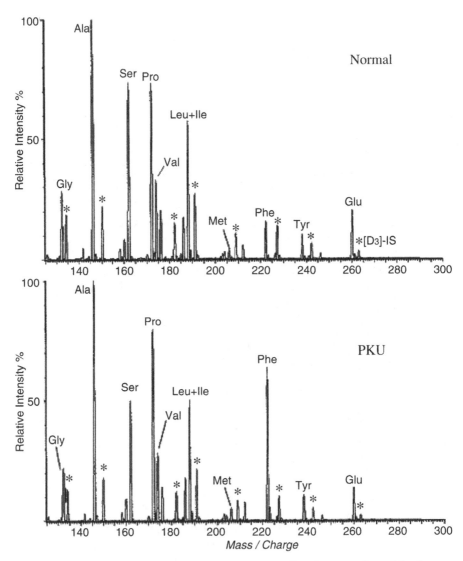

Figure 12.4: Neutral-loss MS–MS amino acid profiles from newborn blood spots showing (top) a healthy subject, and (bottom) a subject tested positive for PKU. Reprinted from [66] with permission. ©1998, American Association for Clinical Chemistry.

3.2   Amino acids

Subsequently, it was demonstrated that the use of ESI–MS–MS also enabled the profiling of amino acids in order to detect amino-acid metabolism deficiencies. This further extended the applicability range in the clinical laboratory. Chace *et al.* [65] reported the diagnosis of phenylketonuria from the quantitative analysis of phenylalanine and tyrosine in neonatal blood spots using FAB in combination with MS–MS. The amino acids were derivatized to their *n*-butyl derivatives. In MS–MS, fragmentation leads to the loss of butyl formate (102 Da), which can be used in a screening based on neutral-loss analysis. Again, this method was modified by the use of ESI–MS–MS on a triple-quadrupole instrument rather than FAB on a double-focussing sector instrument [63, 66]. Comparison of the amino acid profiles for a healthy newborn and a newborn tested positive for PKU is shown in Figure 12.4.

ESI–MS–MS is widely applied in both acylcarnitine and amino-acid profiling in various screening programs throughout the world. Newborn screening and clinical testing for amino acid and acylcarnitine related metabolic disorders was reviewed by Chace and Kalas [67].

4.   Analysis of drugs of abuse

4.1   Amphetamines

Amphetamine derivatives are often misused as stimulants, particularly methylenedioxyamphetamines (MDMA, XTC, 'Ecstasy'). Immunoassays are available for screening, but confirmation of positive results must be performed by MS. The GC–MS requires derivatization. Therefore, LC–MS has been investigated as an alternative.

Amphetamine   3,4-methylenedioxyamphetamine

Bogusz *et al.* [68] reported the LC–APCI–MS analysis of sixteen phenethylamines after SPE. Subsequently, optimization of the mobile-phase composition for the LC–ESI–MS analysis of methylamphetamine and related compounds was described [69]. Methanol in the mobile phase resulted in a slightly higher abundance of the [M+H]$^+$. Formic acid was preferred over other additives. Microporous ultrafiltration was used in the sample pretreatment of urine samples. The intra-assay and inter-assay precision and accuracy in the 100–1000 ng/ml range were 2.3–6.3%, 0.8–5.9%, and 1.6–6%, respectively, for amphetamine in urine.

Clauwaert *et al.* [70] compared the quantitative performance for 3,4-MDMA and related compounds of an existing method based on fluorescence detection to an LC–ESI-MS method on a quadrupole–time-of-flight hybrid (Q–TOF) system. With nearly four decades (2–10,000 pg on-column), the linear dynamic range and the absolute sensitivity of the LC–MS method was superior.

Paramethoxyamphetamine is an illicit designer drug that recently caused a number of lethal intoxications. A method of its analysis together with 3,4-MDMA and related substances was reported by Mortier *et al.* [71], based on LLE of whole blood, urine or post-mortem tissues, narrow-bore LC coupled to an ion-trap MS via a sonic-spray interface (Ch. 5.7.1). The early eluting contamination was directed to waste. The method was validated for all three matrices: precision was better than 17.5% and the accuracy better than 16.2%. Linearity ranging from 10 to 1000 ng/ml in blood and urine and 20 to 2000 ng/g in tissue were obtained.

The determination of amphetamine, methamphetamine, and methylenedioxy analogues in meconium was reported by Pichini *et al.* [72]. The analytes were extracted using 17 mmol/l methanolic HCl, purified by SPE, separated using a linear solvent gradient at pH 9.0, and analysed with ESI-MS in SIM mode. The LOQ was ~0.05 μg/g meconium. The method was applied to newborn screening for fetal exposure to amphetamines.

4.2   Lysergic acid diethylamide (LSD)

LSD analogs and metabolites were used as model compounds by Cai and Henion [73] in the development of on-line immunoaffinity extraction coupled to capillary-column LC–MS–MS. The system consists of a 2.1-mm-ID protein G immunoaffinity column with a noncovalently immobilized antibody, which can be operated at flow-rates of up to 4 ml/min, a short packed-capillary trapping column, and a packed-capillary analytical column, operated at 3.5 μl/min. The method enables low ng/l determination of LSD and related compounds in urine, which is twenty-fold better than previous methods based on SPE and LC–MS–MS [73].

Hoja *et al.* [74] reported a method for LSD and *N*-desmethyl-LSD in urine. The analytes were extracted from urine by means of an Extrelut-3 SPE cartridge. The eluate of the cartridge was evaporated to dryness and reconstituted in mobile phase prior to LC–ESI-MS analysis. LOQ were 0.05 and 0.10 ng/ml, with linearity up to 20 ng/ml. Intra-assay and inter-assay precision at 0.1 ng/ml were better than 9 % and 14%, respectively

De Kanel *et al.* [93] described a quantitative confirmation procedure for LSD and *N*-desmethyl-LSD in blood, serum, plasma, and urine. Fully automated SPE pretreatment and extract evaporation were performed using a Zymark RapidTrace. LC–ESI-MS–MS was performed, using [$^2H_3$]-LSD as IS. The intra-assay and inter-assay precision were 2.2 and 4.4%, respectively. The detection limit was 0.025 ng/ml. Similar results in terms of figures-of-merit were reported by others.

More recent methods not only include *N*-desmethyl-LSD, but also 2-oxo-3-hydroxy-LSD, the main LSD metabolite in urine [76-78]. The metabolite was detected by LC–MS in the analysis of 74 urine samples which were screened positive in a RIA and in which LSD was confirmed by GC–MS. The metabolite was present in urine at considerably higher concentrations than LSD, iso-LSD, and nor-LSD (mean ratio to LSD was 43) [76]. Canezin *et al.* [77] also reported a significant excess of 2-oxo-3-hydroxy-LSD over LSD, but also detected a number of other metabolites by applying neutral-loss analysis: nor-iso-LSD, lysergic acid ethylamide, trioxylated-LSD, lysergic acid ethyl-2-hydroxyethylamide, 13- and 14-hydroxy-LSD and their glucuronide conjugates.

4.3   Morphine and its glucuronides

The determination of the heroin metabolites morphine, morphine-6-glucuronide (M6G), morphine-3-glucuronide (M3G), and 6-monoacetylmorphine (6-MAM) in body fluids is an important application of LC–MS in the toxicology laboratory. In some cases, codeine and codeine-6-glucuronide are determined as well. Several quantitative methods have been reported. However, only a few deal with forensic toxicology, performing the analysis in urine, serum, vitreous humour [79], and autopsy whole blood [79-80].

Morphine   Codeine

In most cases, morphine and its metabolites are determined in plasma for pharmacokinetic studies, *e.g.*, [81-82]. The methodology in these methods is essentially the same. After protein precipitation, the supernatant is subjected to an SPE cartridge. Reversed-phase LC on C_{18}-columns is applied with positive-ion ESI-MS. For forensic applications, the typical analysis time is around 15 min, while shorter run times are achieved in pharmacokinetic studies. Either SIM or SRM is applied. Similar LOQ have been achieved: between 0.5 and 3.8 ng/ml for morphine, 2.3 and 25 ng/ml for M3G, and 0.5 and 12 ng/ml for M6G. Either analogue IS or deuterium-labelled IS is used.

The analysis of morphine and its metabolites in human urine was reported by Murphy and Huestis [83]. After SPE, the analytes were determined by LC–MS in SRM mode. The analysis time is 13 min. Isotopically-labelled IS were applied for each analyte. Good accuracy and precision is achieved in the range 25–4000 ng/ml.

4.4   Cocaine and its metabolites

The analysis of cocaine and its metabolites has been performed by means of GC–MS, but this demands extensive sample pretreatment and derivatization. LC–MS has been evaluated as an alternative. Simultaneous LC–ESI-MS and fluorescence detection of cocaine, benzoylecgonine, and cocaethylene in human hair was described by Clauwaert et al. [84]. The effect of mobile-phase composition and pH on the ESI LC–MS performance of cocaine, ecgonine methyl ester, and benzoylecgonine was investigated by Jeanville et al. [85]. The best sensitivity was achieved with a 1:1 mixture of 60% acetonitrile–40% acetone in 100 mmol/l ammonium acetate. The pH has little influence on the response, suggesting that gas-phase rather than liquid-phase ionization is important.

The analysis of cocaine and some of its metabolites in meconium and human urine was reported [86-88]. The analysis in meconium was directed at finding an appropriate diagnostic marker for neonatal cocaine exposure [86]. For human urine analysis of these basic compounds, the use of a pentafluorophenylpropyl rather than a C_{18} column has been proposed: 12-fold better signals were obtained due to the improved peak shape [88]. In most cases, ESI ionization in combination with SRM is applied. Typical limits of quantification are in the range of 1–5 ng/ml or ng/g.

4.5   Multiresidue methods

For practical use in the toxicological laboratory, multiresidue methods capable of quantifying and confirming drugs of abuse from various compound classes would be highly attractive. Simultaneous determination of amphetamines, morphine, cocaine, and its benzoylecgonine metabolites using a Q–TOF instrument was reported by Mortier et al. [89], while direct injection of 10-µl of urine into an LC–MS–MS system in positive-ion APCI mode and using SRM enabled the analysis of opioids such as morphine, heroin and codeine, cocaine, and metabolites down to the 10–100 ng/ml level [90]. The simultaneous determination of amphetamine and various related compounds, cocaine, codeine, morphine, and their metabolites in preserved oral fluids was reported by Wood et al. [91].

5. Systematic toxicological analysis

With the advent of reliable and robust instrumentation, the potential of LC–MS was also evaluated for use in toxicological and forensic analysis. A large number of review papers have been published, *e.g.*, [92-96]. Systematic toxicological analysis is concerned with the screening of biological samples, especially urine, plasma, and other body fluids, for the presence of medicinal drugs, drugs of abuse, toxins, etc. As far as volatile compounds are concerned, GC–MS is the *golden standard* for identification and quantification. But there is a growing need to extend the applicability range to less volatile compounds, and to simplify sample pretreatment procedures and/or to avoid time-consuming analyte derivatization procedures.

Clinical toxicology is among others concerned with the diagnosis or the definite exclusion of an acute or chronic intoxication. In addition, monitoring of persons addicted to illegal drugs has to be performed. Forensic toxicology is among others concerned with proof of an abuse of illegal drugs or of a murder by poisoning. In both areas, representative sampling is of utmost importance. With respect to sample pretreatment, separation and detection, there will always be the dilemma between broad general unknown screening by one method or the analysis of various compound classes by various methods. In this respect, there is an important difference between GC–MS and LC–MS. While GC–MS in most cases is preferably performed by means of EI, resulting in information-rich mass spectra, LC–MS is performed using a soft-ionization technique. Structural information can in practice only be obtained using MS–MS technology. Obviously, the use of data-dependent acquisition (DDA) can help in this, but these procedures are often not as powerful as needed at trace levels of analytes in complex biological matrices. While extensive mass spectral libraries are available for GC–MS, this is not the case of LC–MS. As these issues have been discussed in more detail elsewhere [92-96], the attention here is focussed at the building and application of mass spectral libraries for LC–MS and LC–MS–MS.

Given the power of a computer search through a mass spectral library for identification by GC–MS using EI, there has been considerable interest in building mass spectral libraries for LC–MS and MS–MS. There are different means to fragment the $[M+H]^+$ by collision-induced dissociation (CID), *e.g.*, in-source CID, CID in the collision cell of a triple-quadrupole or Q–TOF instrument, and CID in an ion-trap instrument. This leads to mass spectra that only partially share common features. In addition, it has been demonstrated that differences in collision energy regimes in triple-quadrupole instruments from different manufacturers can have a distinct influence on the relative abundance and to a lesser extent on the type of fragment ions observed. In the search algorithms, built for the relatively reproducible mass spectra from EI, both the peak position (m/z) and its relative abundance are taken into account. To avoid problems with the poor reproducibility of relative abundances in MS–MS spectra acquired at instruments from different manufacturers, Kienhuis and Geerdink [97] proposed building a library for pesticides where only two levels of relative abundance are discriminated: intense ions are set at 100%, and less intense ions at 50%. Although this is a useful approach, most people prefer a conventional mass spectral library.

Although libraries based on MS–MS in a triple-quadrupole [98] and an ion-trap instrument [99] have been described, in-source CID is applied in most cases, *e.g.*, [100-104]. In this respect, it must be mentioned that an ion-trap can have some distinct advantages in building libraries, because good mass spectral reproducibility between ion-trap instruments can be achieved by the use of normalized collision energy [99]. Recently, Marquet *et al.* [105] evaluated MS–MS on a quadrupole–linear-ion-trap hybrid (Q–LIT) instrument as an alternative to the in-source CID. Promising preliminary results were obtained. Comparison of MS–MS spectra from different instruments was reported by the same group [96].

In-source CID libraries for drugs in general unknown screening for toxicology were developed. In most cases, two or three different settings of the relevant voltage (orifice voltage or declustering potential in Sciex instruments, and fragmentor voltage in Agilent instruments) were used. Weinmann *et al.* [101] investigated haloperidol and paracetamol to act as tuning compounds for the settings of the ion-source parameters. This improved the quality and transferability of libraries between instruments, despite variations in relative abundances that are still observed. Bristow *et al.* [102] compared two different optimization procedures, one based on tuning compounds, the other on attenuation of $[M+H]^+$ signal, but could not decide between the two approaches. Applications of these libraries in toxicological applications have been presented by most authors.

A different approach to systematic toxicological analysis is based on the use of a wide variety of SRM transitions to screen for relevant compounds in blood or urine. Initially, SRM screening for 238 target drugs was developed using a triple-quadrupole instrument. With a dwell time of 25 ms per transition, the total cycle time is 6 s [106]. The method was adopted for a Q–LIT instrument. A survey SRM screening for 301 drugs was performed. When a compound was detected with sufficient signal, an enhanced product-ion analysis was done via DDA. The product-ion MS–MS spectrum was searched against a library [107].

6. References

1.   M. Kinter, *Editorial: Toward broader inclusion of LC–MS in the clinical laboratory*, Clin. Chem., 50 (2004) 1500.

2.   M. Vogeser, *LC–MS–MS - Application in the clinical laboratory*, Clin. Chem. Lab. Med., 41 (2003) 117.

3.   K.C. Dooley, *MS–MS in the clinical chemistry laboratory*, Clin. Biochem., 36 (2003) 471.

4.   R. Bischoff, T.M. Luider, *Methodological advances in the discovery of protein and peptide disease markers*, J. Chromatogr. B, 803 (2004) 27.

5.   F. Streit, U. Christians, H.M. Schiebel, K.L. Napoli, L. Ernst, A. Linck, B.D. Kahan,

K.-F. Sewing, *Sensitive and specific quantification of sirolimus and its metabolites in blood of kidney graft recipients by LC–ESI-MS*, Clin. Chem., 42 (1996) 1417.

6. C. Vidal, G.I. Kirchner, G. Wünsch, K.-F. Sewing, *Automated simultaneous quantification of the immunosuppressants 40-O-(2-hydroxyethyl) rapamycin and cyclosporin in blood with ESI-MSc detection*, Clin. Chem., 44 (1998) 1275.

7. G.I. Kirchner, C. Vidal, W. Jacobsen, A. Franzke, K. Hallensleben, U. Christians, K.-F. Sewing, *Simultaneous on-line extraction and analysis of sirolimus and CsA in blood by LC–ESI-MS*, J. Chomatogr. B, 721 (1999) 285.

8. U. Christians, W. Jacobsen, N, Serkova, L.Z. Benet, C. Vidal, K.-F. Sewing, M.P. Manns, G.I. Kirchner, *Automated, fast and sensitive quantification of drugs in blood by LC–MS with on-line extraction: immunosuppressants*, J. Chomatogr. B, 748 (2000) 41.

9. P.J. Taylor, A. Jones, G.A. Balderson, S.V. Lynch, R.L. Norris, S.M. Pond, *Sensitive, specific quantitative analysis of tacrolimus in blood by LC–ESI-MS–MS*, Clin. Chem., 42 (1996) 279.

10. P.J. Taylor, N.S. Hogan, S.V. Lynch, A.G. Johnson, S.M. Pond, *Improved TDM of tacrolimus by MS–MS*, Clin. Chem., 43 (1997) 2189.

11. P.J. Taylor, C.E. Jones, P.T. Martin, S.V. Lynch, A.G. Johnson, S.M. Pond, *Microscale LC–ESI-MS–MS assay for CsA in blood*, J. Chromatogr. B, 705 (1998) 289.

12. P.J. Taylor, A.G. Johnson, *Quantitative analysis of sirolimus in blood by LC–ESI-MS–MS*, J. Chromatogr. B, 718 (1998) 251.

13. P. Salm, P.J. Taylor, P.I. Pillans, *Analytical performance of microparticle enzyme immunoassay and LC–MS–MS in the determination of sirolimus in whole blood*, Clin. Chem., 45 (1999) 2278.

14. P. Salm, P.J. Taylor, S.V. Lynch, P.I. Pillans, *Quantification and stability of everolimus in human blood by LC–ESI-MS–MS*, J. Chromatogr. B, 772 (2002) 283.

15. A. Volosov, K.L. Napoli, S.J. Soldin, *Simultaneous simple and fast quantification of three major immunosuppressants by LC–MS–MS*, Clin. Biochem., 34 (2001) 285.

16. D.W. Holt, T. Lee, K. Jones, A. Johnston, *Validation of an assay for routine monitoring of sirolimus using LC–MS detection*, Clin. Chem., 46 (2000) 1179.

17. B.G. Keevil, D.P. Tierney, D.P. Cooper, M.R. Morris, *Rapid LC–MS–MS method for routine analysis of CsA over an extended concentration range*, Clin. Chem., 48 (2002) 69.

18. I. Segarra, T.R. Brazelton, N. Guterman, B. Hausen, W. Jacobsen, R.E. Morris, L.Z. Benet, U. Christians, *Development of a LC–ESI-MS assay for the specific and sensitive quantification of the novel immunosuppressive macrolide 40-O-(2-hydroxyethyl)-rapamycin*, J. Chromatogr. B, 720 (1998) 179.

19. L.M. McMahon, S. Luo, M. Hayes, F.L.S. Tse, *High-throughput analysis of everolimus and CsA in whole blood by LC–MS using a semi-automated 96-well SPE system*, Rapid Commun. Mass Spectrom., 14 (2000) 1965.

20. N. Brignol, L.M. McMahon, S. Luo, F.L.S. Tse, *High-throughput semi-automated 96-well LLE and LC–MS analysis of everolimus and CsA in whole blood*, Rapid Commun. Mass Spectrom., 15 (2001) 898.

21. L. Zhou, D. Tan, J. Theng, L. Lim, Y.-P. Liu, K.-W. Lam, *Optimized analytical method for CsA by LC–ESI-MS*, J. Chromatogr. B, 754 (2001) 201.

22. P. Hatsis, D.A. Volmer, *Evaluation of a cyano stationary phase for the determination*

of tacrolimus, sirolimus and CsA in whole blood by LC–MS–MS, J. Chromatogr. B, 809 (2004) 287.

23.   M. Deters, G. Kirchner, K. Resch, V. Kaever, *Simultaneous quantification of sirolimus, everolimus, tacrolimus and CsA by LC–MS*, Clin. Chem. Lab. Med., 40 (2002) 285.

24.   F. Streit, V.W. Armstrong, M. Oellerich, *Rapid LC–MS–MS routine method for simultaneous determination of sirolimus, everolimus, tacrolimus and CsA in whole blood*, Clin. Chem., 48 (2002) 955.

25.   M. Oellerich, V.W. Armstrong, F. Streit, L. Weber, B. Tönshoff, *Immunosuppressive drug monitoring of sirolimus and CsA in pediatric patients*, Clin. Biochem., 37 (2004) 424.

26.   T. Koal, M. Deters, B. Casetta, V. Kaever, *Simultaneous determination of four immunosuppressants by means of high speed and robust on-line SPE–LC–MS–MS*, J. Chromatogr. B, 805 (2004) 215.

27.   U. Ceglarek, J. Lembcke, G.M. Fiedler, M. Werner, H. Witzigmann, J.P. Hauss, J. Thiery, *Rapid simultaneous quantification of immunosuppressants in transplant patients by TFC combined with MS–MS*, Clin. Chim. Acta, 346 (2004) 181.

28.   M.A. Poquette, G.L. Lensmeyer, T.C. Doran, *Effective use of LC–MS in the routine clinical laboratory for monitoring sirolimus, tacrolimus and CsA*, Ther. Drug Monit., 27 (2005) 144.

29.   M. Deters, V. Kaever, G.I. Kirchner, *LC–MS for TDM of immunosuppressants*, Anal. Chim. Acta, 492 (2003) 133.

30.   P.E. Wallemacq, *TDM of immunosuppressant drugs. Where are we?*, Clin. Chem. Med Lab., 42 (2004) 1204.

31.   G.L. Lensmeyer, M.A. Poquette, *TDM of tacrolimus concentrations in blood: semi-automated extraction and LC–MS–ESI-MS*, Ther. Drug Monit., 23 (2001) 239.

32.   D,H, Wilson, D. Sepe, G. Barnes, *Inter-laboratory differences in sirolimus results from six sirolimus testing centers using LC–MS–MS*, Clin. Chim. Acta, 355 (2005) 211.

33.   K. M. L. Crommentuyn, H. Rosing, L. G. A. H. Nan-Offeringa, M. J. X. Hillebrand, A. D. R. Huitema, J. H. Beijnen, *Rapid quantification of HIV protease inhibitors in human plasma by LC–ESI-MS–MS*, J. Mass Spectrom., 38 (2003) 157.

34.   K.M.L. Crommentuyn, H. Rosing, M.J.X. Hillebrand, A.D.R. Huitema, J.H. Beijnen, *Simultaneous quantification of the new HIV protease inhibitors atazanavir and tipranavir in human plasma by LC–ESI-MS–MS*, J. Chromatogr. B, 804 (2004) 359.

35.   V.A. Frerichs, R. DiFrancesco, G.D. Morse, *Determination of protease inhibitors using LC–MS–MS*, J. Chromatogr. B, 787 (2003) 393.

36.   K.M. Rentsch, *Sensitive and specific determination of eight antiretroviral agents in plasma by LC–MS*, J. Chromatogr. B, 788 (2003) 339.

37.   L. Dickinson, L. Robinson, J. Tjia, S. Khoo, D. Back, *Simultaneous determination of HIV protease inhibitors amprenavir, atazanavir, indinavir, lopinavir, nelfinavir, ritonavir and saquinavir in human plasma by LC–MS–MS*, J. Chromatogr. B, 829 (2005) 82.

38.   M. Jemal, S. Rao, M. Gatz, D. Whigan, *LC–MS–MS quantitative determination of the HIV protease inhibitor atazanavir in human PBMC: practical approaches to PBMC preparation and PBMC assay design for high-throughput analysis*, J. Chromatogr. B, 795 (2003) 273.

39. A. Rouzes, K. Berthoin, F. Xuereb, S. Djabarouti, I. Pellegrin, J.L. Pellegrin, A.C. Coupet, S. Augagneur, H. Budzinski, M.C. Saux, D. Breilh, *Simultaneous determination of antiretroviral agents amprenavir, lopinavir, ritonavir, saquinavir and efavirenz in human PBMC by LC–MS*, J. Chromatogr. B, 813 (2004) 209.

40. S. Colombo, A. Beguin, A. Telenti, J. Biollaz, T. Buclin, B. Rochat, L.A. Decosterd, *Intracellular measurements of anti-HIV drugs indinavir, amprenavir, saquinavir, ritonavir, nelfinavir, lopinavir, atazanavir, efavirenz and nevirapine in PBMC by LC–MS–MS*, J. Chromatogr. B, 819 (2005) 259.

41. K.B. Kenney, S.A. Wring, R.M. Carr, G.N. Wells, J.A. Dunn, *Simultaneous determination of lamivudine and zidovudine concentrations in human serum using LC–MS–MS*, J. Pharm. Biomed. Anal., 22 (2000) 967.

42. A.S. Pereira, K.B. Kenney, M.S. Cohen, J.E. Hall, J.J. Eron, R.R. Tidwell, J.A. Dunn, *Simultaneous determination of lamivudine and zidovudine concentrations in human seminal plasma using LC–MS–MS*, J. Chromatogr. B, 742 (2000) 173.

43. J.L. Wiesner, F.C.W. Sutherland, M.J. Smit, G.H. van Essen, H.K.L. Hundt, K.J. Swart, A.F. Hundt, *Sensitive and rapid LC–MS–MS method for the determination of stavudine in human plasma*, J. Chromatogr. B, 773 (2002) 129.

44. RE. Estrela, MC. Salvadori, R.S.L. Raices, G. Suarez-Kurtz, *Determination of didanosine in human serum by on-line SPE–LC–ESI-MS–MS: application to a bioequivalence study*, J. Mass Spectrom., 38 (2003) 378.

45. R.S.L. Raices, M.C. Salvadori, R.E. Estrela, F.R. de Aquino Neto, G. Suarez-Kurtz, *Determination of stavudine in human serum by on-line SPE–LC–ESI-MS–MS: application to a bioequivalence study*, Rapid Commun. Mass Spectrom., 17 (2003) 1611.

46. Y. Huang, E. Zurlinden, E. Lin, X. Li, J. Tokumoto, J. Golden, A. Murr, J. Engstrom, J. Conte, *LC–MS–MS assay for the simultaneous determination of didanosine and stavudine in human plasma, bronchoalveolar lavage fluid, alveolar cells, peripheral blood mononuclear cells, seminal plasma, cerebrospinal fluid and tonsil tissue*, J. Chromatogr. B, 799 (2004) 51.

47. J.D. Moore, G. Valette, A. Darque, X.-J. Zhou, J.-P. Sommadossi, *Simultaneous quantitation of the 5'-triphosphate metabolites of zidovudine, lamivudine, and stavudine in PBMC of HIV infected patients by LC–MS–MS*, J. Am. Soc. Mass Spectrom., 11 (2000) 1134.

48. X. Cahours, T.T. Tran, N. Mesplet, C. Kieda, P. Morin, L.A. Agrofoglio, *Analysis of intracellular didanosine triphosphate at sub-ppb level using LC–MS–MS*, J. Pharm. Biomed. Anal., 26 (2001) 819.

49. E.N. Fung, Z. Cai, T.C. Burnette, A.K. Sinhababu, *Simultaneous determination of Ziagen and its phosphorylated metabolites by ion-pairing LC–MS–MS*, J. Chromatogr. B, 754 (2001) 285.

50. A. Pruvost, F. Becher, P. Bardouille, C. Guerrero, C. Creminon, J.-F. Delfraissy, C. Goujard, J. Grassi, H. Benech, *Direct determination of phosphorylated intracellular anabolites of stavudine by LC–MS–MS*, Rapid Commun. Mass Spectrom., 15 (2001) 1401.

51. F. Becher, A. Pruvost, C. Goujard, C. Guerreiro, J.-F. Delfraissy, J. Grassi, H. Benech, *Improved method for the simultaneous determination of d4T, 3TC and ddI intracellular phosphorylated anabolites in human PBMC using LC–MS–MS*, Rapid Commun. Mass Spectrom., 16 (2002) 555.

52. F. Becher, D. Schlemmer, A. Pruvost, M.-C. Nevers, C. Goujard, S. Jorajuria, T. Brossette, L. Lebeau, C. Créminon, J. Grassi, H. Benech, *Development of a direct assay for measuring intracellular AZT triphosphate in human PBMC*, Anal. Chem., 74 (2002) 4220.

53. F. Becher, A. Pruvost, J. Gale, P. Couerbe, C. Goujard, V. Boutet, E. Ezan, J. Grassi, H. Benech, *A strategy for LC–MS–MS assays of intracellular drugs: application to the validation of the triphosphorylated anabolite of antiretrovirals in PBMC*, J. Mass Spectrom., 38 (2003) 879.

54. H. Pèlerin, S. Compain, X. Duval, F. Gimenez, H. Bénech, A. Mabondzo, *Development of an assay method for the detection and quantification of protease and nNRTI in plasma and in PBMC by LC–UV and MS–MS detection*, J. Chromatogr. B, 819 (2005) 47–57

55. T. Koal, H. Burhenne, R. Römling, M. Svoboda, K. Resch, V. Kaever, *Quantification of antiretroviral drugs in dried blood spot samples by means of LC–MS–MS*, Rapid Commun. Mass Spectrom. 19 (2005) 2995.

56. B. Fan, M.G. Bartlett, J.T. Stewart, *Determination of lamivudine, stavudine and efavirenz in human serum using LC–ESI-MS–MS with ionization polarity switch*, Biomed. Chromatogr., 16 (2002) 383.

57. J. Chi, A.L. Jayewardene, J.A. Stone, F.T. Aweeka, *An LC–MS–MS method for the determination of nevirapine, a nNRTI, in human plasma*, J. Pharm. Biomed. Anal., 31 (2003) 953.

58. M.S. Rashed, *Clinical applications of MS–MS: ten years of diagnosis and screening for inherited metabolic diseases*, J. Chromatogr. B, 758 (2001) 27.

59. D.S. Millington, C.R. Roe, D.A. Maltby, *Application of high resolution FAB and constant B/E ratio linked scanning to the identification and analysis of acylcarnitines in metabolic disease*, Biomed. Mass Spectrom., 11 (1984) 236.

60. D.S. Millington, T.P. Bohan, C.R. Roe, A.L. Yergey, D.J. Liberato, *Valproylcarnitine: a novel drug metabolite identified by FAB and thermospray LC–MS*, Clin. Chim. Acta, 145 (1985) 69.

61. D.L. Norwood, N. Kodo, D.S. Millington, *Application of continuous-flow LC–FAB-MS to the analysis of diagnostic acylcarnitines in human urine*, Rapid Commun. Mass Spectrom., 2 (1989) 269.

62. M.S. Rashed, P.T. Ozand, M.E. Harrison, P.J.F. Watkins, S. Evans, *ESIMS–MS in the diagnosis of organic acidemias*, Rapid Commun. Mass Spectrom., 8 (1994) 129.

63. M.S. Rashed, M.P. Bucknall, D. Little, A. Awad, M. Jacob, M. Alamoudi, M. Alwattar, P.T. Ozand, *Screening blood spots for inborn errors of metabolism by ESI-MS–MS with a microplate batch process and a computer algorithm for automated flagging of abnormal profiles*, Clin. Chem., 43 (1997) 1129.

64. D.H. Chace, S.L. Hillman, J.L.K. van Hove, E.W. Naylor, *Rapid diagnosis of MCAD deficiency: quantitative analysis of octanoylcarnitine and other acylcarnitines in newborn blood spots by MS–MS*, Clin. Chem., 43 (1997) 2106.

65. D.H. Chace, D.S. Millington, N. Terada, S.G. Kahler, C.R. Roe, L.F. Hofman, *Rapid diagnosis of PKU by quantitative analysis for Phe and Tyr in neonatal blood spots by MS–MS*, Clin. Chem., 39 (1993) 66.

66. D.H. Chace, J.E. Sherwin, S.L. Hillman, F. Lorey, G.C. Cunningham, *Use of Phen-to-Tyr ratio determined by MS–MS to improve newborn screening for PKU of early discharge specimens collected in the first 24 hours*, Clin. Chem., 44 (1998) 2405.

67. D.H. Chace, T.A. Kalas, *A biochemical perspective on the use of MS–MS for newborn screening and clinical testing*, Clin. Biochem., 38 (2005) 296.

68. M.J. Bogusz, K.D. Krüger, R.D. Maier, *Analysis of underivatized amphetamines and related phenethylamines with LC–APCI-MS*, J. Anal. Toxicol., 24 (2000) 77.

69. M.-R.S. Fuh, K.-T. Lu, *Determination of methylamphetamine and related compounds in human urine by LC–ESI-MS*, Talanta, 48 (1999) 415.

70. K.M. Clauwaert, J.F. Van Bocxlaer, H.J. Major, J.A. Claereboudt, W.E. Lambert, E.M. Van den Eeckhout, C.H. Van Peteghem, A.P. De Leenheer, *Investigation of the quantitative properties of the Q–TOF-MS with ESI using 3,4-methylenedioxy-methamphetamine*, Rapid Commun. Mass Spectrom., 13 (1999) 1540.

71. K.A. Mortier, R. Dams, W.E. Lambert, E.A. De Letter, S. Van Calenbergh, A.P. De Leenheer, *Determination of paramethoxyamphetamine and other amphetamine-related designer drugs by LC–sonic spray ionization MS*, Rapid Commun. Mass Spectrom., 16 (2002) 865.

72. S. Pichini, R. Pacifici, M. Pellegrini, E. Marchei, J. Lozano, J. Murillo, O. Vall, Ó. García-Algar, *Development and validation of a LC–MS assay for determination of amphetamine, methamphetamine, and methylenedioxy derivatives in meconium*, Anal. Chem., 76 (2004) 2124.

73. J. Cai, J.D. Henion, *On-line immunoaffinity extraction-coupled column capillary LC–MS–MS: trace analysis of LSD analogs and metabolites in human urine*, Anal. Chem., 68 (1996) 72.

74. H. Hoja, P. Marquet, B. Verneuil, H. Lotfi, J.-L. Dupuy, and G. Lachâtre, *Determination of LSD and N-desmethyl-LSD in urine by liquid chromatography coupled to electrospray ionization mass spectrometry*, J. Chromatogr. B, 692 (1997) 329.

75. J. De Kanel, W.E. Vickery, B. Waldner, R.M. Monahan, F.X. Diamond, *Automated extraction of LSD and N-demethyl-LSD from blood, serum, plasma, and urine samples using the Zymark RapidTrace with LC–MS–MS confirmation*, J. Forensic Sci., 43 (1998) 622.

76. G.K. Poch, K.L. Klette, D.A. Hallare, M.G. Manglicmot, R.J. Czarny, L.K. McWhorter, C.J. Anderson, *Detection of metabolites of LSD in human urine specimens: 2-oxo-3-hydroxy-LSD, a prevalent metabolite of LSD*, J. Chromatogr. B, 724 (1999) 23.

77. J. Canezin, A. Cailleux, A. Turcant, P. Le Bouil, P. Harry, P. Allain, *Determination of LSD and its metabolites in human biological fluids by LC–ESI-MS–MS*, J. Chromatogr. B, 765 (2001) 15.

78. S. Stybe Johansen, J. Lundsby Jensen, *LC–MS–MS determination of LSD, iso-LSD, and the main metabolite 2-oxo-3-hydroxy-LSD in forensic samples and application in a forensic case*, J. Chromatogr. B., 825 (2005) 21.

79. M.J. Bogusz , R.-D. Maier, M. Erkens, S. Driessen, *Determination of morphine, M3G, and M6G, codeine, codeine-glucuronide and 6-MAM in body fluids by LC–APCI-MS*, J. Chromatogr. B, 703 (1997) 115.

80. A. Dienes-Nagy, L. Rivier, C. Giroud, M. Augsburger, P. Mangin, *Method for quantification of morphine, M3G, and M6G, codeine, codeine glucuronide and 6-MAM in human blood by LC–ESI-MS for routine analysis in forensic toxicology*, J. Chromatogr. A, 854 (1999) 109.

81. R. Pacifici, S. Pichini, I. Altieri, A. Caronna, A.R. Passa, P. Zuccaro, *LC–ESI-MS*

determination of morphine, M3G, and M6G: application to pharmacokinetic studies, J. Chromatogr. B, 664 (1995) 329.

82.   D. Projean, T. Minh Tu, J. Ducharme, *Rapid and simple method to determine morphine and its metabolites in rat plasma by LC–MS*, J. Chromatogr. B, 787 (2003) 243.

83.   C.M. Murphy, M.A. Huestis, *LC–ESI-MS–MS analysis for the quantification of morphine, M3G, and M6G, codeine,, and codeine-6-β-D-glucuronide in human urine*, J. Mass Spectrom., 40 (2005) 1412.

84.   K.M. Clauwaert, J.F. van Boxclaer, W.E. Lambert, E.G. van den Eeckhout, F. Lemière, E.L. Esmans, A.P. de Leenheer, *Narrow-bore LC in combination with fluorescence and ESI-MS detection for the analysis of cocaine and metabolites in human hair*, Anal. Chem., 70 (1998) 2336.

85.   P.M. Jeanville, E.S. Estapé, I. Torres-Negrón de Jeanville, *The effect of LC eluents and additives on the positive ion responses of cocaine, benzoylecgonine, and ecgonine methyl ester using ESI*, Int. J. Mass Spectrom., 227 (2003) 247.

86.   Y. Xia, P. Wang, M.G. Bartlett, H.M. Solomon, K.L. Busch, *An LC–MS–MS method for the comprehensive analysis of cocaine and cocaine metabolites in meconium*, Anal. Chem., 72 (2000) 764.

87.   P.M. Jeanville, E.S. Estapé, S.R. Needham, M.J. Cole, *Rapid confirmation/ quantitation of cocaine and benzoylecgonine in urine utilizing LC–MS–MS*, J. Am. Soc. Mass Spectrom., 11 (2000) 257.

88.   S.R. Needham, P.M. Jeanville, P.R. Brown, E.S. Estapé, *Performance of a pentafluorophenylpropyl stationary phase for the LC–ESI-MS–MS assay of cocaine and its metabolite ecgonine methyl ester in human urine*, J. Chromatogr. B, 748 (2000) 77.

89.   K.A. Mortier, K.E. Maudens, W.E. Lambert, K.M. Clauwaert, J.F. Van Bocxlaer, D.L. Deforce, C.H. Van Peteghem, A.P. De Leenheer, *Simultaneous, quantitative determination of opiates, amphetamines, cocaine and benzoylecgonine in oral fluid by LC–Q–TOF-MS*, J. Chromatogr. B, 779 (2002) 321.

90.   R. Dams, C.M. Murphy, W.E. Lambert, M.A. Huestis, *Urine drug testing for opioids, cocaine and metabolites by direct injection LC–MS–MS*, Rapid Commun. Mass Spectrom., 17 (2003) 1665.

91.   M. Wood, M. Laloup, M. del Mar Ramirez Fernandez, K.M. Jenkins, M.S. Young, J.G. Ramaekers, G. De Boeck, N. Samyn, *Quantitative analysis of multiple illicit drugs in preserved oral fluid by SPE and LC–MS–MS*, Forensic Sci. Intern., 150 (2005) 227.

92.   H.H. Maurer, *Systematic toxicological analysis procedures for acidic drugs and/or metabolites relevant to clinical and forensic toxicology and/or doping control*, J. Chromatogr. B, 733 (1999) 3.

93.   J.F. van Bocxlaer, K.M. Clauwaert, W.E. Lambert, D.L. Deforce, E.G. van den Eeckhout, A.P. de Leenheer, *LC–MS in forensic toxicology*, Mass Spectrom. Rev., 19 (2000) 165.

94.   M.J. Bogusz, *LC–MS as a routine method in forensic sciences: a proof of maturity*, J. Chromatogr. B, 748 (2000) 3.

95.   D. Thieme, H. Sachs, *Improved screening capabilities in forensic toxicology by application of LC–MS–MS*, Anal. Chim. Acta, 492 (2003) 171.

96.   R. Jansen, G. Lachatre, P. Marquet, *LC–MS–MS systematic toxicological analysis:*

      Comparison of MS–MS spectra obtained with different instruments and settings, Clin. Biochem., 38 (2005) 362.

97.   P.G.M. Kienhuis, R.B. Geerdink, *A mass spectral library based on CI and CID*, J. Chromatogr. A, 974 (2002) 161.

98.   W. Weinmann, M. Gergov, M. Goerner, *MS-MS Libraries with triple quadrupole MS–MS for drug identification and drug screening*, Analusis, 28 (2000) 934.

99.   C. Baumann, M.A. Cintora, M. Eichler, E. Lifante, M. Cooke, A. Przyborowska, J.M. Halket, *A library of API daughter ion mass spectra based on wideband excitation in an ion trap MS*, Rapid Commun. Mass Spectrom., 14 (2000) 349.

100.   P. Marquet, N. Venisse, E. Lacassie, G. Lachâtre, *In-source CID mass spectral libraries for the "general unknown" screening of drugs and toxicants*, Analusis, 28 (2000) 925.

101.   W. Weinmann, M. Stoertzel, S. Vogt, M. Svoboda, A. Schreiber, *Tuning compounds for ESI–in-source CID and mass spectra library searching*, J. Mass Spectrom., 36 (2001) 1013.

102.   A.W.T. Bristow, W.F. Nichols, K.S. Webb, B. Conway, *Evaluation of protocols for reproducible ESI in-source CID on various LC–MS instruments and the development of spectral libraries*, Rapid Commun. Mass Spectrom., 16 (2002) 2374.

103.   H.H. Maurer, C. Kratzsch, T. Kraemer, F.T. Peters, A.A. Weber, *Screening, library-assisted identification and validated quantification of oral antidiabetics of the sulfonylurea-type in plasma by LC–APCI-MS*, J. Chromatogr. B, 773 (2002) 63.

104.   F. Saint-Marcoux, G. Lachâtre, P. Marquet, *Evaluation of an improved general unknown screening procedure using LC–ESI-MS by comparison with GC and LC–DAD*, J. Am. Soc. Mass Spectrom., 14 (2003) 14.

105.   P. Marquet, F. Saint-Marcoux, T.N. Gamble, J.C.Y. Leblanc, *Comparison of a preliminary procedure for the general unknown screening of drugs and toxic compounds using a Q–LIT-MS with a LC–MS reference technique*, J. Chromatogr. B, 789 (2003) 9.

106.   M. Gergov, I. Ojanpera, E. Vuori, *Simultaneous screening for 238 drugs in blood by LC–ESI-MS–MS with SRM*, J. Chromatogr. B, 795 (2003) 41.

107.   C.A. Mueller, W. Weinmann, S. Dresen, A. Schreiber, M. Gergov, *Development of a multi-target screening analysis for 301 drugs using a QTrap LC–MS–MS system and automated library searching*, Rapid Commun. Mass Spectrom., 19 (2005) 1332.

13

LC–MS ANALYSIS OF STEROIDS

1. Introduction

Steroids are compounds with a cyclopenta[a]phenanthrene skeleton (Figure 13.1). Methyl groups can be present at C-10 and C-13. An alkyl side chain may also be present at C-17. Some typical examples of steroid structures, relevant to the present discussion, are given in Figure 13.2.

Steroids are important compounds in many areas in nature. First of all, they are important hormones with a variety of physiological functions in the body, e.g., in relation to the regulation of procreation and a variety of other functions, e.g., in the brain. Characterization and analysis of the steroids present in body fluids and tissues is an important application area of LC–MS. The physiological functions of steroids imply the determination of individual steroids or the monitoring of steroid profiles as a diagnostic tool in clinical studies. At the same time, steroid drugs have been produced to affect the steroid levels or profiles in the body, e.g., contraceptive drugs and various other steroid drugs. Development of new steroid drugs demands for analytical methods to perform metabolic, pharmacokinetic and bioequivalence studies. One of the physiological effects of steroids can be growth promotion. Therefore, various steroids, especially anabolic and corticosteroids, are attractive for both veterinary use and doping in sports. Examples of the use of LC–MS in steroid analysis in some important areas are discussed in Ch. 13.3-5. Regulatory analysis related to the veterinary use of steroids is discussed in Ch. 14.4.

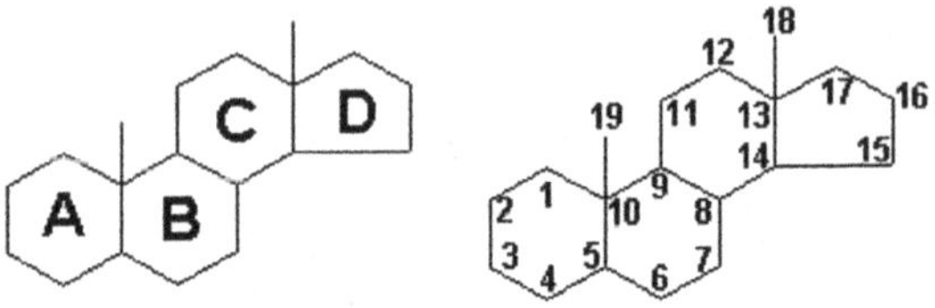

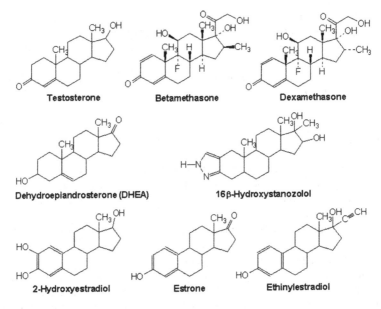

Figure 13.1: General structure of and carbon numbering convention in steroids.

Figure 13.2: Representative structures of some steroids

2. Mass spectrometry of steroids and steroid conjugates

2.1   History

The analysis of steroids is a challenging task. GC–MS has frequently been applied, but requires analyte derivatization [1]. Therefore, over the years most LC–MS interfaces have been tested or applied in the analysis of steroids. As early as 1981, Henion [2] demonstrated the analysis of dexamethasone and cortisone with micro-LC coupled to a capillary-inlet interface. Van der Greef et al. [3] described the quantitation of progesterone in serum using isotope dilution MS in ammonium CI mode using the moving-belt interface. Henion and coworkers [4] described qualitative and quantitative analysis of corticosteroids in equine urine using micro-LC coupled to direct-liquid introduction–MS. Watson *et al.* [5] reported the analysis of 30 steroid mono- and disulfates and 18 steroid glucuronides by negative-ion thermospray (TSP) LC–MS. Esteban *et al.* [6] described the quantitation of serum cortisol by isotope dilution TSP LC–MS. Lindberg *et al.* [7] reported the determination of the glucocorticosteroid budesonide, clinically used in the treatment of asthma and rhinitis, by automated TSP LC–MS. The compound was isolated from human plasma by solid-phase extraction (SPE) and subsequently derivatized to the 21-acetyl derivative, to enhance its ionization efficiency. $[^2H_8]$-Budesonide was used as internal standard (IS). The method was linear over the range 0.30–30 pmol, corresponding to plasma concentrations between 0.10 and 10 nmol/l. Intra-day variation at 0.1 nmol/l was 10–18%. The bias was less than 15% at 0.10 nmol/l and less than 7% at 0.20 nmol/l. The inter-day variation was 9% at 0.30 nmol/l [7]. Polettini *et al.* [8] described the use of various coupled-column approaches for the determination of betamethasone in urine. TSP ionization was used. In LC–LC–MS–MS, the sample throughput was 4.5 samples per hour and the limit of quantification (LOQ) was 1 ng/ml. Watson *et al.* [9] described the analysis of some corticosteroids and corticosteroid esters by LC–MS via a frit fast-atom bombardment interface. Creaser *et al.* [10] described the analysis of synthetic corticosteroids dexamethasone and flumethasone in equine urine after a clean-up by immunoaffinity chromatography coupled on-line to LC–MS via a particle-beam interface (PBI) on an ion-trap system with electron ionization (EI). Recently, Heimark *et al.* [11] performed normal-phase LC–MS via a PBI to characterize ergosterol biosynthetic precursors from their EI spectra. They developed the method to study the ergosterol profiles in *Candida albicans* following treatment with posaconazole.

2.2   Electrospray and APCI of steroids

Nowadays, electrospray ionization (ESI) and atmospheric-pressure chemical ionization (APCI) are the methods-of-choice in steroid analysis by LC–MS, while the potential of atmospheric-pressure photoionization (APPI) has been evaluated as well. The analysis of steroids by LC–MS is challenging, because of their relatively low polarity and proton affinity.

While there were already some reports on the LC–MS analysis of steroids [12-18], Ma and Kim [19] reported a comparative study of the performance of APCI and ESI in the analysis of neutral steroids. Both electrospray and APCI were performed at 0.4 ml/min and a 2.1-mm-ID column. Based on mass spectral data and sensitivity, the steroids investigated were classified into three major groups:

A   Steroids containing a 3-one-Δ^4 group, *e.g.*, testosterone and progesterone.

B   Steroids containing at least one ketone group without conjugation, *e.g.*, estrone and androsterone.

C   Steroids containing only hydroxy groups, *e.g.*, 17β-estradiol.

In APCI, the best performance for steroids was achieved with a methanol–water gradient, containing 1-2% acetic acid. The A-group steroids all showed intense protonated molecules $[M+H]^+$ and only minor peaks due to water loss. From the B-group steroids, only some compounds showed an intact $[M+H]^+$, while loss of 1 or 2 water molecules resulted in abundant ions. For C-group steroids, no intact $[M+H]^+$ were observed. Peaks due to 2 water losses were the most abundant ions for almost all C-group steroids. The response of the A-group steroids was 3–6-fold better than B-group steroids, which in turn was 2-4-fold better than C-group steroids, resulting in detection limits in selected-ion monitoring (SIM) of 50 ± 10 pg for A-group steroids, 1.0 ± 0.3 ng for B-group steroids, and 2.0 ± 0.5 ng for C-group steroids. In ESI, the best performance for A-group and B-group steroids was achieved with a mobile phase of methanol–water, resulting in the observation of sodiated molecules $[M+Na]^+$. The use of acetonitrile or the addition of ammonium acetate had a detrimental effect on the response. Best performance of C-group steroids could be achieved by the addition of acetic acid to the mobile phase.

Spectral characteristics in ESI were similar to those in APCI: primarily intact molecules for A-group steroids, losses of 1 and/or 2 water molecules next to the intact molecules for B-group steroids, and loss of 1 and/or 2 water molecules without observing the intact molecule for C-group steroids. The best response was achieved for A-group steroids, which was in the range of 80-100% of the most sensitive steroids (testosterone and epitestosterone). The response of B-group steroids was 20–45% of the testosterone response, with two exceptions (9% and 85%), while the response of C-group steroids ranged from 1–4% of the testosterone response, again with two exceptions (7% and <0.1%). Detection limits for the A-group steroids were in the range of 5 ± 3 pg.

A chromatogram of 11 steroids (20 pg each), detected as $[M+Na]^+$ in LC–ESI-MS, is shown in Figure 13.3. The observations by Ma and Kim [19] are generally confirmed by others, although it must be taken in account that the source design from different instrument manufacturers appears to influence the choice between ESI and APCI.

Neutral corticosteroids are prone to the formation of acid adducts $[M+RCOO]^-$ in negative-ion mode [20-21]. Abundant acetate adducts are observed for steroids with a relatively acidic hydroxyl group [22]. In negative-ion TSP ionization, Kim et al. [23] observed more abundant acid adducts with decreasing pK_a of the acid. Marwah et al. [24] showed signal enhancement for a variety of steroids like dehydroepiandrosterone (DHEA) and related compounds due to the addition of low concentration of acid, i.e., typically 1–5 mmol/l formic acid, 1–8 mmol/l acetic acid, or 0.05–0.15 mmol/l trifluoroacetic acid, while higher acid concentrations were found to compromise the response. Formic acid was the best choice for the neutral steroids, while acetic acid is preferred for sulfate conjugates. Post-column addition of 10 nmol/l silver nitrate resulted in a ten-fold increase in the response for androst-5-ene-3β,17β-diol. $[M+Ag]^+$ is observed instead of $[M+H-H_2O]^+$ [24].

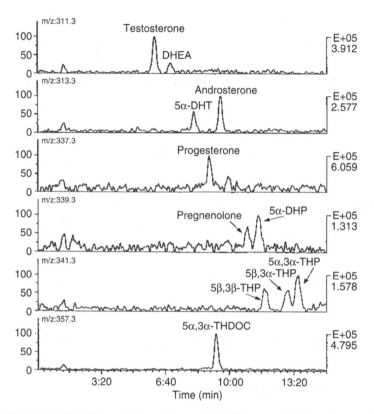

Figure 13.3: LC–ESI-MS analysis of 11 steroids (20 pg each injected) by means of a methanol–water gradient on a 2.1-mm-ID column, operated at a flow-rate of 0.4 ml/min. Reprinted from [19] with permission. ©1997, American Society for Mass Spectrometry.

APPI is promoted as an alternative to APCI in the analysis of relatively nonpolar compounds. Leinonen et al. [25] compared ESI, APCI, and APPI in the analysis of three anabolic steroids, 3'-hydroxystanozolol, 6β-hydroxy-4-chlorodehydromethyl-testosterone, and oxandrolone. All three techniques proved suitable, but ESI was preferred. Optimum mobile phase was a methanol–water gradient containing 5 mmol/l ammonium acetate and 0.01% acetic acid. For all three compounds, intact protonated molecules and some fragments due to consecutive water losses were observed. Significant better performance of APPI compared to ESI and APCI was also reported for corticosteroids by Greig et al. [26].

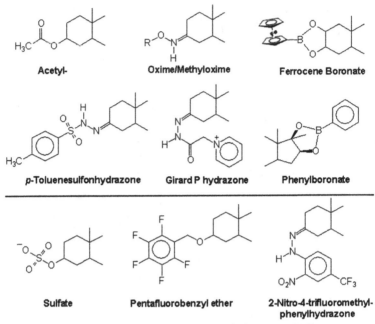

Figure 13.4: Structures of steroid derivatives. Partly based on [36].

2.3   Steroid derivatization

In order to enhance the sensitivity in steroid analysis, a number of derivatization procedures have been proposed. In most cases, the aim of the derivatization is an increase of the proton affinity of the compound, although the introduction of a fixed charge on a quaternary ammonium or a group with high electron affinity has been described as well. An overview of the derivatives generated is given in Figure 13.4. Most derivatization procedures are applied in combination with ESI-MS.

Lindberg *et al.* [7] performed a derivatization of budesonide with acetic anhydride to the 21-acetyl derivative in order to enhance its ionization efficiency in TSP ionization. The same procedure in combination with positive-ion APCI was used by others for budesonide [27], and the catechol estrogens 2- and 4-hydroxy-estrone and 2- and 4-hydroxyestradiol in rat brain [28].

Shimada *et al.* [29] described the conversion pregnenolone and its 3-stearate ester into 3-acetate-20-methyloxime and 20-methyloxime derivatives, respectively. 20-fold improvement in the detection limit was reported by Liu *et al.* [30] as a result of the derivatization of oxosteroids into their oximes. 16β-hydroxystanozolol was derivatized with phenylboronic acid to reduce the fragmentation in MS–MS [31]. Williams *et al.* [32] reported derivatization of 2- and 4-hydroxyestradiol with ferrocene boronic acid to form cyclic boronate esters. A radical cation of the derivative is generated in ESI-MS in a nonaqueous solvent system (10% dichloromethane in acetonitrile with 0.1 mmol/l lithium triflate). Xu *et al.* [33] described the derivatization of 2- and 4-hydroxyestrones to *p*-toluenesulfonhydrazones to significantly enhanced the sensitivity. Johnson *et al.* [34] achieved 10-fold better sensitivity by converting cholesterol into its mono-(dimethylaminoethyl) succinyl ester. Anari *et al.* [35] enhanced the ESI-MS response of ethinyl estradiol by means of a derivatization with dansyl chloride. The most abundant fragment ion in the product ion MS–MS spectrum is the protonated 5-(dimethylamino)naphthalene at *m/z* 171, *i.e.*, not related to the steroid. Griffiths *et al.* [36] demonstrated the derivatization of neutral oxosteroids to Girard P hydrazones. Product-ion MS–MS spectra of the derivatives were structurally informative enabling steroid identification at the sub-pg level.

Chatman *et al.* [37] reported conversion of unconjugated steroids to their sulfate esters to be analysed by negative-ion ESI-MS. Highly selective and sensitive ionization in negative-ion APCI-MS can be achieved by means of the electron-capture process. This requires analytes with a high electron affinity. Singh *et al.* [38] demonstrated 25–100-fold improvement in LOQ of various compounds, *e.g.*, estrone, by means of a pentafluorobenzyl derivatization. Higashi *et al.* [39] found 20-fold enhanced sensitivity in the analysis of 20-oxosteroids in rat brains after derivatization to hydrazones by means of the high electron-affinitive reagent 2-nitro-4-trifluoromethylphenylhydrazine. Even higher sensitivity gain was achieved by derivatization with 2-hydrazino-1-methylpyridine [40]. The same group reviewed the derivatization of neutral steroids for LC–MS [41].

2.4   ESI and APCI of steroid conjugates

Following previous studies with TSP [5], Murray *et al.* [17] showed the analysis of steroid sulfates by microbore LC–MS using an ESI interface. With a 1-mm-ID column and 5-μl injection volume, absolute detection limits in the range of 5 pg were achieved, *i.e.*, a 5-fold improvement relative to previous results. Subsequently, the analysis of steroid sulfates and glucuronides by negative-ion APCI or ESI LC–MS has been described frequently, especially in doping analysis (Ch. 13.3) and clinical applications (Ch. 13.4). A nano-LC–MS system, featuring a 350×0.1-mm-ID capillary column combined with a 50×0.1-mm-ID precolumn, was developed for the analysis of steroid sulfates in plasma [42]. The system was applied to the analysis of neurosteroids in rat brain [43].

Kuuranne *et al.* [44] studied the behaviour of eight anabolic steroid glucuronides in positive-ion and negative-ion ESI and APCI. In positive-ion mode, steroid glucuronides with a 3-keto-Δ^4 group show $[M+H]^+$ due to their high proton affinity, while others show $[M+NH_4]^+$. In negative-ion mode, intense deprotonated molecules $[M–H]^-$ are observed. In product-ion MS–MS, losses of glucuronic acid (176 Da) and one or two water molecules are observed. In negative-ion mode, these diagnostic ions were less intense. Abundance ratios of these ions allow discrimination between the isomeric 5α-nandrolone and 5β-nandrolone glucuronides and between the isomeric testosterone and epitestosterone glucuronides. Nevertheless, the authors concluded that positive-ion ESI-MS–MS seems to be the most promising method for further development of a screening method.

2.5   Fragmentation of steroids in MS–MS

Fragmentation of steroids in MS–MS is frequently applied as part of a screening method, especially in regulatory analysis, where confirmation of identity is generally based on three or four diagnostic product ions. Little attention has been paid to the actual identity of these diagnostic ions. This is obviously due to the complexity of the fragmentation reactions involved.

Williams *et al.* [45] performed an extensive study to elucidate the identity of the fragments ions at *m/z* 97, 109, 121, and 123 observed in the in-source CID and MS–MS spectra of testosterone and its hydroxy analogs. From fragmentation data from various hydroxy isotopically-labelled analogs, the identity could be elucidated. For the fragment ions at *m/z* 109 and 123, detailed fragmentation mechanisms were proposed. The identity of the ion at *m/z* 109 was found to be different from what was earlier proposed. A summary of the results is shown in Figure 13.5, indicating that all four fragments originated from a B-ring cleavage.

This study demonstrates the difficulties faced in identifying fragment ions in the MS–MS spectra of steroids. Therefore, the use of diagnostic ions is more practical in confirmation. In some cases, diagnostic ions do not allow an unambiguous confirmation of identity, especially in the differentiation between isomers.

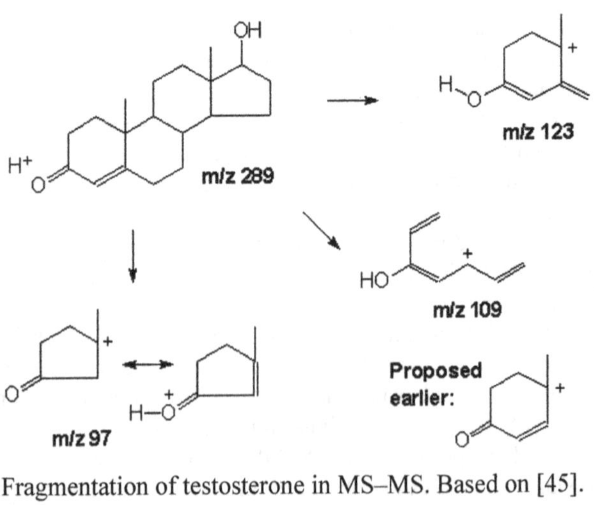

Figure 13.5: Fragmentation of testosterone in MS–MS. Based on [45].

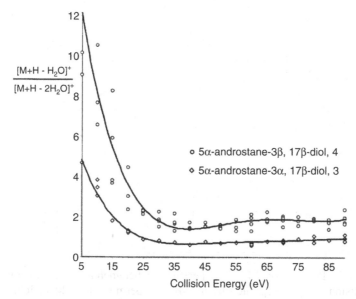

Figure 13.6: Ratio of the relative abundances of $[M+H-H_2O]^+$ and $[M+H-2H_2O]^+$ as a function of the collision energy for 5α-androstane-3β,17β diol and 5α-androstane-3α,17β diol. Reprinted from [46] with permission. ©1998, John Wiley & Sons, Ltd..

Sjöberg and Markides [46] applied the relative abundances of water-loss-related fragment ions in energy-resolved in-source CID or collision-cell CID to differentiate between isomeric hydroxysteroids. As an example, the relative abundance ratio of $[M+H-H_2O]^+$ and $[M+H-2H_2O]^+$ is plotted as a function of the collision energy for 5α-androstane-3β,17β diol and 5α-androstane-3α,17β diol in Figure 13.6.

Differentiation between the stereoisomeric dexamethasone and betamethasone (Figure 13.2) has been investigated in detail [47-49]. While De Wasch *et al.* [47] differentiated from differences in relative abundances, Antignac *et al.* [48] proposed the use of multivariate statistical techniques, based on principal component analysis. Baseline separation between betamethasone, dexamethasone, and various related esters was reported by Arthur *et al.* [49] using a C_8 column and a step gradient.

Antignac *et al.* [20] studied the fragmentation of 14 corticosteroids and metabolites. In the positive-ion mode, subsequent losses of water were observed as well as losses of HF and HCl for fluorinated or chlorinated compounds. At higher collision energies, ring fragmentation is induced. The positive-ion MS–MS spectra of betamethasone acquired under different conditions are shown in Figure 13.7. The negative-ion mode was found to be more specific, showing the loss of formaldehyde from the $[M-H]^-$ for all compounds with a 17α-hydroxyl group. This formaldehyde loss is also observed by others [21, 50-51].

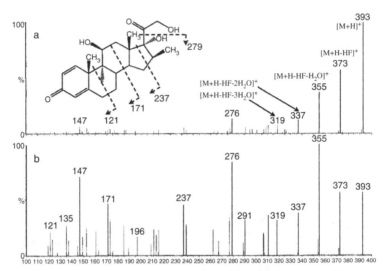

Figure 13.7: Product-ion MS–MS spectrum of betamethasone at (a) low and (b) high collision energy. Reprinted from [20] with permission. ©2000, John Wiley & Sons Ltd..

Characterization of budesonide and structurally related corticosteroids was recently reported by Hou *et al.* [52].

Collections of product-ion MS–MS spectra and/or tables with diagnostic ions have been reported for anabolic steroids [53-55], corticosteroids [56], oxosteroid oximes [30], and anabolic steroid conjugates [44]. The fragmentation of steroids in MS–MS was reviewed by Griffiths [57].

3. Doping analysis in sports

3.1 Racing horses

In early studies, LC–ESI-MS was applied in the analysis of steroid sulfates in equine urine [12], and in the identification of methandrostenolone and stanozolol metabolites in equine and human urine [13-15].

Kim *et al.* [55] developed an LC–MS–MS method, based on positive-ion APCI mode and selected-reaction monitoring (SRM), for the doping analysis of 19-nortestosterone and three of its esters in plasma of race horses. The limits of quantification are 0.16, 0.1, 2.0, and 5.0 ng/ml for 19-nortestosterone, and its phenylpropionate, decanoate, and cyclopentanepropionate ester, respectively, using 2 ml of plasma. The method enabled detection of 19-nortestosterone up to 23 days after an intra-muscular injection of 400 mg of the decanoate ester.

Tang *et al.* [58] reported the analysis of corticosteroids in equine urine after enzymatic hydrolysis, liquid-liquid extraction (LLE) with ethyl acetate, and LC–MS–MS in positive-ion APCI mode. Hydrocortisone, deoxycorticosterone, and 21 synthetic corticosteroids can be analysed at 5 µg/l level in urine within 10 min. As an example, the confirmation of deoxycortone in a urine sample is demonstrated in Figure 13.8.

The determination of corticosteroid beclomethasone dipropionate, the prodrug of beclomethasone used in the treatment of respiratory disorders in horses, in equine plasma and urine was reported by Guan *et al.* [59]. The dosing is only 325 µg per horse, posing challenges to the detection of this drug with restricted use. After LLE with methyl *t*-butyl ether (MTBE), the analytes and its metabolites were analysed by positive-ion LC–ESI–MS. The LOQ was 13 pg/ml for the analyte and between 25 and 50 pg/ml for its three major metabolites, which allowed detection in equine plasma up to 4 h post administration.

The analysis of stanozolol and its metabolites in equine urine was reported by McKinney *et al.* [60]. The analytes were isolated by SPE. Phase II conjugates were acid treated, and stanozolol and the resulting phase I metabolites were analysed by positive-ion ESI–MS on an ion trap. The equine metabolism of stanozolol was investigated as well. Boldenone sulfate and glucuronides were analysed in equine urine by the same group [61].

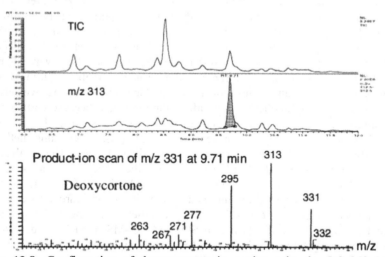

Figure 13.8: Confirmation of deoxycortone in equine urine by LC–MS–MS in positive-ion APCI mode. Reprinted from [58] with permission. ©2001, Elsevier Science BV.

3.2   Human athletes

The detection of too high levels of anabolic steroids, which are physiologically present in the body, presents a particular challenge in doping control. This is certainly true for testosterone. Generally, the illegal use of testosterone can be detected from the ratio between the urinary levels of testosterone and epitestosterone. The physiological ratio is in the range 0.9–1.6. For athletes, a testosterone/epitestosterone ratio in excess of 6 indicates illegal use. The measurement of the ratio from urine samples by GC–MS is complicated by the formation of glucuronide and sulfate conjugates prior to excretion in urine; in GC–MS, deconjugation and derivatization is required. Bean and Henion [62] demonstrated the feasability of using LC–ESI-MS for the determination of sulfate and glucuronide conjugates of testosterone and epitestosterone in human urine without the need for deconjugation and/or derivatization. Their method was further improved by Borts and Bowers [63]. They detected the conjugates in positive-ion LC–ESI-MS–MS in SRM mode. Transitions corresponding to the loss of 176 Da for the glucuronides and either 98 or 80 Da for the sulfates were monitored for both the compounds of interest and their [$^2H_3$]-labelled IS. The IS were necessary to correct for the ion-suppression effect in ESI ionization (Ch. 11.5).

Le Bizec et al. [64] demonstrated the detection of 19-norandrosterone and 19-noretiocholanolone in human urine subsequent to boar consumption using positive-ion LC–ESI-MS–MS. The detection of these steroids is generally used to confirm exogenous administration of 19-nortestosterone (nandrolone). However, a (false) positive result can be obtained from an athlete who ate pork meat. In a subsequent study [65], endogenous 19-norsteroids in boar meat were identified and quantified.

Bévalot et al. [21] analysed corticosteroids in hair for the detection of illegal use by athletes. Using LC–ESI-MS in SIM mode, the detection limit was 0.1 ng/mg from 50 mg powdered hairs. The same method was also applied in urine analysis. Comparison of the results demonstrated the complementarity of the two methods. Urine analysis is only applicable soon after administration, because corticosteroids are rapidly eliminated: some athletes were tested positive for dexa- or betamethasone in hair analysis, but negative in urinalysis. A screening method for nine corticosteroids in doping analysis was validated [66]. The analytes are separated within 20 min after LLE with diethyl ether. The LOQ was 4 ng/ml or better for all nine compounds. Betamethasone could be detected up to 12 days after administration. In order to determine endogenous levels of androgen steroids and their conjugates, Buiarelli et al. [67] developed a method for the analysis of fifteen free and conjugated steroids, based on SPE and LC–MS in SRM mode. The free steroids were analysed in positive-ion ESI-MS, the conjugates in negative-ion mode. Detection limits were between 80 and 100 ng/ml.

An LC–MS method for the urine analysis of tetrahydrogestrinone, a formerly undetectable anabolic steroid, was reported by Catlin et al. [68]. After enzymatic deglucuronidation and LLE with pentane, evaporation, and reconstitution, the extracts were analysed by positive-ion LC–ESI-MS using a mobile phase of 85% acetonitrile in 0.1% acetic acid. The detection limit was ~5 ng/ml.

An overview of doping control analysis during the Summer Olympics of 2004 in Athens was recently reported by Tsivou et al. [69].

4. Clinical studies

In clinical studies, steroid profiling by GC–MS continues to play an important role [1]. However, LC–MS is started to be applied for various special applications. Some results in relation to breast and prostate cancer research, and neurosteroids are discussed below.

4.1   Breast and prostate cancer

Epidemiological studies have suggested that high circulation levels of estrogens may be associated with the development of breast cancer and tumour growth. In order to accurately monitor estrogen levels, Zhang and Henion [70] developed a method to determine five estrogen sulfates in human urine. The method comprises of a C_{18}-SPE step in a 96-well-plate format and isocratic LC–ESI-MS–MS in SRM mode, monitoring the loss of SO_3 from the deprotonated molecule for estriol-3-sulfate, 17β-estradiol-3-sulfate, and estrone-3-sulfate and the fragment HSO_4^- at m/z 97 for 17β-estradiol-3,17-disulfate and 17β-estradiol-17-sulfate. The LOQ is 0.2 ng/ml from 100 μl of human urine. Intra- and inter-assay precision does not exceed 12%. The inter-assay accuracy is 8.6%. The sample throughput is ca. 8 samples per hour. The analysis of seven endogenous ketolic estrogens in a 2.5-ml sample of human urine after hydrolysis of the conjugates, solid-supported LLE with diethyl ether, hydrazone derivatization, and positive-ion LC–ESI-MS was reported by Xu et al. [71]. The LOQ was 0.2 ng in 2.5 ml of urine.

The analysis of 2- and 4-hydroxyestradiol and 2- and 4-hydroxyestrones is relevant in relation to the development of breast and prostate cancer. These compounds have been analysed in urine after hydrolysis by means of β-glucuronidase/sulfatase and extraction by LLE or SPE. Prior to LC–MS analysis, the hydroxyestradiols and hydroxyestrones were derivatized with ferrocene boronic acid to form cyclic boronate esters [32], or with p-toluenesulfonhydrazine to form p-toluenesulfonhydrazones [33], respectively. Product-ion MS–MS enables discrimination between two isomers [32].

Urinary glucuronides of testosterone and dihydrotestosterone from patients with benign prostate hyperplasia were analysed using on-line SPE–LC coupled to ion-trap MS operated in negative-ion ESI-MS [72]. Because DHEA, excreted by the adrenal gland, is converted into testosterone and subsequently into dihydrotestosterone, it can be involved in the growth of prostate cancer. In order to study this in more detail, an LC–MS method was developed for the simultaneous determination of androstenediol 3-sulfate and DHEA sulfate in human serum [73]. The compounds were extracted by SPE using Oasis HLB. The extract was purified by LLE with hexane. Analysis by negative-ion LC–ESI-MS resulted in LOQ of 10 ng/ml for androstenediol 3-sulfate and of 50 μg/ml for DHEA sulfate. This enables the monitoring of these compounds in healthy volunteers and patients with prostate cancer. The analysis of testosterone in rat plasma, after ethyl oxime and acetyl ester derivatization, was reported by Niwa *et al.* [74]

Hydroxyestradiol and hydroxyestrone can be metabolized to glutathione conjugates. Ramanathan *et al.* [75] reported the use of various MS–MS methods for the characterization and discrimination of glutathione conjugates of estrone and estradiol, *i.e.*, low-energy collisions in an ion-trap or a triple-quadrupole instrument, in-source CID and high-energy CID on a four-sector instrument, and metastable-ion decompositions. Both FAB and positive-ion ESI ionization were applied. Various types of fragments were observed. Fragments resulting from cleavages of the peptide backbone were observed by all methods. They revealed that the catechol estrogen is bound at the cysteine moiety of glutathione. Internal fragment ions of the glutathione supported this finding. Fragmentation at the *C–S* bond resulted in thiosteroids. The fourth type of fragment involved oxidation of the steroid ring and reduction of the glutathione moiety (Figure 13.9 for the 2-OHE$_2$-1-SG) and was the most isomer specific fragmentation. These ions were weak at low-energy conditions. All four type of fragment ions were readily observed by in-source CID on the sector instrument. The fragmentation of these isomeric glutathione conjugates was also studied using negative-ion ESI-MS in an ion-trap instrument [76].

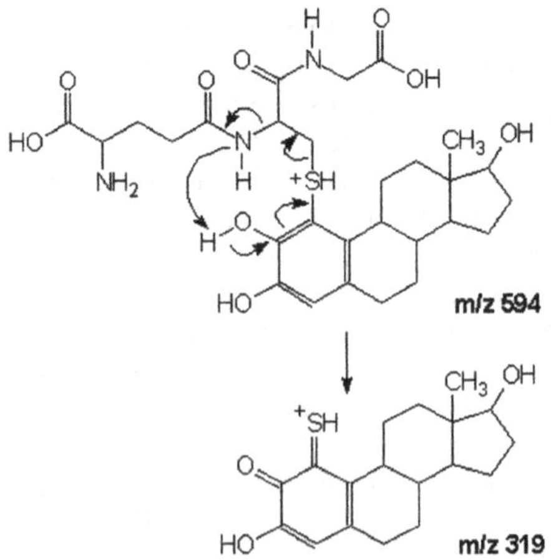

Figure 13.9: Type-4 fragmentation scheme for 2-hydroxy-E$_2$-1-glutathione conjugate. Based on [81].

4.2 Neurosteroids

The discovery of 17- and 20-oxosteroids, the so-called neurosteroids, in rat brain has raised interest in their biological properties. Special attention in this respect has been given to DHEA. Neurosteroids appear to exist in rat brains in the free form, as sulfates, lipoidal esters, and sulfolipids. LC–MS methods based on positive-ion APCI were used to characterize pregnenolone, its sulfate and DHEA in rat brains [77] and to clarify the existence of pregnenolone- and DHEA-3-stearate and -3-palmitate in rat brains [78]. Subsequently, Shimada et al. [29] reported a quantitative method for pregnenolone and its 3-stearate ester using isotopically-labelled IS. Prior to the LC–MS analysis, the steroids and their esters were converted into 3-acetate-20-methyloxime and 20-methyloxime derivatives, respectively. Absolute limits of quantification of 500 pg and 750 pg were reported for pregnenolone and its 3-stearate ester, respectively.

Chatman et al. [37] reported the analysis of DHEA-sulfate in human CSF using negative-ion nano-ESI MS–MS in precursor-ion scanning mode using m/z 97 as common product ion. Griffiths et al. [79] also applied nano-ESI in combination with MS–MS for the identification and detection of neurosteroid sulfates in brain tissue. The nano-ESI was performed on a hybrid magnetic sector–orthogonal acceleration time-of-flight tandem mass spectrometer equipped with a focal plane array detector. Complete structure information could be obtained from 1 ng of steroid sulfate, while detection based on the characteristic sulfate ester fragment ions was possible from 3 pg. An SPE–LC–MS system was developed by the same group [42-43]. This was applied to the analysis of steroid sulfates and unconjugated ketosteroids. Neurosteroids from rat brain tissue were extracted, purified, and separated in two groups by SPE and cation- and anion-exchange LC. The steroid sulfates were analysed by nano-LC–MS directly, while the ketosteroids were first derivatized to their oximes. The presence in the rat brain of pregnenolone, various pregnanolone isomers, progesterone, testosterone, and DHEA was confirmed. The estimated levels were in the range of 0.04–20 ng/g wet brain [43].

The metabolism of DHEA was studied by means of human prostate homogenates [80], as well as human and rat liver microsomal fractions [81-82]. Using human prostate homogenates, androst-5-ene-3β,17β-diol (major component), androst-4-ene-3,17-dione, testosterone, 5α-dihydrotestosterone, androsterone, and 7α-hydroxy-DHEA were identified as metabolites of DHEA [80]. By means of liver microsomal fractions, 16α-hydroxy-DHEA, 7α-hydroxy-DHEA, and 7-oxo-DHEA were identified by Fitzpatrick et al. [81], while in total 19 DHEA metabolites were reported by Marwah et al. [82]. Some of these metabolites were characterized for the first time. Quantitative analysis of 5α-androstane-3β,17β-diol in plasma was recently reported by Reddy et al. [83].

5. Development of steroid drugs

The development of steroid drugs does not significantly differ from the development of any other drug. A number of quantitative bioanalytical studies related to anti-inflammatory steroid drugs are reviewed here.

Budesonide is a glucocorticosteroid used in the treatment of asthma. The analysis of budesonide in plasma after derivatization to the 21-acetyl analogue was reported by Lindberg *et al.* [7] using TSP LC–MS, and by others [17, 84] using LC–APCI-MS. Dimova *et al.* [84] reported the comparison between an SPE-radioimmunoassay (RIA) method and LC–MS. The LOQ of the LC–MS method was 50 pg/ml and more precise. The bioanalysis of budesonide and its two hydroxylated metabolites by negative-ion ESI-MS was reported by Wang *et al.* [85]. The LOQ was 0.1 ng/ml using SPE as sample pretreatment.

Fluticasone propionate is a novel glucocorticoid under investigating for the treatment of asthma and rhinitis. Li *et al.* [15, 27] developed a quantitative method for the analysis of budesonide and fluticasone propionate in human plasma to support a clinical study. The analytes were extracted from plasma by C_{18}-SPE. Budesonide was acetylated to form the 21-acetyl derivatives [7]. The compounds are analysed using LC–MS–MS in positive-ion APCI mode. Linearity was achieved over 0.05–10 ng/ml for budesonide and 0.02–4 ng/ml for fluticasone propionate. Intra- and inter-day precision was better than 14.3%. Plasma concentration profiles have been acquired.

Deflazacourt is an inactive prodrug which after oral administration is rapidly converted to the active 21-hydroxydeflazacort. It has anti-inflammatory activity, and is used in the treatment of rheumatoid arthritis. Ifa *et al.* [87] developed a method for the determination of 21-hydroxydeflazacort in human plasma. The analyte and dexamethasone-21-acetate as the IS are extracted from plasma using a diethyl ether LLE. After evaporation to dryness and reconstitution in mobile phase, the compounds are analysed using LC–MS with positive-ion APCI in SRM mode, monitoring the transitions m/z 400→124 for the analyte and m/z 435→397 for the IS. The linearity was tested in the range of 1–400 ng/ml. The intra-assay precision and accuracy were better than 5.5% and ±7.1%, respectively.

6. References

1. S.A. Wudy, J. Homoki, W.M. Teller, *Clinical steroid analysis by GC–MS*, in: W.M.A. Niessen (Ed.), *Current practice in gas chromatography–mass spectrometry*, 2001, Marcel Dekker Inc., New York, NY.

2. J.D. Henion, *Continuous monitoring of total micro LC eluant by direct liquid introduction LC–MS*, J. Chromatogr. Sci., 19 (1981) 57.

3. J. van der Greef, A.C. Tas, M.A.H. Rijk, M.C. ten Noever de Brauw, M. Höhn, G. Meyerhoff, U. Rapp, *Determination of progesterone by LC–MS using a moving-belt*

interface and isotope dilution, J. Chromatogr., 343 (1985) 397.

4. D.S. Skrabalak, K.K. Cuddy, J.D. Henion, *Quantitative determination of betamethasone and its major metabolite in equine urine by micro-LC–MS*, J. Chromatogr., 341 (1985) 261.

5. D. Watson, G.W. Taylor, S. Murray, *Steroid glucuronide conjugates: Analysis by negative ion TSP LC–MS*, Biomed. Environ. Mass Spectrom., 13 (1986) 65.

6. N.V. Esteban, A.L. Yergey, D.J. Liberato, T. Loughlin, D.L. Loriaux, *Stable isotope dilution method using TSP LC–MS for quantification of daily cortisol production in humans*, Biomed. Environ. Mass Spectrom., 15 (1988) 603.

7. C. Lindberg, A. Blomqvist, J. Paulson, *Determination of (22R,S)budesonide in human plasma by automated TSP LC–MS*, Biol. Mass Spectrom., 21 (1992) 525.

8. A. Polettini, G.M. Bouland, M. Montagna, *Development of a coupled-column LC–MS–MS method for the direct determination of betamethasone in urine*, J. Chromatogr. B, 713 (1998) 339.

9. D.G. Watson, A.G. Davidson, B.I. Knight, *The analysis of some corticosteroids and corticosteroid esters by LC–frit-FAB-MS*, Rapid Commun. Mass Spectrom., 11 (1997) 415.

10. C.S. Creaser, S.J. Feeley, E. Houghton, M. Seymour, *Immunoaffinity chromatography combined on-line with LC–MS for the determination of corticosteroids*, J. Chromatogr. A, 794 (1998) 37.

11. L. Heimark, P. Shipkova, J. Greene, H. Munayyer, T. Yarosh-Tomaine, B. DiDomenico, R. Hare, B.N. Pramanik, *Mechanism of azole antifungal activity as determined by LC–MS monitoring of ergosterol biosynthesis*, J. Mass Spectrom., 37 (2002) 265.

12. L.O.G. Weidolf, E.D. Lee, J.D. Henion, *Determination of boldenone sulfoconjugate and related steroid sulfates in equine urine by LC–MS–MS*, Biomed. Environ. Mass Spectrom., 15 (1988) 283.

13. P.O. Edlund, L. Bowers, J.D. Henion, *Determination of methandrostenolone and its metabolites in equine plasma and urine by coupled-column LC with UV detection and confirmation by MS–MS* J. Chromatogr., 487 (1989) 341.

14. W.M. Mück, J.D. Henion, *LC–MS–MS: its use for the identification of stanozolol and its major metabolites in human and equine urine*, Biomed. Environ. Mass Spectrom., 19 (1990) 37.

15. P.O. Edlund, L. Bowers, J.D. Henion, T.R. Covey, *Rapid determination of methandrostenolone in equine urine by isotope dilution LC–MS*, J. Chromatogr., 497 (1989) 49.

16. F. Komatsu, M. Morioka, Y. Fujita, K. Sugihara, H. Kodama, *Determination of various steroids in normal adrenal glands and adrenocortical tumours by LC–API-MS*, J. Mass Spectrom., 30 (1995) 698.

17. S. Murray, N.B. Rendell, G.W. Taylor, *Microbore LC–ESI-MS of steroid sulphates*, J. Chromatogr. A, 738 (1996) 191.

18. Y.N. Li, B. Tattam, K.F. Brown, J.P. Seale, *Determination of epimers 22R and 22S of budesonide in human plasma by LC–APCI-MS*, J. Chromatogr. B, 683 (1996) 259.

19. Y.-C. Ma, H.-Y. Kim, *Determination of steroids by LC–MS*, J. Am. Soc. Mass Spectrom., 8 (1997) 1010.

20. J.-P. Antignac, B. Le Bizec, F. Monteau, F. Poulain, F. André, *CID of corticosteroids in ESI-MS–MS and development of a screening method by LC–MS–MS*, Rapid

21. F Bévalot, Y. Gaillard, M.A. Lhermitte, G. Pépin, *Analysis of corticosteroids in hair by LC–ESI-MS*, J. Chromatogr. B, 740 (2000) 227.

Commun. Mass Spectrom., 14 (2000) 33.

22. M. Honing, E. van Bockxmeer, D. Beekman, *Adduct formation of steroids in APCI and its relation to structure identification*, Analusis, 28 (2000) 921.

23. Y. Kim, T. Kim, W. Lee, *Detection of corticosteroids in the presence of organic acids in a LC eluent using the negative-ion TSP-MS*, Rapid Commun. Mass Spectrom., 11 (1997) 863.

24. A. Marwah, P. Marwah, H. Lardy, *Analysis of ergosteroids - VIII: Enhancement of signal response of neutral steroidal compounds in LC–ESI-MS analysis by mobile phase additives*, J. Chromatogr. A, 964 (2002) 137.

25. A. Leinonen, T. Kuuranne, R. Kostiainen, *LC–MS in anabolic steroid analysis - optimization and comparison of three ionization techniques: ESI, APCI and APPI*, J. Mass Spectrom., 37 (2002) 693.

26. M.J. Greig, B. Bolaños, T. Quenzer, J.M.R. Bylund, *FT-ICR-MS using APPI for high-resolution analyses of corticosteroids*, Rapid Commun. Mass Spectrom., 17 (2003) 2763.

27. Y.N. Li, B. Tattam, K.F. Brown, J.P. Seale, *Quantification of epimeric budesonide and fluticasone propionate in human plasma by LC–APCI-MS–MS*, J. Chromatogr. B, 761 (2001) 177.

28. K. Mitamura, M. Yatera, K. Shimada, *Studies in neurosteroids. Part XIII. Characterization of catechol estrogens in rat brains using LC–MS–MS*, Analyst, 125 (2000) 811.

29. K. Shimada, Y. Mukai, *Studies on neurosteroids: VII. Determination of pregnenolone and its 3-stearate in rat brains using LC–APCI-MS*, J. Chromatogr. B, 714 (1998) 153.

30. S. Liu, J. Sjövall, W.J. Griffiths, *Analysis of oxosteroids by nano-ESI-MS of their oximes*, Rapid Commun. Mass Spectrom., 14 (2000) 390.

31. M. van de Wiele, K. de Wasch, J. Vercammen, D. Courtheyn, H. den Brabander, S. Impens, *Determination of 16β-hydroxystanozolol in urine and faeces by LC–MS^n*, J. Chromatogr. A, 904 (2000) 203.

32. D. Williams, S. Chen, M.K. Young, *Ratiometric analysis of the ferrocene boronate esters of 2- and 4-hydroxyestradiol by ESI-MS–MS*, Rapid Commun. Mass Spectrom., 15 (2001) 182.

33. X. Xu, R.G. Ziegler, D.J. Waterhouse, J.E. Saavedra, L.K. Keefer, *Stable isotope dilution LC–ESI-MS method for endogenous 2- and 4-hydroxyestrones in human urine*, J. Chromatogr. B, 780 (2002) 315.

34. D.W. Johnson, H.J. ten Brink, C. Jakobs, *A rapid screening procedure for cholesterol and dehydrocholesterol by ESI-MS–MS*, J. Lipid Res., 42 (2001) 1699.

35. M.R. Anari, R. Bakhtiar, B. Zhu, S. Huskey, R.B. Franklin, D.C. Evans, *Derivatization of ethinylestradiol with dansyl chloride to enhance ESI: Application in its trace analysis in rhesus monkey plasma*, Anal. Chem., 74 (2002) 4136.

36. W.J. Griffiths, S. Liu, G. Alvelius, J. Sjovall, *Derivatisation for the characterisation of neutral oxosteroids by ESI and MALDI-MS–MS: the Girard P derivative*, Rapid Commun. Mass Spectrom., 17 (2003) 924.

37. K. Chatman, T. Hollenbeck, L. Hagey, M. Vallee, R. Purdy, F. Weiss, G. Siuzdak, *Nano-ESI-MS and precursor ion monitoring for quantitative steroid analysis and*

attomole sensitivity, Anal. Chem., 71 (1999) 2358.
38. G. Singh, A. Gutierrez K. Xu, A.I. Blair, *LC–ECNI-MS: Analysis of pentafluorobenzyl derivatives of biomolecules and drugs in the attomole range*, Anal. Chem., 72 (2000) 3007.
39. T. Higashi, N. Takido, K. Shimada, *Detection and characterization of 20-oxosteroids in rat brains using LC-ECNI-MS after derivatization with 2-nitro-4-trifluoromethyl-phenylhydrazine*, Analyst, 128 (2003) 130.
40. T. Higashi, A. Yamauchi, K. Shimada, *2-Hydrazino-1-methylpyridine: a highly sensitive derivatization reagent for oxosteroids in LC–ESI-MS*, J. Chromatogr. B, 825 (2005) 214.
41. T. Higashi, K. Shimada, *Derivatization of neutral steroids to enhance their detection characteristics in LC–MS*, Anal. Bioanal. Chem., 378 (2004) 875.
42. S. Liu, W.J. Griffiths, J. Sjövall, *Capillary LC–ESI-MS for analysis of steroid sulfates in biological samples*, Anal. Chem., 75 (2003) 791.
43. S. Liu, J. Sjövall, W.J. Griffiths, *Neurosteroids in rat brain: extraction, isolation, and analysis by nanoscale LC–ESI-MS*, Anal. Chem., 75 (2003) 5835.
44. T. Kuuranne, M. Vahermo, A. Leinonen, R. Kostiainen, *ESI and APCI-MS–MS behavior of eight anabolic steroid glucuronides*, J. Am. Soc. Mass Spectrom., 11 (2000) 722.
45. T.M. Williams, A.J. Kind, E. Houghton, D.W. Hill, *ESI-CID of testosterone and its hydroxy analogs*, J. Mass Spectrom., 34 (1999) 206.
46. P.J.R. Sjöberg, K.E. Markides, *Energy-resolved CID APCI-MS of constitutional and stereo steroid isomers*, J. Mass Spectrom., 33 (1998) 872.
47. K. De Wasch, H.F. De Brabander, M. Van de Wiele, J. Vercammen, D. Courtheyn, S. Impens, *Differentiation between dexamethasone and betamethasone in a mixture using multiple MS*, J. Chromatogr. A, 926 (2001) 79.
48. J.-P. Antignac, B. Le Bizec, F. Monteau, F. Andre, *Differentiation of betamethasone and dexamethasone using LC–ESI-MS and multivariate statistical analysis*, J. Mass Spectrom., 37 (2002) 69.
49. K.E. Arthur, J.-C. Wolff, D.J. Carrier, *Analysis of betamethasone, dexamethasone and related compounds by LC–ESI-MS*, Rapid Commun. Mass Spectrom., 18 (2004) 678.
50. D.A. Volmer, J.P.M. Hui, *Rapid determination of corticosteroids in urine by combined SPME and LC–MS*, Rapid Commun. Mass Spectrom., 11 (1997) 1926.
51. R. Draisci, C. Marchiafava, L. Palleschi, P. Cammarata, S. Cavalli, *Accelerated solvent extraction and LC–MS–MS quantitation of corticosteroid residues in bovine liver*, J. Chromatogr. B, 753 (2001) 217.
52. S. Hou, M. Hindle, and P.R. Byron, *Chromatographic and mass spectral characterization of budesonide and a series of structurally related corticosteroids using LC–MS*, J. Pharm. Biomed. Anal., 39 (2005) 196.
53. M. Fiori, E. Pierdominici, F. Longo, G. Brambilla, *Identification of main corticosteroids as illegal feed additives in milk replacers by LC–APCI-MS*, J. Chromatogr. A, 807 (1998) 219.
54. P.E. Joos, M. van Ryckeghem, *LC–MS–MS of some anabolic steroids*, Anal. Chem., 71 (1999) 4701.
55. J. Y. Kim, M.H. Choi, S.J. Kim, B.C. Chung, *Measurement of 19-nortestosterone and its esters in equine plasma by LC–MS–MS*, Rapid Commun. Mass Spectrom., 14

(2000) 1835.
56. I. Mikšík, M. Vylitová, J. Pácha, Z. Deyl, *Separation and identification of corticosterone metabolites by LC–ESI-MS*, J. Chromatogr. B, 726 (1999) 59.
57. W.J. Griffiths, *MS–MS in the study of fatty acids, bile acids, and steroids*, Mass Spectrom. Rev., 22 (2003) 81.
58. P.W. Tang, W.C. Law, T.S.M. Wan, *Analysis of corticosteroids in equine urine by LC–MS*, J. Chromatogr. B, 754 (2001) 229.
59. F. Guan, C. Uboh, L. Soma, A. Hess, Y. Luo, D.S. Tsang, *Sensitive LC–MS–MS method for the determination of beclomethasone dipropionate and its metabolites in equine plasma and urine*, J. Mass Spectrom., 38 (2003) 823.
60. A.R. McKinney, C.J. Suann, A.J. Dunstan, S.L. Mulley, D.D. Ridley, A.M. Stenhouse, *Detection of stanozolol and its metabolites in equine urine by LC–ESI-ion trap-MS*, J. Chromatogr. B, 811 (2004) 75.
61. F. Pu, A.R. McKinney, A.M. Stenhouse, C.J. Suann, M.D. McLeod, *Direct detection of boldenone sulfate and glucuronide conjugates in horse urine by ion trap LC–MS*, J. Chromatogr. B, 813 (2004) 241.
62. K.A. Bean, J.D. Henion, *Direct determination of anabolic steroid conjugates in human urine by combined liquid chromatography and tandem mass spectrometry*, J. Chromatogr. B, 690 (1997) 65.
63. D.J. Borts, L.D. Bowers, *Direct measurement of urinary testosterone and epitestosterone conjugates using LC–MS–MS*, J. Mass Spectrom., 35 (2000) 50.
64. B. Le Bizec, I. Gaudin, F. Monteau, F. Andre, S. Impens, K. De Wasch, H. De Brabander, *Consequence of boar edible tissue consumption on urinary profiles of nandrolone metabolites. I. MS detection and quantification of 19-norandrosterone and 19-noretiocholanolone in human urine*, Rapid Commun. Mass Spectrom., 14 (2000) 1058.
65. K. De Wasch, B. Le Bizec, H. De Brabander, F. André, S. Impens, *Consequence of boar edible tissue consumption on urinary profiles of nandrolone metabolites. II. Identification and quantification of 19-norsteroids responsible for 19-norandrosterone and 19-noretiocholanolone excretion in human urine*, Rapid Commun. Mass Spectrom., 15 (2001) 1442.
66. K. Deventer, F.T. Delbeke, *Validation of a screening method for corticosteroids in doping analysis by LC–MS–MS*, Rapid Commun. Mass Spectrom., 17 (2003) 2107.
67. F. Buiarelli, F. Coccioli, M. Merolle, B. Neri, A. Terracciano, *Development of a LC–MS–MS method for the identification of natural androgen steroids and their conjugates in urine samples*, Anal. Chim. Acta., 526 (2004) 113.
68. D.H. Catlin, M.H. Sekera, B.D. Ahrens, B. Starcevic, Y.-C. Chang, C.K. Hatton, *Tetrahydrogestrinone: discovery, synthesis, and detection in urine*, Rapid Commun. Mass Spectrom., 18 (2004) 1245.
69. M. Tsivou, N. Kioukia-Fougia, E. Lyris, Y. Aggelis, A. Fragkaki, X. Kiousi, Ph. Simitsek, H. Dimopoulou, I.-P. Leontiou, M. Stamou, M.-H. Spyridaki, C. Georgakopoulos, *An overview of the doping control analysis during the Olympic Games of 2004 in Athens, Greece*, Anal. Chim. Acta, 555 (2006) 1.
70. H. Zhang and J.D. Henion, *Quantitative and qualitative determination of estrogen sulfates in human urine by LC–MS–MS using 96-well technology*, Anal. Chem., 71 (1999) 3955.
71. X. Xu, L.K. Keefer, D.J. Waterhouse, J.E. Saavedra, T.D. Veenstra, R.G. Ziegler,

Measuring seven endogenous ketolic estrogens simultaneously in human urine by LC–MS, Anal. Chem., 76 (2004) 5829.

72. M.H. Choi, J.N. Kim, and B.C. Chung, *Rapid LC–ESI-MS–MS assay for urinary testosterone and dihydrotestosterone glucuronides from patients with benign prostate hyperplasia*, Clin. Chem., 49 (2003) 322.

73. K. Mitamura, Y. Nagaoka, K. Shimada, S. Honma, M. Namiki, E. Koh, A. Mizokami, *Simultaneous determination of androstenediol 3-sulfate and DHEA sulfate in human serum using isotope diluted LC–ESI-MS*, J. Chromatogr. B, 796 (2003) 121.

74. M. Niwa, N. Watanabe, H. Ochiai, K. Yamashita, *Determination of testosterone concentrations in rat plasma using LC–APCI-MS combined with ethyl oxime and acetyl ester derivatization*, J. Chromatogr. B, 824 (2005) 258.

75. R. Ramanathan, K. Cao, E. Cavalieri, M.L. Gross, *MS methods for distinguishing structural isomers of glutathione conjugates of estrone and estradiol*, J. Am. Soc. Mass Spectrom., 9 (1998) 612.

76. E. Rathahao, A. Page, I. Jouanin, A. Paris, L. Debrauwer, *LC coupled to negative ESI ion trap MS for the identification of isomeric glutathione conjugates of catechol estrogens*, Int. J. Mass Spectrom., 231 (2004) 119.

77. K. Shimada, Y. Mukai, K. Yago, *Studies on neurosteroids. VII. Characterization of pregnenolone, its sulfate and DHEA in rat brains using LC–MS*, J. Liq. Chromatogr., 21 (1998) 765.

78. K. Shimada, Y. Mukai, A. Nakajima, Y. Naka, *Studies on neurosteroids. VI. Characterization of fatty acid esters of pregnenolone and DHEA in rat brains using LC–MS*, Anal. Commun. 34 (1997) 145.

79. W.J. Griffiths, S. Liu, Y. Yang, R.H. Purdy, J. Sjövall, *Nano-electrospray tandem mass spectrometry for the analysis of neurosteroid sulphates*, Rapid Commun. Mass Spectrom., 13 (1999) 1595.

80. K. Mitamura, T. Nakagawa, K. Shimada, M. Namiki, E. Koh, A. Mizokami, S. Honma, *Identification of DHEA metabolites formed from human prostate homogenate using LC–MS and GC–MS*, J. Chromatogr. A, 961 (2002) 97.

81. J.L. Fitzpatrick, L.S. Ripp, N.B. Smith, W.M. Pierce Jr., R.A. Prough, *Metabolism of DHEA by cytochrome P450 in rat and human liver microsomal fractions*, Arch. Biochem. Biophys., 389 (2001) 278.

82. A. Marwah, P. Marwah, H. Lardy, *Ergosteroids - VI. Metabolism of DHEA by rat liver in vitro: an LC–MS study*, J. Chromatogr. B, 767 (2002) 285.

83. D.S. Reddy, L. Venkatarangan, B. Chien, K. Ramu, *A LC–MS–MS assay of the androgenic neurosteroid 3α-androstanediol in plasma*, Steroids, 70 (2005) 879.

84. H. Dimova, Y. Wang, S. Pommery, H. Moellmann, G. Hochhaus, *SPE–RIA vs LC–MS for measurement of low levels of budesonide in plasma*, Biomed. Chromatogr., 17 (2003) 14.

85. Y. Wang, Y. Tang, H. Moellmann, G. Hochhaus, *Simultaneous quantification of budesonide and its two metabolites, 6-hydroxybudesonide and 16-hydroxyprednisolone, in human plasma by LC–ESI-MS–MS*, Biomed. Chromatogr., 17 (2003) 158.

86. D.R. Ifa, M.E. Moraes, M.O. Moraes, V. Santagada, G. Caliendo, G. de Nucci, *Determination of 21-hydroxydeflazacort in human plasma by LC–APCI-MS–MS. Application to bioequivalence study*, J. Mass Spectrom., 35 (2000) 440.

14

LC–MS IN FOOD SAFETY ANALYSIS

1. Introduction

In this chapter, the role of LC–MS in various aspects of food safety analysis is discussed. LC–MS is important in the quantification and confirmation of identity of food contaminants, *e.g.*, residues of pesticides and related substances in fruit and vegetables (Ch. 7.7), antibiotic and antibacterial compounds, other veterinary drugs, anabolic steroids, and various toxins in feed and food. In addition, attention is paid to the analysis of heterocyclic aromatic amines in food.

An important issue in food safety analysis is the criteria for the confirmation of identity. In a study on the analysis of betalactam antibiotics, Heller and Ngoh [1] apply four criteria for spectral comparison:
- the occurrence of the expected precursor m/z,
- the presence of at least two diagnostic product ions above a minimum threshold,
- correct ion ratios for these diagnostic ions, and
- the ability to account for extraneous signals as chemical noise or peaks appearing in control samples.

Table 14.1: LC–MS of antibiotic and antibacterial compounds			
Compound class	Examples	ESI	APCI
Aminoglycosides	streptomycin A, gentamicin	×	
Amphenicols	chloramphenicol		
Betalactams	penicillin G, amoxicillin, cefadroxil, cephadrine	×	
Fluoroquinolones	sarafloxacine, ciprofloxacine	×	×
Macrolides	erythromycin, tylosin,	×	×
Nitrofurans	nitrofurazone, furazolidone		×
Nitroimidazoles	metronidazole, ronidazole	×	×
Polyether ionophores	monensin A, salinomycin	×	
Sulfonamides	sulfamethazine, sulfamoxole	×	×
Tetracyclines	tetracycline, oxytetracycline	×	×

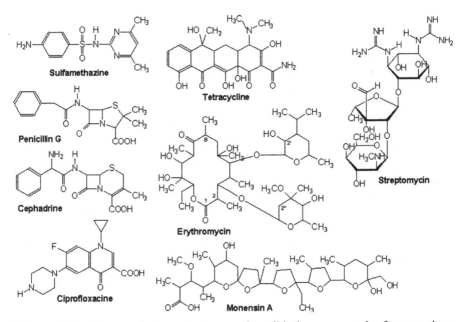

Figure 14.1: Representative structures of antibiotic compounds from various compound classes.

These criteria very much meet the ones recently established in the EU guidelines [2]. These criteria are based on the principle of "identification points". One point is earned for each ion in the mass spectrum and/or for the precursor ion in a product-ion MS–MS spectrum, and one-and-a-half point for a product ion in the MS–MS spectrum. For the confirmation of Group A (illegal) compounds four identification points are required, while for the confirmation of Group B (legal, but with maximum residue level, MRL) compounds three identification points are required. In addition, ion ratios must be within 20–30% of the ratios in the reference spectrum.

2. Mass spectrometry of antibiotics

LC–MS plays an important role in the confirmation of identity of the veterinary residues of antibiotics in animal food products for human consumption [3-6]. LC with UV or UV-DAD detection is often applied in quantitative determination, while confirmation of identity is performed by LC–MS. GC–MS is applicable to a limited number of compound classes after derivatization. The LC–MS characterization of various classes of antibiotics is discussed in this section, while residue analysis is discussed in Ch. 14.3. An overview of important classes of antibiotic and antibacterial compounds is given in Table 14.1. Typical examples of various compound classes are shown in Figure 14.1. Only a limited number of compound classes is discussed here.

2.1   Sulfonamides

Sulfonamides, N-derivatives of 4-amino-benzene sulfonamide, are synthetic broad-spectrum antibacterial compounds. They are widely used as veterinary drugs for prophylactic and therapeutic purposes, but they also show growth-promoting activity. Sulfonamides can be analysed in LC–MS using both electrospray ionization (ESI) and atmospheric-pressure chemical ionization (APCI) in either positive-ion or negative-ion mode [7]. Protonated molecules [M+H]$^+$ are observed next to sodium adducts [M+Na]$^+$ in positive-ion mode. The fragmentation scheme in MS–MS$^+$ is illustrated in Figure 14.2a. The main fragmentation occurs in the sulfonamide function on either side of the SO_2 group with charge retention on either side. For sulfamethazine (Figure 14.2b), this leads to two pairs of complement fragments, *i.e.*, m/z 92 and 186, and m/z 156 and 124. Other fragments are due to the loss of SO from m/z 156, leading to m/z 108, and due to the loss of H_2 and SO_2 from [M+H]$^+$, *i.e.*, at m/z 213 for sulfamethazine. The fragments at m/z 92, 156, and 108 are group-specific fragments. They occur at these m/z for all sulfonamides. The other three ions, *i.e.*, m/z 124, 186, and 213, are compound-specific fragments, depending on the identity of the heterocyclic ring. In ion-trap MS–MS spectra of sulfonamides [8], an additional peak is observed, *i.e.*, at m/z 204 for sulfamethazine, attributed to a solvent adduct [m/z 186 + H_2O]$^+$.

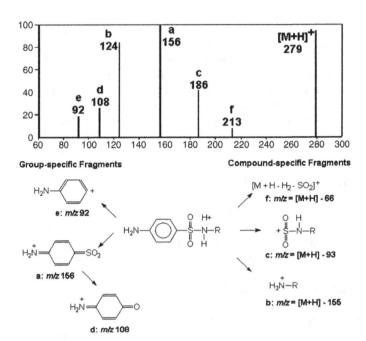

Figure 14.2: General fragmentation of sulfonamides and product-ion MS–MS spectrum of sulfamethazine.

The fragmentation pathway of sulfonamides has recently been investigated in more detail [9]. In negative-ion mode, the fragmentation is similar, except a loss of SO_2 rather than H_2SO_2 is observed [10].

There are some pairs of isomeric sulfonamides, which cannot be unambiguously identified from these fragments. Discrimination between these isomers can be achieved either by two stages of MS–MS in an ion-trap instrument or by means of a combination of in-source collision-induced dissociation (CID) to generate the compound-specific RNH_3^+ and subsequent MS–MS on this ion in a triple-quadrupole instrument [11].

2.2   (Fluoro)quinolones

Quinolone and fluoroquinolone antibiotics are a group of highly-potent, synthetic antibiotics, derived from 3-quinolone carboxylic acid They are used as broad-spectrum antibiotics in the treatment of both human and veterinary diseases.

Fluoroquinolones are detected as $[M+H]^+$ in positive-ion mode after either ESI or APCI. In MS–MS at low collision energy, the major fragments are due to losses of H_2O and CO_2, whereas at higher collision energy other fragments are observed [12].

2.3 Tetracyclines

Tetracyclines (TC) are broad-spectrum antibacterial compounds against both Gram-positive and Gram-negative bacteria. They can be analysed in LC–MS using positive-ion ESI or APCI. $[M+H]^+$ are observed. $[M+Na]^+$ are observed in mobile phases containing in excess of 0.5 mmol/l of Na^+. At higher analyte concentrations, proton-bound dimers $[2M+H]^+$ are observed, especially at higher ESI voltages [13].

Kamel et al. [14] studied the fragmentation of TC in MS–MS. At low collision energy, the initial fragmentation of oxy-TC involves the loss of either ammonia, or water, or both. H/D-exchange was used to show that the water loss involves the hydroxy-group at the 6 position and is initiated by an internal proton transfer from the initial site of protonation at the dimethylamine function, and not due to a charge-remote fragmentation mechanism. At higher collision energies, more extensive fragmentation is achieved. Stepwise fragmentation in multistage MS–MS in an ion-trap instrument demonstrated that the fragmentation can be considered as subsequent losses of small neutrals like water and CO in addition to retro-Diels-Alder cleavages. Bruno et al. [15] reported the formation of intense methanol adducts of the $[M+H–NH_3]^+$.

2.4 Betalactam antibiotics

Betalactam antibiotics comprise several classes of compounds, among which the cephalosporins and the penicillins are most important. Both classes contain bulky side chains attached to the 7-aminocephalosporanic acid or 6-aminopenicillanic acid nuclei, respectively. The betalactams have limited stability, especially in organic solvents like methanol and acetonitrile, which may hamper accurate trace analysis.

Under thermospray LC–MS conditions, thermally-induced hydrolytic ring-opening of the betalactam ring was observed, followed by the loss of CO_2 [16]. This is also observed in APCI. Betalactams are generally analysed in positive-ion or negative-ion LC–ESI-MS.

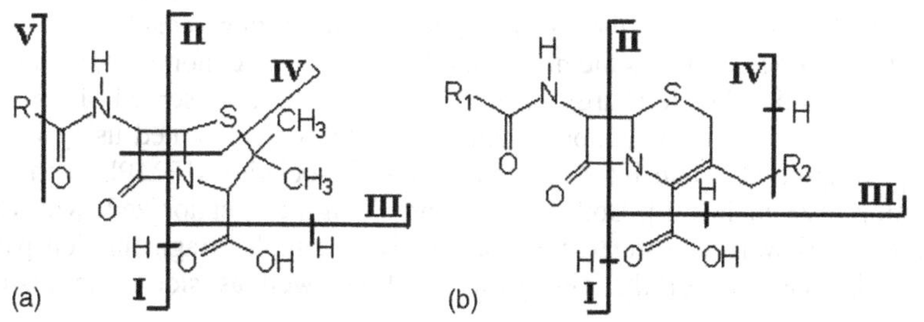

Figure 14.3: General fragmentation scheme of (a) penicillins and (b) cephalosporins. Based on [1].

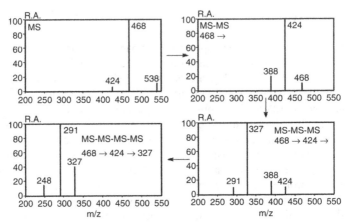

Figure 14.4: Negative-ion multi-stage ion-trap product-ion MS–MS of dicloxacillin. Based on [17].

Fragmentation of betalactams was studied using ion-trap MS–MS. In positive-ion mode, a number of group-specific fragments can be observed (Figure 14.3) [1]. For individual compounds, additional side-chain related fragments may be observed. In negative-ion mode, a more stepwise fragmentation is observed [17-18]. The negative-ion ion-trap multi-stage MS–MS spectra of dicloxacillin are shown in Figure 14.4.

2.5   Aminoglycosides

Aminoglycosides are broad spectrum antibiotics consisting of two to four linked sugars, with a substituted deoxystreptamine as a central unit. Accurate characterization of aminoglycosides is important because small changes in the structure can influence their biological activity. They are labile molecules, demanding for a soft ionization technique: ESI is the method of choice. McLaughlin et al. [20-21] first reported the LC–MS analysis of aminoglycosides. The hydrate $[M+H+H_2O]^+$ at m/z 351 was the most abundant ion for spectinomycin, while for the other compounds double-charge ions, $[M+2H]^{2+}$, were observed as the most abundant ions. The fragmentation of aminoglycosides was studied using CID in a triple-quadrupole [19] and infrared multiphoton dissociation (IRMPD) and CID in an ion-trap instrument [22]. Both $[M+H]^+$ and alkali-metal cationized (with Li^+ and Na^+) molecules were investigated, the latter giving simpler fragmentation patterns. In general, cleavages of the glycosidic bonds as well as side-chain losses are observed.

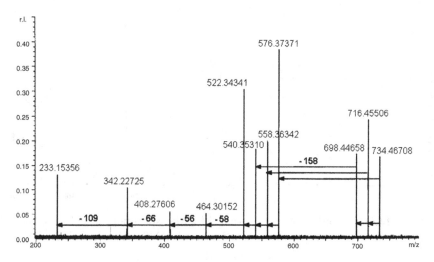

Figure 14.5: High-resolution FT-ICR-MS–MS spectrum of erythromycin A. Reprinted from [24] with permission. ©1999, John Wiley and Sons Ltd.

2.6 Macrolide antibiotics

Macrolide antibiotics consist of a 12-, 14-, or 16-membered macrocyclic lactones to which amino- and deoxy-sugar groups are attached. They are produced by various *Streptomyces* strains. Under ESI conditions, the mass spectrum of erythromycin A shows the [M+H]$^+$ at m/z 734, while some fragmentation can be induced by in-source CID, especially the loss of water (m/z 718) and of cladinose (m/z 576). In MS–MS, the same type of fragmentation is observed [23]. Multi-stage MS–MS in ion-trap and FT-ICR systems [24-25] were used to study the fragmentation of erythromycins in more detail. The high-resolution FT-ICR MS–MS of erythromycin A is shown in Figure 14.5. The fragmentation involves various neutral losses, *e.g.*, water and cladinose, from the [M+H]$^+$, and ring opening followed by losses of 58 and 56 Da. Ring opening of the macrocyclic lactone aglycone was studied in more detail [26].

2.7 Chloramphenicol

The analysis of the broad-spectrum antibiotics chloramphenicol (CAP) has recently attained considerable attention. CAP has been used to treat food-producing animals from the 1950s until it was banned by the EU in 1994. The minimum required performance limit for analytical methods for CAP within the EU is 2 µg/kg.

In most cases, CAP is analysed in negative-ion ESI [27-29], although results with APCI [30] and APPI [31] were reported as well.

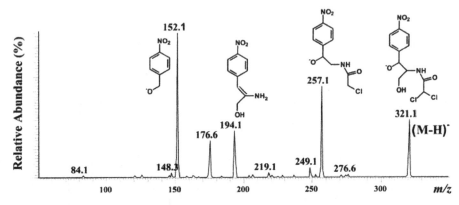

Figure 14.6: Negative-ion product-ion MS–MS spectrum of chloramphenicol. Reprinted from [28] with permission, and adapted. ©2003, Elsevier Science BV.

[M–H]⁻ was detected at m/z 321 in all cases, except one where M⁻˙ was detected at m/z 322 [27]. In selected-reaction monitoring (SRM), the transitions m/z 321→152, 194, and/or 257 are monitored in most cases. These fragments are explained in the negative-ion product-ion MS–MS spectrum, shown in Figure 14.6.

3.   Residue analysis of antibiotics

3.1   Sulfonamides

Typical maximum residue limit (MRL) for sulfonamides is 100 μg/kg in meat and 10 μg/kg in milk. Sulfonamide residues are of concern because of the possible development of antibiotic resistance and their potential carcinogenic properties.

A multiresidue method for 21 sulfonamides in milk was described by Volmer [32]. Separation of all compounds was achieved in only 6 min on a 50×4.0-mm-ID ODS-AQ column (3 μm) using a fast gradient program (10–45% acetonitrile in 0.1% aqueous formic acid in 7 min). The column was run at a flow-rate of 1 ml/min. A post-column split was used to deliver 90 μl/min to ESI. A three-step analytical strategy is adopted:

- Pre-screening and possibly confirmation by precursor-ion scan and SRM experiments using the group-specific ions at m/z 156, 108, and 92 (Ch. 14.2.1).
- Quantitation of identified target compounds using selected-ion monitoring (SIM) at [M+H]⁺.
- If needed, further confirmation by time-scheduled SRM using compound-specific ions.

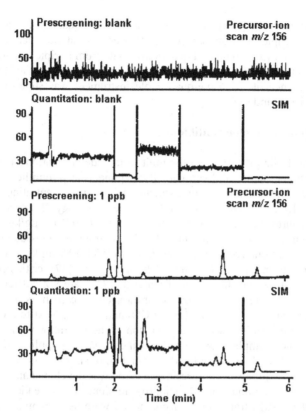

Figure 14.7: Results of the pre-screening and quantitation of sulfonamides at 1 µg/kg level in milk using LC–ESI-MS–MS. Peak identification: 5=sulfadiazine, 6=sulfapyridine, 10=sulfamethazine, 20=sulfisoxazole, and 23= sulfaquinoxaline. Reprinted from [32] with permission. ©1996, John Wiley and Sons Ltd.

The method detection limit in the SIM quantitation procedure was between 0.2 and 0.5 µg/kg for most compounds in milk. The results of the pre-screening and the quantitation of 5 sulfonamides, *i.e.*, sulfadiazine, sulfapyridine, sulfamethazine, sulfisoxasole, and sulfaquinoxaline, spiked at 1 ppb, are shown in Figure 14.7.

A variety of other multiresidue methods for sulfonamides have been reported, based on the use of SRM with two or three product ions on a triple quadrupole, or full-scan MS–MS on an ion-trap instrument. Some examples of the current state-of-the art are the quantification and confirmation of sulfonamides in swine urine (16 µg/l on a triple-quadrupole and 24 µg/l on an ion-trap instrument, [33]), in bovine kidney (5–13.5 µg/kg) on a triple-quadrupole instrument after on-line extraction and sample clean-up [34], in eggs (50 µg/kg) using an ion-trap instrument [8], in bovine liver and kidney (5–14 µg/kg using SIM on a single quadrupole, and 1–8 µg/kg using SRM on a triple-quadrupole instrument) after hot-water extraction [35].

Bartolucci et al. [33] compared quantitative data from a triple-quadrupole and an ion-trap instrument. Both instruments were applicable in residue analysis. The overall performance of the triple-quadrupole instrument was superior in terms of linearity, precision, and sensitivity.

3.2   (Fluoro)quinolone antibiotics

Doerge and Bajic [36] reported the multiresidue determination of the quinolone antibiotics oxolinic acid, nalidixic acid, flumequine, and piromidic acid in catfish muscle tissue. The analytes were extracted by a preconcentrating liquid-liquid extraction (LLE) and analysed via a heated-nebulizer APCI interface. MS and MS–MS procedures were compared. Detection down to 0.8-1.7 µg/kg was achieved using SIM with in-source CID on the $[M+H]^+$ and fragment ions due to the loss of water and CO_2. Alternatively, detection down to 0.08-0.16 µg/kg was achieved in SRM, monitoring the loss of water. Schilling et al. [37] criticized this SRM procedure, indicating that the losses of water and CO_2 are not sufficiently specific for confirmation. In the confirmation of sarafloxacin in catfish tissue, they used an additional fragment ion, due to the loss of CO_2 and C_2H_5N. Next to the detection of four ions, confirmation is based on comparison of the ion ratios with corresponding values in reference samples. Later on, Delepine et al. [38] applied SIM on the in-source generated $[M+H-H_2O]^+$ in the determination of six floxacine analogues in pig muscle tissue at 7.5 µg/kg. A multiresidue method was developed and validated [39-40], enabling the determination of 11 (fluoro)quinolones in swine kidney down to 10 µg/kg level. Two SRM transitions per compound were used. Similar methods were described for residue analysis of quinolones and fluoroquinolones in fish tissue and seafood [41], and in eggs [42].

3.3   Tetracyclines

Blanchflower et al. [43] reported the analysis of TC, oxy-TC, and chlor-TC in pig muscle and kidney tissue by after extraction in a glycine-HCl buffer and solid-phase extraction (SPE) clean up of the extracts. Gradient LC–APCI-MS was performed with 10-90% acetonitrile in water with 0.04% heptafluorobutyric acid, 10 mmol/l oxalic acid, and 10 µmol/l EDTA. Detection limits were 10 µg/kg in muscle and 20 µg/kg in kidney. Ion ratio measurements, e.g., on m/z 410, 426, and 445 for tetracycline, were performed for confirmation.

Bruno et al. [15] followed a similar approach to extract TC from milk or eggs. A Carbograph 4 SPE cartridge was used prior to analysis by LC–ESI-MS. Time-scheduled SIM is performed using $[M+H]^+$, $[M+H-NH_3]^+$, $[M+H-NH_3+CH_3OH]^+$, or structure-specific fragments. A chromatogram for a milk sample spiked with 25 µg/kg of TC is shown in Figure 14.8.

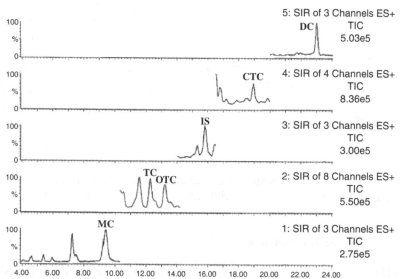

Figure 14.8: LC–ESI-MS chromatogram in time-scheduled SIM mode for the analysis of milk spiked with 25 μg/kg of tetracyclines. Peaks: MC=minocycline, TC=tetracycline, OTC=oxytetracycline, CTC=chlortetracycline, and DC=doxycycline. Reprinted from [15] with permission. ©2002, John Wiley and Sons Ltd.

Estimated quantification limits are 2–9 μg/kg in milk and 2–19 μg/kg in eggs.

Cherlet *et al.* [44] reported the determination of TC, chlor-TC, oxy-TC, and doxycycline and their 4-epimers, in edible pig tissue. The compounds are extracted with a sodium succinate solution. After protein precipitation with trifluoroacetic acid (TFA) and paper filtration, the extract was purified on an Oasis HLB SPE cartridge, prior to LC–ESI-MS analysis. The limit of quantification (LOQ) was 50 μg/kg, which is 50% of the MRL in the EU. A similar method was applied by Andersen *et al.* [45] for the analysis in shrimp and whole milk.

The quantitative analysis of chlor-TC is hampered by its keto-enol tautomerism and epimerization [46]. Protein precipitation of pg or chicken plasma was performed at 56°C with TFA, which forces chlor-TC in its keto-form. Analysis by LC–MS now provides the total chlor-TC content. The method was validated according to EU criteria.

Chlor-TC was also analysed in swine plasma [47]. After acetonitrile liquid extraction, the compound was analysed with an isocratic mobile phase and a C_8-column. Linear weighted calibration was performed in the range of 20–2000 ng/ml. The method was validated.

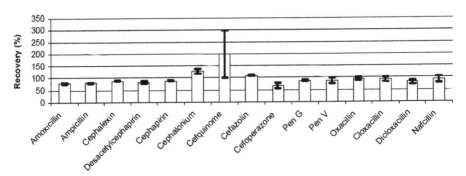

Figure 14.9: Evaluation of matrix effects in the analysis of betalactam antibiotics in bovine kidney. Reprinted from [55] with permission, ©2004. Elsevier BV.

3.4 Betalactam antibiotics

Initial results in the application of LC–ESI-MS in the multiresidue analysis of betalactam antibiotics in bovine milk were reported by Voyksner *et al.* [48-50]. First, the determination of penicillin G, ampicillin, amoxicillin, cloxacillin, and cephapirin by LC–ESI-MS using a 2-mm-ID column and a 70:1 post-column split was described [48]. The estimated detection limit in the simultaneous determination of these 5 compounds in bovine milk was 100 µg/kg. Preliminary research was also reported on the use of a 320-µm-ID packed capillary column. Subsequently, a fast micro-LC method based on perfusion LC and combined with ultrasonically-assisted ESI-MS was developed for the analysis of 6 betalactam antibiotics, *i.e.*, penicillin G, ampicillin, amoxicillin, cephapirin, cloxacillin, and ceftiofur, at 10-µg/kg level in milk [49]. Ultrafiltration through a 10 kDa cutoff filter is used for initial sample cleanup. On-column focussing of 75-µl sample is applied. The total chromatographic analysis time, including the on-column focussing, fast perfusion LC on a 320-µm-ID packed capillary column operated at 50 µl/min, and reequilibration after gradient elution, was only 13 min, while with a conventional packed microcapillary column the analysis time was 40 min. All 6 compounds could be analysed within 6 min. Linear calibration was possible in the range of 10 to 1500 µg/kg in milk. Control samples, spiked at 30 or 300 µg/kg, were measured to within 15% of their spiked value. These methods were subsequently applied as a multiresidue method for various betalactam antibiotics in bovine milk [50].

Heller and Ngoh [1] described the use of full-scan LC–MS–MS using an ion-trap for the confirmation of 7 betalactam antibiotics in bovine milk at 10 µg/kg level. They compared full-scan LC–MS–MS spectra obtained for different sample conditions, *i.e.*, standard and extracts of both fortified and control milk using C_{18} or Oasis HLB SPE. Similar studies for betalactam antibiotics in milk have been reported by others, demonstrating detection limits in the range of 1 to 25 µg/kg [51] and 0.1 to 1.1 µg/kg [52] using SRM on triple quadrupole instruments. Makeswaran *et al.* [53] reported the analysis of residues ceftiofur and its conjugated metabolites in milk. De Baere *at al.* [54] reported the quantitative analysis using SRM of amoxycillin and its major metabolites amoxycilloic acid and amoxycillin piperazine-2',5-dione in animal tissue using ampicillin as IS and ion-pair LC with 9.6 mmol/l pentafluoropropionic acid (PFPA). The quantification limit was 25 µg/kg, which corresponds to 50% of the MRL in the EU.

Becker *et al.* [55] reported the residue analysis of 15 penicillins and cephalosporins in bovine muscle, kidney, and milk. They paid special attention to the evaluation of matrix effects by comparison of the response in solvent and matrix-matched standards. Their results for kidney tissue are shown in Figure 14.9. Cefquinome was found to give erroneous results in kidney tissue, unless the standard addition method was applied to compensate for matrix effects. All 15 components were determined in one run by time-scheduled SRM. The early eluting compounds were analysed in positive-ion mode, the later eluting in negative-ion mode.

3.5   Aminoglycosides

McLaughlin *et al.* [20-21] pioneered the LC–MS determination of amino-glycosides. Spectinomycin, hygromycin B, streptomycin, and dihydrostreptomycin were analysed in bovine tissues by ion-pair LC using the volatile ion-pairing agents PFPA or heptafluorobutyric acid (HFBA) at a concentration of 5-20 mmol/l in combination with ESI-MS. SIM detection in bovine kidney extracts spiked at 20 mg/kg was demonstrated. The regulatory tolerance levels are 20 µg/kg for hygromycin B and in the range of 0.1–0.5 mg/kg for the others [20]. Subsequently, the method was adapted. Neomycin B and the gentamicin C complex were added to the target compounds. An improved sample pretreatment method was implemented. SRM was performed on three product ions [21]. This enabled the determination of all compounds except spectinomycin in bovine kidney tissue at or below the regulatory tolerance level. Spectinomycin showed poor recovery in the sample pretreatment.

For the analysis of spectinomycin in milk [56], acid protein precipitation and analyte extraction was combined with SPE, ion-pair LC with HFBA as ion-pair agent, and multi-stage ESI-MS–MS on an ion-trap instrument. $[M+H+H_2O]^+$ at m/z 351 was fragmented to yield $[M+H]^+$ at m/z 333 in the first stage of MS–MS and further fragmentation of $[M+H]^+$ in a second stage of MS–MS. The estimated confirmation limit for spectinomycin was 100 µg/kg milk. Further improvement of this method, especially with respect to variability of the ion-trap performance, is required to meet regulatory demands.

Several diagnostic ions, generated by fragmentation of $[M+H]^+$, were used in confirmation and quantitation of neomycin and gentamicin [57]. Matrix-matched standards had to be used to avoid matrix effects. Neomycin and gentamicin could be confirmed in milk at the 300 µg/kg level. More recently, the same group reported the analysis of gentamicin in bovine plasma, urine, milk, and kidney biopsy samples [58]. With TFA in the mobile phase, the various gentamicin components co-eluted and were determined in total. The LOQ was 3.3, 4.5, and 3.8 ng/ml in plasma, milk, and urine, respectively.

3.6   Macrolides

Pleasance *et al.* [23] described residue analysis of erythromycin A and its metabolites in salmon tissue using LC–ESI-MS. Detection limits of erythromycin A in salmon tissue were below 10 µg/kg in SIM and below 50 µg/kg in SRM, while confirmatory full-scan LC–MS or LC–MS-MS was achieved at the 0.5- and 1-mg/kg level, respectively. Next to erythromycin A, a variety of metabolites were detected, *e.g.*, anhydroerythromycin and N-desmethylerythromycin.

Dubois *et al.* [59] determined the macrolides tylosin, tilmicoson, spiramycin, josamycin, and erythromycin in swine and bovine muscle, kidney and liver tissue, in bovine milk, and in hen eggs, using roxithromycin as IS. The method involves extraction in a Tris buffer, protein precipitation, SPE clean-up on a Oasis HLB cartridge, and LC–MS–MS analysis in SRM mode. All analytes were confirmed by four ions with an ion-ratio reproducibility ranging from 2.4 to 15%. The sample throughput is 50 samples per analyst per day. Draisci *et al.* [60] developed a confirmatory method for tylosin, tilmicosin, and erythromycin in bovine muscle, liver, and kidney. The quantification limits were 30, 20, and 50 µg/kg in muscle, 40, 150, and 50 µg/kg in liver, and 40, 150, 80 µg/kg in kidney for tylosin, tilmicoson, and erythromycin, respectively. Horie *et al.* [61] reported the multiresidue determination of erythromycin, oleandromycin, litasamycin, josamycin, mirosamycin, spiramycin, tilmicoson, and tylosin in meat and fish. The LOQ was 10 µg/kg in positive-ion LC–ESI-MS in SIM mode.

3.7   Chloramphenicol

CAP can be analysed using GC–MS after derivatization. Comparison between GC–MS and LC–MS was reported by Impens *et al.* [27] and Gantverg *et al.* [30]. The latter demonstrated that LC–MS is superior, proving a detection limit of 0.02 µg/kg compared with 2µg/kg after GC–MS. Screening for CAP residues in rainbow trouts using GC–MS and subsequent confirmation by LC–MS was reported by Santos *et al.* [62]. The decision limit (CCα) and detection capability (CCβ) were 0.27 and 0.45 µg/kg, respectively.

Mottier, Guy, and coworkers [28-29] reported the LC–ESI-MS analysis of CAP in meat (chicken, turkey, pork, and beef) and seafood (fish, shrimps) and in milk powder. Liquid extraction was used, followed by an SPE clean-up. CCα and CCβ in meat and fish samples were 0.01 µg/kg and 0.02 µg/kg, respectively, and in milk powder 0.02 µg/kg and 0.04 µg/kg, respectively.

Takino *et al.* [31] applied APPI instead of ESI. CAP was extracted from flatfish or young yellowtail fish using liquid extraction with acetonitrile and sample clean-up by LLE. APPI gave better LOQ. Detection limits were 0.27 and 0.1 μg/kg in two fish species.

Bogusz *et al.* [63] reported the determination of CAP in chicken meat, seafood, and honey. The total run time was 7 min, and the detection limit 0.1 μg/kg in meat. A method for the determination of the glucuronide of CAP was also developed, but due to the lack of a reference standard, this could not be fully validated.

Kaufman and Butcher [64] reported the use of ultra-performance LC (1.7-μm particles) in the residue analysis of CAP in kidney tissue and honey. Compared to a conventional LC column (5 μm), 3–4-fold overall improvement in S/N was achieved. CCα was 0.011 μg/kg for kidney, which is significant between the MRPL.

4. Residue analysis of steroids

Because of their growth-promotion properties, the veterinary use of steroids, especially corticosteroids and anabolic steroids, is attractive. Within the EU, the use of growth-promoting substances (thyreostats, anabolics, and beta-agonist) in animal fattening is forbidden. Considerable effort has been put in the development in appropriate methods for tracing the use of illegal steroid hormones [6, 65].

General aspects of the LC–MS analysis of steroids are discussed in detail in Ch. 13. In this section, attention is paid to veterinary residue analysis of steroids. LC–MS has been used at regulatory level in tracing the use of illegal hormones via the analysis of body fluids, *e.g.*, urine, serum, or plasma, and animal tissue, *e.g.*, meat, kidney, liver.

The sample-pretreatment method applied in the regulatory analysis of steroids to some extent depends on the matrix. Urine is first submitted to an enzymatic hydrolysis step in order to convert the steroid conjugates into free steroids. The next step for all matrices is a liquid extraction or LLE, generally followed by a clean-up of the extract by SPE. The sample is then injected into the LC–MS system. In most studies, an IS is added prior to the sample pretreatment. This IS can be an analogue, but more recently isotopically-labelled IS are used more frequently.

With respect to MS analysis, SRM is applied most frequently, using up to four diagnostic ions, although the use of full-scan ion-trap MS–MS has been described as well. Antignac *et al.* [66] compared three acquisition methods for the determination of corticosteroids by negative-ion electrospray MS–MS: full-scan product ion mode, a neutral-loss analysis mode monitoring the loss of acetic acid and formaldehyde from [M+CH$_3$COO]⁻, and SRM monitoring the same losses. The SRM mode was found to be 10-fold more sensitive than the neutral-loss mode, which in turn was 10-fold more sensitive than the full-scan MS–MS mode.

With respect to the analysis of anabolic steroids, methods described are the analysis of 36 anabolic steroids in kidney fat, faeces, and urine [67], the analysis of the anabolic steroids 17β-19-nortestosterone, 17β-testosterone and progesterone and their major metabolites 17α-19-nortestosterone and 17α-testosterone in bovine serum and urine [68], the determination of the anabolic steroid trenbolone and its metabolite in bovine urine [69], the analysis of the acetyl gestagens megestrol acetate, medroxyprogesterone acetate, chlormadinone acetate and melengestrol acetate from bovine kidney fat [70], the confirmatory analysis of 17α- and 17β-boldenone and androsta-1,4-diene-3,17-dione in bovine urine [71-72], and in bovine faeces, feed and skin swab [72]. A number of methods were reported discussing the analysis of stanozolol and its major metabolite 16β-hydroxystanozolol in cattle urine [73-76] or faeces [73]. Typical LOQ for the anabolic steroids are in the range of 30–200 ng/l in urine, and 0.1–1 µg/kg in faeces and various tissue materials.

For the analysis of corticosteroids, some of the methods described are the determination of corticosteroids in feed and urine [77], the analysis of dexamethasone and betamethasone in bovine liver [78], a multiresidue method for the quantification and confirmation of five corticosteroids in urine [79], and the analysis of seven corticosteroids in bovine urine [80]. The LOQ for corticosteroids in urine are generally around 1 µg/l.

Some approaches reported to improve the sample pretreatment methods and replace the combination of liquid extraction and purification by SPE are:
• Preparative LC fractionation of the kidney fat, faeces, and urine extracts into six fractions prior to the LC–MS analysis of 36 anabolic steroids [67].
• Automated supercritical fluid extraction (SFE) for the isolation of the acetyl gestagenic steroids megestrol acetate, medroxyprogesterone acetate, chlormadinone acetate, and melengestrol acetate from bovine kidney fat [70].
• Accelerated solvent extraction (ASE) to achieve rapid extraction of dexamethasone and betamethasone from bovine liver [78].

Stolker *et al.* [77] compared various combinations of sample pretreatment (immunoaffinity chromatography (IAC), tandem SPE, and SPE) for LC–MS in the determination of corticosteroids in feed and urine. For the analysis in feed, SPE with LC–MS is preferred, while either SPE or IAC in combination with LC–MS is preferred for urine analysis.

One of the disadvantages of the current strategies in screening for the illegal veterinary use of steroids is that the approach is target-compound oriented. Nielen *et al.* [81] proposed the use of a bioassay to direct the search for any growth promoting agent. This necessitates the utilization of a LC–MS instrument with excellent full-spectrum sensitivity, *e.g.*, a Q–TOF instrument. In search for estrogen residues in calf urine, a rapid reporter gene bioassay was used for screening for estrogen activity. The urine was enzymatically deconjugated and pretreated using SPE. The positively screened urine samples were analysed by gradient LC in parallel with a bioactivity screening with a 20-s time resolution and ESI-MS analysis using a Q–TOF instrument. The method enables detection of estrogen activity and identification of unknown estrogens in urine at the 1–2 ng/l level.

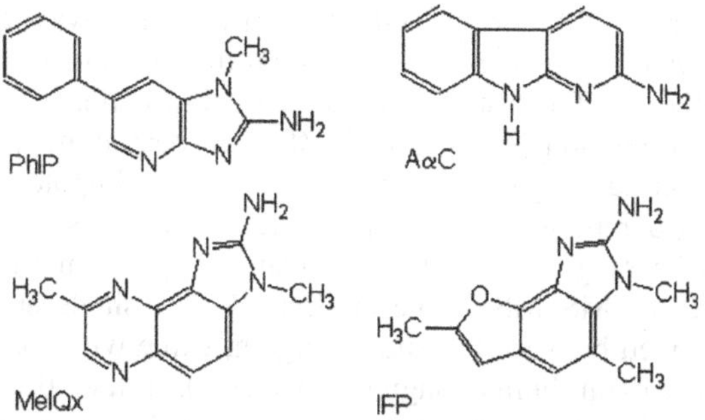

Figure 14.10: Structures of four of the most abundant HAA in cooked food. PhIP is 2-amino-1-methyl-6-phenylimidazo[4,5-*b*]pyridine, AαC is 2-amino-9*H*-pyrido[2,3-*b*]indole, MeIQx is 2-amino-3,8-dimethylimidazo[4,5-*f*]quinoxaline, and IFP is 2-amino-1,6-dimethylfuro[3,2-*e*]imidazo[4,5-*b*]pyridine.

5. Heterocyclic aromatic amines

Heterocyclic aromatic amines (HAA) are carcinogenic compounds that may occur in food. They are probably formed during cooking processes by the pyrolysis of amino acids and proteins. Since the mid 1990s, LC–MS plays a role in the analysis and characterization of these HAA. Both ESI-MS and APCI-MS are applied. A recent special issue of *Journal of Chromatography B* [82] emphasizes the analytical challenges related to HAA. The structures of the most abundant HAA in cooked food are shown in Figure 14.10.

Method development for the LC–MS analysis of HAA in meat was performed by the group of Galceran [83-90]. Initially, complex multi-step extraction and sample pretreatment methods were applied, *e.g.*, liquid extraction from meat, clean-up by solid-supported LLE, and ion-exchange SPE. Later on, different SPE procedures for sample clean-up were evaluated [85]. A combination of solid-supported-LLE and SPE on C$_{18}$ material was most efficient in selective elution of the polar and less polar HAA. LC was first performed on a 100×1-mm-ID C$_{18}$ column (5 μm) with 50% acetonitrile in 5 mmol/l aqueous ammonium acetate (pH 6.7) at a flow-rate of 50 μl/min. ESI-MS was applied in SIM mode, enabling detection of HAA at 1–6 μg/kg in beef extracts [83].

In subsequent studies, gradient elution was performed with an acetonitrile gradient in 50 mmol/l aqueous ammonium acetate (pH 4.7–5.7) on a 4.6-mm-ID C$_{18}$ or C$_8$ column in combination with APCI-MS [84, 86]. In the most recent studies, ESI-MS was used again, often in combination with 1–2.1-mm-ID columns [88-90]. Six different reversed-phase materials were evaluated [89]. The TSK Gel Semi-Micro ODS-80TS proved to show best performance. While initially SIM was applied [83-84] in combination with in-source CID, later MS–MS on ion-trap [86-

89], triple quadrupole [88], or Q–TOF [90] instruments was applied. Structure elucidation of HAA by multistage MS^n on an ion-trap instrument was reported [86-87]. The use of the Q–TOF instrument enabled confirmation of identity based on accurate-mass determination, and resolving MS–MS spectral interpretation issues. A comparison between a single-quadrupole, a triple-quadrupole, and an ion-trap instrument was also reported [88]. Typical detection limits in lyophilized meat extracts were in the range of 1.4–9.0 µg/kg in full-spectrum MS and of 0.1–3.6 µg/kg in full-spectrum MS–MS on the ion-trap. For the single-quadrupole MS, the detection limits ranged between 0.1 and 1.7 µg/kg, while with the triple quadrupole in SRM mode detection limits ranging between 0.01 and 0.1 µg/kg could be achieved. Interestingly, different compounds showed the lowest limits with these instruments, *i.e.*, MeIQ in full-spectrum MS, and Trp-P-1 in full-spectrum MS–MS on the ion-trap, 7.8-dimethyl-MEIQx in SIM on the single-quadrupole, and PhIP in SRM on the triple-quadrupole instrument. The triple-quadrupole instrument in SRM mode generally showed better precision, especially intra-day precision [88].

Method development for the determination of HAA in meat (extracts) was also reported by Holder *et al.* [91] using LLE for clean-up, and by Guy *et al.* [92] using SPE for clean-up. Both groups apply APCI on a triple-quadrupole instrument operated in SRM mode. Guy *et al.* [92] also reported the use of the neutral-loss analysis mode, based on the loss of the methyl radical, to screen for additional HAA in cooked meat, and found two components rarely reported. Various slightly modified methods were reported by others. An interlaboratory study in the analysis of HAA in food products was also reported [93], clearly recommending LC–MS as the method of choice.

All methodology was developed in order to determine HAA concentrations in cooked foods or to assess human exposure. In addition, the formation of HAA upon cooking of meat was studied [94]. It appears that HAA are especially abundant in well-done grilled chicken meat. Widely different HAA concentrations can be found in different preparations of different meat types, *e.g.*, concentration ranges between 0.045 and 45,500 µg/kg [92]. Typical amounts in cooked meat, as indicated by Skog [94], are 35 µg/kg for PhIP, 10 µg/kg for MeIQx and IFP, and 20 µg/kg for AαC. With respect to human exposure, Knize *et al.* [95] reported that after eating a well-done chicken meal 8-fold variation in the total amount of PhIP metabolites were found in the urine of volunteers, and 20-fold variation in the relative amounts of the various metabolites.

6. Toxins

LC–MS plays an important role in the characterization and analysis of toxins. It often concerns labile and complex molecules, not amenable to GC–MS. LC–MS is especially applied in the analysis of:

- Neurotoxins produced by cyanobacteria (blue-green algae), such as saxitoxin and analogues, and anatoxin a [96].
- Hepatotoxins also produced by cyanobacteria, *e.g.*, the cyclic heptapeptide microcystins and nodularins [97].
- Various peptide and protein toxins, such bacterial toxins like cholera and botulinum, *e.g.*, [98], venoms from spiders, frogs, snakes.
- Phycotoxins originate from microalgae and accumulate in filter-feeding shellfish and have caused various syndromes after eating contaminated shellfish.
- Various groups of mycotoxins produced by *Fusarium* strains, such as trichothecenes, fumonisins, and ochratoxin A. A database containing 474 fungal metabolites was compiled by Nielsen and Smedsgaard [99]

6.1   Trichothecene mycotoxins

Trichothecene mycotoxins are a group of sesquiterpenoid mycotoxins produced by fungi from the *Fusarium* family. There are four types: Type-B such as nivalenol differs from Type-A such as diacetoxyscirpenol by the presence of the keto-group in the C8 position. Type-C has an additional epoxide group, and Type-D are macrocyclic trichothecenes. Human and animal toxicoses by these toxins have been due to the consumption of contaminated grain. The structure of some of the Type-A and Type-B compounds is shown in Table 14.2.

The determination of Type-A and Type-B trichothecenes in wheat and structure elucidation by means of multi-stage positive-ion LC–APCI-MS^n on an ion-trap instrument was reported [100]. The analytes were liquid extracted from wheat. After SPE clean-up, the extract was separated on a 125×2-mm-ID C_{18} column with a linear gradient of 25–98% methanol in water at a flow-rate of 250 µl/min. Confirmation of identity was done by retention time and fragmentation pattern in MS^n, while quantitation was based on peak areas in the mass chromatograms of $[M+H]^+$ or of abundant fragments. Typical LOQ range from 10 to 100 µg/kg in wheat.

Razzazi-Fazeli *et al.* [101] reported the determination of nivalenol and deoxynivalenol, Type-B trichothecenes, in wheat using the negative-ion APCI-MS in SIM mode. LC was performed on a 250×4.6-mm-ID C_{18} column (5 µm) with a 82:9:9 water–acetonitrile–methanol mobile phase at a flow-rate of 1 ml/min. The LOQ was 40–50 µg/kg in wheat. For the analysis of six Type-A trichothecenes [102], this method was modified in various ways: a 2-mm-ID column was applied, operated at 0.3 ml/min, and positive-ion APCI-MS. The LOQ ranged from 50 to 85 µg/kg. This modified method was applied to the determination of six Type-B trichothecenes in pig urine and maize was reported [103]. The LOQ ranged from 25 to 150 µg/kg.

Similar methods were subsequently reported by others [104-106]. Dall'Asta *et al.* [105] explored the addition of sodium chloride to the mobile phase in order to enhance positive-ion ESI-MS by adduct formation. The LOQ reported ranged from 20 to 50 µg/kg.

| Table 14.2: Structures of some type-A and type-B trichothecene mycotoxins |

Compound	M	R_1	R_2	R_3	R_4	R_5
T-2 toxin	466	OH	OAc	OAc	H	i-C$_4$H$_9$COO
HT-2 toxin	424	OH	OH	OAc	H	i-C$_4$H$_9$COO
Diacetoxyscirpenol	366	OH	OAc	OAc	H	H
T-2 Tetraol	298	OH	OH	OH	H	OH
Triacetoxyscirpenol	408	OAc	OAc	OAc	H	H
Deoxynivalenol	296	OH	H	OH	OH	=O (α)
Nivalenol	312	OH	OH	OH	OH	=O (α)

By means of a systematic comparison and optimization, Laganà et al. [104] concluded negative-ion ESI-MS showed better performance than APCI for Type-B trichothecenes. Enhanced selectivity and inherent better detection limits by means of SRM compared to SIM were also demonstrated: the LOQ ranged from 1.5 to 10 µg/kg in maize using ESI-MS–MS [104] or from 0.3 to 3.8 µg/kg using APCI-MS–MS [106].

6.2   Other mycotoxins

Zearalenone (ZON) is another mycotoxin produced by *Fusarium* species. It can infect wheat and maize. The analysis of ZON was reported by several groups [106-108]. Rosenberg et al. [107] reported the analysis of ZON in food and feed, using LC separation on a C$_{18}$ column with 40% acetonitrile in water and positive-ion APCI-MS. The LOQ was 0.12 µg/kg in maize, which is a 50-fold improvement over fluorescence detection. They also compared 100 and 20 mm long columns. Sample clean-up was only required when using the 20 mm column. Pressurized liquid extraction was applied for the isolation of ZON from wheat and corn prior to LC–ESI-MS in the negative-ion mode [108]. The LOQ was 12–15 µg/kg.

Berthiller et al. [106] reported the detection of ZON next to eight trichothecene mycotoxins using positive-ion APCI-MS in SRM mode. The LOQ was 3.2 µg/kg in maize.

Zearalenone   Ochratoxin A

Ochratoxin A (OTA) is a mycotoxin produced by *Aspergillus* and *Penicillium* species. It has nephrotoxic and nephrocarcinogenic properties and is suspected as a possible cause of a chronic kidney disease in south-eastern Europe. The LC–MS analysis of OTA was described by various groups [109-113]. Positive-ion ESI-MS is the method of choice. SRM with a triple-quadrupole instrument [109-111] or an ion-trap instrument [112-113] is applied. Consecutive reaction monitoring (CRM) in MS^3 on an ion-trap instrument was reported as well [113]. Gradient elution LC is applied with an acidic mobile phase, *e.g.*, 0.05% TFA. Becker *et al.* [109] compared positive-ion and negative-ion ESI-MS. OTA could be detected in wheat, beer, and coffee down to 0.01 µg/kg. Lau *et al.* [110] detected OTA down to 0.5 µg/l in human plasma. In order to exclude matrix effects, three different quantitation procedures were compared: standard addition, IS, and external standard. The dechlorinated analogue ochratoxin B (OTB) was used as IS. External standardization often leads to erroneous results. The IS method requires an additional injection in order to assure the absence of OTB in the sample, while in the standard addition methods a relatively large sample amount is required. De Saeger *et al.* [111] reported the analysis of OTA in kidney samples. The method was validated according to EU regulations [2]. CCα and CCβ after a strong-anion-exchange sample clean-up were 0.11 and 0.25 µg/kg. OTB was used as internal standard. Lindenmeier *et al.* [112] reported the analysis of OTA after immunoaffinity clean-up. A $[D_5]$-labelled IS was synthesized and applied. The LOQ was 1.4 µg/kg. OTA was analysed in wheat flour, coffee, liquorice, beer, wine, and some spices. In most foods, OTA could not be detected, but it was detected at concentrations up to 3.3 µg/kg in mulled wine, up to 1.8 µg/kg in nutmeg powder, and up to 29.8 µg/kg in raisins.

6.3   Marine biotoxins

Marine biotoxins are produced by marine organisms. They have led to numerous cases of seafood poisoning, resulting in various syndromes, such as diarrhetic (DSP), paralytic (PSP), and amnesic shellfish poisoning (ASP).

A major contribution in initial method development for the LC–MS analysis of marine biotoxins was made by Quilliam and coworkers at the Institute for Marine Biosciences of the National Research Council of Canada. Quilliam [114] reviewed the early development in this field. Initially, the attention was focussed on three compounds and their derivatives: domoic acid, saxitoxin, okadoic acid.

Domoic acid   Saxitoxin

Okadaic acid

More recently, other compound classes were studied as well, such as yessotoxins, dinophysistoxins, pectenotoxins, and azaspiracids.

Domoic acid and its derived products are involved in ASP. These compounds show [M+H]$^+$ and [M–H]$^-$ in positive-ion and negative-ion ESI-MS, respectively. In MS–MS, structure-informative fragmentation is observed. In an early study [115], domoic acid was detected at 37 μg/ml in a heavily contaminated mussel tissue extract. In a more recent study [116], using CRM in an ion-trap MS^3 experiment, detection limits as low as 8 ng/ml was achieved in various marine biological samples including scallops.

Saxitoxin and its derived products are involved in PSP. Saxitoxin shows a [M+H]$^+$ in positive-ion mode, while no significant response is obtained in the negative-ion mode. In an early study [115], a detection limit of 30 ng/ml was established in column-bypass injections of standard solution. In a more recent study [117], the estimated on-column detection limit for saxitoxin was 30 ng/ml as well.

Okadaic acid and its derived products (dinophysistoxins and pectenotoxins) are involved in DSP. In positive-ion ESI-MS, okadaic acid shows an abundant [M+H]$^+$ and no significant fragmentation. In negative-ion mode, [M–H]$^-$ is observed. In an early study using gradient-elution LC–ESI-MS in SIM mode, okadaic acid was detected down to 40 μg/kg in mussel tissue [118]. In a more recent study [119], okadaic acid, dinophysistoxins-1, and pectenotoxin-6 were analysed as [M–H]$^-$ in negative-ion ESI-MS and subsequently quantified in SIM mode in bivalves. The compounds were detected in scallops collected at Mutsu Bay in Japan, but not in mussels collected from the same site. Between 40 and 450 μg/kg okadaic acid was found in midgut glands of scallops, and between 400 and 1600 µg/kg of esterified dinophysistoxin-1. Okadaic acid, dinophysistoxins, and pectenotoxins were also analysed in marine phytoplankton, *Dinophysis acuta*, and mussels, *Mytilus edulis*, collected along the southwest coast of Ireland [120]. Okadaic acid could be analysed in shellfish extracts down to 7 ng/ml. DSP toxin profiles were made in mussels and phytoplankton in Ireland and Norway. Concentration ranges in *M. edulis* were for Ireland 0.07–8.2 µg/g for okadaic acid, and 0.3–15 µg/g for dinophysistoxin-2, while dinophysistoxin-1 was not detected, and for Norway 0.59–1.7 µg/g for okadaic acid, and 12-26 µg/g for dinophysistoxin-1, while dinophysistoxin-2 was not detected in the Norwegian mussels.

Yessotoxin (YTX) belongs to the DSP compounds. The LC–MS analysis of YTC was described by Draisci *et al.* [121]. In negative-ion ESI-MS the compound shows an intense peak due to $[M-2Na+H]^-$ at *m/z* 1141, in addition to $[M-Na]^-$ at *m/z* 1163 and $[M-H]^-$ at *m/z* 1185. The fragmentation of YTX in MS–MS was also studied. YTX was detected in Italian shellfish samples collected in 1997. The samples were earlier tested negative by the official mouse bioassay. A two-step procedure for the detection of YTX and related DSP toxins, including okadaic acid, dinophysistoxins, and pectenotoxins, in total twelve compounds, in mussels and phytoplankton was proposed by the same Italian group, in collaboration with three other research groups from Ireland and Japan [122]. The samples were first analysed in negative-ion mode, to detect all twelve compounds in one run using SIM. Positive samples were subsequently analysed in positive-ion mode in order to differentiate between the structurally related toxins from the dinophysistoxin and pectenotoxin groups. Toxin profiling in samples collected in various countries was reported as well. Quantitative determination of ten DSP toxins in muscle and scallops was reported by Goto *et al.* [123]. Detection limits were 5 and 10 ng/g for okadaic acid, 10 and 20 ng/g for dinophysistoxin-1, and 40 and 80 ng/g for pectenotoxins and YTX in the muscle and digestive glands of scallops, respectively. Nano-LC–MS using a Q–TOF instrument for the determination of YTX in marine phytoplankton was recently reported by Cañás *et al.* [124]. Standard injection of YTX enabled detection down to 0.75 ng/ml. YTX was detected in phytoplankton cells of *Protoceratium reticulatum* in the range of 0.01–0.02 ng/cell.

Azaspiracid (AZA-1) and related compounds were involved in a number of recent human intoxifications, causing DSP-like symptoms. The toxic syndrome is called azaspiracid poisoning (AZP). The LC–MS analysis of AZA-1 was reported by Draisci *et al.* [125]. Isocratic elution with 85% acetonitrile in 0.03% aqueous TFA from a 1.0-mm-ID C_{18} column at 30 µl/min was performed. $[M+H]^+$ was observed at *m/z* 842. In MS–MS, three subsequent water losses were observed, next to several other minor fragments. The instrumental detection limit was 20 pg, which is considerably better than the conventional mouse bioassay (2.8 µg). Lehane *et al.* [126] reported the analysis of AZA-1 and four related compounds in shellfish down to 0.05 µg/ml. The same group compared various SPE methods [127]. They developed a CRM procedure with an ion-trap instrument for the determination of azaspiracids in shellfish [128], enabling detection of 0.8 ng/ml of AZA-1, equivalent to 0.37 ng/g shellfish tissue. Later on, this method was modified to include the detection of hydroxy analogues as well. Ten azaspiracids could be detected with this method [129].

In a series of papers, the group of Volmer [130-132] studied the analysis of azaspiracid biotoxins. Ultrafast and/or high-resolution LC of azaspiracids on monolithic LC columns was evaluated [130]. Chromatograms of five azaspiracids on a 100-mm and a 700-mm monolithic column are shown in Figure 14.11. Fragmentation of azaspiracids in MS–MS on ion-trap and triple-quadrupole instruments was studied as well [131]. The interpretation was confirmed using accurate-mass data from a Q–TOF instrument. Validation of a quantitative method for AZA-1 was also reported [132]. The LOQ was 5 and 50 pg/ml extract using a triple-quadrupole in SRM mode and an ion-trap instrument, respectively.

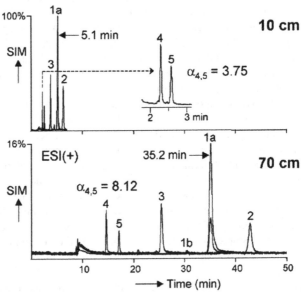

Figure 14-11: Ultrafast or high-resolution separations of five azaspiracids using LC–ESI-MS on monolithic columns of various lengths. Flow-rate is 1 ml/min for both columns. Reprinted from [130] with permission. ©2002, John Wiley and Sons Ltd.

7. **References**

1. D.N. Heller, M.A. Ngoh, *ESI and ion trap MS–MS for the confirmation of seven betalactam antibiotics in bovine milk*, Rapid Commun. Mass Spectrom., 12 (1998)

2031.

2. SANCO 2002/657/EC (Commission Decision of 12 August 2002 implementing Council Directive 96/23/EC concerning the performance of analytical methods and the interpretation of results).

3. W.M.A. Niessen, *Analysis of antibiotics by LC–MS*, J. Chromatogr. A, 812 (1998) 53.

4. D.G. Kennedy, R.J. McCracken, A. Cannavan, S.A. Hewitt, *LC–MS in antibiotic residue analysis in meat and milk*, J. Chromatogr. A, 812 (1998) 77.

5. A. Di Corcia, M. Nazzari, *LC–MS methods for analyzing antibiotic and antibacterial agents in animal food products*, J. Chromatogr. A, 974 (2002) 53.

6. A.A.M. Stolker, U.A.Th. Brinkman, *Analytical strategies for residue analysis of veterinary drugs and growth-promoting agents in food-producing animals*, J. Chromatogr. A, 1067 (2005) 15.

7. S. Pleasance, P. Blay, M.A. Quilliam, G. O'Hara, *Determination of sulfonamides by LC, UV-DAD and ESI-MS–MS with application to cultured salmon flesh*, J. Chromatogr., 558 (1991) 155.

8. D.N. Heller, M.A. Ngoh, D. Donoghue, L. Podhorniak, H. Righter, M.H. Thomas, *Identification of incurred sulfonamide residues in eggs: methods for confirmation by LC–MS–MS and quantitation by LC–UV detection*, J. Chromatogr. B, 774 (2002) 39.

9. K. Klagkou, F. Pullen, M. Harrison, A. Organ, A. Firth, G.J. Langley, *Fragmentation pathways of sulphonamides under ESI-MS–MS conditions*, Rapid Commun. Mass Spectrom., 17 (2003) 2373.

10. Z. Wang, C.E.C.A. Hop, M.-S. Kim, S.-E.W. Huskey, T.A. Baillie, Z. Guan, *The unanticipated loss of SO_2 from sulfonamides in CID*, Rapid Commun. Mass Spectrom., 17 (2003) 81.

11. K.P. Bateman, S.J. Locke, D.A. Volmer, *Characterization of isomeric sulfonamides using CE coupled with nano-ESI quasi-MS–MS–MS*, J. Mass Spectrom., 32 (1997) 297.

12. P.A. D'Agostino, J.R. Hancock, L.R. Provost, *ESI-MS characterization of fluoro-quinolone antibiotics: norfloxacin, enoxacin, ciprofloxacin and ofloxacin*, Rapid Commun. Mass Spectrom., 9 (1995) 1038.

13. A.M. Kamel, P.R. Brown, B. Munson, *ESI-MS of TC, oxy-TC, chloro-TC, minocycline, and methacycline*, Anal. Chem., 71 (1999) 968.

14. A.M. Kamel, H.G. Fouda, P.R. Brown, B. Munson, *MS characterization of TC by ESI, H/D exchange, and multiple stage MS*, J. Am. Soc. Mass Spectrom., 13 (2002) 543.

15. F. Bruno, R. Curini, A. Di Corcia, M. Nazzari, M. Pallagrosi, *An original approach to determining traces of TC antibiotics in milk and eggs by SPE and LC–MS*, Rapid Commun. Mass Spectrom., 16 (2002) 1365.

16. R.D. Voyksner, K.L. Tyczkowska, A.L. Aronson, *Development of analytical methods for some penicillins in bovine milk by ion, paired chromatography and confirmation by TSP MSy*, J. Chromatogr., 567 (1991) 389.

17. S. Rabbolini, E. Verardo, M. Da Col, A.M. Gioacchini, P. Traldi, *Negative ion ESI-MS–MS in the structural characterization of penicillins*, Rapid Commun. Mass Spectrom., 12 (1998) 1820.

18. S. Tenconi, L. de Filippo, M. Da Col, A.M. Gioacchini, P. Traldi, *ESI-MS in the structural characterization of cephalosporins*, J. Mass Spectrom., 34 (1999) 268.

19. P. Hu, E.K. Chess, S. Brynjelsen, G. Jabubowski, J. Melchert, R.B. Hammond, T.D. Wilson, *CID of aminocyclitol-aminoglycoside antibiotics and their application in the identification of a new compound in tobramycin samples*, J. Am. Soc. Mass Spectrom.,

11 (2000) 200.

20. L.G. McLaughlin, J.D. Henion, *Determination of aminoglycoside antibiotics by ion pair RPLC coupled with pulsed amperometry and ESI-MS*, J. Chromatogr., 591 (1992) 195.

21. L.G. McLaughlin, J.D. Henion, P.J. Kijak, *Multi-residue confirmation of aminoglycoside antibiotics and bovine kidney by LC–ESI-MS–MS*, Biol. Mass Spectrom., 23 (1994) 417.

22. B.J. Goolsby, J.S. Brodbelt, *Analysis of protonated and alkali metal cationized aminoglycoside antibiotics by CID and IRMPD in the quadrupole ion trap*, J. Mass Spectrom., 35 (2000) 1011.

23. S. Pleasance, J. Kelly, M.D. LeBlanc, M.A. Quilliam, R.K. Boyd, D.D. Kitts, K. McErlane, R.M. Bailey, D.H. North, *Determination of erythromycin A in salmon tissue by LC–ESI-MS*, Biol. Mass Spectrom., 21 (1992) 675.

24. P.J. Gates, G.C. Kearney, R. Jones, P.F. Leadlay, J. Staunton, *Structural elucidation studies of erythromycins by ESI-MS*, Rapid Commun. Mass Spectrom., 13 (1999) 242.

25. G.C. Kearney, P.J. Gates, P.F. Leadlay, J. Staunton, R. Jones, *Structural elucidation studies of erythromycins by ESI-MS–MS II*, Rapid Commun. Mass Spectrom., 13 (1999) 1650.

26. M. Roddis, P. Gates, Y. Roddis, J. Staunton, *Structural elucidation studies on 14- and 16-membered macrolide aglycones by accurate-mass ESI sequential MS*, J. Am. Soc. Mass Spectrom., 13 (2002) 862.

27. S. Impens, W. Reybroeck, J. Vercammen, D. Courtheyn, S. Ooghe, K. De Wasch, W. Smedts, H. De Brabander, *Screening and confirmation of CAP in shrimp tissue using ELISA in combination with GC–MS^2 and LC–MS^2*, Anal. Chim. Acta, 483 (2003) 153.

28. P. Mottier, V. Parisod, E. Gremaud, P.A. Guy, R.H. Stadler, *Determination of the antibiotic CAP in meat and seafood products by LC–ESI-MS–MS*, J. Chromatogr. A, 994 (2003) 75.

29. P.A. Guy, D. Royer, P. Mottier, E. Gremaud, A. Perisset, R.H. Stadler, *Quantitative determination of CAP in milk powders by isotope dilution LC–MS–MS*, J. Chromatogr. A, 1054 (2004) 365.

30. A. Gantverg, I. Shishani, M. Hoffman, *Determination of CAP in animal tissues and urine. LC–MS–MS versus GC–MS*, Anal. Chim. Acta, 483 (2003) 125.

31. M. Takino, S. Daishima, T. Nakahara, *Determination of CAP residues in fish meats by LC–APPI-MS*, J. Chromatogr. A, 1011 (2003) 67.

32. D.A. Volmer, *Multiresidue determination of sulfonamide antibiotics in milk by short-column LC–ESI-MS–MS*, Rapid Commun. Mass Spectrom., 10 (1996) 1615.

33. G. Bartolucci, G. Pieraccini, F. Villanelli, G. Moneti, A. Triolo, LC–MS–MS *quantitation of sulfamethazine and its metabolites: direct analysis of swine urine by triple-quadrupole and ion-trap MS*, Rapid Commun. Mass Spectrom., 14 (2000) 967.

34. N. Van Eeckhout, J. Castro Perez, C. Van Peteghem, *Determination of eight sulfonamides in bovine kidney by LC–MS–MS with on-line extraction and sample clean-up*, Rapid Commun. Mass Spectrom., 14 (2000) 2331.

35. S. Bogialli, R. Curini, A. Di Corcia, M. Nazzari, M. Sergi, *Confirmatory analysis of sulfonamide antibacterials in bovine liver and kidney: extraction with hot water and LC coupled to a single- or triple-quadrupole MS*, Rapid Commun. Mass Spectrom., 17 (2003) 1146.

36. D.R. Doerge, S. Bajic, *Multiresidue determination of quinolone antibiotics using*

LC–APCI-MS–MS, Rapid Commun. Mass Spectrom., 9 (1995) 1012.
37. J.B. Schilling, S.P. Cepa, S.R. Menacherry, L.T. Bavda, B.M. Heard, B.L. Stockwell, *LC–MS–MS for the confirmation of sarafloxacin in catfish tissue*, Anal. Chem., 68 (1996) 1905.
38. B. Delepine, D. Hurtaud-Pessel, P. Sanders, *Simultaneous determination of six quinolones in pig muscle by LC–APCI-MS*, Analyst, 123 (1998) 2743.
39. G. van Vyncht, A. Janosi, G. Bordin, B. Toussaint, G. Maghuin-Rogister, E. De Pauw, A.R. Rodriguez, *Multiresidue determination of (fluoro)quinolone antibiotics in swine kidney using LC–MS–MS*, J. Chromatogr. A, 952 (2002) 121.
40. B. Toussaint, G. Bordin, A. Janosi, A.R. Rodriguez, *Validation of a LC–MS–MS method for the simultaneous quantification of 11 (fluoro)quinolone antibiotics in swine kidney*, J. Chromatogr. A, 976 (2002) 195.
41. L. Johnston, L. Mackay, M. Croft, *Determination of (fluoro)quinolones in fish tissue and seafood by LC–ESI-MS–MS detection*, J. Chromatogr. A, 982 (2002) 97.
42. M.J. Schneider, D.J. Donoghue, *Multiresidue determination of fluoroquinolone antibiotics in eggs using LC–fluorescence–MS*, Anal. Chim. Acta, 483 (2003) 39.
43. W.J. Blanchflower, R.J. McCracken, A.S. Haggan, D.G. Kennedy, *Confirmatory assay for the determination of TC, oxy-TC, chlor-TC and its isomers in muscle and kidney using LC–MS*, J. Chromatogr. B, 692 (1997) 351.
44. M. Cherlet, M. Schelkens, S. Croubels, P. De Backer, *Quantitative multiresidue analysis of TC and their 4-epimers in pig tissues by positive-ion LC–ESI-MS*, Anal. Chim. Acta, 492 (2003) 199.
45. W.C. Andersen, J.E. Roybal, S.A. Gonzales, S.B. Turnipseed, A.P. Pfenning, L.R. Kuck, *Determination of TC residues in shrimp and whole milk using LC with UV detection and residue confirmation by MS*, Anal. Chim. Acta, 529 (2005) 145.
46. M. Cherlet, S. Croubels, P. De Backer, *Quantitative determination of chlor-TC content in animal plasma at controlled keto-enol tautomerism by LC–ESI-MS–MS*, J. Chromatogr. A, 1102 (2006) 116.
47. F. Beaudry, J.R.E. del Castillo, *Determination of chlor-TC in swine plasma by LC–ESI-MS–MS*, Biomed. Chromatogr., 19 (2005) 523.
48. R.F. Straub, R.D. Voyksner, *Determination of penicillin G, ampicillin, amoxicillin, cloxacillin and cephapirin by LC–ESI-MS*, J. Chromatogr., 647 (1993) 167.
49. R. Straub, M. Linder, R.D. Voyksner, *Determination of β-lactamresidues in milk using perfusion LC combined with ultrasonic nebulization ESI-MS*, Anal. Chem., 66 (1994) 3651.
50. K.L. Tyczkowska, R.D. Voyksner, R.F. Straub, A.L. Aronson, *Simultaneous multiresidue analysis of betalactam antibiotics in bovine milk by LC with UV detection and confirmation by ESI-MS*, J. AOAC Int., 77 (1994) 1122.
51. E. Daeseleire, H. De Ruyck, R. Van Renterghem, *Confirmatory assay for the simultaneous detection of penicillins and cephalosporins in milk using LC–MS–MS*, Rapid Commun. Mass Spectrom., 14 (2000) 1404.
52. S. Riediker, R.H. Stadler, *Simultaneous determination of five betalactam antibiotics in bovine milk using LC–ESI-MS–MS*, Anal. Chem., 73 (2001) 1614.
53. S. Makeswaran, I. Patterson, J. Points, *An analytical method to determine conjugated residues of ceftiofur in milk using LC–MS–MS*, Anal. Chim. Acta, 529 (2005) 151.
54. S. De Baere, M. Cherlet, K. Baert, P. De Backer, *Quantitative analysis of amoxycillin and its major metabolites in animal tissues by LC–ESI-MS–MS*, Anal. Chem., 74

(2002) 1393.

55. M. Becker, E. Zittlau, M. Petz, *Residue analysis of 15 penicillins and cephalosporins in bovine muscle, kidney and milk by LC–MS–MS*, Anal. Chim. Acta, 520 (2004) 19.

56. M.C. Carson, D.N. Heller, *Confirmation of spectinomycin in milk using ion-pair SPE and LC–ESI ion trap MS*, J. Chromatogr. B, 718 (1998) 95.

57. D.N. Heller, S.B. Clark, H.F. Righter, *Confirmation of gentamicin and neomycin in milk by weak cation-exchange extraction and ESI ion trap MS–MS*, J. Mass Spectrom., 35 (2000) 39.

58. D.N. Heller, J.O. Peggins, C.B. Nochetto, M.L. Smith, O.A. Chiesa, K. Moulton, *LC–MS–MS measurement of gentamycin in bovine plasma, urine, milk and biopsy samples taken from kidneys of standing animals*, J. Chromatogr. B, 821 (2005) 22.

59. M. Dubois, D. Fluchard, E. Sior, Ph. Delahaut, *Identification and quantification of five macrolide antibiotics in several tissues, eggs and milk by LC–ESI-MS–MS*, J. Chromatogr. B, 753 (2001) 189.

60. R. Draisci, L. Palleschi, E. Ferretti, L. Achene, A. Cecilia, *Confirmatory method for macrolide residues in bovine tissues by micro-LC–MS–MS*, J. Chromatogr. A, 926 (2001) 97.

61. M. Horie, H. Takegami, K. Toya, H. Nakazawa, *Determination of macrolide antibiotics in meat and fish by LC–ESI-MS*, Anal. Chim. Acta, 492 (2003) 187.

62. L. Santos, J. Barbosa, M.C. Castilho, F. Ramos, C.A.F. Ribeiro, M.I.N. de Silveira, *Determination of CAP residues in rainbow trouts by GC–MS and LC–MS–MS*, Anal. Chim. Acta, 529 (2005) 249.

63. M.J. Bogusz, H. Hassan, E. Al-Enazi, Z. Ibrahim, M. Al-Tufail, *Rapid determination of CAP and its glucuronide in food products by negative-ion LC–ESI-MS–MS*, J. Chromatogr. B, 807 (2004) 343.

64. A. Kaufman, P. Butcher, *Quantitative LC–MS–MS determination of CAP residues in food using sub-2 μm particulate LC columns for sensitivity and speed*, Rapid Commun. Mass Spectrom., 19 (2005) 3694.

65. B. Le Bizec, P. Marchand, D. Maume, F. Monteau, F. André, *Monitoring anabolic steroids in meat-producing animals. Review of current hyphenated MS techniques*, Chromatographia, S59 (2004) S3.

66. J.-P. Antignac, B. Le Bizec, F. Monteau, F. Poulain, F. André, *CID of corticosteroids in ESI-MS–MS and development of a screening method by LC–MS–MS*, Rapid Commun. Mass Spectrom., 14 (2000) 33.

67. P.E. Joos, M. van Ryckeghem, *LC–MS–MS of some anabolic steroids*, Anal. Chem., 71 (1999) 4701.

68. R. Draisci, L. Pallkeschi, E. Ferretti, L. Lucentini, P. Cammarata, *Quantitation of anabolic hormones and their metabolites in bovine serum and urine by LC–MS–MS*, J. Chromatogr. A, 870 (2000) 511.

69. F. Buiarelli, G.P. Cartoni, F. Coccioli, A. De Rossi, B. Neri, *Determination of trenbolone and its metabolite in bovine fluids by LC–MS–MS*, J. Chromatogr. B, 784 (2003) 1.

70. A.A.M. Stolker, P.W. Zoontjes, P.L.W.J. Schwillens, P.R. Kootstra, L.A. van Ginkel, R.W. Stephany, U.A.Th. Brinkman, *Determination of acetyl gestagenic steroids in kidney fat by automated SFE and LC ion-trap MS*, Analyst, 127 (2002) 748.

71. R. Draisci, L. Palleschi, E. Ferretti, L. Lucentini, F. delli Quadri, *Confirmatory analysis of 17β-boldenone, 17α-boldenone and androsta-1,4-diene-3,17-dione in*

bovine urine by LC–MS–MS, J. Chromatogr. B, 789 (2003) 219.

72. M.W.F. Nielen, P. Rutgers, E.O. van Bennekom, J.J.P. Lasaroms, J.A. van Rhijn, *Confirmatory analysis of 17β-boldenone, 17β-boldenone and androsta-1,4-diene-3,17-dione in bovine urine, faeces, feed and skin swab samples by LC–ESI-MS–MS*, J. Chromatogr. B, 801 (2004) 273.

73. M. van de Wiele, K. de Wasch, J. Vercammen, D. Courtheyn, H. den Brabander, S. Impens, *Determination of 16β-hydroxystanozolol in urine and faeces by LC–MS*, J. Chromatogr. A, 904 (2000) 203.

74. R. Draisci, L. Palleschi, C. Marchiafava, E. Ferretti, F. delli Quadri, *Confirmatory analysis of residues of stanozolol and its major metabolite in bovine urine by LC–MS–MS*, J. Chromatogr. A, 926 (2001) 69.

75. C. Van Poucke, C. Van Peteghem, *Development and validation of a multi-analyte method for the detection of anabolic steroids in bovine urine with LC–MS–MS*, J. Chromatogr. B, 772 (2002) 211.

76. M.W.F. Nielen, H. Hooijerink, M.L. Essers, J.J.P. Lasaroms, E.O. van Bennekom, L. Brouwer, *Value of alternative sample matrices in residue analysis for stanozolol*, Anal. Chim. Acta, 483 (2003) 11.

77. A.A.M. Stolker, P.L.W.J. Schwillens, L.A. van Ginkel, *Comparison of different LC methods for the determination of corticosteroids in biological matrices*, J. Chromatogr. A, 893 (2000) 55.

78. R. Draisci, C. Marchiafava, L. Palleschi, P. Cammarata, S. Cavalli, *ASE and LC–MS–MS quantitation of corticosteroid residues in bovine liver*, J. Chromatogr. B, 753 (2001) 217.

79. M.J. O'Keeffe, S. Martin, L. Regan, *Validation of a multiresidue LC–MS–MS method for the quantitation and confirmation of corticosteroid residues in urine, according to the proposed SANCO 1085 criteria for banned substances*, Anal. Chim. Acta, 483 (2003) 341.

80. E. Sangiorgi, M. Curatolo, W. Assini, E. Bozzoni, *Application of neutral loss mode in LC–MS for the determination of corticosteroids in bovine urine*, Anal. Chim. Acta, 483 (2003) 259.

81. M.W.F. Nielen, E.O. van Bennekom, H.H. Heskamp, J.A. van Rhijn, T.F.H. Bovee, L.A.P. Hoogenboom, *Bioassay-directed identification of estrogen residues in urine by LC–ESI-Q–TOF-MS*, Anal. Chem., 76 (2004) 6600.

82. S. Knasmüller, M. Murkovic, W. Pfau, G. Sontag, *HAA – still a challenge for scientists*, J. Chromatogr. B, 802 (2004) 1.

83. M.T. Galceran, E. Moyano, L. Puignou, P. Pais, *Determination of HAA by pneumatically assisted LC–ESI-MS*, J. Chromatogr. A, 730 (1996) 185.

84. P. Pais, E. Moyano, L. Puignou, M.T. Galceran, *LC–APCI-MS as a routine method for the analysis of mutagenic amines in beef extracts*, J. Chromatogr. A, 778 (1997) 207.

85. F. Toribio, L. Puignou, M.T. Galceran, *Evaluation of different clean-up procedures for the analysis of HAA in a lyophilized meat extract*, J. Chromatogr. A, 836 (1999) 223.

86. F. Toribio, E. Moyano, L. Puignou, M.T. Galceran, *Ion-trap MS–MS for the determination of HAA in food*, J. Chromatogr. A, 948 (2002) 267.

87. F. Toribio, E. Moyano, L. Puignou, M.T. Galceran, *Multistep MS of HAA in a quadrupole ion trap mass analyser*, J. Mass Spectrom., 37 (2002) 812.

88. E. Barceló-Barrachina, E. Moyano, L. Puignou, M.T. Galceran, *Evaluation of different LC–ESI-MS systems for the analysis of HAA*, J. Chromatogr. A, 1023 (2004) 67.

89. E. Barceló-Barrachina, E. Moyano, L. Puignou, M.T. Galceran, *Evaluation of RP columns for the analysis of HAA by LC–ESI–MS*, J. Chromatogr. B, 802 (2004) 45.
90. E. Barceló-Barrachina, E. Moyano, M.T. Galceran, *Determination of HAA by LC–Q–TOF-MS*, J. Chromatogr. A, 1054 (2004) 409.
91. C.L. Holder, S.W. Preece, S.C. Conway, Y.M. Pu, D.R. Doerge, *Quantification of HAA in cooked meats using isotope dilution LC–APCI-MS–MS*, Rapid Commun. Mass Spectrom., 11 (1997) 1667.
92. P.A. Guy, E. Gremaus, J. Richoz, R.J. Turesky, *Quantitative analysis of mutagenic HAA in cooked meat using LC–APCI-MS–MS*, J. Chromatogr. A, 883 (2000) 89.
93. F.J. Santos, E. Barcelo-Barrachina, F. Toribio, L. Puignou, M.T. Galceran, E. Persson, K. Skog, C. Messner, M. Murkovic, U. Nabinger, A. Ristic, *Analysis of HAA in food products: interlaboratory studies*, J. Chromatogr. B, 802 (2004) 69.
94. K. Skog, *Problems associated with the determination of HAA in cooked foods and human exposure*, Food Chem. Toxicol., 40 (2002) 1197.
95. M.G. Knize, K.S. Kulp, M.A. Malfatti, C.P. Salmon, J.S. Felton, *LC–MS–MS method of urine analysis for determining human variation in carcinogen metabolism*, J. Chromatogr. A, 914 (2001) 95.
96. M.A. Quilliam, *The role of chromatography in the hunt for red tide toxins*, J. Chromatogr. A, 1000 (2003) 527.
97. M. Maizels, W.L. Budde, *A LC–MS method for the determination of cyanobacteria toxins in water*, Anal. Chem., 76 (2004) 1342.
98. B.L.M. van Baar, A.G. Hulst, A.L. de Jong, E.R.J. Wils, *Characterisation of botulinum toxins type C, D, E, and F by MALDI and ESI-MS*, J. Chromatogr. A, 1035 (2004) 97.
99. K.F. Nielsen, J. Smedsgaard, *Fungal metabolite screening: database of 474 mycotoxins and fungal metabolites for dereplication by standardised LC–UV–MS methodology*, J. Chromatogr. A, 1002 (2003) 111.
100. U. Berger, M. Oehme, F. Kuhn, *Quantitative determination and structure elucidation of type A- and B-trichothecenes by LC–ion trap MS^n*, J. Agric. Food Chem., 47 (1999) 4240.
101. E. Razzazzi-Fazeli, J. Böhm, W. Luf, *Determination of nivalenol and deoxynivalenol in wheat using LC–MS with negative ion APCI*, J. Chromatogr. A, 854 (1999) 45.
102. E. Razzazi-Fazeli, B. Rabus, B. Cecon, J. Böhm, *Simultaneous quantification of A-trichothecene mycotoxins in grains using LC–APCI-MS*, J. Chromatogr. A, 967 (2002) 129.
103. E. Razzazi-Fazeli, J. Böhm, K. Jarukamjorn, J. Zentek, *Simultaneous determination of major B-trichothecenes and the de-epoxy-metabolite of deoxynivalenol in pig urine and maize using LC–MS*, J. Chromatogr. B, 796 (2003) 21.
104. A. Laganà, R. Curini, G. D'Ascenzo, I. De Leva, A. Faberi, E. Pastorini, *LC–MS–MS for the identification and determination of trichothecenes in maize*, Rapid Commun. Mass Spectrom., 17 (2003) 1037.
105. C. Dall'Asta, S. Sforza, G. Galaverna, A. Dossena, R. Marchelli, *Simultaneous detection of type A and type B trichothecenes in cereals by LC–ESI-MS using NaCl as cationization agent*, J. Chromatogr. A, 1054 (2004) 389.
106. F. Berthiller, R. Schuhmacher, G. Buttinger, R. Krska, *Rapid simultaneous determination of major type A- and B-trichothecenes as well as ZON in maize by LC–MS–MS*, J. Chromatogr., 1062 (2005) 209.
107. E. Rosenberg, R. Krska, R. Wissiack, V. Kmetov, R. Josephs, E. Razzazi-Fazeli, M.

Grasserbauer, *LC–APCI-MS as a new tool for the determination of mycotoxin ZON in food and feed*, J. Chromatogr. A, 819 (1998) 277.
108. L. Pallaroni, C. von Holst, *Determination of ZON from wheat and corn by pressurized liquid extraction and LC–ESI-MS*, J. Chromatogr. A, 993 (2003) 39.
109. M. Becker, P. Degelmann, M. Herderich, P. Schreier, H.-U. Humpf, *Column LC–ESI-MS–MS for the analysis of OTA*, J. Chromatogr. A, 818 (1998) 260.
110. B. P.-Y. Lau, P. M. Scott, D. A. Lewis, S. R. Kanhere, *Quantitative determination of OTA by LC–ESI-MS*, J. Mass Spectrom., 35 (2000) 23.
111. S. De Saeger, F. Dumoulin, C. Van Peteghem, *Quantitative determination of OTA in kidneys by LC–MS*, Rapid Commun. Mass Spectrom., 18 (2004) 2661.
112. M. Lindenmeier, P. Schieberle, M. Rychlik, *Quantification of OTA in foods by a stable isotope dilution assay using LC–MS–MS*, J. Chromatogr. A, 1023 (2004) 57.
113. I. Losito, L. Monaci, F. Palmisano, G. Tantillo, *Determination of OTA in meat products by LC–ESI-MSⁿ*, Rapid Commun. Mass Spectrom., 18 (2004) 1965.
114. M.A. Quilliam, *LC–MS in seafood toxins*, in: D. Barceló (Ed.), *Applications of LC–MS in Environmental Chemistry*, 1996, Elsevier Science, Amsterdam, Ch. 10, p. 415.
115. M.A. Quilliam, B.A. Thomson, G.J. Scott, K.W.M. Siu, *ESI-MS of marine neurotoxins*, Rapid Commun. Mass Spectrom., 3 (1989) 145.
116. A. Furey, M. Lehane, M. Gillman, P. Fernandez-Puente, K.J. James, *Determination of domoic acid in shellfish by LC with ESI and multiple MS–MS*, J. Chromatogr. A, 938 (2001) 167.
117. C. Dell'Aversano, G.K. Eaglesham, M.A. Quilliam, *Analysis of cyanobacterial toxins by HILIC–MS*, J. Chromatogr. A, 1028 (2004) 155.
118. S. Pleasance, M.A. Quilliam, J.C. Marr, *ESI-MS of marine toxins. IV. Determination of DSP toxins in mussel tissue by LC–MS*, Rapid Commun. Mass Spectrom., 6 (1992) 121.
119. T. Suzuki, T. Yasumoto, *LC–ESI-MS of DSP toxins okadaic acid, dinopysistoxin-1 and pectenotoxin-6 in bivalves*, J. Chromatogr. A, 874 (2000) 199.
120. P.F. Puente, M.J.F. Saez, B. Hamilton, M. Lehane, H. Ramstad, A. Furey, K.J. James, *Rapid determination of polyether marine toxins using LC–MSⁿ*, J. Chromatogr. A, 1056 (2004) 77.
121. R. Draisci, L. Giannetti, L. Lucentini, E. Ferretti, L. Palleschi, C. Marchiafava, *Direct identification of YTX in shellfish by LC–MS–MS*, Rapid Commun. Mass Spectrom., 12 (1998) 1291.
122. R. Draisci, L. Palleschi, L. Giannetti L. Lucentini, K.J. James, A.G. Bishop, M. Sataka, T. Yasumoto, *New approach to the direct detection of known and new DSP toxins in mussels and phytoplankton by LC–MS*, J. Chromatogr. A, 847 (1999) 213.
123. H. Goto, T. Igarashi, M. Yamamoto, M. Yasuda, R. Sekiguchi, M. Watai, K. Tanno, T. Yasumoto, *Quantitative determination of marine toxins associated with DSP by LC–MS*, J. Chromatogr. A, 907 (2001) 181.
124. I.R. Cañás, B. Hamilton, M.F. Amandi, A. Furey, K.J. James, *Nano-LC–Q–TOF-MS for the determination of YTX in marine phytoplankton*, J. Chromatogr. A, 1056 (2004) 253.
125. R. Draisci, L. Palleschi, E. Ferretti, A. Furey, K.J. James, M. Sataka, T. Yasumoto, *Development of a method for the identification of AZA-1 in shellfish by LC–MS–MS*, J. Chromatogr. A, 871 (2000) 13.
126. M. Lehane, A. Braña-Magdalena, C. Moroney, A. Furey, K.J. James, *LC with ESI ion*

trap MS for the determination of five azaspiracids in shellfish, J. Chromatogr. A, 950 (2002) 139.

C. Moroney, M. Lehane, A. Braña-Magdalena, A. Furey, K.J. James, *Comparison of SPE methods for the determination of azaspiracids in shellfish by LC–ESI-MS*, J. Chromatogr. A, 963 (2002) 353.

A. Furey, A. Braña-Magdalena, M. Lehane, C. Moroney, K.J. James, M. Satake, T. Yasumoto, *Determination of azaspiracids in shellfish using LC–ESI-MS–MS*, Rapid Commun. Mass Spectrom., 16 (2002) 238.

M. Lehane, M.J. Fidalgo Saez, A.Braña Magdalena, I. Ruppén Cañas, M. Díaz Sierra, B. Hamilton, A. Furey, K.J. James, *LC–MS^n for the determination of ten azaspiracids, including hydroxyl analogues in shellfish*, J. Chromatogr. A, 1024 (2004) 63.

D.A. Volmer, S. Brombacher, B. Whitehead, *Studies on azaspiracid biotoxins. I. Ultrafast high-resolution LC–MS separations using monolithic columns*, Rapid Commun. Mass Spectrom., 16 (2002) 2298.

S. Brombacher, S. Edmonds, D.A. Volmer, *Studies on azaspiracid biotoxins. II. Mass spectral behavior and structural elucidation of azaspiracid analogs*, Rapid Commun. Mass Spectrom., 16 (2002) 2306.

P.K.S. Blay, S. Brombacher, D.A. Volmer, *Studies on azaspiracid biotoxins. III. Instrumental validation for rapid quantification of AZA-1 in complex biological matrices*, Rapid Commun. Mass Spectrom., 17 (2003) 2153.

15

LC–MS ANALYSIS OF PLANT PHENOLS

1. Introduction

The study of natural products in plant extracts is an interesting challenge to LC–MS. Generally, the relevant compounds must be detected as minor components in complex mixtures. A combination of LC separation, especially to resolve isomeric structures, and MS detection is needed. Furthermore, structural information is needed for the identification and dereplication of the unknown plant constituents. Because of the complexity of the sample pretreatment procedures involved in the isolation, MS in most cases is the only applicable spectrometric technique; too much of a purified component would be needed for IR and NMR analysis. On-line analysis in relatively crude samples is obligatory for the detection of minor constituents. When electrospray ionization (ESI) or atmospheric-pressure chemical ionization (APCI) are applied for analyte ionization, structural information must be obtained by application of collision-induced dissociation (CID), either via in-source CID or preferably via MS–MS or MS^n. LC–MS and LC–MS–MS have proved to be extremely successful in this area.

Whereas LC–MS has been applied in the detection and characterization of other plant natural products, it was decided to focus on plant phenols, especially flavonoids. A general classification of plant phenols is given in Table 15.1. The application of MS to plant phenol characterization was reviewed [1].

413

Table 15.1: Classification of plant phenols [1].		
Class	**Structure**	**Example**
Phenols	C_6	Catechol, resorcinol
Phenolic acid	C_6C_1	Gallic acid, salicylic acid
Hydroxycinnamic acids	C_6C_3	Ferulic acid, chlorogenic acid
Lignans	$(C_6C_3)_2$	
Coumarins	C_6C_3	Umbelliferone
Chromones	C_6C_3	Eugenin
Flavonoids	$C_6C_3C_6$	Quercetin, cyanidin, catechin

2. Mass spectrometry of flavonoids

2.1   General structure

Flavonoids are widespread secondary plant metabolites. These plant phenols show many biological and physiological effects and may serve as chemotaxonomic markers. Flavonoids are present in plants as flavonoid aglycones, flavonoid O-glycosides, flavonoid C-glycosides, and/or flavonoid O,C-glycosides. The basic structures of the six main classes of flavonoids are shown in Figure 15.1. Various substituents can be present at the A and B ring, *e.g.*, hydroxy-, methoxy-, prenyl-, isoprenyl-. In total, more than 1500 different flavonoid aglycones and more than 6500 flavonoids have been reported. Common O-glycosylation positions are C7 in flavones, isoflavones, flavonones, and flavonols, and C3 in flavonols and anthocyanidins. Common C-glycosylation positions are C6 and C8 in flavones. The molecular masses of flavonoids can be calculated from residue masses of the various building blocks. Some of the relevant residue masses are summarized in Table 15.2. The mass is the sum of residue masses of all the groups in the molecule.

Flavonoids show antioxidative activity. They can interfere with free radical producing systems and act as scavengers of oxidizing free radicals and nitric oxide. They have multiple molecular targets, such as cyclo-oxygenase, estrogen receptors, and tumours. They are used in the prevention of cancer, atherosclerosis, and coronary heart disease. The best sources for flavonoids in the Western diet are onions, tea, and red wine. *In vitro* screening methods have been developed to evaluate the antioxidant activity of plant phenols and related compounds, *e.g.*, the ferric reducing/antioxidant power (FRAP) assay, the β-carotene–linoleic acid model system (β-CLAMS), the oxygen radical absorption capacity (ORAC) method, the Troloc equivalent antioxidant capacity (TEAC) test, and a luminol photochemiluminescence (PCL) method [2]. Some of these tests involve reactions with stable free radicals such as 2,2-diphenyl-1-picrylhydrazyl (DPPH•) and 2,2'-azinobis-(3-ethylbenzothiazoline-6-sulfonic acid (ABTS•, used in the TEAC test).

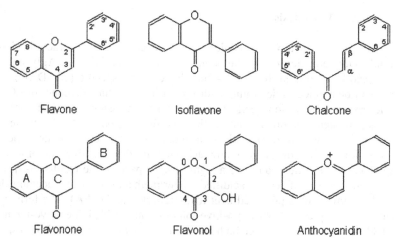

Flavone   Isoflavone   Chalcone

Flavonone   Flavonol   Anthocyanidin

Figure 15.1: Basic structures of the six main classes of flavonoids. Ring nomenclature for all classes is indicated for flavonone. Carbon numbering and C-ring bonds for all classes except chalcones is indicated for flavone and flavonol, respectively.

Table 15.2: Residue masses of flavonoid building blocks			
Flavonoid building block		**Glycosides**	
Anthocyanidin	207	Ara, Xyl	132
Chalcone	208	Rha	146
(Iso) Flavone	222	Glu, Gal	162
Flavonone	224	GluA	176
Flavonol	240	**Acyl substituents**	
Catechin	290	Acetyl	58
Substituents		Malonyl	86
Hydroxy	16	Benzoyl	104
Methyl	14	Coumaroyl	146
Prenyl	68	Galloyl	152

LC–MS plays a significant role in the analysis, dereplication, identification, and structural characterization of flavonoids. The MS analysis of flavonoid aglycones and flavonoid glycosides was reviewed by Cuyckens and Claeys [3].

2.2   LC–MS analysis

LC–MS is the method of choice in the analysis and characterization of flavonoids. Derivatization of the flavonoids is required for GC–MS. Aglycones can also be readily identified from UV-photodiode array spectra (UV-DAD). Shift reagents can be applied to determine hydroxy-positions in the aglycone. In LC–MS, both electrospray ionization (ESI) and atmospheric-pressure chemical ionization (APCI) are frequently applied. Some authors prefer ESI, others APCI. In the same way, either positive-ion or negative-ion mode is preferred, although the best sensitivity for flavonoids is generally achieved in the negative-ion mode. The mobile-phase composition can have a significant impact on the results [3-6]. From the sensitivity point of view, acidified water/methanol or water/acetonitrile is preferred. Formic acid is preferred over trifluoroacetic acid (TFA). The formic acid concentration is typically 0.5% in positive-ion mode and 0.1% in negative-ion mode. Basic additives negatively affect both the analyte retention and the ESI response. Acetic acid slightly increases ESI response, but reduces analyte retention on the C_8 and C_{18} columns, which are most frequently utilized.

Rauha et al. [5] compared the ionization efficiency for flavonoids with ESI, APCI, and atmospheric-pressure photoionization (APPI) with nine different mobile-phase compositions in both positive-ion and negative-ion mode. The mobile-phase composition can have distinct influence on the response. Best response was achieved with 0.4% formic acid (pH 2.3) for positive-ion ESI and APCI, with 10 mmol/l ammonium acetate adjusted to pH 4.0 for negative-ion ESI and APCI, and with 5 mmol/l ammonium acetate in APPI.

De Rijke et al. [6] compared positive-ion and negative-ion ESI and APCI on triple-quadrupole and ion-trap instruments. The best response for fifteen isoflavone, flavonone, and flavonone glycoside model compounds was achieved using negative-ion APCI and methanol/aqueous ammonium formate (adjusted to pH 4) as eluent.

Brodbelt et al. [7-8] showed enhanced detection of flavonoids (two orders of magnitude higher intensity in positive-ion mode, and one-and-a-half order in negative-ion mode) by metal complexation with transition metal ions such as Cu^{2+} and Co^{2+} and 2,2'-bipyridine or 4,7-diphenyl-1,10-phenanthroline as an auxiliary neutral ligand. This results in the formation of [M (Flavonoid – H) Ligand]$^+$ ions.

2.3   Information in mass spectra

Next to protonated or deprotonated molecules, [M+H]$^+$ and [M–H]$^-$, various adducts ions may be observed, especially sodium adducts [M+Na]$^+$ in positive-ion ESI-MS, as well as proton-bound dimers [2M+H]$^+$. Often, more fragmentation due to in-source fragmentation is observed in the positive-ion mode than in the negative-ion mode.

In their comparison of ESI, APCI, and APPI, Rauha *et al.* [5] applied five model compounds, *i.e.*, the flavanol catechin, the flavonol isorhamnetin, the *C*-glycosidic flavone vitexin, the 3-*O*-glycosidic flavonol isoquercitrin, and the 3',7-*O*-diglycoside of the flavone luteolin. For both aglycones, $[M+H]^+$ was the base peak. No significant fragmentation was observed. For the glycosides, $[M+H]^+$ was observed as well, but next to various fragment ions, *e.g.*, due to the loss of water and of $C_4H_8O_4$ for vitexin, the loss of dehydroglucose ($C_6H_{10}O_5$, 162 Da) for isoquercitrin, and the loss of dehydroglucose or two dehydroglucose units for luteolin-3',7-*O*-diglycoside (see also Ch. 15.2.6).

In negative-ion mode, $[M–H]^-$ was observed, *i.e.*, as the base peak for the aglycones in ESI and APCI, and for all model compounds in APPI. The loss of a methyl radical was observed for isorhamnetin. For vitexin, a peak due to the loss of $C_4H_8O_4$ was the base peak in negative-ion ESI and APCI. For luteolin-3',7-*O*-diglycoside, the loss of glucose resulted in the base peak in negative-ion ESI and APCI, and in an intense peak in APPI. In negative-ion APCI, the base peak for isoquercitrin was found at *m/z* 300 (also observed in negative-ion ESI), indicating a radical fragment.

2.4 Aglycone fragmentation in positive-ion MS–MS

The fragmentation of flavonoid aglycones was studied in detail using fast-atom bombardment (FAB) and low-energy CID by the group of Claeys [3, 9-11]. Their interpretation was applied and confirmed in many later studies. Wolfender *et al.* [12] tabulated positive-ion and negative-ion MS–MS spectra, acquired at various collision energies using quadrupole–time-of-flight hybrid (Q–TOF) and ion-trap instruments, for various flavones, flavonols, flavonones, flavanes, and the flavanol catechin. Differences in relative abundances of the various fragments are observed when comparing spectra from different instruments. This is illustrated in the product-ion MS–MS spectra of isorhamnetin, acquired by ion-trap and Q–TOF instruments at two different collision energies (Figure 15.2).

Losses of small molecules and/or radicals from $[M+H]^+$ are observed in the positive-ion mode: losses of H_2O, CO, and $H_2C=C=O$, and combinations thereof. The loss of a methyl radical $CH_3\bullet$ is characteristic for *O*-methylated isoflavones, flavones, and flavonols [13]. Loss of C_4H_8 indicates the presence of prenyl substituents (at 3-, 6-, and/or 8-position) [14]. Kuhn *et al.* [15] demonstrated the differentiation between isomeric flavones and isoflavones by means of positive-ion MS–MS in an ion-trap instrument. Isoflavones show an abundant fragment ion due to the loss of 2 CO molecules from the C-ring, while flavones only show a single CO loss.

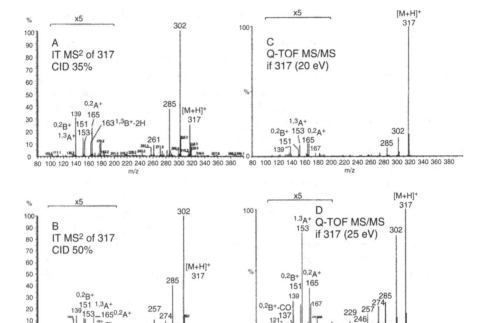

Figure 15.2: Product-ion MS–MS os isorhamnetin acquired by (a) ion-trap at an energy of 35%, (b) ion-trap at 50%, (c) Q–TOF with 20eV collision energy, and (d) Q–TOF with 25 eV collision energy. Reprinted from [12] with permission. ©2000, EDP Sciences.

Cleavages of the C-ring (refer to ring nomenclature in the flavonone structure in Figure 15.1) are more relevant in terms of structure elucidation. Nomenclature for the resulting fragments was proposed by the group of Claeys [3, 9-11] and is now widely accepted. The fragment ions are indicated as A or B, depending on the ring the charge is retained at. The two bonds cleaved in the C-ring (refer to bond numbering in the flavonol structure in Figure 15.1) are indicated as superscripts. Often, both A and B ions for the same C-ring cleavage are observed in the spectrum. In addition, losses of H_2O and CO from the A- and B-fragments may occur as well.

The most important C-ring fragmentation, often leading to the most abundant fragment ions, is the 1,3-cleavage, corresponding to a retro-Diels-Alder (RDA) fragmentation in the C-ring. The $^{1,3}A^+$-fragment ion is observed for all flavonoid classes. Together with the mass of the intact flavonoid aglycone, the $^{1,3}A^+$ and $^{1,3}B^+$ fragments provide information on the substituents on the A and B ring (Table 15.3).

Table 15.3: Positive-ion retro-Diels-Alder fragments of some flavonoids					
	Substituents				
Example	A-ring	B-ring	$[M+H]^+$	$^{1,3}A^+$	$^{1,3}B^+$
Flavone	none	none	223	121	103
Chrysin	5,7-OH	none	255	153	103
7,4'-Dihydroxyflavone	7-OH	4'-OH	255	137	119
Baicalein	5,6,7-OH	none	271	169	103
Apigenin	5,7-OH	4'-OH	271	153	119
Genkwanin	5-OH, 7-OCH₃	4'-OH	285	167	119
Biochanin A	5,7-OH	4'-OCH₃	285	153	133

The RDA cleavage is the most important C-ring fragmentation pathway for flavanones, dihydroxyflavonols, flavones, flavonols, and flavanols. In methoxy-substituted flavonoids, fragment ions due to RDA fragmentation are less abundant. In prenyl flavonoids, the characteristic A-fragment is observed as $^{1,3}A^+ - C_4H_8$ [14].

Other C-ring cleavages may be observed for specific groups of flavonoids. For flavonols, the 0,2-cleavage is a common pathway: $^{0,2}A^+$, $^{0,2}A^+ - CO$, $^{1,4}A^+ + 2H$, and $^{1,3}B^+ - 2H$ are found. For flavones, $^{1,3}B^+$, $^{0,4}B^+$, $^{0,4}B^+ - H_2O$ are observed. At higher collision energy, the abundance of $^{0,2}A^+$ and $^{1,3}B^+$ decreases for flavonols and flavones, respectively, while the abundance of the corresponding ions, *i.e.*, $^{0,2}B^+$ and $^{1,3}A^+$, increases.

March and Miao [16] performed a detailed study on the fragmentation of the flavonol kaempferol, using accurate-mass determination with a Q–TOF instrument. The data were consistent with previous findings. Next to the common C-ring fragmentations, leading to $^{1,4}H^+ + 2H$ and the pairs $^{1,3}A^+$ and $^{1,3}B^+ - 2H$, and $^{0,2}B^+$ and $^{0,2}A^+$, losses of C_2H_2O, CHO•, CO, and H_2O were observed.

Identification of flavonoids by ion-trap MS–MS, operated with normalized collision energy and wideband activation, and subsequent library searching in a laboratory-built MS–MS library (300 entries) was reported by Lee *et al.* [17].

2.5 Aglycone fragmentation in negative-ion MS–MS

Positive-ion MS–MS is applied more frequently in structure elucidation and dereplication of flavonoids. However, the negative-ion mode is considered more sensitive. Different fragmentation behaviour is observed, providing complementary information. Negative-ion MS–MS of flavonoids was investigated in detail by various groups [18-21]. Again, the RDA fragmentation of the C-ring, generating $^{1,3}A^-$ and $^{1,3}B^-$, is the most important fragmentation pathway. The $^{1,3}A^-$ and $^{1,3}B^-$ fragments have been reported for most flavonoids. The $^{1,3}A^-$-ion often is the major fragment ion. The $^{1,3}B^-$ is characteristic for isoflavones. Lower collision energies are needed when hydroxyl groups are present at the B-ring.

The 0,3-cleavage of the C-ring, leading to $^{0,3}A^-$ and $^{0,3}B^-$ fragment ions, is observed for isoflavones and flavones. Low-abundance fragments resulting from the 0,4-cleavage can be observed, especially for isoflavones. A 1,2-cleavage, especially the $^{1,2}A^-$ fragment, was observed for 3',4'-dihydroxyflavonols [18].

Losses of small neutrals such as H_2O and CO are observed in the negative-ion mode as well. Again, the loss of $CH_3\bullet$ can be observed from methoxylated flavonoids, as was studied in detail by Justesen [22-23]. [M–H]$^-$ of monomethyl-flavonoids show the loss of $CH_3\bullet$ and subsequent losses of CO or HCO$\bullet$, while polymethyl-flavonoids show losses of $CH_3\bullet$ and subsequent losses of $CH_3\bullet$, or $CH_3\bullet$ and CO. It is not clear whether the loss of two CH3$\bullet$ radicals was checked by accurate-mass determination. Alternatively, the loss of 30 Da could be explained by the loss of $H_2C=O$, which is frequently observed in the fragmentation of polymethoxylated compounds. Isomeric methoxylated flavonoids showed different fragmentation behaviour, but standards were still needed for determination of the methoxy-position. The product-ion MS–MS spectra of rhamnetin and isorhamnetin are shown in Figure 15.3. No loss of the prenyl substituent is observed for prenyl flavonoids in negative-ion mode [14].

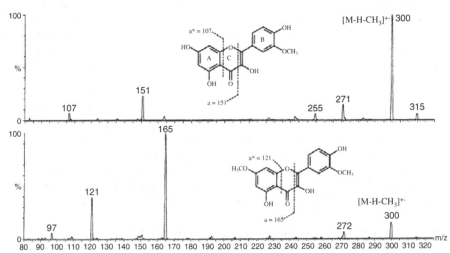

Figure 15.3: Negative-ion product-ion MS–MS spectra of rhamnetin and isorhamnetin. Reprinted from [23] with permission. ©2000, Elsevier Science BV.

Using accurate-mass determination with a Q–TOF instrument, March and Miao [16] studied the fragmentation of deprotonated kaempferol. The fragments observed were apparently not consistent with the above discussion. Accurate-mass assisted interpretation indicated that most fragments can be explained from consecutive losses of small neutrals, especially H_2O, $OH\bullet$, CO, CH_2O, and C_2H_2O.

2.6 Fragmentation of flavonoid glycosides

In annotating glycoside fragmentation, the nomenclature rules proposed by Domon and Costello [24] are followed (Ch. 20.3.2 and Figure 20.2). Discrimination between O-glycosidic, C-glycosidic, and O,C-glycosidic flavonoids can be made from the MS or MS–MS spectrum. The O-glycosides show the loss of a dehydrosugar unit, $e.g.,$, loss of 162 Da, by proton rearrangement at the glycosidic bond, while C-glycosides only show cross-ring cleavages. The loss of $C_4H_8O_4$ (120 Da) was observed for the C-glycoside vitexin, whereas the loss of dehydroglucose ($C_6H_{10}O_5$, 162 Da) was found for the O-glycoside isoquercitrin [5]. Negative-ion ESI-MS–MS spectra of sixteen flavonoid C- and O-glycosides were tabulated by Sánchez $et\ al.$ [20].

In the MS–MS spectra of flavonoid-O-diglycosides, fragment ions may be observed from subsequent cleavages of the interglycosidic bonds with charge retention on the part containing the aglycone (Y-ions). In some cases, the corresponding B-ion is observed as well. Because the loss of glycoside residues corresponds to specific neutral losses, the neutral-loss analysis mode in a triple-quadrupole instrument can be applied to screen for glycosides in crude extracts of plant material [25-28].

Whereas in positive-ion mode only even-electron fragments are observed from interglycosidic cleavages, in negative-ion mode radical cleavage of flavonoid-O-glycosides into a odd-electron aglycone product ion are observed [29-31]. This is correlated to the antioxidant activity of the flavonoids. The abundance of the radical fragment relative to the even-electron fragment increases and decreases with an increasing number of hydroxy-groups at the B-ring for 3-O-glycosides and 7-O-glycosides, respectively. No radical fragment ions were observed for flavanone and dihydrochalcone glycosides, indicating the 2,3-double bond is involved in the stabilization of the radical aglycone product ion. A combination of an even-electron and an odd-electron aglycone product ion was also observed upon fragmentation of deprotonated isoquercitrin 3-O-glycoside in a triple-quadrupole [5] and genistein-7-O-glucoside in a Q–TOF instrument [32].

A potentially confusing observation in the characterization of flavonoid-O-glycosides is the occurrence of internal glucose residue loss, indicated as a Y* ion [33-35]. It was found to strongly depend on the aglycone type: it was more pronounced with flavonones. It may yield erroneous glycan sequence information. On the other hand, the internal glucose residue loss may be help to differentiate between the isomeric O-diglycosides containing neohesperidose (rhamnosyl-

($\alpha 1\rightarrow 2$)-glucose) or rutinose (rhamnosyl-($\alpha 1\rightarrow 6$)-glucose). Initial differentiation can be made, because with a neohesperidoside a more abundant Y_0 fragment is observed, whereas with a rutinoside the Y_1 fragment is more abundant. The internal glucose residue loss is observed for both flavonone-7-O analogues, and only for flavonol-3-O-rutinoside [35]. As an example, the MS–MS spectra of two isomeric flavonoid-7-O-diglycosides are shown in Figure 15.4. Brodbelt *et al.* [8, 36] demonstrated that (1$\rightarrow$2) and (1$\rightarrow$6) disaccharide isomers can also be differentiated from the MS–MS spectra of complexes of flavonoids with Cu^{2+} or Co^{2+} and 2,2'-bipyridine or 4,7-diphenyl-1,10-phenanthroline (Ch. 15.2.2).

Internal glucose residue loss is not observed in the fragmentation of $[M+Na]^+$ and $[M–H]^-$. Differentiation between neohesperidose (1$\rightarrow$2 linkage) and rutinose (1$\rightarrow$6 linkage) is possible in the negative-ion mode [34].

In *C*-glycosides, the major fragmentation pathway involves cross-ring cleavages (Ch. 20.3.2). Differentiation between 6- and 8-*C*-glycosylation is the most important issue. Water loss from both $[M+H]^+$ and $[M–H]^-$ is more predominant for 6-*C*- than for 8-*C*-glycosides. Unambiguous differentiation is possible by studying the patterns of small-molecule losses from the $^{0,2}X^+$ fragment ion at *m/z* of $[M+H–120]^+$ in an ion-trap MS^3 experiment. The dereplication process is outlined in detail [37]. In *C*-diglycosides with glycan residues with different mass, the loss of the sugar at the C_6-position shows more abundant fragment ions than that at the C_8-position.

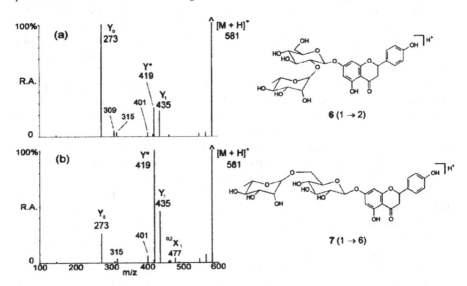

Figure 15.4: MS–MS spectra of (a) naringenin-7-O-neohesperidoside and (b) naringenin-7-O-rutinoside, showing the Y* peak due to internal glucose residue loss. Reprinted from [35] with permission.©2001, John Wiley and Sons, Ltd.

In the fragmentation of 6-*C*-arabinose-8-*C*-glucose-chrysin in negative-ion MS–MS with an ion-trap instrument, Wu *et al.* [21] found that the [M–H–90]⁻ fragment due to the $^{0,2}_5X^-$ fragment was more abundant than the [M–H–120]⁻, due to the $^{0,2}_6X^-$ fragment.

Data-dependent acquisition (DDA) on an ion-trap instrument was applied in combination with in-source CID for the analysis of *C*-glycosylflavone-*O*-glycosides in plant extracts [38]. DDA controlled switching between a survey MS scan and MS–MS allowed automated LC–MS analysis of plant extracts. Differentiation between 6- and 8-*C*-glycosylation is made via the in-source CID generation of $^{0,1}X^-$ ions and subsequent MS–MS analysis.

2.7 Acetylated flavonoid glycosides

Acetylation of the glycoside, *e.g.*, at the 6-position of a hexose moiety, occurs as well (Table 15.2). Next to molecular mass information from the MS spectrum, characteristic acyl-related product ions can be observed in low-energy MS–MS spectra using [M+H]⁺and [M+Na]⁺ precursor ions. Often, the loss of the acyl group is observed after proton rearrangement, *e.g.*, the loss of 86 Da for malonyl, 152 Da for galloyl, and 146 Da for coumaroyl. In addition, the loss of the acylhexose may be observed, *e.g.*, the loss of 248 Da for malonylhexose, and 314 Da for galloylhexose.

2.8 LC–NMR in structure elucidation of flavonoids

Next to LC–MS–MS, nuclear magnetic resonance spectroscopy (NMR) is an important tool in the structure elucidation of flavonoids and other phenolic plant constituents. Therefore, when it became available in the late 1990s, on-line LC–NMR started to be applied in conjunction with LC–MS in this area [39-41].

3. LC–MS analysis of flavonoids in plant material

In this section, some studies in the LC–MS analysis of flavonoids and flavonoid glycosides in plant material are briefly reviewed. The discussion is not meant to be comprehensive, neither in terms of literature coverage, nor in terms of plant species studied.

Sample handling strategies for the determination of plant phenols in food and plant material have been reviewed [42]. Attention was paid to hydrolysis, liquid extraction, solid-phase extraction (SPE), and supercritical-fluid extraction.

3.1   Flavonoids in plant materials

The research group of Hostettmann significantly contributed to the current widespread use of LC–MS in the analysis, characterization, and dereplication of plant phenols including flavonoids. Initially, thermospray and continuous-flow FAB were applied. ESI and APCI were implemented in the mid 1990s [39]. The group evaluated both Q–TOF and multistage MS^n with an ion-trap instrument for the dereplication of flavonoids in crude plant extracts [12], and pioneered the application of on-line LC–NMR in this field [39].

A number of studies reported the utilization of LC–MS in the analysis of flavonoids in red clover, *Trifolium pratense*, and related species. He *et al.* [43] applied positive-ion LC–ESI-MS to red clover extracts. They detected 11 flavonoids, including the isoflavonoids genistin, ononin, daidzein, sissotrin, formononetin, and biochanin A and the flavonol isoquercitrin. The same group reported the detection of 20 flavonoid glycoside malonates, some of which were never reported before [44]. Flavones are the major flavonoids in the flowers, and isoflavones in the leaves. Using two-dimensional SPE and positive-ion LC–ESI-MS, Klejdus *et al.* [45] identified 14 isoflavone glycoside malonates and six acetyl glycosides in red clover. Most of the malonates and all acetyl analogues were reported for the first time. In another study, the analysis and identification of 31 isoflavones in red clover, including aglycones, glycosides, and glycoside malonates, was reported [46]. Several of these isoflavones were detected for the first time in related species such as white clover, *T. repense*, and alsike clover, *T. hybridum*. The concentrations of 10 isoflavones, determined in acid-hydrolysed red clover extracts, ranged from 6 to 3125 ng/ml based on LC–MS detection and from 24 to 12500 ng/ml based on LC–UV detection. De Rijke *et al.* [47] reported the analysis of flavonoids in red clover and related species (*T. dubium*, *T. repens*, and *L. corniculatus*) using UV detection for general screening, classification, and determination of total flavonoid content, fluorescence detection for isoflavones, and negative-ion APCI-MS for dereplication. Flavonoid glucoside malonates are the main constituents in red and white clover; flavonoid (di)glucosides were found in the other two species.

26 Flavonoid aglycones were identified from leaf surfaces of three species of *Chrysothamnus*, collected in Oregon (US) [13]. Among the compounds identified were many methoxy- and dimethoxy-flavonoids. It was shown that the flavonoid profiles have diagnostic value for taxonomy at species level, but no evidence was found in favour of inclusion of *Chrysothamnus* into *Ericameria*, as proposed from morphological similarities.

Hvattum [48] reported the analysis of various phenolic compounds in *Rosa canina* using LC–DAD, negative-ion LC–ESI-MS, MS–MS on [M–H]⁻, and MS–MS on fragments generated by in-source CID. In this way, the aglycones catechin and quercetin, and various glycosides of quercetin, taxifolin, and eriodictyol were identified as well as some other plant phenol compounds.

Qu *et al.* [25] reported the use of the neutral-loss analysis mode (common loss of 176 Da) in the LC–MS screening for flavonoid *O*-glucuronides in extracts of *Erigeron breviscapus*. This revealed the presence of three precursors at *m/z* 445, 461, and 477, presumably due to apigenin 7-*O*-glucuronide, scutellarin, and quercetin 3-*O*-glucuronide. Confirmation was made by comparison with standards. Flavonoid glucuronides and their aglycones were determined using selected-reaction monitoring (SRM).

Tolonen and Uusitalo [49] developed a method for the fast screening of total flavonoid content in extracts of plants and food products using LC–MS with polarity switching on a TOF instrument. The good full-spectrum sensitivity of the TOF enabled sensitive detection in full-spectrum mode. The acquired chromatograms were automatically searched for [M+H]$^+$ and [M–H]$^-$ of flavonoids and a variety of their glycosides. Quantitation was performed using quercetin, quercitrin, rutin, and kuromanine as external and dextromethorphan as internal standard. The LOQ ranged from 0.2 to 10 μg/ml extract for anthocyanins and from 0.2 to 4 μg/ml for other flavonoids. The method was applied to the determination of flavonoid content of *Hypericum perforatum* and *Rhodolia rosea*, as well as to various food products.

3.2 Flavonoids in medicinal plants

St. John's Wort is a traditional medicine with various biological effects, which currently is especially applied as antidepressant. It is an extract of *H. perforatum*. LC–MS has been frequently applied in the characterization of its constituents. Next to hypericin, pseudohypericin, hyperforin, and adhyperforin (Ch. 15.7), a number of flavonoids have been observed. Brolis *et al.* [50] applied both thermospray and negative-ion ESI-MS to hot-methanolic extracts. Flavonol-*O*-glycosides like rutin, hyperoside, and quercitrin were found. Hansen *et al.* [40] described the structure elucidation of flavonoids and other constituents in an extract of *H. perforatum* using LC–NMR and LC–MS. Next to the already known constituents, two new components were tentatively identified, *i.e.*, quercetin-arabinoside and quercetin-galacturonide. Silva *et al.* [51] correlated antioxidant activity, based on an *in vitro* TEAC assay using DPPH•, and phenol content for extracts of *H. perforatum*. LC–DAD–MS was applied for the full characterization of the extracts. Kaempferol 3-rutinoside and rutin-acetyl were identified in these extracts for the first time.

4. LC–MS analysis of flavonoids in food products

Numerous health benefits have been reported or claimed for flavonoids and isoflavones, such as in cancer prevention, cardiovascular effects, and relief of menopausal syndromes. However, concerns have been raised about potential adverse effects, such as enhancement of reproductive organs and anti-thyroid effects. In order to evaluate potential beneficial or hazardous health effects from flavonoids, more knowledge is needed about the flavonoid content of various food products. In addition, intervention studies are performed to systematically investigate correlations between flavonoid intake and health effects. LC–MS is an important analytical tool in supporting such studies.

4.1   Flavonoid profiling

Justesen *et al.* [52] developed a method for the quantitative screening of flavones, flavonols, and flavonones in foods. The compounds were analysed in negative-ion APCI-MS as [M–H]⁻ of their aglycones after acid hydrolysis. The average concentrations of quercetin, kaempferol, hesperetin, naringenin. myricetin, apigenin, and luteolin ranging from 0.1 to 185 mg per 100 gr fresh weight were determined in thirty freeze-dried fruits, vegetables, and beverages. The same group [23] characterized flavonoids in extracts of fresh herbs such as dill, lovage, mint, parsley, and thyme. Various flavonoid glycosides were detected in the herb extracts, *e.g.*, glucosides of quercetin, isorhamnetin, and kaempferol in chives, rhamno-glucosides of quercetin and luteolin in lovage, luteolin glycoside, apigenin-acetyl-diglucoside, diosmetin-acetyl-apiosylglucoside, and rosmarinic acid in oregano. Subsequently, they calculated the contribution to the flavonoid uptake of the consumption of some traditional Danish dishes containing herbs [53]. The five major flavonoids, *i.e.*, apigenin, isorhamnetin, kaempferol, luteolin, and quercetin, were analysed after acid hydrolysis in herb extracts using negative-ion LC–APCI-MS.

Häkkinen and Auriola [54] studied the flavonoid contents of various types of Finnish berries by a combination of LC–ESI-MS, LC–UV-DAD, and GC–MS for identification of the sugar residues. Quercetin was found in all berry species. In addition, myricetin and kaempferol were found in some species. Hexose-, deoxyhexose–hexose-, and pentose-flavonols were most abundant. The presence of rhamnose, glucose, and galactose was identified GC–MS. The method was applied to assess the health-promoting effects of the flavonoids as antioxidants and anticarcinogens

Various reports were published on flavonoid profiling in tomatoes or tomato-based products. Stewart *et al.* [55] investigated the occurrence of flavonols in 20 varieties of tomato fruit and 10 commonly consumed tomato-based food products. Both aglycones and glycosides were observed, especially quercetin and kaempferol glycosides such as rutin. The glycosides are present for 99% in the tomato skin. The total flavonol content varied between 1.3 and 22.2 µg/g of fresh weight. In contrast to fresh tomatoes, free flavonols were found in significant amounts in tomato-based products. Le Gall *et al.* [56] characterized flavonoid glycosides in the flesh of genetically modified tomato, *Lycopersicon esculentum*. Rutin, kaempferol-3-*O*-rutinoside, and naringenin chalcone were detected in tomato for the first time. In addition, kaempferol-3,7-di-*O*-glucoside, kaempferol-3-*O*-rutinoside-7-*O*-glucoside, two dihydrokaempferol-*O*-hexosides, kaempferol-3-*O*-glucoside, and naringenin-7-

O-glucoside were detected. The total flavonoid content was 10-fold higher than the control, and kaempferol glycosides accounted for 60% of this.

A validated method for the determination of 11 flavonoids, *i.e.*, the flavone aglycones apigenin and chrysin, the flavonol aglycones kaempferol, galangin, and quercetin, the flavanone 7-O-rutinosides eriocitrin, hesperidin, and narirutin, and the flavanone 7-O-neohesperidosides naringin, neoeriocitrin, and neohesperidin, in food extracts was developed by Careri *et al.* [57]. The absolute detection limits ranged from 0.08 to 0.4 ng. The method was applied to the analysis of an orange juice sample. The concentrations of the six compounds found ranged from 0.14 mg naringin to 76.9 mg hesperidin per 100 g juice.

More than 40 plant phenols were detected in artichoke waste, using negative-ion LC-ESI-MS [26]. The major compounds were caffeoylquinic and dicaffeoylquinic acids, luteolin glucuronide, luteolin galactoside, quercetin, and various quercetin glycosides. Various neutral-loss MS–MS experiments were used, monitoring the loss of 162 Da characteristic of dehydrohexose loss, and 176 Da, characteristic of dehydroglucuronic acid loss. Precursor-ion analysis mode was applied to screen for glycosides of particular aglycones, *e.g.*, using the product ion m/z 301 for quercetin, m/z 285 for luteolin, and m/z 269 for apigenin related compounds. Parejo *et al.* [58] identified more than 40 water-soluble phenols in fennel, *Foeniculum vulgare*, after liquid extraction, fractionation, and negative-ion LC–ESI–MS. 27 of these, including various hydrocinnamic acid derivatives and flavonoid glycosides, were not reported in fennel before. Vrhovsek *et al.* [59] studied the phenol contents in 41 apple samples, representing 8 of the most popular varieties cultivated in western Europe. Three hydroxycinnamates, three flavanol, four anthocyanins, three dihydrochacones, and six quercetin-3-O-glycosides were monitored. Proanthocyanidin oligomers were quantified as a group. In total, 21 compounds were determined. The total phenol content range from 0.66 mg/g in Fuji to 2.12 mg/g in Renetta. Catechins and proanthocyanidins account for 71–90% of the phenols. Sánchez *et al.* [27] reported the qualitative analysis of phenols in apple pomace, a waste product of the apple juice industry. After extraction, the sample was fractionated in 13 fractions, which were analysed by LC–MS, applying not only product-ion MS–MS, but neutral-loss and precursor-ion analysis mode as well. Among the 60 compounds identified, there were flavonoid aglycones, a wide variety of flavonoid glycosides, and cinnamic and benzoic acid derivatives.

4.2　　Flavonoids and isoflavones in soybean products

There has been special interest in the analysis of flavonoids and isoflavones like genistein and daidzein in soybeans and derived soy products. As part of an on-going nutritional study, the isoflavone aglycone and glycoside content of 24 high-soy food products and 15 low-soy food products was determined using LC–MS [60]. Daidzin, daidzein, 6"-O-malonyldaidzin, 6"-O-acetyldaidzin, genistein, genistin, 6"-O-malonylgenistin, 6"-O-acetylgenistin, glycitin, glycitein, 6"-O-malonylglycitin, and

6"-*O*-acetylglycitin were measured in positive-ion LC–APCI-MS using selected-ion monitoring (SIM).

Boué *et al.* [61] reported the identification of three flavone aglycones and two flavone glycosides in soybean pods. Two other flavone glycosides were tentatively identified. The compounds detected are 7,4'-dihydroxyflavone-7-*O*-glucoside, luteolin-7-*O*-glucoside, apigenin-7-*O*-glucoside, 7,4'-dihydroxyflavone, apigenin-7-*O*-glucoside-6"-*O*-malonate, luteolin, and apigenin.

Klejdus *et al.* [62] modified and optimized an accelerated solvent extraction (ASE) method for the determination of isoflavones in various soy products. The modified method enabled more accurate and precise results. The concentrations of daidzin and genistin in dry soybean foods ranged between 30 and 60 μg/g, and 60 and 130 μg/g, respectively.

4.3   Flavonoids and catechins in tea

Catechins are a group of phenolic anthocyanidins (Figure 15.1), abundantly contained in *Camellia sinensis*, the Chinese green tea. Catechin is a flavanol. The main components are epicatechin (EC), epigallocatechin (EGC), and their 3-gallate analogues (ECG and EGCG). These compounds are considered to protect against cancer, and inflammatory and cardiovascular diseases. This effect has been attributed to their anti-oxidative activities by scavenging free radicals. The determination of tea catechins by LC–UV, LC–MS, and other techniques was reviewed by Dalluge and Nelson [63].

Catechin and related compounds were analysed by Poon *et al.* [64] using direct infusion of crude tea extracts into ESI-MS. Various related compounds were identified from their negative-ion MS–MS spectra, *e.g.*, EGC, EGCG, and ECG. Zeeb *et al.* [65] reported the identification of 12 catechins in green and black teas. Eight major catechins were found in the LC–MS chromatogram, while the monitoring of the catechin-specific RDA fragment at *m/z* 139 enabled the identification of an additional four minor components. In another study, Miketova *et al.* [66] identified two minor components and one new compound in green tea. In the two minor components, the gallic acid was 3"-*O*-methylated in ECG and EGCG. The new component was very similar to 3"-*O*-methyl-ECG, except that one of the two hydroxy-substituents was missing in either the A- or the B-ring.

The determination of the antioxidant activity and phenolic profile of various tea and herbal infusions were recently reported. Atoui *et al.* [67] applied TEAC and luminol PCL assays to assess the antioxidant activity, and positive-ion LC–DAD–ESI-MS to assess the phenolic profile in 9 different teas, including Greek mountain tea, mint, chamomile, black Ceylon, and Chinese green tea. About 60 different flavonoid and phenolic acid derivatives were identified.

5. LC–MS analysis of flavonoids in body fluids

Given the various biological effects ascribed to flavonoids, questions arise on bioavailability. Therefore, there is a need for analytical methods for the analysis of flavonoid aglycones and glycosides in human biofluids. In addition, the usefulness of flavonoid as a biomarker for the intake of fruit and vegetables must be evaluated. MS methods for the determination of flavonoids in biological samples were reviewed by Prasain *et al.* [68].

5.1 Flavonoids in human body fluids

Nielsen *et al.* [69] applied a coupled-column setup (Ch. 1.4.6) prior to APCI-MS detection for the identification and quantitation of flavonoids in human urine. Urine was collected from female subjects with highly controlled diets, especially with respect to the flavonoid intake (from fruits, berries, and vegetables). The method enabled detection over a wide range (2.5–1000 ng/ml). Only aglycones were detected in urine, no glycosides. A significant increase in the flavonoid excretion is observed with the flavonoid-rich diet. In a subsequent study [70], they studied the relation between the intake of fruits and vegetables and the urinary excretion of flavonoids. Twelve relevant flavonoids were monitored in urine samples collected from subjects prior and after an intervention study, *i.e.*, from subjects that were on their habitual diet and from subjects that were on diets either high or low in fruits and vegetables. Significant differences in urinary flavonoid excretion could be established between the three groups of urine samples. Therefore, flavonoids can be used as useful biomarkers to study their possible health protective effects.

In order to assess the systemic availability of quercetin glycosides after the intake of food or herbal medicinal products, Wittig *et al.* [71] developed a method for the determination of quercetin and its glucuronide metabolites in human plasma. Five different glucuronides of quercetin were detected in human plasma after consumption of fried onions, but no quercetin or its glycosides. In another study, Mullen *et al.* [72] reported 23 mixed sulfate, methyl, glucuronide, and glucose derivatives of quercetin in both urine and plasma of human volunteers 1 h after ingestion of lightly fried red onions. In this study, glycosides of both quercetin and isorhamnetin were detected in plasma. In order to assess quercetin exposure in an onion intervention study, Franke *et al.* [73] developed an LC–DAD–MS method for the determination of the dietary phytoestrogens *O*-desmethylangolensin, genistein, dihydrogenistein, daidzein, dihydrodaidzein, enterodiol, enterolactone, glycitein, hesperetin, naringenin, and quercetin in human plasma, serum, and urine. The method is based on negative-ion LC–MS in SRM mode.

High-throughput quantitation of the soy isoflavones genistein, daidzein, and equol in blood from human and animal studies was developed by Twaddle *et al.* [74] using positive-ion LC–ESI-MS in SIM or SRM mode. Samples were processed in 96-well plate format: protein precipitation, enzymatic deconjugation by *H. pomatia* glucuronidase (only when the total isoflavone content was determined), and sulfatase S3009, followed by sample clean-up on 96-well parallel SPE. As internal standards, [D$_4$]-genistein, [D$_3$]-daidzein, and [D$_4$]-equol were applied. The estimated LOQ with 100 µl of plasma or serum was 30 nmol/l for genistein and daidzein using SIM or 5 nmol/l using SRM. The method was validated.

5.2   Phytoestrogenic activity of flavonoids

Phytoestrogens are plant constituents with estrogenic activity. They are found in various plants consumed in the human diet, *e.g.*, in soybeans. A special issue of *Journal of Chromatography B* devoted to analytical methods for the analysis of phytoestrogens was published in 2002 [75], containing various reviews, *e.g.*, on tools for *in vitro* and *in vivo* phytoestrogen activity tests, bioavailability, and thyropid peroxidase inhibition. Analytical methods for the determination of phytoestrogens were reviewed by Wang *et al.* [76].

In order to assess phytoestrogenic activity, the soy isoflavones genistein and daidzein were analysed in rat serum samples using on-line SPE on a restricted-access column (RAM, Ch. 1.5.6) [77]. The use of the on-line RAM column enabled rapid and robust analysis of the rat serum samples after enzymatic deconjugation and centrifugation. When the rats were fed with a standard diet, average genistein and daidzein levels were 0.62 and 0.25 µmol/l in female rat serum, and 0.35 and 0.20 µmol/l in male rat serum. The method is applied to assess the impact of diet on the results of animal bioassays for reproductive toxicity and carcinogenicity testing.

The estrogenic activity of the flavonoid aglycones apigenin, kaempferol, genistein, and equol was investigated in immature female mice by Breinholt *et al.* [78]. Administration of genistein and equol resulted in an increase in the overall uterine concentration of the estrogen receptor alpha (ERα). While equol was almost completely recovered in the urine, biochanin A, genistein, and daidzein were extensively metabolized to hydroxylated and dihydroxylated metabolites, some of which showed even greater *in vitro* estrogenic potency.

6. Anthocyanidins and related compounds

Anthocyanins are intensely coloured plant pigments, consisting of glycosylated anthocyanidins. MS analysis of phenols, procyanidins, and anthocyanins in grapes and wines was reviewed by Flamini [79]. Some studies on the LC–MS analysis of anthocyanidins are briefly reviewed here.

Piovan *et al.* [80] reported the characterization of anthocyanins in both flower material and *in vitro* cultures of *Catharanthus roseus*. The anthocyanidins malvidin, petunidin, and hirsutidin were identified from both sources. Next to known aglycones, three 3-*O*-glycosides as well as three 3-*O*-(6-*O*-*p*-coumaroyl)glycosides of petunidin, malvidin, and hirsutidin were identified. Significant differences in the quantitative profile between the three sources were observed.

Revilla *et al.* [81] identified 30 anthocyanin analogues, including 3-*O*-glycosides and 6-coumaroyl-3-*O*-glycosides of delphinidin, cyanidin, petunidin, peonidin, and malvidin, 6-acetyl-3-glycosides of malvidin and peonidin, in extracts of red grape skin and in red wines. The combination of UV-DAD and MS detection proved to be essential for identification.

The characterization of eight anthocyanins in red raspberry fruit was reported by Mullen *et al.* [82] using positive-ion LC–APCI-MS. The compounds detected were the 3-sophoroside, 3-(2-glucosylrutinoside), 3-glucoside, and 3-rutinoside analogues of cyanidin and pelargonidin. The same group also reported the detection of ellagitannins and conjugates of ellagic acid in raspberry fruits [83].

The screening of red and black raspberries, highbush blueberries, and grapes for anthocyanins was reported by Tian *et al.* [28]. They applied a combination of precursor-ion and neutral-loss analysis mode, product-ion MS–MS, and SRM. The neutral-loss analysis enabled discrimination between glucosides/galactosides and arabinosides.

Procyanidins are plant phenolic compounds found in several plant species. They may be present as monomers or as oligomers (tannins). Two linkage types are possible: in A-type procyanidins both C–C and ether bonds are formed, while in the B-type only C–C bonds are formed, either via 4→6 or 4→8 links. The LC–MS of B-type procyanidins is reported most frequently.

The identification of procyanidins up to heptamers or even decamers in cocoa and chocolate was reported by various groups [84-85], using negative-ion ESI-MS. Tannins of polymeric proanthocyanidins with up to 27 units were detected in grape stems [86]. Gu *et al.* [87] reported the screening of 88 different types of food including fruits, beans, and nuts for proanthocyanidins using NPLC and RPLC and negative-ion LC–MS. Both oligomeric and polymeric proanthocyanidins were detected. An average DP of 47.9 was found in blackcurrent. The same group [88] studied the fragmentation of the proanthocyanidin oligomers and identified sequences through diagnostic ions derived from quinone methide cleavages of the interflavan bond. Novel heterogenous proanthocyanidins containing (epi)afzelechin as subunits were also found in pinto beans.

In their studies on phenols in apples (Ch. 15.4.1), Vrhovsek *et al.* [59] monitored four anthocyanins. In addition, the proanthocyanidin oligomers were quantified as a group. Catechins and proanthocyanidins account for 71–90% of the phenols in apples.

7. Related plant phenolic compounds

Next to flavonoids and related substances, plant material contains a wide variety of other phenolic components (Table 15.1), such as chlorogenic acid, caffeoylquinic acid, and analogues. LC–MS has also been widely applied in the characterization of such compounds, often in conjunction with flavonoids. In most cases, the negative-ion mode of either ESI or APCI is applied [39].

Another group of plant phenols is related to St. John's Wort, an extract of *H. perforatum*. Brolis *et al.* [50] reported the identification of chlorogenic acid, hypericin and pseudohypericin, and hyperforin and adhyperforin in St. John's Wort. The same compounds were also reported by Hansen *et al.* [40] using LC–NMR and LC–MS, and by Pirker *et al.* [89]. The latter group reported the simultaneous determination of hypericin and hyperforin from St. John's Wort extract in human plasma and serum. LLE with 70% hexane in ethyl acetate was applied to extract the analytes from plasma or serum. The method was applied to measure time dependent concentrations after ingestion of a St. John's Wort extract capsule. Silva *et al.* [51] correlated antioxidant activity and phenol content of *H. perforatum* extracts.

8. References

1.   D. Ryan, K. Robards, P. Prenzler, M. Antolovich, *Applications of MS to plant phenols*, Trends Anal. Chem., 18 (1999) 362.

2.   R. Tsao, Z, Deng, *Separation procedures for naturally occurring antioxidant phytochemicals*, J. Chromatogr. B, 812 (2004) 85.

3.   F. Cuyckens, M. Claeys, *MS in the structural analysis of flavonoids*, J. Mass Spectrom., 39 (2004) 1.

4.   F. Cuyckens, M. Claeys, *Optimization of a LC method based on simultaneous ESI-MS and UV-DAD for analysis of flavonoid glycosides*, Rapid Commun. Mass Spectrom., 16 (2002) 2341.

5.   J.-P. Rauha, H. Vuorela, R. Kostiainen, Effect of eluent on the ionization efficiency of flavonoids by ESI, APCI, and APPI-MS, J. Mass Spectrom., 36 (2001) 1269.

6.   E. de Rijke, H. Zappey, F. Ariese, C. Gooijer, U.A.Th. Brinkman, *LC with APCI and ESI-MS of flavonoids with triple-quadrupole and ion-trap instruments*, J. Chromatogr. A, 984 (2003) 45.

7.   M. Satterfield, J.S. Brodbelt, *Enhanced detection of flavonoids by metal complexation and ESI-MS*, Anal. Chem., 72 (2000) 5898.

8.   M. Pikulski, J.S. Brodbelt, *Differentiation of flavonoid glycoside isomers by using metal complexation and ESI-MS*, J. Am. Soc. Mass Spectrom., 14 (2003) 1437.

9.   Y.L. Ma, Q.M. Li, H. Van den Heuvel, M. Claeys, *Characterization of flavone and flavonol aglycones by CID MS–MS*, Rapid Commun. Mass Spectrom., 11 (1997) 1357.

10. Y. L. Ma, H. Van den Heuvel, M. Claeys, *Characterization of 3-methoxyflavones using FAB and CID MS–MS*, Rapid Commun. Mass Spectrom., 13 (1999) 1932.

11. F. Cuyckens, Y.L Ma, G. Pocsfalvi, M. Claeys, *Tandem mass spectral strategies for the structural characterization of flavonoid glycosides*, Analusis, 28 (2000) 888.

12. J.-L. Wolfender, P. Waridel, K. Ndjoko, K.R. Hobby, H.J. Major, K. Hostettmann, *Evaluation of Q–TOF-MS–MS and multiple-stage IT-MS^n for the dereplication of flavonoids and related compounds in crude plant extracts*, Analusis, 28 (2000) 895.

13. J.F. Stevens, E. Wollenweber, M. Ivancic, V.L. Hsu, S. Sundberg, M.L. Deinzer, *Leaf surface flavonoids of Chrysothamnus*, Phytochem., 51 (1999) 771.
14. J.F. Stevens, M. Ivancic, V.L. Hsu, M.L. Deinzer, *Prenylflavonoids from Humulus lupulus*, Phytochem., 44 (1997) 1575.
15. F. Kuhn, M. Oehme, F. Romero, E. Abou-Mansour, R. Tabacchi, *Differentiation of isomeric flavone/isoflavone aglycones by ion trap MS^2 and a double neutral loss of CO*, Rapid Commun. Mass Spectrom., 17 (2003) 1941.
16. R.E. March, X.-S. Miao, *A fragmentation study of kaempferol using ESI-Q–TOF-MS at high mass resolution*, Int. J. Mass Spectrom., 231 (2004) 157.
17. J.S. Lee, D.H. Kim, K-H. Liu. T.K. Oh, C.H. Lee, *Identification of flavonoids using LC with ESI and ion-trap MS–MS with an MS–MS library*, Rapid Commun. Mass Spectrom., 19 (2005) 3539.
18. N. Fabre, I. Rustan, E. de Hoffmann, J. Quetin-Leclercq, *Determination of flavone, flavonol, and flavanone aglycones by negative ion LC–ESI ion trap MS*, J. Am. Soc. Mass Spectrom., 12 (2001) 707.
19. R.J. Hughes, T.R. Croley, C.D. Metcalfe, R.E. March, *A MS–MS study of selected characteristic flavonoids*, Int. J. Mass Spectrom., 210/211 (2001) 371.
20. F. Sánchez-Rananeda, O. Jáuregui, I. Casals, C. Andrés-Lacueva, M. Izquierdo-Pulido, R.M. Lamuela-Raventós, *LC–ESI-MS–MS study of the phenolic composition of cocoa (Theobroma cacao)*, J. Mass Spectrom., 38 (2003) 35.
21. W. Wu, C. Yan, L. Li, Z. Liu, S. Liu, *Studies on the flavones using LC–ESI-MS–MS*, J. Chromatogr. A, 1048 (2004) 213.
22. U. Justesen, *CID of deprotonated methoxylated flavonoids, obtained by ESI-MS*, J. Mass Spectrom., 36 (2001) 169.
23. U. Justesen, *Negative APCI low-energy CID-MS for the characterization of flavonoids in extracts of fresh herbs*, J. Chromatogr. A, 902 (2000) 369.
24. B. Domon, C.E. Costello, *A systematic nomenclature for carbohydrate fragmentation in FAB–MS–MS spectra of glycoconjugates*, Glyconconj. J., 5 (1988) 397.
25. J. Qu, Y. Wang, G. Loa, Z. Wu, *Identification and determination of glucuronides and their aglycones in Erigeron brevicapus by LC–MS–MS*, J. Chromatogr. A, 928 (2001) 155.
26. F. Sánchez-Rabaneda, O. Jáuregui, R.M. Lamuela-Raventós, J. Bastida, F. Viladomat, C. Codina, *Identification of phenolic compounds in artichoke waste by LC–MS–MS*, J. Chromatogr. A, 1008 (2003) 57.
27. F. Sánchez-Rabaneda, O. Jáuregui, R.M. Lamuela-Raventós, F. Viladomat, J. Bastida, C. Codina, *Qualitative analysis of phenolic compounds in apple pomace using LC–MS–MS*, Rapid Commun. Mass Spectrom., 18 (2004) 553.
28. Q. Tian, M.M. Giusti, G.D. Stoner, S.J. Schwartz, *Screening for anthocyanins using LC–ESI-MS–MS with precursor-ion analysis, product-ion analysis, common-neutral-loss analysis and SRM*, J. Chromatogr. A, 1091 (2005) 72.
29. E. Hvattum, D. Ekeberg, *Study of the collision-induced radical cleavage of flavonoid glycosides using negative ESI-MS–MS*, J. Mass Spectrom., 38 (2003) 43.
30. F. Cuyckens, M. Claeys, *Determination of the glycosylation site in flavonoid-mono-O-glycosides by CID of ESI-generated deprotonated and sodiated molecules*, J. Mass Spectrom., 40 (2005) 364.

31. R.E. March, E.G. Lewars, C.J. Stadey, X.-S. Miao, Z. Zhao, C.D. Metcalfe, *A comparison of flavonoid glycosides be ESI-MS–MS*, Int. J. Mass Spectrom., 248 (2006) 61.

32. R.E. March, X.-S. Miao, C.D. Metcalfe, M. Stobiecki, L. Marczak, *A fragmentation study of an isoflavone glycoside, genistein-7-O-glucoside, using ESI-Q–TOF-MS at high mass resolution*, Int. J. Mass Spectrom., 232 (2004) 171.

33. Y.L. Ma, I. Verdernikova, H. van den Heuvel, M. Claeys, *Internal glucose residue loss in protonated O-diglycosyl flavonoids upon low-energy CID*, J. Am. Soc. Mass Spectrom., 11 (2000) 136.

34. F. Cuyckens, R. Rozenberg, E. de Hoffmann, M. Claeys, *Structure characterization of flavonoid O-diglycosides by positive and negative nano-ESI ion trap MS*, J. Mass Spectrom., 36 (2001) 1203.

35. Y.L. Ma, F. Cuyckens, H. van den Heuvel, M. Claeys, *MS methods for the characterization and differentiation of isomeric O-diglycosyl flavonoids*, Phytochem. Anal., 12 (2001) 159.

36. M. Satterfield, J.S. Brodbelt, *Structural characterization of flavonoid glycosides by CID of metal complexes*, J. Am. Soc. Mass Spectrom., 12 (2001) 537.

37. P. Waridel, J.-L. Wolfender, K. Ndjoko, K.R. Hobby, H.J. Major, K. Hostettmann, *Evaluation of Q–TOF-MS and ion-trap multiple-stage MS for the differentiation of C-glycosidic flavonoid isomers*, J. Chromatogr. A, 926 (2001) 29.

38. G.C. Kite, E.A. Porter, F.C. Denison, R.J. Grayer, N.C. Veitch, I. Butler, M.S.J. Simmonds, *Data-directed scan sequence for the general assignment of C-glycosylflavone O-glycosides in plant extracts by LC ion trap MS*, J. Chromatogr. A, 1104 (2006) 123.

39. J.-L. Wolfender, S. Rodriguez, K. Hostettmann, *LC coupled to MS and NMR for the screening of plant constituents*, J. Chromatogr. A, 794 (1998) 299.

40. S.H. Hansen, A.G. Jensen, C. Cornett, I. Bjørnsdottir, S. Taylor, B. Wright, I.D. Wilson, *LC on-line coupled to high-field NMR and MS for structure elucidation of constituents of Hypericum perforatum*, Anal. Chem., 71 (1999) 5235.

41. H.B. Xiao, M. Krucker, K. Putzbach, K. Albert, *Capillary LC–microcoil ^{1}H NMR and LC–ion trap-MS for on-line structure elucidation of isoflavones in Radix astragali*, J. Chromatogr. A, 1067 (2005) 135.

42. D. Tura, K. Robards, *Sample handling strategies for the determination of biophenols in food and plants*, J. Chromatogr. A, 975 (2002) 71.

43. X.-G. He, L.-Z. Lin, L.-Z. Lian, *Analysis of flavonoids from red clover by LC–ESI-MS*, J. Chromatogr. A, 755 (1996) 127.

44. L.-Z. Lin, X.-G. He, M. Lindenmaier, J. Yang, M. Cleary, S.-X. Qiu, G.A. Cordelle, *LC–ESI-MS Study of the flavonoid glycoside malonates of red clover (Trifolium pratense)*, J. Agric. Food Chem., 48 (2000) 354.

45. B. Klejdus, D. Vitamvásová-Štěerbová, V. Kubáň, *Identification of isoflavone conjugates in red clover (T. pratense) by LC–MS after two-dimensional SPE*, Anal. Chim. Acta, 450 (2001) 81.

46. Q. Wu, M. Wang, J.E. Simon, *Determination of isoflavones in red clover and related species by LC–UV–MS detection*, J. Chromatogr. A, 1016 (2003) 195.

47. E. de Rijke, H. Zappey, F. Ariese, C. Gooijer, U.A.Th. Brinkman, *Flavonoids in Leguminosae: Analysis of extracts of T. pratense L ., T. dubium L ., T. repens L., and L. corniculatus L. leaves using LC with UV, MS and fluorescence detection*, Anal.

Bioanal. Chem., 378 (2004) 995.
48. E. Hvattum, *Determination of phenolic compounds in rose hip (Rosa canina) using LC coupled to ESI-MS–MS and DAD*, Rapid Commun. Mass Spectrom., 16 (2002) 655.
49. A. Tolonen, J. Uusitalo, *Fast screening method for the analysis of total flavonoid content in plants and foodstuffs by LC–ESI-TOF-MS with polarity switching*, Rapid Commun. Mass Spectrom., 18 (2004) 3113.
50. M. Brolis, B. Gabetta, N. Fuzzati, R. Pace, F. Panzeri, F. Peterlongo, *Identification by LC–DAD–MS and quantification by LC–UV absorbance of active constituents of H. perforatum*, J. Chromatogr. A, 825 (1998) 9.
51. B.A. Silva, F. Ferreres, J.O. Malva, A.C.P. Dias, *Phytochemical and antioxidant characterization of H. perforatum alcoholic extracts*, Food Chem., 90 (2005) 157.
52. U. Justesen, P. Knuthsen, T. Leth, *Quantitative analysis of flavonols, flavones and flavonones in fruits, vegetables and beverages by LC with DAD and MS detection*, J. Chromatogr. A, 799 (1998) 101.
53. U. Justesen, P. Knuthsen, *Compositions of flavonoids in fresh herbs and calculation of flavonoid intake by the use of herbs in traditional Danish dishes*, Food Chem., 73 (2001) 245.
54. S. Häkkinen, S. Auriola, *LC with ESI-MS and UV-DAD in identification of flavonol aglycones and glycosides in berries*, J. Chromatogr. A, 829 (1998) 91.
55. A.J. Stewart, S. Bozonnet, W. Mullen, G.I. Jenkins, M.E.J. Lean, A. Crozier, *Occurrence of flavonols in tomatoes and tomato-based products*, J. Agric. Food Chem., 48 (2000) 2663.
56. G. Le Gall, M.S. Dupont, F.A. Mellon, A.L. Davis, G.J. Collins, M.E. Verhoeyen, I.J. Colquhoun, *Characterization and content of flavonoid glycosides in genetically modified tomato (Lycopersicon esculentum) fruits*, J. Agric. Food Chem., 51 (2003) 2438.
57. M. Careri, L. Elviri, A. Mangia, *Validation of a LC–ESI-MS method for the analysis of flavanones, flavones and flavonols*, Rapid Commun. Mass Spectrom., 13 (1999) 2399.
58. I. Parejo, O. Jauregui, F. Sánchez-Rabaneda, F. Viladomat, J. Bastida, C. Codina, *Separation and characterization of phenolic compounds in fennel (Foeniculum vulgare) using negative-ion LC–ESI-MS–MS*, J. Agric. Food Chem., 52 (2004) 3679.
59. U. Vrhovsek, A. Rigo, D. Tonon, F. Mattivi, *Quantitation of polyphenols in different apple varieties*, J. Agric. Food Chem., 52 (2004) 6532.
60. H. Wiseman, K. Casey, D.B. Clarke, K.A. Barnes, E. Bowey, *Isoflavone aglycon and glucoconjugate content of high- and low-soy U.K. foods used in nutritional studies*, J. Agric. Food Chem., 50 (2002) 1404.
61. S.M. Boué, C.H. Carter-Wientjes, B.Y. Shih, T.E. Cleveland, *Identification of flavone aglycones and glycosides in soybean pods by LC–MS–MS*, J. Chromatogr. A, 991 (2003) 61.
62. B. Klejdus, R. Mikelova, V. Adam, J. Zehnalek, J. Vacek, R. Kizek, V. Kuban, *LC–MS determination of genistin and daidzin in soybean food samples after ASE with modified content of extraction cell*, Anal. Chim. Acta., 517 (2004) 1.
63. J.J. Dalluge, B.C. Nelson, *Determination of tea catechins*, J. Chromatogr. A, 881 (2000) 411.
64. G.K. Poon, *Analysis of catechins in tea extracts by LC–ESI-MS*, J. Chromatogr. A,

794 (1998) 63.
65. D.J. Zeeb, B.C. Nelson, K. Albert, J.J. Dalluge, *Separation and identification of twelve catechins in tea using LC–APCI-MS*, Anal. Chem., 72 (2000) 5020.
66. P. Miketova, K.H. Schram, J. Whitney, M. Li, R. Huang, E. Kerns, S. Valcic, B.N. Timmermann, R. Rourick, S. Klohr, *MS–MS studies of green tea catechins. Identification of three minor components in the polyphenolic extract of green tea*, J. Mass Spectrom., 35 (2000) 860.
67. A.K. Atoui, A. Mansouri, G. Boskou, P. Kefalas, *Tea and herbal infusions: Their antioxidant activity and phenolic profile*, Food Chem., 89 (2005) 27.
68. J.K. Prasain, C.-C. Wand, S. Barnes, *MS methods for the determination of flavonoids in biological matrices*, Free Rad. Biol. Med., 37 (2004) 1324.
69. S.E. Nielsen, R. Freese, C. Cornett, L.O. Dragsted, *Identification and quantification of flavonoids in human urine samples by column-switching LC–APCI-MS*, Anal. Chem., 72 (2000) 1503.
70. S.E. Nielsen, R. Freese, P. Kleemola, M. Mutanen, *Flavonoids in human urine as biomarkers for intake of fruits and vegetables*, Cancer Epidemiol. Biomark. Prev., 11 (2002) 459.
71. J. Wittig, M. Herderich, E.U. Graefe, M. Veit, *Identification of quercetin glucuronides in human plasma by LC–MS–MS*, J. Chromatogr. B, 753 (2001) 237.
72. W. Mullen, A. Boitier, A.J. Stewart, A. Crozier, *Flavonoid metabolites in human plasma and urine after the consumption of red onions: analysis by LC with DAD and full scan MS–MS detection*, J. Chromatogr. A, 1058 (2004) 163.
73. A.A. Franke, L.J. Custer, L.R. Wilkens, L. Le Marchand, A.M.Y. Nomura, M.T. Goodman, L.N. Kolonel, *LC–DAD–MS analysis of dietary phytoestrogens from human urine and blood*, J. Chromatogr. B, 777 (2002) 45.
74. N.C. Twaddle, M.I. Churchwell, D.R. Doerge, *High-throughput quantification of soy isoflavones in human and rodent blood using LC–ESI-MS–MS detection*, J. Chromatogr. B, 777 (2002) 139.
75. M. Metzler and G. Sontag, *Foreword*, J. Chromatogr. B, 777 (2002) 1.
76. C.-C. Wang, J.K. Prasain, S. Barnes, *Review of the methods used in the determination of phytoestrogens*, J. Chromatogr. B, 777 (2002) 3.
77. D.R. Doerge, M.I. Churchwell, K.B. Delclos, *On-line sample preparation using RAM in the analysis of the soy isoflavones, genistein and daidzein, in rat serum using LC–ESI-MS*, Rapid Commun. Mass Spectrom., 14 (2000) 673.
78. V. Breinholt, A. Hossaini, G.W. Svendsen, C. Brouwer, S.E. Nielsen, *Estrogenic activity of flavonoids in mice. The importance of estrogen receptor distribution, metabolism and bioavailability*, Food Chem. Toxicol., 38 (2000) 555.
79. R. Flamini, *MS in grape and wine chemistry: Part I: Polyphenols*, Mass Spectrom. Rev., 22 (2003) 218.
80. A. Piovan, R. Filippini, D. Favretto, *Characterization of the anthocyanins of Catharanthus roseus (L.) G. Don in vivo and in vitro by ESI ion trap MS*, Rapid Commun. Mass Spectrom., 12 (1998) 361.
81. I. Revilla, S. Pérez-Magariño, M.L. González-SanJosé, S. Beltrán, *Identification of anthocyanin derivatives in grape skin extracts and red wines by LC–DAD–MS*, J. Chromatogr. A, 847 (1999) 83.
82. W. Mullen, M.E.J. Lean, A. Crozier, *Rapid characterization of anthocyanins in red raspberry fruit by LC–MS*, J. Chromatogr. A, 966 (2002) 63.

83. W. Mullen, T. Yokota, M.E.J. Lean, A. Crozier, *Analysis of ellagitannins and conjugates of ellagic acid and quercetin in raspberry fruits by LC–MS*[n], Phytochem., 64 (2003) 617.

84. J.F. Hammerstone, Sh.A. Lazarus, A.E. Mitchell, R. Rucker, H.H. Schmitz, *Identification of procyanidins in cocoa (Theobroma cacao) and chocolate using LC–MS*, J. Agr. Food Chem., 47 (1999) 490.

85. J. Wollgast, L. Pallaroni, M.-E. Agazzi, E. Anklam, *Analysis of procyanidins in chocolate by RPLC–ESI-MS–MS detection*, J. Chromatogr. A, 926 (2001) 211.

86. J.-M. Souquet, B. Labarbe, C. Le Guernevé, V. Cheynier, M. Moutounet, *Phenolic composition of grape stems*, J. Agric. Food Chem., 48 (2000) 1076.

87. L. Gu, M.A. Kelm, J.F. Hammerstone, G. Beecher, J. Holden, D Haytowitz, R.L. Prior, *Screening of foods containing proanthocyanidins and their structural characterization using LC–MS–MS and thiolytic degradation*, J. Agric. Food Chem., 51 (2003) 7513.

88. L. Gu, M.A. Kelm, J.F. Hammerstone, Z. Zhang, G. Beecher, J. Holden, D. Haytowitz, R.L. Prior, *LC–ESI-MS studies of proanthocyanidins in foods*, J. Mass Spectrom., 38 (2003) 1272.

89. R. Pirker, C.W. Huck, G.K. Bonn, *Simultaneous determination of hypericin and hyperforin in human plasma and serum using LLE, LC and LC–MS–MS*, J. Chromatogr. B, 777 (2002) 147.

APPLICATIONS: BIOMOLECULES

16

LC–MS ANALYSIS OF PROTEINS

1. Introduction

Until 1988, the mass spectrometric analysis of peptides and proteins was difficult. Some results were achieved using (continuous-flow) fast-atom bombardment (FAB) and ^{252}Cf plasma desorption. The major breakthrough in the characterization of proteins by mass spectrometry (MS) is due to the introduction of matrix-assisted laser desorption/ionization (MALDI) and electrospray ionization (ESI) in 1988. Currently, peptides and proteins form the compound class most intensively studied by MS. This is primarily due to the prominent role ESI-MS and MALDI-MS play in the field of proteomics.

The impact of MS, and within this book especially of ESI-MS, in protein characterization is difficult to review, because of the multitude of papers appearing, both in journals with analytical orientation and in journals with specialized biochemical orientation. This chapter should be considered as a technology overview rather than a comprehensive review of current applications of ESI-MS in peptide and protein characterization.

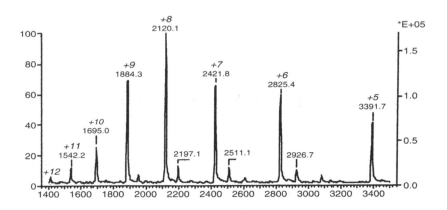

Figure 16.1: Thermospray spectrum of myoglobin. Reprinted from [7] with permission. ©1990, Wiley & Sons Ltd.

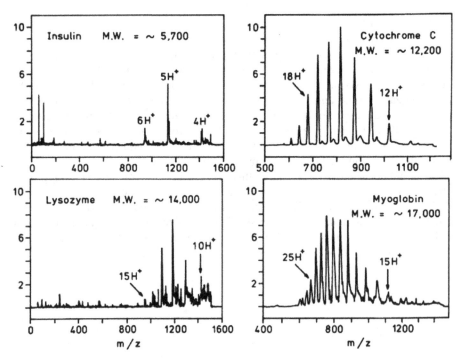

Figure 16.2: First electrospray mass spectra of multiple-charge peptide and protein ions. Reprinted from [6] with permission. ©1989, American Chemical Society.

2. Electrospray ionization of peptides and proteins

2.1   History

The analysis of peptides and proteins has played a role in the development of LC–MS from the early 1980s. Yu *et al.* [1] reported the analysis of a *N*-acetyl-*N,O,S*-permethylated octapeptide derived from the *C*-terminus of glucagon as early as 1984, using a moving-belt interface and isobutane chemical ionization. Stroh *et al.* [2] applied FAB and the moving-belt interface to the structure elucidation of antibiotic leucinostatin peptides and a tryptic digest of antiamoebin I.

Amino acids and peptides were among the test compounds used by Vestal and coworkers [3] in the development of thermospray (TSP, Ch. 4.7). Mass spectra of 6 nmol of underivatized renin substrate (1757 Da) were obtained with single-, double-, and triple-charge ions and without significant fragmentation. The potential of TSP was further explored by Pilosof *et al.* [4] with the analysis of the α-melanocyte stimulating hormone (1665 Da, 4 nmol analysed) and glucagon (3483 Da, 2 nmol analysed). The abundance of the multiple-charge ions appeared to be correlated to the solution pH. Following the introduction of ESI-MS [5-6], Straub and Chan [7] reported series of multiple-charge ions from TSP of 100 pmol of proteins, such as myoglobin, carbonic anhydrase II, and trypsinogen. The major change compared to previous experiments was replacing 0.1 mol/l ammonium acetate by 5% acetic acid. The spectrum for myoglobin (16,951 Da) is given in Figure 16.1.

The development of the continuous-flow FAB interface by Caprioli *et al.* [8] was primarily directed at the analysis of peptides (Ch. 4.6). Continuous-flow FAB provided reduced suppression of hydrophilic peptides in peptide mixtures. It was widely applied in the field of peptide characterization and the analysis of proteolytic digests, *e.g.*, in the analysis of a tryptic digest of bovine ribonuclease B before and after treatment with N-glycanase [9]. A single injection of 100 pmol provided *ca.* 70% sequence coverage.

A major breakthrough was achieved by Fenn *et al.* [5-6] using ESI-MS in 1988. They demonstrated the generation of multiple-charge ions from large proteins by ESI-MS, enabling their detection and molecular-mass determination with a relatively inexpensive quadrupole mass analyser. Some of their early mass spectra are shown in Figure 16.2.

2.2   Molecular-weight determination

Fenn *et al.* [5-6] demonstrated the ability to perform accurate molecular-weight determination (generally within 0.01%) of proteins up to 30 kDa. Because of the low sample consumption, the excellent sensitivity, and the ease of operation, this method was rapidly implemented in the biochemistry and biotechnology toolbox [10-13].

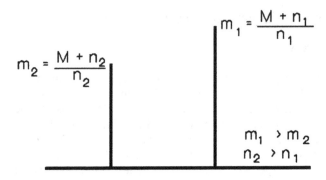

Figure 16.3: Principle of molecular-mass determination from the ion envelope of multiple-charge protein ions.

Molecular-weight determination from an ion envelope of multiple-charge ions can be done by using the averaging algorithm [6, 14]. This algorithm assigns a number of charges to the peaks in the ion envelope and then averages the resulting values for M to give an estimate of the molecular weight. If adjacent peaks in the ion envelope differ by only one charge and charging is due to protonation (see Figure 16.3), the average molecular weight M of the protein can be calculated from:

$$\text{\textit{If:}} \quad n_1 = n_2 + 1 \quad \text{\textit{Then:}} \quad n_2 = \frac{m_1 - 1}{m_2 - m_1}$$

$$M = n_2 \, (m_2 - 1)$$

Next to this simple and straightforward approach, which can only be applied to relatively simple spectra with good signal-to-noise ratio (S/N), algorithms for the deconvolution or transformation of ESI-MS mass spectra of proteins were introduced [14] They use all the information available in the spectrum and provide good accuracies if other than protonated adducts are present. The software is readily available from the instrument manufacturers.

Maximum-entropy based algorithms for transformation of ESI-MS mass spectra were proposed by Reinhold and Reinhold [15] and Ferrige *et al.* [16-17]. The algorithm detects patterns of series of multiple-charge ions in a complex spectrum. An entropy measure signals the presence of a specified pattern, the ion envelope of multiple-charge ions, in the mass spectra [15]. Alternatively, commercial maximum-entropy techniques, well established in radio astronomy and photography, were adapted to the deconvolution of ion envelopes [16-17]. The software, commercially available from various MS instrument manufacturers, is especially useful in the interpretation of mass spectra generated from mixtures of proteins.

A recent application of the maximum-entropy deconvolution is the interpretation of ESI-MS spectra from the giant, extracellular hexagonal bilayer hemoglobin from the leech *Nephelopsis oscura*, consisting of three monomer globin chains, five non-globin linker chains, and two subunits, ranging between 16 and 33 kDa [18].

2.3 Ionization mechanism and charge-state distribution

The ionization mechanism of ESI is discussed in Ch. 6.3. Specific topics related to the ionization and charge-state distribution of proteins are briefly discussed here. In principle, the maximum number of charges on a protein depends on the number of protonation or deprotonation sites in the protein. In positive-ion mode, the maximum charge state is determined by number of basic sites, *i.e.*, *N*-terminus plus the number of basic amino acids (Lys, Arg, His). A good correlation between the number of basic sites and the maximum number of charges was suggested by Smith *et al.* [10]. Steric hindrance, *e.g.*, due to disulfide bridges, may limit the actual maximum number of charges observed. However, it may also be argued that protein folding results in a hydrophobic core with the hydrophilic groups, *i.e.*, the protonation sites, being exposed to solution. For lysozyme, it has been shown that after reduction of the four disulfide bridges the maximum number of charges in the spectrum increases from 10 to 14 [19]. Based on experiments with multiple-charge sodium adducts of polyethylene glycols, Fenn [20-21] argued that an additional constraint is that in the gas-phase ion the bonding energy of one charge at any site on the molecule should be equal to or exceed the electrostatic Coulomb repulsion energy by all other charges on the molecule. Ashton *et al.* [22] observed a maximum number of charges of 27 for subtilisin (*Carlsberg*), while only 19 basic amino acids are present in the sequence. In the data collected by Smith *et al.* [10], they found similar behaviour for quite a number of proteins. A possible explanation would be protonation of other amino acids, *e.g.*, Trp, Pro, Tyr, and Glu.

Apart from protonation in the positive-ion mode and deprotonation in the negative-ion mode, various other adducts may be formed, *e.g.*, with Na^+ or K^+. This broadens the individual peaks in the ion envelope.

In a fundamental discussion [23], Fenn actually reversed the arguments related to the charge states of desorbed molecules. The number of protons found on a desorbing protein is determined by the number density of net protons on the droplet surface, and not by the number of protonated basic groups in the neutral bulk solution. Since the protons on the droplet surface are expected to be approximately equidistant, determined by the forces of Coulomb repulsion, the number of charges attached to a protein depends on the number of charges it can span at the surface. When the surface charges move closer upon solvent evaporation, the analyte molecule is able to span a larger number of protons. Actual charging of the analyte molecule requires the presence of a protonation site in the vicinity of the charge at the droplet surface. This means that only one of two closely positioned potential charge sites on a molecule will be charged, because of the equidistant charge distribution. This concept does not completely explain the observed mass spectra, because the desorption rate of charged analyte molecules increases upon proceeding solvent evaporation, *i.e.*, upon increasing surface field.

To illustrate the complex behaviour, the ESI-MS spectra of gramicidin S (1141 Da) and cyclosporin A (1202 Da), two cyclic peptides of similar size, were studied as a function of analyte concentration [23]. Over the concentration range studied, the abundance ratio M^{2+}/M^+ was higher for gramicidin than for cyclosporin. This can be explained from the more solvophobic nature of cyclosporin: it desorbs substantially faster at lower fields and thus at lower charge densities. From these observations, the situation with more complex molecules can be extrapolated, and the bell-shaped distribution of multiple-charge ions can be explained. Molecules with a small number of charges start to desorb from the droplet at a relatively early stage of the droplet lifetime. The desorption rate is not very high, due to a number of factors including the charge at the molecule, the desorption rate, and the surface concentration. The molecules with higher charge states show an increasing abundance. At higher charge states, *i.e.*, after continued solvent evaporation, the ion desorption from the droplet surface is more efficient. The molecules with higher charge states have a shorter lifetime on the surface as well as a higher surface concentration due to an increasing number of locations and orientations to span multiple charges on the droplet surface. The process of increasing abundance of higher charged molecules proceeds until the supplies of charge and/or analyte species become depleted. Consequently, the desorption rate of ions in the highest charge states, and thus their abundance, must decrease. Similarly, the effect of protein conformation can readily be understood. A more compact configuration leads to a smaller two-dimensional projection of the molecule at the droplet surface and will allow a smaller number of charges to be spanned. The ion desorption rate is thus strongly determined by the time course of the droplet, *i.e.*, its evaporation rate. A rapid evaporation of the droplets should lead to charge distributions biassed to the high charge states, whereas slow evaporation leads to an increase in abundance of low charge states. It also follows that the lower charge states are thus desorbed from the droplet at an earlier stage in the droplet evaporation process than the higher charge states.

An alternative model for ESI was developed by the group of Siu [24-25]. They correlated the ion envelope of a protein with the predicted distribution of charge states of the protein in solution. The latter depends on pH and pK_a values of the acidic and basic amino acids in the protein. From the good correlation obtained, they conclude that the ion envelope nicely reflects the abundances of the preformed ions of the protein in solution. Based on further experiments, they postulate that the droplet evaporation is not that important and that the ESI-MS spectrum results from ions directly emitted from the Taylor cone [25-26].

In subsequent studies by others, it was experimentally demonstrated that this direct correlation between solution chemistry and ion envelope is not generally observed for all proteins. For instance, Kelly *et al.* [19] studied the positive-ion and negative-ion mass spectra of horse heart myoglobin (isoelectric point, pI 7), hen egg white lysozyme, and bovine α-lactalbumin, as a function of the solution pH between 3 and 10. At pH 3.5, both positive-ion and negative-ion mass spectra of myoglobin were observed. The ion current in the positive-ion mode was 100× larger than in the negative-ion mode. From the isoelectric point, one would not expect to observe a negative-ion spectrum, at least when only evaporation of preformed ions in solution is involved in ESI. At pH 10, again both positive-ion and negative-ion mass spectra were observed. While the intensity of the negative-ion mass spectrum was about equal to that observed at pH 3.5, the intensity of the positive-ion spectrum was 10× less. Myoglobin contains 33 basic and 22 acidic sites; the maximum charge state observed are 23 in positive-ion and 20 in negative-ion mode.

Fenn [23] indicated that the model of Siu [24-25] passes over the fact that ion desorption does not take place from a neutral bulk solution, but from a highly charged droplet. The probably drastic changes in the solution conditions during the droplet evaporation were studied by (among others) Gatlin and Tureček [27]. They investigated the pH changes in the sprayed solution by means of the pH-dependent dissociation of $Fe^{2+}(bpy)_3$ and $Ni^{2+}(bpy)_3$ complexes. They concluded that a 10^3–10^4-fold increase in the $[H_3O^+]$ concentration takes place upon solvent evaporation. Mansoori et al. [28] measured the pH of collected sprayed solutions and observed only small changes in the solution pH (less than 1 pH unit) (see also Ch. 6.3.5).

The mechanism of ESI continues to be a fascinating topic, which continues to raise research interest and efforts, as for instance indicated by the observation of higher charge states with enhanced abundance via the addition of glycerol or m-nitrobenzyl alcohol to the ESI solvent [29].

3. LC–MS of intact proteins

In the early years, considerable attention was paid to extensive characterization and optimization of ESI parameters. Optimum experimental conditions were investigated, instrumentation was improved, and the limits of the new technique were explored. Developments in ESI-MS instrumentation are discussed in Ch. 5.4–5. These developments resulted in the widespread availability of easy-to-use, dedicated instruments for ESI-MS from all major MS instrument manufacturers.

3.1 Direct infusion experiments

Studies on the analysis and characterization of proteins by ESI-MS generally involve the direct infusion of a protein solution. This enables the determination of the molecular weight of the protein. Initially, the influence of a variety of experimental parameters, e.g., solvent composition, temperature, and influence of nozzle-skimmer potential, was studied for model proteins [10-13]. Given the diverse nature and properties of proteins, general statements on optimum solvent conditions are difficult to make.

With respect to the solvent composition, many additives commonly applied in biochemistry, such as phosphate buffers and phosphoric acid, surfactants, and chaotropes, are not suitable to ESI-MS (Ch. 6.6.3). They must be replaced by volatile alternatives. In addition, ESI-MS is preferentially performed in mixtures of methanol or acetonitrile and water, a solvent that is not generally applied in biochemistry. Organic solvent may induce conformational changes and/or protein denaturation, which in turn influences the charge-state distribution. Proteins are generally analysed in positive-ion mode. Acidic conditions are most favourable (Ch. 6.3.1). Trifluoroacetic acid (TFA), formic acid, and acetic acid are used as additives. Counterions in solution may cause a shift in the charge-state distribution to higher m/z because of ion-pair formation and neutralization of the positive charges [30]. TFA is especially notorious in this respect. Increasing concentrations of ammonium acetate also results in a slight shift to higher m/z, probably due to competition for protons between the basic sites in the protein and the ammonia.

In order to solubilize many proteins, especially membrane proteins, surfactants and/or chaotropes like urea or guanidine are required. These additives are generally not applicable in ESI-MS. The influence of surfactants on the ESI-MS response of proteins was studied in detail [31-33]. Most cationic, Zwitterionic, and nonionic surfactants can be applied in concentrations up to 0.01%. At higher concentrations, significant signal suppression is observed. Most anionic surfactants, $e.g.$, sodium dodecylsulfate (SDS), show significant signal suppression even at 0.01%. Typical background ions resulting from the presence of various detergents were tabulated [31]. Maximum tolerable concentrations were established for some surfactants [32].

Recommended conditions for ESI-MS of proteins comprise samples that are free of surfactants, chaotropes, polyethylene glycol, and salts, especially nonvolatile salts (5 mmol/l of sodium chloride already hampers analysis). Buffers like phosphate, TRIS, and HEPES cannot be used. Volatile buffers, $e.g.$, 1-10 mmol/l ammonium carbonate or acetate, can be applied. The first choice of solvent is 50% water/acetonitrile containing 0.1–1% formic acid. Other water/organic mixtures have been successfully applied, $e.g.$, with methanol, 2-propanol, or chloroform/methanol. Pure water can be applied, but better response and stability is achieved in the presence of as little as 5–10% organic solvent. Samples containing significant amounts of salts must be desalted, $e.g.$, using solid-phase extraction (SPE) on a reversed-phase cartridge or a ZipTip® (Ch. 17.4.2).

Instrumental parameters also influence the observed charge-state distribution. An increasing nozzle-skimmer potential difference (NSP) results in a shift towards higher m/z due to charge-stripping [22], $i.e.$, collision-induced gas-phase proton exchange between protein and solvent constituents present in this relatively high-pressure region. Countercurrent gas may have a similar influence.

A wide variety of studies have been performed to determine the molecular weight of intact proteins. At an early stage, Feng and Konishi [34] explored the upper mass limit of ESI-MS in the analysis of antibodies and other large glycoproteins. They determined the molecular weight of human complement component C4 at 196,863 ± 29 Da. More recently, van Berkel *et al.* [35] passed this limit by showing the 47+-ion of the octamer of vanillyl-alcohol oxidase (~510 kDa) and the 72+-ion of its dimer as well (mass accuracy better than 0.15%!).

Direct-infusion ESI-MS provides molecular-weight determination of intact proteins and can therefore be useful in rapid characterization of recombinant proteins [36], protein isomers and conformers [37], and isoenzymes [38].

3.2   Reversed-phase LC–MS of proteins

Despite the limited tolerance to nonvolatile additives, ESI-MS is a powerful tool for protein and peptide characterization by direct infusion, especially when static nano-ESI is used ([39], Ch. 17.2). However, in LC–MS, the situation is more complex, because the solvent composition must be a compromise between optimum conditions for LC separation and ESI. The recent review by García [40] and the references therein are useful additional reading on this topic.

The most widely applied LC method for the separation of peptides and proteins is reversed-phase LC (RPLC). A typical mobile-phase composition is an acetonitrile–water gradient with a fixed concentration of TFA (typically 0.05–0.5%). TFA acts as an ion-pairing agent enhancing the retention of peptides and proteins, but also masks secondary interactions with the silica-based stationary phase. TFA is a volatile additive, but due to its ion-pairing properties and effect on the surface tension, it may significantly suppress the ESI response in positive-ion mode.

Apffel *et al.* [41] proposed the use of a post-column sheath liquid of propionic acid–2-propanol (3:1, v/v) to reduce the detrimental effect of TFA in the mobile phase. This approach may result in 10–100-fold improvement in S/N, but is only effective with peptides and smaller proteins.

Huber and Premstaller [42] studied the influence of 0.1–0.5% TFA, formic acid, and acetic acid on the separation and the detectability of eight proteins ranging in molecular weight between 14 and 80 kDa (ribonuclease A (RIB), cytochrome *c* (CYT), lysozyme (LYS), trypsin (TRY), myoglobin (MYO), α-lactalbumin (LALB), β-lactoglobulin B (LAC B), and carbonic anhydrase (CAH)). They used a 1-mm-ID column packed with a highly-crosslinked octadecylated poly(styrene–divinylbenzene) (PS–DVB) copolymer (2.3-μm particles), *i.e.*, a non-silica-based material. Substitution of TFA by formic acid resulted in a 35–160-fold improvement in the detection limits at the expense of only 32–104% increase in peak width at half height. The separation of the protein mixture under optimized conditions is shown in Figure 16.4. In a subsequent study [43], they studied the LC conditions in more detail. Best chromatographic separation, but poorest MS response was achieved with mobile phases containing heptafluorobutyric acid (HFBA), while best S/N for proteins was obtained in 0.1% formic acid. However, 0.05% TFA is the best compromise between LC and MS performance.

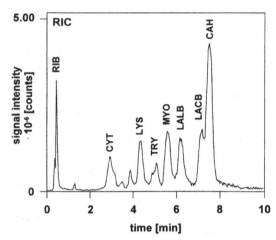

Figure 16.4: LC–MS chromatogram of eight proteins (see text, 0.7–1.9 pmol each). Mobile-phase gradient: 10–75% B in A in 10 min, with (A) 0.50% formic acid, acetonitrile–water (15:85), and (B) 0.50% formic acid, acetonitrile–water (60:40). Column is 60×1-mm-ID PS–DVB-C$_{18}$ (2.3-μm particles), flow-rate 50 μl/min, and temperature 90°C. Reprinted from [42] with permission. ©1999, Elsevier Science.

The influence of various mobile-phase additives on separation and detection of proteins after RPLC–MS was systematically studied by García *et al.* [44]. First, the response of myoglobin, cytochrome *c*, and bovine serum albumin (BSA) in ESI-MS was evaluated by column-bypass injections. The best response was achieved using 0.2% formic acid, followed by 0.3% acetic acid, 10 mmol/l ammonium formate (pH 3), and 50 mmol/l ammonium bicarbonate (pH 9). Poor responses were achieved with TFA, and 10 mmol/l ammonium formate or acetate (pH 6). Low additive concentrations were favourable, except for ammonium bicarbonate. Separation of these proteins in RPLC could only be achieved using TFA, formic and acetic acid as additive. Formic acid showed poor recovery of the proteins from the column, while TFA resulted in signal suppression. Therefore, the use of acetic acid was preferred.

Decreasing the TFA content of the mobile phase is a useful strategy to reduce signal suppression effects. To this end, a mixture of 0.02% TFA and 0.5% acetic acid was used by Clarke *et al.* [45]. Another approach would be modification of the silica-based RPLC material in order to reduce the secondary interactions, as is achieved in for instance PepMap materials, or the use of polymeric materials. In that respect, the advent of monolithic columns for peptide and protein separation is an interesting development [46-47]. Only a small decrease in separation efficiency on the monolithic column was observed when the TFA concentration was reduced from 0.2% to 0.05%.

3.3   Other LC modes in LC–MS of proteins

Next to RPLC, there are other LC modes that can be applied in protein separation, *e.g.*, size-exclusion (SEC), ion-exchange (IEC), affinity (AfC), and immobilized metal-ion affinity chromatography (IMAC). However, in most cases high concentrations of nonvolatile salts have to be used to achieve the elution of proteins. All modes have been applied as the first separation step in two-dimensional (2D) LC systems for the separation of peptides (Ch. 17.5.4 and Ch. 18.3.2).

SEC provides a separation primarily based on molecular size and would be a highly attractive separation method for proteins. High salt concentrations are needed to overcome secondary interactions between proteins and the stationary phase. In addition, the separation efficiency of SEC is limited. Direct coupling of SEC and ESI-MS has been reported for the study of protein-drug interactions (Ch. 9.4).

IEC can be performed with strong or weak cation or anion exchange materials. In peptide and protein separation, strong cation exchange (SCX) is used most often. The use of LC–LC (SCX and RPLC) in combination with ESI-MS on a time-of-flight (TOF) instrument was reported for the study of protein complexes. A dual-RPLC column setup is used: One column is gradient eluted to the MS, while peptides are eluted from the SCX in the next salt concentration step to the other column [48].

The use of perfusion columns (Ch. 1.4.1) was also evaluated to speed up the LC separation of proteins prior to ESI-MS [49]. Five-fold faster analysis was reported. Due to the narrow chromatographic peaks (5–10 s), the number of protein mass spectra available for transformation is limited.

3.4   Selected applications

The LC–MS analysis of intact proteins is not widely applied. This is due to the poor efficiency of RPLC in the separation of proteins, being large molecules. Limited sensitivity can be achieved, partly due to the ion envelope of multiple-charge ions in the MS spectrum. Selected applications show the possibilities and the limitation of direct LC–MS of intact proteins.

Metallothionein (MT) is a polymorphic nonenzymatic metal-binding protein (6–7 kDa), involved in the detoxification of trace metals like Cd^{2+} and Zn^{2+}. The various isoforms can be separated by RPLC and detected by ESI-MS [50-51]. The chromatogram of horse kidney MT-α and MT-β and mass spectra taken at the apex of the peaks are shown in Figure 16.5. In the spectrum, the occurrence of mixed Cd–Zn–MT complexes is observed.

The fingerprinting of gliadins and glutenins in various wheat flours by LC–MS shows complex traces with more than 20 peaks. The two major components, *i.e.*, γ_2- and γ_3-gliadin, were identified through *N*-terminal sequencing and accurate-mass determination [52]. The components were proposed as markers to detect traces of wheat in gluten-free food preparations.

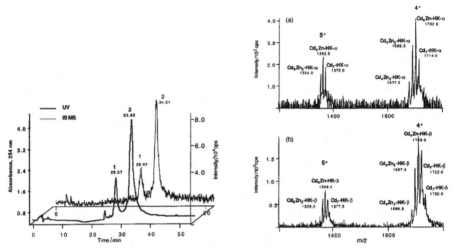

Figure 16.5: Chromatogram and mass spectra for a 500 µg/ml solution of horse kidney metallothionein, detected by UV and ESI-MS. Mass spectra are taken at the apex of peak 1 (a) and peak 2 (b). Reprinted from [50] with permission. ©1998, Royal Chemical Society.

White *et al.* [53] reported open-access LC–MS (*cf.* Ch. 9.2) for the rapid characterization of proteins. The system allows web-based sample submission and registration, automated data processing, interpretation, and report generation. The submitted amino-acid sequence is used to set parameters for the transformation of the mass spectrum into the protein molecular weight and to evaluate whether the submitted protein is actually detected. Data acquisition is performed on a TOF instrument in order to achieve sufficient mass accuracy.

For the analysis of specific proteins from complex biological samples, multiple pretreatment and isolation steps are required. Some of these steps can be performed on-line with the LC–MS.

Lacey *et al.* [54] reported the on-line combination of immunoaffinity chromatography (IAC), a short C$_4$ trapping column, and ESI-MS for the determination of transferrin (Trf) isoforms. Carbohydrate-deficient transferrin isoforms are related to congenital disorders of glycosylation (CDG), which may also occur with chronic alcohol abuse. A 25-µl serum sample is vortex-mixed with 100 µl water. A 5-µl injection is performed onto a 20×1-mm-ID IAC column containing anti-human-Trf immobilized at a POROS-aldehyde medium. A flow of 100 µl/min of PBS (pH 7.4) is applied for 2 min. The IAC column is eluted for 2 min at 100 µl/min with a 0.1 mol/l glycine buffer (set at pH 2.5 with acetic acid) to a 4×2-mm-ID C$_4$ cartridge. After washing to remove excess salt and buffer components, the Trf isoforms are eluted with 5 ml/l acetic acid in 0.2 g/l TFA–methanol–acetonitrile

(5:48:48 by volume) at 100 µl/min into ESI-MS and detected. The complete procedure takes 9.5 min per sample. The averaged mass spectrum of the Trf peak, eluting at ~7 min, is transformed. The intact dioligosaccharide tetraantennary Trf is observed at 79,561 Da, while in serum of CDG patients additional peaks are detected at 75,145 and 77,353 Da, representing losses of one or two oligosaccharide antennae (~2208 Da each).

Zhou and Johnston [55] reported protein characterization by capillary isoelectric focussing (CIEF) on-line coupled to RPLC–MS. Direct coupling of CIEF to ESI-MS is limited by interferences by the ampholytes. Inserting RPLC in-between can help removing these interferences. CIEF is performed in combination with a microdialysis membrane-based cathodic cell to remove the ampholyte and to collect protein fractions by stop-and-go CIEF prior to transfer to a 5×0.3-µm-ID C_{18} trapping column and RPLC separation on a 50×0.3-µm-ID C_4 column. The separation is performed using an acetonitrile–water gradient (0.1% acetic acid). ESI-MS is performed on a quadrupole–TOF hybrid (Q–TOF) instrument.

Feng et al. [56] reported high-throughput protein characterization for chemically modified or recombinant protein products. After preparative LC, the compounds are measured with a system consisting of eight parallel 2.1-mm-ID POROS columns connected to a eight-channel multiplexed ESI source on a TOF instrument (Ch. 5.5.3). Automated maximum-entropy software (Ch. 16.2.2) is applied to accurately determine the molecular weight of the proteins.

4. Characterization of proteins

The approaches described in the previous section enable the molecular-mass determination of intact proteins, generally with an accuracy better than 0.01%. Further structural characterization of the protein requires determination of possible post-translational modifications (PTM) as well as the amino acid sequence. In addition, issues related to tertiary and quaternary structure of the protein, the presence of cofactors, etc., may be relevant. LC–MS plays an important role in the primary and secondary structural characterization of proteins, i.e., in terms of amino-acid sequencing and PTM. The procedure generally involves chemical or enzymatic treatment of the intact protein, acquisition of a peptide map or peptide mass fingerprint by either direct infusion (nano-)ESI-MS or RPLC–MS, and the amino-acid sequencing of individual peptides by means of product-ion MS–MS. Further experiments may be needed in relation to PTM, as outlined in more detail in Ch. 19.

With respect to the use of ESI-MS, this approach was pioneered by the group of Henion [57-58]. The intact protein is digested using trypsin into smaller peptides under neutral pH conditions at 37°C. Trypsin selectively cleaves the protein at the C-terminal side of Lys or Arg (Ch. 17.4.4). The tryptic digest is separated by RPLC–MS. ESI-MS generates abundant double-charge ions for the tryptic peptides.

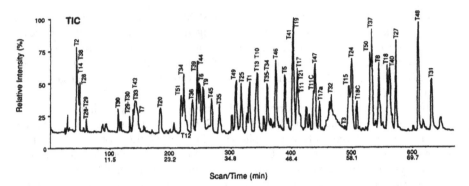

Figure 16.6: Reversed-phase electrospray LC–MS tryptic peptide map of recombinant tissue plasminogen activator. Reprinted from [60] with permission, ©1991, American Chemical Society.

These can be efficiently fragmented by collision-induced dissociation (CID) in tandem mass spectrometry (MS–MS). The procedure of enzymatic digestion and MS–MS fragmentation of the tryptic peptides was already well established for FAB, e.g., from the excellent work of Hunt et al. [59] in this area. Interpretation of the data reveals the amino-acid sequence of the individual tryptic peptides in the digest mixture. Nomenclature rules have been established to unambiguously annotate the various fragments formed under different fragmentation conditions. Fragmentation of peptides and interpretation of the MS–MS is discussed in detail in Ch. 17.6.1.

In the early LC–MS–MS studies, CID is performed with low-energy collisions by means of triple-quadrupole instruments. Under these conditions, series of N-terminal b-ions and C-terminal y-ions result from cleavages at the peptide bond and charge retention on either side, although double-charge tryptic peptide ions tend to favour fragmentation towards more abundant y-ions. From these ladders of sequence ions in the MS–MS spectrum, the amino-acid sequence of the peptide may be derived. This is the bottom-up protein identification approach (Ch. 18.3.1).

Impressive examples from the early days are the characterization of recombinant DNA derived tissue plasminogen activator (rt-PA) [60], and the characterization of a mixture of lobster digestive cysteine proteinases (23.4 kDa) [61]. The rt-PA is a glycoprotein with 527 amino acids and a molecular weight of 59,008 Da, excluding 3–9 kDa due to glycosylation at two or three sites [60]. The RPLC–MS chromatogram of its tryptic peptide map is shown in Figure 16.6.

Numerous similar studies have been performed along these lines. Impressive results have been reported. Some examples are briefly reviewed here.

The identification of endogenous peptides recognized by the human class I major histocompatibility complex (MHC) encoded human leucocyte antigen (HLA-A2.1) by microcapillary LC–MS–MS was pioneered by the group of Hunt [62]. The peptides are involved in the recognition of host cells by the human immune system and are also involved in the rejection of tissue transplants. Class I molecules can associate with thousands of different endogenous peptides. After fractionation by RPLC, individual fractions still contain 50–100 peptides, of which only one provides the specific immune response. The study involves the detection and identification of these specific nonapeptides at femtomole level. The progress in this field was reviewed by De Jong [63].

The characterization of the sequence and the heterogeneity of a humanized antibody against the interleukin 2 receptor, using of direct infusion ESI-MS, LC–MS, and MS–MS analysis of the disulfide-reduced and trypsin-digestion antibody, was studied by Lewis *et al.* [64].

The structural characterization of the protein and carbohydrate components of the anti-respiratory syncytial virus monoclonal antibody, using LC–MS for peptide mapping and selective identification of glycopeptides, Edman degradation and MS–MS for sequencing of selected peptides, MALDI-MS for molecular-weight determination of the intact protein and the characterization of endo- and exoglycosidase digests of isolated glycopeptides to identify glycosylation sites and carbohydrate structures, was investigated by Roberts *et al.* [65].

5. Protein conformation

In cells of higher organisms, the functioning of proteins is frequently determined by larger associates of several subunits. Therefore, one has to study these protein associates or complexes. An important step in these studies is the detection of the intact associates by ESI-MS [66].

In early optimization studies of ESI-MS of proteins, solvent-induced shifts of the ion envelope in the mass spectrum were observed. These early findings were correlated to solvent-induced conformational changes of the proteins [67-68]. Solvent-induced denaturation of the protein results in the exposure to the bulk solvent of a larger number of basic amino acids, which in turn leads to an increase of the charge at the protein and a shift of the ion-envelope. Hydrogen/deuterium (H/D) exchange was used by Katta and Chait [68-69] to probe conformational changes in bovine ubiquitin induced by the addition of methanol to aqueous acidic solutions. Similar studies have been done for other proteins and protein complexes by others. Topics in MS studies of protein conformation are reviewed by Kaltashov and Eyles [70]. An important tool in these studies is H/D exchange [71].

These observations in turn led to the observation of noncovalent protein complexes. The noncovalent binding between the immunosuppressive binding protein FKBP and the immunosuppressive agents FK506 and rapamycin, probed by electrospray MS, was reported by Ganem *et al.* [72] in 1991. This report and the next one on enzyme–substrate interactions with hen egg-white lysozyme [73] attracted considerable attention. The study of noncovalent interactions by ESI-MS became an important research topic. In subsequent years, many reports on noncovalent complexes between proteins and proteins, oligonucleotides, and drugs were published. The topic is reviewed by Loo [74]. The detection of noncovalent complexes by ESI-MS plays an important role in drug discovery studies (Ch. 9.4) and in the study of protein-protein interactions (Ch. 18.5).

An important tool in the study of protein conformation and noncovalent protein complexes is the on-line combination of ion-mobility spectrometry (IMS) and MS. The IMS–MS instruments consists of an ESI source with related ion optics, a drift tube, and a mass spectrometer [75-76]. Quadrupole and TOF-MS instruments have been applied most frequently. In an IMS instrument, ions drift through a buffer gas under the influence of a weak uniform electric field. The IMS separation of ions is based on differential mobility of ions related to shape and charge state. Within a particular charge state, compact ions show a higher mobility than more extended structures, because they experience fewer collisions. In this way, conformation differences between ions can be discovered. Compact ions have a smaller collision cross section.

IMS–MS can be applied to resolve different conformations of a protein, as first demonstrated by Clemmer for the 7+-ions of cytochrome c [77]. This can be combined with H/D exchange of different conformations in the gas phase of the drift tube [78]. It also shows great promise in the analysis of protein complexes, as demonstrated by a recent study by Loo *et al.* [79]. They performed IMS–MS to characterize the structure of the noncovalent 28-subunit 20S protein complex from *Methanosarcina thermophila* and rabbit. This allowed confirmation of the stoichiometries of the 192 kDa α_7-ring and the intact 690 kDa $\alpha_7\beta_7\beta_7\alpha_7$ complexes. The results suggest that the structure of large protein complexes is preserved in the gas-phase upon desolvation.

IMS–MS has also been applied to study protein aggregation into insoluble fibrils or plaques, which seems to be important in relation to neurodegenerative diseases like Alzheimer, Parkinson, and Creutzfeldt-Jacob disease [76, 80]. The two major alloforms of the naturally occurring amyloid β-proteins (Aβ) are the 42-residue form (Aβ42) and the more-abundant 40-residue form (Aβ40). Aβ42 fibrillizes faster, is more neurotoxic, and forms other oligomer populations than Aβ40. This oligomer formation seems to be linked to Alzheimer disease. IMS–MS is found to be an important tool in these investigations.

6. Conclusions and perspectives

Most studies discussed in this chapter are focussed around early developments in the field of protein characterization. In Ch. 17, the characterization of proteins by LC–MS is discussed in more detail, introducing the technologies which are at the basis of current proteomics research (Ch. 18) and the characterization of PTM (Ch. 19). As such, this is the prelude to more recent developments in the mid-1990s.

7. References

1. T.J. Yu, H.A. Schwartz, S.A. Cohen, P. Vouros, B.L. Karger, *Sequence analysis of derivatized peptides by LC–MS*, J. Chromatogr., 301 (1984) 425.
2. J.G. Stroh, J. Carter Cook, R.M. Milberg, L. Brayton, T. Kihara, Z. Huang, K.L. Rinehart, Jr., I.A.S. Lewis, *On-line LC–FAB-MS*, Anal. Chem., 57 (1985) 985.
3. C.R. Blakley, M.L. Vestal, *TSP interface for LC–MS*, Anal. Chem., 55 (1983) 750.
4. D. Pilosof, H.-Y. Kim, D.F. Dyckes, M.L. Vestal, *Determination of nonderivatized peptides by TSP LC–MS*, Anal. Chem., 56 (1984) 1236.
5. C.K. Meng, M. Mann, J.B. Fenn, Presented at the 36th ASMS Conference on Mass Spectrometry and Allied Topics, June 5-10, 1988, San Francisco, CA, p. 771.
6. M. Mann, C.K. Meng, J.B. Fenn, *Interpreting mass spectra of multiply charged ions*, Anal. Chem., 61 (1989) 1702.
7. K. Straub and K. Chan, *Molecular weight determination of proteins from multiply charged ions using thermospray ionization mass spectrometry*, Rapid Commun. Mass Spectrom., 4 (1990) 267.
8. R.M. Caprioli, T. Fan, J.S. Cottrell, *Continuous-flow sample probe for FAB-MS*, Anal. Chem., 58 (1986) 2949.
9. K. Mock, J. Firth, J.S. Cottrell, *Application of on-line HPLC–FAB MS to the analysis of enzymatic digests of ribonuclease B*, Org. Mass Spectrom., 24 (1989) 591.
10. R.D. Smith, J.A. Loo, C.G. Edmonds, C.J. Barinaga, H.R. Udseth, *New developments in biochemical MS: ESI*, Anal. Chem., 62 (1990) 882.
11. M. Mann, *ESI: Its potential and limitations as an ionization method for biomolecules*, Org. Mass Spectrom., 25 (1990) 575.
12. S.A. Carr, M.E. Hemling, M.F. Bean, G.D. Roberts, *Integration of MS in analytical biotechnology*, Anal. Chem., 63 (1991) 2802.
13. R.D. Smith, J.A. Loo, R.R. Ogorzalek Loo, M. Busman, H.R. Udseth, *Principles and practice of ESI-MS for large polypeptides and proteins*, Mass Spectrom. Rev., 10 (1991) 359.
14. T.R. Covey, R.F. Bonner, B.I. Shushan, J.D. Henion, *The determination of protein, oligonucleotide and peptide molecular weights by ESI-MS*, Rapid Commun. Mass Spectrom., 2 (1988) 249.
15. B.B. Reinhold, V.N. Reinhold, *ESI-MS: Deconvolution by an entropy-based algorithm*, J. Am. Soc. Mass Spectrom., 3 (1992) 207.
16. A.G. Ferrige, M.J. Seddon, S. Jarvis, *Maximum entropy deconvolution in ESI-MS*, Rapid Commun. Mass Spectrom., 5 (1991) 374.
17. A.G. Ferrige, M.J. Seddon, B.N. Green, S.A. Jarvis, J. Skilling, *Disentangling ESI spectra with maximum entropy*, Rapid Commun. Mass Spectrom., 6 (1992) 707.
18. B.N. Green, S.N. Vinogradov, *An ESI-MS study of the subunit structure of the giant hemoglobin from the leech Nephelopsis oscura*, J. Am. Soc. Mass Spectrom., 15 (2004) 22.
19. M.A. Kelly, M.M. Vestling, C.C. Fenselau, P.B. Smith, *ESI analysis of proteins: A comparison of positive-ion and negative-ion mass spectra at high and low pH*, Org. Mass Spectrom., 27 (1992) 1143.
20. J.B. Fenn, *Ion formation from charged droplets: roles of geometry, energy, and time*, J. Am. Soc. Mass Spectrom., 4 (1993) 524.
21. S.F. Wong, C.K. Meng, J.B. Fenn, *Multiple charging in ESI of poly(ethylene glycols)*,

J. Phys. Chem., 92 (1988) 546.

22. D.S. Ashton, C.R. Beddell, D.J. Cooper, B.N. Green, R.W.A. Oliver, *Mechanism of production of ions in ESI-MS*, Org. Mass Spectrom., 28 (1993) 721.

23. J.B. Fenn, *Ion formation from charged droplets: roles of geometry, energy, and time*, J. Am. Soc. Mass Spectrom., 4 (1993) 524.

24. R. Guevremont, K.W.M. Siu, J.C.Y. Le Blanc, S.S. Berman, *Are the ESI mass spectra of proteins related to their aqueous solution chemistry?*, J. Am. Soc. Mass Spectrom., 3 (1992) 216.

25. K.W.M. Siu, R. Guevremont, J.C.Y. Le Blanc, R.T. O'Brien, S.S. Berman, *Is droplet evaporation crucial in the mechanism of ESI-MS*, Org. Mass Spectrom., 28 (1993) 579.

26. R. Guevremont, J.C.Y. Le Blanc, K.W.M. Siu, *ESI-MS - Ethylene glycol as a solvent and its effects on ion desorption*, Org. Mass Spectrom., 28 (1993) 1345.

27. C.L. Gatlin, F. Tureček, *Acidity determination in droplets formed by electrospraying methanol–water solutions*, Anal. Chem., 66 (1994) 712.

28. B.A. Mansoori, D.A. Volmer, R.K. Boyd, *'Wrong-way-around' ESI of amino acids*, Rapid Commun. Mass Spectrom., 11 (1997) 1120.

29. A.T. Iavarone, J.C. Jurchen, E.R. Williams, *Supercharged protein and peptide ions formed by ESI*, Anal. Chem., 73 (2001) 1455.

30. U.A. Mirza, B.T. Chait, *Effects of anions on the positive ion ESI mass spectra of peptides and proteins*, Anal. Chem., 66 (1994) 2898.

31. R.R. Ogorzalek Loo, N. Dales, P.C. Andrews, *Surfactant effects on protein structure examined by ESI-MS*, Protein Sci, 3 (1994) 1975.

32. J. Funk, X. Li, T. Franz, *Threshold values for detergents in protein and peptide samples for MS*, Rapid Commun. Mass Spectrom., 19 (2005) 2986.

33. Y. Ishihama, H. Katayama, N. Asakawa, *Surfactants usable for ESI-MS*, Anal. Biochem., 287 (2000) 45.

34. R. Feng and Y. Konishi, *Analysis of antibodies and other large glycoproteins in the mass range of 150,000–200,000 Da by electrospray ionization mass spectrometry*, Anal. Chem., 64 (1992) 2090.

35. W.J.H. van Berkel, R.H.H. van den Heuvel, C. Versluis, A.J.R. Heck, *Detection of intact megaDalton protein assemblies of vanillyl-alcohol oxidase by MS*, Protein Sci., 9 (2000) 435.

36. A. Van Dorsselaer, F. Bitsch, B.N. Green, S.A. Jarvis, P. Lepage, R. Bischoff, H.V.J. Kolbe, C. Roitsch, *Application of ESI-MS to the characterization of recombinant proteins up to 44 kDa*, Biomed. Environ. Mass Spectrom., 19 (1990) 692.

37. T.W. Muit, M.J. Williams, S.B.H. Kent, *Detection of synthetic protein isomers and conformers by ESI-MS*, Anal. Biochem., 224 (1995) 100.

38. P. Rouimi, L. Debrauwer, J. Tulliez, *ESI-MS as a tool for characterization of glutathione S-transferase isoenzymes*, Anal. Biochem., 229 (1995) 304.

39. M.S. Wilm, M. Mann, *Analytical properties of the nano-ESI ion source*, Anal. Chem., 68 (1996) 1.

40. M.C. García, *The effect of the mobile phase additives on sensitivity in the analysis of peptides and proteins by LC–ESI-MS*, J. Chromatogr. B, 825 (2005) 111.

41. A. Apffel, S. Fischer, G. Goldberg, P.C. Goodley, F.E. Kuhlmann, *Enhanced sensitivity for peptide mapping with LC–ESI-MS in the presence of signal suppression due to TFA-containing mobile phases*, J. Chromatogr. A, 712 (1995) 177.

42. C.G. Huber, A. Premstaller, *Evaluation of volatile eluents and electrolytes for LC–ESI-MS and capillary electrophoresis–ESI-MS of proteins. I. LC*, J. Chromatogr. A, 849 (1999) 161.

43. W. Walcherm, H. Toll, A. Ingendoh, C.G. Huber, *Operational variables in LC–ESI-MS of peptides and proteins using PS-DVB monoliths*, J. Chromatogr. A, 1053 (2004) 107.

44. M.C. García, A.C. Hogenboom, H. Zappey, H. Irth, *Effect of the mobile phase composition on the separation and detection of intact proteins by RPLC–ESI-MS*, J. Chromatogr. A, 957 (2002) 187.

45. N.J. Clarke, F.W. Crow, S. Younkin, S. Naylor, *Analysis of in vivo-derived amyloid-b polypeptides by on-line 2D-LC–MS*, Anal. Biochem., 298 (2001) 32.

46. A. Premstaller, H. Oberacher, W. Walcher, A.M. Timperio, L. Zolla, J.-P. Chervet, N. Cavusoglu, A. van Dorsselaer, C.G. Huber, *LC–ESI-MS using monolithic capillary columns for proteomic studies*, Anal. Chem., 73 (2001) 2390.

47. W. Walcher, H. Oberacher, S. Troiani, G. Hölzl, P. Oefner, L. Zolla, C.G. Huber, *Monolithic capillary columns for LC–ESI-MS in proteomic and genomic research*, J. Chromatogr. B, 782 (2002) 111.

48. H. Liu, S.J. Berger, A.B. Chakraborty, R.S. Plumb, S.A. Cohen, *Multidimensional LC coupled to ESI-TOF-MS as an alternative to 2D gels for the identification and analysis of complex mixtures on intact proteins*, J. Chromatogr. B, 782 (2002) 267.

49. J.F. Banks, Jr., *Separation and analysis of proteins by perfusion liquid chromatography and electrospray ionization mass spectrometry*, J. Chromatogr. A, 691 (1995) 325.

50. H. Chassaigne, R. Lobinski, *Characterization of horse kidney metallothionein isoforms by ESI-MS and RPLC–ESI-MS*, Analyst, 123 (1998) 2125.

51. H. Chassaigne, R. Lobinski, *Polymorphism and identification of metallothionein isoforms by RPLC with on-line ESI-MS detection*, Anal. Chem., 70 (1998) 2536.

52. G. Mamone, P. Ferranti, L. Chianese, L. Scafuri, F. Addeo, *Qualitative and quantitative analysis of wheat gluten proteins by LC and ESI-MS*, Rapid Commun. Mass Spectrom., 14 (2000) 897.

53. W.L. White, C.D. Wagner, J.T. Hall, E.E. Chaney, B. George, K. Hofmann, L.A.D. Miller, J.D. Williams, *Protein open-access LC–MS*, Rapid Commun. Mass Spectrom., 19 (2005) 241.

54. J.M. Lacey, H.R. Bergen, M.J. Magera, S. Naylor, J.F. O'Brien, *Rapid determination of transferrin isoforms by IAC and ESI-MS*, Clin. Chem., 47 (2001) 513.

55. F. Zhou, M.V. Johnston, *Protein characterization by on-line CIEF, RPLC, and MS*, Anal. Chem., 76 (2004) 2734.

56. B. Feng, M.S. McQueney, T.M. Mezzasalma, J.R. Slemmon, *An integrated ten-pump, eight-channel, LC–MS system for automated high-throughput analysis of proteins*, Anal. Chem., 73 (2001) 5691.

57. E.C. Huang, J.D. Henion, *LC–MS and LC–MS–MS determination of protein tryptic digest*, J. Am. Soc. Mass Spectrom., 1 (1990) 158.

58. T.R. Covey, E.C. Huang, J.D. Henion, *Structural characterization of protein tryptic peptides via LC–MS and CID of their doubly charged molecular ions*, Anal. Chem., 63 (1991) 1193.

59. D.F. Hunt, J.R. Yates, III, J. Shabanowitz, S. Winston, C.R. Hauer, *Protein sequencing by MS–MS*, Proc. Natl. Acad. Sci. USA, 83 (1986) 6233.

60. V. Ling, A.W. Guzzetta, E. Canova-Davis, J.T. Stults, W.S. Hancock, T.R. Covey,
     B.I. Shushan, *Characterization of the tryptic map of recombinant DNA derived tissue
     plasminogen activator by LC–ESI-MS*, Anal. Chem., 63 (1991) 2909.
61. P. Thibault, S. Pleasance, M.V. Laycock, R.M. MacKay, R.K. Boyd,
     *Characterization of a mixture of lobster digestive cysteine proteinases by ESI-MS and
     tryptic mapping with LC–MS and LC–MS–MS*, Int. J. Mass Spectrom. Ion Processes,
     111 (1991) 317.
62. R.A. Henderson, A.L. Cox, K. Sakaguichi, E. Appella, J. Shabanowitz, D.F. Hunt,
     V.H. Engelhard, *Direct identification of an endogeneous peptide recognized by
     multiple HLA-A2.1 specific cytotoxic T cells*, PNAS, 90 (1993) 10275.
63. A.P.J.M. de Jong, *Contribution of MS to contemporary immunology*, Mass Spectrom.
     Rev., 17 (1998) 311.
64. D.A. Lewis, A.W. Guzzetta, W.S. Hancock, *Characterization of humanized anti-
     TAC, an antibody directed against the interleukin 2 receptor, using ESI-MS by direct
     infusion, LC–MS, and MS–MS*, Anal. Chem., 66 (1994) 585.
65. G.D. Roberts, W.P. Johnson, S. Burman, K.R. Anumula, S.A. Caw, *An integrated
     strategy for structural characterization of the protein and carbohydrate components
     of monoclonal antibodies: application to anti-respiratory syncytial virus MAb*, Anal.
     Chem., 67 (1995) 3613.
66. A.J.R. Heck, R.H.H. van den Heuvel, *Investigation of intact protein complexes by
     MS*, Mass Spectrom. Rev., 23 (2004) 368.
67. J.A. Loo, R.R. Ogorzalek Loo, H.R. Udseth, C.G. Edmonds, R.D. Smith, *Solvent-
     induced conformational changes of polypeptides probed by ESI-MS*, Rapid Commun.
     Mass Spectrom., 5 (1991) 101.
68. V. Katta, B.T. Chait, *Conformational changes in proteins probed by H/D exchange
     ESI-MS*, Rapid Commun. Mass Spectrom., 5 (1991) 214.
69. V. Katta, B.T. Chait, *H/D exchange ESI-MS: a method for probing protein
     conformational changes in solution*, J. Am. Chem. Soc., 115 (1993) 6317.
70. I.A. Kaltashov, S.J. Eyles, *Studies of biomolecular conformations and conformational
     dynamics by MS*, Mass Spectrum. Rev., 21 (2002) 37.
71. X. Yan, J. Watson, P.S. Ho, M.L. Deinzer, *MS approaches using ESI charge states
     and H/D exchange for determining protein structures and their conformational
     changes*, Mol Cell Proteomics, 3 (2004) 10.
72. B. Ganem, Y.-T. Li, J.D. Henion, *Detection of noncovalent receptor-ligand
     complexes by MS*, J. Am. Chem. Soc., 113 (1991) 6294.
73. B. Ganem, Y.-T. Li, J.D. Henion, *Observation of non-covalent enzyme-substrate and
     enzyme-product complexes by ESI-MS*, J. Am. Chem. Soc., 113 (1991) 7818.
74. J.A. Loo, *ESI-MS: a technology for studying noncovalent macromolecular
     complexes*, Int. J. Mass Spectrom., 200 (2000) 175.
75. D.E. Clemmer, M.F. Jarrold, *Ion mobility measurements and their applications to
     clusters and biomolecules*, J. Mass Spectrom., 32 (1997) 577.
76. T. Wyttenbach, E.S. Baker, S.L. Bernstein, A. Frezoco, J. Gidden, D. Liu, M.T.
     Bowers, *The IMS–MS method and its application to duplex formation of
     oligonucleotides and aggregation of proteins*, in: A.E. Ashcroft, G. Brenton, J.J.
     Monaghan (Eds.), Advances in Mass Spectrometry, Volume 16, 2004, Elsevier
     Science, p. 189.
77. D.E. Clemmer, R.R. Hudgins, M.F. Jarrold, *Naked protein conformations:*

cytochrome c in the gas phase, J. Am. Chem. Soc., 117 (1995) 10141.

78. S. J. Valentine, D. E. Clemmer, *H/D Exchange levels of shape-resolved cytochrome c conformers in the gas phase*, J. Am. Chem. Soc., 119 (1997) 3558.

79. J.A. Loo, B. Berhane, C.S. Kaddis, K.M. Wooding, Y. Xie, S.L. Kaufman, I.V. Chernushevich, *ESI-MS and IMS analysis of the 20S proteasome complex*, J. Am. Soc. Mass Spectrom., 16 (2005) 998.

80. S.L. Bernstein, T. Wyttenbach, A. Baumketner, J.-E. Shea, G. Bitan, D.B. Teplow, M.T. Bowers, *Amyloid β-protein: Monomer structure and early aggregation states of Aβ42 and its Pro^{19} alloform*, J. Am. Chem. Soc., 127 (2005) 2075.

17

LC–MS ANALYSIS OF PEPTIDES
ENABLING TECHNOLOGIES

1. Introduction

Soon after the introduction of electrospray ionization (ESI) and matrix-assisted laser desorption ionization (MALDI) for mass spectrometry (MS) in 1988, these techniques were intensively applied to the characterization of peptides and proteins. After initial studies, discussed in Ch. 16, significant new developments started in the mid-1990s. Step-by-step developments of technologies and procedures of protein characterization brought us to where we stand today. Nowadays, high-throughput liquid chromatography–mass spectrometry (LC–MS) analysis of tryptic digests and multidimensional LC–MS of digests of the complete proteome of a bacteria or of human plasma are routinely performed in many laboratories. Interpretation of product-ion MS–MS spectra of individual tryptic peptides has been replaced by automated database searching of multidimensional data sets. Currently, ESI-MS and MALDI-MS play a prominent role in the field of proteomics. The advent and development of these enabling technologies are highlighted in this chapter. This chapter is a technology overview rather than a comprehensive review of current applications of LC–MS of peptides.

2. Nanoelectrospray

Early ESI-MS studies involving proteins were generally done using flow-rates of 1–10 µl/min [1]. Because often only limited amounts of protein are available, reduction of the flow-rate is advantageous. Andrén et al. [2] demonstrated ESI-MS of peptides with flow-rates of 0.2–6 µl/min using microelectrospray (micro-ESI) needles with an internal diameter of 50 µm and an emitter tip diameter of 10–20 µm. With the zeptomole per microliter sensitivity achieved, they clearly illustrated one of the major advantages of reducing the flow-rate in ESI-MS.

Subsequently, nano-ESI was introduced [3]. Initially, the nano-ESI needles were made of gold-coated pulled glass capillaries with an emitter tip diameter of 1–3 µm. The needle is filled with 0.2–2 µl of liquid and positioned 1–2 mm from the ion sampling orifice in the ESI source. With a needle potential of ~1–2 kV, an electrospray can be generated with a flow-rate of ~20 nl/min.

Besides the improvements in sensitivity and the reduction of sample consumption, nano-ESI shows several other advantages over conventional ESI [4]. Stable performance of ESI-MS with aqueous samples is easier to achieve. There is greater tolerance towards salts and buffers, almost allowing physiological conditions to be used, and enhanced performance for certain compound classes, e.g., noncovalent protein complexes. These effects can be explained in terms of the reduced number of fissions to off-spring droplets and improved surface-to-volume ratio, resulting from the smaller initial droplet size in nano-ESI relative to conventional ESI.

Nano-ESI can be performed via direct infusion of a sample deposited directly in the needle. This static sample analysis allows prolonged continuous spraying (as long as 45 min). This results in improved signal-to-noise ratio (S/N) by signal averaging and/or the ability to perform a wide variety of MS and MS–MS experiments, e.g., time-consuming experiments like sequencing a multitude of components in complex (tryptic) peptide mixture by multi-stage MS–MS fragmentation in an ion-trap instrument [5] or by data-dependent acquisition (DDA) on a quadrupole–time-of-flight hybrid (Q–TOF) instrument [6].

Alternatively, nano-ESI is used in continuous-flow mode by injecting samples in a mobile-phase flow or in on-line combination with nano-LC (Ch. 17.5.2) and other low-flow separation methods (Ch. 6.6.2 and Ch. 17.5.5–6). As such, nano-ESI plays an important role in the progress in the field of protein characterization.

A versatile approach to nano-ESI is the use of a microchip containing an array of 10×10 nano-ESI emitter tips, typically 8 µm ID. A sample-handling robot performs continuous infusion of samples from a 96-well or 384-well plate using a new emitter tip at the chip for each individual sample. This system is commercially available from Advion BioSciences and has found wide application. An example is the identification of proteins excised from a 2D gel electrophoresis gel and in-gel digested prior to infusion [7]. Implementation of ZipTip®-based sample pretreatment strategies in combination with these nano-ESI microchips was reported as well [8].

3. Proteins from gel electrophoresis

One- and two-dimensional gel electrophoresis (1D- or 2D-GE) is an important tool in the separation and isolation of intact proteins [9]. In 1D-GE, the proteins are separated in a sodium dodecylsulfate poly(acrylamide) gel (SDS-PAGE). The separation is according to molecular weight. In 2D-GE, the proteins are first separated by isoelectric point (pI, isoelectric focussing, IEF), and next by molecular weight. 2D-GE is considered to be the most powerful tool in protein separation. Nevertheless, the technique suffers from problems: it is labour-intensive, analysis time is long, and the reproducibility poor. Furthermore, hydrophobic proteins do not behave well in the first IEF step and tend to form broad bands.

After the separation, the proteins are visualized by staining using Coomassie Brilliant Blue or silver. This represents another limitation of 2D-GE: staining results in a relatively narrow dynamic range, i.e., the relative difference between most and least abundant protein detected. Low-abundance proteins may be completely obscured by high-abundance proteins like albumin and immunoglobulins. For visualization, ~10 femtomoles of protein is needed. Procedures have been reported for the selective removal of the high-abundance proteins prior to 2D-GE [10-11].

(Electro)blotting techniques have been developed to transfer specific proteins from the gel to a nitrocellulose or polyvinylidenefluoride (PVDF) membrane for further handling prior to MS analysis, e.g., cutting spots of interest for extraction and in-solution digestion, or on-membrane enzymatic digestion. Electroelution into a buffer solution can also be applied. Alternatively, the part of the gel containing the protein of interest is excised from the gel, the staining is removed, and in-gel digestion is performed. After digestion, the peptides are extracted from the gel prior to MS analysis. Impressive results have been achieved with this procedure, e.g., in applying either MALDI-MS or nano-ESI-MS in combination with database searching. An illustrative example is the identification of the yeast proteome [12].

Because Coomassie staining is a slow process and it may interfere in MS analysis, silver staining has been proposed as an alternative [13]. Silver staining improves the detection limit from ~50 ng to ~1–10 ng It is easier to remove as it does not act by specific binding to the protein.

The presence of high SDS concentrations in the excised gel spots is not favourable, because SDS may significantly suppress the signal in ESI-MS. Various tools were developed to remove SDS prior to LC–MS (Ch. 17.4.2). Alternatively, an acid-labile surfactant (ALS) can be applied instead of SDS [14].

Despite its general applicability and separation power, 2D-GE does not provide a global picture of the proteome, even apart from the dynamic range issue. Some classes of proteins are excluded, e.g., very small or very large proteins (>100 kDa), very acidic or very basic proteins, i.e., with pI<3.5 or pI>9.5, and highly hydrophobic proteins, such as membrane proteins. It has been estimated that these restrictions apply to 50% of the proteins expressed.

Another issue related to the use of GE is the possible formation of acrylamide adducts with Cys residues [15]. The covalent modification of Cys by acrylamide leads to a shift in the residue mass (from 103 to 174 Da, *cf.* Table 17.1).

4. Sample pretreatment

Procedures have been reported to selectively enrich samples for specific proteins and/or to fractionate proteomes [16]. This may involve fractionation of cells in subcellular fractions, *e.g.*, organelles, or isolation of protein complexes and/or specific protein enrichment strategies. Strategies for revealing low-abundance proteins have been reviewed [17]. In addition to prefractionation procedures, applied prior to LC–MS or 2D-LC–MS, stable-isotope labelling to some extent facilitates selection of specific peptides within the mass spectrometer (Ch. 18.4.1–2).

4.1   Selective enrichment of proteins

Specific protein enrichment strategies are based on protein-interaction and/or (immuno)affinity methods. This often involves immobilized affinity tags [18].
• Avidin and streptavidin loaded affinity columns (AfC) are applied for selective enrichment of biotinylated proteins and peptides (*e.g.*, Ch. 18.4.1–2).
• Immobilized monoclonal or polyclonal antibodies have been used to selectively enrich specific proteins or protein complexes (Ch. 18.4.5 and 18.5.1).
• Multiple polyclonal IAC has been applied for the selective removal of high-abundance proteins such as human serum albumin and immunoglobulins, from human plasma samples [10-11]. The approach has been extended to immunodepletion of transferrin, haptoglobin, and antitrypsin as well and commercialized [19]. The effectiveness of some commercial kits in plasma and cerebrospinal fluid was evaluated by Ramström *et al.* [20].
• Immobilized metal-ion affinity chromatography (IMAC) is a powerful technique for the selective enrichment of phosphopeptides (Fe^{3+} and Ga^{3+}, Ch. 18.3.2), or His-containing peptides (Cu^{2+}, Ni^{2+}).
• Lectins like concanavalin A and wheat germ agglutinin are used for selective glycoprotein enrichment [21-22].

4.2   Desalting and solvent switching

Given the limited tolerance of ESI-MS to nonvolatile additives, considerable effort has been put in developing technologies for sample purification. Various approaches have been described:
• Off-line solid-phase extraction (SPE) procedures, based on cartridges, Empore disks, or ZipTip® procedures, are frequently applied for clean-up of peptide mixtures prior to ESI-MS or LC–MS analysis. This was evaluated by Erdjument-Bromage *et al.* [23].

- Off-line desalting by SPE on Waters Oasis® HLB material in 96-well-plate format was reported by Gilar *et al.* [24].
- On-line SPE–ESI-MS: Injection onto a short trapping column, washing away of the salts and/or SDS, and elution of the proteins into the ESI source [25-26]. An integrated device for on-column enrichment and desalting via an exit vent was reported as well (valveless SPE–LC) [27].
- Injection of the sample onto an analytical column, from which the buffer is washed away before the LC gradient is started [28]. This procedure also allows relatively large injection volumes by on-column focussing.
- On-line microdialysis desalting of protein solutions [29] or in-between LC and ESI-MS [30-31].

4.3   Derivatization

Peptide and protein derivatization strategies have been applied to improve the chromatographic properties, enhance the response in ESI-MS, change the fragmentation characteristics in MS–MS, or facilitate data interpretation, *e.g.*, in *de novo* sequencing (Ch. 17.6.3). A variety of derivatization strategies aiming at the implementation of stable-isotope labels for quantitative proteomics is developed (Ch. 18.4.1–2). The widely applied ICAT labelling strategy (Ch. 18.4.1) implements an affinity tag enabling the selective enrichment of particular peptides from complex mixtures. An example of peptide derivatization to enhance the chromatographic properties is the *N*-terminal benzoyl formation of hydrophilic peptides [32].

4.4   Enzymatic digestion

Although a variety of (specific) proteolytic enzymes are available, trypsin is applied in most studies. Trypsin cleaves the protein at the *C*-terminal side of Lys or Arg, unless the next *C*-terminal amino acid is Pro. Other missed cleavage sites occur as well. Tryptic peptides are often indicated with Tx, where x is the number of the tryptic peptide, counted from the *N*-terminal side of the protein. The reaction can be performed in various ways: homogeneous digestion, with trypsin immobilized on beads, or by means of a trypsin immobilized enzyme reaction (IMER).

Homogeneous digestion is performed in solution at pH 7.5, using a phosphate or ammonium bicarbonate buffer, and 37 °C, and takes 6–24 hours. In order to limit the autodigestion of trypsin, the trypsin-to-substrate ratio must be kept low. For many studies, reduction and alkylation of the protein must be performed prior to proteolytic digestion. Reduction of the Cys–Cys disulfide bridges in the protein by means of β-mercaptoethanol or dithiothreitol (DTT) results in a better accessibility of the protease. Subsequent alkylation using iodoacetamide is required to protect the reactive Cys-SH groups. The influence of organic solvents like methanol, acetone, 2-propanol, and acetonitrile in aqueous–organic solvent mixtures on protein digestion rate and efficiency was evaluated as well [33].

For complex mixtures, proteins must be solubilized in a denaturing and reducing solvent, *e.g.*, 8 mol/l urea, 10 mmol/l DTT, and a TRIS buffer. After alkylation, the mixture is diluted and subjected to overnight trypsin digestion.

In-gel digestion is a widely applied technique in off-line coupling of GE and either nano-ESI [12-13] or LC–MS [34] (Ch. 17.3). Tools have been developed for interfacing in-gel digestion and capillary LC–MS, *e.g.*, electroextraction and trapping on a strong cation exchange (SCX) column [35], on-line in-gel digestion in a cartridge in stop-flow mode and subsequent solvent extraction and concentration on a SPE trapping column, prior to LC–MS analysis [36].

The presence of SDS significantly reduces the efficiency of tryptic digestion. In a comparison between SDS and ALS, it was found that SDS results in a 80% reduction of the trypsin activity, while with ALS only 1% reduction is observed [37]. During in-solution digestion, it acts as a mild protein denaturant, thereby enhancing the digestion. It is compatible with ESI-MS as well.

Digestion can also be achieved using a trypsin IMER, where trypsin is immobilized to a solid support, *e.g.*, macroporous silica [38], on POROS material (Porozyme IMER) [39-40], a PVDF membrane in a microreactor [41], or silica-based [42] or porous polymer monoliths [43-45].

A trypsin IMER has several advantages over homogeneous digestion. It allows a larger enzyme-to-substrate ratio. This results in a enhanced digestion efficiency. Reduced and alkylated proteins from *E. coli* have been digested within 20 min using a trypsin column operated at elevated temperatures [46]. Moreover, the enzyme on an IMER can be used repeatedly. Generally, the denaturation rate is lower. As an IMER can be part of a multidimensional system, less manual handling reduces the risks of sample contamination.

The trypsin IMER can be applied either in off-line [38] or on-line mode [47] with ESI-MS, or with SPE–LC–MS [42, 45], or as part of a multidimensional on-line system (Ch. 17.5.4). The on-line digestion by means of a trypsin IMER can be performed either in flow-through mode or in stop-flow mode [46]. Systems consisting of a trypsin IMER and an SPE–LC–MS system were reported for on-line digestion and peptide mapping of lactate dehydrogenase [42].

Although trypsin digestion is applied in the majority of the applications, alternative proteolytic enzymes can be used as well. Mostly, specific proteases are applied that show predictable cleavage specificity, *e.g.*, LysC which cleaves at the *C*-terminal side of Lys unless it is next to Pro, AspN which cleaves at the *N*-terminal side of Asp, or chymotrypsin, which cleaves at the *C*-terminal of Trp, Tyr, or Phe, unless they are next to Pro. Digests with multiple proteolytic enzymes were tested to enhance the sequence coverage in LC–MS–MS on an ion-trap instrument [48-49]. For some applications, such as membrane proteins (Ch. 18.3.6), nonspecific proteolytic enzymes are preferred, *e.g.*, proteinase K or pepsin, with cleavage after hydrophobic amino acids, and Phe, Met, Leu, or Trp, respectively. This may result in better sequence coverage. Finally, chemical treatment with cyanogen bromide can be applied.

5. Liquid-phase separations

Despite the power of 2D-GE for protein separations, the use of liquid-phase separations (LC and capillary electrophoresis (CE)) is an attractive alternative. They show flexibility, higher speed, ease of automation, and easy hyphenation to MS. Enzymatic digest samples in protein characterization or proteomics studies pose significant challenges to the analytical separation technologies. Limited amounts of very complex samples of labile compounds, present in a wide range of abundances (>6 orders of magnitude), must be analysed. The analytes involved span a wide range of properties in terms of hydrophobicity and hydrophilicity, solubility, and molecular weight. The samples often contain significant amounts of salts, buffers, detergents, and chaotropes, either to achieve solubilization or due to previous sample handling, e.g., 2D-GE and enzymatic digestion.

5.1   Reversed-phase LC–MS of peptides

The most widely applied LC method for the separation of peptides is reversed-phase LC (RPLC). Most of the issues discussed in relation to RPLC of proteins (Ch. 16.3.2) also apply to peptides. The solvent composition must be a compromise between optimum conditions for LC separation and ESI. The recent review by García [50] and the references therein provides useful additional reading.

A typical mobile-phase composition is an acetonitrile–water gradient with a fixed concentration of trifluoroacetic acid (TFA), formic, or acetic acid (typically 0.05–0.5%). TFA acts as an ion-pairing agent and masks secondary interactions with the silica-based stationary phase. TFA may significantly suppress the ESI response in positive-ion mode. To avoid this, either formic acid is preferred or a mixture of 0.02% TFA and 0.5% acetic acid can be used. Some silica-based RPLC materials can be used with a lower TFA concentration (PepMap®). Alternatively, poly(styrene–divinylbenzene) polymeric materials (PS–DVB) can be applied. With a monolithic PS–DVB column, only a small decrease in separation efficiency on the monolithic column was observed when the TFA concentration was reduced from 0.2% to 0.05% [51].

5.2   Nano-LC

In early RPLC–MS studies on tryptic digests, typically 1-mm-ID columns were used. Further column miniaturization led to the introduction and use of packed microcapillary columns (typically 250–320 μm ID) and nano-LC columns (typically 75–150 μm ID). Nano-LC is nowadays routinely applied in proteomics studies.

Nanoscale LC for LC–MS was pioneered by Deterding et al. [52] using a coaxial continuous-flow fast-atom bombardment interface. Nano-LC was commercialized by LC Packings, both with respect to column technology and instrumentation [53]. Nowadays, nano-LC is available from several instrument and column manufacturers.

Packed microcapillary and nano-LC columns are ideally suited for sample-limited analysis. They are operated at low flow-rates, *i.e.,* 50 µl/min for a 1-mm-ID and 0.3 µl/min for a 75-µm-ID column, and are thus readily compatible with micro-ESI and nano-ESI interfacing (Ch. 17.2). The typical column length is between 50 and 150 mm (3–5-µm particles). Longer columns, up to 800 mm, can show superior resolution in the analysis of complex samples, as demonstrated by Shen *et al.* [54], but this also leads to prolonged analysis times.

The volume that can be injected onto a packed microcapillary or nano-LC column is very limited, *e.g.,* less than 0.1 µl for a 100-µm-ID column. This seriously compromises the achievable concentration detection limits, unless on-column preconcentration would be performed. However, for dilute sample solution, the injection volume is restricted by external peak broadening, and not by column loadability (typically ~50–200 µg/g of porous packing material). Therefore, on-line SPE can be applied for sample preconcentration.

On-line SPE–nano-LC is for instance described by Van der Heeft *et al.* [55], using 30–50 µl injections onto a 7×0.1-mm-ID precolumn at 10 µl/min and subsequent elution with an acetonitrile–water gradient (0.1 mol/l acetic acid) to a 250×0.1-mm-ID analytical column, operated at 0.5 µl/min (see Figure 1.3). To reduce memory effects in subsequent injections, the use of two parallel precolumns has been proposed [56]. Meiring *et al.* [57] reviewed instrumental considerations and design features for nano-LC–MS. Flow-split systems are applied to generate the required low-flow solvent gradients.

In order to reduce peak broadening in the nano-ESI needle attached to the nano-LC column, the use of packed needles has been promoted. Gatlin *et al.* [58] reports the use of 100-µm-ID fused-silica in-needle packed columns (10-µm particles) with a laser-pulled tip with a diameter of ~2 µm. Figeys and Aebersold [59] reported the use of such in-needle RPLC columns for the LC–MS analysis of tryptic digests in combination with a microfluidic device to generate a nanoflow solvent gradient via electroosmotic flows.

Further developments of nano-LC comprise further reduction of column inner diameter and the use of smaller particles to enhance the separation efficiency. Some examples of these developments are the use of a 150-µm-ID in-needle column with a 0.3–0.5-µm-ID laser-pulled emitter tip, packed with 1-µm particles, and operated at flow-rates of <50 nl/min [60], and the use of long 15–75-µm-ID nano-LC columns, packed with 3-µm particles, and operated at flow-rates as low as ~20 nl/min [54].

Monolithic PS–DVB columns have advantages in the separation of tryptic digests, both in terms of speed and in their ability to perform efficient separation at low TFA concentrations [51, 61]. For instance, a 12-min separation of the tryptic digest of bovine catalase was demonstrated, using a 60×0.2-mm-ID column [61]. Moore *et al.* [62] reported the generation of monolithic PS–DVB columns inside a 150-µm-ID nano-ESI-needle (5–10 µm ID emitter tip), and the application to tryptic digests from proteins isolated from silver-stained GE. Low-attomole detection of tryptic digests was reported with the use of 20-µm-ID PS–DVB [63] and silica-based [64] monolithic columns in combination with nano-ESI-MS. This is a 20-fold improvement relative to the 75-μm-ID nano-LC columns. The principles and applications of monolithic silica columns were reviewed by Rieux *et al.* [65].

Next to the many advantages of on-line LC separation instead of direct-infusion ESI-MS, one of the major disadvantages is the restriction in the time available for the MS experiments during development of the chromatogram. All MS and MS–MS experiments must be finished within the time scale of the chromatographic peak. A useful approach to solve this problem is 'peak parking': during the elution of a peak of special interest the flow-rate through the column is greatly reduced to enabling longer MS acquisition times [66-67].

5.3   Reducing analysis time in LC

Various approaches have been investigated to shorten the analysis time of peptide separation by RPLC. Packed microcapillary perfusion columns (Ch. 1.4.1) for LC–MS analysis of tryptic digests were evaluated by Kassel *et al.* [68]. The 320-μm-ID perfusion column was operated at 45 μl/min, which is ~10× higher than normal. Compared to a 180-μm-ID RPLC column, operated at 2 μl/min, the analysis time for a tryptic digest and the peak width were reduced from 30 min and 5–10 s to 10 min and 20–45 s. The use of nonporous silica particles also allows a peptide separation within a few minutes [69]. The narrow peaks achieved in such fast separations challenge the speed of the mass spectrometer in the acquisition of MS and MS–MS spectra for the eluting tryptic peptides. The use of a TOF-MS would be attractive for fast LC separations.

Various examples have also been described of dual columns systems: while the analytes are eluted from one column, the other column is regenerated. For example, two 50×0.3-mm-ID columns, operated at 4 μl/min, are used for the separation of tryptic digests within 17 min per sample, which corresponds to a sample throughput of ~80 samples/24 hr [70].

Reduction of particle size also contributes to a shorter analysis time. In that respect, recent developments in ultra-performance RPLC, featuring a commercially available system for LC with 1.7-μm particles, are of great interest [71]. RPLC–MS with 0.8-μm particles was investigated by Shen *et al.* [72]. The results demonstrate the compromise between chromatographic resolution and analysis time: in 50 min ~1000 proteins of *Shewanella oneidensis* could be identified from ~4000 peptides, in 20 min ~550 proteins from ~1800 peptides, or in 8 min ~250 proteins from ~700 peptides.

5.4   Multidimensional LC

There are two general approaches to two-dimensional LC, *i.e.*, a heart-cut approach as applied in coupled-column LC (Ch. 1.4.6), and the comprehensive LC×LC approach. In LC×LC, all subsequent fractions of the first LC separation are transferred to a second-dimension, preferentially providing an orthogonal separation, *i.e.*, with another phase system.

Initial results with LC×LC–MS in the analysis of proteins were reported by Opiteck *et al.* [73-74], using either ion-exchange (IEC) or size-exclusion (SEC) chromatography in the first dimension and RPLC in the second dimension. Protein mapping similar to 2D-GE was demonstrated, but with inferior efficiency.

SEC×RPLC–MS was also applied to tryptic digests [75-76]. After the void volume of the SEC (120 min), the flow-rate was reduced from 0.29 to 0.1 ml/min. For the next 200 min, a fraction was transferred every 5 min from the SEC to one of the two parallel RPLC columns for the second-dimension separation.

The most widely used 2D-LC or better LC×LC approach is IEC×RPLC, where a salt-step gradient is applied to elute the tryptic peptides of a complete or prefractionated proteome from a SCX column onto the second-dimension RPLC separation. Apart from the good sorption properties of SCX for positively-charged peptides, it also allows washing away neutral and negatively-charged sample constituents (SDS, urea, DTT), and provides strong sorption of trypsin. The procedure was first developed using a biphasic in-needle column and named multidimensional protein identification technology (MudPIT) [77]. Later on, the system was modified into a three-column setup and the use of valve-switching technology [78]. Commercial systems have become available [79-80] (Figure 17.1). Collection of fractions from the SCX column and subsequent off-line analysis by RPLC–MS is a viable approach as well [81]. SCX×RPLC is discussed in more detail in Ch. 18.3.2.

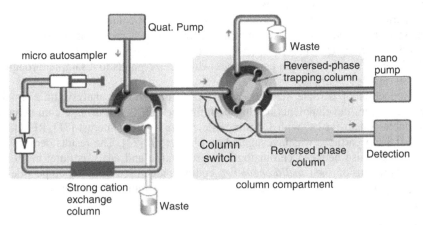

Figure 17.1: SCX×RPLC system for rapid analysis of tryptic digests of complex protein samples. The step gradient over the SCX elutes peptide fractions to the RPLC trapping column. After valve-switching, the trapping column is eluted to the gradient RPLC–MS system. Reprinted from [79] with permission. ©2003, Elsevier B.V.

A multidimensional separation system may also include an IMER for on-line digestion (Ch. 17.4.4). Not-too-complex mixtures allow direct injection into the IMER and subsequent SPE–LC–MS analysis, whereas for more complicated samples the IMER is applied as part of a multidimensional system. This field is pioneered by the group of Regnier [38, 46, 82-83]. Their systems consist of a trypsin IMER, affinity selection of specific peptides from the tryptic digest, and subsequent analysis of these peptides by RPLC–MS. Affinity selection of the specific so-called signature peptides can be based on lectin AfC for glycopeptides [82], IMAC for phosphorylated peptides [39], or covalent chromatography on a thiopropyl Sepharose affinity resin with subsequent reduction and alkylation for Cys-containing peptides [83]. A system based on SEC of intact proteins, digestion by means of an IMER, and RPLC–MS of the tryptic peptides was also reported by the same group [46]. More recently, two multidimensional LC–MS systems with an IMER were reported. Carol *et al.* [84] reported the on-line combination of SEC, IMER, and SPE–LC–MS for the analysis of protein biotoxins, whereas Hoos *et al.* [85] reported the use of IAC, IMER, and SPE–LC–MS for the targeted quantitative analysis of protein drugs.

5.5 Microfluidic chips

In recent years, impressive progress has been made in the development of microchips for sample pretreatment (SPE, dialysis, tryptic digestion) and/or separation by either LC or CE. The on-line combination of microchip devices to ESI-MS has been reviewed [86-87]. Some examples are given here.

A microchip device with an attached nano-ESI emitter tip was developed to facilitate the introduction of tryptic digests by means of nano-ESI [88-89]. Instead of off-line filling of the nano-ESI needle, the sample is transferred from a vial on the chip to the nano-ESI needle by electroosmosis. Detection limits of 2 fmol/μl were achieved for fibrinopeptide A (1699 Da). Further developments enabled sequential automated analysis of protein digests by ESI-MS [90]. On-chip sample pretreatment and desalting by either sample stacking via polarity switching or SPE prior to on-chip CE was described by Li *et al.* [91], and applied to the identification of 2D-GE separated proteins from *Haemophilus influenzae* using a Q–TOF instrument.

The most complete approach is a microfluidic chip with integrated SPE, a nano-LC column, and a nano-ESI emitter tip, as has recently become commercially available [92]. Typical dimensions are a 40 nl SPE enrichment column, 43 or 150 mm × 75 μm ID nano-LC column, both packed with 5 μm particles. Sample loading is performed with 4 μl/min, LC analysis with 300 nl/min. The nano-ESI emitter has a tip diameter of 10 μm ID.

5.6   Capillary electrophoresis–MS

Given the important role of GE in protein separations (Ch. 17.3), an on-line combination of CE and MS would be a logical choice. CE–MS with ESI has been under development for many years [93-96]. CE is also an important technique in the development of microchip technology [86].

There are two ways of coupling CE and MS. The most robust and widely used approach is the use of a sheath liquid surrounding the separation capillary, ensuring proper electrical contact, essential for stable ESI performance. Alternatively, a sheathless approach can be taken, where the end of the capillary acts as a micro-ESI or nano-ESI needle. In either case, good results are generally easier to obtain when the ESI needle is at ground potential.

Only a small injection volume can be applied to a CE separation capillary. This seriously limits the achievable concentration detection limits. Although various approaches have been developed to circumvent this problem, it effectively limits the use of CE–MS in real-life applications, although the technique has demonstrated its potential in several specific areas, *e.g.*, the study of protein glycosylation [97]. On-line SPE–CE removes the problems with the small injection volume. In a direct comparison between SPE–CE–MS and SPE–LC–MS on the same ion-trap instrument [98], LC–MS was 5-fold more sensitive, but CE–MS provided better sequence coverage, especially for larger tryptic peptides.

Next to CE, on-line coupling of capillary IEF and MS is attractive. For protein characterization, CIEF–MS on a Fourier-transform ion-cyclotron resonance mass spectrometer (FT-ICR-MS) instrument was pioneered by the group of Smith [99-100]. They demonstrated the high-resolution analysis of *E. coli* proteins, revealing >400 proteins (2–100 kDa) from an injection of only ~300 ng.

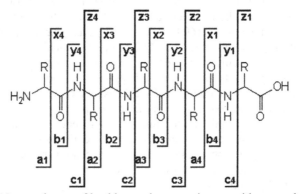

Figure 17.2: Nomenclature of backbone cleavages in a peptide. y- and c-ions should be annotated as y'' and c'', because of the proton rearrangement involved in their generation. For more details [103-104].

6. Identification of peptides and proteins

6.1   Fragmentation of peptides

The general procedure of protein identification involves digestion of the intact protein using trypsin and ESI-MS–MS analysis of the resulting peptides. This procedure was already well established for FAB [101]. The double-charge tryptic peptide ions generated in ESI-MS can be efficiently fragmented by collision-induced dissociation (CID) in MS–MS. Interpretation of the data reveals the amino-acid sequence of the individual tryptic peptides in the digest mixture [102]. Nomenclature rules have been established to unambiguously annotate the various fragments formed [103-104] (Figure 17.2). Fragmentation mechanisms and interpretation strategies have been reviewed for both positive-ion [102, 105] and negative-ion [106] mode.

In early LC–MS–MS studies, CID was performed with low-energy collisions in triple-quadrupole instruments. Under these conditions, series of N-terminal a- and b-ions and C-terminal y-ions result from cleavages at the peptide bond and charge retention on either side. Double-charge tryptic peptide ions tend to favour fragmentation towards more abundant y-ions than b-ions. In high-energy CID, the other backbone fragment ions (c-, x-, and z-ions) can occur as well as amino-acid side-chain cleavages leading to d-, v-, and w-ions. From the ladders of sequence ions in the MS–MS spectrum, the amino-acid sequence of the peptide may be derived. This is the bottom-up protein identification approach (Ch. 18.3.1).

For a known sequence, the sequence ions of b- and y-type can be predicted using the equations summarized in Table 17.1 and the residue masses of amino acids given in Table 17.2. Interpretation problems arise because some peptide bonds appear to be more difficult to cleave than others, resulting in missing sequence ions. In addition, certain amino acids show unique fragmentation characteristics (Glu, Asp) or influence the fragmentation pattern (Pro), and some amino acids are isomers and isobars. Finally, problems may arise from the poor S/N due to limited sample amounts. Fragmentation of larger peptides (>3 kDa) is generally not successful, except by means of FT-ICR-MS (see also Ch. 17.7.3 and Ch. 18.3.5).

Table 17.1: Prediction of peptide sequence ions			
From **N**-terminal (–NH$_2$)		From **C**-terminal (–COOH)	
b_1	$M_{residue} + H$	y_1	$M_{residue} + 19$
b_2	$b_1 + M_{residue}$	y_2	$y_1 + M_{residue}$
b_{n-1}	$[M+H]^+ - 18 - M_{residue}$	y_{n-1}	$[M+H]^+ - M_{residue}$
b_n	$[M+H]^+ - 18$	y_n	$[M+H]^+$

Table 17.2: Monoisotopic residue masses for DNA-encoded Amino Acids			
Amino Acid		Monoisotopic Residue Mass (Da)	Immonium Ion (m/z)
G	Gly	57	57.0215	30
A	Ala	71	71.0371	44
S	Ser	87	87.0320	60
P	Pro	97	97.0528	70
V	Val	99	99.0684	72
T	Thr	101	101.0477	74
C	Cys	103	103.0092	76
L	Leu	113	113.0841	86
I	Ile	113	113.0841	86
N	Asn	114	114.0429	87
D	Asp	115	115.0269	88
Q	Gln	128	128.0586	101
K	Lys	128	128.0950	101
E	Glu	129	129.0426	102
M	Met	131	131.0405	104
H	His	137	137.0589	110
F	Phe	147	147.0684	120
R	Arg	156	156.1011	129
Y	Tyr	163	163.0633	136
W	Trp	186	186.0793	159

The step-wise fragmentation in multi-stage ion-trap MS–MS results in different fragmentation [107]. Moreover, peptide adduct ions with metal cations (Li^+, Na^+, but also transition metal cations) can be fragmented in ion-trap MS–MS. This open new directions towards enhanced response in ESI-MS and/or altered fragmentation characteristics.

Other fragmentation techniques have been introduced [108]. Some of these, *e.g.*, sustained off-resonance irradiation (SORI) and infrared multiphoton dissociation (IRMPD), provide similar fragmentation as in CID, *i.e.*, preferential backbone cleavages at the peptide amide bond (b- and y-ions). Others like electron-capture dissociation (ECD) [109-110] induce different fragmentation reactions, *i.e.*, the formation of c- and z'-ions due to cleavage of $N–C_\alpha$ bonds.

6.2   Database searching

An important development in high-throughput protein identification is the introduction of protein database searching [111]. After separation on 1D- or 2D-GE, the proteins were blotted onto a membrane and enzymatically digested after reduction and alkylation. The tryptic peptide mixture is analysed by MALDI-MS to achieve a peptide map or peptide mass fingerprint (PMF). The m/z information of the peptides is used to search the protein database, $e.g.$, the Protein Identification Resource (PIR) database [112-114]. If the mass of just 4–6 tryptic peptides is accurately measured (between 0.1 and 0.01%), a useful database search can be performed.

The procedure was further refined by the use of MS–MS data [115]. Although the complete interpretation of the peptide MS–MS spectrum is often hindered by poor S/N and missing sequence ions, a partial sequence of a peptide can be derived. The partial sequence divides the peptides in three parts: m_1 of the unknown N-terminal end, the identified partial sequence, and m_2 of the unknown C-terminal end. These peptide sequence tags are used together with PMF data to search the database. These approaches were developed into the web-based software tools (PeptideSearch and MS-tag), and subsequently in second-generation programs like MOWSE [116], ProFound [117], and MASCOT [118].

At the same time, the group of Yates [119-123] developed software to correlate uninterpreted ion-trap MS–MS spectra of peptides with protein sequence and DNA databases. The protein sequence database is searched to identify sequences within a mass tolerance of 1 Da of the precursor mass. A cross correlation function is used to evaluate the similarity between the MS–MS data and the predicted sequence ions for the peptides found in the database. Later on, the procedure is refined to enable searching of spectra of modified peptides, $e.g.$, carboxymethylated Cys, oxidized Met, phosphorylated Ser, Thr, or Tyr [121]. The spectra are searched against the OWL database containing protein sequence information from other databases like GenBank, Swiss-Prot, and PIR. A further extension of the database search procedure involves searching in cDNA and expressed sequence tag (EST) databases [122], generated by sequencing of reverse transcribed messenger RNA, as takes place within the framework of the Human Genome Project. In this way, automated identification of proteins can be achieved by taking advantage of the available information in protein and nucleotide databases [34]. The software was commercialized as SEQUEST and continues to be developed, refined, and evaluated. It was tested for protein identification in digested mixtures of proteins [123] and as such is the basis of the shotgun protein identification approach (Ch. 18.3.2). Other improvements and modifications are filtering of spectra with poor quality prior to the search [124], preprocessing of MS–MS data [125], and deriving partial sequence tags from MS–MS spectrum of peptides with unknown post-translational modifications (PTM) (GutenTag) [126]. Searches of MS–MS data can also be performed via MASCOT [118].

Comparison shows that peptide mass fingerprinting is rapid but is not necessarily conclusive, whereas SEQUEST is usually conclusive but slow [89].

Next to these programs, a variety of other tools have been reported, especially for spectrum interpretation and sequence alignment in *de novo* sequencing, *e.g.*, Lutefisk [127], PEAKS [128], and OpenSea [129]. Commercial implementations from MS instrument manufacturers have also been developed.

Software tools for database searching and available protein sequence databases for protein identification were reviewed [130-131]. Liska and Shevchenko [132] reviewed approaches to study proteomes with nonsequenced genomes via sequence similarity database searching.

6.3   *De novo* sequencing

The strategy for protein identification outlined so far involves protein isolation, tryptic digestion, and subsequent ESI-MS, LC–MS and/or MS–MS analysis to obtain PMF data, sequence tags, and information-rich MS–MS spectra, which can be searched against protein and DNA databases. Though extremely powerful, the approach does not always lead to a satisfactory result, simply because the protein detected was not (yet) included in the database. In those cases, *de novo* sequencing [115] of the protein or some of its specific peptides is required. This requires interpretation and measurement strategies along the lines outlined by (among others) the group of Hunt in the 1980s [100]. In order to enhance the confidence in sequence assignment, the methyl esterified peptides are analysed by MS–MS as well. Stable isotopic labelling has also been applied to simplify the interpretation of the MS–MS spectra [101, 104].

De novo sequencing seldom leads to a complete sequence because in most cases the y- and b-ion series are not completely present, some sequence ions might be lost in the noise, and/or ambiguities exist due to additional peaks (neutral losses of CO, H_2O, NH_3, etc.), and isobaric and isomeric amino acids (Table 17.2).

Therefore, and also to compare the results with existing information, one may still want to search the database with the results obtained by *de novo* sequencing. The approach to follow is a BLAST search (Basic Local Alignment Search Tool, [133-134]). Sequence alignment is a powerful way to compare novel sequences with previously characterized sequences and search for similarities. The BLAST search is much slower than the database searches outlined in Ch. 17.6.2. A nice example is the *de novo* identification of the bacterium *Shewanella putrefaciens* proteome by 2D-GE, in-gel digestion, direct infusion nano-ESI, and nano-LC–MS [135]. The results were BLAST searched against the incomplete genome of the bacterium.

Enzymatic digestion of proteins in 1:1 $^{16}O/^{18}O$ water has been applied to facilitate *de novo* sequencing [136]. The labelling results in split peaks with a 2 Da difference for the y-ions, enabling easy discrimination between peaks due to b- and y-ions. Other stable isotope labelling methods (Ch. 18.4.2) involving modification of either the *N*- or the *C*-terminal can assist in *de novo* sequencing as well.

7. Mass spectrometry

7.1 Mass analysers

The developments in protein characterization by MS have become possible by the continuous developments in mass analyser technology, resulting in improvements in both performance and possibilities. Next to the triple-quadrupoles, that were used in the first few years, other instruments became available. The ion-trap mass spectrometer provides significantly better full-spectrum sensitivity, both in MS and MS–MS operation. In addition, it enables multiple stages of MS–MS, to some extent providing stepwise fragmentation of the precursor ion [137]. With slower scan rates, it also provides enhanced mass resolution. The Q–TOF instrument provided excellent full-spectrum sensitivity, both in MS and MS–MS operation, as well as more accurate m/z determination (typically <5 ppm) due to the high mass resolution (in excess of 10,000 FWHM) achievable in the TOF analyser [89, 138]. The ion-trap and Q–TOF instruments rapidly became the MS workhorses in protein and proteomics studies. Recently, quadrupole–linear-ion-trap hybrid (Q–LIT) instruments have been introduced. These combine triple-quadrupole fragmentation with ion-trap ion accumulation and full-spectrum sensitivity (Ch. 2.4.5) [139].

7.2 Data-dependent acquisition

Data-dependent switching between a survey-MS mode and the product-ion MS–MS mode (Ch. 2.4.2) in the LC–MS analysis of tryptic digest on a triple-quadrupole instrument was pioneered by Stahl *et al.* [140]. The MS–MS spectra obtained were correlated with a protein sequence database by using the SEQUEST program. DDA (also called SmartSelect, or Information-Dependent Acquisition, IDA) on ion-trap, Q–LIT, and Q–TOF instruments have become important tools in high-throughput protein characterization.

With ion-trap instruments, DDA often involves a 'TriplePlay' experiment: survey MS, switching to slow enhanced resolution scan ('ZoomScan') when the peak intensity exceeds the threshold value, and subsequent product-ion MS–MS. During acquisition, averaging of multiple scans may be applied. An important issue in DDA is the duty cycle of the instrument. Factors affecting the performance of DDA in ion-trap instruments were systematically evaluated [141]. Averaging of (at least) three MS–MS spectra was required to get good-quality data. MS–MS data may be acquired for multiple precursor ions within one MS spectrum within ~10 s. This means that ~1000 sequencing attempts may have been acquired in a 30-min LC run. This number will even increase when the chromatographic speed is reduced. Dynamic exclusion utilities in the acquisition software have been applied to increase the number of precursor ions subjected to MS–MS. Ion-trap DDA has been utilized, for instance, in integrated systems enabling unattended LC–MS–MS and automated database searching for the identification of in-gel digested and extracted proteins, *e.g.*, 90 proteins from the yeast *Saccharomyces cerevisiae* [34].

The Q–LIT enables relatively fast switching in DDA, and provides structure informative MS–MS spectra with fragments over a wide range, and enhanced full-spectrum sensitivity [139]. An example of its advanced DDA possibilities is discussed in Ch. 19.3.1. In a Q–LIT instrument, fragmentation can be achieved in a conventional (quadrupole) collision cell. An additional stage of MS–MS can be performed after accumulation of the product ions in the LIT itself.

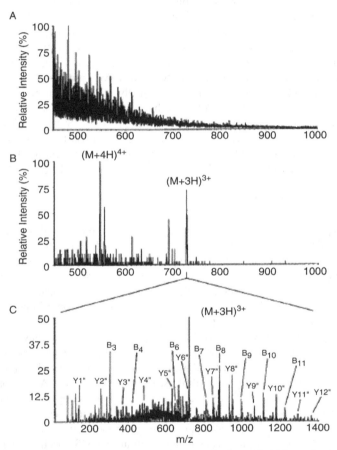

Figure 17.3: Application of precursor-ion analysis MS–MS analysis mode to screening for peptides in complex mixtures. (A) Spectrum of a synthetic peptide, hidden in the noise, (B) precursor-ion MS–MS spectrum using the Leu/Ile-immonium ion at *m/z* 86 as common product ion to detect the peptide, (C) product-ion MS–MS spectrum of the synthetic peptide based on the [M+3H]$^{3+}$-ion. Reprinted from [144] with permission. ©1996. American Chemical Society.

This MS–MS^2 approach was found to be helpful in solving ambiguities in peptide identification by means of further fragmentation of especially *C*-terminal fragments [142]. Correlation of fragment information from ion-trap MS–MS and MS–MS^2 can assist in the *de novo* interpretation of the MS–MS spectra [143].

Direct infusion nano-ESI-MS of complex peptide mixtures is a powerful tool in protein characterization, but is often hindered by poor S/N of lower-level peptides present. The use of precursor-ion MS–MS analysis (Ch. 2.4.1) with immonium ions facilitates the determination of the peptide *m/z* and enables further characterization by product-ion MS–MS [144]. The approach is illustrated in Figure 17.3.

Lehmann [145] reported the use of combined in-source CID and precursor-ion analysis, using either *m/z* 147 or 175 as common product-ion, and neutral-loss analysis (28 Da) to generate MS–MS product-ion spectra showing only y- or b-ions, respectively, which are easier to interpret.

7.3 Fourier-transform ion-cyclotron resonance MS

In the early days of ESI, the potential of fragmenting multiple-charge protein ions by in-source CID or MS–MS has been investigated [146]. A major difficulty in the interpretation of these spectra lies in establishing the charge state of the fragment ions. It was soon realised that high-resolution MS in an FT-ICR-MS (Ch. 2.4.6) can have substantial benefits, especially in charge-state determination [147]. Significant instrumental developments were required in the coupling to ESI (Ch. 5.8.3) and in the operation and measurement protocols for FT-ICR-MS [148-149].

This greatly expanded the potential of FT-ICR-MS in protein identification. Unit-mass resolution for the 112-kDa chondroitinase I and II proteins and within 3-Da mass accuracy are clear examples of this [150]. Unit-mass resolution can be achieved for proteins with a molecular weight in kDa of 10×B, the magnetic field strength. Therefore, investments have been done for higher magnetic fields. A 100% sequence coverage was obtained for a tryptic digest of bovine serum albumin by signal averaging of 100 spectra [151]. In the averaged spectrum, 71 isotope distributions were found that were within 2 ppm error limits of the expected tryptic fragments. Next to resolving power, excellent sensitivity was demonstrated, *e.g.*, the detection of *ca.* 7 amol of carbonic anhydrase in a crude extract from human blood [152], and 30 zmol of cytochrome C or myoglobin [153].

Despite these impressive results, on-line LC–MS on a FT-ICR-MS could only be achieved after modification of the operation procedures by the advent of external ion accumulation [154]. This adapts the FT-ICR-MS duty cycle to the LC time scale. In addition, it solves space-charging effects in the FT-ICR cell by controlling the number of ions that enter the cell. Impressive early examples of on-line LC–FT-ICR-MS were reported by Emmett *et al.* [155] and Martin *et al.* [156]. These features are now routinely available in modern commercial FT-ICR-MS instruments, which are in fact hybrid systems featuring either a quadrupole [157] or a Q–LIT hybrid [158] as the first stage and the FT-ICR-MS as the second stage.

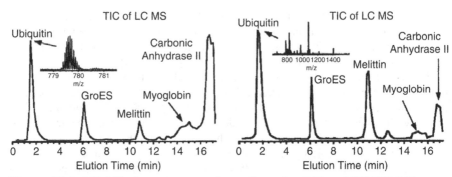

Figure 17.4: LC–MS chromatograms from alternating MS and IRMPD-MS scans for the separation of five known proteins by LC–ESI-FT-ICR-MS. Reprinted from [162] with permission. ©1999, American Chemical Society.

Initial results with the fragmentation of multiple-charge protein ions in FT-ICR-MS were achieved, either by in-source CID [159], sustained off-resonance irradiation (SORI) [160], or infrared multiphoton dissociation (IRMPD) [161], and later on with electron-capture dissociation [109-110].

Li *et al.* [162] reported the use of on-line LC–ESI-FT-ICR-MS for the identification of intact proteins in mixtures by alternating scans: a short list of candidate proteins is generated from a database search by accurate-mass determination (±0.1 Da at 10 kDa), and IRMPD is applied to generate fragment ions and sequence tags for each protein, enabling further protein identification from the short list. Typical results are shown in Figure 17.4.

FT-ICR-MS enables the identification of unknown proteins with limited enzymatic cleavage and eventually in-cell fragmentation via a variety of ion-fragmentation techniques and high-resolution m/z determination of the fragment ions. This is the top-down protein identification approach [163] (Ch. 18.3.5). The introduction of ECD was an important development in this respect, as it allows in-cell fragmentation of the complete protein [110].

The high resolving power of FT-ICR-MS in the characterization of peptides and proteins can still be exploited when the ion fragmentation is performed prior to the introduction of ions into the FT-ICR cell. This can be achieved by in-source CID or in the first-stage mass analyser of the commercial hybrid systems [157-158]. In-source CID has some interesting additional features. Because of the extremely low pressure in the FT-ICR cell (>10^{-7} Pa), a series of nozzle-skimmer devices is applied in the vacuum interface (*cf.* Ch. 5.8.3). In-source CID can be performed in each stage, thereby enabling improved fragmentation of intact proteins [164].

If the mass of a single peptide in a tryptic digest of a protein can be measured with sufficiently high mass accuracy (~1 ppm), it can be used as an accurate mass tag (AMT) and an effective marker for unambiguous protein identification [165].

Peptides with a mass higher than 2500 Da are more likely to be suitable as an AMT than smaller ones. Only 10% of the peptides with masses below 1000 Da were found to be suitable. FT-ICR-MS has to be applied to acquire these AMT. MS–MS can be used to fill the gaps in the case of ambiguous measurements and/or to validate the AMT. This opens ways to high-throughput analysis for proteomics.

8. Conclusions and perspectives

The discussion in this chapter focussed at developments in the field of protein characterization, and provided an overview of the technology developed to enable proteomics studies (Ch. 18). The strategies outline above have been applied to increasingly complex samples. Some examples are the detection and identification of human leucocyte antigen peptides related to the major histocompatibility complex [166], and the unattended identification of 90 proteins from the yeast *Saccharomyces cerevisiae* by means of an integrated workstation for LC–MS–MS under DDA and with database searching [34].

9. References

1. R.D. Smith, J.A. Loo, R.R. Ogorzalek Loo, M. Busman, H.R. Udseth, *Principles and practice of ESI-MS for large polypeptides and proteins*, Mass Spectrom. Rev., 10 (1991) 359.
2. P.E. Andrén, M.R. Emmett, R.M. Caprioli, *Micro-ESI: Zeptomole/attomole per microliter sensitivity for peptides*, J. Am. Soc. Mass Spectrom., 5 (1994) 867.
3. M.S. Wilm, M. Mann, *Analytical properties of the nano-ESI ion source*, Anal. Chem., 68 (1996) 1.
4. M. Karas, U. Bahr, T. Dülcks, *Nano-ESI-MS: addressing analytical problems beyond routine*, Fres. J. Anal. Chem., 366 (2000) 669.
5. R. Körner, M. Wilm, K. Morand, M. Schubert, M. Mann, *Nano-ESI combined with a quadrupole ion trap for the analysis of peptides and protein digests*, J. Am. Soc. Mass Spectrom., 7 (1996) 150.
6. E. Hoyes, S.J. Gaskell, *Automatic function switching and its usefulness in peptide and protein analysis using direct infusion micro-ESI Q–TOF MS*, Rapid Commun. Mass Spectrom., 15 (2001) 1802.
7. S. Zhang, C.K. van Pelt, J.D. Henion, *Automated chip-based nano-ESI-MS for rapid identification of proteins separated by 2D-GE*, Electrophoresis, 24 (2003) 3620.
8. J.G. Williams, K.B. Tomer, *Disposable chromatography for a high-throughput nano-ESI-MS and nano-ESI-MS–MS platform*, J. Am. Soc. Mass Spectrom., 15 (2004) 1333.
9. D.E. Garfin, *2D-GE: an overview*, Trends Anal.Chem., 22 (2003) 263.
10. L.F. Steel, M.G. Trotter, P.B. Nakajima, T.S. Mattu, G. Gonye, T. Block, *Efficient and specific removal of albumin from human serum samples*, Mol. Cell. Proteomics, 2 (2003) 262.

11. C. Greenough, R.E. Jenkins, N.R. Kitteringham, M. Pirmohamed, B.K. Park, S.R. Pennington, *Method for the rapid depletion of albumin and immunoglobulin from human plasma*, Proteomics, 4 (2004) 3107.

12. A. Shevchenko, O.N. Jensen, A.V. Podtelejnikov, F. Sagliocco, M. Wilm, O. Vorm, P. Mortesen, A. Shevchenk, H. Boucherie, M. Mann, *Linking genome and proteome by MS: large-scale identification of yeast proteins from 2D-GE*, PNAS, 93 (1996) 14440.

13. A, Shevchenko, M. Wilm, O. Vorm, M. Mann, *MS sequencing of proteins from silver-stained polyacrylamide gels*, Anal. Chem., 68 (1996) 850.

14. F. Meng, B.J. Cargile, S.M. Patrie, J.R. Johnson, S.M. McLoughlin, N.L. Kelleher, *Processing complex mixtures of intact proteins for direct analysis by MS*, Anal. Chem., 74 (2002) 2923.

15. M. Garzotti, L. Rovatti, M. Hamdan, *On-line LC–ESI-MS to investigate acrylamide adducts with cysteine residues: implications for polyacrylamide GE separations of proteins*, Rapid Commun. Mass Spectrom., 12 (1998) 484.

16. T. Stasyk, L.A. Huber, *Zooming in: Fractionation strategies in proteomics*, Proteomics, 4 (2004) 3704.

17. A. Ahmed, G.E. Rice, *Strategies for revealing lower abundance proteins in two-dimensional protein maps*, J. Chromatogr. B, 815 (2005) 39.

18. W.-C. Lee, K.H. Lee, *Applications of AfC in proteomics*, Anal. Biochem., 324 (2004) 1.

19. J. Martosella, N. Zolotarjova, H. Liu, G. Nicol, B.E. Boyes, *RPLC prefractionation of immunodepleted human serum proteins to enhance MS identification of lower-abundant proteins*, J. Proteome Res., 4 (2005) 1522.

20. M. Ramström, C. Hagman, J.K. Mitchell, P.J. Derrick, P. Håkansson, J. Bergquist, *Depletion of high-abundant proteins in body fluids prior to LC–FT-ICR-MS*, J. Proteome Res., 4 (2005) 410.

21. R. Qiu, F.E. Regnier, *Use of multidimensional lectin AfC in differential glycoproteomics*, Anal. Chem., 77 (2005) 2802.

22. M. Madera, Y. Mechref, M.V. Novotny, *Combining lectin microcolumns with high-resolution separation techniques for enrichment of glycoproteins and glycopeptides*, Anal. Chem., 77 (2005) 4081.

23. H. Erdjument-Bromage, M. Lui, L. Lacomis, A. Grewal, R.A. Annan, D.E. McNulty, S.A. Carr, P. Tempst, *Examination of micro-tip RPLC extraction of peptide pools for MS analysis*, J. Chromatogr. A, 826 (1998) 167.

24. M. Gilar, A. Belenky, B.H. Wang, *High-throughput biopolymer desalting by SPE prior to MS analysis*, J. Chromatogr. A, 921 (2001) 3.

25. I. Kay, A.I. Mallet, *Use of an on-line LC trapping column for the purification of protein samples prior to ESI-MS analysis*, Rapid Commun. Mass Spectrom., 7 (1993) 744.

26. J.P.C. Vissers, J.-P. Chervet, J.-P. Salzmann, *SDS removal from tryptic digest samples for on-line capillary LC–ESI-MS*, J. Mass Spectrom., 31 (1996) 1021.

27. L.J. Licklider, C.C. Thoreen, J. Peng, S.P. Gygi, *Automation of nanoscale microcapillary LC–MS–MS with a vented column*, Anal. Chem., 74 (2002) 3076.

28. J. Roboz, Q. Yu, A. Meng, R. van Soest, *On-line buffer removal and fraction selection in gradient capillary LC prior to ESI-MS of peptides and proteins*, Rapid Commun. Mass Spectrom., 8 (1994) 621.

29.   Q. Wu, C. Liu, R.D. Smith, *On-line microdialysis desalting for ESI-MS of proteins and peptides*, Rapid Commun. Mass Spectrom., 10 (1996) 835.
30.   C. Liu, S.S. Verma, *Direct coupling of ionic LC with ESI-MS utilizing a microdialysis junction interface*, J. Chromatogr. A, 835 (1999) 93.
31.   S. Canarelli, I. Fisch, R. Freitag, *On-line microdialysis of proteins with high-salt buffers for direct coupling of ESI-MS and LC*, J. Chromatogr. A, 948 (2002) 139.
32.   S. Julka, F.E. Regnier, *Benzoyl derivatization as a method to improve retention of hydrophilic peptides in tryptic peptide mapping*, Anal. Chem., 76 (2004) 5799.
33.   W.K. Russell, Z.-Y. Park, D.H. Russell, *Proteolysis in mixed organic-aqueous solvent systems: applications for peptide mass mapping using MS*, Anal. Chem., 73 (2001) 2682.
34.   A. Ducret, I. van Oostveen, J.K. Eng, J.R. Yates, III, R. Aebersold, *High throughput protein characterization by automated RPLC–ESI-MS–MS*, Protein Sci., 7 (1998) 706.
35.   A.T. Timperman, R. Aebersold, *Peptide electroextraction for direct coupling of in-gel digestion with capillary LC–MS–MS for protein identification and sequencing*, Anal. Chem., 72 (2000) 4115.
36.   N.F.C. Visser, H. Lingeman, K.W. Li, H. Irth, *Quantitative transfer of in-gel digest products to LC–ESI-MS using on-line coupled extraction*, Chromatographia, 61 (2005) 433.
37.   Y.-Q. Yu, M. Gilar, P.J. Lee, E.S.P. Bouvier, J.C. Gebler, *Enzyme-friendly, MS compatible surfactant for in-solution enzymatic digestion of proteins*, Anal. Chem., 75 (2003) 6023.
38.   M.T. Davis, T.D. Lee, M. Ronk, S.A. Hefta, *Microscale IMER columns for peptide mapping by LC–MS analysis*, Anal. Biochem., 224 (1995) 235.
39.   L. Riggs, C. Sioma, F. Regnier, *Automated signature peptide approach for proteomics*, J. Chromatogr. A, 924 (2001) 359.
40.   G. Vecchione, B. Casetta, R. Santacroce, M. Margaglione, *A comprehensive on-line digestion–LC–MS–CID–MS approach for the characterization of human fibrinogen*, Rapid Commun. Mass Spectrom., 15 (2001) 1383.
41.   J.W. Cooper, J. Chen, Y. Li, and C.S. Lee, *Membrane-based nanoscale proteolytic reactor enabling protein digestion, peptide separation, and protein identification using MS*, Anal. Chem., 75 (2003) 1067.
42.   J. Samskog, D. Bylund, S.P. Jacobsson, and K.E. Markides, *Miniaturized on-line proteolysis–capillary LC–MS for peptide mapping of lactate dehydrogenase*, J. Chromatogr. A, 998 (2003) 83.
43.   D.S. Peterson, T. Rohr, F. Svex, J.M.J. Fréchet, *High-throughput peptide mass mapping using a microdevice containing trypsin immobilized on a porous polymer monolith coupled to MALDI-TOF and ESI- TOF-MS*, J. Proteome Res., 1 (2002) 563.
44.   A.K. Palm, M.V. Novotny, *Analytical characterization of a facile porous polymer monolithic trypsin microreactor enabling peptide mass mapping using MS*, Rapid Commun. Mass Spectrom., 18 (2004) 1374.
45.   E. Calleri, C. Temporini, E. Perani, C. Stella, S. Rudaz, D. Lubda, G. Mellerio, J. -L. Veuthey, G. Caccialanza, G. Massolini, *Development of a bioreactor based on trypsin immobilized on monolithic support for the on-line digestion and identification of proteins*, J. Chromatogr. A, 1045 (2004) 99.
46.   S. Wang, F.E. Regnier, *Proteolysis of whole cell extracts with immobilized enzyme*

columns as part of multidimensional chromatography, J. Chromatogr. A, 913 (2001) 429.

47. J. Křenková, Z. Bilková, F. Foret, *Characterization of a monolithic immobilized trypsin microreactor with on-line coupling to ESI-MS*, J. Sep. Sci., 28 (2005) 1675.

48. M.P. Washburn, D.A. Wolters, J.R. Yates, III, *Large-scale analysis of the yeast proteome by MudPIT*, Nat. Biotechnol., 19 (2001) 242.

49. G. Choudhary, S.-L. Wu, P. Shieh, W.S. Hancock, *Multiple enzymatic digestion for enhanced sequence coverage of proteins in complex proteomic mixtures using capillary LC with ion trap MS–MS*, J. Proteome Res., 2 (2003) 59.

50. M.C. García, *The effect of the mobile phase additives on sensitivity in the analysis of peptides and proteins by LC–ESI-MS*, J. Chromatogr. B, 825 (2005) 111.

51. W. Walcher, H. Oberacher, S. Troiani, G. Hölzl, P. Oefner, L. Zolla, C.G. Huber, *Monolithic capillary columns for LC–ESI-MS in proteomic and genomic research*, J. Chromatogr. B, 782 (2002) 111.

52. L.J. Deterding, M.A. Moseley, K.B. Tomer, J.W. Jorgenson, *Nanoscale separations combined with MS–MS*, J. Chromatogr. 554 (1991) 73.

53. J.P. Chervet, M. Ursem, J.P. Salzmann, *Instrumental requirements for nanoscale LC*, Anal. Chem., 68 (1996) 1507.

54. Y. Shen, R. Zhao, S.J. Berger, G.A. Anderson, N. Rodrigues, R.D. Smith, *High-efficiency nanoscale LC coupled on-line with MS using nano-ESI for proteomics*, Anal. Chem., 74 (2002) 4235.

55. E. van der Heeft, G.J. ten Hove, C.A. Herberts, H.D. Meiring, C.A.C.M. van Els, A.P.J.M. de Jong, *A microcapillary column switching LC–ESI-MS system for the direct identification of peptides presented by major histocompatibility complex class I molecules*, Anal. Chem., 70 (1998) 3742.

56. H. Schaefer, J.-P. Chervet, C. Bunse, C. Joppich, H.E. Meyer, K. Marcus, *A peptide preconcentration approach for nano-LC to diminish memory effects*, Proteomics, 4 (2004) 2541.

57. H.D. Meiring, E. van der Heeft, G.J. ten Hove, A.P.J.M. de Jong, *Nanoscale LC–MS$^{(n)}$: technical design and applications to peptide and protein analysis*, J. Sep. Sci., 25 (2002) 557.

58. C.L. Gatlin, G.R. Kleemann, L.G. Hays, A.J. Link, J.R. Yates, III, *Protein identification at the low femtomole level from silver-stained gels using a new fritless ESI interface for LC–micro-ESI- and nano-ESI-MS*, Anal. Biochem., 263 (1998) 93.

59. D. Figeys, R. Aebersold, *Nanoflow solvent gradient delivery from a microfabricated device for protein identifications by ESI-MS*, Anal. Chem., 70 (1998) 3721.

60. T. Natsume, Y. Yamauuchi, H. Nakayama, T. Shinkawa, M. Yanagida, N. Takahashi, T. Isobe, *A direct nanoflow LC–MS–MS system for interaction proteomics*, Anal. Chem., 74 (2002) 4725.

61. A. Premstaller, H. Oberacher, W. Walcher, A.M. Timperio, L. Zolla, J.-P. Chervet, N. Cavusoglu, A. van Dorsselaer, C.G. Huber, *LC–ESI-MS using monolithic capillary columns for proteomic studies*, Anal. Chem., 73 (2001) 2390.

62. R.E. Moore, L. Licklider, D. Schumann, T.D. Lee, *A microscale ESI interface incorporating a monolithic, PS–DVB support for on-line LC–MS–MS analysis of peptides and proteins*, Anal. Chem., 70 (1998) 4879.

63. A.R. Ivanov, L. Zang, B.L. Karger, *Low-attomole ESI-MS and MS–MS analysis of protein tryptic digests using 20-µm-i.d. PS–DVB monolithic capillary columns*, Anal.

Chem., 75 (2003) 5306.
64. Q. Luo, Y. Shen, K.K. Hixson, R. Zhao, F. Yang, R.J. Moore, H.M. Mottaz, R.D. Smith, *Preparation of 20-μm-i.d. silica-based monolithic columns and their performance for proteomics analyses*, Anal. Chem., 77 (2005) 5028.
65. L. Rieuz, H. Niederländer, E. Verpoorte, R. Bischoff, *Silica monolithic columns: Synthesis, characterization and applications to the analysis of biological molecules*, J. Sep. Sci., 28 (2005) 1628.
66. M.T. Davis, T.D. Lee, *Variable flow LC–MS–MS and the comprehensive analysis of complex digest mixtures*, J. Am. Soc. Mass Spectrom., 8 (1997) 1059.
67. J.P.C. Vissers, R.K. Blackburn, M.A. Moseley, *A novel interface for variable flow nanoscale LC–MS–MS for improved proteome coverage*, J. Am. Soc. Mass Spectrom., 13 (2002) 760.
68. D.B. Kassel, B. Shushan, T. Sakuma, J.-P. Salzmann, *Evaluation of packed capillary perfusion column HPLC–MS–MS for the rapid mapping and sequencing of enzymatic digests*, Anal. Chem., 66 (1994) 236.
69. J.F. Banks, E.E. Gulcicek, *Rapid peptide mapping by LC on nonporous silica with on-line ESI-TOF-MS*, Anal. Chem., 69 (1997) 3973.
70. D.C. Delnisky, K.D. Greis, *Capillary LC–MS with column switching for rapid identification of proteins from 2D-GE*, J. Proteome Res., 1 (2002) 279.
71. M.I. Churchwell, N.C. Twaddle, L.R. Meeker, D.R. Doerge, *Improving LC–MS sensitivity through increases in chromatographic performance: Comparisons of UPLC–ES–MS–MS to HPLC–ES–MS–MS*, J. Chromatogr. B, 825 (2005) 134-143.
72. Y. Shen, R.D. Smith, K.K. Unger, D. Kumar, D. Lubda, *Ultrahigh-throughput proteomics using fast RPLC separations with ESI-MS–MS*, Anal. Chem., 77 (2005) 6692.
73. G.J. Opiteck, K.C. Lewis, J.W. Jorgenson, R.J. Anderegg, *Comprehensive on-line LC–LC–MS of proteins*, Anal. Chem., 69 (1997) 1518.
74. G.J. Opiteck, S.M. Ramirez, J.W. Jorgenson M.A. Moseley, III, *Comprehensive 2D-LC for the isolation of overexpressed proteins and proteome mapping*, Anal. Biochem., 258 (1998) 349.
75. G.J. Opiteck, J.W. Jorgenson, R.A. Anderegg, *2D SEC×RPLC coupled to MS for the analysis of peptides*, Anal. Chem., 69 (1997) 2283.
76. G.J. Opiteck, J.W. Jorgenson, M.A. Moseley, III, R.J. Anderegg, *2D microcolumn LC coupled to single quadrupole MS for the elucidation of sequence tags and peptide mapping*, J. Microcol. Sep., 10 (1998) 365.
77. D.A. Wolters, M.P. Washburn, J.R. Yates, III, *An automated MudPIT for shotgun proteomics*, Anal. Chem., 73 (2001) 5683.
78. M.T. Davis, J. Beierle, E.T. Bures, M.D. McGinley, J. Mort, J.H. Robinson, C.S. Spahr, W. Yu, R. Luethy, S.D. Patterson, *Automated LC–LC–MS–MS platform using binary IEC and gradient RPLC for improved proteomic analyses*, J. Chromatogr. B, 752 (2001) 281.
79. E. Nägele, M. Volmer, P. Hörth, *2D nano-LC–MS system for applications in proteomics*, J. Chromatogr. A, 1009 (2003) 197.
80. G. Mitulović, C. Stingl, M. Smoluch, R. Swart, J.-P. Chervet, I. Steinmacher, C. Gerner, K. Mechtler, *Automated, on-line 2D nano LC–MS–MS for rapid analysis of complex protein digests*, Proteomics, 4 (2004) 2545.
81. J. Peng, J.E. Elias, C.C. Thoreen, L.J. Licklider, S.P. Gygi, *Evaluation of*

multidimensional LC coupled with MS–MS for large-scale protein analysis: the yeast proteome, J. Proteome Res., 2 (2003) 43.

82.   J. Ji, A. Chakraborty, M. Geng, X. Zhang, A. Amini, M. Bina, F. Regnier, *Strategy for qualitative and quantitative analysis in proteomics based on signature peptides*, J. Chromatogr. B, 745 (2000) 197.

83.   S. Wang, X. Zhang, F. Regnier, *Quantitative proteomics strategy involving the selection of peptides containing both cysteine and histidine from tryptic digests of cell lysate*, J. Chromatogr. A, 949 (2002) 153.

84.   J. Carol, M.C.J.K. Gorseling, C.F. de Jong, H. Lingeman, C.E. Kientz, B.L.M. van Baar, H. Irth, *Determination of denatured proteins and biotoxins by on-line SEC–digestion–LC–ESI-MS*, Anal. Biochem., 346 (2005) 150.

85.   J.S. Hoos, H. Sudergat, J.-P. Hoelck, M. Stahl, J.S.B. de Vlieger, W.M.A. Niessen, H. Lingeman, H. Irth, *Selective quantitative bioanalysis of proteins in biological fluids by on-line IAC–protein digestion–LC–MS*, J. Chromatogr. B, 830 (2006) 262.

86.   N. Lion, T.C. Rohner, L. Dayon, I.L. Arnaud, E. Damoc, N. Youhnovski, Z.-Y. Wu, C. Roussel, J. Josserand, H. Jensen, J.S. Rossier, M. Przybylski, H.H. Girault, *Microfluidic systems in proteomics*, Electrophoresis, 24 (2003) 3533.

87.   W.-C. Sung, H. Makamba, S.-H. Chen, *Chip-based microfluidic devices coupled with ESI-MS*, Electrophoresis, 26 (2005) 1783.

88.   D. Figeys, Y. Ning, R. Aebersold, *A microfabricated device for rapid protein identification by micro-ESI ion trap MS*, Anal. Chem., 69 (1997) 3153.

89.   D. Figeys, C. Lock, L. Taylor, R. Aebersold, *Microfabricated device coupled with an ESI-Q–TOF-MS: protein identifications based on enhanced-resolution MS and MS–MS data*, Rapid Commun. Mass Spectrom., 12 (1998) 1435.

90.   D. Figeys, S.P. Gygi, G. McKinnon, R. Aebersold, *An integrated microfluidics–MS–MS system for automated protein analysis*, Anal. Chem., 70 (1998) 3728.

91.   J. Li, C. Wang, J.F. Kelly, D.J. Harrison, P. Thibault, *Rapid and sensitive separation of trace level protein digests using microfabricated devices coupled to a Q–TOF-MS*, Electrophoresis, 21 (2000) 198.

92.   H. Yin, K. Killeen, R. Brennen, D. Sobek, M. Werlich, T. van der Goor, *Microfluidic chip for peptide analysis with an integrated HPLC column, sample enrichment column, and nano-ESI tip*, Anal. Chem., 77 (2005) 527.

93.   P. Schmitt-Kopplin, M. Frommberger, *CE–MS: 15 years of developments and application*, Electrophoresis, 24 (2003) 3837.

94.   J. Hernández-Borges, C. Neusüß, A. Cifuentes, M. Pelzing, *On-line CE–MS for the analysis of biomolecules*, Electrophoresis, 25 (2004) 2257.

95.   D.C. Simpson, R.D. Smith, *Combining CE with MS for applications in proteomics*, Electrophoresis, 26 (2005) 1291.

96.   C.W. Klampfl, *Recent advances in the application of CE with MS detection*, Electrophoresis, 27 (2006) 3.

97.   A. Zamfir, J. Peter-Katalinić, *CE–MS for glycoscreening in biomedical research*, Electrophoresis, 25 (2004) 1949.

98.   M. Pelzing, C. Neusüß, *Separation techniques hyphenated to ESI-MS in proteomics: CE versus nano-LC*, Electrophoresis, 26 (2005) 2717.

99.   L. Yang, C.S. Lee, S.A. Hofstadler, L. Paša-Tolić, R.D. Smith, *CIEF–ESI-FT-ICR-MS for protein characterization*, Anal. Chem., 70 (1998) 3235.

100. P.K. Jensen, L. Paša-Tolić, G.A. Anderson, J.A. Horner, M.S. Lipton, J.E. Bruce,

R.D. Smith, *Probing proteomes using CIEF–ESI-FT-ICR-MS*, Anal. Chem., 71 (1999) 2076.

101. D.F. Hunt, J.R. Yates, III, J. Shabanowitz, S. Winston, C.R. Hauer, *Protein sequencing by MS–MS*, PNAS, 83 (1986) 6233.

102. I.A. Papayannopoulos, *The interpretation of CID tandem mass spectra of peptides*, Mass Spectrom. Rev., 14 (1995) 49.

103. P. Roepstorff, J. Fohlmann, *Proposal for a common nomenclature for sequence ions in mass spectra of peptides*, Biomed. Mass Spectrom., 11 (1984) 601.

104. K. Biemann, *Contributions of MS to peptide and protein structure*, Biomed Mass Spectrom., 16 (1988) 99.

105. B. Paizs, S. Suhai, *Fragmentation pathways of protonated peptides*, Mass Spectrom. Rev., 24 (2005) 508.

106. J.H. Bowie, C.S. Brinkworth, S. Dua, *CID of the [M–H]⁻ parent anions of underivatized peptides: an aid to structure determination and some unusual negative ion cleavages*, Mass Spectrom. Rev., 21 (2002) 87.

107. R.W. Vachet, K.L Ray, G.L. Glish, *Origin of product ions in the MS–MS spectra of peptides in a quadrupole ion trap*, J. Am. Soc. Mass Spectrom., 9 (1998) 341.

108. L. Sleno, D.A. Volmer, *Ion activation methods for MS–MS*, J. Mass Spectrom., 30 (2004) 1091.

109. R.A. Zubarev, D.M. Horn, E.K. Fridriksson, N.L. Kelleher, N.A. Kruger, M.A. Lewis, B.K. Carpenter, F.W. McLafferty, *ECD for structural characterization of multiply charged protein cations*, Anal. Chem., 72 (2000) 563.

110. R.A. Zubarev, *Reactions of polypeptide ions with electrons in the gas phase*, Mass Spectrom. Rev., 22 (2003) 57.

111. W.J. Henzel, T.M. Billeci, J.T. Stults, S.C. Wong, C. Grimley, C. Watanabe, *Identifying proteins from 2D-GE by molecular mass searching of peptide fragments in protein sequence databases*, PNAS, 90 (1993) 5011.

112. M. Mann, P. Højrup, P. Roepstorff, *Use of MS molecular weight information to identify proteins in sequence databases*, Biol. Mass Spectrom., 22 (1993) 338.

113. E. Mørtz, O. Vorm, M. Mann, P. Roepstorff, *Identification of proteins in polyacrylamide gels by MS peptide mapping combined with database search*, Biol. Mass Spectrom., 23 (1994) 249.

114. O.N. Jensen, A.V. Podtelejnikov, M. Mann, *Identification of the components of simple protein mixtures by high-accuracy peptide mass mapping and database searching*, Anal. Chem., 69 (1997) 4741.

115. M. Mann, M. Wilm, *Error-tolerant identification of peptides in sequence databases by peptide sequence tags*, Anal. Chem., 66 (1994) 4390.

116. D.D.J. Pappin, P. Højrup, A.J. Bleasby, *Rapid identification of proteins by PMF*, Curr. Biol., 3 (1993) 327.

117. W. Zhang, B.T. Chait, *ProFound – an expert system for protein identification using MS PMF information*, Anal. Chem., 72 (2000) 2482.

118. D.N. Perkins, D.J. Pappin, D.M. Creasy, J.S. Cottrell, *Probability-based protein identification by searching sequence databases using MS data*, Electrophoresis, 20 (1999) 3551.

119. J.R. Yates, III, S. Speicher, P.R. Griffin, T. Hunkapiller, *Peptide mass maps: A highly informative approach to protein identification*, Anal. Biochem., 214 (1993) 397.

120. J.K. Eng, A.L. McCormack, J.R. Yates, III, *An approach to correlate tandem mass*

spectral data of peptides with amino acid sequences in a protein database, J. Am. Soc. Mass Spectrom., 5 (1994) 976.

121.   J.R. Yates, III, J.K. Eng, A.L. McCormack, D. Schieltz, *Methods to correlate tandem mass spectra of modified peptides to amino acid sequences in the protein database*, Anal. Chem., 67 (1995) 1426.

122.   J.R. Yates, III, J.K. Eng, A.L. McCormack, *Mining genomes: Correlating tandem mass spectra of modified and unmodified peptides to sequences in nucleotide databases*, Anal. Chem., 67 (1995) 3202.

123.   A.L. McCormack, D.M. Schieltz, B. Goode, S. Yang, G. Barnes, D. Drubin, J.R. Yates, III, *Direct analysis and identification of proteins in mixtures by LC–MS–MS and database searching at the low-femtomole level*, Anal. Chem., 69 (1997) 767.

124.   R.E. Moore, M.K. Young, T.D. Lee, *Method for screening peptide fragment ion mass spectra prior to database searching*, J. Am. Soc. Mass Spectrom., 11 (2000) 422.

125.   B. Ma, K. Zhang, C. Hendrie, C. Liang, M. Li, A. Doherty-Kirby, G. Lajoie, *PEAKS: powerful software for peptide de novo sequencing by MS–MS*, Rapid Commun. Mass Spectrom., 17 (2003) 2337.

126.   D.L. Tabb, A. Saraf, J.R. Yates, III, *GutenTag: High-throughput sequence tagging via an empirically derived fragmentation model*, Anal. Chem., 75 (2003) 6415.

127.   J.A. Taylor, and R.S. Johnson, *Implementation and uses of automated de novo peptide sequencing by MS*, Anal. Chem., 73 (2001) 2594.

128.   G.K. Taylor, Y.-B. Kim, A.J. Forbes, F. Meng, R. McCarthy, N.L. Kelleher, *Web and database software for identification of intact proteins using "top down" MS*, Anal. Chem., 75 (2003) 4081.

129.   B.C. Searle, S. Dasari, M. Turner, A.P. Reddy, D. Choi, P.A. Wilmarth, A.L. McCormack, L.L. David, S.R. Nagalla, *High-throughput identification of proteins and unanticipated sequence modifications using a mass-based alignment algorithm for MS–MS de novo sequencing results*, Anal. Chem., 76 (2004) 2220.

130.   A.J. Liska, A. Shevchenko, *Combining MS with database interrogation strategies in proteomics*, Trends Anal. Chem., 22 (2003) 291.

131.   R. Apweiler, A. Bairoch, C.H. Wu, *Protein sequence databases*, Curr. Opin. Chem. Biol., 8 (2004) 76.

132.   A.J. Liska, A. Shevchenko, *Expanding the organismal scope of proteomics: Cross-species protein identification by MS and its applications*, Proteomics, 3 (2003) 19.

133.   S.F. Altschul, W. Gish, W. Miller, E.W. Myers, D.J. Lipman, *Basic local alignment search tool*, J. Mol. Biol., 215 (1990) 403.

134.   S.F. Altschul, D.J. Lipman, *Protein database searches for multiple alignments*, PNAS, 87 (1990) 5509.

135.   B. Devreese, F. Vanrobaeys, J. Van Beeumen, *Automated nanoflow LC–MS–MS identification of proteins from Shewanella putrefaciens separated by 2D-GE*, Rapid Commun Mass Spectrom., 15 (2001) 50.

136.   A. Shevchenko, I. Chernushevich, W. Ens, K.G. Standing, B. Thomson, M. Wilm, M. Mann, *Rapid 'de novo' peptide sequencing by a combination of nano-ESI, isotopic labeling and a Q–TOF-MS*, Rapid Commun. Mass Spectrom., 11 (1997) 1015.

137.   K.R. Jonscher, J.R. Yates, III, *The quadrupole ion trap mass spectrometer – A small solution to a big challenge*, Anal. Biochem., 244 (1997) 1.

138.   R.K. Blackburn, M.A. Moseley, III, *Q–TOF-MS: A powerful new tool for protein identification and characterization*, Am Pharm. Rev., 2 (1999) 49.

139. J.C.Y. Le Blanc, J.W. Hager, A.M.P. Ilisiu, C. Hunter. F. Zhong, I. Chu, *Unique scanning capabilities of a new hybrid LIT-MS (Q TRAP) used for high sensitivity proteomics applications*, Proteomics, 3 (2003) 859.

140. D.C. Stahl, K.M. Swiderek, M.T. Davis, T.D. Lee, *Data-controlled automation of LC–MS–MS analysis of peptide mixtures*, J. Am. Soc. Mass Spectrom., 7 (1996) 532.

141. B.R. Wenner, B.C. Lynn, *Factors that affect ion trap data-dependent MS–MS in proteomics*, J. Am. Soc. Mass Spectrom., 15 (2004) 150.

142. J.V. Olsen, M. Mann, *Improved peptide identification in proteomics by two consecutive stages of MS fragmentation*, PNAS, 101 (2004) 13417.

143. Z. Zhang, J.S. McElvain, *De novo peptide sequencing by 2D fragment correlation MS*, Anal. Chem., 72 (2000) 2337.

144. M. Wilm, G. Neubauer, M. Mann, *Parent ion scans of unseparated peptide mixtures*, Anal. Chem., 68 (1996) 527.

145. W.D. Lehmann, *Single series peptide fragment ion spectra generated by two-stage CIDn in a triple quadrupole*, J. Am. Soc. Mass Spectrom., 9 (1998) 606.

146. J.A. Loo, C.G. Edmonds, R.D. Smith, *MS–MS of very large molecules: Serum albumin sequence information from multiply charged ions formed by ESI*, Anal. Chem., 63 (1991) 2488.

147. K.D. Henry, J.P. Quinn, F.W. McLafferty, *High-resolution ESI-MS of large molecules*, J. Am. Chem. Soc., 113 (1991) 5447.

148. A.G. Marshall, *Milestones in FT-ICR-MS technique development*, Int. J. Mass Spectrom., 200 (2000) 331.

149. R.M.A. Heeren, A.J. Kleinnijenhuis, L.A. McDonnell, T.H. Mize, *A mini-review of MS using high-performance FT-ICR-MS methods*, Anal. Bioanal. Chem., 378 (2004) 1048.

150. N.L. Kelleher, M.W. Senko, M.M. Siegel, F.W. McLafferty, *Unit resolution mass spectra of 112 kDa molecules with 3 Da accuracy*, J. Am. Soc. Mass Spectrom., 8 (1997) 380.

151. J.E. Bruce, G.A. Anderson, J. Wen, R. Harkewicz, R.D. Smith, *High-mass-measurement accuracy and 100% sequence coverage of enzymatically digested bovine serum albumin from an ESI-FTICR mass spectrum*, Anal. Chem., 71 (1999) 2595.

152. G.A. Valaskovic, N.L. Kelleher, F.W. McLafferty, *Attomole protein characterization by CE–MS*, Science, 273 (1996) 1199.

153. M.E. Belov, M.V. Gorshkov, H.R. Udseth, G.A. Anderson, R.D. Smith, *Zeptomole-sensitivity ESI–FT-ICR-MS of proteins*, Anal. Chem., 72 (2000) 2271.

154. M.W. Senko, C.L. Hendrickson, M.R. Emmett, S.D.-H. Shi, A.G. Marshall, *External accumulation of ions for enhanced ESI-FT-ICR-MS*, J. Am. Soc. Mass Spectrom., 8 (1997) 970.

155. M.R. Emmett, F.M. White, C.L. Hendrickson, S.D.H. Shi, A.G. Marshall, *Application of micro-ESI LCy techniques to FT-ICR-MS to enable high-sensitivity biological analysis*, J. Am Soc. Mass Spectrom., 9 (1998) 333.

156. S.E. Martin, J. Shabanowitz, D.F. Hunt, J.A. Marto, *Subfemtomole MS and MS–MS peptide sequence analysis using nano-HPLC micro-ESI-FT-ICR-MS*, Anal. Chem., 72 (2000) 4266.

157. S.M. Patrie, J.P. Charlebois, D. Whipple, N.L. Kelleher, C.L. Hendrickson, J.P. Quinn, A.G. Marshall, B. Mukhopadhyay, *Construction of a hybrid quadrupole–FT-*

ICR-MS for versatile MS–MS above 10 kDa, J. Am. Soc. Mass Spectrom., 15 (2004) 1099.

158. J.E.P. Syka, J.A. Marto, D. L. Bai, S. Horning, M.W. Senko, J.C. Schwartz, B. Ueberheide, B. Garcia, S. Busby, T. Muratore, J. Shabanowitz, D.F. Hunt, *Novel LIT–FT-ICR-MS: Performance characterization and use in the comparative analysis of histone H3 post-translational modifications*, J. Proteome Res., 3 (2004) 621.

159. J.A. Loo, J.P. Quinn, S.I. Ryu, K.D. Henry, M.W. Senko, F.W. McLafferty, *High-resolution μ of large biomolecules*, PNAS, 89 (1992) 286.

160. M.W. Senko, J.P. Speir, F.W. McLafferty, *Collision activation of large multiply charged ions using FT-ICR-MS*, Anal. Chem., 66 (1994) 2801.

161. D.P. Little, J.P. Speir, M.W. Senko, P.B. O'Connor, F.W. McLafferty, *IRMPD of large multiply charged ions for biomolecule sequencing*, Anal. Chem., 66 (1994) 2809.

162. W. Li, C.L. Hendrickson, M.R. Emmett, A.G. Marshall, *Identification of intact proteins in mixtures by alternated capillary LC–ESI and LC–ESI-IRMPD-FT-ICR-MS*, Anal. Chem., 71 (1999) 4397.

163. N.L. Kelleher, H.Y. Lin, G.A. Valaskovic, D.J. Aaserud, E.K. Fridriksson, F.W. McLafferty, *Top down versus bottom up protein characterization by tandem high-resolution MS*, J. Am. Chem. Soc., 121 (1999) 806.

164. H. Zhai, X. Han, K. Breuker, F.W. McLafferty, *Consecutive ion activation for top down MS: Improved protein sequencing by nozzle–skimmer dissociation*, Anal. Chem., 77 (2005) 5777.

165. T.P. Conrads, G.A. Anderson, T.D. Veenstra, L. Paša-Tolić, R.D. Smith, *Utility of accurate mass tags for proteome-wide protein identification*, Anal. Chem., 72 (2000) 3349.

166. R.A. Henderson, A.L. Cox, K. Sakaguichi, E. Appella, J. Shabanowitz, D.F. Hunt, V.H. Engelhard, *Direct identification of an endogenous peptide recognized by multiple HLA-A2.1 specific cytotoxic T cells*, PNAS, 90 (1993) 10275.

18

LC–MS IN PROTEOMICS

1. Introduction

In the previous chapter (Ch. 17), key technologies in the LC–MS analysis of peptides and proteins were introduced and discussed. The technologies discussed can be considered as the enabling technologies, at least from the LC–MS point of view, to the current developments in the field of proteomics.

Important enabling technologies and concepts are:

- The group of Hunt [1] pioneered the identification of trace-level peptides in complex biological samples by means of microcapillary reversed-phase RPLC–MS–MS in the analysis of endogenous human class I major histocompatibility complex (MHC) encoded human leucocyte antigen (HLA-A2.1) peptides.
- The group of Mann [2-3] pioneered in nanoelectrospray (nano-ESI), enabling the identification of proteins separated by two-dimensional gel electrophoresis (2D-GE), digested with trypsin, and infused by nano-ESI-MS–MS.
- Database searching with m/z data from peptide mass fingerprints (PMF) or uninterpreted MS–MS spectra was pioneered by several groups [4-6].
- Integrated workstations were developed enabling autosampler injection of protein digests into a nano-LC column, coupled to ESI-MS–MS, operated with data-dependent acquisition (DDA), and providing automated database searching using the SEQUEST algorithm [7], as demonstrated in the unattended identification of 90 proteins from the yeast *Saccharomyces cerevisiae*. Integrated

systems are available based on triple-quadrupole instruments, and especially ion-trap and quadrupole–time-of-flight hybrid (Q–TOF) instruments.

- Multidimensional protein identification (MudPIT) systems were introduced by the group of Yates [7-8]. MudPIT is based on LC×LC–MS–MS with a strong cation exchange (SCX) column as the first dimension, RPLC as the second dimension, ESI-MS–MS under DDA operation, and subsequent SEQUEST database searching.
- Protein characterization by means of LC–MS or direct-infusion on FT-ICR-MS instruments was realized by the efforts of several groups [9-12].
- Stable isotope-labelled affinity tags have been introduced [13] to enable comparative quantitative proteomics.

Some of the challenges of proteomics research by means of LC–MS were nicely illustrated by Julka and Regnier [25]. The proteome of an average cell contains ~10,000 proteins. If each protein would be digested into 30–50 peptides, the tryptic digest contains $3–5\times10^5$ peptides. The typical peak capacity of a RPLC column is ~400 and no more than 20–40 peptides can be analysed simultaneously. This means that the analytical capacity of LC–MS is at best 2×10^4 peptides, which is more than an order of magnitude lower than ideally required. An additional problem is that the protein concentrations span a very wide range, e.g., 10^6 in *Escherichia coli* and ~10^9 in human, while the dynamic range of most mass spectrometers is ~10^4. Given the limited peak capacity of the LC–MS system, co-elution of peptides from low-abundance and high-abundance proteins is highly likely to occur, which among others may lead to signal suppression of the low-abundance peptide.

A large number of review papers were published on the role of MS in proteomics or specific application areas within the field. Some of these papers were especially useful in preparing this chapter and are recommended as further reading [14-18]. The book by Liebler [19] provides a concise overview. The discussion in this chapter provides a technology-oriented review of developments in relation to LC–MS, and is certainly not a comprehensive overview of proteomics.

2. Proteomics, a concise overview

Proteomics is the study of the proteome, the expression of the genome in cellular and extracellular proteins, *i.e.*, involving the entire complement of proteins expressed by a particular cell, organism, or tissue. It comprises a transformation of protein chemistry and structure biology, where the structure and function of individual proteins are studied, into systems biology, where partial peptide sequencing is performed in combination with database matching to identify proteins in complex mixtures. The focus has moved to understand the cellular function of proteins. This means that next to identification of which proteins are present, other questions are relevant, *e.g.*, related to functional modification of proteins and sets of apparently related proteins, and to up- and down-regulation of concentrations of specific proteins in cells as a function of cellular state. This implies that next to qualitative aspects, one also has to focus on quantitative proteomics.

The major application areas of proteomics involve:

- Mining the proteome, *i.e.*, identify as many proteins in a sample as possible.
- Protein-expression profiling, *i.e.*, identify proteins in a sample as a function of the cellular state (life cycle development, differentiation, disease state).
- Protein-protein interactions and protein network mapping, *i.e.*, studying the interaction between proteins in relation to their function in the living cell.
- Mapping of protein (post-translational) modifications, which determine targeting, structure, function, and turnover of proteins (Ch. 19).

An alternative classification would indicate the areas of profiling, functional, and structural proteomics, dealing with differential expression of proteins, function of individual proteins and/or protein complexes in the living cell, and elucidation of the tertiary structure of proteins in the cell. In this classification, the role of MS would be considered from a different perspective.

Within this text, the emphasis is on analytical proteomics, *i.e.*, the development and application of (LC–MS-based) analytical technologies needed to enable the biologist to perform his or her studies in profiling, functional, and structural proteomics, and not so much on the results achieved. Analytical requirements in proteome analysis are sensitivity, chromatographic, and in some instances mass spectrometric, resolution, sample throughput, and confidence in protein identification. The emergence and evolution of fundamental technologies and concepts required in proteomics studies are to some extent discussed in Ch. 17. To some extent, because next to ESI-MS and LC–MS discussed in detail in this text, matrix-assisted laser desorption ionization (MALDI-MS) is also of utmost importance in this field. Some important technologies are briefly discussed here.

2.1   Peptide mass fingerprinting

When a mixture of peptides, generated by the enzymatic digestion of a protein or protein mixture, is directly analysed without prior separation by means of nano-ESI-MS or MALDI-MS, the resulting mass spectrum is called a peptide map or peptide mass fingerprint (PMF). The m/z values of the peptides detected can be searched against protein databases, *e.g.*, using the PeptideSearch, MASCOT, or similar software [4-6, 20] (Ch. 17.6.2). The quality of the search results is strongly determined by number of peptide m/z included in the search and especially the mass accuracy with which the peptide map was acquired [21]. The quality of PMF can be improved by performing additional searching using peptide sequence tags.

Due to the different ionization characteristics, complementary data may be obtained from ESI and MALDI. However, the peptide map is much easier to obtain by MALDI-MS. Therefore, PMF using MALDI-MS is often the method-of-choice, also because advantage can be taken from the good mass accuracy of the TOF mass analyser. Robotic workstations have been developed for automated cutting of protein spots from gels, destaining, in-gel digestion, peptide extraction, and subsequent transfer of the peptide mixture to a MALDI target. Instruments have been developed that automatically perform high-throughput measurement of the peptide maps contained at different spots on the MALDI target, automated data processing, and database searching for protein identification [22]. PFM was reviewed and critically evaluated by Henzel *et al.* [23].

2.2   Peptide sequence analysis

A complementary approach to protein identification is based on peptide sequencing analysis (PSA) by collision-induced dissociation (CID) in MS–MS. For proteomics studies, the *de novo* sequencing of MS–MS spectra is too time-consuming. Therefore, algorithms have been developed to either provide automated MS–MS spectrum interpretation, or perform protein identification by means of a SEQUEST database search [6-7]. In the latter case, the precursor m/z values are matched against a virtual digestion of all the proteins in the database. Sequence ions are predicted for the peptides that match the precursor m/z values, and compared with the measured MS–MS data. A correlation score is calculated for each match.

The SEQUEST database search algorithm (Ch. 17.6.2) is a powerful tool in protein identification, although like any statistical tool it relies on the judgments of the user to critically evaluate its results. Problems may especially arise, when unanticipated modifications are present in the protein investigated.

The potential of using other peptide properties to support the identification was evaluated. Parameters such as retention time [24-25] or isoelectric point (pI) [104, 153] were evaluated for this purpose.

A nice application of this approach is the automated identification of amino-acid sequence variations resulting from single nucleotide polymorphisms (SNP), which is relevant in the detection and identification of hemoglobin variants [28]. Interestingly, these types of unanticipated variations are not effectively handled by the SEQUEST algorithm. Therefore, the SEQUEST program was modified. Isolated, dehemed, reduced, and alkylated hemoglobins were subjected to a combination proteolytic digestion, to obtain a complex peptide mixture with multiple sequence overlaps. By means of DDA, LC–MS–MS spectra were acquired for the five most abundant ions in each scan. In this way, amino-acid changes in six hemoglobin variants were correctly identified.

2.3   Accurate-mass, affinity, or sequence tags

With sufficiently high specificity in MS detection, which may result from either accurate-mass measurement or PSA by MS–MS, often one (or only a few) peptides from a protein tryptic digest is sufficient for unequivocal identification of this protein. This implies that strategies can be developed, in which tryptic peptides with specific properties are selectively isolated from the complex protein mixture prior to introduction in RPLC–MS. This is the signature peptide approach, promoted by the group of Regnier [29-30]. The isolation of peptides labelled with an isotope-coded affinity tag (ICAT) using avidin AfC (Ch. 18.4.1) is an example of this approach.

The accurate mass tag (AMT) approach, promoted by the group of Smith [31-33], is the mass spectrometric counterpart of the signature peptide approach, where the high resolving power of FT-ICR-MS is applied to perform accurate-mass determination (<1 ppm) of one peptide to identify the complete protein. Selection of a peptide as AMT is based on potential tags derived from MS–MS experiments. The subsequent use of these AMT in FT-ICR-MS avoids the need to perform MS–MS and will therefore improve sensitivity and throughput of the method. In a latter study, the AMT approach was extended by inclusion of normalized chromatographic retention time (NET) data [24-25]. For more complex proteomes, mass accuracies better than 1 ppm and NET within 0.01 are required.

Both approaches are applied in both proteome mining (Ch. 18.3), protein-expression profiling (Ch. 18.4), and PTM-mapping (Ch. 19).

A third approach, pioneered by the group of Liebler [19, 34], involves a pattern recognition software called scoring algorithm for spectral analysis (SALSA) to search specific sequence motifs in the MS–MS data. Potential applications envisaged for SALSA include identification of specific protein modifications, *e.g.*, PTM, identifications of peptides with common sequences, *e.g.*, wild-type and mutant forms, and targeted analysis of isoforms and conformers in complex samples.

3. Mining the proteome

Proteome mining involves the identification of as many proteins of a proteome as possible. However, the proteome of a cell or organism changes with the developmental stage or state. In higher organisms, different genes are expressed in different tissues. This poses even more challenges to proteomics studies. The aim is thus mining the proteome at a specific state (and perhaps comparing it to other states, see Ch. 18.4). Important aspects related to selection and subsequent collection of the proteins from the tissue, organisms or cell are not discussed here. The focus is on the measurement strategy. The two central technologies in proteome mining are PMF (Ch. 18.2.1) and PSA (Ch. 18.2.2). With respect to PMF, a strategy involving MALDI-MS is often the method-of-choice, because of its speed, simplicity, and robustness [22]. Interestingly, developments in the last few years, especially with respect to LC×LC–MS in shotgun proteomics (Ch. 18.3.2), have resulted in more focus on LC–MS in mining studies. This is partly due to the added value of PSA involving MS–MS in protein identification. In that respect, MALDI-MS recently started to regain territory as a result of the introduction of commercial TOF–TOF-MS workstations. In addition, MALDI and atmospheric-pressure MALDI sources have become available for ion-trap and Q–TOF instruments.

In an inter-laboratory comparison of PMF based on MALDI-MS and PSA based on LC–MS–MS, a total of 162 2D-GE-separated protein spots from *Methanococcus jannaschii* and *Pyrococcus furiosus* were studied [35]. PSA matched 100% of the gel spots, and PMF 97%. Multiple proteins were detected in 9% and 50% of the gel spots by PMF and PSA, respectively. PSA provided better sequence coverage than PMF.

The complementary nature of MALDI-MS and ESI-MS–MS can also be exploited to increase the proteome coverage via PSA [36]. A fraction (20%) of the column effluent of a nano-LC column was sent to ESI-MS–MS, while the rest was fractionated onto MALDI spots for automated off-line LC–MALDI-MS and MS–MS analysis. About 63% overlap in protein identified from a digest of mammalian mitochondrial ribosomes was observed, but unique proteins were also identified by either technique.

The power of the currently available methods for proteome mining can be illustrated with data reported on the identification of ~800 proteins from 50 million *E. coli* K12 cells using multiple fractionation, RPLC–MS–MS, and various ways of database searching [37]. With MASCOT searching, 754 proteins were identified from 1326 peptides and 2167 MS–MS spectra, which means on average 1.75 peptides and 2.87 spectra per protein. About 17% of the *E. coli* K12 proteome was covered in this way.

Protein identification in a proteome by tryptic digestion and RPLC–MS relies on the ability of trypsin to achieve efficient digestion of the proteins in the mixture, the ability of RPLC to separate the resulting peptides, and deliver them to nano-ESI-MS for mass analysis. The peptides in a complex mixture widely differ in their physicochemical properties, which in turn influence both LC and nano-ESI-MS performance. Data were compared for experimentally detected and *in-silico* predicted peptides from three different complex protein mixtures [38]. The peptides detected actually form only a small subset of the peptides present.

The focus in this text is on the application of LC–MS technology. Three general approaches to proteome mining can be discriminated, which are related to the three general approaches to protein identification: bottom-up, top-down, and shotgun. The latter essentially is a bottom-up approach, but is discussed separately.

3.1   Bottom-up protein identification

The bottom-up approach very much resembles classical protein identification strategies. The proteins in the proteome are first separated by 2D-GE (Ch. 17.3), or in some cases by SCX, size-exclusion (SEC), or affinity (AfC) chromatography. Specific proteins are excised from the gel, blotted, or electroeluted. The protein is digested, and the digest is analysed by LC–MS. The LC separation involves either RPLC with microcapillary or nano-LC columns (Ch. 17.5.2), or 2D-LC with typically SEC or SCX in the first dimension and RPLC in the second (Ch. 17.5.4). Alternatively, the sample may be introduced via either direct-infusion nano-ESI (Ch. 17.2), CE–MS (Ch. 17.5.6), or a microfluidic device coupled to MS (Ch. 17.5.5).

2D-GE continues to be an important tool in protein separation. In-gel digestion of excised protein spots and extraction of the tryptic digests can be performed.

The high resolving power of FT-ICR-MS can readily be exploited in bottom-up protein identification. A nice example is the identification of high-abundant proteins in a tryptic digest of human plasma without any prior separation. The 2745 peaks in the spectrum could be reduced to 1165 isotopic clusters and 669 unique masses, 82 of which matched tryptic fragments of albumin (93% sequence coverage) and 16 others transferrin (41%) [39]. The same group showed that a theoretically predicted retention time of a tryptic peptide can be applied as an additional protein identification tool, next to its accurate mass acquired in LC–FT-ICR-MS. [40-41]

3.2   Shotgun protein identification: on-line LC×LC–MS

The shotgun protein identification approach can be considered as a modification of the bottom-up approach. In conventional bottom-up approaches, the proteins in a mixture are first separated prior to their tryptic digestion. In the shotgun approach, a mixture of proteins is digested with a proteolytic enzyme to produce a complex mixture of peptides. This mixture is analysed by LC–MS, generally operated under DDA with switching between survey-MS and product-ion MS–MS mode (Ch. 17.7.2). The large number of MS–MS spectra acquired in this way are then searched against protein or expressed sequence tags (EST) databases [42] to identify the proteins present in the mixture.

This approach was pioneered by the group of Yates [43]. Using standard mixtures of proteins, they established the proteins in the mixtures with 30-fold difference in molar quantity can still be successfully identified. Initially, the method was applied to protein mixtures obtained by immunoaffinity precipitation or similar specific isolation procedures based on protein interaction (Ch. 17.4.1). In another study, *E. coli* periplasmic proteins were identified by the same approach [44]. The protein fraction was enriched using anion-exchange chromatography (AEC). Part of each fraction was separated by GE. The proteins were detected by silver staining. The other part was enzymatically digested and analysed by LC–MS–MS, operated under DDA. The resulting MS–MS spectra were searched against the *E. coli* sequence database. In GE, ~160 bands were observed, whereas ~80 proteins were identified by LC–MS.

Subsequently, this approach was modified by the implementation of a biphasic column, featuring a 2D-LC separation on an SCX followed by RPLC [8]. The column is an in-needle 100-µm-ID fritless column (Ch. 17.5.2), consisting of 100 mm with 5-µm C_{18} and 40 mm with 5-µm SCX particles (see Figure 18.1a). The system was called multidimensional protein identification technology (MudPIT). After digestion, the protein mixture is injected onto the column. The tryptic peptides are eluted from the SCX column by means of a salt step gradient (ten 2-min steps with between 25 and 250 mmol/l ammonium acetate). A 3-min wash step with 5% acetonitrile in 0.5% aqueous acetic acid is followed by an 80 min RPLC gradient with each elution step of the SCX column. The column effluent is introduced into an ESI-MS–MS system operated under DDA. In this way, 288 proteins of *S. cerevisiae* were identified, using data from 666 peptides via SEQUEST database searching. Even better results (1484 proteins identified) were obtained with RPLC fractionation in five fractions prior to the MudPIT procedure [8, 45].

The MudPIT procedure is a powerful approach to comprehensive proteome mining. In a comparison of the biphasic MudPIT approach (Figure 18.1a) with either 1D-RPLC–MS or triphasic MudPIT (Figure 18.1b), 26, 55, and 62 proteins were identified from a complex protein mixture associated with bovine brain microtubules using the 1D, 2D, and 3D approach, respectively [46]. The 3D approach enabled the identification of some hydrophilic peptides that were missed by the 2D approach. Even a tetraphasic MudPIT with longer columns was described [47].

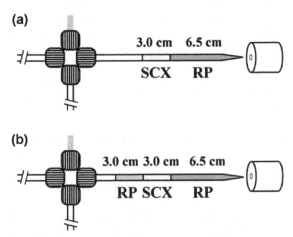

Figure 18.1: Setup of (a) biphasic and (b) triphasic MudPIT. Reprinted from [46] with permission. ©2002, Elsevier Science BV.

More recently, attention was also paid to the potential use of predicted retention times to strengthen the confidence in protein identification [38, 40-41, 48-49]. It also emphasizes the still restricted perspective we have on the proteome by means of the current technology. Whereas this is not (yet) a major problem in proteome mining, it may significantly influence the results in protein-expression profiling (Ch. 18.4) and/or protein-related biomarker discovery (Ch. 18.6).

The MudPIT approach is not without problems. A fully automated procedure as described requires (almost) continuous MS data acquisition for more than 15 hours, resulting in several thousands of mass spectra that put high demands on computers and bioinformatics. The procedure is limited by the ability of the mass spectrometer to rapidly switch between MS and MS–MS analysis under DDA control. With current technology, certainly not all peptides present can be analysed with both modes (Ch. 17.7.2). The power of the approach is greatly enhanced by performing a protein prefractionation, either by RPLC [8, 45] or by means of a protein-specific enrichment technique such as AfC (Ch. 17.4.1).

The MudPIT approach was modified in order to perform LC×LC in two separated columns (SCX and RPLC), which are connected via switching valves and a short RP trapping column [50]. This allows intermediate enrichment and desalting of the fractions eluted from the SCX prior to RPLC–MS analysis. The application of such a system in combination with a Q–TOF instrument has been applied to identification of the *Caenorhabditis elegans* proteome [51]. In total, 1616 proteins were identified, including 110 secreted/targeted proteins and 242 transmembrane proteins. Many of the ~5400 peptides identified in the course of this study contained a PTM such as acetylation and phosphorylation. This type of system has become commercially available as a fully-automated workstation [52-53] (Figure 17.1).

Another modification of the LC×LC–MS approach is the use of ultra-high-efficiency RPLC columns [54-55]. The system consists of a 800×0.32-μm-ID SCX column (packed with 3-μm polysulfoethyl aspartamide-bonded silica), a 40-mm×75-μm-ID RPLC trapping column, and a 800-mm×30-μm-ID RPLC column. The system was applied to the identification of proteins in the proteome of *Deinococcus radiodurans* [54], and of human plasma proteins [55]. In that case, more than 800 proteins, *i.e.*, both low-abundance cytokines and high-abundance proteins, were identified from in total 365 μg plasma.

Alternatively, high-speed RPLC was reported by means of a short 50-μm-ID column packed with 0.8-μm porous C_{18}-modified silica particles in combination with TOF (0.2-s acquisition) or FT-ICR-MS (0.3–0.6 s acquisition) [56]. Applications in proteome mining of *Shewanella oneidensis* were shown: ~600 proteins were identified from ~2000 peptides within 3 min.

3.3   Shotgun protein identification: off-line prefractionation

It was argued that off-line SCX fractionation using a shallow gradient and the collection of small effluent fraction significantly increases the overall peak capacity of the system [57]. The reduced complexity of the fractions to be analysed using nano-LC–MS facilitates a shorter analysis time in this step. Off-line prefractionation prior to shotgun LC–MS–MS analysis is frequently reported.

Yet another modification involves the use of SCX fraction collection and subsequent off-line analysis of these fractions by RPLC–MS. An impressive result of this approach is the detection of ~7537 unique peptides and the identification of ~1504 proteins in the yeast proteome [58]. This means that in this case about the same number of proteins was identified as in a previous study [45]. Detailed data evaluation showed that 858 proteins were identified in both studies, while 607 and 646 unique proteins were found [58].

In another study, related to the *E. coli* K12 proteome, by another group [37], the tryptic digest was first fractionated in 40 fractions by RPLC. Next, Met-containing peptides in these fractions were oxidized by H_2O_2. These modified proteins were separated from the bulk unmodified proteins by RPLC and fractionated in 8 fractions. These fractions were subsequently analysed by nano-LC–MS–MS on a Q–TOF instrument operated under DDA. After the first RPLC run, an exclusion list was generated from peptides identified in a MASCOT search of the data, and the sample was then rerun. Following this procedure, 754 proteins were identified, which is about 34% of the expressed *E. coli* K12 proteome.

Prefractionation of proteins in an *Arabidopsis thaliana* leaf extract by means of fast-performance LC prior to SCX×RPLC–MS on a monolithic RPLC column allowed the identification of 1032 unique proteins in 4 mg of a protein plant leaf tissue extract [59]. Fractionation of human plasma proteins by on-line SCX and RPLC was performed prior to tryptic digestion [60]. The resulting 30 fractions were digested and analysed by LC–MS on a quadrupole–linear-ion-trap (Q–LIT) instrument. Using this approach, 1292 proteins were identified, some of which are know to be present at <10 ng/ml.

Fractionation by capillary isoelectric focussing (CIEF) prior to LC–MS was reported by several groups (Ch. 17.5.6). Retention time from the RPLC and pI-data from the CIEF separation were applied to validate potentially identified peptides from the SEQUEST search [27]. The initial number of 7629 identified peptides was reduced to 1837 identified and 1130 likely hits by pI and retention time validation.

3.4   Shotgun protein identification: FT-ICR-MS

FT-ICR-MS is a superior tool in bottom-up and shotgun protein identification. This is due to its extreme resolving power and accurate-mass determination capabilities as well as the availability of a variety of pre-ICR-cell or in-ICR-cell peptide fragmentation technologies (Ch. 17.7.3). An illustrative bottom-up example is the analysis of digests of carbonic anhydrase (29 kDa). FT-ICR-MS identifies 64 peptides in the chymotryptic digest (95% sequence coverage) and 17 peptides as well as 23 unassigned masses in the Lys-C digest (95% sequence coverage) [9].

Automated DDA operation for switching between in-cell MS and MS–MS in LC–FT-ICR-MS has been illustrated with *D. radiodurans* whole cell digests [62]. Alternatively, the AMT approach (Ch. 18.2.3) may be applied for the identification of expressed proteins by FT-ICR-MS. In a preliminary study, AMT were identified for >60% of the potentially expressed proteins in *D. radiodurans* [33]. Further results on the *D. radiodurans* proteome were reported elsewhere [63].

Alternating scanning between MS and IRMPD-induced MS–MS in LC–FT-ICR-MS enables both PMF and PSA in one run [64]. The method was applied to the identification of a rat liver diacetyl-reducing enzyme.

An alternative method to peptide fragmentation is a hybrid system, involving either a triple-quadrupole type of a Q–LIT-type first stage of mass analysis and an FT-ICR-MS second stage of mass analysis. This decouples fragmentation and accurate-mass determination and results in a more user-friendly instrument (Ch. 17.7.3). More elaborate use of these commercial instruments can be anticipated.

Improving the mass accuracy also reduces the number of peptides needed for an unequivocal protein identification (AMT, Ch. 18.2.3). This may enable shotgun protein identification in complex mixtures without prior peptide separation. A nice example of AMT involves Cys-alkylation with 2,4-dichlorobenzyliodoacetamide of the peptides in a tryptic digest of a yeast lysate. This incorporates chlorine in the Cys-containing tryptic fragments. The Cys-containing fragments are readily recognized in FT-ICR-MS from the Cl-isotope pattern and mass defect [31].

3.5   Top-down protein identification

Resolving power, mass accuracy, and flexibility in protein fragmentation permitted the use of FT-ICR-MS in a different protein identification strategy as well. In the top-down strategy, the mass of the intact protein is accurately determined by ESI-FT-ICR-MS first. Then, the protein is subjected to mild digestion conditions, *e.g.*, by Lys-C or cyanogen bromide cleavage. Upon infusion of the resulting mixture containing only a limited number of relatively large peptides (5–36 kDa), accurate-mass determination of the fragments is performed. The masses of the peptides sum up to the mass of the protein. In this way, 100% coverage was achieved for carbonic anhydrase [9].

As such, the top-down strategy is a tool to recognize errors in the DNA-derived sequence, to identify (post-translational) modifications, or to evaluate modifications among complementary sets of the protein, *e.g.*, in order to confirm the identity of a recombinant product. Further study with digestion and/or MS–MS then allows identification of the modification and its location.

In-ICR-cell fragmentation of intact proteins was found to yield useful sequence information [64] (Ch. 17.7.3). This was applied in sequencing six gene products (7–74 kDa) of the *E. coli* thiamin biosynthetic operon by multistage MS–MS [65]. First, the multiple-charge protein ions are fragmented into a number of large peptide fragments, covering the complete sequence, Then, these peptide fragments are fragmented further to obtain information on their amino acid sequence. This revealed several discrepancies between the determined sequences for these six proteins and the DNA-derived sequences.

The top-down strategy avoids a disadvantage of the bottom-up strategy: in principle, it verifies the complete DNA-predicted sequence of the protein, as both the intact proteins and the sum of the fragments are measured.

Top-down identification is only applicable to isolated proteins or not-too-complex protein mixtures. For more complex protein mixtures, prefractionation is obligatory, as gas-phase isolation of proteins then is no longer an option. Software tools have been developed for automated isolation and fragmentation of intact proteins and subsequent database identification. They have been tested with mixtures of protein standards and cytosolic protein mixtures of *M. jannaschii* [66]. During acquisition, the software recognizes charge states, generates isolation waveforms, and fragments ions by IRMPD. Enhanced performance in top-down sequencing of intact proteins by FT-ICR-MS can be achieved by applying multi-stage in-source CID in the vacuum interface [67].

Off-line continuous GE, and RPLC were applied to generate fractions which were subsequently analysed by nano-ESI-FT-ICR-MS for the detection of diverse PTM in a lysate of *S. cerevisiae* [68]. In total, 117 gene products were identified with 100% sequence coverage, revealing among others 26 acetylation spots.

Bottom-up/shotgun and top-down procedures can be considered complementary, especially when dealing with PTM, where the shotgun approach provides information on the amino-acid sequence and the top-down approach on nature and position of the PTM. Integration of top-down and bottom-up protein identification procedures has also been reported, *e.g.*, in the proteome analysis of the Gram-negative γ-proteobacterium *S. oneidensis* [69]. A total of 868 proteins were identified. With the same combined strategy, the 70S ribosome of *Rhodopseudomonas pallustris* was characterized [70]. The method enabled protein identification, assignment of PTM, and discrimination between isoforms.

A combined approach was also applied in the analysis of β-casein and the epidermal growth factor receptor using a hybrid instrument with a Q–LIT front end and an FT-ICR-MS back-end [71]. The protein was digested with Lys-C to get a small number of peptides. The increased hydrophobicity of the larger peptide fragments is advantageous when retention time shifts due to polar PTM occur. First, an FT-ICR-MS survey scan (mass accuracy < 2 ppm) is performed to characterize the higher charge states. MS–MS and MS–MS^2 is performed in the Q–LIT at higher scan speeds, compatible with LC separation. Combination of the data provided

>95% sequence coverage of the two phosphorylated proteins with at sub-picomole level.

The potential of a top-down strategy based on the generation of peptide-sequence tags from intact proteins by CID in a Q–TOF instrument was evaluated as well [72]. In most cases, *N*- and *C*-terminal fragments and some stretches of consecutive product ions from the protein interior were observed. In a Q–TOF-based top-down approach, in-source CID of the intact protein and subsequent Q–TOF MS–MS of the resulting fragments was applied. The resulting sequence tags were database searched using PeptideSearch [73].

The potential of CID, gas-phase ion-molecule, and ion-ion reactions of multiple-charge ions of intact proteins in an ion-trap in top-down protein characterization has been evaluated as well [74]. Following this approach, top-down identification of a previously unknown, modified protein in a protein mixture derived from *S. cerevisiae* was reported [75].

3.6 Identification of membrane proteins

Membrane proteins are composed of alternating hydrophobic and hydrophilic regions. Characterization of membrane proteins is important, given their role in cell signalling, cell–cell interactions, and transport across the membrane. Membrane proteins are also important targets in drug discovery. They pose significant challenge to protein identification strategies based on 2D-GE and/or LC–MS via bottom-up or shotgun strategies. Due to their hydrophobicity and limited solubility, they require special strategies for isolation and handling. Often, the use of surfactants and/or high concentrations of chaotropes like urea or guanidine is required. This is not readily compatible with ESI-MS. Given the challenges involved, a few selected applications of the identification of membrane proteins are discussed here. General procedures and early results are discussed in more detail elsewhere [76].

A first step in the LC–MS characterization is the solubilization of the membrane proteins, which may be achieved by chemical and/or (multi-)enzymatic digestion in combination with the use of surfactants, organic solvents, and/or acids. High pH treatment and nonspecific proteinase K digestion has also been proposed [77].

For shotgun-LC–MS identification, selective prefractionation by means of AfC is a powerful tool. Biotinylation of cell-surface proteins and subsequent streptavidin-AfC enabled a 1600-fold and 400-fold enrichment of plasma membrane proteins over proteins from mitochondria and endoplasmic reticulum, respectively [78-79]. Because of the selective isolation procedure, only 3% of the 918 unique proteins identified were related to mitochondria and endoplasmic reticulum proteins, and ~25% were either plasma membrane proteins or membrane-associated proteins.

Similar technology was applied in the global analysis of the membrane subproteome of the bacterium *Pseudomonas aeruginosa* [80]. Biotinylation of the Cys-containing proteins and selective AfC enrichment of low-abundance membrane proteins was performed prior to nano-LC–MS–MS on a 650×0.15-mm-ID column in combination with a ion-trap instrument and SEQUEST database searching. In total 768 proteins were identified, 333 of which had at least one transmembrane domain.

3D-LC–MS (RPLC×SCX×RPLC) was applied to three fractions of the yeast proteome, *i.e.*, the soluble proteins, the urea-solubilized peripheral membrane proteins, and SDS-solubilized membrane proteins [81]. In total, 3019 proteins were identified with on average 5.5 peptides per protein. From the 1221 proteins known to contain two or more transmembrane domains, 495 were identified.

4. Protein-expression profiling

An important topic in functional proteomics is the evaluation of differences in protein expression in related sets of samples, *e.g.*, cells in healthy and disease state, or disease state with or without treatment. This means measuring and comparing the concentrations of sets of proteins in two samples, *i.e.*, identification of the relevant proteins in both samples and then comparison of their levels.

Conventional methods of protein-expression profiling are based on imaging of 2D-GE and comparison of protein spot intensities. Powerful software tools have been developed for this. However, the approach suffers from problems with precision and reproducibility, and a too limited dynamic range. 2D-GE provides ~3 orders of magnitude, while at least 6 orders are required, in human even ~10 orders.

The alternative approach involving LC–MS (or MALDI-MS) involves selective labelling of proteins, enzymatic digestion, AfC selection of labelled peptides, and measurement of the labelled peptides by alternating LC–MS and LC–MS–MS for simultaneous relative quantification and protein identification.

4.1   Labelling of proteins: ICAT

In order to enable evaluation of differences in protein expression in two related sets of samples by means of LC–MS, the concept of isotope-coded affinity tagging (ICAT) was introduced by Gygi *et al.* [13]. An ICAT label consists of three parts: the biotin affinity part, a linker containing either only H (light label) or D_8 (heavy label), and a Cys-specific reactive group. The procedure of comparative quantitative analysis is as follows (Figure 18.2). The reduced proteins of two cell states are labelled with ICAT, one state with the light label, the other with the heavy label. The two samples are mixed, eventually fractionated, and digested with trypsin. Avidin AfC is applied to selectively isolate the ICAT labelled, *i.e.*, Cys containing, peptides from the entire tryptic digest. The isolated peptides are analysed by LC–MS with DDA. Each peptide yields two peaks in the mass spectrum, differing 8 Da (or a multiple thereof, if more than one Cys is present in the peptide). The relative abundance indicates the relative expression level of the protein.

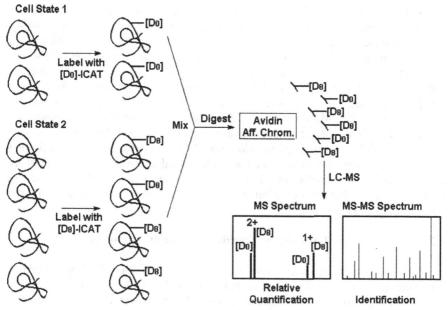

Figure 18.2: Principle of ICAT labelling for quantitative proteomics.

The MS–MS data, acquired simultaneously, are applied to identify the proteins by database searching. The presence of Cys is relatively rare, thereby greatly reducing the complexity of the peptide mixture and the risk of ion suppression. Of the 293 potential tryptic peptides for glyceraldehyde-3-phosphate dehydrogenase, only 44 contain at least one Cys residue [13]. Protein without Cys are missed in the analysis, and most likely not all Cys-residues in a reduced protein will be captured in the ICAT labelling.

Initial results with ICAT labelling were shown for different protein expression in the yeast *S. cerevisiae* grown on either ethanol or galactose as carbon source [13]. The potential of MudPIT (Ch. 18.3.2) for the quantitative proteomics was evaluated by means of mixing soluble portions of the *S. cerevisiae* proteome grown on either natural or ^{15}N-enriched media in various ratios (1:1 to 10:1) [82]. Reproducibility in quantitative peptide analysis by means of metabolic labelling of *S. cerevisiae* was evaluated by Washburn *et al.* [83].

ICAT labelling kits are commercially available. In a modified procedure, two separate measurement steps are performed to enhance the sample throughput [84]. In the first step, LC–TOF-MS is applied for quantification and triggering of collection fractions containing up- or down-regulated peptides. In the second step, the fractions are identified via direct-infusion nano-ESI-Q–TOF-MS.

With isotopically-labelled compounds, one has to pay attention to heavy-isotope effects, that may result in different retention times for light- and heavy-ICAT labelled peptides, especially in the case of D-labelling [14, 85]. The position of the D-label in the molecule is important as well: the least isotopic retention-time shift is observed, when the D-labels are positioned at a polar group with little interaction with the hydrophobic RPLC packing. The importance of the isotope effect in data interpretation should not be underestimated. Labelling with ^{13}C causes no observable retention-time shifts. More recently, ICAT labels have been introduced containing $^{13}C_9$ instead of D_8 in order to reduce the isotope effects (cICAT) [86-87].

Lee *et al.* [88] discussed optimization of RPLC for quantitative proteomics. They especially paid attention to the fact that both a good chromatographic peak shape must be achieved to enable reliable quantification and the peak capacity must be maximized to enable identification with sufficient proteome coverage. Gradient range and slope were evaluated.

4.2   Labelling of proteins or peptides: Alternative methods

There are a number of alternative ways to achieve isotopic labelling of peptides for quantitative purposes. One can start at the protein level, like in ICAT:

- Protein digestion in $H_2^{18}O$ to incorporate ^{18}O in the C-terminal carboxylic acid group [89-90]
- Metabolic labelling is applicable to organisms that are grown on defined media, like yeast, bacteria, and cell cultures. Cells grown in ^{15}N or ^{13}C enriched media are compared to cells grown in natural media [91-92]. In inverse ^{15}N metabolic labelling, four instead of two sample pools are generated: both cell states are grown on natural and ^{15}N-enriched media to facilitate data interpretation [93].
- Metabolic labelling by growing organisms on media enriched in an isotopically-labelled amino acid (SILAC) [94-95] or by feeding them ^{15}N-labelled bacteria or yeast [96]. Impressive results were obtained for the nematode *C. elegans* after metabolic labelling via ^{15}N-labelled *E. coli*. Several down- and up-regulated proteins were found in a wild-type and a germline deficient *glp-4* mutant [96].
- Tandem mass tags (TMT) consist of a CH_3/CD_3-labelled substituent linked to a guanidino group, a fragmentation-enhancing group, a mass-normalization group, and a reactive functionality [97]. If linked to a peptide, the TMT fragment in MS–MS to provide a high-abundance low-mass fragment ion, containing the CH_3/CD_3-label, as well as sequence ions from the peptide. The mass-normalization group ensures that the mass of the CH_3-substituent is balanced to ensure co-elution of the labelled and nonlabelled peptides in RPLC. Modified TMT have been commercialized as the iTRAQ labelling technique.

Whereas the isotope-labelling can be performed at the protein level, labelling can also be performed after tryptic digestion of the proteome:

- Acetylation of *N*-terminal amines by means of *N*-hydroxysuccinimide (or its D_3 analogue) [29] or succinic anhydride (or its $^{13}C_4$ analogue), or esterification of *C*-terminal carboxylic acids and of Asp and Glu [98].
- Solid-phase labelling using immobilized acid-labile isotope-coded extractants (ALICE) [99] or photo-cleavable isotope-coded tags [82]. A thiol-reactive group with an isotope tag is immobilized onto a solid support via an acid- or photo-cleavable linker. Cys-containing peptides are captured. After washing, the labelled peptides are released by cleaving the linker.
- *N*-terminal labelling with phenyl isocyanate, which is a fast and quantitative reaction also applied during Edman degradation [100]. An alternative reagent is 4-sulfophenyl isothiocyanate [101].

Alternatives to stable-isotope labelling have also been proposed. *N*-terminal and Lys labelling with either *S*-methyl thioacetimidate and *S*-methyl thiopropionimidate (mass difference 14 Da) provides labelling in every part of the digest and improves ionization efficiency [102]. In mass-coded abundance tagging (MCAT) [103], the proteome digest from one of the two cell states is subject to a Lys-guanidination, while the other is not (mass difference of 42 Da). A disadvantage of MCAT is the absence of a label-selective purification. Labelling by means of mass-defect tags, containing chlorine or bromine, yields a readily resolvable mass difference between tagged and nontagged species in FT-ICR-MS [31, 104].

Labelling strategies for proteins and peptides haven recently been extensively reviewed, *e.g.*, [14, 105]. Specialized software applications have been developed for data processing [106].

4.3   Isolation of labelled peptides

Quantitative proteomics requires global approaches to the proteome, as prefractionation of the proteome complicates quantification due to distribution of proteins over various fractions and difficulties in determining the recovery of proteins in complex analytical procedures. However, after labelling and tryptic digestion, the tryptic peptides relevant to the study are preferentially selectively isolated from the very complex digest of the proteome. In the ICAT procedure, this is performed by avidin AfC (Ch. 18.4.1). Alternative strategies are applied, in the liquid phase based on signature peptides, or in FT-ICR-MS.

The signature peptide approach can be applied to significantly reduce the complexity of the protein digests. The use of a combined selection of Cys-containing peptides by covalent chromatography (capturing with thiopropyl Sepharose) and His-containing peptides by Cu^{2+}-loaded immobilized metal-ion affinity chromatography (IMAC) was applied to detect up-regulated proteins from plasmid bearing *E. coli* that had been induced with isopropyl β-thiogalacto-pyranoside [107].

The potential of AMT in FT-ICR-MS was already discussed [31-33]. If an AMT can be isotope-coded, the same strategy can be applied to quantitative proteomics. One of the problems of FT-ICR-MS in quantitative proteomics is the space charging constraint of the FT-ICR cell. Data-dependent selective ejection of highly abundant ions and prolonged accumulation of low-abundant ions in the external quadrupole, prior to ion transfer to the FT-ICR cell, helps in extending the dynamic range of the method [108].

4.4   Targeted quantitative analysis

In the previous sections, comparative quantitative proteomics is discussed. This is aiming at relative quantification of protein levels in various cell states. In some applications, absolute rather than relative quantification of proteins is relevant, *e.g.*, as an alternative to immunoaffinity methods for quantitative bioanalysis of protein drugs. This requires different strategies.

Optimization of the experimental parameters, *e.g.*, sample pH, cone voltage, collision energy, comparison of positive-ion and negative-ion ESI-MS, and of selected-ion monitoring (SIM) and selected-reaction monitoring (SRM), was reported for the quantitative analysis of human insulin (5808 Da) [109]. Best performance was achieved in SIM at the 6+ ion (*m/z* 969). SRM would probably be more appropriate for analysis in plasma. In a subsequent study [110], they applied negative-ion LC–ESI-MS–MS in SRM mode for the determination of the urinary concentration of the C-peptide (3020 Da), the connecting peptide in proinsulin.

Various methods were evaluated for the targeted proteomics of human growth hormone (hGH) in human plasma [111]. hGH was spiked in plasma 10-fold above natural level (~16 pg/µl). Initially, the full plasma proteome was reduced, alkylated, and digested prior to LC–MS via DDA on an ion-trap instrument. hGH could be identified from its T_1 peptide. Next, the plasma proteome was fractionated by RPLC and GE prior to digestion and LC–MS analysis. hGH could be identified with higher confidence. Finally, an LC×LC–MS approach was applied, which enabled hGH identification from five tryptic peptides. An important conclusion was that hGH could be detected in a complex sample at the low femtomole level among proteins that were 40,000× more abundant. The results show that a multidimensional approach may be taken for targeted proteomics and quantitative protein bioanalysis.

Absolute quantitative analysis of recombinant Kringle5 (rK5, 10,464 Da) of plasminogen protein in human plasma by LC–MS on a triple-quadrupole instrument was described [112]. An analogue with lower molecular weight was selected as internal standard (IS). The protein analytes were extracted from plasma. For rK5, the transition *m/z* 1495.5→1462.9 (7+) was monitored by SRM. The IS did not show fragmentation, but monitoring the transition *m/z* 1371.7→1371.7 (7+) showed reduced background interference. With sodium heparin anticoagulated human plasma, good linearity was achieved in the range of 50 ng/ml to 10 µg/ml. The relative standard deviations (RSD) for the individual standard levels ranged from 2.6 to 15.6%.

The use of isotope-labelled synthetic peptides as IS was proposed for the absolute quantification of proteins in protein expression studies [113]. If necessary, these synthetic peptides can be covalently modified to be applied as IS in for instance phosphopeptide quantification. This approach was applied in the quantitative analysis of two glutathione S-transferase isoforms in human liver cytosol by LC–MS in SRM mode [114]. A series of pilot experiments were performed to select the most suitable IS peptides for four human plasma proteins (hemopexin, α1 antichymotrypsin, interleukin-6, and tumor necrosis factor-α) [115]. Rabbit polyclonal antibodies were raised against these selected peptides and immobilized on POROS supports. These IAC columns were applied to achieve a 120-fold enrichment of the antigen peptide from digested human plasma proteins. The peptides and its IS were measured by LC–MS in SIM or SRM mode. The methods appears to be a tailor method for targeted protein analysis.

The use of tryptic peptides derived from related proteins, $e.g.$, horse myoglobin in the analysis of human myoglobin, as IS in targeted quantitative analysis has been proposed as well [116].

5. Protein-protein interactions

In cells of higher organisms, the functioning of proteins is frequently determined by larger associates, of either more or less identical subunits like in hemoglobin (homomers) or in complexes containing a number of different proteins (heteromers). This means that in functional proteomics one actually has to study these protein associates or complexes. Important early steps in these studies are the detection of the intact associate [117] and the identification of the individual proteins involved. LC–MS plays only a limited role in this area.

The technique of choice in the detection and isolation of protein complexes is immunoprecipitation [15], whereas in some cases centrifugation or free-flow electrophoresis can be applied for the purification of cellular organelles. Immunoprecipitation permits the purification of specific proteins against which an antibody has been raised. Affinity-based methods have been developed for a variety of organelles. When applied to the isolation of protein complexes, the antibody should target one of the proteins in the complex. The antibody–protein complex precipitates from solution. Alternatively, immobilization of the primary antibody to agarose (beads) can be applied. Multi-step procedures have been developed as well.

In the identification step, the protein complex is released from the antibody prior to enzymatic digestion, especially when antibody immobilization was used, or the antibody–protein complex is digested. The proteins in the complex are then identified by means of PMF and/or PSA using LC–MS or MALDI-MS analysis.

An illustrative example is the identification of the *Saccharomyces cerevisiae* ribosome using LC–MS–MS and SEQUEST database searching [118]. The eukaryotic 80S ribosome, comprising of a small 40S subunit predicted to contain 32 proteins, and a large 60S subunit predicted to contain 46 subunits, is one of the largest macromolecular complexes in the cell. Using the 2D-LC MudPIT approach (Ch. 18.3.2), 95 unique proteins were identified in an isolated ribosomal fraction, 90 of which were related to ribosomal gens. These represent 75 of the 78 predicted ribosomal proteins. Off-line prefractionation of digests by cation- or anion-exchange columns prior to RPLC–MS for the characterization of yeast ribosome protein complexes was reported by another group [119].

A slightly modified approach was applied in the identification of protein complexes in *Schizosaccharomyces pombe* [120]. The major modification involves the use of a tandem affinity purification [121]. To a protein of interest, an epitope is appended that contains two different affinity tags separated by a specific protease cleavage site. This enables isolation of the complex under milder conditions.

From the analytical point-of-view, developments in this field comprise further improvements in the chromatographic separation efficiency, as for instance demonstrated by the use of a triphasic MudPIT approach [46], and of in-needle columns packed with 1-μm particles, and operated at flow-rates of <50 nl/min [122].

6. Biomarker discovery

A biomarker is a characteristic that is objectively measured and evaluated as an indicator of normal biologic processes, pathogenic processes, or pharmacologic responses to a therapeutic intervention [123]. In proteomics, such a biomarker could be a protein or peptide (native or modified), or a pattern of peaks. Biomarkers are proposed to identify unique molecular signatures related to certain (disease) states.

While significant efforts have been undertaken, the progress of bringing biomarkers into clinical application is limited due to the lack of the required sensitivity, specificity, and validation of protein biomarker tests for clinical application such as diagnosis, prognosis, or monitoring. What it takes to perform biomarker discovery is nicely outlined by LaBaer [124] in the introduction to a special issue of *Journal of Proteome Research* on biomarkers. Requirements for proteomics-based tools to be applicable as *in vitro* clinical diagnostic tests were discussed by Vitzthum *et al.* [125]. Efforts to develop multi-disciplinary programs to biomarker discovery, development, and validation have been reported [123, 126].

As far as protein-related biomarkers are considered, an important tool in biomarker discovery is quantitative protein-expression profiling (Ch. 18.4). Mixtures of nonlabelled and stable-isotopically-labelled peptides derived from two different states, *e.g.*, a healthy and disease state, or with and without treatment, are analysed to search for up- or down-regulated protein concentrations. One may question whether approaches based on either the LC–MS or the MALDI-MS and surface-enhanced laser desorption ionization (SELDI-MS) are most suitable for biomarker discovery [127-128].

Obviously, the systems biology must go deeper than just quantification of the change. Apart from intra-individual differences in response, resulting from differences in nutritional habits, natural cycles, and genetic differences between individuals, it must evaluate and investigate how the up- or down-regulated proteins are associated with the onset or progression of the disease or treatment. Most likely a complex interrelated pattern is beyond the measured changes. The changes occur in a time-related pattern of changes in signalling, transcription, translation, PTM, and/or metabolic alterations. Validation of the observed changes in protein concentrations or in interrelated patterns is a difficult and lengthy process, requiring large patient populations.

As discussed in relation to protein phosphorylation (Ch. 19.3), substoichiometric changes in PTM such as phosphorylation may occur, demanding for a strategy to enable temporal monitoring of differential phosphorylation as a function of (potentially unknown) external or internal stimuli. In fact, some PTM may be disease-related, but others may not. Alternatively, several disease states are related to single amino-acid polymorphism. The state difference may not necessarily be expressed at protein level, but rather in different concentration(s) of small molecules. This is where metabolomics comes into play.

In the past few years, significant research efforts have been devoted to the discovery of biomarkers related to disease states, especially various forms of cancer and neurological diseases. The review by Journet and Ferro [129] on lysosomes and cancer provides a more application-oriented state-of-the-art perspective. A variety of review papers summarize the state of the art from technology point-of-view [127-128, 130]. Existing technologies such as 2D-GE, multidimensional LC–MS, and FT-ICR-MS are available. New technologies appear such as tissue imaging with direct MALDI-MS [131]. Apart from the technological innovations, significant input from bioinformatics is needed to a successful biomarker discovery. Whereas apparently all the technology needed is actually available, significant progress and development is still needed between the early development of a technology prototype and real-life, reliable application of that technology in a validated clinical setting.

Significant efforts and progress are to be expected in this area in the years to come.

7. References

1.   R.A. Henderson, A.L. Cox, K. Sakaguichi, E. Appella, J. Shabanowitz, D.F. Hunt, V.H. Engelhard, *Direct identification of an endogeneous peptide recognized by multiple HLA-A2.1 specific cytotoxic T cells*, PNAS, 90 (1993) 10275.

2. M. Wilm, A. Shevchenko, T. Houthaeve, S. Breit, L. Schweigerer, T. Fotsis, M. Mann, *Femtomole sequencing of proteins from polyacrylamide gels by nanoESI-MS*, Nature, 379 (1996) 466.

3. A. Shevchenko, O.N. Jensen, A.V. Podtelejnikov, F. Sagliocco, M. Wilm, O. Vorm, P. Mortesen, A. Shevchenk, H. Boucherie, M. Mann, *Linking genome and proteome by MS: large-scale identification of yeast proteins from 2D gels*, PNAS, 93 (1996) 14440.

4. W.J. Henzel, T.M. Billeci, J.T. Stults, S.C. Wong, C. Grimley, C. Watanabe, *Identifying proteins from 2D gels by molecular mass searching of peptide fragments in protein sequence databases*, PNAS, 90 (1993) 5011.

5. M. Mann, P. Højrup, P. Roepstorff, *Use of MS molecular weight information to identify proteins in sequence databases*, Biol. Mass Spectrom., 22 (1993) 338.

6. J.K. Eng, A.L. McCormack, J.R. Yates, III, *An approach to correlate tandem mass spectral data of peptides with amino acid sequences in a protein database*, J. Am. Soc. Mass Spectrom., 5 (1994) 976.

7. A. Ducret, I. van Oostveen, J.K. Eng, J.R. Yates, III, R. Aebersold, *High throughput protein characterization by automated RPLC–MS–MS*, Protein Sci., 7 (1998) 706.

8. D.A. Wolters, M.P. Washburn, J.R. Yates, III, *An automated MudPIT for shotgun proteomics*, Anal. Chem., 73 (2001) 5683.

9. N.L. Kelleher, H.Y. Lin, G.A. Valaskovic, D.J. Aaserud, E.K. Fridriksson, F.W. McLafferty, *Top down versus bottom up protein characterization by tandem high-resolution MS*, J. Am. Chem. Soc., 121 (1999) 806.

10. A.G. Marshall, *Milestones in FT-ICR-MS technique development*, Int. J. Mass Spectrom., 200 (2000) 331.

11. B. Bogdanov, R.D. Smith, *Proteomics by FT-ICR-MS: top down and bottom up*, Mass Spectrom. Rev., 24 (2005) 168.

12. R.M.A. Heeren, A.J. Kleinnijenhuis, L.A. McDonnell, T.H. Mize, *A mini-review of MS using FT-ICR-MS methods*, Anal. Bioanal. Chem., 378 (2004) 1048.

13. S.P. Gygi, B. Rist, S.A. Gerber, F. Tureček, M.H. Gelb, R. Aebersold, *Quantitative analysis of complex protein mixtures using ICAT*, Nat. Biotechnol., 17 (1999) 994.

14. S. Julka, F. Regnier, *Quantification in proteomics through stable isotope coding: A review*, J. Proteome Res., 3 (2004) 350.

15. J.R. Yates, III, *MS and the age of proteome*, J. Mass Spectrom., 33 (1998) 1.

16. N.L. Kelleher, *From primary structure to function: biological insights from large-molecule mass spectra*, Chem Biol., 7 (2000) R37.

17. J. Peng, S.P. Gygi, *Proteomics: the move to mixtures*, J. Mass Spectrom., 26 (2001) 1083.

18. J. Reinders, U. Lewandrowski, J. Moebius, Y. Wagner, A. Sickmann, *Challenges in MS-based proteomics*, Proteomics, 4 (2004) 3686.

19. D.C. Liebler, *Introduction to proteomics*, 2002, Human Press, Totowa, NJ.

20. A.J. Liska, A. Shevchenko, *Combining MS with database interrogation strategies in proteomics*, Trends Anal. Chem., 22 (2003) 291.

21. K.R. Clauser, P. Baker, A.L. Burlingame, *Role of accurate mass measurement (±10 ppm) in protein identification strategies employing MS or MS–MS and database searching*, Anal. Chem., 71 (1999) 2871.

22. B. Thiede, W. Höhenwarter, A. Krak, J. Mattow, M. Schmid, F. Schmidt, P.R. Jungblut, *Peptide mass fingerprinting*, Methods, 35 (2005) 237.

23. W.J. Henzel, C. Watanabe, J.T. Stults, *Protein identification: The origins of PMF*, J. Am. Soc. Mass Spectrom., 14 (2003) 931.

24. E.F. Strittmatter, P.L. Ferguson, K. Tang, R.D. Smith, *Proteome analyses using accurate mass and elution time peptide tags with capillary LC–TOF-MS*, J. Am. Soc. Mass Spectrom., 14 (2003) 980.

25. A.D. Norbeck, M.E. Monroe, J.N. Adkins, K.K. Anderson, D.S. Daly, R.D. Smith, *The utility of accurate mass and LC elution time information in the analysis of complex proteomes*, J. Am. Soc. Mass Spectrom., 16 (2005) 1239.

26. B.J. Cargile, J.L. Stephenson, Jr., *An alternative to MS–MS: pI and accurate mass for the identification of peptides*, Anal. Chem., 76 (2004) 267.

27. M. Heller, M. Ye, P.E. Michel, P. Morier, D. Stalder, M.A. Jünger, R. Aebersold, F. Reymond, J.S. Rossier, *Added value for MS–MS shotgun proteomics data validation through IEF of peptides*, J. Proteome Res., 4 (2005) 2273.

28. C.L. Gatlin, J.K. Eng, S.T. Cross, J.C. Detter, J.R. Yates, III, *Automated identification of amino acid sequence variations in proteins by HPLC–micro-ESI-MS–MS*, Anal. Chem., 72 (2000) 757.

29. J. Ji, A. Chakraborty, M. Geng, X. Zhang, A. Amini, M. Bina, F. Regnier, *Strategy for qualitative and quantitative analysis in proteomics based on signature peptides*, J. Chromatogr. B, 745 (2000) 197.

30. L. Riggs, C. Sioma, F. Regnier, *Automated signature peptide approach for proteomics*, J. Chromatogr. A, 924 (2001) 359.

31. D.R. Goodlett, J.E. Bruce, G.A. Anderson, B. Rist, L. Paša-Tolić, O. Fiehn, R.D. Smith, R. Aebersold, *Protein identification with a single accurate mass of a Cys-containing peptide and constrained database searching*, Anal. Chem., 72 (2000) 1112.

32. T.P. Conrads, G.A. Anderson, T.D. Veenstra, L. Paša-Tolić, R.D. Smith, *Utility of AMT for proteome-wide protein identification*, Anal. Chem., 72 (2000) 3349.

33. R.D. Smith, G.A. Anderson, M.S. Lipton, L. Paša-Tolić, Y. Shen, T.P. Conrads, T.D. Veenstra, H.R. Udseth, *An AMT strategy for quantitative and high-throughput proteome measurements*, Proteomics, 2 (2002) 513.

34. D.C. Liebler, B.T. Hansen, S.W. Davey, L. Tiscareno, D.E. Mason, *Peptide sequence motif analysis of MS–MS data with the SALSA algorithm*, Anal. Chem., 74 (2002) 203.

35. H. Lim, J. Eng, J.R. Yates, III, S.L. Tollaksen, C.S. Giometti, J.F. Holden, M.W.W. Adams, C.I. Reich, G.J. Olsen, L.G. Hays, *Identification of 2D-gel proteins: A comparison of MALDI-TOF PMF to μLC–ESI-MS–MS*, J. Am. Soc. Mass Spectrom., 14 (2003) 957.

36. W.M. Bodnar, R.K. Blackburn, J.M. Krise, and M.A. Moseley, *Exploiting the complementary nature of LC–MALDI-MS–MS LC–ESI-MS–MS for increased proteome coverage*, J. Am. Soc. Mass Spectrom., 14 (2003) 971.

37. K. Gevaert, J. Van Damme, M. Goethals, G.R. Thomas, B. Hoorelbeke, H. Demol, L. Martens, M. Puype, A. Staes, J. Vandekerckhove, *Chromatographic isolation of Met-containing peptides for gel-free proteome analysis: Identification of more than 800 Escherichia coli proteins*, Mol Cell Proteomics 1 (2002) 896.

38. T. Le Bihan, M.D. Robinson, I.I. Stewart, D. Figeys, *Definition and characterization of a "trypsinosome" from specific peptide characteristics by nano-HPLC–MS–MS and in silico analysis of complex protein mixtures*, J. Proteome Res., 3 (2004) 1138.

39. M. Palmblad, M. Wetterhall, K. Markides, P. Håkansson, J. Bergquist, *Analysis of enzymatically digested proteins and protein mixtures using a 9.4 Tesla FT-ICR-MS*, Rapid Commun. Mass Spectrom., 14 (2000) 1029.

40. M. Palmblad, M. Ramström, K.E. Markides, P. Håkansson, J. Bergquist, *Prediction of chromatographic retention and protein identification in LC–MS*, Anal. Chem., 74 (2002) 5826.

41. M. Palmblad, M. Ramstrom, C.G. Bailey, S.L. McCutchen-Maloney, J. Bergquist, L.C. Zeller, *Protein identification by LC–MS using retention time prediction*, J. Chromatogr. B, 803 (2004) 131.

42. R. Apweiler, A. Bairoch, C.H. Wu, *Protein sequence databases*, Curr. Opin. Chem. Biol., 8 (2004) 76.

43. A.L. McCormack, D.M. Schieltz, B. Goode, S. Yang, G. Barnes, D. Drubin, J.R. Yates, III, *Direct analysis and identification of proteins in mixtures by LC–MS–MS and database searching at the low-femtomole level*, Anal. Chem., 69 (1997) 767.

44. A.J. Link, E. Carmack, J.R, Yates, III, *A strategy for the identification of proteins localized to subcellular spaces: application to E. coli periplasmic proteins*, Int. J. Mass Spectrom. Ion Processes, 160 (1997) 303.

45. M.P. Washburn, D.A. Wolters, J.R. Yates, III, *Large-scale analysis of the yeast proteome by MudPIT*, Nat. Biotechnol., 19 (2001) 242.

46. W.H. McDonald, R. Ohi, D.T. Miyamoto, T.J. Mitchison, J.R. Yates, III, *Comparison of three directly coupled LC–MS–MS strategies for identification of proteins from complex mixtures: single-dimension LC–MS–MS, 2-phase MudPIT, and 3-phase MudPIT*, Int. J. Mass Spectrom., 219 (2002) 245.

47. A.W. Guzzetta, A.S. Chien, *A double-vented tetraphasic continuous column approach to MudPIT analysis on long capillary columns demonstrates superior proteomic coverage*, J. Proteome Res., 4 (2005) 2412.

48. E.F. Strittmatter, L.J. Kangas, K. Petritis, H.M. Mottaz, G.A. Anderson, Y. Shen, J.M. Jacobs, D.G. Camp, II, R.D. Smith, *Application of peptide LC retention time information in a discriminant function for peptide identification by MS–MS*, J. Proteome Res., 3 (2004) 760.

49. Y. Wang, J. Zhang, X. Gu, X.-M. Zhang, *Protein identification assisted by the prediction of retention time in LC–MS–MS*, J. Chromatogr. B, 826 (2005) 122.

50. M.T. Davis, J. Beierle, E.T. Bures, M.D. McGinley, J. Mort, J.H. Robinson, C.S. Spahr, W. Yu, R. Luethy, S.D. Patterson, *Automated LC–LC–MS–MS platform using binary IEC and gradient RPLC for improved proteomic analyses*, J. Chromatogr. B, 752 (2001) 281.

51. K.G. Mawuenyega, H. Kaji, Y. Yamauchi, T. Shinkawa, H. Saito, M. Taoka, N. Takahashi, T. Isobe, *Large-scale identification of C. elegans proteins by multidimensional LC–MS–MS*, J. Proteome Res., 2 2003) 23.

52. E. Nägele, M. Volmer, P. Hörth, *2D-LC–MS system for applications in proteomics*, J. Chromatogr. A, 1009 (2003) 197.

53. G. Mitulović, C. Stingl, M. Smoluch, R. Swart, J.-P. Chervet, I. Steinmacher, C. Gerner, K. Mechtler, *Automated, on-line 2D nano LC–MS–MS for rapid analysis of complex protein digests*, Proteomics, 4 (2004) 2545.

54. Y. Shen, N. Tolić, C. Masselon, L. Paša-Tolić, D.G. Camp, II, K.K. Hixson, R. Zhao, G.A. Anderson, R.D. Smith, *Ultrasensitive proteomics using high-efficiency on-line micro-SPE–nano-LC–nano-ESI-MS and MS–MS*, Anal. Chem., 76 (2004) 144.

55. Y. Shen, J.M. Jacobs, D.G. Camp, II, R. Fang, R.J. Moore, R.D. Smith, W. Xiao, R.W. Davis, R.G. Tompkins, *Ultra-high-efficiency SCX–RPLC–MS–MS for high dynamic range characterization of the human plasma proteome*, Anal. Chem., 76 (2004) 1134.

56. Y. Shen, E.F. Strittmatter, R. Zhang, T.O. Metz, R.J. Moore, F. Li, H.R. Udseth, R.D. Smith, K.K. Unger, D. Kumar, D. Lubda, *Making broad proteome protein measurements in 1-5 min using high-speed RPLC separations and high-accuracy mass measurements*, Anal. Chem., 77 (2005) 7763.

57. M. Vollmer, P. Hörth, E. Nägele, *Optimization of 2D off-line LC–MS separations to improve resolution of complex proteomic samples*, Anal. Chem., 76 (2004) 5180.

58. J. Peng, J.E. Elias, C.C. Thoreen, L.J. Licklider, S.P. Gygi, *Evaluation of LC×LC–MS–MS for large-scale protein analysis: the yeast proteome*, J. Proteome Res., 2 (2003) 43.

59. S. Wienkoop, M. Glinski, N. Tanaka, V. Tolstikov, O. Fiehn, W. Weckwerth, *Linking protein fractionation with multidimensional monolithic RPLC–MS enhances protein identification from complex mixtures even in the presence of abundant proteins*, Rapid Commun. Mass Spectrom., 18 (2004) 643.

60. W.-H. Jin, J. Dai, S.-J. Li, Q.-C. Xia, H.-F. Zou, R. Zeng, *Human plasma proteome analysis by multidimensional LC prefractionation and LIT-MS identification*, J. Proteome Res., 4 (2005) 613.

61. L. Li, C.D. Masselon, G.A. Anderson, L. Paša-Tolić, S.-W. Lee, Y. Shen, R. Zhao, M.S. Lipton, T.P. Conrads, N. Tolić, R.D. Smith, *High-throughput peptide identification from protein digests using DDA multiplexed tandem FT-ICR-MS coupled with capillary LC*, Anal. Chem., 73 (2001) 3312.

62. M.S. Lipton, L. Paša-Tolić, G.A. Anderson, D.J. Anderson, D.L. Auberry, J.R. Battista, M.J. Daly, J. Fredrickson, K.K. Hixson, H. Kostandarithes, C. Masselon, L.M. Markillie, R.J. Moore, M.F. Romine, Y. Shen, E. Stritmatter, N. Tolić, H.R. Udseth, A. Venkateswaran, K.-K. Wong, R. Zhao, R.D. Smith, *Global analysis of the D. radiodurans proteome by using AMT*, PNAS, 99 (2002) 11049.

63. T. Kosaka, T. Yoneyama-Takazawa, K. Kubota, T. Matsuola, I. Sato, T. Sasaki, Y. Tanaka, *Protein identification by PMF and PSA with alternating scans of nano-LC–IRMPD-FT-ICR-MS*, J. Mass Spectrom., 38 (2003) 1281.

64. W. Li, C.L. Hendrickson, M.R. Emmett, A.G. Marshall, *Identification of intact proteins in mixtures by alternated capillary LC–ESI and LC–ESI-IRMPD-FT-ICR-MS*, Anal. Chem., 71 (1999) 4397.

65. N.L. Kelleher, S.V. Taylor, D. Grannis, C. Kinsland, H.-J. C., T.P. Begley, F.W. McLafferty, *Efficient sequence analysis of the six gene products (7-74 kDa) from the E. coli thiamin biosynthetic operon by high-resolution MS–MS*, Protein Sci., 7 (1998) 1796.

66. J.R. Johnson, F. Meng, A.J. Forbes, B.J. Cargile, N.L. Kelleher, *FT-ICR-MS for automated fragmentation and identification of 5-20 kDa proteins in mixtures*, Electrophoresis, 23 (2003) 3217.

67. H. Zhai, X. Han, K. Breuker, F.W. McLafferty, *Consecutive ion activation for top down MS: Improved protein sequencing by nozzle–skimmer dissociation*, Anal. Chem., 77 (2005) 5777.

68. F. Meng, Y. Du, L.M. Miller, S.M. Patrie, D.E. Robinson, N.L. Kelleher, *Molecular-level description of proteins from S. cerevisiae using quadrupole FT hybrid MS for*

top down proteomics, Anal. Chem., 76 (2004) 2852.

69. N.C. VerBerkmoes, J.L. Bundy, L. Hauser, K.G. Asano, J. Razumovskaya, F. Larimer, R.L. Hettich, J.L. Stephenson, Jr., *Integrating "top-down" and "bottom-up" MS approaches for proteomic analysis of S. oneidensis*, J. Proteome Research, 1 (2002) 239.

70. M.B. Strader, N.C. VerBerkmoes, D.L. Tabb, H.M. Connelly, J.W. Barton, B.D. Bruce, D.A. Pelletier, B.H. Davison, R.L. Hettich, F.W. Larimer, G.B. Hurst, *Characterization of the 70S ribosome from R. palustris using an integrated "top-down" and "bottom-up" MS approach*, J. Proteome Res., 3 (2004) 965.

71. S.-L. Wu, J. Kim, W.S. Hancock, B. Karger, *Extended range proteomic analysis (ERPA): A new and sensitive LC–MS platform for high sequence coverage of complex proteins with extensive PTM-comprehensive analysis of beta-casein and epidermal growth factor receptor (EGFR)*, J. Proteome Res., 4 (2005) 1155.

72. J.F. Nemeth-Cawley, J.C. Rouse, *Identification and sequencing of intact proteins via CID and Q–TOF-MS*, J. Mass Spectrom., 37 (2002) 270.

73. J.M. Ginter, F. Zhou, M.V. Johnston, *Generating protein sequence tags by combining cone and conventional CID in a Q–TOF-MS*, J. Am. Soc. Mass Spectrom., 15 (2004) 1478.

74. G.E. Reid, S.A. McLuckey, *'Top down' protein characterization via MS–MS*, J. Mass Spectrom., 27 (2002) 663.

75. R. Amunugama, J.M. Hogan, K.A. Newton, S.A. McLuckey, *Whole protein dissociation in a quadrupole ion trap: Identification of an a priori unknown modified protein*, Anal. Chem., 76 (2004) 720.

76. C.C. Wu, J.R. Yates, III, *The application of MS to membrane proteomics*, Nat. Biotechnol., 21 (2003) 262.

77. C.C. Wu, M.J. MacCoss, K.E. Howell, J.R. Yates, III, *A method for the comprehensive proteomic analysis of membrane proteins*, Nat. Biotechnol., 21 (2003) 532.

78. Y. Zhao, W. Zhang, M.A. White, Y. Zhao, *Capillary CE–MS analysis of proteins from affinity-purified plasma membrane*, Anal. Chem., 75 (2003) 3751.

79. Y. Zhao, W. Zhang, Y. Kho, Y. Zhao, *Proteomic analysis of integral plasma membrane proteins*, Anal. Chem., 76 (2004) 1817.

80. J. Blonder, M.B. Goshe, W. Xiao, D.G. Camp, II, M. Wingerd, R.W. Davis, R.D. Smith, *Global analysis of the membrane subproteome of P. aeruginosa using LC–MS–MS*, J. Proteome Res., 3 (2004) 434.

81. J. Wei, J. Sun, W. Yu, A. Jones, P. Oeller, M. Keller, G. Woodnutt, J.M. Short, *Global proteome discovery using an online 2D-LC–MS–MS*, J. Proteome Res., 4 (2005) 801.

82. M.P. Washburn, R. Ulaszek, C. Deciu, D.M. Schieltz, J.R. Yates, III, *Analysis of quantitative proteomic data generated via MudPIT*, Anal. Chem., 74 (2002) 1650.

83. M.P. Washburn, R.R. Ulaszek, J.R. Yates, III, *Reproducibility of quantitative proteomic analyses of complex biological mixtures by MudPIT*, Anal. Chem., 75 (2003) 5054.

84. T.J. Griffin, D.K.M. Han, S.P. Gygi, B. Rist, H. Lee, R. Aebersold, K.C. Parker, *Toward a high-throughput approach to quantitative proteomic analysis: expression-dependent protein identification by MS*, J. Am. Soc. Mass Spectrom., 12 (2001) 1238.

85. R. Zhang, F.E. Regnier, *Minimizing resolution of isotopically coded peptides in*

comparative proteomics, J. Proteome Res., 1 (2002) 139.

86. K.C. Hansen, G. Schmitt-Ulms, R.J. Chalkley, J. Hirsch, M.A. Baldwin, A.L. Burlingame, *MS analysis of protein mixtures at low levels using cleavable ^{13}C-ICAT and multidimensional LC*, Mol Cell Proteomics, 2 (2003) 299.

87. L.-R. Yu, T.P. Conrads, T. Uo, H.J. Issaq, R.S. Morrison, T.D. Veenstra, *Evaluation of the acid-cleavable ICAT reagents: application to camptothecin-treated cortical neurons*, J. Proteome Res., 3 (2004) 469.

88. H. Lee, E.C. Yi, B. Wen, T.P. Reily, L. Pohl, S. Nelson, R. Aebersold, D.R. Goodlett, *Optimization of microcapillary RPLC for quantitative proteomics*, J. Chromatogr. B, 803 (2004) 101.

89. I.I. Stewart, T. Thomson, D. Figeys, *^{18}O Labelling: a tool for proteomics*, Rapid Commun. Mass Spectrom., 15 (2001) 2456.

90. Y.K. Wang, Z. Ma, D.F. Quinn, E.W. Fu, *Inverse ^{18}O labeling MS for the rapid identification of marker/target proteins*, Anal. Chem., 73 (2001) 3742.

91. Y. Oda, K. Huang, F.R. Cross, D. Cowburn, B.T. Chait, *Accurate quantitation of protein expression and site-specific phosphorylation*, PNAS, 96 (1999) 6591.

92. T.P. Conrads, K. Alving, T.D. Veenstra, M.E. Belov, G.A. Anderson, D.J. Anderson, M.S. Lipton, L. Paša-Tolić, H.R. Udseth, W.B. Chrisler, B.D. Thrall, R.D. Smith, *Quantitative analysis of bacterial and mammalian proteomes using a combination of cysteine affinity tags and 15N-metabolic labeling*, Anal. Chem., 73 (2001) 2132.

93. Y.K. Wang, Z. Ma, D.F. Quinn, E.W. Fu, *Inverse ^{15}N-metabolic labelling–MS for comparative proteomics and rapid identification of protein markers/targets*, Rapid Commun. Mass Spectrom., 16 (2002) 1389.

94. T.D. Veenstra, S. Martinović, G.A. Anderson, L. Paša-Tolić, R.D. Smith, *Proteome analysis using selective incorporation of isotopically labeled amino acids*, J. Am. Soc. Mass Spectrom., 11 (2000) 78.

95. S.-E. Ong, B. Blagoev, I. Kratchmarova, D.B. Kristensen, H. Steen, A. Pandey, M. Mann, *SILAC as a simple and accurate approach to expression proteomics*, Mol. Cell. Prot., 1 (2002) 376.

96. J. Krijgsveld, R.F. Ketting, T. Mahmoudi, J. Johansen, M. Artal-Sanz, C.P. Verrijzer, R.H.A. Plasterk, A.J.R. Heck, *Metabolic labelling of C. elegans and D. melanogaster for quantitative proteomics*, Nat. Biotechnol., 21 (2003) 927.

97. A. Thompson, J. Schäfer, K. Kuhn, S. Kienle, J. Schwarz, G. Schmidt, T. Neumann, C. Hamon, *TMT: a novel quantification strategy for comparative analysis of complex protein mixtures by MS–MS*, Anal. Chem., 75 (2003) 1895.

98. D.R. Goodlett, A. Keller, J.D. Watts, R. Newitt, E.C. Yi, S. Purvine, J.K. Eng, P. von Haller, R. Aebersold, E. Kolker, *Differential stable isotope labeling of peptides for quantitation and de novo sequence derivatization*, Rapid Commun. Mass Spectrom., 15 (2001) 1214.

99. Y. Qiu, E.A. Sousa, R.M. Hewick, J.H. Wang, *ALICE: A class of reagents for quantitative MS analysis of complex protein mixtures*, Anal. Chem., 74 (2002) 4969.

100. D.E. Mason, D.C. Liebler, *Quantitative analysis of modified proteins by CE–MS–MS of peptides labelled with phenyl isocyanate*, J. Proteome Res., 2 (2003) 265.

101. Y.H. Lee, H. Han, S.-B. Chang, S.-W. Lee, *Isotope-coded N-terminal sulfonation of peptides allows quantitative proteomic analysis with increased de novo peptide sequencing capability*, Rapid Commun. Mass Spectrom., 18 (2004) 3019.

102. C. Hagman, M. Ramström, P. Håkansson, J. Bergquist, *Quantitative analysis of*

tryptic protein mixtures using ESI-FT-ICR-MS, J. Proteome Res., 3 (2004) 587.

103.   G. Cagney, A. Emili, *De novo peptide sequencing and quantitative profiling of complex protein mixtures using MCAT*, Nat. Biotechnol., 20 (2002) 163.

104.   M.P. Hall, S. Ashrafi, I. Obegi, R. Petesch, J.N. Peterson, L.V. Schneider, *'Mass defect' tags for biomolecular MS*, J. Mass Spectrom., 38 (2003) 809.

105.   A. Leitner, W. Lindner, *Current chemical tagging strategies for proteome analysis by MS*, J. Chromatogr. B, 813 (2004) 1.

106.   B.D. Halligan, R.Y. Slyper, S.N. Twigger, W. Hicks, M. Olivier, A.S. Greene, *ZoomQuant: An application for the quantitation of stable isotope labeled peptides*, J. Am. Soc. Mass Spectrom., 16 (2005) 302.

107.   S. Wang, X. Zhang, and F. Regnier, *Quantitative proteomics strategy involving the selection of peptides containing both cysteine and histidine from tryptic digests of cell lysate*, J. Chromatogr. A, 949 (2002) 153.

108.   L. Paša-Tolić,, R. Harkewicz, G.A. Anderson, M. Tolić, Y. Shen, R. Zhao, B. Thrall, C. Masselon, R.D. Smith, *Increased proteome coverage for quantitative peptide abundance measurements based upon high performance separations and DREAMS FT-ICR-MS*, J. Am. Soc. Mass Spectrom., 13 (2002) 954.

109.   C. Fierens, D. Stöckl, L.M. Thienpoint, A.P. De Leenheer, *Strategies for determination of insulin with ESI-MS–MS: implications for other analyte proteins?*, Rapid Commun. Mass Spectrom., 15 (2001) 1433.

110.   C. Fierens, D. Stöckl, D. Baetens, A.P. De Leenheer, L.M. Thienpoint, *Application of a C-peptide ESI–isotope dilution–LC–MS–MS measurement procedure for the evaluation of five C-peptide immunoassays for urine*, J. Chromatogr. B, 792 (2003) 249.

111.   S.-L. Wu, H. Amato, R. Biringer, G. Choudhary, P. Shieh, W.S. Hancock, *Target proteomics of low-level proteins in human plasma by LC–MS^n: Using human growth hormone as a model system*, J. Proteome Res., 1 (2002) 459.

112.   Q.C. Ji, E.M. Gage, R. Rodila, M.S. Chang, T.A. El-Shourbagy, *Method development for the concentration determination of a protein in human plasma utilizing 96-well solid-phase extraction and LC–MS–MS detection*, Rapid Commun. Mass Spectrom., 17 (2003) 794.

113.   S.A. Gerber, J. Rush, O. Stemman, M.W. Kirschner, S.P. Gygi, *Absolute quantification of proteins and phosphoproteins from cell lysates by MS–MS*, PNAS, 100 (2003) 6940.

114.   F. Zhang, M.J. Bartels, W.T. Stott, *Quantitation of human glutathione S-transferases in complex matrices by LC–MS–MS with signature peptides*, Rapid Commun. Mass Spectrom., 18 (2004) 491.

115.   N.L. Anderson, N.G. Anderson, L.R. Haines, D.B. Hardie, R.W. Olafson, T.W. Pearson, *MS quantitation of peptides and proteins using stable isotope standards and capture by anti-peptide antibodies (SISCAPA)*, J. Proteome Res., 3 (2004) 235.

116.   B.M. Mayr, O. Kohlbacher, K. Reinert, M. Sturm, C. Gröpl, E. Lange, C. Klein, C.G. Huber, *Absolute myoglobin quantitation in serum by combining 2D-LC–ESI-MS and novel data analysis algorithms*, J. Proteome Res., 5 (2006) 414.

117.   A.J.R. Heck, R.H.H. van den Heuvel, *Investigation of intact protein complexes by MS*, Mass Spectrom. Rev., 23 (2004) 368.

118.   A.J. Link, J.K. Eng, D.M. Schieltz, E. Carmack, G.J. Mize, D.R. Morris, B.M. Garvik, J.R. Yates, III, *Direct analysis of protein complexes using MS*, Nat.

Biotechnol, 17 (1999) 676.
119. Y. Wagner, A. Sickmann, H.E. Meyer, G. Daum, *Multidimensional nano-LC for analysis of protein complexes*, J. Am. Soc. Mass Spectrom., 14 (2001) 1003.
120. K.L. Gould, L. Ren, A,S. Feoktistova, J.L. Jennings, A.J. Link, *Tandem affinity purification and identification of protein complex components*, Methods, 33 (2004) 239.
121. G. Rigaut, A. Schevchenko, B. Rutz, M. Wilm, M. Mann, B. Séraphin, *A generic protein purification method for protein complex characterization and proteome exploration*, Nat. Biotechnol., 17 (1999) 1030.
122. T. Natsume, Y. Yamauuchi, H. Nakayama, T. Shinkawa, M. Yanagida, N. Takahashi, T. Isobe, *A direct nano-LC–MS–MS system for interaction proteomics*, Anal. Chem., 74 (2002) 4725.
123. S. Srivastava, R.-G. Srivastava, *Proteomics in the forefront of cancer biomarker discovery*, J. Proteome Res., 4 (2005) 1098.
124. J. LaBaer, *So, you want to look for biomarker?*, J. Proteome Res., 4 (2005) 1053.
125. F. Vitzthum, F. Behrens, N.L. Anderson, J.H. Shaw, *Proteomics: From basic research to diagnostic application. A review of requirements and needs*, J. Proteome Res., 4 (2005) 1086.
126. R. Aebersold, L. Anderson, R. Caprioli, B. Druker, L. Hartwell, R. Smith, *Perspective: A program to improve protein biomarker discovery for cancer*, J. Proteome Res., 4 (2005) 1104.
127. Z. Xiao, DaRue Prieta, T.P. Conrads, T.D. Veenstra, H.J. Issaq, *Proteomic patterns: their potential for disease diagnosis*, Mol. Cell. Endocrin., 230 (2005) 95.
128. R. Bischoff, T.M. Luider, *Methodological advances in the discovery of protein and peptide disease markers*, J. Chromatogr. B, 803 (2004) 27.
129. A. Journet, M. Ferro, *The potentials of MS-based subproteomic approaches in medical science: The case of lysosomes and breast cancer*, Mass Spectrom. Rev., 23 (2004) 393.
130. W. Kolch, C. Neusüß, M. Pelzing, H. Mischak, *CE–MS as a powerful tool in clinical diagnosis and biomarker discovery*, Mass Spectrom. Rev., 24 (2005) 959.
131. P. Chaurand, S.A. Schwartz, R.M. Caprioli, *Assessing protein patterns in disease using imaging MS*, J. Proteome Res., 3 (2004) 245.

19

LC–MS FOR IDENTIFICATION OF
POST-TRANSLATIONAL MODIFICATIONS

1. Introduction

The proteome consists of the expressed proteins in a cell, tissue, or organism. Whereas one gene in principle encodes for one protein, the expressed protein may occur as a population of heterogenous species in various modification states, involving among others phosphorylation and glycosylation. In fact, more than 200 modification types have been recorded, while new ones are being discovered. These modified forms of one protein, often indicated as post-translation modifications (PTM), may be present at different locations in the cell and in widely different concentrations. PTM are known to determine targeting, structure, function, and turnover of proteins. They play an important role in the regulatory actions with respect to the biological activity of the protein. PTM result from a variety of mechanisms, involving both enzymatic and nonenzymatic reactions in the cell. They may also be due to interactions between proteins and xenobiotics. The analysis of a wide variety of protein adducts was reviewed [1].

The discussion in this chapter concentrates on liquid chromatography–mass spectrometry (LC–MS) related technologies to detect and identify protein phosphorylation and glycosylation, The focus is on general concepts and technologies rather than on applications.

2. General considerations

From an analytical point of view, PTM greatly affect the complexity of the proteome: a particular amino-acid sequence may be found in a wide variety of species with widely different abundances in the cell. Both issues demand efficient (multidimensional) separation technologies capable of selective enrichment of the lower abundance products. With respect to MS analysis, the different PTM species may either have different molecular mass (heterogeneity) or may be isomeric. Unlike in protein identification (Ch. 18.3), generally higher sequence coverage is needed. If a protein is enzymatically digested in 8 peptides, two of which may be phosphorylated, *e.g.*, T5 and T7, the protein itself can already be identified from a limited number of tryptic peptides. Perhaps, only T1, T4, and T8 are readily observed in the peptide map. However, to be able to pinpoint the phosphorylation, detection and characterization of the phosphorylated fragments is required. Unfortunately, a PTM may adversely influence the detectability of the peptide in electrospray ionization (ESI-MS) or matrix-assisted laser desorption ionization (MALDI-MS). This is certainly true for the two most widely studied PTM, *i.e.*, phosphorylation and glycosylation. In addition, the presence of a PTM may significantly alter the fragmentation by collision-induced dissociation (CID) in MS–MS. The bond between the peptide and the modification is often easier to cleave than the peptide backbone bonds. As a result, the MS–MS spectrum of a glycosylated peptide may show fragments due to internal fragmentation within the glycosylation chain, which completely overwhelm the amino-acid sequence ions.

In order to characterize the modified protein, both the position(s) and the nature of the modification must be elucidated. While the nature of a PTM can be very simple, *e.g.*, with phosphorylation or acetylation, it may also be very complex, *e.g.*, with glycosylation. For an isolated modified protein, the strategy may involve:

- Determination of the molecular mass of the intact protein, both with its modifications and without, after enzymatic or chemical removal. From the mass difference, the identity of the modification may be elucidated. MALDI-MS is generally preferred in this step. Searchable databases of protein modifications are available, *e.g.*, [2]. The heterogeneity of many modified proteins prevents an accurate-mass determination of the intact protein by ESI-MS, unless high-resolution MS is applied. In addition, some modifications are prone to formation of multiple sodium adducts in ESI-MS, which further hinders data interpretation and adequate transformation of the ion envelope of multiple-charge ions.
- Peptide mass fingerprinting (PMF) of tryptic digests of both the modified and the unmodified protein (complementary peptide mapping). By careful comparison of the two spectra, *m/z* shifts can be found, from which the identity of the modification may be elucidated, as well as the tryptic fragment(s) that are actually modified. When the amino-acid sequence of the protein is known (and validated), the position of the modification may be known. For example,

phosphorylation primarily occurs at Ser, Thr, or Tyr. The ability to predict post-translational glycosylation and phosphorylation of proteins from the amino acid sequence is reviewed [3]. Thus, if only one of the target amino acids is in the modified peptide, the position is known as well. Obviously, additional study is required in many cases.

- Peptide sequence analysis (PSA) of the peptides that contains a PTM. If the amino-acid sequence of the protein is not known or the position is not unambiguously determined in PMF, sequence analysis of the modified peptides is necessary. Strategies for shotgun identification (Ch. 18.3.2–4) of post-translation modifications are reviewed by Cantin and Yates [4].

- Further structure elucidation of the modification itself, if applicable. This is especially important in the case of glycosylation, as this may involve extremely complex and/or heterogenous structures (Ch. 20).

Shotgun and top-down procedures can be considered complementary, especially when dealing with PTM, where the shotgun approach provides information on the amino-acid sequence and the top-down approach on nature and position of the PTM.

Within proteomics, PTM must be studied within a complete proteome, *i.e.*, a complex mixture of both modified and unmodified proteins. In that case, either liquid-phase or MS tools must be applied for the selective isolation of modified proteins or the enzymatic peptides derived from these.

Identification of the PTM is just one step. In terms of functional proteomics, elucidation of the dynamics of modification and the role in the physiological process is important. This involves comparative quantitative proteomics along the lines discussed for protein-expression profiling (Ch. 18.4).

3. Protein phosphorylation

Protein phosphorylation is important in many aspects of cellular growth and development as well as in a number of other regulatory tasks and signal transduction pathways within the cell. Phosphorylated proteins are generated in the cell by phosphotransferase enzymes, known as protein kinases, while another group of enzymes, known as phosphatases, catalyse the removal of specific phosphate groups. Phosphorylation is sometimes considered as a molecular switch in biological systems. Phosphorylation occurs by two classes of kinases, one involved in phosphorylation of Ser and Thr to pSer and pThr, the other of Tyr to pTyr. A large number of kinases has been identified and for many the consensus sequence for recognition and phosphorylation have been identified. Other types of protein phosphorylation, *e.g.*, N-phosphates at Arg, Lys, or His, acylphosphates at Asp or Glu, and S-phosphates at Cys, are not discussed in this text.

Conventional methods for the study of protein phosphorylation rely on ^{32}P radioactive labelling, 2D-GE protein mapping, and Edman degradation. Early studies in LC–MS characterization of protein phosphorylation involve MS–MS analysis of modified tryptic peptide to determine the phosphorylation site by complementary peptide mapping, e.g., [5-7]. In the LC–MS analysis of tryptic and V8-protease digests of a phosphorylated (pp19) and nonphosphorylated (p19) 19-kDa cytosolic protein, two sets of ions with a phosphate-characteristic mass difference of 80 Da were observed. Sequence analysis of the relevant peptides by MS–MS showed that phosphorylation occurs at Ser-25 and Ser-38 [7].

In an early study, concerning the phosphorylation of α-casein [8], the various steps of the LC–MS procedure are nicely outlined and illustrated. Molecular-weight determination by transformation of the ESI-MS ion envelope (Ch. 16.2.2) of the intact protein (both phosphorylated and non-phosphorylated) revealed that eight phosphate groups are present. From the known amino-acid sequence, it can be derived that phosphorylation takes place at Ser in the tryptic fragments T7, T8, and T15. In the LC–MS chromatograms, these peptides are found as well as the T14/15 peptide due to a missed cleavage. Significant differences in chromatographic retention times were observed between some phosphorylated and non-phosphorylated peptides, as expected from the influence phosphorylation has on the hydrophobicity of the peptide. The same tryptic peptides were also detected in a spectrum acquired in neutral-loss analysis mode (Ch. 19.3.1) by direct-infusion ESI-MS or LC–MS–MS, and by combining negative-ion LC–ESI-MS data from high in-source CID in selected-ion monitoring (SIM, m/z 79) and full-spectrum analysis (Ch. 19.3.1). Finally, product-ion MS–MS was applied for peptide sequence analysis of the phosphopeptide.

3.1   MS screening for phosphopeptides

Screening for phosphorylation by MS is based on specific fragmentation of the phosphorylated peptides. pSer and pThr readily lose phosphoric acid by β-elimination, resulting in a loss of 98 Da. This does not occur in pTyr. In negative-ion mode, phosphate-related fragments are generated, i.e., m/z 63, 79, and 97, corresponding to PO_2^-, PO_3^-, and $H_2PO_4^-$, respectively. The fragment at m/z 97 can not be used in screening, because it is isobaric with the $CF_3C=O^-$ fragment of TFA, frequently applied in LC–MS of peptide digests (Ch. 17.5.1)

Covey et al. [9] proposed the use of the neutral-loss analysis mode (common neutral loss of 98 Da) for the selective screening of tryptic digests of phosphorylated proteins, in order to selectively detect phosphopeptides in the digest. In multiple-charge ions with n charges, the loss is 98/n Da. Whereas in triple-quadrupole MS–MS a 98-Da is not observed for pTyr, it does occur in an ion-trap [10].

Huddleston et al. [11] reported the selective detection of phosphopeptides in complex mixtures by LC–MS. They analyse the tryptic digest in negative-ion ESI-

MS with both low or high in-source CID settings, in full-spectrum analysis (m/z 59–99 and 400–2000) or in SIM mode (m/z 63 and 79), respectively. In-source CID results in the formation of diagnostic ions for phosphorylation at m/z 63 and 79. The same group [12] reported a similar method, based on precursor-ion analysis (m/z 79 as the common product ion) in negative-ion mode direct-infusion ESI-MS–MS from a high-pH solvent. An illustration of the power of this method is given in Figure 19.1. Despite the 10× lower signal intensity in the precursor-ion mode, the S/N is superior to negative-ion or positive-ion full-spectrum analysis. The method enables low femtomole detection of phosphopeptides. A similar procedure for the detection of phosphorylation in intact proteins was reported by others [13].

Both these methods were further developed, taking advantage of the advanced features in data acquisition in later generations of LC–MS instruments. Negative-ion MS–MS data for phosphopeptides do not always yield useful peptide sequence information [14]. The screening based on the detection of diagnostic phosphate ions in the negative-ion mode is best combined with PSA in positive-ion ESI-MS–MS.

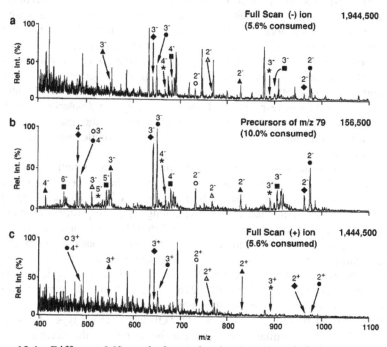

Figure 19.1: Different MS analysis modes in the direct-infusion nano-ESI-MS analysis of an α-casein tryptic digest (2 μl of 250 fmol/μl solution at pH 10 loaded). Mono- (▲,●), di- (◆,△), tri- (★), and pentaphosphorylated (■) peptides were detected. Reprinted from [12] with permission. ©1996, Academic Press Inc.

Allen *et al.* [15] reported mixed-mode scanning for phosphopeptide detection during LC–MS, *i.e.*, alternating acquisition of SIM *m/z* 79 in the negative-ion mode and full-spectrum data in positive-ion mode. Annan *et al.* [16] refined their earlier procedure [11-12] into a three-step procedure involving (1) screening of proteolytic digests in a split-flow LC–MS system with negative-ion ESI-MS, high in-source CID, and SIM on *m/z* 79, and fractionation of the column effluent, (2) molecular-mass determination of phosphopeptides in these fractions by direct-infusion negative-ion nano-ESI-MS in precursor-ion analysis mode (*m/z* 79 as the common product ion), and (3) PSA by direct-infusion positive-ion nano-ESI-MS–MS. A 20-fold improvement in sensitivity was realized by subsequent miniaturization of this system to nanoscale LC [17].

The use of RPLC prior to nano-ESI-MS analysis by the strategies outlined above is especially important for the study of low-abundance phosphoproteins in complex proteomes. This results in conflicts in mobile-phase composition: the negative-ion mode (for *m/z* 79 detection via in-source CID) provides the best response from high-pH solvents, whereas in the positive-ion mode low-pH solvents are favourable. A comparison of low-pH and high-pH mobile phases in the LC–MS analysis of phosphopeptides was reported by Beck *et al.* [18]. At high pH, ~3× better overall response is achieved in *m/z* 79 screening. A similar study was reported by Tholey *et al.* [19] using a monolithic poly(styrene–divinylbenzene) (PS–DVB) copolymer column. The gain in detectability in LC–MS at high pH depended on the degree of phosphorylation: 5- to 5000-fold improvements were reported for mono- and tetra-phosphorylated peptides, respectively. In order to totally avoid polarity switching for phosphopeptide detection, Alverdi *et al.* [20] proposed the use of neutral-loss analysis for monitoring the loss of HPO_3 (80 Da for single-charge ions) in the positive-ion mode.

While these structure-specific screening methods were initially developed on triple-quadrupole instruments, similar strategies have been reported for quadrupole–time-of-flight (Q–TOF) hybrid instruments [21]. The same group also proposed an alternative method for pTyr screening based on positive-ion nano-ESI-MS and the precursor-ion analysis mode (*m/z* 216.043 as the common product ion, corresponding to the immonium ion of pTyr) [21]. This approach allows detection of synthetic phosphopeptides at low fmol/µl level, and required ~100 fmol of an in-gel digestion protein. Because of the possible presence of isobaric b_2 ions (Asn/Thr, Ser/Lys, and Ser/Gln at *m/z* 216.098, 216.135, and 216.098, respectively), this method can only successfully be applied at high-resolution MS instruments.

Comparison of triple-quadrupole and Q–TOF use in these type of studies was reported [22]. While similar sensitivity can be achieved, the selectivity of the Q–TOF is better due to its high resolving power, thereby minimizing isobaric interferences. Bateman *et al.* [23] reported a novel mode of operation of a Q–TOF instrument for precursor-ion analysis of phosphoproteins. The first-stage quadrupole is operated in radiofrequency-only mode, serving as a wide-band ion-transmission device to the quadrupole collision cell. Alternating, a high and a low collision energy is applied. Using advanced data-dependent acquisition (DDA), both precursor-ion m/z-216 data for pTyr, neutral-loss 98-Da data for pSer and pThr, and full-spectrum MS–MS data are acquired during one LC–MS run of a tryptic-digest.

The enhanced DDA and faster scanning capabilities and good sensitivity of a quadrupole–linear-ion-trap hybrid (Q–LIT) instrument make this instrument highly suitable for the analysis of phosphopeptides [24]. Two different DDA experiments have been described for on-line LC–MS of phosphopeptides. The first one comprises of precursor-ion analysis (m/z 79) in negative-ion mode (2 s), a polarity switch (0.7 s), an enhanced-resolution scan to determine the charge state of the precursor (0.5 s), an enhanced-product-ion scan (0.4 s), a polarity switch (0.7 s), and addition of the precursor ion to the exclusion list. The second one performs a neutral-loss analysis (49 Da, double charge assumed) (2 s), an enhanced-resolution scan to determine the charge state of the precursor (0.5 s), an enhanced-product-ion scan (0.4 s), and addition of the precursor ion to the exclusion list. Each loop is repeated during the entire chromatographic run time. These types of procedures were applied in phosphorylation mapping of various protein kinases [25].

For application with triple-quadrupole and especially Q–LIT instruments, a selected-reaction monitoring (SRM) procedure was developed for the sensitive and selective detection of phosphopeptides in proteomes with known amino-acid sequences [26]. A list of SRM transitions of potential phosphopeptides is generated for all expected tryptic peptides in the mixture with Ser, Thr, or Tyr and for double- and triple-charge ions in the mass range of m/z 400–1600. The number of transitions included is limited by the maximum cycle time of ~10 s, which assures that peptides in a 30 s wide peak are at least analysed twice. The procedure was applied to the cell cycle regulatory protein Cyclin B from *Schizosaccharomyces pombe*.

In an alternative strategy, also using DDA in a Q–LIT instrument, both MS–MS and MS–MS^2 scan were acquired [27]. The loss of phosphoric acid in the MS–MS scan automatically triggered the acquisition of an MS–MS^2 scan. Both MS–MS and MS–MS^2 data were applied for site-specific quantification of phosphorylation.

Another highly sensitive tool for phosphopeptide detection (0.1 fmol) in LC–MS is the selective detection of a ^{31}P by means of inductively-coupled plasma (high-resolution) MS (ICP-MS). This approach was pioneered by the group of Lehmann *et al.* [28]. Parallel runs of LC–ICP-MS and LC–ESI-MS allow a selective detection of the phosphopeptides by alignment of the chromatograms of the tryptic digest.

While all these strategies rely on specific properties of phosphopeptides in MS analysis, a more global approach involving shotgun protein identification strategies and SEQUEST database searching (Ch. 18.3.2) is demonstrated by the group of Yates for protein complexes and lens tissue proteins [29].

Due to its high resolving power and its abilities to accurate-mass determination, Fourier-transform ion-cyclotron resonance MS (FT-ICR-MS) has become an important tool in proteomics (Ch. 18.3.4–5). Its potential in phosphoprotein characterization has been evaluated as well. A nice example is the characterization of protein kinase C phosphorylation, studied by LC–MS on an FT-ICR-MS instrument equipped with a dual-ESI source for internal calibration, and external ion accumulation in a linear octapole ion trap [30]. Four previously unidentified phosphorylated peptides were detected in positive-ion mode, and five known ones were observed in negative-ion ESI experiments.

To this end, different ion fragmentation tools have been characterized with respect to phosphopeptide fragmentation, *e.g.*, electron-capture dissociation (ECD) [31] and infrared multiphoton dissociation (IRMPD) [32]. An application of ECD in PTM analysis is the top-down protein characterization (Ch. 18.3.5) of carbonic anhydrase [33]. IRMPD is applied in the study on protein kinase C phosphorylation [30]. Both ECD and IRMPD were applied in a subsequent nano-ESI–FT-ICR-MS study on protein kinase A phosphorylation [34]. Combined ECD and IRMPD for multistage MS–MS in FT-ICR-MS was applied for phosphopeptide characterization [35]. ECD provides backbone cleavages (c- and z•-ions) without H_3PO_4 loss, whereas in IRMPD the loss of H_3PO_4 is prominent and only a few backbone cleavages (b- and y-ions) are observed (*cf.* Ch. 17.6.1).

The potential of IRMPD in ion-trap fragmentation of phosphopeptides was also evaluated [36]. Due to their greater photon absorption efficiency, IRMPD provides selective dissociation of phosphopeptides.

3.2   Liquid-phase selection of phosphopeptides

Selective isolation of phosphopeptides from a complex peptide mixture can be achieved using Ga^{3+}- or Fe^{3+}-loaded immobilized metal-ion affinity chromatography (IMAC) [37]. Phosphopeptides are selectively sorbed from a low-pH solvent (pH <3), and can be released from the IMAC column by means of a high-pH solvent (pH 10.5). The selectivity of IMAC in phosphopeptide isolation from complex tryptic digests can be improved by the formation of methyl esters, because highly acidic peptides show affinity to the IMAC column as well [38-39]. The use of strong-cation exchange chromatography (SCX) prior to IMAC was found to reduce the complexity of the IMAC purified sample and to provide better coverage of the monophosphorylated peptides [40]. IMAC cannot only be applied off-line. It can also be applied on-line with ESI-MS [41] or with RPLC–MS [42].

Selective detection of Tyr-phosphorylation in LC–MS was achieved by means of an on-line microreactor with immobilized tyrosine phosphatase β. Comparison of the chromatograms of the digests with or without the reactor reveals the presence of phosphopeptides [43].

Selective isolation of phosphopeptides from tryptic digests using a dual-precolumn setup (titanium oxide and RPLC) prior to RPLC–MS was reported by Pinkse *et al.* [44]. The tryptic digest is injected onto two precolumns in series: phosphopeptides are selectively trapped by the TiO_2-column, whereas nonphosphorylated peptides are trapped by the RP-column. The RP column is first eluted under acidic conditions to a nano-RPLC–MS system. Subsequently, the phosphopeptides are eluted and transferred under basic conditions.

A continuous-flow post-column ligand-exchange reaction of phosphopeptides with fluorescent Fe^{3+}–methylcalcein blue (MCB) coupled to ESI-MS has been proposed as alternative means to selectively detect phosphopeptides in tryptic digests [45]. The system detects the presence of a phosphopeptide in the LC effluent as a peak in the MCB reporter trace, measured by SIM. Because not all phosphopeptide reacts with Fe^{3+}, the free phosphopeptide can be simultaneously characterized by MS or MS–MS. Using conventional-flow ESI-MS (100 µl/min), detection limits were 2 pmol/µl.

3.3 Derivatization and labelling of phosphopeptides

In order to enhance the response in ESI-MS and to modify the fragmentation characteristics of phosphopeptides, various derivatization strategies have been proposed. Given their regulatory role, quantitative proteomics of phosphorylated proteins is an important topic. To this end, various labelling strategies have been reported. Various strategies applicable to nonphosphorylated peptides (Ch. 18.4.1–2) have been adapted for phosphopeptides, such as the use of isotopically-labelled amino acids in cell cultures (SILAC). Some additional and phosphopeptide-specific strategies are:

- Tris[(2,4,6-trimethoxyphenyl)phosphonium] acetylation [46].
- Conversion in a dimethylamine-containing sulfenic acid derivative, enabling a precursor-ion analysis with m/z 122 as the common product ion, corresponding to 2-dimethylaminoethanesulfoxide [47].
- β-elimination of phosphate from pSer and pThr by means of $Ba(OH)_2$ followed by selective Michael addition of isotope-coded ethanethiol [48-51]. This converts pSer and pThr in cysteic and β-methylcysteic acid, respectively. Michael addition with isotope-coded methyl amine was also reported [52].
- Extension of the above procedure using ethanedithiol, enabling subsequent biotinylation to obtain a phosphoprotein isotope-coded affinity tag (PhIAT) [53-54]. This enables phosphopeptide isolation by avidin affinity chromatography.

A combination of SILAC and IMAC strategies were applied in the quantitative phosphoproteomics of the yeast pheromone signalling pathway [55]. More than 700 phosphopeptides were identified, 139 of which were at least twofold up- or down-regulated in response to mating pheromone.

In protein-expression profiling using labelling strategies of phosphorylated proteins, care must be taken in the selection of an appropriately labelled peptide for quantification. PTM of proteins results in a variety of isoforms. Only peptides common to all isoforms of a protein can be used for quantification. Since most likely not all PTM of a protein are known, redundant peptide analysis must be performed to establish whether all peptides derived from that specific protein all show up- or down-regulation. If not, the peptide in question is associated with a protein containing PTM.

3.4   Selected applications

The identification of the phosphorylation sites in neurofilament proteins NF-L, NF-M, and NF-H found in the neuronal cytoskeleton by nano-ESI-MS was reported [56]. Phosphorylation is an important factor in the stability of the neurofilaments. Some of the special technologies outlined above were applied during these studies, *e.g.*, precursor-ion MS–MS analysis and IMAC. In this study, as well as others [57-58], the characterization of phosphorylated proteins separated by 2D-GE (Ch. 17.3) was performed, using LC–MS on an ion-trap instrument [57], or using precursor-ion analysis in combination with nano-ESI [58]. In the latter study, a existing procedure for in-gel digestion and subsequent extraction of tryptic peptides was modified by the selection of a packing material developed for oligonucleotides (POROS OligoR3) and by adjusting the pH [58]. The methods were applied to assign at least 12 phosphorylation sites to tryptic peptides derived from human double-stranded RNA-activated protein kinase expressed in a mutant strain of the yeast *Saccharomyces cerevisiae* [57], and to phosphorylation mapping in α-casein and chicken Src-ΔU protein expressed in *S. pombe* [58].

Ficarro *et al.* [38] reported an integrated strategy for phosphoprotein analysis in cell lysate proteomes of *S. cerevisiae*, using trypsin digestion of the proteins, methylation of the carboxylic acids, IMAC isolation of phosphopeptides from the digest, LC–MS using an ion-trap instrument operated under DDA, and SEQUEST database searching (Ch. 17.6.2). More than 1000 phosphopeptides were detected, and 216 peptide sequences defining 383 phosphorylation sites were determined. The same strategy was also applied in studying the reversible tyrosine phosphorylation pathways in human cells [59].

Based on a comprehensive strategy for the LC–MS characterization of membrane proteins [60], the analysis of in-vivo phosphorylated plasma membrane proteins from *Arabidopsis thaliana* was reported [40]. Phosphopeptide isolation was performed using SCX–IMAC, characterization with MALDI-MS and nano-LC–MS on a Q–TOF instrument. Six of the identified sequences originated from different isoforms of plasma membrane H^+-ATPase, among which two new sites at the regulatory *C*-terminus.

The identification of 374 different proteins in the purified post-synaptic density of central excitatory synapses by LC–MS–MS was reported [61]. 13 phosphorylated sites were detected at 8 proteins. They also determined the molar concentration and relative stoichiometries of various scaffold proteins in the post-synaptic density.

Comprehensive mapping of protein phosphorylation of the murine circadian protein period 2 was performed using a mix of low-specificity proteases, phosphopeptide enrichment onto a 50-μm-ID TiO$_2$ SPE column, and nano-LC–MS with a 25-μm-ID RPLC column, operated at 25 nl/min [62]. The protease mix enabled detection of 21 phosphorylation sites, whereas only 6 were revealed via trypsin digestion.

Numerous other applications have recently been reported, taking advantages of the technologies here described for phosphopeptide analysis and more general proteomics approaches (Ch. 18).

4. Protein glycosylation

Glycosylation of proteins is an important regulatory aspect of the biological function of proteins. Glycosylation of proteins plays an important role in the biological activity of proteins. By estimation, 60–90% of all mammalian proteins are glycosylated. As such, they serve a role in recognition in cell-cell and cell-molecule interactions, *e.g.*, in fertilization, cell growth, cell-cell adhesion, immune response, viral replication, parasitic infection, and inflammation.

The discussion is split in two parts. In this section, screening strategies and characterization of glycoproteins is discussed [63-65], while a more general discussion on glycan and oligosaccharide structure and analysis follows in Ch. 20.

Glycosylation may occur in various ways. The oligosaccharide can be *N*-linked at an Asn or *O*-linked at an Ser or Thr in the protein.

In mammals, *N*-glycosylation may take place at an Asn in the consensus sequence: Asn – X – Ser/Thr, with X any amino acid except Pro. A common pentasaccharide core structure is attached to Asn, *i.e.*,

{Man (α1-6), Man (α1-3)} Man (β1-4) GlcNAc (β1-4) GlcNAc

This core can be extended to form (1) a high mannose type, where several Man residues are attached to the core, (2) a complex type, where up to five antennae of Gal–GlcNAc and capped with sialic acids are attached to the core, or (3) a hybrid type with a combination of (α1-6)-linked Man and (α1-3)-linked complex antennae are attached to the core.

The *O*-linked glycans are attached to Ser or Thr. There is no consensus sequence and no common core structure, although a number of mainly α-GalNAc-links have been recognized. Typical extensions consist of Gal, Fuc, sialic acid, and GlcNAc.

There can be considerable heterogeneity in the length and type of the glycan side chains at a particular glycosylation site in a glycoprotein, because the enzyme reactions do not invariably go to completion. Different glycosylated variants of a protein are called glycoforms. This further contributes to the complexity of the analysis, because the 'pure' glycoprotein actually consists of a complex mixture of glycoforms, which may significantly differ in mass.

Glycosylation can have a distinct influence on the characterization of the glycoprotein, as it may limit the efficiency of the proteolytic digestion, it may significantly alter the ionization characteristics of the glycoprotein or glycopeptides, and it may obscure the amino-acid sequence information, because sugar losses can be more dominant in the MS–MS spectrum than amino acid losses. Therefore, amino-acid sequence information is generally acquired after chemical or enzymatic removal of the glycan. This can be done in various ways. By hydrazinolysis, both N-linked and O-linked glycans can be released. Alternatively, N-linked glycans can be selectively removed via treatment with the peptide N-glycosidase F (PNGase F) in $H_2{}^{18}O$, which converts the Asn–Glycan into an Asp with a incorporated ^{18}O [66]. O-linked glycans can be selectively removed as alditols via a base-catalysed β-elimination with ammonium hydroxide at 45°C, which converts Ser–Glycan into dehydroalanine and Thr–Glycan into dehydrobutyric acid. Alternatively, O-glycans can be removed enzymatically by means of an O-glycanase, although the currently available enzymes are highly specific. In subsequent PSA, these modified amino acids can be recognized and used to indicate the position of the original glycan.

Orlando and Yang [63] described a three-step procedure for the characterization of glycoproteins:

• The molecular mass of the intact glycoprotein is determined, in most cases using MALDI-MS, because this suffers less from the heterogeneity in the glycans.

• The peptide backbone is enzymatically digested and the resulting glycopeptides are detected and isolated. This can be done in a number of ways, *e.g.*, by comparative mapping using LC–MS of the glycoprotein digest prior to and after a PNGase F treatment, revealing N-linked glycans only [67-68], by searching for diagonal ladder peaks in two-dimensional maps or contour plots of the LC–MS chromatogram [69-70], or by mean of diagnostic sugar oxonium marker ions [67-68].

• Finally, the primary structure of the glycans must be determined. To this end, the mass of each glycan side chain is determined. A search through the public-domain Complex Carbohydrate Structure Database (CCSD, www.ccrc.uga.edu) provides a short list, which is applied in the differentiation between potential structures by using endo- and exoglycosidase digestion and MS and MS–MS analysis.

4.1   Glycoprotein detection using diagnostic oxonium ions

The detection of glycopeptides in a digest mixture of a glycoprotein is an important step in the characterization of the glycoprotein. Carr *et al.* [67-68] reported an elegant method, based on the formation by CID of oxonium fragments, especially HexNAc$^+$ at *m/z* 204 and HexHexNAc$^+$ at *m/z* 366. These fragments can be generated by in-source CID, enabling the differentiation between glycopeptides and other peptides in a single-quadrupole MS, or in the collision cell of a triple-quadrupole instrument, enabling precursor-ion analysis for the selective detection of the glycopeptides. The method has been frequently applied, *e.g.*, by performing scan-wise switching between full-spectrum analysis at a low cone voltage and SIM *m/z* 204 at a high cone voltage [64]. An example of precursor-ion analysis, using *m/z* 204 as common product ion, in the screening of *N*-linked glycans from recombinant human thrombomodulin [71] is shown in Figure 19.2.

This method was extended to include fragments related to sialic acid, *i.e.*, at *m/z* 274 and 292, as well as some other fragments [72]. The applicability of precursor-ion analysis using *m/z* 204, 366, or 163 as the common product ions was demonstrated in the analysis of 2-AMAC-derivatized *N*-linked glycans [73]. The generation of oxonium ions at *m/z* 204, 292, and 366 by in-source CID and subsequent ion-trap MS analysis of glycopeptides in the digest of human α-1 acid glycoprotein and a therapeutic monoclonal antibody was also reported [74].

In order to reduce the complexity of the spectra acquired by precursor-ion analysis (*m/z* 204 as common product ion) in the analysis of complex glycan mixtures without prior LC separation, precursor-ion analysis was performed with larger ions, *e.g.*, with common product ions like *m/z* 1935, 1406, and 1396 [70]. This method was applied in the characterization of over thirty complex *N*-linked glycan structures from the scrapie-associated prion protein PrP[Sc].

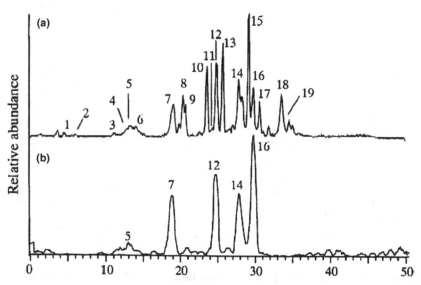

Figure 19.2: Detecting glycopeptides in the chromatogram of a tryptic digest of recombinant human thrombomodulin. (a) Total-ion-current chromatogram, and (b) precursor-ion analysis *m/z* 204 chromatogram. Reprinted from [71] with permission. ©2002, Elsevier Science BV.

4.2   Glycoprotein characterization

Strategies for the characterization of glycoproteins using LC–MS en MALDI-MS have been reviewed [63-65, 76].

Carr *et al.* [67-68] reported methods to differentiate between *N*-linked and *O*-linked oligosaccharides using the selective removal of *N*-linked oligosaccharides by PNGase F. The method was applied to the 42-kDa bovine fetuin glycoprotein, containing three *N*-linked and at least three *O*-linked glycans [67], and to the T11 fragment of recombinant tissue plasminogen activator (rt-PA) [68]. The structural characterization of a recombinant reshaped human monoclonal antibody against respiratory syncytial virus by means of an integrated strategy was reported by the same group [77]. In the strategy proposed, LC–MS was applied in peptide mapping and selective detection of the glycopeptides, Edman degradation and MS–MS for peptide sequencing, and MALDI-MS for the characterization of endo- and exoglycosidase digest of isolated glycopeptides. In excess of 99% of the heavy- and light-chain amino acid sequence was verified. Glycosylation was only found at Asn[296] of the heavy chain. The various glycoforms consisted of biantennary core fucosylated oligosaccharides.

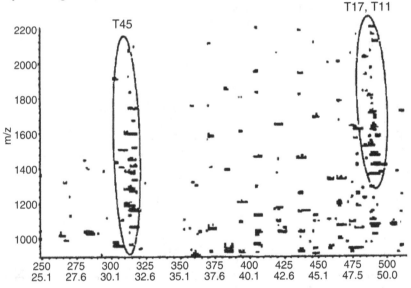

Figure 19.3: Contour plot (*m/z*–retention time) of rt-PA showing the series of negatively sloping ions characteristic for glycoproteins. Three glycopeptides were detected: the complex-type glycopeptides T45 and T17, and the high-mannose type glycopeptide T11. Reprinted from [69] with permission. ©1993, American Chemical Society.

A series of negatively sloping ions in a contour plot (m/z–retention time, Figure 19.3) was used to characterize glycopeptides in a tryptic digest of rt-PA [69]. The complex-type glycopeptides T45 and T17 elute at 31 and 48 min, respectively. T17 co-elutes with the high-mannose type glycopeptide T11.

After on-line enrichment on lectin affinity columns, the glycosylation pattern of *Desmodus* salivary plasminogen activator was characterized using selective enzymatic deglycosylation with PNGase F or *O*-glycosidase to remove sialic acid and *O*-linked glycans, LC–MS, and MALDI-MS [70]. The sample was first digested with the Lys-C. Part of the sample was deglycosylated. Comparison of the peptide maps acquired using LC–MS enables the identification of the *N*-linked glycosylation sites and the *N*-linked glycans. No *O*-linked glycans were found.

Similar strategies were applied by others, *e.g.*, in the evaluation of site heterogeneity and glycan identification of bovine α_1-acid glycoprotein [78], and in the characterization of *N*-linked glycans in recombinant pectate lysate [79] and in pectin methylesterase [80] from the fungus *Aspergillus niger*, using exoglycosidase digestion to confirm possible glycans indicated by the CCSD. Multidimensional LC–MS involving lectin affinity LC is also reported by others [81-82].

Nano-ESI-MS on a Q–TOF-MS was applied to characterize *O*-glycosylation in Mucin 2 membrane proteins [83]. Sequencing of 8 *O*-glycolylated peptides was performed. The MS–MS spectra showed intense B-type fragment ions of the oligosaccharides as well as diagnostic amino-acid sequence-specific b- and y-type ions, enabling determination of both the glycosylation site and the oligosaccharide sequence. This approach was also applied by the same group in the determination of glycosylation sites in *O*-fucosylated glycopeptides from thrombospondin-1 [84].

Karlsson *et al.* [85] reported the structural characterization of fifty different neutral *O*-linked glycans ranging from di- to nonasaccharides from salivary mucin MUC5B, using negative-ion LC–MS and MS–MS. Primary sequence information could be derived from Y- and Z-ions. In addition, some linkage information could be deduced from additional fragmentation pathways.

Triple-quadrupole, ion-trap, and Q–TOF instruments were compared in MS and MS–MS with product-ion, neutral-loss, and precursor-ion analysis modes for the profiling of the Asn^{56}-*N*-glycan of the α-unit of ovine luteinizing hormone [86]. More than 20 glycoforms were detected and identified. The Q–TOF allowed the lowest sample consumption, but all three instruments contributed to the profiling study.

Profiling of *O*-linked glycan from mucin was based on the diagnostic ions m/z 408 and 390 for Gal(β1-3)GalNAc sequences [87]. The core type, and the position and linkage of fucose residues was determined. Differentiation between isomeric structures and linkage types were derived from ^{0,2}A-type cross-ring cleavages in the negative-ion mode on a Q–TOF-MS.

Characterization of the protein glycosylation of *Dolichos biflorus* lectin was reported by Wuhrer *et al.* [88]. The protein was digested by means of unspecific proteolysis into 2–8 amino-acid peptides. The mixture was analysed on a nanoscale Amide-80 column with an acetonitrile–0.5% aqueous formic acid mobile-phase gradient (hydrophilic interaction chromatography, Ch. 1.4.5). The ion-trap MS–MS spectra were dominated by cleavages of glycosidic bonds, while peptide backbone fragmentation was obtained in MS–MS^2.

While the discussion in this text focusses on the use of LC–MS, on-line capillary electrophoresis–MS is another important tool in glycopeptide and glycoprotein analysis, as shown by the data reviewed by Zamfir and Peter-Katalinić [89].

4.3   Towards glycomics

The analysis and characterization of glycoproteins by LC–MS and related techniques is an important, but difficult topic. Analogous to proteomics, protein glycosylation analysis is evolving towards glycomics [90-91], indicating the moving from just characterization to a systems-biology oriented approach. Carbohydrate microarrays and surface-plasmon resonance are important tools to study glycoprotein interactions. With respect to the role of MS, MALDI-MS appears to be more important than ESI-MS. Nevertheless, the potential role of LC–MS in this area has been critically discussed and reviewed [92-93]. LC–MALDI-MS via fraction collection strategies (Ch. 5.9) may be important in that respect.

FT-ICR-MS is rapidly becoming another important tool. Froesch et al. [94] provided a challenging view on future oligosaccharide characterization in their study on a glycoconjugate mixture containing O-glycosylated sialylated peptides from urine of a patient suffering from a hereditary N-acetylhexosaminidase deficiency (Schindler's disease). Sample introduction into the ESI-MS source of an FT-ICR-MS instrument is performed using commercial multi-sprayer microchip device (Ch. 17.5.5). MS–MS is achieved using sustained off-resonance irradiation collision-induced dissociation (SORI-CID, Ch. 2.4.6). They consider this approach as a new platform for advanced glycomics. Glycomics of tear fluids using MALDI-MS on an FT-ICR-MS was reported by An et al. [95]. Combined ECD and IRMPD for multistage MS–MS in FT-ICR-MS was applied for the characterization of N-glycosylated peptides [35]. IRMPD is found to give extensive cleavages of glycosidic bonds, while ECD favours peptide backbone cleavages.

5. References

1.   M. Törnqvist, C. Fred, J. Haglund, H. Helleberg, B. Paulsson, P. Rydberg, *Protein adducts: quantitative and qualitative aspects of their formation, analysis and applications*, J. Chromatogr. B, 778 (2002) 279.

2.   D.M. Creasy, J.S. Cottrell, *Unimod: Protein modifications for MS*, Proteomics, 4 (2004) 1534.

3. N. Blom, T. Sicheritz-Pontén, R. Gupta, S. Gammeltoft, S. Brunak, *Prediction of post-translational glycosylation and phosphorylation of proteins from the amino acid sequence*, Proteomics, 4 (2004) 1633.

4. G.T. Cantin, J.R. Yates, III, *Strategies for shotgun identification of PTM by MS*, J. Chromatogr. A, 1053 (2004) 7.

5. A.J. Rossomando, J. Wu, H. Michel, J. Shabanowitz, D.F. Hunt, M.J. Weber, T.W. Sturgill, *Identification of Tyr-185 as the site of Tyr autophosphorylation of recombinant mitogen-activated protein kinase p42^{mapk}*, PNAS, 89 (1992) 5779.

6. K. Palczewski, J. Buczylko, P. Van Hooser, S.A. Carr, M.J. Huddleston, J.W. Crabb, *Identification of the autophosphorylation sites in rhodopsin kinase*, J. Biol. Chem., 267 (1992) 18991.

7. J.E. Labdon, E. Nieves, U.K. Schubart, *Analysis of phosphoprotein p19 by LC–MS. Identification of two Pro-directed Ser phosphorylation sites and a blocked amino terminus*, J. Biol. Chem., 267 (1992) 3506.

8. A.P. Hunter, D.E. Games, *Chromatographic and MS methods for the identification of phosphorylation sites in phosphoproteins*, Rapid Commun. Mass Spectrom., 8 (1994) 559.

9. T. Covey, B. Shushan, R. Bonner, W. Schröder, F. Hucho, in: H. Jörnvall, J.O. Höög, A.M. Gustavsson (Eds), *Methods in protein sequence analysis*, 1991, Birkhäuser Press, Basel, pp. 249.

10. S.C. Moyer, R.J. Cotter, A.S. Woods, *Fragmentation of phosphopeptides by atmospheric-pressure MALDI and ESI ion trap mass spectrometry*, J. Am. Soc. Mass Spectrom., 13 (2002) 274.

11. M.J. Huddleston, R.S. Annan, M.F. Bean, S.A. Carr, *Selective detection of phosphopeptides in complex mixtures by LC–ESI-MS*, J. Am. Soc. Mass Spectrom., 4 (1993) 710.

12. S.A. Carr, M.J. Huddleston, R.S. Annan, *Selective detection and sequencing of phosphopeptides at the femtomole level by MS*, Anal. Biochem., 239 (1996) 180.

13. G. Neubauer, M. Mann, *Parent ion scans of large molecules*, J. Mass Spectrom., 32 (1997) 94.

14. M. Busman, K.L. Schey, J.E. Oatis, Jr., D.R. Knapp, *Identification of phosphorylation sites in phosphopeptides by positive and negative mode electrospray ionization–tandem mass spectrometry*, J. Am. Soc. Mass Spectrom., 7 (1996) 243.

15. M. Allen, J. Anacleto, R. Bonner, P. Bonnici, B. Shushan, L. Nuwaysir, *Characterization of protein digests using novel mixed-mode scanning with a single quadrupole instrument*, Rapid Commun. Mass Spectrom., 11 (1997) 325.

16. R.S. Annan, M.J. Huddleston, R. Verma, R.J. Deshaies, S.A. Carr, *A multidimensional ESI-MS-based approach to phosphopeptide mapping*, Anal. Chem., 73 (2001) 393.

17. F. Zappacosta, M.J. Huddleston, R.L. Karcher, V.I. Gelfand, S.A. Carr, R.S. Annan, *Improved sensitivity for phosphopeptide mapping using capillary LC and micro-ESI-MS: comparative phosphorylation site mapping from gel-derived proteins*, Anal. Chem., 74 (2002) 3221.

18. A. Beck, M. Deeg, K. Moeschel, E.K. Schmidt, E.D. Schleicher, W.Voelter, H.U. Häring, R. Lehmann, *Alkaline LC–ESI-skimmer-CID-MS for phosphopeptide screening*, Rapid Commun. Mass Spectrom., 15 (2001) 2324.

19.   A. Tholey, H. Toll, C.G. Huber, *Separation and detection of phosphorylated and nonphosphorylated peptides in LC–MS using monolithic columns and acidic or alkaline mobile phases*, Anal. Chem., 77 (2005) 4618.

20.   V. Alverdi, F. Di Pancrazio, G. Lippe, C. Pucillo, B. Casetta, G. Esposito, *Determination of protein phosphorylation sites by MS: a novel ESI-based method*, Rapid Commun. Mass Spectrom., 19 (2005) 3343.

21.   H. Steen, B. Küster, M. Fernandez, A. Pandey, M. Mann, *Detection of Tyr phosphorylated peptides by precursor ion scanning Q–TOF-MS in positive-ion mode*, Anal. Chem., 73 (2001) 1440.

22.   H. Steen, B. Küster, M. Mann, *Q–TOF versus triple-quadrupole MS for the determination of phosphopeptides by precursor ion scanning*, J. Mass Spectrom., 36 (2001) 782.

23.   R.H. Bateman, R. Carruthers, J.B. Hoyes, C. Jones, J.L. Langridge, A. Millar, J.P.C. Vissers, *A novel precursor ion discovery method on a hybrid Q–TOF-MS for studying protein phosphorylation*, J. Am. Soc. Mass Spectrom., 13 (2002) 792.

24.   J.C.Y. Le Blanc, J.W. Hager, A.M.P. Ilisiu, C. Hunter. F. Zhong, I. Chu, *Unique scanning capabilities of a new Q–LIT-MS (Q TRAP) used for high sensitivity proteomics applications*, Proteomics, 3 (2003) 859.

25.   B.L. Williamson, J. Marchesem, N.A. Morrice, *Automated identification and quantification of protein phosphorylation sites by LC–MS on a hybrid triple Q–LIT-MS*, Mol. Cell. Proteomics, 5 (2006) 337.

26.   R.D. Unwin, J.R. Griffiths, M.K. Leverentz, A. Grallert, I.M. Hagan, A.D. Whetton, *SRM to identify sites of protein phosphorylation with high sensitivity*, Mol. Cell. Proteomics, 4 (2005) 1134.

27.   F. Wolschin, U. Lehmann, M. Glinski, W. Weckwerth, *An integrated strategy for identification and relative quantification of site-specific protein phosphorylation using LC coupled to MS^2/MS^3*, Rapid Commun. Mass Spectrom., 19 (2005) 3626.

28.   M. Wind, M. Edler, N. Jakubowski, M. Linscheid, H. Wesch, W.D. Lehmann, *Analysis of protein phosphorylation by capillary LC coupled to element MS with ^{31}P detection and to ESI-MS*, Anal. Chem., 73 (2001) 29.

29.   M.J. MacCoss, W.H. McDonald, A. Saraf, R. Sadygov, J.M. Clark, J.J. Tasto, K.L. Gould, D. Wolters, M. Washburn, A. Weiss, J.I. Clark, J.R. Yates, III, *Shotgun identification of protein modifications from protein complexes and lens tissue*, PNAS, 99 (2002) 7900.

30.   M.J. Chalmers, J.P. Quinn, G.T. Blakney, M.R. Emmett, H. Mischak, S.J. Gaskell, A.G. Marshall, *LC–FT-ICR-MS characterization of protein kinase C phosphorylation*, J. Proteome Res., 2 (2003) 373.

31.   A. Stensballe, O.N. Jensen, J.V. Olsen, K.F. Haselmann, R.A. Zubarev, *ECD of singly and multiply phosphorylated peptides*, Rapid Commun. Mass Spectrom., 14 (2000) 1793.

32.   J.W. Flora, D.C. Muddiman, *Selective, sensitive, and rapid phosphopeptide identification of enzymatic digests using ESI-FT-ICR-MS with IRMPD*, Anal. Chem., 73 (2002) 2305.

33.   S. K. Sze, Y. Ge, H. Oh, F.W. McLafferty, *Top-down MS of a 39-kDa protein for characterization of any PTM to within one residue*, PNAS, 99 (2002) 1774.

34. M.J. Chalmers, K. Håkansson, R. Johnson, R. Smith, J. Shen, M.R. Emmett, A.G. Marshall, *Protein kinase A phosphorylation characterized by tandem FT-ICR-MS*, Proteomics, 4 (2004) 970.
35. K. Håkansson, M.J. Chalmers, J.P. Quinn. M.A. McFarland, C.L. Hendrickson, A.G. Marshall, *Combined ECD and IRMPD for multistage MS–MS in a FT-ICR-MS*, Anal. Chem., 75 (2003) 3256.
36. M.C. Crowe J.S. Brodbelt, *IRMPD and CID of peptides in a quadrupole ion trap with selective IRMPD of phosphopeptides*, J. Am. Soc. Mass Spectrom., 15 (2004) 1581.
37. V. Gaberc-Porekar, V. Menart, *Perspectives of IMAC*, J. Biochem. Biophys. Methods, 49 (2001) 335.
38. S.B. Ficarro, M.L. McCleland, P.T. Stukenberg, D.J. Burke, M.M. Ross, J. Shabanowitz, D.F. Hunt, F.M. White, *Phosphoproteome analysis by MS and its application to Saccharomyces cerevisiae*, Nat. Biotechnol., 20 (2002) 301.
39. L.M. Brill, A.R. Salomon, S.B. Ficarro, M. Mukherji, M. Stettler-Gill, E.C. Peters, *Robust phosphoproteomic profiling of Tyr phosphorylation sites from human T cells using IMAC and MS–MS*, Anal. Chem., 76 (2004) 2763.
40. T.S. Nühse, A. Stensballe, O.N. Jensen, S.C. Peck, *Large-scale analysis of in vivo phosphorylated membrane proteins by IMAC and MS*, Mol. Cell. Proteomics, 2 (2003) 1234.
41. L.M. Muwaysir, J.T. Stults, *ESI-MS of phosphopeptides isolated from on-line IMAC*, J. Am. Soc. Mass Spectrom., 4 (1993) 662.
42. J.D. Watts, M. Affolter, D.L. Krebs, R.L. Wange, L.E. Samelson, R. Aebersold, *Identification by ESI-MS of the sites of Tyr phosphorylation induced in activated Jurkat T cells on the protein Tyr kinase ZAP-70*, J. Biol. Chem., 269 (1994) 29520.
43. L.N. Amankwa, K. Harder, F. Jirik, R. Aebersold, *High-sensitivity determination of Tyr-phosphorylated peptides by on-line enzyme reactor and ESI-MS*, Protein Sci., 4 (1995) 113.
44. M.W.H. Pinkse, P.M. Uitto, M.J. Hilhorst, B. Ooms, A.J.R. Heck, *Selective isolation at the femtomole level of phosphopeptides from proteolytic digests using 2D-nano-LC–ESI-MS–MS and titanium oxide precolumns*, Anal. Chem., 76 (2004) 3935.
45. J.G. Krabbe, H. Lingeman, W.M.A. Niessen, H. Irth, *Ligand-exchange detection of phosphorylated peptides using LC–ESI-MS*, Anal. Chem., 75 (2003) 6853.
46. N. Sadagopan, M. Malone, J.T. Watson, *Effect of charge derivatization in the determination of phosphorylation sites in peptides by ESI-CID-MS–MS*, J. Mass Spectrom., 34 (1999) 1279.
47. H. Steen, M. Mann, *A new derivatization strategy for the analysis of phosphopeptides by precursor ion scanning in positive ion mode*, J. Am. Soc. Mass Spectrom., 13 (2002) 996.
48. W. Weckwerth, L. Willmitzer, O. Fiehn, *Comparative quantification and identification of phosphoproteins using stable isotope labeling and LC–MS*, Rapid Commun. Mass Spectrom., 14 (2000) 1677.
49. M.P. Molloy, P.C. Andrews, *Phosphopeptide derivatization signatures to identify Ser and Thr phosphorylated peptides by MS*, Anal. Chem., 73 (2001) 5387.
50. W. Li, R.A. Boykins, P.S. Backlund, G. Wang, H.-C. Chen, *Identification of pSer and pThr as cysteic acid and β-methylcysteic acid residues in peptides by MS–MS sequencing*, Anal. Chem., 74 (2002) 5701.

51. D.J. McCormick, M.W. Holmes, D.C. Muddiman, B.J. Madden, *Mapping sites of protein phosphorylation by MS utilizing a chemical-enzymatic approach: characterization of products from α-S1 casein phosphopeptides*, J. Proteome Res., 4 (2005) 424.

52. M. Adamczyk, J.C. Gebler, J. Wu, *Identification of phosphopeptides by chemical modification with an isotopic tag and ion trap MS*, Rapid Commun. Mass Spectrom., 16 (2002) 999.

53. M.B. Goshe, T.P. Conrads, E.A. Panisko, N.H. Angell, T.D. Veenstra, R.D. Smith, *Phosphoprotein ICAT approach for isolating and quantifying phosphopeptides in proteome-wide analyses*, Anal. Chem., 73 (2001) 2578.

54. M.B. Goshe, T.D. Veenstram E.A. Panisko, T.P. Conradts, N.H. Angell, R.D. Smith, *Phosphoprotein ICAT: Application to the enrichment and identification of low-abundance phosphoprotein*, Anal. Chem., 74 (2002) 607.

55. A. Gruhler, J.V. Olsen, S. Mohammed, P. Mortensen, N.J. Færgemann, M. Mann, O.N. Jensen, *Quantitative phosphoproteomics applied to the yeast pheromone signaling pathway*, Mol. Cell. Proteomics, 4 (2005) 310.

56. J.C. Betts, W.P. Blackstock, M.A. Ward, B.H. Anderton, *Identification of phosphorylation sites on neurofilament proteins by nano-ESI-MS*, J. Biol. Chem., 272 (1997) 12922.

57. X. Zhang, C.J. Herring, P.R. Romano, J. Szczepanowska, H. Brzeska, A.G. Hinnebusch, J. Qin, *Identification of phosphorylation sites in proteins separated by polyacrylamide GE*, Anal. Chem., 70 (1998) 2050.

58. G. Neubauer, M. Mann, *Mapping of phosphorylation sites of gel-isolated proteins by nano-ESI-MS–MS: potentials and limitations*, Anal. Chem., 71 (1999) 235.

59. A.R. Salomon, S.B. Ficarro, L.M. Brill, A. Brinker, Q.T. Phung, C. Ericson, K. Sauer, A. Brock, D.M. Horn, P.G. Schultz, E.C. Peters, *Profiling of Tyr phosphorylation pathways in human cells using MS*, PNAS, 100 (2003) 443.

60. C.C. Wu, M.J. MacCoss, K.E. Howell, J.R. Yates, III, *A method for the comprehensive proteomic analysis of membrane proteins*, Nat. Biotechnol., 21 (2003) 532.

61. J. Peng, M.J. Kim, D. Cheng, D.M. Duong, S.P. Gygi, M. Sheng, *Semiquantitative proteomic analysis of rat forebrain postsynaptic density fractions by MS*, J. Biol. Chem., 279 (2004) 21003.

62. A. Schlosser, J.T. Vanselow, A. Kramer, *Mapping of phosphorylation sites by a multi-protease approach with specific phosphopeptide enrichment and nano-LC–MS–MS analysis*, Anal. Chem., 77 (2005) 5243.

63. R. Orlando, Y. Yang, *Analysis of glycoproteins*, in: B.S. Larsen and C.N. McEwen, *MS of biological materials*, 2nd Ed., Marcel Dekker Inc., NY, (1998), 215.

64. P.M. Rudd, R.A. Dwek, *Rapid, sensitive sequencing of oligosaccharides from glycoproteins*, Curr. Opinion Biotechnol., 8 (1997) 488.

65. W. Morelle, J.-C. Michalski, *The MS analysis of glycoproteins and their glycan structure*, Curr. Anal. Chem., 1 (2005) 29.

66. J. Suzuki-Sawada, Y. Umeda, A. Kondo, I. Kato, *Analysis of oligosaccharides by on-line LC–ESI-MS*, Anal. Biochem., 207 (1992) 203.

67. S.A. Carr, M.J. Huddleston, M.F. Bean, *Selective identification and differentiation of N- and O-linked oligosaccharides in glycoproteins by LC–MS*, Protein Sci., 2 (1993)183.

68.   M.J. Huddleston, M.F. Bean, S.A. Carr, *Collisional fragmentation of glycopeptides by ESI LC–MS and LC–MS–MS: Methods for selective detection of glycopeptides in protein digests*, Anal. Chem. 65 (1993) 877.
69.   A.W. Guzzetta, L.J. Basa, W.S. Hancock, B.A. Keyt, W.F. Bennett, *Identification of carbohydrate structures in glycoprotein PMF by the use of LC–MS with selected ion extraction with special reference to tissue plasminogen activator and a glycosylation variant produced by site directed mutagenesis*, Anal. Chem., 65 (1993) 2953.
70.   A. Apffel, J.A. Chakel, W.S. Hancock, C. Souders, T. M'Timkulu, E. Pungor, Jr., *Application of LC–ESI-MS and MALDI-TOF-MS in combination with selective enzymatic modifications in the characterization of glycosylation patterns in single-chain plasminogen activator*, J. Chromatogr. A, 732 (1996) 27.
71.   S. Itoh, N. Kawasaki, M. Ohta, T. Hayakawa, *Structural analysis of a glycoprotein by liquid chromatography-mass spectrometry and liquid chromatography with tandem mass spectrometry: Application to recombinant human thrombomodulin*, J. Chromatogr. A, 978 (2002)141.
72.   I. Mazsaroff, W. Yu, B.D. Kelly, J.E. Vath, *Quantitative comparison of global carbohydrate structure of glycoproteins using LC–MS and in-source fragmentation*, Anal. Chem., 69 (1997) 2517.
73.   J. Charlwood, J. Langridge, P. Camilleri, *Structural characterisation of N-linked glycan mixtures by precursor ion scanning and MS–MS analysis*, Rapid Commun. Mass Spectrom., 13 (1999) 1522.
74.   B. Sullivan, T.A. Addona, S.A. Carr, *Selective detection of glycopeptides on ion-trap MS*, Anal. Chem., 76 (2004) 3112.
75.   M.A. Ritchie, A.C. Gill, M.J. Deery, K. Lilley, *Precursor-ion scanning for detection and structural characterization of heterogenous glycopeptide mixtures*, J. Am. Soc. Mass Spectrom., 13 (2002) 1065.
76.   J. Zaia, *MS of oligosaccharides*, Mass Spectrom. Rev., 23 (2004) 161.
77.   G.D. Roberts, W.P. Johnson, S. Burman, K.R. Anumula, S.A. Carr, *An integrated strategy for structural characterization of the protein and carbohydrate components of monoclonal antibodies: Application to anti-respiratory syncytial virus MAb*, Anal. Chem., 67 (1995) 3613.
78.   A.P. Hunter, D.E. Games, *Evaluation of glycosylation site heterogeneity and selective identification of glycopeptides in proteolytic digest of bovine α_1-acid glycoprotein by MS*, Rapid Commun. Mass Spectrom., 9 (1995) 42.
79.   J. Colangelo, V. Licon, J. Benen, J. Visser, C. Bergmann R. Orlando, *Characterization of the N-linked glycosylation site of recombinant pectate lyase*, Rapid Commun. Mass Spectrom., 13 (1999) 2382.
80.   M.E. Warren, H. Kester, J. Benen, J. Colangelo, J. Visser, C. Bergmann, R. Orlando, *Studies in the glycosylation of wild-type and mutant forms of Aspergillus niger pectin methylesterase*, Carbohydr. Res., 337 (2002) 803.
81.   R. Qiu, F.E. Regnier, *Use of multidimensional lectin affinity chromatography in differential glycoproteomics*, Anal. Chem., 77 (2005) 2802.
82.   M. Madera, Y. Mechref, M.V. Novotny, *Combining lectin microcolumns with high-resolution separation techniques for enrichment of glycoproteins and glycopeptides*, Anal. Chem., 77 (2005) 4081.
83.   K. Alving, H. Paulsen, J. Peter-Katalinic, *Characterization of O-glycosylation sites in MUC2 glycopeptides by nano-ESI-Q–TOF-MS,*. J. Mass Spectrom., 34 (1999) 395.

84.   B. Maček, J. Hofsteenge, J. Peter-Katalinic, *Direct determination of glycosylation sites in O-fucosylated glycopeptides using nano-ESI-Q–TOF-MS*, Rapid Commun. Mass Spectrom., 15 (2001) 771.

85.   N.G. Karlsson, B.L. Schulz, N.H. Packer, *Structural determination of neutral O-linked oligosaccharide alditols by negative-ion LC–ESI-MS"*, J. Am. Soc. Mass Spectrom., 15 (2004) 659.

86.   H. Jiang, H. Desaire, V.Y. Butnev, G.R. Bousfield, *Glycoprotein profiling by ESI-MSy*, J. Am. Soc. Mass Spectrom., 15 (2004) 750.

87.   C. Robbe, C. Capon, B. Coddeville, J.-C. Michalski, *Diagnostic ions for the rapid analysis by nano-ESI-Q–TOF-MS of O-glycans from human mucins*, Rapid Commun. Mass Spectrom., 18 (2004) 412.

88.   M. Wuhrer, C.A.M. Koeleman, C.H. Hokke, A.M. Deelder, *Protein glycosylation analyzed by normal-phase nano-LC–MS of glycopeptides*, Anal. Chem., 77 (2005) 886.

89.   A. Zamfir, J. Peter-Katalinić, *CE–MS for glycoscreening in biomedical research*, Electrophoresis, 25 (2004) 1949.

90.   D.M. Ratner, E.W. Adams, M.D. Disney, P.H. Seeberger, *Tools for glycomics: Mapping interactions of carbohydrates in biological systems*, ChemBioChem, 5 (2004) 1375.

91.   R. Raman, S. Raguram, G. Venkataraman, J.C. Paulson, R. Sasisekharan, *Glycomics: an intergrated systems approach to structure-function relationships of glycan*, Nature Methods, 2 (2005) 817.

92.   M. Wuhrer, A.M. Deelder, C.H. Hokke, *Protein glycosylation analysis by LC–MS*, J. Chromatogr. B, 825 (2005) 124.

93.   M.V. Novotny, Y. Mechref, *New hyphenated methodologies in high-sensitivity glycoprotein analysis*, J. Sep Sci., 28 (2005) 1956.

94.   M. Froesch, L.M. Bindila, G. Baykut, M. Allen, J. Peter-Katalinic, A.D. Zamfir, *Coupling of fully automated chip ESI to FT-ICR-MS for high-performance glycoscreening and sequencing*, Rapid Commun. Mass Spectrom., 18 (2004) 3084.

95.   A.J. An, M. Ninonuevo, J. Aguilan, H. Liu, C.B. Lebrilla, L.S. Alvarenga, M.J. Mannis, *Glycomics analyses of tear fluid for the diagnostic detection of ocular rosacea*, J. Proteome Res., 4 (2005) 1981.

20

LC–MS ANALYSIS OF OLIGOSACCHARIDES

1. Introduction

Electrospray ionization (ESI), and to a lesser extent atmospheric-pressure chemical ionization (APCI), has given a major impetus to the analysis and characterization of biomolecules like peptides, proteins, oligosaccharides, lipids, and nucleotides. The state of the art of the LC–MS analysis of oligosaccharides is discussed in this chapter. The focus in the discussion is on technology rather than on detailed results with particular glycans investigated. Examples of various approaches are discussed, without intending to be comprehensive. The focus is predominantly on oligosaccharides derived from glycoproteins.

Oligosaccharides serve many purposes in a biological system, e.g., as important building blocks, as key players in energy metabolism, and as mediators in many cellular events. In these functions, they act as monomeric sugar units, simple or complex oligosaccharides and/or polysaccharides, and either alone or as constituents of glycoproteins, glycolipids, or other glycoconjugates.

The current interest in the characterization of oligosaccharides is predominantly related to glycoproteins. Glycosylation of proteins plays an important role in the biological activity of proteins (Ch. 19.4). Next to glycoproteins, there is interest in the characterization of oligosaccharides, e.g., plant cell wall polysaccharides and oligosaccharides derived thereof, and a variety of glycosylated species that may occur in living systems, e.g., Phase-II metabolites of drugs (Ch. 10.5), glycosides of flavonoids and related plant phenols (Ch. 15).

Table 20.1: Residue masses of some monomeric sugar units			
Type	Mass	CH₃	Example
Hex	162.0528	3	Galactose (Gal), Glucose (Glc), Mannose (Man)
dHex	146.0579	2	Fucose (Fuc), Rhamnose (Rha)
HexNAc	203.0794	3	N-Acetyl-galactose (GalNAc), N-Acetyl-glucose (GlcNAc)
Neu5Ac	291.0954	5	Sialic Acid
Pent	132.0423	2	Xylose (Xyl)
GluA	176.0321	2*	Glucuronic Acid (GluA or Glu)

Branching reduces the number of CH_3 by 1 per branch.

2   Structure of oligosaccharides

The structural complexity of oligosaccharides goes far beyond that of proteins and nucleotides [1-4]. Because with nucleotides or amino acids only linear chains with just one linkage type are formed, a tetramer of nucleotides or amino acids with four different monomeric subunits can have only 24 different structures, while in excess of 10,000 different structures are possible for a tetrasaccharide. The various stereochemical centres, the anomeric centre, the multiple linkage types, and the ability to form branched structures all contribute to this structural diversity.

A monomeric sugar unit has several structural characteristics:
* There is an anomeric carbon, resulting from the equilibrium between the acyclic form and the four cyclic forms of a sugar monomer in solution. These cyclic forms are comprised of a five-membered furanose ring and the generally more abundant six-membered pyranose ring. Mutorotation around the anomeric centre converts the α-form into the β-form.
* The other C-atoms in the ring are stereochemical centres, four in a hexose, leading to sixteen different hexose structures. The stereochemistry at C5 determines the D- or L-configuration. In mammalian glycoproteins, the D-configuration is most common.
* The monomeric units can be linked via a range of glycosidic linkage type, *e.g.*, 1→2, 1→3, 1→4, 1→6. Because each monomeric unit can be involved in more than one linkage, branched structures may be formed. The terminus with a free anomeric centre is called the reducing terminus. The linkage to the protein or any other aglycone is made via this reducing terminus.

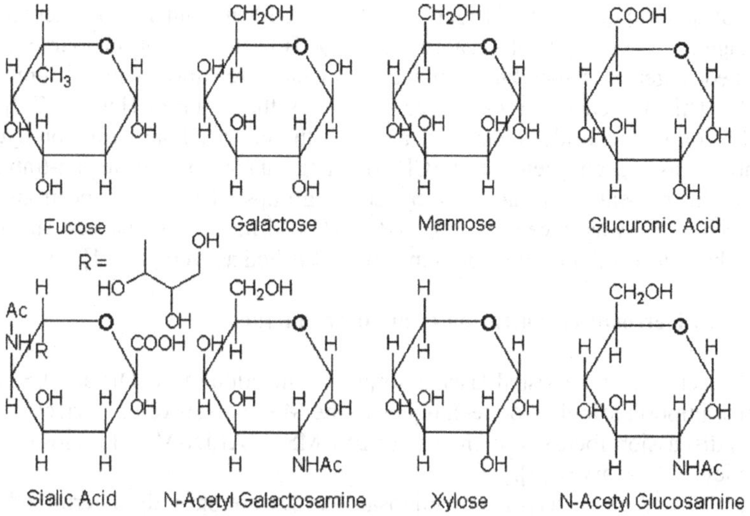

Figure 20.1: Structures of some important monomeric sugar units.

Although the theoretical number of possible oligosaccharide structures is very high, only a limited number of these actually occur in biological systems.

The molecular mass of a oligosaccharide can be calculated from residue masses of individual sugar units in the molecule from $M = 18 + \Sigma$ (Residue masses of monomeric units). Residue masses of some important monomeric units are given in Table 20.1. Relevant sugar monomer structures are shown in Figure 20.1.

3. Mass spectrometry of oligosaccharides

An important step in the characterization of oligosaccharides may be the compositional analysis with respect to the type of monomeric units present. This is mostly done using methylation of the monomers after methanolysis or hydrolysis of the oligosaccharides and subsequent GC–MS analysis [5]. Permethylation of intact oligosaccharides can also be applied prior to ESI-MS analysis, in most cases using dimethylsulfoxide/NaOH and methyliodide [6]. The number of methyl substituents added to each sugar unit in an oligosaccharide is indicated in Table 20.1.

Initially studies in the MS and MS–MS analysis of oligosaccharides were performed using analyte ionization by fast-atom bombardment (FAB) on a sector instrument, enabling both high-energy and low-energy collision-induced dissociation (CID). In the early 1990s, ESI-MS on triple-quadrupole instruments was increasingly used. In the mid 1990s, ion-trap instruments enabling multistage MS–MS became available. Further progress was made in the late 1990s with the advent of quadrupole–time-of-flight hybrid (Q–TOF) instrument. In order to mimic multistage MS–MS in a triple-quadrupole or Q–TOF instrument, in-source CID can be applied to generate specific fragment ions, which are subsequently fragmented in MS–MS. This approach is for instance applied by the group of Harvey [7-8]. More recently, oligosaccharide characterization with Fourier-transform ion-cyclotron resonance mass spectrometers (FT-ICR-MS) and quadrupole–linear-ion-trap hybrid (Q–LIT) instruments was adapted by some groups. The mass spectrometry of oligosaccharides was recently reviewed [3-4], while review papers specific on glycan characterization of glycoprotein were published as well [1-2, 4].

3.1   Electrospray ionization of oligosaccharides

At present, matrix-assisted laser desorption ionization (MALDI) and ESI are the two most important ionization techniques in the MS analysis of oligosaccharides [1-4]. This discussion focusses on the use of ESI-MS. MALDI-MS of oligosaccharides was reviewed by Harvey [9].

The ESI-MS characteristics of oligosaccharides are strongly determined by the identity. The ESI-MS of oligosaccharides is complicated by their hydrophilicity, which limits their surface activity in the ESI droplets. Derivatization strategies to reduce their hydrophilicity, *e.g.*, permethylation or introducing chromophores or fluorophores (Ch. 20.3.5), generally result in an improved response. Permethylated oligosaccharides can be analysed as protonated molecules or as sodium adducts in positive-ion ESI-MS.

Alternatively, the use of nano-ESI results in a significantly improved sensitivity [10]. In the conventional ESI-MS analysis of mixtures of peptides and oligosaccharides, the response of the more hydrophilic oligosaccharides is often suppressed in favour of the peptides, whereas in nano-ESI comparable responses were obtained for oligosaccharides and peptides.

Protonated molecules $[M+H]^+$ are mainly observed for oligosaccharides containing *N*-acetyl groups or after thorough desalting. Oligosaccharides without *N*-atoms are prone to adduct formation with sodium or potassium, *i.e.*, $[M+Na]^+$ or $[M+K]^+$, unless either formic acid to produce $[M+H]^+$, or other alkali salts are added. In the initial LC–ESI-MS studies on oligosaccharides, detection was mainly based on $[M+Na]^+$, lithiated $[M+Li]^+$, or ammoniated $[M+NH_4]^+$ molecules in the positive-ion mode [11]. Most oligosaccharides can also be analysed in the negative-ion mode as deprotonated molecules $[M-H]^-$. Oligosaccharides containing sialic acid were best detected in the negative-ion mode, while the response of neutral oligosaccharides was greatly enhanced by the addition of 10 mmol/l sodium or ammonium acetate to the mobile phase [12]. Formation of $[M-H]^-$ of neutral oligosaccharides after loss of HF and acetic acid from their $[M+F]^-$ and $[M+acetate]^-$ ions was studied by Jiang and Cole [13]. The asialo oligosaccharides from *N*-linked glycans provided good response in ESI-MS, although they were very susceptible to in-source CID without giving specific structural information [14]. Compared to the asialo analogues, an oligosaccharide with just one sialic acid group yielded a 10-fold lower response, while an analogue with two sialic acid was not detected at all.

Additionally, experiments have been reported with adduct formation of oligosaccharides with a variety of divalent metal ions, e.g., with Ca^{2+} and Mg^{2+} [15-17], with Co^{2+}, Cu^{2+}, and Mn^{2+} [16-17], and with Ag^+ [18]. Post-column addition of metal-ion solution by means of a triaxial probe was also described [19]. The rationale for the application of divalent cations is response enhancement and/or induction of specific fragmentation reactions. Co^{2+} and Ca^{2+} appear to be most effective in this respect.

3.2   MS–MS of oligosaccharides

The structural characterization of oligosaccharides by MS–MS is based on specific fragmentation reactions that may occur. Initially, FAB and high-energy MS–MS on sector instruments were applied [20-21]. Domon and Costello [20] have proposed a systematic nomenclature for the fragments generated in MS–MS of oligosaccharides (Figure 20.2). A_i, B_i, and C_i represent ions containing a nonreducing terminal sugar unit, whereas X_j, Y_j, and Z_j labels are used to designate fragments still containing the reducing sugar unit or the aglycone. The subscripts indicate the cleavage position relative to the terminus, while superscripts indicate cross-ring cleavages (with A and X ions). The most frequently observed ions are B and Y ions, corresponding to cleavages of the glycosidic bond at the nonreducing terminal side. With a terminal hexose, the protonated species are found at m/z 163 for the B_1^+ and at $[M+H-162]^+$ for Y_{n-1}^+. In addition, branches are indicated by α, β, ... in the order of decreasing mass. Additional fragments due to secondary fragmentations are indicated as D and E ions.

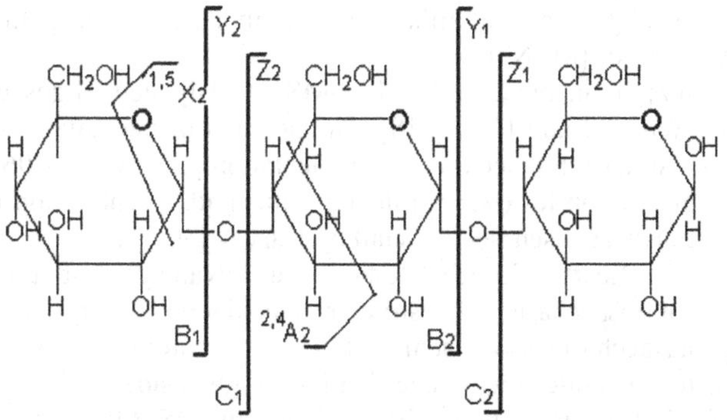

Figure 20.2: Nomenclature of carbohydrate fragmentation, as proposed by Domon and Costello [20].

MS–MS is frequently applied in the structure elucidation of oligosaccharides. Hofmeister et al. [22] demonstrated that the cross-ring fragments generated in FAB-MS–MS of [M+Li]$^+$ of disaccharides can be applied to differentiate between various glycosidic linkages. Later on, Asam and Glish [23] demonstrated similar results can be obtained using MS–MS in an ion-trap instrument. In general, cross-ring cleavages are more readily observed by fragmentation of [M+Li]$^+$, [M+Na]$^+$, or [M–H]$^-$, than by fragmentation of [M+H]$^+$.

Harvey [24] studied positive-ion ESI-MS–MS of neutral N-linked glycans in a Q–TOF instrument using [M+Na]$^+$, [M+H]$^+$, or [M+Li]$^+$ as precursor ion. [M+Li]$^+$ was especially useful in structure elucidation. The fragmentation of [M+Na]$^+$ required four times more energy than the corresponding [M+H]$^+$. While fragmentation of [M+H]$^+$ mainly resulted in B- and Y-type glycosidic fragments, additional information from cross-ring cleavages was obtained by MS–MS on the [M+Na]$^+$ or [M+Li]$^+$.

Chai et al. [25] studied negative-ion ESI-MS–MS of neutral underivatized non-, mono-, di-, and trifucosylated oligosaccharides. Fragmentation of [M–H]$^-$ resulted in fragments due to losses from the nonreducing terminal, providing series of C-ions, enabling sequence determination. In addition, partial linkage information could be derived from D-ions due to cleavage of 3-linked *GlcNAc* or *Glc*, or ^{0,2}A-ions due to cleavage of 4-linked *GlcNAc* or *Glc*. In a later study, this method was extended by fragmenting both [M–H]$^-$ and [M–2H]$^{2-}$ to obtain more detailed branching information [26].

An issue of considerable concern in the structure elucidation of oligosaccharides is the occurrence of internal residue losses, observed in high-energy MS–MS spectra of [M+H]$^+$ of glycosylated plant phenols (*cf.* Ch. 15.2.6). It was demonstrated that such internal residue losses can occur in low-energy MS–MS of [M+H]$^+$ as well, but that they do not occur from [M+Na]$^+$ or [M–H]$^-$ [27-28]. Harvey et al. [29] reported internal residue losses in MS–MS spectra of [M+H]$^+$ of 2-aminobenzamide-derivatized O-linked glycans and milk sugars involving fucose migration. It was not observed in MS–MS of [M+Na]$^+$.

The fragmentation in negative-ion ESI-MS of sialylated oligosaccharides was investigated by Wheeler and Harvey [8]. The position of the sialic acid, ($\alpha2{\rightarrow}3$) or ($\alpha2{\rightarrow}6$), had a distinct influence on the fragmentation pattern: m/z 306 was found to be a useful diagnostic ion for ($\alpha2{\rightarrow}6$)-linked sialic acid. Negative-ion fragmentation of N-linked glycans was discussed in detail by Harvey [30-32].

The group of Leary [33-35] studied the advantages of transition metal coordination of oligosaccharides for structure elucidation. Determination of linkage positions in pentasaccharides was achieved via Co^{2+} coordination [33]. Metal ligand derivatization to generate diethylenetriamine N-glycoside M^{2+} complexes was studied as well [34]. Unambiguous differentiation by MS–MS of the stereoisomers glucose, galactose, mannose, and talose was demonstrated using diastereomeric diethylenetriamine N-glycoside Zn^{2+} complexes. Metal ligand derivatization with Co^{2+}, Ni^{2+}, Cu^{2+}, and Zn^{2+} as metal ions was also applied to sialic oligosaccharides

[35]. The formation of diethylenetriamine N-glycoside M^{2+} complexes suppressed the loss of sialic acid in MS–MS. Therefore, linkage information could be obtained via cross-ring cleavages.

3.3 Multistage MS–MS using ion-trap instruments

As indicated in Ch. 20.3.1, permethylation of oligosaccharides can be a useful procedure to enhance the LC separation and/or enable structure elucidation. A number of studies report the fragmentation of permethylated oligosaccharides in MS^n in an ion-trap instrument.

Viseux *et al.* [36-37] reported the MS–MS analysis of protonated permethylated oligosaccharides, initially in a triple-quadrupole instrument [36] and later also in an ion-trap instrument [37]. The spectra were characterized by series of B-ions and less intense Y-ions, enabling identification of the monomer sequence. Secondary fragmentation of *HexNAc* oligomers results in the elimination of the substituent from the C3-position, *i.e.*, either a sugar unit when located at the C3-position, or methanol, when the sugar is located at the C4-position, thus enabling differentiation between branching locations. The use of MS^n on the ion trap further extended the possibilities [37]. Fragmentation patterns of unknown subunit fragments obtained from a larger oligosaccharide can be compared to those of isobaric reference compounds. This enables more straightforward identification: although the theoretical number of possible oligosaccharide structures is very high, only a limited number of these actually occur in biological systems. These findings were confirmed by others [38-40].

Weiskopf *et al.* [38-39] also characterized permethylated linear oligosaccharides and N-linked glycans using an ion-trap. They identified two isobaric oligosaccharides from chicken ovalbumin (*HexNAc$_5$Hex$_5$*) by performing MS^n on specific fragments in the MS^2 spectrum [32]. The use of MS^3 and MS^4 facilitates the generation of cross-ring fragmentation, necessary to obtain information on linkage and branching in large oligosaccharides. This feature was demonstrated by the elucidation of N-acetyllactosamine and sialyl-N-acetyllactosamine oligosaccharide antennae from biantennary glycans [39]. Sheeley and Reinhold [40] also found that methylation amplifies cross-ring and double glycosidic cleavages, important in assessing linkage and branching information. They concluded that, compared to triple quadrupole instruments, the ion trap provides additional advantages in oligosaccharide characterization, especially due to the MS^n possibilities.

An additional advantage of permethylation of oligosaccharides was demonstrated by Delaney and Vouros [41]. These compounds can be efficiently separated by reversed-phase LC (RPLC), including differentiation between α and β anomers.

Most results with ion-trap instrument were obtained using three-dimensional ion traps. A Q–LIT was recently introduced as well. Such a system shows great promise in oligosaccharide characterization. It provides collision-cell CID with ion-trap sensitivity but without low m/z cut-off, CID in the LIT giving MS^3 possibilities, and the various powerful triple-quadrupole analysis modes, *e.g.*, precursor-ion analysis. Sandra *et al.* [42] reported first results with a Q–LIT instrument in oligosaccharide characterization, demonstrating these features.

Strategies for carbohydrate sequencing by means of multistage MS–MS have recently been outlined in detail by the group of Reinhold [43-45].

3.4   Structural characterization of oligosaccharides

While a limited number of applications of actual on-line LC–MS in the characterization of oligosaccharides have been described (Ch. 20.4), numerous studies describe the utilization of liquid sample introduction into an ESI-MS system for the characterization of (mixtures of) oligosaccharides, eventually after prefractionation or preparative LC. Some examples in the characterization of oligosaccharides not derived from glycoproteins are highlighted in this section. Structural characterization of glycans from glycoproteins is discussed in Ch. 19.4.2.

Brüll *et al.* [46] developed a method for the structure elucidation of oligosaccharides derived from plant cell wall polysaccharides. The oligosaccharide mixtures were fractionated using size-exclusion chromatography (SEC) and high-performance anion-exchange chromatography (HPAEC). After neutralization and drying, the sample was per-*O*-acetylated. The derivatives were isolated and desalted by dissolution in dichloromethane and aqueous partitioning. The per-*O*-acetylated oligosaccharides were analysed using ESI-MS–MS and MALDI-MS.

Pfenninger *et al.* [47-48] reported the structure elucidation of underivatized neutral human milk oligosaccharides using negative-ion nano-ESI and multistage MS–MS on an ion-trap instrument. In MS–MS, C-ion fragmentation involving sugar losses from the reducing end as well as cross-ring cleavages were observed. The technology enabled the identification in human milk of three isomeric fucosylated lacto-*N*-hexaoses, which were not described before.

3.5   Derivatization of oligosaccharides

Because of the lack of a chromophore, oligosaccharides are difficult analytes in conventional LC procedures. They can only be detected via amperometry or refractive index detection. Many derivatization procedures have been developed directed at the introduction of chromophores for UV detection or a fluorophore for fluorescence detection [49]. Often, derivatization also facilitates the separation of the derivatized oligosaccharides by means of RPLC. Because oligosaccharides are also not very efficiently ionized by ESI, various derivatization procedures are directed at improving the response in ESI and/or at directing the fragmentation in MS–MS. The mass increments due to some of the derivatization reactions are given in Table 20.2.

Table 20.2: Mass increment in oligosaccharide mass due to derivatization		
Label	**Abbreviation**	**Increment (Da)**
2-Aminopyridine	PA	78
2-Aminoacridone	2-AMAC	194
4-aminobenzoic ethyl ester	ABEE	149
4-aminobenzoic butyl ester	ABBE	177
1-phenyl-3-methyl-5-pyrazolone	PMP	330

A variety of reagents have been described for reductive amination, a labelling reaction based on the attachment of an amino group at the carbonyl group in a reducing sugar under mild acidic conditions. This yields an acid-labile Schiff base (imine derivative), which is subsequently reduced to a stable secondary amine. Frequently applied reagents are 2-aminopyridine (PA), 2-aminoacridone (2-AMAC), and 4-aminobenzene derivatives such as 4-aminobenzoic ethyl ester (ABEE) or aminobenzoic butyl ester (ABBE).

Derivatization with 2-aminopyridine (PA) is frequently applied to enable UV or fluorescence detection of oligosaccharides. The reduced polarity of the PA-derivatives enables RPLC [50]. It is also frequently applied in LC–MS studies. PA-derivatization of periodate-treated glycan-alditols, as released by β-elimination from glycoproteins, was reported by Morelle et al. [51]. Takegawa et al. [52-53] reported structural and quantitative analysis of isomeric PA-derivatized oligosaccharides derived from complex-type N-glycans of IgG by RPLC–MS^n on an ion-trap instrument. An MS^n mass spectral library for PA-derivatized oligosaccharides was generated as well [53]. It was applied in the assignment of isomeric glycans from IgG. The use of the library allowed identification of coeluting isomeric glycans.

Derivatization with 2-aminoacridone (2-AMAC) results in a hydrophobic fluorescent derivative. It was applied to enable RPLC separation or fractionation of derivatized oligosaccharides [54]. Fragmentation of 2-AMAC-derivatized oligosaccharides resulted in series of Y-ions, including 2-AMAC as aglycone as well as some low-m/z B-ions [54-55]. Charlwood et al. [55] reported the MS–MS spectra of examples of all three common types of N-glycans, acquired on a quadrupole–time-of-flight hybrid (Q–TOF) instrument. As an example, the MS–MS spectrum of a 2-AMAC-derivatized high-mannose glycan with Man 8 is shown in Figure 20.3.

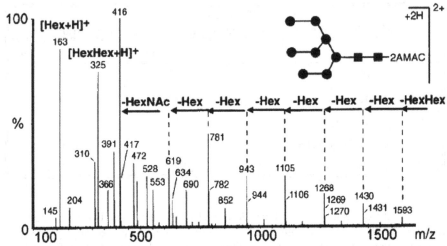

Figure 20.3: ESI–MS–MS spectrum of a 2-AMAC-derivatized high-mannose glycan with Man 8. Reprinted from [55] with permission. ©1999, John Wiley and Sons Ltd.

Derivatization with *p*-aminobenzene derivatives, such as *p*-aminobenzoic ethyl ester (ABEE) or butyl ester (ABBE), has also been described for both RPLC and UV detection. Li and Her [56] reported the formation of glycosylamines after derivatization with ABEE. MS–MS fragmentation of derivatized disaccharides resulted in specific fragment ions in the negative-ion mode, depending on the linkage, *i.e.*, *m/z* 368 and 410 were observed for (β1→2), *m/z* 218, 260, and 398 for (β1→3), *m/z* 263 and 290 for (β1→4), and *m/z* 235, 251, and 266 for (β1→6) disaccharides.

Harvey [7] studied the ESI–MS of various reductively aminated *N*-linked glycan derivatives. [M+Na]$^+$ ions were most abundant. Best response was obtained for ABBE-derivatized oligosaccharides. In MS–MS of [M+Na]$^+$, [M+H]$^+$, and [M+Li]$^+$ on a Q–TOF instrument, mostly glycosidic cleavages were observed leading to sequence-informative B- and Y-ions. Linkage revealing cross-ring cleavages were observed as well. The fragmentation pattern was not significantly influenced by the type of derivatization agent.

Next to reductive amination, other procedures for derivatization of the reducing end are reported. One of the reagents applied is 1-phenyl-3-methyl-5-pyrazolone (PMP) [57-59]. The response of PMP-derivatized oligosaccharides in ESI–MS is better than that of PA-derivatized oligosaccharides. In positive-ion ESI–MS with in-source CID, the PMP-derivatized neutral oligosaccharides generate predominantly Y-ions, while N-acetylated oligosaccharides generate both B and C, and Y and Z-ions [59]. Saba *et al.* [14] compared the ESI–MS response of native asialo and sialylated *N*-linked glycans with that of PMP-derivatized compounds. The PMP-asialo sugar did not provide enhanced response, but showed more specific fragmentation by in-source CID. The PMP-sialylated glycans provided a 100-fold improvement in response and yielded clean and informative spectra.

Suzuki *et al.* [57] systematically compared the responses of some oligosaccharides derivatized with PA, ABEE, PMP, 4-methoxyphenyl-3-methyl-5-pyrazolone, 2-aminoethanethiol, and 2-aminobenzenethiol in continuous-infusion ESI-MS and frit-FAB-MS. In ESI-MS, the PMP derivatives gave the best results. Reductively aminated oligosaccharides can be analysed as $[M+H]^+$ or $[M+Na]^+$ in positive-ion ESI-MS and as $[M–H]^-$ in negative-ion ESI-MS.

Various other derivatization strategies have been reported, *e.g.*, *N*-(2-diethylamino)ethyl-4-aminobenzamide [7], 2-aminonaphthalene trisulfone (ANTS) [60], 2-amino-5-bromopyridine [61]. With the latter label, isotope tagging of the oligosaccharides is achieved, enabling easy access to sequence information through the diagnostic twin peaks.

4. LC–MS analysis of oligosaccharides

Although results have been reported using RPLC, HPAEC is the method-of-choice for the efficient separation of oligosaccharides.

4.1   High-performance anion-exchange chromatography

HPAEC analysis of sugar oligomers is based on anion-exchange of sugar anions generated at high pH, typically pH 13, on Dionex CarboPac columns. The solvent system consists of gradients of up to 1 mol/l sodium acetate in 0.1 mol/l aqueous sodium hydroxide. The increasing sodium-acetate concentration acts as a displacer, inducing the subsequent elution of oligosaccharides with higher degree of polymerisation (DP). While in a particular gradient program, the oligomers with increasing DP from a homologous series elute with increasing sodium-acetate concentration, there is no straightforward relation between elution of a particular DP and that concentration. Within a given gradient program, large differences may exist in the elution of various series of oligomers, *e.g.*, Xyl-oligomers elute at much lower sodium-acetate concentration than Ara-oligomers. For mixed oligomers, consisting of various sugar residues, it is impossible to predict the elution order.

On-line HPAEC–MS was first reported by Simpson *et al.* [62] using a thermospray (TSP) interface. An anion micromembrane suppressor (AMMS) was applied to achieve a post-column exchange of sodium ions (up to ~0.3 mol/l) for hydronium ions. The same approach was applied by Niessen *et al.* [63-64] using TSP ionization in the characterization of plant-cell-wall oligosaccharides, and by the group of Henion [11-12] using ESI-MS in the analysis of *N*-linked glycans from glycoproteins.

Later on, the AMMS was replaced by an anion self-regeneration suppressor (ASRS). This device is based on hydronium-ion generation by continuous electrolysis of water. It provides more efficient salt removal (up to ~0.6 mol/l) and is easier in use. The application of the ASRS for LC–ESI–MS coupling was reported by Wunschel *et al.* [65] in their study on the quantitative analysis of neutral and acidic sugar monomers in whole bacterial cell hydrolysates. Torto *et al.* [66] reported the on-line monitoring of the hydrolysis of wheat starch using on-line microdialysis introduction into a HPAEC–MS system. The same approach was used in monitoring the enzymatic degradation of polysaccharides in two *Phaseolus* bean varieties [67]. A capillary Nafion cation-exchange desalter was recently described by Bruggink *et al.* [68] for use in combination with a 0.38-mm-ID HPAEC-column in the analysis of an urine sample from an G_{M1}-gangliosidosis patient.

4.2   Reversed-phase chromatography

Efficient RPLC of oligosaccharides on conventional C_{18} materials is difficult, unless the oligosaccharides are derivatized prior to the separation (Ch. 20.3.5). Suzuki-Sawada *et al.* [69] reported the analysis of 16 PA-derivatized complex-type biantennary oligosaccharides derived from IgG. The combination of RPLC (partial) separation and m/z determination by ESI-MS enables two-dimensional mapping of PA-oligosaccharides [52-53]. At present, more than 400 PA-oligosaccharides have been mapped.

The use of a 150×2.1- mm-ID RPLC column with a 0–15% gradient of acetonitrile in 0.12% formic acid was applied for the efficient desalting of underivatized sialylated and neutral oligosaccharides [70]. The oligosaccharides eluted as a mixture and could be analysed as $[M+H]^+$.

In the analysis of glycoprotein digests, the glycopeptides can be analysed by RPLC using water–acetonitrile gradients, containing trifluoroacetic acid (TFA). This is demonstrated in the characterization of glycopeptides derived from ribonuclease B, ovomucoid, fetuin, and asialofetuin [71]. The same approach was demonstrated by Ohta *et al.* [72] using a Vydac C_{18} column for the selective glycopeptide mapping of recombinant human erythropoietin (EPO). Later on, it was demonstrated that this approach can be applied to differentiate EPO from different sources, revealing differences in acetylation, sialylation, and sulfation at each glycosylation site [73].

Graphitized-carbon columns (GCC) exhibit properties similar to RPLC columns in the separation of native oligosaccharides, but they are able to retain less hydrophobic oligosaccharides and to separate closely related oligosaccharides without derivatization. Kawasaki *et al.* [74-75] demonstrated the potential of these columns in the LC–MS analysis of enzymatically released oligosaccharides from RNase B. As an example, the base-peak chromatogram of RNase B oligosaccharide alditols is shown in Figure 20.4. Further structure elucidation of some of the oligosaccharides was performed using LC–MS–MS.

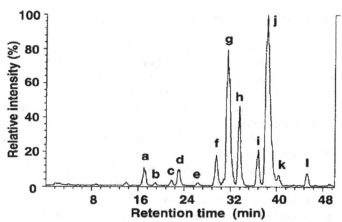

Figure 20.4: Base-peak chromatogram of RNase B oligosaccharide alditols, acquired using a graphitized carbon column in LC–MS. Peak annotation: (a) *Man₉GlcNAc*, (b) and (c) *Man₈GlcNAc*, (d) *Man₇GlcNAc*, (e) *Man₆GlcNAc*, (f) *Man₇GlcNAc*, (g) *Man₆GlcNAc*, (h) *Man₈GlcNAc*, (i) *Man₇GlcNAc*, (j) *Man₅GlcNAc*, (k) *Man₆GlcNAc*, and (l) *Hex₆HexNAc₂*. Reprinted from [74] with permission. ©1999, Academic Press/Elsevier BV.

In the next paper [75], the separation power of GCC for underivatized oligosaccharides was utilized in combination with LC–MS and MS–MS to characterize a wide variety of *N*-linked sialylated oligosaccharides enzymatically released from EPO. In a subsequent study [72], using a C_{18} column for peptide separation, 4 glycopeptides and 9 peptides were found after Glu-C digestion. Each glycopeptide contained many glycoforms. 20, 16, and 22 *N*-linked glycans were identified at Asn[24], Asn[38], and Asn[83], respectively, while 2 *O*-linked glycans were observed at Ser[126].

The combined use of MS–MS and both RPLC–MS for the characterization of glycopeptides and GCC-LC–MS for the characterization of *O*- and *N*-linked glycans is a powerful approach in the structure analysis of glycoproteins, as demonstrated for EPO [71-72, 75] and recombinant human thrombomodulin [76]. Karlsson *et al.* [77] recently reported the use of a 100×150-μm-ID nano-LC GCC column for the negative-ion LC–MS characterization of neutral and acidic *O*- and *N*-linked glycans, *e.g.*, 4 sialylated *O*-linked glycans from membrane proteins of ovarian cancer tissue. 10-fold improved detection limits were achieved using the 150-μm-ID column, operated at 0.6 μl/min, compared to the 300-μm-ID column, operated at 6 μl/min. A 100-fold increase in the absolute response for the [M–nH]ⁿ⁻ ions was achieved.

Permethylated oligosaccharides can be efficiently separated by RPLC, including α and β anomers [41]. This enables characterization of the oligosaccharides of a recombinant glycoprotein soluble CD4 with varying sialic acid content.

4.3   Other modes of chromatography

Hydrophilic interaction chromatography (HILIC, Ch. 1.4.5, often called normal-phase LC) can be an interesting tool in the separation of oligosaccharides. Saba *et al.* [60] evaluated different derivatization strategies and LC modes in combination with ESI-MS for the characterization of ANTS- and PMP-derivatized glycans from ovalbumin. HILIC was compared to RPLC. In HILIC, an amino column was used with a gradient of acetonitrile water (adjusted to pH 3.5 with formic acid). Better separation efficiency in either RPLC or HILIC was achieved with PMP-derivatives. More than 20 different glycoforms of PGNase-F released *N*-glycans of ovalbumin were detected. Wuhrer *et al.* [78] reported nano-HILIC–MS using a 100×75-µm-ID amide column and a 0–20% acetonitrile gradient in 50 mmol/l aqueous formic acid. Detection of fmol amounts of 2-aminobenzamide-derivatized *N*-glycans from keyhole limpet hemocyanin, asialofetuin, and RNase B was demonstrated.

The separation of ANTS-derivatized glycans released from bovine RNase B, bovine fetuin, and chicken ovalbumin was reported by means of ion-pair RPLC with trimethylammonium as ion-pairing agent [79].

Deery *et al.* [80] reported the on-line coupling of SEC and ion-trap MS for the detection of sodiated polysaccharides with masses in excess of 9 kDa. Multiple-charge ions with up to five sodium ions were observed. On-line SEC–MS^n allowed the determination of linkage information in permethylated arabinogalactan oligomers.

5. Conclusion and perspectives

The characterization of oligosaccharides is a fascinating research area, which is due to the complexity of the structure and the wide variety of technologies applied and required to solve the relevant analytical problems. In fact, the discussion in this chapter focussed on a limited number of relevant aspects of oligosaccharide characterization, *i.e.*, mainly involving glycans from glycoproteins, following the discussions in Ch. 19.4. Several other relevant groups of oligosaccharides have been analysed and characterized by LC–MS and ESI-MS technologies, including sulfated glycosamineglycans like heparin.

Modern LC–MS technology and especially liquid introduction via nano-ESI has become an important tool in the structure elucidation of oligosaccharides, not only for glycans from glycoproteins, but for a wide variety of other oligosaccharides as well. The multi-platform approach, where triple-quadrupole, linear and three-dimensional ion-trap, and Q–TOF instruments are applied to solve the analytical problems, is becoming more and more important. FT-ICR-MS is rapidly becoming another important tool in oligosaccharide characterization [81-82].

6. References

1. R. Orlando, Y. Yang, *Analysis of glycoproteins*, in: B.S. Larsen, C.N. McEwen, *Mass spectrometry of biological materials*, 2nd Ed., Marcel Dekker Inc., NY, (1998), 215.

2. P.M. Rudd, R.A. Dwek, *Rapid, sensitive sequencing of oligosaccharides from glycoproteins*, Curr. Opinion Biotechnol., 8 (1997) 488.

3. J. Zaia, *MS of oligosaccharides*, Mass Spectrom. Rev., 23 (2004) 161.

4. W. Morelle, J.-C. Michalski, *MS analysis of glycoproteins and their glycastructures*, Curr Anal. Chem., 1 (2005) 29.

5. R.K. Merkle, I. Poppe, *Carbohydrate composition analysis of glycoconjugates by GLC–MS*, Methods Enzymol., 230 (1994) 1.

6. I. Ciucanu, F. Kerek, *A simple and rapid method for the permethylation of carbohydrates*, Carbohydr. Res., 131 (1984) 209.

7. D.J. Harvey, *ESI-MS and fragmentation of N-linked carbohydrates derivatized at the reducing terminus*, J. Am. Soc. Mass Spectrom., 11 (2000) 900.

8. S.F. Wheeler, D.J. Harvey, *Negative ion MS of sialylated carbohydrates: discrimination of N-acetylneuraminic acid linkages by MALDI-TOF and ESI-TOF MS*, Anal. Chem., 72 (2000) 5027.

9. D.J. Harvey, *MALDI-MS of carbohydrates and glycoconjugates*, Int. J. Mass Spectrom., 226 (2003) 1.

10. U. Bahr, A. Pfenninger, M. Karas, B. Stahl, *High-sensitivity analysis of neutral underivatized oligosaccharides by nano-ESI-MS*, Anal. Chem,, 69 (1997) 4530.

11. J.J. Conboy, J.D. Henion, *HPAEC–MS for the determination of carbohydrates*, Biol. Mass Spectrom., 21 (1992) 397.

12. K.L. Duffin, J.K. Welply, E. Huang, J.D. Henion, *Characterization of N-linked oligosaccharides by ESI-MS–MS*, Anal. Chem., 64 (1992) 1440.

13. Y. Jiang, R.B. Cole, *Oligosaccharide analysis using anion attachment in negative mode ESI-MS*, J. Am Soc. Mass Spectrom., 16 (2005) 60.

14. J.A. Saba, X. Shen, J.C. Jamieson, H. Perreault, *Effect of PMP labeling on the fragmentation behavior of asialo and sialylated N-linked glycans under ESI conditions*, Rapid Commun. Mass Spectrom., 13 (1999) 704.

15. A. Fura, J.A. Leary, *Differentiation of Ca^{2+}-coordinated and Mg^{2+}-coordinated branched trisaccharide isomers – An ESI-MS–MS study*, Anal. Chem., 65 (1993) 2805.

16. E.M. Sible, S.P. Brimmer, J.A. Leary, *Interaction of first row transition metals with alpha 1-3, alpha 1-6 mannotriose and conserved trimannosyl core oligosaccharides: A comparative ESI study of doubly and singly charged complexes*, J. Am. Soc. Mass Spectrom., 8 (1997) 32.

17. D.J. Harvey, *Ionization and CID of N-linked and related carbohydrates using divalent cations*, J. Am. Soc. Mass. Spectrom 12 (2001) 926.

18. D.J. Harvey, *Ionization and fragmentation of N-linked glycans as silver adducts by ESI-MS*, Rapid Commun. Mass Spectrom., 19 (2005) 484.

19. M. Kohler, J.A. Leary, *LC–MS–MS of carbohydrates with postcolumn addition of metal chlorides using a triaxial ESI probe*, Anal. Chem., 67 (1995) 3501.

20. B. Domon, C.E. Costello, *A systematic nomenclature for carbohydrate fragmentations in FAB-MS–MS spectra of glycoconjugates*, Glycoconjugate J. 5 (1988) 397.

21. R. Orlando, C.A. Bush, C. Fenselau, *Structural-analysis of oligosaccharides by MS–MS: CID of sodium adduct ions*, Biomed. Environ. Mass Spectrom., 19 (1990) 747.

22. G.E. Hofmeister, Z. Zhou, J.A. Leary, *Linkage position determination in lithium-cationized disaccharides: MS–MS and semiempirical calculations*, J. Am. Chem. Soc., 113 (1991) 5964.

23. M.R. Asam, G.L. Glish, *MS–MS of alkali cationized polysaccharides in a quadrupole ion trap*, J. Am. Soc. Mass Spectrom., 8 (1997) 987.

24. D.J. Harvey, *CID of underivatized N-linked carbohydrates ionized by ESI*, J. Mass Spectrom., 35 (2000) 1178.

25. W. Chai, V. Piskarev, A.M. Lawson, *Negative-ion ESI-MS of neutral underivatized oligosaccharides*, Anal. Chem., 73 (2001) 651.

26. W. Chai, A.M. Lawson, V. Piskarev, *Branching pattern and sequence analysis of underivatized oligosaccharides by combined MS–MS of singly and doubly charged molecular ions in negative-ion ESI-MS*, J. Am. Soc. Mass Spectrom., 13 (2002) 670.

27. L.P. Brüll, W. Heerma, J.E. Thomas-Oates, J. Haverkamp, V. Kovácik, P. Kovác, *Loss of internal 1-6 substituted monosaccharide residues from underivatized and per-O-methylated trisaccharides*, J. Am. Soc. Mass Spectrom., 8 (1997) 43.

28. L.P. Brüll, V. Kovácik, J.E. Thomas-Oates, W. Heerma, J. Haverkamp, *Sodium-cationized oligosaccharides do not appear to undergo 'internal residue loss' rearrangement processes on MS–MS*, Rapid Commun. Mass Spectrom., 12 (1998) 1520.

29. D.J. Harvey, T.S. Mattu, M.R. Wormald, L. Royle, R.A. Dwek, P.M. Rudd, *"Internal residue loss": Rearrangements occurring during the fragmentation of carbohydrates derivatized at the reducing terminus*, Anal. Chem., 74 (2002) 734.

30. D.J. Harvey, *Fragmentation of negative ions from carbohydrates: Part 1. Use of nitrate and other anionic adducts for the production of negative ion ESI spectra from N-linked carbohydrates*, J. Am. Soc. Mass Spectrom., 16 (2005) 622.

31. D.J. Harvey, *Fragmentation of negative ions from carbohydrates: Part 2. Fragmentation of high-mannose N-linked glycans*, J. Am. Soc. Mass Spectrom., 16 (2005) 631.

32. D.J. Harvey, *Fragmentation of negative ions from carbohydrates: Part 3. Fragmentation of hybrid and complex N-linked glycans*, J. Am. Soc. Mass Spectrom., 16 (2005) 647.

33. S. König, J.A. Leary, *Evidence for linkage position determination in cobalt coordinated pentasaccharides using ion trap MS*, J. Am. Soc. Mass Spectrom., 9 (1998) 1125.

34. S.P. Gaucher, J.A. Leary, *Stereochemical differentiation of mannose, glucose, galactose, and talose using zinc(II) diethylenetriamine and ESI ion trap MS*, Anal. Chem., 70 (1998) 3009.

35. M.D. Leavell, J.A. Leary, *Stabilization and linkage analysis of metal-ligated sialic acid containing oligosaccharides*, J. Am. Soc. Mass Spectrom., 12 (2001) 528.

36. N. Viseux, E. de Hoffmann, B. Domon, *Structural analysis of permethylated oligosaccharides by ESI-MS–MS*, Anal. Chem., 69 (1997) 3193.

37. N. Viseux, E. de Hoffmann, B. Domon, *Structural assignment of permethylated oligosaccharide subunits using sequential MS–MS*, Anal. Chem., 70 (1998) 4951.

38. A.S. Weiskopf, P. Vouros, D.J. Harvey, *Characterization of oligosaccharide*

composition and structure by quadrupole ion trap MS, Rapid Commun. Mass Spectrom., 11 (1997) 1493.

39. A.S. Weiskopf, P. Vouros, D.J. Harvey, *ESI-ion trap MS for structural analysis of complex N-linked glycoprotein oligosaccharides*, Anal. Chem., 70 (1998) 4441.

40. D.M. Sheeley, V.N. Reinhold, *Structural characterization of carbohydrate sequence, linkage, and branching in a quadrupole ion trap MS: Neutral oligosaccharides and n-linked glycans*, Anal. Chem., 70 (1998) 3053.

41. J. Delaney, P. Vouros, *LC ion trap MS analysis of oligosaccharides using permethylated derivatives*, Rapid Commun. Mass Spectrom., 15 (2001) 325.

42. K. Sandra, B. Devreese, J. Van Beeumen, I. Stals, M. Claeyssens, *The Q-trap MS, a novel tool in the study of protein glycosylation*, J. Am. Soc. Mass Spectrom., 15 (2004) 413.

43. D. Ashline, S. Singh, A. Hanneman, V. Reinhold, *Congruent strategies for carbohydrate sequencing. 1. Mining structural details by MS^n*, Anal. Chem., 77 (2005) 6250.

44. H. Zhang, S. Singh, V.N. Reinhold, *Congruent strategies for carbohydrate sequencing. 2. FragLib: An MS^n spectral library*, Anal. Chem., 77 (2005) 6263.

45. A.J. Lapadula, P.J. Hatcher, A.J. Hanneman, D.J. Ashline, H. Zhang, V.N. Reinhold, *Congruent strategies for carbohydrate sequencing. 3. OSCAR: An algorithm for assigning oligosaccharide topology from MS^n data*, Anal. Chem., 77 (2005) 6271.

46. L. Brüll, M. Huisman, H.A. Schols, A.G.J Voragen, G. Critchley, J.E. Thomas-Oates, J. Haverkamp, *Rapid molecular mass and structural determination of plant cell wall-derived oligosaccharides using HPAEC and MS*, J. Mass Spectrom., 33 (1998) 713.

47. A. Pfenninger, M. Karas, B. Finke, B. Stahl, *Structural analysis of underivatized neutral human milk oligosaccharides in the negative ion mode by nano-ESI-MS^n (Part 1: methodology)*, J. Am. Soc. Mass Spectrom., 13 (2002) 1331.

48. A. Pfenninger, M. Karas, B. Finke, B. Stahl, *Structural analysis of underivatized neutral human milk oligosaccharides in the negative ion mode by nano-ESI-MS^n (Part 2: application to isomeric mixtures)*, J. Am. Soc. Mass Spectrom., 13 (2002) 1341.

49. F.N. Lamari, R. Kuhn, N.K. Karamanos, *Derivatization of carbohydrates for chromatographic, electrophoretic and MS structure analysis*, J. Chromatogr. B, 793 (2003) 15.

50. S.A. Carr, M.J. Huddleston, M.F. Bean, *Selective identification and differentiation of N- and O-linked oligosaccharides in glycoproteins by LC–MS*, Protein Sci., 2 (1993) 183.

51. W. Morelle, J. Lemoine, G. Strecker, *Structural analysis of O-linked oligosaccharide-alditols by ESI-MS–MS after mild periodate oxidation and derivatization with 2-PA*, Anal. Biochem., 259 (1998) 16.

52. Y. Takegawa, S. Ito, S. Yoshioka, K. Deguchi, H. Nakagawa, K. Monde, S.-I. Nishimura, *Structural assignment of isomeric 2-PA-derivatized oligosaccharides using MS^n spectral matching*, Rapid Commun. Mass Spectrom., 18 (2004) 385.

53. Y. Takegawa, K. Deguchi, S. Ito, S. Yoshioka, A. Sano, K. Yoshinari, K. Kobayashi, H. Nakagawa, K. Monde, S.-I. Nishimura, *Assignment and quantification of 2-PA derivatized oligosaccharide isomers coeluted on RPLC–MS by MS^n spectral library*, Anal. Chem., 76 (2004) 7294.

54. G. Okafo, L. Burrow, S.A. Carr, G.D. Roberts, W.Johnson, P. Camilleri, *A*

coordinated HPLC, CE, and MS approach for the analysis of oligosaccharide mixtures derivatized with 2-AMAC, Anal. Chem., 68 (1996) 4424.

55. J. Charlwood, J. Langridge, D. Tolson, H. Birrell, P. Camilleri, *Profiling of 2-AMAC derivatised glycans by ESI-MS*, Rapid Commun. Mass Spectrom., 13 (1999) 107.

56. D.T. Li, G.R. Her, *Structural analysis of chromophore-labeled disaccharides and oligosaccharides by LC–ESI-MS*, J. Mass Spectrom., 33 (1998) 644.

57. S. Suzuki, K. Kakehi, S. Honda, *Comparison of the sensitivities of various derivatives of oligosaccharides in LC–MS with FAB and ESI interfaces*, Anal. Chem., 68 (1996) 2073.

58. X. Shen, H. Perreault,, *Characterization of carbohydrates using a combination of derivatization, LC and MS detection*, J. Chromatogr. A, 811 (1998) 47.

59. X. Shen, H. Perreault, *ESI-MS of PMP derivatives of neutral and N-acetylated oligosaccharides*, J. Mass Spectrom., 34 (1999) 502.

60. J.A. Saba, X. Shen, J.C. Jamieson, H. Perreault, *Investigation of different combinations of derivatization, separation methods and ESI-MS for standard oligosaccharides and glycans from ovalbumin*, J. Mass Spectrom., 36 (2001) 563.

61. M. Li, J.A. Kinzer, *Structural analysis of oligosaccharides by a combination of ESI-MS and bromine isotope tagging of reducing-end sugars with 2-amino-5-bromopyridine*, Rapid Commun. Mass Spectrom., 17 (2003) 1462.

62. R.C. Simpson, C.C. Fenselau, M.R. Hardy, R.R. Townsend, Y.C. Lee, R.J. Cotter, *Adaptation of a TSP LC–MS interface for use with HPAEC of carbohydrates*, Anal. Chem., 62 (1990) 248.

63. W.M.A. Niessen, R.A.M. van der Hoeven, J. van der Greef, H.A. Schols, G. Lucas-Lokhorst, A.G.J. Voragen, C. Bruggink, *HPAEC–TSP-MS in the analysis of oligosaccharides*, Rapid Commun. Mass Spectrom., 6 (1992) 474.

64. H.A. Schols, M. Mutter, A.G.J. Voragen, W.M.A. Niessen, R.A.M. van der Hoeven, J. van der Greef, *The use of combined HPAEC–TSP-MS in the structural analysis of pectic oligosaccharides*, Carbohydr. Res., 261 (1994) 335.

65. D.S. Wunschel, K.F. Fox, A. Fox, M.L. Nagpal, K. Kim, G.C. Stewart, M. Shahgholi, *Quantitative analysis of neutral and acidic sugars in whole bacterial cell hydrolysates using HPAEC–ESI-MS–MS*, J. Chromatogr. A, 776 (1997) 205.

66. N. Torto, A. Hofte, R.A.M. van der Hoeven, U.R. Tjaden, L. Gorton, G. Marko-Varga, C. Bruggink, J. van der Greef, *Microdialysis introduction HPAEC–ESI-MS for monitoring of on-line desalted carbohydrate hydrolysates*, J. Mass Spectrom., 33 (1998) 334.

67. H. Okatch, N. Torto, J. Amarteifio, *Characterization of carbohydrate hydrolysates from legumes using enzymatic hydrolysis, microdialysis sampling and HPAEC–ESI-MS*, J. Chromatogr. A, 992 (2003) 67.

68. C. Bruggink, M. Wuhrer, C.A.M. Koeleman, V. Barreto, Y. Liu, C. Pohl, A. Ingendoh, C.H. Hokke, A.M. Deelder, *Oligosaccharide analysis by capillary-scale HPAEC with on-line ion-trap MS*, J. Chromatogr. B, 829 (2005) 136.

69. J. Suzuki-Sawada, Y. Umeda, A. Kondo, I. Kato, *Analysis of oligosaccharides by on-line LC–ESI-MS*, Anal. Biochem., 207 (1992) 203.

70. L. Huang, R.M. Riggin, *Analysis of nonderivatized neutral and sialylated oligosaccharides by ESI-MS*, Anal. Chem., 72 (2000) 3539.

71. J.J. Conboy, J.D. Henion, *The determination of glycopeptides by LC–MS with CID*, J. Am. Soc. Mass Spectrom., 3 (1992) 804.

72. M. Ohta, N. Kawasaki, S. Hyuga, M. Shyuga, T. Hayakawa, *Selective glycopeptide mapping of erythropoietin by on-line LC–ESI-MS*, J. Chromatogr. A, 910 (2001) 1.

73. M. Ohta, N. Kawasaki, S. Itoh, T. Hayakawa, *Usefulness of glycopeptide mapping by LC–MS in comparability assessment of glycoprotein products*, Biologicals, 30 (2002) 235.

74. N. Kawasaki, M. Ohta, S. Hyuga, O. Hashimoto, T. Hayakawa, *Analysis of carbohydrate heterogeneity in a glycoprotein using LC–MS–MS*, Anal. Biochem., 269 (1999) 297.

75. N. Kawasaki, M. Ohta, S. Hyuga, M. Shyuga, T. Hayakawa, *Application of LC–MS–MS to the analysis of the site-specific carbohydrate heterogeneity in erythropoietin*, Anal. Biochem., 285 (2000) 82.

76. S. Itoh, N. Kawasaki, M. Ohta, T. Hayakawa, *Structural analysis of a glycoprotein by LC–MS–MS: Application to recombinant human thrombomodulin*, J. Chromatogr. A, 978 (2002)141.

77. N.G. Karlsson, N.L. Wilson, H.-J. Wirth, P. Dawes, *Negative ion GCC nano-LC–MS increases sensitivity for glycoprotein oligosaccharide analysis*, Rapid Commun. Mass Spectrom., 18 (2004) 2282.

78. M. Wuhrer, C.A.M. Koeleman, C.H. Hokke, A.M. Deelder, *Nano-scale LC–MS of 2-aminobenzamide-labeled oligosaccharides at low femtomole sensitivity*, Int. J. Mass Spectrom., 232 (2004) 51.

79. L.A. Gennaro, D.J. Harvey, P. Vouros, *Ion-pairing RPLC–ion trap MS for the analysis of negatively charged, derivatized glycans*, Rapid Commun. Mass Spectrom., 17 (2003) 1528.

80. M.J. Deery, E. Stimson, C.G. Chappell, *SEC–MS applied to the analysis of polysaccharides*, Rapid Commun. Mass Spectrom., 15 (2001) 2273.

81. M. Froesch, L.M. Bindila, G. Baykut, M. Allen, J. Peter-Katalinic, A.D. Zamfir, *Coupling of fully automated chip ESI to FT-ICR-MS for high-performance glycoscreening and sequencing*, Rapid Commun. Mass Spectrom., 18 (2004) 3084.

82. Y. Park, C.B. Lebrilla, *Application of FT-ICR-MS to oligosaccharides,* Mass Spectrom. Rev., 24 (2005) 232.

21

LC–MS ANALYSIS OF LIPIDS AND PHOSPHOLIPIDS

1. Introduction

The analysis of fatty acids is typically performed using GC–MS after derivatization, *e.g.*, to methyl or trimethylsilyl esters. The GC–MS analysis of triacylglycerides (TAG) is more complicated. With the advent of LC–MS, the potential of LC–MS in the analysis of fatty acids, triglycerides, and related substances has been evaluated. LC–MS opens additional application areas in lipid analysis, as it enables the characterization of intact phospholipids without derivatization. The analysis and characterization of lipids is discussed in this chapter, in order to provide insight in the possibilities and limitations of LC–MS in this area. The LC–MS analysis of lipids was reviewed by Byrdwell [1]. MS–MS in the analysis of lipids was reviewed by Griffiths [2]. Electrospray ionization (ESI) MS of phospholipids was reviewed by Pulfer and Murphy [3].

2. Fatty acid analysis

Reversed-phase LC (RPLC) in the LC–MS analysis of fatty acids and hydroxy fatty acids was pioneered by the group of Kusaka [4-5]. Positive-ion atmospheric-pressure chemical ionization (APCI) was applied for analyte ionization. Because methyl esters did not give sufficient sensitivity, various amide derivatives were analysed. The best results were achieved for *N-n*-propyl amide derivatives.

In later studies, other derivatization agents were proposed to enhance the response of fatty acids in ESI-MS and/or to control their fragmentation in MS–MS. Some of the derivatives reported are *N*-methyl-2-alkylimidazoline [6], dimethyl-aminoethyl [7], and alkyltrimethylaminoethyl ester iodides [8]. The latter strategy was applied in the screening of very long chain fatty acids in the diagnosis of peroxisomal disorders [8-9], using column-bypass injection in ESI-MS and selected-reaction monitoring (SRM) for quantitation. In another study from the same group [10], pristanic acid, phytanic acid, and other very long chain fatty acids were separated on a 150×2.1-mm-ID C_8 column (5 μm) using a gradient of 80–100% acetonitrile in water (with 0.1% trifluoroacetic acid). The general procedures in these studies are very similar to the ones applied in neonatal screening for acylcarnitines (Ch. 12.3.1). Another approach for the analysis of very long chain fatty acids is the use of reversed-phase LC–MS on a C_{18} column using a stepwise gradient of partly miscible solvents, methanol/water, and methanol/*n*-hexane [11]. This method allowed quantitation in the low-pg range.

Direct negative-ion LC–ESI-MS analysis of free fatty acids in chocolate without derivatization on a C_{18} column was described by Perret *et al.* [12]. A gradient of 95–100% methanol in water (0.25 mmol/l formic acid) was used with post-column addition of a 40-mmol/l methanolic ammonia. The LOQ was 150 ng/g.

An important issue in fatty acid analysis is the structural characterization, especially in terms of positions of double-bond, hydroxy, and other groups. Although results with low-energy negative-ion ESI-MS–MS was described [13], the MS–MS analysis of lithiated lithium salts $[M–H+2Li]^+$ is the method of choice. Determination of the double-bond position relies on charge-remote fragmentation [2, 14].

3. Triacylglycerols

3.1   Chromatographic separation

The LC–MS analysis of TAG using APCI for analyte ionization was pioneered by Byrdwell and Emken [15]. After nonaqueous RPLC with a propionitrile/hexane solvent gradient, positive-ion APCI was performed. Protonated molecules $[M+H]^+$ and diacylglycerol (DAG) fragments $[M+H–RCOOH]^+$ were observed. The ratio between $[M+H]^+$ and DAG fragment increased with increasing saturation. $[M+H]^+$ and DAG fragments were also reported by Mottram and Evershed [16-17], next to less abundant $[RC≡O]^+$ acylium ions. They also studied relative abundances of the DAG fragments to differentiate between positional isomers. The loss of the *sn-2* fatty acids is less likely to occur, leading to less abundant fragment ions.

Laakso and Voutilainen [18] demonstrated the applicability of silver-loaded ion-exchange (Ag^+) columns in the LC–MS analysis of TAG. Isomers of α- and γ-linolenic acid from seed oils were separated and subsequently characterized by LC–APCI-MS. Schuyl *et al.* [19] applied Ag^+-column LC–ESI-MS for the characterization of TAG in terms of degree of unsaturation, the position of the most unsaturated fatty acid, and the carbon number (CN). Compared to non-aqueous RPLC, information on fatty acid position and CN is found more readily. A typical chromatogram is shown in Figure 21.1. In order to solve mobile-phase incompatibility problems between the toluene-hexane/toluene-ethyl acetate solvent system applied at the Ag^+-column and ESI-MS, a post-column addition of 0.1 mmol/l methanolic sodium acetate was applied. Under these conditions, $[M+Na]^+$ of the TAG were observed without fragmentation. Ag^+-LC–MS has become an important tool in TAG analysis and characterization. Dugo *et al.* [20] recently reported the fractionation of complex TAG mixtures using non-aqueous RPLC and subsequent analysis of these fractions using Ag^+-LC–APCI-MS in the positive-ion mode.

Nagy *et al.* [11] reported the use of RPLC–MS on a C_{18} column using a stepwise gradient of partly miscible solvents, methanol/water and methanol/*n*-hexane, for the analysis of TAG in blood and/or plant oil.

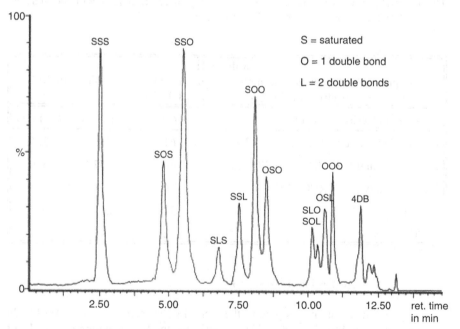

Figure 21.1: Total-ion-chromatogram of an interesterified palm oil, after Ag^+-LC–ESI-MS. Reprinted from [19] with permission. ©1998, Elsevier Science BV.

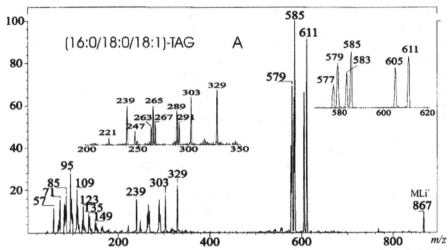

Figure 21.2: Product-ion MS–MS spectrum of [M+Li]⁺ of (16:0 / 18:0 / 18:1)-TAG. The fragments are explained in Table 21.1 Reprinted from [22] with permission. ©1999, American Society for Mass Spectrometry.

Table 21.1: Interpretation of the most informative fragments in the product-ion MS–MS spectrum of [M+Li]⁺ of (16:0 / 18:0 / 18:1)-TAG			
Fragments Resulting from:	**P (16:0)**	**S (18:0)**	**O (18:1)**
Loss of Fatty Acid ([M+Li–RCOOH]⁺)	611	583	585
Loss of Lithium Salts ([M+H–RCOOLi]⁺)	605	577	579
Lithiated Fatty Acid [RCOOLi]⁺	263	291	289
[RCO]⁺	239	267	265
[RCO]⁺ – 18	221	249	247

3.2   Characterization by MS–MS

Considerable attention was paid to the MS–MS characterization of TAG [2]. In an excellent study, Stroobant *et al.* [21] investigated the fragmentation of [M–H]⁻ of TAG, generated by desorption chemical ionization. A reversible rearrangement due to an internal Claisen condensation is proposed to explain the lower abundance of the DAG fragment resulting from the loss of the *sn-2* fatty acid. Hsu and Turk [22] discussed the MS–MS analysis of [M+Li]$^+$ of TAG. As an example, the product-ion MS–MS spectrum of (16:0 / 18:0 / 18:1)-TAG is shown in Figure 21.2. The structurally most informative fragments are explained in Table 21.1.

Hvattum [23] reported similar fragmentation patterns in the MS and MS–MS spectra of [M+NH$_4$]$^+$ of TAG, *i.e.*, formation of [M+NH$_4$]$^+$, [M+NH$_4$–RCOONH$_4$]$^+$, [RCO]$^+$, and [RCO – 18]$^+$.

3.3   Quantitative analysis

In addition to structural characterization, quantitative analysis of the various species is important in many application areas. Due to the effect of unsaturation on the relative abundance of the [M+H]$^+$ and mass discrimination effects in favour of the DAG fragment ion in a quadrupole mass analyser, determination of response factors for the various species is required for an accurate quantitative analysis [24]. Therefore, limiting the number of standards that are required in the analysis of complex TAG mixtures is an important topic in quantitation. From a comparison of four different approaches, Byrdwell *et al.* [24] concluded that the most accurate method involved the calculation of response factors from contributions from individual fatty acids to the TAG response. This approach was applied with good results by others as well [17].

Quantitative evaluation of TAG regioisomers in palm oil and cocoa butter was reported by Fauconnet *et al.* [25] using nonaqueous RPLC on a C$_{18}$ column, a 20–40% acetonitrile/chloroform gradient, and positive-ion APCI-MS. Characterization of TAG and their regioisomers is performed on basis of regression curves of mixed regioisomers (AAB and ABA) with different ratios at three different concentration levels. From the [M+H–RCOOH]$^+$ DAG fragment ions of two regioisomers, an regioisomeric impurity factor was calculated. The method enabled the accurate determination (±3%) of the proportion of each regioisomer in an AAB/ABA pair within the working range of 10–1000 µg/ml.

3.4   Selected applications

Autooxidation products of triolein, trilinolein, and trilinolenin were analysed by nonaqueous RPLC with a C$_{18}$ column and a acetonitrile/dichloromethane gradient and positive-ion APCI-MS [26]. Among others, mono- and dihydroxyperoxides and different types of mono- and diepoxides were observed. In a subsequent study [27], two parallel mass spectrometers, one running ESI-MS, the other APCI-MS, were applied in the TAG analysis of canola oil. ESI-MS especially showed advantages over APCI-MS in the analysis of TAG oxidation products, because it yielded a more abundant intact [M+NH$_4$]$^+$ than APCI-MS did. The ability of ESI-MS to reduce fragmentation of TAG also played an important role in the analysis of regioisomers of short-chain TAG, as reported by Kalo *et al.* [28]. LC–ESI-MS was also applied in the characterization of the TAG content in Malaysian cocoa butter, reported by

Segall *et al.* [29]

In order to study lipid absorption, TAG were determined in lymph samples from rats, which either received a structured lipid (8:0/18:1/8:0 or 8:0/18:2/8:0) or safflower oil [30]. Nonaqueous RPLC–APCI-MS was applied with a gradient of acetonitrile–isopropanol/hexane and post-column addition of 50 mmol/l ammonium acetate. The TAG were detected as $[M+NH_4]^+$ and DAG fragment ions.

Mottram and Evershed [31] studied the TAG composition of bovine milk using nonaqueous RPLC–APCI-MS with propionitrile as mobile phase. Prefractionation of the fat using thin-layer or size-exclusion chromatography was necessary because of the complexity of the mixture. Some fractions were also analysed using GC–MS.

Holčapek *et al.* [32] reported the LC–APCI-MS characterization of TAG and DAG in plant-oil samples from almond, amaranth, Brazil-nut, corn, evening primrose, hazelnut, linseed, macadamia, palm, pistachio, poppy-seed, rapeseed, soy bean, sunflower, *Dracocephalum moldavica*, and *Silybum arianum* using non-aqueous RPLC with an ethanol/acetonitrile gradient system and two C_{18} columns in series. The fragmentation of TAG in ion-trap MS^n was investigated as well.

4. Metabolites of arachidonic acid

Metabolites of arachidonic acid (5,8,11,14-eicosatetraenoic acid, (20:4ω6) $CH_3(CH_2)_3(CH_2CH=CH)_4(CH_2)_3COOH$) play important roles in many physiological and pathological processes such as stimulation of muscle activity, lowering of blood pressure, inflammatory responses, allergy and other immunological processes. There are several classes of eicosanoids, *i.e.*, cyclo- and lipoxygenase metabolites of arachidonic acid, including prostaglandins (PGs), hydroxy-eicosatetraenoic acids, and leukotrienes (LTs) (Figure 21.3).

The LC–MS analysis of these compounds is briefly reviewed in this section. Balazy [33] reviewed the analysis of eicosanoids in terms of targeted lipidomics.

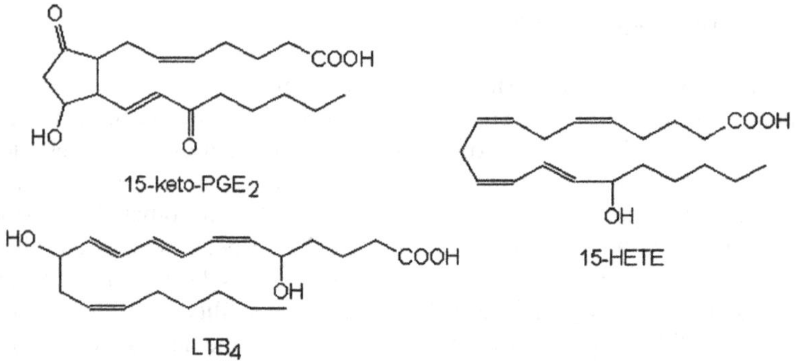

Figure 21.3: Structures of some typical arachidonic metabolites.

4.1   Hydroxyeicosatetraenoic acids

Lipid hydroeicosatetraenoic (HETE) and hydroperoxyeicosatetraenoic acids (HPETE) are important products of both enzymatic processes and autooxidation of polyunsaturated fatty acids. They appear to exhibit biological activity in disease processes and in ageing.

MacMillan and Murphy [34] reported the analysis of HPETE by means of negative-ion ESI-MS–MS. Hydroperoxides like 5-, 12-, and 12-HPETE ($C_{20:4}$) , 9- and 13-hydroperoxyoctadecadienoic acid (HPODE, $C_{18:2}$), and 5,12-dihydroperoxy-tetraenoic acid (5,12-diHPETE, $C_{20:3}$) all gave abundant carboxylate anions [M–H]⁻. By in-source CID, various fragments can be formed, *e.g.*, due to (a) the loss of water from [M–H]⁻, (b) vinylic cleavage of the bond adjacent to the hydroperoxide group, and (c) cleavage of the double bond allylic to the hydroperoxide group accompanied by the loss of water. These fragment ions are thus indicative of the position of the hydroperoxide group in the molecule. Good full-spectrum response was obtained with a direct 1-μl injection of 100 pg (0.3 pmol) of 15-HPETE (flow-rate 250 μl/min). The identity of the various fragments generated in MS–MS of [M–H]⁻ was elucidated by the use of $^{18}O_2$- and D_8-labelled compounds. Some of the major fragments are similar to those obtained by in-source CID.

Based on the strategies developed for HPETE and leukotriene-related metabolites, Wheelan *et al.* [35] studied the negative-ion ESI-MS and MS–MS spectra of dihydroxyeicosatrienoic acids (diHET), dihydroxyeicosatetraenoic acids (diHETE), and the triHETE lipoxins A_4 and B_4. Fragmentation in MS–MS allowed the determination of the hydroxy positions in the various molecules.

Griffiths *et al.* [36] described the analysis of HETE and diHETE by ESI-MS–MS of [M–H]⁻ on a sector–time-of-flight hybrid instrument equipped with a focal-plane detector. The HETE and diHETE could be analysed in underivatized form at the pmol level. The MS–MS spectra allow differentiation of isomeric eicosanoids.

Next to these more fundamental studies on the structural characterization of eicosanoids, LC–MS was applied in the qualitative and/or quantitative analysis of eicosanoids in biological systems. Some selected examples are briefly reviewed.

Kerwin and Torvik [37] reported the identification of various HETE generated by incubation of fatty acids with soybean lipoxygenase. In the negative-ion ESI-MS–MS of 2-hydroxy compounds, the loss of 46 Da (HCOOH and not C_2H_5OH) rather than 44 Da (CO_2) was observed. The compounds formed in the incubate were 5-, 8-, 9-, 11-, 12-, and 15-HETE.

In order to study the metabolism of arachidonic acid by the cytochrome P450 complex, Nithipatikom *et al.* [38] developed a negative-ion LC–ESI-MS method for the quantitative determination of 5,6-, 8,9-, 11,12-, and 14,15-epoxyeicosatrienoic acids (EET), their corresponding diHET, and 20-HETE. The analytes and their deuterium-labelled internal standards were measured as [M–H]⁻ in selected-ion monitoring (SIM) mode. The detection limit was 1 pg (2-mm-ID column).

Powell *et al.* [39] reported the quantitative analysis of 5-oxo-6,8,11,14-eicosatetraenoic acid (5-oxo-ETE), which is considered to be an important mediator in asthma. The compound and its $[D_4]$-analogue as internal standard were analysed in biological fluids using negative-ion LC–ESI-MS in SRM mode.

Suzuki *et al.* [40] reported the urinary analysis of 12-HETE using RPLC–MS in SRM mode, and the chiral analysis of 12-HETE using a column-switching setup, in which the 12-HETE containing fraction from the first RPLC column was heartcut and transferred to a Chiral CD-Ph column and separated using isocratic elution with methanol–water–acetic acid (65:35:0.02). 12(*S*)-HETE was found to be the major enantiomer.

Lee *et al.* [41] developed a method for the quantitative analysis of enantiomers and regioisomers of fatty acids, hydroxy fatty acids, and related substances, including several HETEs and prostaglandins. The method is based on pre-column derivatization to pentafluorobenzyl derivatives and subsequent LC–MS analysis in an organic mobile phase using electron-capture negative-ionization (ECNI). The method is applied to lipidomic profiling, for instance of rat epithelial cells.

4.2   Prostaglandins

Initially, the LC–MS analysis of prostaglandins was performed using thermospray ionization (Ch. 4.7). Post-column methylation of the prostaglandins [42] and conversion into diethyl amino ethyl (DEAE) derivatives [43] was proposed to improve the response. Considerable attention has been paid to the structural characterization of prostaglandins (PG), leukotrienes (LT), and related compounds using ESI-MS–MS, especially by the group of Murphy [44-46].

Wheelan *et al.* [44] described the negative-ion ESI-MS and MS–MS analysis of LTB_4 and 16 of its metabolites, among which were glutathionyl- and cysteinylglycyl-conjugates. LTB_4 and its metabolites were analysed as $[M–H]^-$. Major fragment ions in MS–MS resulted from charge-remote α-hydroxy fragmentation. The fragmentation was studied by the use of ^{18}O- and *D*-labelled compounds. All 16 LTB_4 related metabolites showed unique product-ion mass spectra. The MS and MS–MS spectra of the cysteinyl eicosanoids LTC_4 and FOG_7 were interpreted in detail by Hevko and Murphy [45] using deuterated analogues,.

Dickinson and Murphy [46] reported the MS–MS characterization of LTA_4, PGH_2 and PGI_2, reactive intermediates in the biosynthesis of PG and LT. Negative-ion MS–MS spectra were acquired on triple-quadrupole, ion-trap, and quadrupole–time-of-flight hybrid (Q–TOF) instruments. The fragmentation was studied in detail. These compounds must be analysed at high pH using an XTerra LC column with a mobile phase of 10 mmol/l triethylamine in 25% (65:35) acetonitrile/methanol in water.

Oda *et al.* [47] reported the quantitative determination using LC–ESI-MS of thromboxane B_2 (TXB_2), PGE_2, and LTB_4 in whole blood after a two-step liquid-liquid extraction (LLE). The extracted sample was analysed with on-line solid-phase extraction–LC–MS (SPE–LC–MS) using a 10×4.0-mm-ID C_8 SPE column and a 150×0.7-mm-ID C_{18} column operated at a flow-rate of 20 μl/min. Linear calibration plots over the range 30–10000 pg/ml were obtained for all three components. Good intra-day precision, with an RSD generally better than 10%, and acceptable inter-day precision, with an RSD better than 20%, and good accuracies were achieved.

Margalit *et al.* [48] reported the simultaneous quantitative analysis of 14 eicosanoids in biological samples using column-bypass negative-ion ESI-MS in SRM mode. Detection limits ranged from 0.5 pg for TXB_2 to 10 pg for 6-keto $PGF_{1\alpha}$.

Sajiki and Kakimi [49] reported the LC–MS analysis of various eicosanoids in red algae, *Gracilaria asiatica*, including PGE_2, 15-keto-PGE_2, and 8-HETE as major components, and PGA_2, and LTB_4 as minor ones. The RPLC separation was achieved using a gradient of water–formic acid (40:20:40:0.1) and acetonitrile–methanol–water–formic acid (40:20:20:0.08). ESI-MS enabled differentiation between the co-eluting 8- and 12-HETE.

Nithipatikom *et al.* [50] reported the positive-ion LC–MS analysis of the prostaglandins 6-keto-$PGF_{1\alpha}$, PGD_2, PGE_2, $PGF_{2\alpha}$, and PGJ_2 in cultured cells. The compounds were detected at 1 pg concentration using SIM.

Murphey *et al.* [51] reported a negative-ion LC–MS method in SRM mode for the determination of urinary metabolites of PGE_2, especially 11α-hydroxy-9,15-dioxo-2,3,4,5-tetranor-prostane-1,20-dioic acid (PGE-M), after methyloxime derivatization. The PGE-M concentration is 2-fold higher in urine from healthy males than from healthy females. Higher urinary PGE-M concentrations are found in the urine from patients suffering lung cancer.

Montuschi *et al.* [53] described the negative-ion LC–ESI-MS analysis of LTB_4 in exhaled breath consendate (EBC), a noninvasive method to collect airway secretions. In comparison between APCI-MS and ESI-MS in both positive-ion and negative-ion mode, negative-ion ESI-MS was found to provide the best results. The LOQ was 100 pg/ml. The method was applied to compare LTB_4 concentrations in EBC of healthy and asthmatic patients, and to evaluate the effect of anti-inflammatory therapy.

5. Phospholipids

Phospholipids are hydrophobic molecules present in all living organisms. They are applied as building blocks of cellular membranes, but serve several other functions as well. The most important classes of phospholipids in eukaryotic cells are the sphingomyelins (SM, Figure 21.4) and glycerophospholipids (GPL, Table 21.2), while phosphoglycolipids are found in prokaryotic cells. Fast-atom bombardment ionization (FAB) first enabled the use of MS and MS–MS for the structural characterization of phospholipids. ESI-MS further facilitated this. ESI-MS characterization of phospholipids was reviewed by Pulfer and Murphy [3].

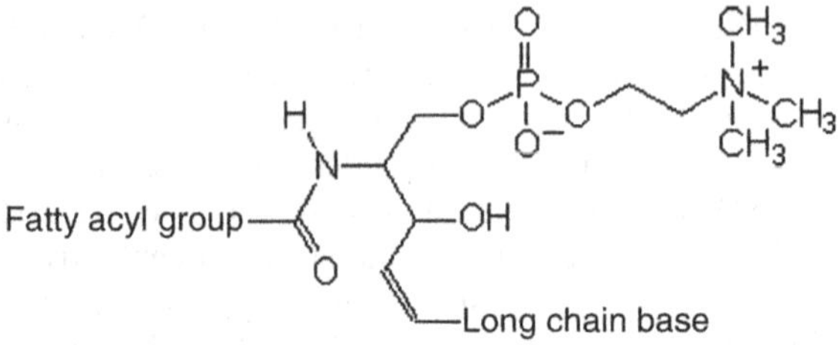

Figure 21.4: General structure of sphingomyelins. Each individual compound is characterized by a combination of a fatty acyl group (FA) and a long chain base (LCB), *e.g.*, a $C_{13}H_{27}$ alkyl chain.

Table 21.2: General structure of glycerophospholipids		
	Substituent	**Compound class**
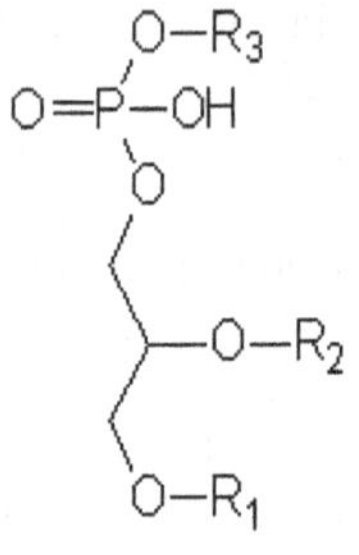	Choline	Glycerophosphocholine (GPCho)
	Ethanolamine	Glycerophosphoethanolamine (GPEtn)
	Inositol	Glycerophosphoinositol (GPIns)
	Serine	Glycerophosphoserine (GPSer)
	Glycerol	Glycerophosphoglycerol (GPGro)
	H	Glycerophosphatidic acid (GPA)
	Fatty acyl	(*sn-2*)
	Fatty ether	(*sn-1*)
	Fatty vinyl ether	
	Fatty acyl	

5.1   Sphingomyelins

Karlsson *et al.* [53] performed positive-ion NPLC–ESI-MS for the molecular-mass determination of SM. The compounds were separated on a diol column using a gradient of 63–80% hexane in 1-propanol (containing triethylamine and formic acid). The SM were detected as $[M+H]^+$. By in-source CID, fragmentation to m/z 184 corresponding to phosphorylcholine $[(HO)_2PO_2(CH_2)_2N(CH_3)_3]^+$ takes place. Further structure elucidation was performed in LC–APCI-MS mode, where the SM generated ceramide ions $[M+H–183]^+$, which were subsequently fragmented in MS–MS. Compounds with at least 36 different masses ranging from 673 tot 815 Da were detected in bovine milk. Among the 12 most abundant ions, at least 25 different combinations of FA and LCB groups (Figure 21.4) were found. Both saturated and unsaturated LCB and FA were detected next to hydroxy FA. Structural characterization of SM was discussed by Hsu and Turk [54].

5.2   Chromatography of glycerophospholipids

GPL can be separated using either normal-phase (NP) LC or RPLC. NPLC allows the separation of phospholipids by class, i.e., first the neutral lipids, followed by GPEtn, GPIns, and GPA, while GPCho elute later.

Kim et al. [55] pioneered the LC–MS analysis of phospholipids. They reported nonaqueous RPC using a C_{18} column and a gradient from 12% water in hexane to 12% methanol in hexane containing 0.5% ammonium hydroxide. Only 1% of the effluent is transferred to the ESI-MS. The analytes were detected as $[M+H]^+$ or $[M+Na]^+$ under these conditions. The best response was obtained for GPCho, followed by GPEtn, while the response of GPSer was twenty-fold less. Vernooij et al. [56] also reported RPLC–MS of GPCho, applied as excipients in pharmaceutical products. A mobile phase of acetonitrile–methanol–triethylamine (550:1000:25 by weight) was used on a C_{18} column to separate according to fatty-acid chain length and degree of unsaturation. A similar method for the analysis of GPCho and SM was reported by Isaac [57] using capillary LC on a C_{18} column with a mobile phase of methanol–tetrahydrofuran–water (80:15:5) containing 0.1% formic acid. The method was applied in GPCho profiling in human brain extracts.

Byrdwell [58] reported the use of two LC systems in parallel, equipped with UV-VIS and evaporative light-scattering detection as well as ESI-MS and APCI-MS for the analysis of SM, GPL, and plasmologen in human plasma and porcine lens extracts. A complex nonaqueous quaternary mobile-phase system was applied on a amino column. Hvattum et al. [59] reported the quantitative analysis of (16:0)(18:0)GPSer in human blood, using a gradient of 100–40% chloroform in 25% ammonia in methanol with 0.2% formic acid (pH 5.3) on a diol column. The LOQ in negative-ion LC–MS in SRM mode was 1.2 ng. The endogenous concentration ranged from 1.7 to 3.1 µg/ml. In a later study [60], group separation of GPGro, GPCho, GPEtn, lyso-GPCho, GPIns, and GPSer was achieved using NPLC on a diol column with a mobile-phase gradient of 95–20% chloroform in methanol with 0.1% formic acid, 0.05% ammonia to achieve pH 5.3, and 0.05% triethylamine. A typical chromatogram is shown in Figure 21.5. Detection limits varied between 0.1 and 5 ng. The method allowed the detection of 17 not previously described disaturated GPLs in blood. NPLC–ESI-MS was also applied in the GPL analysis in blood mononuclear cells in relation to cystic fibrosis [61]. Nonaqueous RPLC–ESI-MS on a pellicular C_8 column was applied to detect and identify GPCho, GOEtn, GPA, and lyso-GPCho in human bronchoalveolar lavage fluid (BALF) in the study of chronic obstructive pulmonary disease and bronchiolitis obliterans [62]. The identification was performed using ion-trap MS–MS. GPL were extracted from BALF by LLE. Detection limits ranged from 0.05–0.25 ng injected, depending on the class studied.

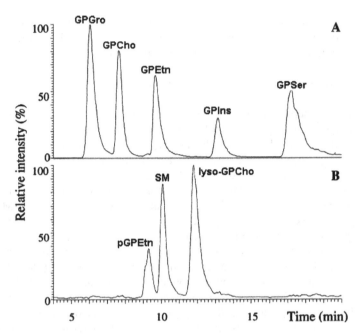

Figure 21.5: Negative-ion LC–ESI-MS analysis of a phospholipid mixture. (A) Summed mass chromatograms of model compounds of GPGro (*m/z* 777.6), GPCho (*m/z* 834.6), GPEtn (*m/z* 716.5), GPIns (*m/z* 885.5), and GPSer (*m/z* 762.5). (B) Summed mass chromatograms of plasmalogen GPEtn (*m/z* 750.6), SM (*m/z* 747.5), and lyso-GPCho (*m/z* 508.3). Reprinted from [60] with permission. ©2001 Elsevier Science BV.

5.3   Characterization by MS–MS of glycerophospholipids

Smith *et al.* [63] reported the negative-ion ESI-MS and MS–MS analysis of bacterial GPL. All GPL generated [M–H]⁻, except GPCho that showed [M–15]⁻. Generally occurring fragment ions from the [M–H]⁻ precursor are [RCH₂COO]⁻, [M–RCH₂COOH]⁻, and [M–RCHCO]⁻. The latter two ions were not observed for GPSer. The method was applied at GPL profiling in four different bacterial species. Hvattum *et al.* [64] studied the fragmentation mechanisms involved in the MS–MS of [M–H]⁻ from GPLs. GPEtn and GPCho showed a preferential loss of the *sn-2* group at low collision energy and of the *sn-1* group at high collision energy. No preferential loss of either the *sn-1* or *sn-2* group was observed for GPA, GPSer, or GPIns, while GPA is fragmenting considerably easier than the other two.

Hsu and Turk [65-72] reported a number of mechanistic studies on the fragmentation of various classes of GPL. They studied the fragmentation of [M–H]⁻ or [M+Li]⁺ of GPCho [65], GPEtn [66], GPA [67], diacyl-GPEtn [68], GPIns,

GPIns-4-biphosphate, and GPIns-4,5-biphosphate [69], GPGro [71], and GPCho [72]. The fragmentation of GPLs in an ion-trap instrument was investigated by Larsen et al. [73] and compared to that in triple-quadrupole instruments. Ho and Huang [74] indicated that fragmentation patterns of GPCho may differ between triple-quadrupole and ion-trap instruments. They proposed the use of trifluoroacetic acid adducts of GPCho-potassium salts as precursor-ions in negative-ion ESI-MS in an ion-trap instrument. A preferential loss of the *sn-1* group was observed under these conditions. In a later study [75], the same group investigated the ion-trap MS–MS of metal-ion adducts and complexes of GPLs, including adducts with Li^+, Na^+, K^+, Sr^{2+}, Ba^{2+}, and first-row transition metal ions. MS–MS spectra of Co^{2+}-complexes revealed structural information of both lipid class and regiospecificity of the two fatty acid substituents in GPEtn, GPGro, and GPSer. Moe *et al.* [76] studied unsaturated FA substituents of GPA, GPSer, GPIns, and GPGro after conversion into 1,2-d-hydroxy derivatives by OsO_4. Low-energy MS–MS can then be applied to determine the double-bond positions.

5.4 Lipidomics

Various strategies, based on multiple precursor-ion analysis and metabolic labelling with stable isotopes, were applied in the quantitative profiling of endogenous GPL in *Escherichia coli* using a Q–TOF instrument [77]. Specific common fragment ions were applied for various compound classes, *i.e.*, *m/z* 184 for protonated GPCho and SMs, *m/z* 148 for lithiated GPEtn, *m/z* 192 for lithiated GPSer, and *m/z* 196 and 241 for deprotonated GPEtn and GPIns, respectively.

Wang *et al.* [78-80] explored the potential of a quadrupole–linear-ion-trap hybrid (Q–LIT) instrument in the structural characterization of GPL in human blood. Data-dependent acquisition (DDA) was applied to identify GPL in human blood. Survey precursor-ion and neutral-loss analysis with direct-injection ESI-MS was applied to detect specific head groups of various GPL classes. DDA-triggered enhanced resolution and product-ion MS–MS was applied to identify the various components in each class. The methodology was applied for plasma GPL metabolic profiling in the search for biomarkers of type-2 diabetes mellitus [80]. Similar strategies were applied by Taguchi *et al.* [81]. Automated high-throughput quantitative analysis of GPL in biological samples was developed by Hermansson *et al.* [82]. This type of research is moving towards lipidomics [83], comprehensive liquid content analysis in systems biology.

6. References

1. W.C. Byrdwell, *APCI-MS for analysis of lipids*, Lipids, 36 (2001) 327.
2. W.J. Griffiths, *MS–MS in the study of fatty acids, bile acids, and steroids*, Mass Spectrom. Rev., 22 (2003) 81.

3. M. Pulfer, R.C. Murphy, *ESI-MS of phospholipids*, Mass Spectrom. Rev., 22 (2003) 332.
4. T. Kusaka, M. Ikeda, H. Nakano, Y. Numajiri, *LC–MS of fatty acids as their anilides*, J. Biochem., 104 (1988) 495.
5. M. Ikeda, T Kusaka, *LC–MS of hydroxy and non-hydroxy fatty acids as amide derivatives*, J. Chromatogr., 575 (1992) 197.
6. W. Vetter, W. Meister, G. Oesterhelt, *2-Alkylimidazoline derivative to control fatty acid fragmentation upon electron impact and ESI*, J. Mass Spectrom., 33 (1998) 461.
7. D.W. Johnson, *Dimethylaminoethyl esters for trace, rapid analysis of fatty acids by ESI-MS–MS*, Rapid Commun. Mass Spectrom., 13 (1999) 2388.
8. D.W. Johnson, *Alkyldimethylaminoethyl ester iodides for improved analysis of fatty acids by ESI-MS–MS*, Rapid Commun. Mass Spectrom., 14 (2000) 2019.
9. D.W. Johnson, M.-U. Trinh, *Analysis of isomeric long-chain hydroxy fatty acids by MS–MS: application to the diagnosis of long-chain 3-hydroxyacyl CoA dehydrogenase deficiency*, Rapid Commun. Mass Spectrom., 17 (2003) 171.
10. D.W. Johnson, M.-U. Trinh, T. Oe, *Measurement of plasma pristanic, phytanic and very long chain fatty acids by LC–ESI-MS–MS for the diagnosis of peroxisomal disorders*, J. Chromatogr. B, 798 (2003) 159.
11. K. Nagy, A. Jakab, J. Fekete, K. Vékey, *An LC–MS approach for analysis of very long chain fatty acids and other apolar compounds on octadecyl-silica phase using partly miscible solvents*, Anal. Chem., 76 (2004) 1935.
12. D. Perret, A. Gentili, S. Marchese, M. Sergi, L. Caporossi, *Determination of free fatty acids in chocolate by LC–MS–MS*, Rapid Commun. Mass Spectrom., 18 (2004) 1989.
13. J.L. Kerwin, A.M. Wiens, L.H. Ericsson, *Identification of fatty acids by ESI-MS–MS*, J. Mass Spectrom., 31 (1996)184.
14. F.-F. Hsu, J. Turk, *Distinction among isomeric unsaturated fatty acids as lithium adducts by ESI-MS using low energy CID on a triple stage quadrupole instrument*, J. Am. Soc. Mass Spectrom., 10 (1999) 600.
15. W.C. Byrdwell, E.A. Emken, *Analysis of TAG using APCI-MS*, Lipids, 30 (1995) 173.
16. H.R. Mottram, R.P. Evershed, *Structure analysis of TAG positional isomers using APCI-MS*, Tetrahedron Lett., 37 (1996) 8593.
17. H.R. Mottram, S.E. Woodbury, R.P. Evershed, *Identification of TAG positional isomers present in vegetable oils by LC–APCI-MS–MS*, Rapid Commun. Mass Spectrom., 11 (1997) 1240.
18. P. Laakso, P. Voutilainen, *Analysis of TAG by Ag⁺-LC–APCI-MS–MS*, Lipids, 31 (1996) 1311.
19. P.J.W. Schuyl, T. de Joode, M.A. Vasconcellos, G.S.M.J.E. Duchateau, *Ag⁺-LC–ESI-MS of TAG*, J. Chromatogr. A, 810 (1998) 53.
20. P. Dugo, O. Favoino, P.Q. Tranchida, G. Dugo, L. Mondello, *Off-line coupling of non-aqueous RP and Ag⁺-LC–MS for the characterization of rice oil TAG positional isomers*, J. Chromatogr. A, 1041 (2004) 135.
21. V. Stroobant, R. Rozenberg, el M. Bouabse, E. Deffensem, E. de Hoffmann, *Fragmentation of conjugate bases of esters derived from multifunctional alcohols including TAG*, J. Am. Soc. Mass Spectrom., 6 (1995) 498.
22. F.F. Hsu, J. Turk, *Structural characterization of TAG as lithiated adducts by ESI-MS using low-energy CID on a triple stage quadrupole instrument*, J. Am. Soc. Mass

Spectrom., 10 (1999) 587.

23. E. Hvattum, *Analysis of TAG with non-aqueous RPLC and ESI-MS–MS*, Rapid Commun. Mass Spectrom., 15 (2001) 187.

24. W.C. Byrdwell, E.A. Emken, W.E. Neff, R.O. Adlof, *Quantitative analysis of TAG using APCI-MS*, Lipids, 31 (1996) 919.

25. L. Fauconnot, J. Hau, J.-M. Aeschlimann, L.-B.Fay, F. Dionisi, *Quantitative analysis of TAG regioisomers in fats and oils using RPLC–APCI-MS*, Rapid Commun. Mass Spectrom., 18 (2004) 218.

26. W.E. Neff, W.C. Byrdwell, *Characterization of model TAG autoxidation products via LC–APCI-MS*, J. Chromatogr. A, 818 (1998) 169.

27. W.C. Byrdwell, W.E. Neff, *Dual parallel ESI and APCI-MS, MS-MS and MS-MS-MS for the analysis of TAG and triacylglycerol oxidation products*, Rapid Commun. Mass Spectrom., 16 (2002) 300.

28. P. Kalo, A. Kemppinen, V. Ollilainen, A. Kuksis, *Analysis of regioisomers of short-chain TAG by NPLC–ESI-MS–MS*, Int. J. Mass Spectrom., 229 (2003) 167.

29. S.D. Segall, W.E. Artz, D.S. Raslan, V.P. Ferra, J.A. Takahashi, *Analysis of triacylglycerol isomers in Malaysian cocoa butter using LC–MS*, Food Res. Int., 38 (2005) 167.

30. H. Mu, C.-E. Hoy, *Application of LC–APCI-MS in identification of lymph TAG*, J. Chromatogr. B, 748 (2000) 425.

31. H.R. Mottram, R.P. Evershed, *Elucidation of the composition of bovine milk fat TAG using LC–APCI-MS*, J. Chromatogr. A, 926 (2001) 239.

32. M. Holčapek, P. Jandera, P. Zderadička, L. Hrubá, *Characterization of triacylglycerol and diacylglycerol composition of plant oils using LC–APCI-MS*, J. Chromatogr. A, 1010 (2003) 195.

33. M. Balazy, *Eicosanomics: target lipidomics of eiconasoids in biological systems*, Prostaglandins Lipid Mediators, 73 (2004) 173.

34. D.K. MacMillan, R.C. Murphy, *Analysis of lipid hydroperoxides and long-chain conjugated keto-acids by negative-ion ESI-MS*, J. Am. Soc. Mass Spectrom., 6 (1995) 1190.

35. P. Wheelan, J.A. Zirrolli, R.C. Murphy, *ESI and low energy MS–MS of polyhydroxy unsaturated fatty acids*, J. Am. Soc. Mass Spectrom., 7 (1996) 140.

36. W.J. Griffiths, Y.Yang, J. Sjövall, J.Å. Lindgren, *ESI-CID of mono-, di- and trihydroxylated lipoxygenase products, including LTB and lipoxins*, Rapid Commun. Mass Spectrom., 10 (1996) 183.

37. J.L. Kerwin, J.J. Torvik, *Identification of monohydroxy fatty acids by ESI-MS–MS*, Anal. Biochem., 237 (1996) 56.

38. K. Nithipatikom, A.J. Grall, B.B. Holmes, D.R. Harder, J.R. Falck, W.B. Campbell, *LC–ESI-MS analysis of cytochrome P450 metabolites of arachidonic acid*, Anal. Biochem., 298 (2001) 327.

39. W.S. Powell, D. Boismenu, S.P. Khanapure, J. Rokach, *Quantitative analysis of 5-oxo-6,8,11,14-ETE by ESI-MS using a deuterium-labeled internal standard*, Anal. Biochem., 295 (2001) 262.

40. N. Suzuki, T. Hishinuma, T. Saga, J. Sato, T. Toyota, J. Goto, M Mizugaki, *Determination of urinary 12(S)-HETE by LC–MS–MS with column-switching technique: sex difference in healthy volunteers and patients with diabetes mellitus*, J. Chromatogr. B, 783 (2003) 383.

41. S.H Lee, M.V. Williams, R.N. DuBois, I.A. Blair, *Targeted lipidomics using ECNI-MS*, Rapid Commun. Mass Spectrom., 17 (2003) 2168.

42. R.D. Voyksner, E.D. Busch, *Determination of PG and other metabolites of arachidonic acid by thermospray LC–MS using post column derivatization*, Biomed. Environ. Mass Spectrom., 14 (1987) 213.

43. R.D. Voyksner, E.D. Busch, D. Brent, *Derivatization to improve thermospray LC–MS sensitivity for the determination of PB and TXB₂*, Biomed. Environ. Mass Spectrom., 14 (1987) 523.

44. P. Wheelan, J.A. Zirrolli, R.C. Murphy, *Negative ion ESI-MS–MS structural characterization of LTB₄ and LTB₄-derived metabolites*, J. Am. Soc. Mass Spectrom., 7 (1996) 129.

45. J.M. Hevko, R.C. Murphy, *ESI-MS–MS of cysteinyl eicosanoids: LLTC₄ and FOG₇*, J. Am. Soc. Mass Spectrom., 12 (2001) 763.

46. J.S. Dickinson, R.C. Murphy, *MS analysis of LTA₄ and other chemically reactive metabolites of arachidonic acid*, J. Am. Soc. Mass Spectrom., 13 (2002) 1227.

47. Y. Oda, N. Mano, N. Asakawa, *Simultaneous determination of TXB₂, PGE₂ and LTB₄ in whole blood by LC–MS*, J. Mass Spectrom., 30 (1995) 1671.

48. A. Margalit, K.L. Duffin, P.C. Isakson, *Rapid quantitation of a large scope of eicosanoids in two models of inflammation: development of an ESI-MS–MS method and application to biological studies*, Anal. Biochem., 235 (1996) 73.

49. J. Sajiki, H. Kakimi, *Identification of eicosanoids in the red algae, Gracilaria asiatica, using LC–MS–ESI-MS*, J. Chromatogr. B, 795 (1998) 227.

50. K. Nithipatikom, N.D. Laabs, M.A. Isbell, W.B. Campbell, *LC–MS determination of cyclooxygenase metabolites of arachidonic acid in cultured cells*, J. Chromatogr. B, 785 (2003) 135.

51. L.J. Murphey, M.K. Williams, S.C. Sanchez, L.M. Byrne, I. Csiki, J.A. Oates, D.H. Johnson, J.D. Morrowa, *Quantification of the major urinary metabolite of PGE₂ by a LC–MS assay: determination of cyclooxygenase-specific PGE₂ synthesis in healthy humans and those with lung cancer*, Anal. Biochem., 334 (2004) 266.

52. P. Montuschi, S. Martello, M. Felli, C. Mondino, M. Chiarotti, *Ion trap LC–MS–MS analysis of LTB₄ in exhaled breath condensate*, Rapid Commun. Mass Spectrom., 18 (2004) 2723.

53. A.Å. Karlsson, P. Michélsen, G. Odham, *Molecular species of SM determination by LC–MS with ESI and APCI*, J. Mass Spectrom., 33 (1998) 1192.

54. F.-F. Hsu, J. Turk, *Structural determination of SM by ESI-MS–MS*, J. Am. Soc. Mass Spectrom., 11 (2000) 437.

55. H.-Y. Kim, T.-C. Wang, Y.-C. Ma, *LC–MS of GPL using ESI*, Anal. Chem., 66 (1994) 3977.

56. E.A.A.M. Vernooij, J.F.H.M. Browers, J.J. Kettenes-van den Bosch, D.J.A. Crommelin, *RPLC–ESI-MS determination of acyl chain positions in GPL*, J. Sep. Sci., 25 (2002) 285.

57. G. Isaac, D. Bylund, J.-E. Månsson, K.E. Markides, J. Bergquist, *Analysis of GPCho and SM molecular species from brain extracts using capillary LC–ESI-MS*, J. Neurosci. Meth., 128 (2003) 111.

58. W.C. Byrdwell, *Dual parallel MS for analysis of SM, GPL and plasmologen molecular species*, Rapid Commun. Mass Spectrom., 12 (1998) 256.

59. E. Hvattum, Å. Larsen, S. Uran, P.M. Michelsen, T. Skotland, *Specific detection and*

quantification of palmitoyl-stearoyl-GPSer in human blood using NPLC–ESI-MS, J. Chromatogr. B, 716 (1998) 47.

60. S. Uran, Å. Larsen, P.B. Jacobsen, T. Skotland, *Analysis of GPL species in human blood using NPLC–ESI ion-trap MS–MS*, J. Chromatogr. B, 758 (2001) 265.

61. M. Malavolta, F. Bocci, E. Boselli, N.G. Frega, *NPLC–ESI-MS–MS analysis pf GPL molecular species in blood mononuclear cells: application to cystic fibrosis*, J. Chromatogr. B, 810 (2004) 173.

62. B. Barroso, R. Bischoff, *LC–MS analysis of GPL and lyso-GPL in human bronchoalveolar lavage fluid*, J. Chromatogr. B, 814 (2005) 21.

63. P.B.W. Smith, A.P. Snyder, C.S. Harden, *Characterization of bacterial GPL by ESI-MS–MS*, Anal. Chem., 67 (1995) 1824.

64. E. Hvattum, G. Hagelin, Å. Larsen, *Study of mechanisms involved in the CID of carboxylate anions from GPL using negative ion ESI-MS–MS*, Rapid Commun. Mass Spectrom., 12 (1998) 1405.

65. F.-F. Hsu, A. Bohrer, J. Turk, *Lithiated adducts of GPCho lipids to identify by ESI-MS–MS*, J. Am. Soc. Mass Spectrom., 9 (1998) 516.

66. F.-F. Hsu, J. Turk, *Characterization of GPEtn as a lithiated adduct by triple quadrupole ESI-MS–MS*, J. Mass Spectrom., 35 (2000) 596.

67. F.-F. Hsu, J. Turk, *Charge-driven fragmentation processes in diacyl GPA upon low-energy CID. A mechanistic proposal*, J. Am. Soc. Mass Spectrom., 11 (2000) 797.

68. F.-F. Hsu, J. Turk, *Charge-remote and charge-driven fragmentation processes in diacyl GPEtn upon low-energy CID: A mechanistic proposal*, J. Am. Soc. Mass Spectrom., 11 (2000) 892.

69. F.-F. Hsu, J. Turk, *Characterization of GPIns, GPIns-4-phosphate, GPIns-4,5-biphosphate by ESI-MS–MS: A mechanistic study*, J. Am. Soc. Mass Spectrom., 11 (2000) 986.

70. F.-F. Hsu, J. Turk, *Structural determination of SM as lithiated adducts by ESI-MS using low-energy CID on a triple stage quadrupole instrument*, J. Am. Soc. Mass Spectrom., 12 (2001) 61.

71. F.-F. Hsu, J. Turk, *Studies on GPGro with triple quadrupole ESI-MS–MS: fragmentation processes and structural characterization*, J. Am. Soc. Mass Spectrom., 12 (2001) 1036.

72. F.-F. Hsu, J. Turk, *ESI-MS–MS studies on GPCho: The fragmentation processes*, J. Am. Soc. Mass Spectrom., 14 (2003) 352.

73. Å. Larsen, S. Uran, P. B. Jacobsen, T. Skotland, *CID of GPL using ESI ion-trap MS*, Rapid Commun. Mass Spectrom., 15 (2001) 2393.

74. Y.-P. Ho, P.-C. Huang, *A novel structural analysis of GPCho as TFA/K⁺ adducts by ESI ion trap MS–MS*, Rapid Commun. Mass Spectrom., 16 (2002) 1582.

75. Y.-P. Ho, P.-C. Huang, K.-H. Deng, *Metal ion complexes in the structural analysis of GPL by ESI-MS–MS*, Rapid Commun. Mass Spectrom., 17 (2003) 114.

76. M.K. Moe, T. Anderssen, M.B. Strøm, E. Jensen, *Total structure characterization of unsaturated acidic GPL provided by vicinal di-hydroxylation of FA double bonds and negative ESI-MS*, J. Am. Soc. Mass Spectrom., 16 (2005) 46.

77. K. Ekroos, I.V. Chernushevich, K. Simons, A. Shevchenko, *Quantitative profiling of GPL by multiple precursor ion scanning on a Q–TOF-MS*, Anal. Chem., 74 (2002) 941.

78. C. Wang, S. Xie, J. Yang, Q. Yang, G. Xu, *Structural identification of human blood*

GPL using LC–Q–LIT-MS, Anal. Chim. Acta., 525 (2004) 1.

79. C. Wang, J. Yang, P. Gao, X. Lu, G. Xu, *Identification of GPL structures in human blood by direct-injection Q–LIT-MS*, Rapid Commun. Mass Spectrom., 19 (2005) 2443.

80. C. Wang, H. Kong, Y. Quan, J. Yang, J. Gu, S, Yang, G. Xu, *Plasma GPL metabolic profiling and biomarkers of type 2 diabeter mellitus based on LC–ESI-MS and multivariate statistical analysis*, Anal. Chem., 77 (2005) 4108.

81. R. Taguchi, T. Houjou, H. Nakanishi, T. Yamazaki, M. Ishida, M. Imagawa, T. Shimizu, *Focused lipidomics by MS–MS*, J. Chromatogr. B, 625 (2005) 26.

82. M. Hermansson, A. Uphoff, R. Käkelä, P. Somerharju, *Automated quantitative analysis of complex lipidomes by LC–MS*, Anal. Chem., 77 (2005) 2166.

83. X. Han, R.W. Gross, *Shotgun lipidomics: ESI-MS analysis and quantitation of cellular lipidomes directly from crude extracts of biological samples*, Mass Spectrom. Rev., 24 (2005) 367.

LC–MS ANALYSIS OF NUCLEIC ACIDS

1. Introduction

LC–MS only plays a modest role in the analysis and characterization of oligonucleotides and related compounds. Unlike in proteomics, the role of LC–MS in DNA sequencing is not very significant. While in the mid 1990s some researchers expected a significant role of MS–MS in DNA sequencing in relation to the Human Genome Project, this expectation was not made true, because DNA sequencers rather than mass spectrometers were applied. Nevertheless, LC–MS is applied in the analysis and characterization of nucleic acids and related compounds in a variety of areas. Like with other biomolecules, two different MS approaches are applied in solving the analytical problems at hand, *i.e.*, matrix-assisted laser desorption/ ionization (MALDI) or electrospray ionization (ESI). The focus in this discussion is on ESI-MS and LC–MS technology development rather than on the results achieved.

Nucleosides and nucleotides were important model compounds in the development of the thermospray ionization [1-2]. Currently, the most widely applied LC–MS technique in the analysis of oligonucleotides and other nucleic acid derived compounds is ESI-MS. Covey *et al.* [3] was the first to show the applicability of ESI-MS to the analysis of oligonucleotides. They showed a spectrum of a 14-mer oligonucleotide. A complex ion envelope of multiple-charge negative ions $[M-nNa]^{n-}$ and $[M-nNa+mH]^{(n-m)-}$ is observed. Stults and Marsters [4] reported the use of ion-exchange sample pretreatment to replace sodium by ammonium and showed a spectrum of, among others, a 77-mer ($dA_{23}dC_{17}dG_{29}dT_8$).

The LC–MS analysis of nucleic acids was reviewed by Huber and Oberacher [5].

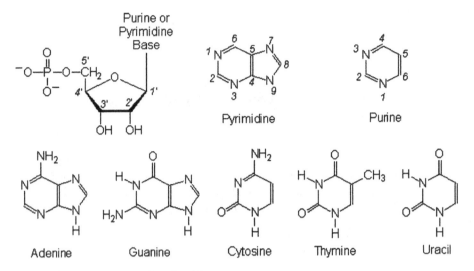

Figure 22.1: Structures of nucleotides.

1.1   Nucleosides and nucleotides

A nucleic acid or nucleotide consists of three parts: a nucleobase, a (deoxy) ribose moiety, and a phosphate group. In a nucleoside, the phosphate group is removed. Next to the common nucleobases guanine, uracil, cytosine, adenine, and thymine (Figure 18.1), various modified bases is known. The deoxyribonucleosides are indicated as dG, dU, dC, dA, and dT, respectively, or dN in general. The nucleic acid shown in Figure 22.1 is a ribonucleotide (in RNA, with uracil instead of thymine). In deoxyribonucleosides, 2'-OH is replaced by 2'-H. Linkage of the sugar to the nucleobase proceed at the N-9-position for pyrimidine bases and at the N-1-position for purine bases.

Next to nucleotides in DNA and RNA, related compounds may be present in living systems. For instance, adenosine mono-, di-, and triphosphate (AMP, ADP, and ATP) play an important role in energy management of many enzymatic reactions.

1.2   Oligonucleotides

In an oligonucleotide, the monomeric nucleic acid units are linked via a (3'-5')-phosphodiester linkage between the (deoxy)riboses to linear chains, the so-called single-stranded oligonucleotides. In solution, these molecules show a rather linear conformation, with both the sugar-phosphates and the nucleobase exposed to solution. An oligonucleotide is written like 5'-ACGA-3' or d(ACGA).

In nature, complementary single-stranded oligonucleotide chains combine to double-stranded DNA, as a result of hydrogen binding between the A–T pairs and the G–C pairs. The typical conformation is the double-helix structure, where the sugar-phosphates are facing outwards to solution and the more hydrophobic nucleobases facing inwards.

2. General considerations

2.1   LC analysis of nucleosides and nucleotides

Nucleosides can be analysed by conventional reversed-phase (RP) LC with a buffered mobile phase. The separation of nucleotides is somewhat complicated by the dissociation of the phosphate groups and *H/Na* exchange at these sites. As indicated for the phosphorylated anabolites of nucleoside reverse transcriptase inhibitors (NRTI, Ch. 13.2.4), ion-pair RPLC using *N,N*-dimethylhexylamine (DMHA) can be applied to both reduce adduct formation and obtain sufficient retention [6-7]. Alternatively, enzymatic dephosphorylation of the nucleotides prior to LC–MS analysis can be performed.

2.2   ESI-MS of nucleosides and nucleotides

In ESI-MS, the ionization characteristics of nucleosides are mainly determined by the nucleobase. The compounds are best analysed as $[M+H]^+$ in the positive-ion mode. Some $[M+Na]^+$ may be observed as well. In low-energy collision-induced dissociation (CID) in tandem mass spectrometry (MS–MS), the bond between the sugar and the purine or pyrimidine base is the weakest link. The fragmentation results in proton rearrangement and charge retention on the base, resulting in a characteristic loss of dehydrodeoxyribose (116 Da) or dehydroribose (132 Da).

For nucleotides, the situation is more complex. Nucleotides dissociate to polyanions in aqueous solution, which can be readily analysed as $[M-H]^-$ in negative-ion mode. Acidic protons in the sugar-phosphate backbone may be exchanged for mono and divalent cations from solution, *e.g.*, Na^+, K^+, Mg^{2+}, and NH_4^+. This actually disperses the signal for a particular nucleic acid over several peaks with different *m/z*. Desalting and/or ion-pairing agents like DMHA may be applied to reduce these problems.

The major fragmentation route of $[M-H]^-$ in MS–MS involves the loss of the sugar with charge retention on the deprotonated base for dNMP, and the loss of the nucleoside or nucleoside monophosphate for the dNDP and dNTP, respectively, resulting in a fragment ion at *m/z* 159 due to a deprotonated pyrophosphate [6].

2.3   LC analysis of oligonucleotides

LC–MS of oligonucleotides is most frequently performing with ion-pair RPLC. Retention of oligonucleotides in conventional RPLC is very limited. In addition, the ESI-MS response is not very high in the typical RPLC mobile phase of 100 mmol/l ammonium acetate and up to only 10% acetonitrile [8]. Triethylammonium (TEA) salts are applied as ion-pairing agents. The oligonucleotides are primarily separated according to chain length, while secondary interactions may result in sequence-dependent retention, especially for single-stranded oligonucleotides. Size-exclusion LC is mainly used for desalting or prefractionation, while the salt gradients with high concentrations of nonvolatile salts applied in anion-exchange and mixed-mode LC are not amenable to LC–MS.

Huber and Krajete [9] promoted the use of ion-pair RPLC on capillary columns packed with 2.3-μm micropellicular octadecylated poly(styrene–divinylbenzene) (PS–DVB) particles and solvent gradients of acetonitrile in 50 mmol/l TEA-bicarbonate (TEABC). Post-column addition of an acetonitrile sheath resulted in stable ESI-MS conditions with good response for the oligonucleotides. A typical chromatogram is shown in Figure 22.2. In a separate study [10], they compared various sheath liquids, *e.g.*, isopropanol, methanol, and acetonitrile, and various additives to these liquids. Acetonitrile gave best performance. They also showed LC on PS–DVB monolithic columns [11-12]. In this way, significant improvement of the chromatographic efficiency was achieved. As an example, the separation of double-stranded DNA fragments from a restriction digest of the pBR322 plasmid is shown in Figure 22.3.

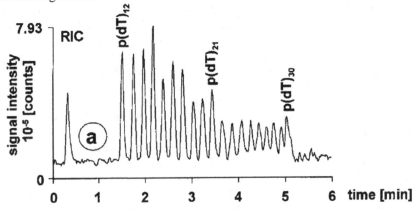

Figure 22.2: LC–MS reconstructed total-ion chromatogram of a separation of a homologous series of oligonucleotides $p(dT)_n$ with n between 12 and 30 on a micropellicular octadecylated PS–DVB column using a solvent gradient of acetonitrile in 50 mmol/l TEABC. Reprinted from [9] with permission. ©1999, American Chemical Society.

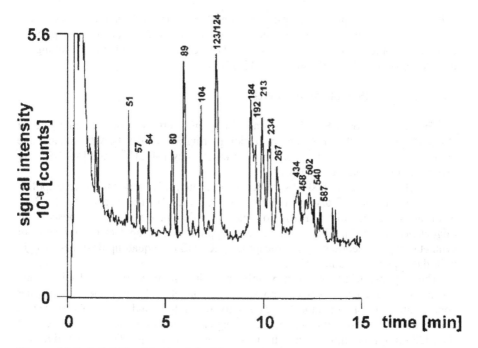

Figure 22.3: LC–MS analysis of double-stranded DNA fragments from a restriction digest of the pBR322 plasmid (180 fmol of each fragment) on a monolithic PS–DVB column using a solvent gradient of acetonitrile in 50 mmol/l TEABC. Reprinted from [11] with permission. ©2000, American Chemical Society.

2.4   Desalting sample pretreatment strategies

In the ESI-MS analysis of oligonucleotides, the adduct formation with Na^+, K^+, Mg^{2+}, and NH_4^+ and exchange of acidic protons may result in very complicated mass spectra with poor signal-to-noise ratios (S/N). It may seriously compromise the mass determination of especially larger nucleotides. Sample pretreatment procedures directed at the reduction of the cation concentration are therefore of utmost importance in nucleic acid analysis by LC–MS.

Desalting can be achieved by: precipitation with ethanolic ammonium acetate [4], solid-phase extraction (SPE), microdialysis [13], Fe^{3+}-loaded immobilized metal affinity chromatography (IMAC) [14], cation-exchange columns (SCX) [15-16], or combinations thereof [17-18]. The SCX procedure can be applied on-line with direct infusion of the effluent into ESI-MS [15-16]. The mobile phase applied is similar to the one applied in LC–MS, *i.e.*, 50% acetonitrile in 10 mmol/l aqueous TEA.

Ion-pairing agents like TEA also suppress the adduct formation [9, 16, 19].

Willems *et al.* [20] proposed on-column desalting using 1,2-diaminocyclohexane-tetraacetic acid (CDTA), ammonium acetate, formic acid, or acetic acid prior to chromatographic separation on a 300-μm-ID microcapillary column. Desalting by the addition of EDTA was also reported [21].

2.5   ESI-MS of oligonucleotides

The negative-ion ESI-MS spectra of oligonucleotides show an envelope of $[M–nH]^{n-}$ ions [3]. The presence of an ion-pairing agent like TEABC reduces the number of charge states observed. In most cases, only $[M–2H]^{2-}$ and $[M–3H]^{3-}$ are observed in the presence of TEA [9].

In early negative-ion ESI-MS studies, halogenated solvent additives were applied to reduce the risk of discharge formation (Ch. 6.3.2). The use of 1,1,1,3,3,3,-hexafluoro-2-propanol (HFIP) to the mobile phase was proposed for oligonucleotides [22]. This is for instance applied in the rapid characterization of synthetic oligonucleotides by microcapillary LC–MS on a quadrupole–time-of-flight hybrid (Q–TOF) instrument [20].

The response in ESI-MS can be influenced in a number of ways. Bleicher and Bayer [8] reported that better response for oligonucleotides is achieved at higher concentrations of organic modifier, especially acetonitrile, and at high pH. However, both conditions are unfavourable in LC. Greig and Griffey [19] studied the effects of the addition of micromolar concentrations of organic bases like TEA, piperidine, and imidazole. The co-addition of imidazole and TEA provided best results. Post-column addition of 0.1 mol/l imidazole in acetonitrile to a acetonitrile/0.1 mol/l TEA acetate gradient was applied to improve the response in ESI-MS [23]. Heating the sample and the ion source was also found to reduce adduct formation [21, 24].

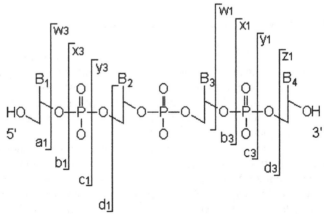

Figure 22.4: Nomenclature for designating MS–MS fragments of oligonucleotides.

The addition of TEA is frequently applied in oligonucleotide LC–MS analysis. Huber and Krajete [14] showed the advantages of TEABC in combination with post-column addition of an acetonitrile sheath. Further improvement in ESI-MS response can be achieved by replacing the TEABC by butyldimethylammonium bicarbonate (BdMABC) [25]. In that case, a somewhat higher acetonitrile concentration in the mobile phase can be used, which results in a significant improvement of the sensitivity.

With proper calibration, excellent mass accuracy with relative mass deviations better than 0.01% can be achieved, as for instance demonstrated by Oberacher et al. [25] for $p(dA)_n$ with n between 40 and 60. Typical detection limits for $p(dT)_{16}$ are 104 fmol in full-spectrum analysis and 710 amol in selected-ion monitoring (SIM) with TEABC as ion-pairing agent [9]. Similar detection limits can be achieved for double-stranded DNA fragments.

2.6 Fragmentation by MS–MS

DNA sequencing by MS was reviewed by Murray [26]. There are three general sequencing strategies: (1) mass analysis of Sanger sequencing reaction products, (2) ladder sequencing, and (3) gas-phase fragmentation in MS–MS.

The Sanger method is based on the mass analysis of a series of oligonucleotides with different strand lengths, produced as complementary strands by DNA polymerase in the presence of 2',3'-dideoxy chain terminating agents. This approach imposes very high demands to the mass resolution at very high m/z. Given the formation of multiple-charge oligonucleotide ions and the adduct formation in ESI-MS, MALDI-MS would then be the most realistic choice.

In the ladder sequencing method, the oligonucleotide is enzymatically digested to subsequently remove either 3'- or 5'-terminal residues. The sequence can be derived from the mass change after each cleavage. Ladder sequencing was demonstrated using either MALDI-MS or ESI-MS [27].

Gas-phase fragmentation of the oligonucleotide anions by CID in MS–MS was pioneered by McLuckey et al. [28] using an ion-trap instrument. It is readily compatible with LC–ESI-MS. They also proposed the nomenclature, commonly applied to designate DNA fragments in MS–MS (Figure 22.4).

The first fragmentation step involves the loss of a nucleobase (indicated as –B, adenine is lost most easily [28]), followed by the cleavage of the 3'-C–O phosphodiester bond at the nucleotide that lost the base. As a result, (a–B)- and w-ions are predominantly observed. The resulting series of 3'-terminal and 5'-terminal ions show characteristic mass differences, from which the sequence can be read. However, these MS–MS spectra are quite complex with a huge variety of fragment peaks. Ni et al. [29] described a procedure for the interpretation of oligonucleotide MS–MS spectra and developed a computer algorithm to derive the sequence from the m/z values observed in the MS–MS spectrum. Oberacher et al. [30-31] developed an algorithm to compare the observed m/z values from an unknown sequence with *m/z* values predicted from the fragmentation of known sequences. This approach is especially powerful in quality control of synthetic oligomers and to localize point mutations (Ch. 22.4.1). Later on, this software was adapted to enable *de novo* sequencing as well [32]. Other software tools were reported as well [33].

Vrkic *et al.* [34] studied the fragmentation reactions of [M+H]$^{+}$ of all 64 protonated oligodeoxynucleotide trimers and of the 16 isomeric mixed-base tetramers in ion-trap MS–MS. Similar to negative-ion MS–MS, the first loss involves a nucleobase. The relative abundance of the resulting fragment depends on the nucleobase (C≈G>A>>T) and its position (5'>3'>internal). Further MS^{3} experiments on the [M+H–B$_n$H]$^{+}$ fragment ion result in 3'-C–O phosphodiester cleavage to w- and (a–B)- ions.

Premstaller *et al.* [35-36] compared the performance of triple-quadrupole, ion-trap, and sector instruments in the analysis of oligonucleotides. For smaller oligonucleotides (8-mer) the ion-trap was more sensitive than the other two, but for larger analytes (24-mer) there is no difference anymore. Best day-to-day reproducibility was achieved with the triple-quadrupole instrument. All three instruments enabled the sequencing of a 5-mer oligonucleotide without problems. For *de-novo* sequencing of a 10-mer, the triple quadrupole gave better results, while for larger oligonucleotides the interpretation of the ion-trap MS^n spectra was more straightforward.

Given the complexity of the oligonucleotide MS and MS–MS spectra, the application of high-resolution instruments like Fourier-transform ion-cyclotron resonance mass spectrometers (FT-ICR-MS) is beneficial. While some initial results were reported in the mid-1990s, *e.g.*, [37-38], the frequent utilization of FT-ICR-MS in oligonucleotide characterization is more recent (Ch. 22.3.3).

3. Selected applications on oligonucleotides

3.1   Quality control of synthetic oligonucleotides

Quality control of oligonucleotides synthesized by solid-phase synthesis is an important application of LC–MS. With a coupling efficiency of 98–99% per synthesis cycle, the maximum yield of a 32-mer oligonucleotide is only 52–72% [5]. Contamination of the target sequence is observed with a number of failure sequences or partially protected sequences [9, 16]. Ion-pair RPLC–MS is an excellent tool for such studies. Oberacher *et al.* [30-31] developed an algorithm for comparative sequencing, enabling characterization of oligonucleotides up to 60-mers via on-line SCX desalting and direct infusion into the ion-trap ESI-MS system. Ni and Chan [39] reported enhanced and more reliable sequence verification by the use of a Q–TOF instrument, enabling charge-state determination and more accurate mass determination.

3.2 Antisense oligonucleotides

In antisense oligonucleotide therapy, the biosynthesis of particular proteins is inhibited by the use of short pieces of synthetic (modified) oligonucleotides (15–30-mers) that bind to messenger-RNA. The therapy shows potential in the treatment of viral infections and cancer. Synthetic modifications, *e.g.*, a phosphorothioate backbone or methylation of the phosphonate backbone, must provide resistance to cleavage by nucleases. Reliable quality control of these compounds is required. Baker *et al.* [40] reported the use of positive-ion ESI-MS for the characterization of an antisense methylphosphonate product (18-mer). MS–MS on $[M+H]^{5+}$ is used for sequence verification. Griffey *et al.* [41] reported the characterization of *in vivo* metabolism of antisense oligonucleotide products in pig kidney. By the use of LC separation with a methanol gradient in aqueous TEA–HFIP and sequencing with negative-ion MS–MS in an ion-trap instrument, a pattern of nuclease degradation was revealed. Lotz *et al.* [42] reported MS–MS sequence verification of oligophosphorothioates on a triple-quadrupole instrument.

3.3 Polymerase chain reaction products

The polymerase chain reaction (PCR) is an important tool in DNA research, as it allows amplification of DNA sequences. PCR analysis involves isolation of the template DNA from the sample matrix, amplification of the targeted DNA region, and determination of the molecular weight of the products. The latter is conventionally done using agarose gel electrophoresis. In order to speed up and improve the accuracy of the molecular-weight determination, the applicability of MS was evaluated. Muddiman *et al.* [17] reported the characterization of PCR products from the genome of various *Bacillus* species using negative-ion ESI on an FT-ICR-MS instrument. Molecular-weight determination of 89-base pair (bp) and 114-bp oligonucleotide portion were demonstrated. In a subsequent paper, the same group differentiated 89-bp PCR products differing by a single nucleotide only [43]. Subsequently, increasingly larger PCR products were characterized by FT-ICR-MS technology. Muddiman *et al.* [44] reported a precision of ±27 Da (87 ppm) in the molecular-weight determination of a 500-bp PCR product of 309 kDa.

Molecular-weight determination and evaluation of sequence modifications of PCR products can be performed using either single- or double-stranded oligonucleotides. The locations of the modifications can only be determined from MS–MS fragmentation of single-stranded oligonucleotides. Null *et al.* [45] reported the use of the lambda exonuclease DNA repair enzyme to selectively digest one DNA strand without affecting the other. They applied this approach to intact PCR products from the Human Tyrosine Hydroxylase gene (HUMTHO1). Another approach was proposed by Oberacher *et al.* [25]. Ion-pair RPLC was applied for the separation of single- and double-stranded PCR products, enabling further on-line MS–MS characterization of the purified product.

Obviously, FT-ICR-MS for routine molecular-weight determination is rather expensive. Krahmer *et al.* [46] explored the potential of ESI-MS on a quadrupole analyser. Both nucleotide substitutions and insertion/deletion can be detected in PCR products with up to 62 base pairs.

3.4   Single nucleotide polymorphism

With the completion of the Human Genome Project in sequencing the human DNA, new research themes emerge, such as the correlation between genotype and phenotype. Single nucleotide polymorphisms (SNP) are singly-base changes occurring at a specific position in a genome, thus leading to different sequence alternatives (alleles) in the individuals within a population. An allele must have a frequency higher than 1%, otherwise it is considered a mutation. On average, one SNP is found in every 500–1000 bases in humans. Considerable attention is paid to the development of high-throughput SNP genotyping methods, which would enable the characterization of genes involved in complex human diseases like cancer. MS plays a role in genotyping SNP, especially MALDI-MS [47]. Some studies report the use of ESI-MS and/or LC–MS. In fact, the characterization of PCR products discussed above is an example of SNP detection. Various tools for PCR product detection as well as sequence verification of synthetic oligonucleotides can be applied in SNP studies as well.

Krahmer *et al.* [48] reported the use of quadrupole ESI-MS and ion-trap ESI-MS–MS for the identification of SNP in the PCR products of the Pro and Arg variants of the tumour suppressor protein p53 gene. The 69-bp Arg variant is isomeric with the Pro variant. A 43-bp fragment, created by means of restriction enzyme digestion, could be sequenced and characterized.

An important tool in SNP genotyping is polymerase-mediated single nucleotide primer extension (SNuPE). Zhang *et al.* [49] reported the use of ESI-MS as part of a SNuPE-based Survivor assay. The procedure involves PCR amplification of the genomic DNA, purification of the resulting DNA, reaction of the single-strand oligonucleotide with an SNP primer, extension of the primer by a single dideoxynucleotide (ddNTP), and subsequent ESI-MS–MS analysis for the detection of free ddNTP. The assay offers the detection of just four ddNTP for any SNP without the need for labelling. In a subsequent paper, further simplification and validation of the approach was described [50]. Combined sample preparation and analysis takes only 2 min per sample.

Walters, Muhammad *et al.* [24, 51] reported the characterization of SNP in PCR products using ESI-MS on either a quadrupole [24] or a Q–TOF instrument [51] from relatively small mass shifts in the PCR products. While the transversion of C–G (40 Da) can still be detected by the quadrupole, the enhanced resolution of the Q–TOF was required for the detection of A–T transversion (9 Da).

The group of Huber [21, 52-53] demonstrated the applicability of the ion-pair RPLC–MS methodology developed (Ch. 21.2.3) in genotyping of SNP.

4. LC–MS analysis of modified nucleosides

Urinary excretion of modified nucleosides, originating from transfer-RNA, may be used as a biomarker for tumours and AIDS. Dudley *et al.* [54-57] reported method development for the analysis of urinary nucleosides by LC–MS. Initially, LC–MS conditions were optimized [54]. In positive-ion ESI-MS, detection limits were achieved ranging from 7 pmol for tubercidin to 110 pmol for uridine. Next, a comparison was made between GC–MS, LC–MS on an ion-trap instrument, and capillary LC–MS on a triple-quadrupole instrument [55]. These methods proved complementary rather than that just one could be selected as optimal. Therefore, in the next study [56], all three techniques were applied to identify the unexpected 5'-deoxycytidine in the urine of a patient suffering with head and neck cancer. In another study [57], they demonstrated the detection of dA, 1-methyl-dA, xanthosine, *N*-1-methyl-dG, *N*-2-methyl-dG, *N*-2,*N*-2-dimethyl-dG, *N*-2,*N*-2,*N*-7-trimethyl-dG, inosine, and 1-methylinosine in urine samples from various cancer patients.

4.1 Urinary analysis of oxidized nucleobases

LC–MS is also applied in the quantitative analysis of oxidized nucleobases, which serve as biomarkers for *in vivo* oxidative stress. An important target compound is 8-oxo-7,8-dihydro-2'-deoxyguanosine (8-OH-dG). Ravanat *et al.* [58] reported the determination of 8-OH-dG in cellular liver DNA and urine using positive-ion LC–MS. Detection limits of 5 pmol in SIM and 20 fmol in SRM were reported. Hua *et al.* [59] compared positive-ion and negative-ion ESI-MS in the analysis of oxidized deoxynucleosides, *i.e.*, 8-OH-dG, 8-OH-dA, dT-glycol, and 5-hydroxy-methyl-dU. The two modes showed similar S/N, except for dT-glycol, which could be measured 100× more sensitive in negative-ion mode. Renner *et al.* [60] applied SPE and LC–MS for the detection of 8-OH-dG in urine. The detection limit was 0.2 ng/ml (7 fmol absolute). Pietta *et al.* [61] reported a detection limit of 1 ng/ml for 8-OH-dG in human urine using LC–APCI-MS in SRM mode. Hu *et al.* [62] compared LC–ESI-MS and an immunoassay for the quantitative analysis of urinary 8-OH-dG in the urine of workers occupationally exposed to polycyclic aromatic hydrocarbons. A good correlation was found between the results of the two methods. However, LC–MS showed a significant difference in the urinary levels of exposed and control subjects. This was not detected by the immunoassay. Sabatini *et al.* [63] reported routine quantitation of 8-OH-dG in urine using SPE and micro-LC–ESI-MS in SRM mode. Matrix effects were evaluated. The LOQ was 0.2 ng/ml.

5. LC–MS analysis of DNA adducts

One of the possible causes for the development of cancer is the modification of the DNA base after exposure to a chemical carcinogen, resulting in a DNA adduct. MS plays an important role in the detection and especially the structure elucidation of DNA adducts. In this section, analytical strategies and selected examples of the use of LC–MS in the analysis of DNA adducts are reviewed, focussing on developments reported after the publication of the reviews by Apruzzese and Vouros [64] and Esmans et al. [65] in 1998. Koc and Swenberg [66] published a more recent review (2002) on the MS quantitation of DNA adducts.

The general strategy for the isolation and characterization of DNA adducts involves the extraction of DNA from the biological sample, hydrolysis and enzymatic digestion and dephosphorylation, SPE and/or LC cleanup in order to remove unmodified nucleosides, and LC–MS–MS analysis of the modified nucleosides [64]. The major challenge is lowering the detection levels in LC–MS. In 1998, the detection limits were several orders of magnitude higher than those obtained by ^{32}P-post-labelling. In addition, carcinogen-modified oligonucleotides have been analysed as well.

Tretyakova et al. [67] reported the quantitative analysis of adducts of 1,3-butadiene epoxides with dA and dG in DNA. The butadiene metabolites 3,4-epoxy-1-butene, diepoxybutane, and 3,4-epoxy-1,2butanediol were found to react with dG at the N-7-position and dA at the N-1-, N-3-, N-6-, and N-7-positions. Quantitative analysis of the modified nucleobases was performed by positive-ion LC–ESI-MS in SRM mode.

Lemière et al. [68] studied isomeric phenylglycidyl ether adducts of dG and dGMP. From MS–MS data, the formation of adducts at the N-7- and N-2-position was proposed, while NMR proved the N-2-adduct was actually adducted at the N-1-position. In subsequent studies from this group [69-70], the conventional LC was replaced by capillary LC and even nano-LC providing better absolute detection limits. In order not to compromise the concentration detection limits, as the capillary or nano-LC system allow smaller injection volumes, sample injection was done at a short SPE column, which was switched in-line with the LC column for elution and LC–MS analysis. This approach was applied in the analysis of mephalan adducts of dAMP in calf thymus hydrolysates. A solution of 1.1 nmol/l mephalan–dAMP adduct was detected with a S/N of 8 with the capillary LC column and with a S/N ratio of 22 with the nano-LC, indicating a 2.5-fold improvement. Considerable attention was also paid to the structure elucidation of the various adducts formed [69-72], also in an in vivo study in rats [72].

Siethoff et al. [73] reported the quantitative bioanalysis of nucleotides from DNA modified by styrene oxide by a combination of LC with ESI-MS and inductively coupled plasma mass spectrometry (ICP-MS). The LC–ICP-MS system was applied for phosphorous detection. This helped to evaluate response factors of various adducts in LC–ESI-MS, which were found to be almost identical. The detection limit for the styrene oxide adducts was 20 pg, using a 5 μg DNA sample (corresponds to 14 adducts in 10^8 bases).

Gangl et al. [74] reported a 100-fold improvement in the detection of in vivo formed DNA adducts derived from the food-derived 2-amino-3-methylimidaz[4,5-f] quinoline (IQ, one of the heterocyclic aromatic amines, Ch. 14.5) by the application of capillary LC in combination with micro-ESI-MS. As a result, the detection limit approaches 1 adduct in 10^9 nucleobases using 500 μg DNA. In a subsequent study [75], this technology was applied to the quantitative analysis of the IQ–dG adduct in rat liver samples in a dose–response study. The major adduct (C8-IQ–dG) could be detected at 17.5 fmol in 300 μg of liver DNA (corresponding to 2 adducts in 10^8 nucleobases).

Roberts et al. [76] reported the detection of etheno-dC adducts in crude DNA hydrolysates on the order to 5 adducts in 10^8 nucleobases using 100 μg DNA. The use of on-line affinity LC (Ch. 1.4.2) allowed a 100-fold improvement in detection limits compared to a conventional C_{18}-based SPE approach.

There is considerable interest in the analysis of DNA adducts formed by the reaction of hydroxylated metabolic products of estrogens with DNA. The reactivity of estradiol-2,3-quinone towards dG and dC was studied by Van Aerden et al. [77]. Several adducts were characterized, including a new estrogen–dC adduct. Embrechts et al. [78] applied nano-LC (300-μm-ID) coupled to nano-ESI-MS (300 nl/min) for the detection of adducts with 4-hydroxyequilenin. Different isomeric adducts were found with dA, dC, and dG, but not with dT. A SRM detection limit of 197 fg for an equilenin–dG was reported. In a subsequent study, the same group identified 4-hydroxyequilenin–DNA adducts as well as DNA adducts with 4-hydroxy-estradiol or 4-hydroxy-estrone in human breast tumour tissue [79]. Regio- and stereo-selectivity in the linkage of 2-hydroxyestradiol to dG was investigated by Debrauwer et al. [80].

6. References

1. C.R. Blakley, J.J. Carmody, M.L. Vestal, *LC–MS for analysis of nonvolatile samples*, Anal. Chem., 52 (1980) 1636.
2. R.D. Voyksner, *Characteristics of ion evaporation ionization in thermospray LC–MS*, Org. Mass Spectrom., 22 (1987) 513.
3. T.R. Covey, R.F. Bonner, B.I. Shushan, J.D. Henion, *The determination of protein, oligonucleotide and peptide molecular weights by ESI-MS*, Rapid Commun. Mass Spectrom., 2 (1988) 249.
4. J.T. Stults, J.C. Marsters, *Improved ESI of synthetic oligonucleotides*, Rapid Commun. Mass Spectrom., 5 (1991) 359.
5. C.G. Huber, H. Oberacher, *Analysis of nucleic acids by on-line LC–MS*, Mass Spectrom. Rev., 20 (2001) 310.
6. A. Pruvost, F. Becher, P. Bardouille, C. Guerrero, C. Creminon, J.-F. Delfraissy, C. Goujard, J. Grassi, H. Benech, *Direct determination of phosphorylated intracellular*

anabolites of stavudine (d4T) by LC–MS–MS, Rapid Commun. Mass Spectrom., 15 (2001) 1401.

7. R. Tuytten, F. Lemière, W. Van Dongen, E. L. Esmans, H. Slegers, *Short capillary ion-pair LC–ESI-MS–MS for the simultaneous analysis of nucleoside mono-, di- and triphosphates*, Rapid Commun. Mass Spectrom., 16 (2002) 1205.

8. K. Bleicher, E. Bayer, *Various factors influencing the signal intensity of oligonucleotides in ESI-MS*, Biol. Mass Spectrom., 23 (1994) 320.

9. C.G. Huber, A. Krajete, *Analysis of nucleic acids by capillary ion-pair RPLC coupled to negative-ion ESI-MS*, Anal. Chem., 71 (1999) 3730.

10. C.G. Huber, A. Krajete, *Sheath liquid effects in capillary LC–ESI-MS of oligonucleotides*, J. Chromatogr. A, 870 (2000) 413.

11. A. Premstaller, H. Oberacher, C.G. Huber, *LC–ESI-MS of single- and double-stranded nucleic acids using monolithic capillary columns*, Anal. Chem., 72 (2000) 4386.

12. G. Hölzl, H. Oberacher, S. Pitsch, A. Stutz, C.G. Huber, *Analysis of biological and synthetic ribonulceaic acids by LC–MS using monolithic capillary columns*, Anal. Chem., 77 (2005) 673.

13. C. Liu, Q. Wu, A.C. Harms, R.D. Smith, *On-line microdialysis sample cleanup for ESI-MS of nucleic acid samples*, Anal. Chem., 68 (1996) 3295.

14. R. Tuytten, F. Lemière, W. Van Dongen, H. Slegers, R.P. Newton, E.L. Esmans, *Investigation of the use of IMAC for the on-line sample clean up and pre-concentration of nucleotides prior to their determination by ion pair LC–ESI-MS: a pilot study*, J. Chromatogr. B, 809 (2004) 189.

15. C.G. Huber, M.R. Buchmeiser, *On-line cation exchange for suppression of adduct formation in negative-ion ESI-MS of nucleic acids*, Anal. Chem., 70 (1998) 5288.

16. C.G. Huber, A. Krajete, *Comparison of direct infusion and on-line LC–ESI-MS for the analysis of nucleic acids*, J. Mass Spectrom., 35 (2000) 870.

17. D.C. Muddiman, D.S. Wunschel, C. Liu, L. Paša-Tolić, K.F. Fox, A. Fox, G.A. Anderson, R.D. Smith, *Characterization of PCR products from bacilli using ESI-FT-ICR-MS*, Anal. Chem.68 (1996) 3705.

18. A.P. Null, L.T. George, D.C. Muddiman, *Evaluation of sample preparation techniques for mass measurements of PCR products using ESI-FT-ICR-MS*, J. Am. Soc. Mass Spectrom., 13 (2002) 338.

19. M. Greig, R.H. Griffey, *Utility of organic bases for improved ESI-MS of oligonucleotides*, Rapid Commun. Mass Spectrom., 9 (1995) 97.

20. A.V. Willems, D.L. Deforce, W.E. Lambert, C.H.V. Peteghem, J.F. Van Bocxlaer, *Rapid characterization of oligonucleotides by capillary nano-LC–ESI-Q–TOF-MS*, J. Chromatogr. A, 1052 (2004) 93.

21. H. Oberacher, W. Parson, G. Holzl, P.J. Oefner, C.G. Huber, *Optimized suppression of adducts in polymerase chain reaction products for semi-quantitative SNP genotyping by LC–MS*, J. Am. Soc. Mass Spectrom., 15 (2004) 1897.

22. A. Apffel, J.A. Chakel, S. Fischer, K. Lichtenwalter, W.S. Hancock, *Analysis of oligonucleotides by LC–ESI-MS*, Anal. Chem., 69 (1997) 1320.

23. K. Deguchi, M. Ishikawa, T. Yokokura, I. Ogata, S. Ito, T. Mimura, C. Ostrander, *Enhanced mass detection of oligonucleotides using RPLC–ESI ion-trap MS*, Rapid Commun. Mass Spectrom., 16 (2002) 2133.

24. J.J. Walters, W. Muhammad, K.F. Fox, A. Fox, D. Xie, K.E. Creek, L. Pirisi,

Genotyping SNP using intact PCR products by ESI-MS, Rapid Commun. Mass Spectrom., 15 (2001) 1752.

25. H. Oberacher, W. Parson, R. Mühlmann, C.G. Huber, *Analysis of PCR products by on-line LC–MS for genotyping of polymorphic short tandem repeat loci*, Anal. Chem., 73 (2001) 5109.

26. K.K. Murray, *DNA sequencing by MS*, J. Mass Spectrom., 31 (1996) 1205.

27. N.I. Taranenko, S.L. Allman, V.V. Golovlev, N.V. Taranenko, N.R. Isola, C.H. Chen, *Sequencing DNA using MS for ladder detection*, Nucl. Acids Res., 26 (1998) 2488.

28. S.A. McLuckey, G.J. van Berkel, G.L. Glish, *MS–MS of small, multiply charged oligonucleotides*, J. Am. Soc. Mass Spectrom., 3 (1992) 60.

29. J. Ni, S.C. Pomerantz, J. Rozenski, Y. Zhang, J.A. McCloskey, *Interpretation of oligonucleotide mass spectra for determination of sequence using ESI-MS–MS*, Anal. Chem., 68 (1996) 1989.

30. H. Oberacher, B. Wellenzohn, C.G. Huber, *Comparative sequencing of nucleic acids by LC–MS–MS*, Anal. Chem., 74 (2002) 211.

31. H. Oberacher, W. Parson, P.J. Oefner, B.M. Mayr, C.G. Huber, *Applicability of MS–MS to the automated comparative sequencing of long-chain oligonucleotides*, J. Am. Soc. Mass Spectrom., 15 (2004) 510.

32. H. Oberacher, B.M. Mayr, C.G. Huber, *Automated de novo sequencing of nucleic acids by LC–MS–MS*, J. Am. Soc. Mass Spectrom., 15 (2004) 32.

33. J. Rozenski, J.A. McCloskey, *SOS: a simple interactive program for ab initio oligonucleotide sequencing by MS*, J. Am. Soc. Mass Spectrom., 13 (2002) 200.

34. A.K. Vrkic, R.A.J. O'Hair, S. Foote, G.E. Reid, *Fragmentation reactions of all 64 protonated trimer and 16 mixed base tetramer oligodeoxynucleotides via MS–MS in an ion-trap*, Int. J. Mass Spectrom., 194 (2000) 145.

35. A. Premstaller, K.-H. Ongania, C.G. Huber, *Factors determining the performance of triple quadrupole, quadrupole ion trap and sector field mass spectrometers in ESI-MS–MS of oligonucleotides. 1. Comparison of performance characteristics*, Rapid Commun. Mass Spectrom., 15 (2001) 1045.

36. A. Premstaller, C.G. Huber, *Factors determining the performance of triple quadrupole, quadrupole ion trap and sector field mass spectrometers in ESI-MS–MS of oligonucleotides. 1. Comparison of performance characteristics*, Rapid Commun. Mass Spectrom., 15 (2001) 1053.

37. X. Cheng, D.C. Gale, H.R. Udseth, R.D. Smith, *Charge state reduction of oligonucleotide negative ions from ESI*, Anal. Chem., 67 (1995) 586.

38. D.P. Little, F.W. McLafferty, *Infrared photodissociation of non-covalent adducts of electrosprayed nucleotide ions*, J. Am. Soc. Mass Spectrom., 9 (1996) 209.

39. J. Ni, K. Chan, *Sequence verification of oligonucleotides by ESI-Q–TOF-MS*, Rapid Commun. Mass Spectrom., 15 (2001) 1600.

40. T.R. Baker, T. Keough, R.L.M. Dobson, T.A. Riley, J.A. Hasselfield, P.E. Hesselberth, *Antisense DNA oligonucleotides I: the use of ESI-MS–MS for the sequence verification of methylphosphonate oligodeoxyribo-nucleotides*, Rapid Commun. Mass Spectrom., 7 (1993) 190.

41. R.H. Griffey, M.J. Greig, H.J. Gaus, K. Liu, D. Monteith, M. Winniman, L.L. Cummins, *Characterization of oligonucleotide metabolism in vivo via LC–ESI-MS–MS with a quadrupole ion trap MS*, J. Mass Spectrom., 32 (1997) 305.

42. R. Lotz, M. Gerster, E. Bayer, *Sequence verification of oligodeoxynucleotides and oligophosphorothioates using ESI-MS–MS*, Rapid Commun. Mass Spectrom., 12 (1998) 389.

43. D.S. Wunschel, D.C. Muddiman, K.F. Fox, A. Fox, R.D. Smith, *Heterogeneity in Bacillus cereus PCR products detected by ESI-FT-ICR-MS*, Anal. Chem., 70 (1998) 1203.

44. D.C. Muddiman, A.P. Null, J.C. Hannis, *Precise mass measurement of a double-stranded 500 base-pair (309 kDa) polymerase chain reaction product by negative ion ESI-FT-ICR-MS*, Rapid Commun. Mass Spectrom., 13 (1999) 1201.

45. A.P. Null, J.C. Hannis D.C. Muddiman, *Preparation of single-stranded PCR products for ESI-MS using the DNA repair enzyme lambda exonuclease*, Analyst, 125 (2000) 619.

46. M.T. Krahmer, Y.A. Johnson, J.J. Walters, K.F. Fox, A. Fox, M. Nagpal, *ESI-MS analysis of model oligonucleotides and PCR products: determination of base substitutions, nucleotide additions/deletions, and chemical modifications*, Anal. Chem., 71 (1999) 2893.

47. J. Tost, I.G. Gut, *Genotyping SNP by MS*, Mass Spectrom. Rev., 21 (2003) 388.

48. M.T. Krahmer, J.J. Walters, K.F. Fox, A. Fox, K.E. Creek, L. Pirisi, D.S. Wunschel, R.D. Smith, D.L. Tabb, J.R. Yates, III, *MS for identification of SNP and MS–MS for discrimination of isomeric PCR products*, Anal. Chem., 72 (2000) 4033.

49. S. Zhang, C.K. Van Pelt, G.A. Schultz, *ESI-MS-based genotyping: an approach for identification of SNP*, Anal. Chem., 73 (2001) 2117.

50. S. Zhang, C.K. Van Pelt, X. Huang, G.A. Schultz, *Detection of SNP using ESI-MS: validation of a one-well assay and quantitative pooling studies*, J. Mass Spectrom., 37 (2002) 1039.

51. W.T. Muhammad, K.F. Fox, A. Fox, W. Cotham, M. Walla, *ESI-Q–TOF-MS and quadrupole MS for genotyping SNP in intact PCR products in K-ras and p53*, Rapid Commun. Mass Spectrom., 16 (2002) 2278.

52. B. Berger, G. Holzl, H. Oberacher, H. Niederstatter, C.G. Huber, W. Parson, *SNP genotyping by on-line LC–MS in forensic science of the Y-chromosomal locus M9*, J. Chromatogr. B, 782 (2002) 89.

53. H. Oberacher, C.G. Huber, *Genotyping of SNP by LC–ESI-MS*, Anal. Bioanal. Chem., 376 (2003) 292.

54. E. Dudley, S. El-Sharkawi, D. E. Games, R.P. Newton, *Analysis of urinary nucleosides. I. Optimisation of LC–ESI-MS*, Rapid Commun. Mass Spectrom., 14 (2000) 1200.

55. E. Dudley, F. Lemière, W. Van Dongen, J.I. Langridge, S. El-Sharkawi, D.E. Games, E.L. Esmans, R.P. Newton, *Analysis of urinary nucleosides. II. Comparison of MS methods for the analysis of urinary nucleosides*, Rapid Commun. Mass Spectrom., 15 (2001) 1701.

56. E. Dudley, F. Lemière, W. Van Dongen, J.I. Langridge, S. El-Sharkawi, D.E. Games, E.L. Esmans, R.P. Newton, *Analysis of urinary nucleosides. III. Identification of 5'-deoxycytidine in urine of a patient with head and neck cancer*, Rapid Commun. Mass Spectrom., 17 (2003) 1132.

57. E. Dudley, F. Lemière, W. Van Dongen, R. Tuytten, S. El-Sharkawi, A.G. Brenton, E.L. Esmans, R.P. Newton, *Analysis of urinary nucleosides. IV. Identification of urinary purine nucleosides by LC–ESI-MS*, Rapid Commun. Mass Spectrom., 18

(2004) 2730.

58. J.-L. Ravanat, B. Duretz, A. Guiller, T. Douki, J. Cadet, *Isotope dilution LC–ESI-MS–MS assay for the measurement of 8-OH-dG in biological samples*, J. Chromatogr. B, 715 (1998) 349.

59. Y. Hua, S.B. Wainhaus, Y. Yang, L. Shen, Y. Xiong, X. Xu, F. Zhang, J.L. Bolton, R.B. van Breemen, *Comparison of negative and positive ion LC–ESI-MS–MS analysis of oxidized deoxynucleosides*, J. Am. Soc. Mass Spectrom., 12 (2001) 80.

60. T. Renner, G. Scherer, *Fast quantification of the urinary marker of oxidative stress 8-OH-dG using SPE and LC–MS–MS*, J. Chromatogr. B, 738 (2000) 311.

61. P.G. Pietta, P. Simonetti, C. Gardana, S. Cristoni, L. Bramati, and P.L. Mauri, *LC–APCI-MS–MS analysis of urinary 8-OH-dG*, J. Pharm. Biomed. Anal., 32 (2003) 657.

62. C.-W. Hu, M.-T. Wu, M.-R. Chao, C.-H. Pan, C.-J. Wang, J.A. Swenberg, K.-Y. Wu, *Comparison of analyses of urinary 8-OH-dG by isotope-dilution LC–ESI-MS–MS and by enzyme-linked immunosorbent assay*, Rapid Commun. Mass Spectrom., 18 (2004) 505.

63. L. Sabatini, A. Barbieri, M. Tosi, A. Roda, F.S. Violante, *A method for routine quantitation of urinary 8-OH-dG based on SPE and micro-LC–ESI-MS–MS*, Rapid Commun. Mass Spectrom., 19 (2005) 147.

64. W. Apruzzese, P. Vouros, *Analysis of DNA adducts by capillary methods coupled to MS: a perspective*, J. Chromatogr. A, 794 (1998) 97.

65. E.L. Esmans, D. Broes, I. Hoes, F. Lemière, K. Vanhoutte, *LC–MS in nucleoside, nucleotide and modified nucleotide characterization*, J. Chromatogr. A, 794 (1998) 109.

66. H. Koc, J.A. Swenberg, *Applications of MSy for quantitation of DNA adducts*, J. Chromatogr. B, 778 (2002) 323.

67. N.Yu. Tretyakova, S.-Y. Chiang, V.E. Walker, J.A. Swenberg, *Quantitative analysis of 1,3-butadiene-induced DNA adducts in vivo and in vitro using LC–ESI-MS–MS*, J. Mass Spectrom., 33 (1998) 363.

68. F. Lemière, K. Vanhoutte, T. Jonckers, R. Marek, E.L. Esmans, M. Claeys, E. Van den Eeckhout, H. Van Onckelen, *Differentiation between isomeric phenylglycidyl ether adducts of 2-deoxyguanosine and 2-deoxyguanosine-5-monophosphate using LC–ESI-MS–MS*, J. Mass Spectrom., 34 (1999) 820.

69. I. Hoes, E.L. Esmans, *Analysis of melphalan adducts of 2'-deoxynucleotides in calf thymus DNA hydrolysates by capillary LC–ESI-MS–MS*, J. Chromatogr. B, 736 (1999) 43.

70. I. Hoes, W. Van Dongen, F. Lemière, E.L. Esmans, D. Van Bockstaele, Z.N. Berneman, *Comparison between capillary and nano LC–ESI-MS for the analysis of minor DNA-melphalan adducts*, J. Chromatogr. B, 748 (2000) 197.

71. B. Van den Driessche, F. Lemière, W. Van Dongen, E.L. Esmans, *Structural characterization of melphalan modified 2'-oligodeoxynucleotides by miniaturized LC–ES I-MS–MS*, J. Am. Soc. Mass Spectrom., 15 (2004) 568.

72. B. Van den Driessche, F. Lemière, W. Van Dongen, A. Van der Linden, E.L. Esmans, *Qualitative study of in vivo melphalan adduct formation in the rat by miniaturized column-switching LC–ESI-MS*, J. Mass Spectrom., 39 (2004) 29.

73. C. Siethoff, I. Feldmann, N. Jakubowski, M. Linscheid, *Quantitative determination of DNA adducts using LC–ESI-MS and LC–high resolution ICP-MS*, J. Mass

Spectrom., 34 (1999) 421.

74.   E.T. Gangl, R.J. Turesky, P. Vouros, *Detection of in vivo formed DNA adducts at the part-per-billion level by capillary LC–ESI-MS*, Anal. Chem., 73 (2001) 2397.

75.   J.R. Soglia, R.J. Turesky, A. Paehler, P. Vouros, *Quantification of the heterocyclic aromatic amine DNA adduct N-(deoxyguanosin-8-yl)-2-amino-3-methylimidazo[4,5-f]quinoline in livers of rats using capillary LC–ESI-MS: A dose–response study*, Anal. Chem., 73 (2001) 2819.

76.   D.W. Roberts, M.I. Churchwell, F.A. Beland, J.-L. Fang, D.R. Doerge, *Quantitative analysis of etheno-2'-deoxycytidine DNA adducts using on-line immunoaffinity chromatography coupled with LC–ESI-MS–MS detection*, Anal. Chem., 73 (2001) 303.

77.   C. Van Aerden, L. Debrauwer, J.C. Tabet, A. Paris, *Analysis of nucleoside– estrogen adducts by LC–ESI-MS–MS*, Analyst, 123 (1998) 2677.

78.   J. Embrechts, F. Lemière, W. Van Dongen, E.L. Esmans, *Equilenin-2'-deoxynucleoside adducts: analysis with nano-LC–ESI-MS–MS*, J. Mass Spectrom., 36 (2001) 317.

79.   J. Embrechts, F. Lemière, W. Van Dongen, E.L. Esmans, P. Buytaert, E. Van Marck, M. Kockx, A. Makar, *Detection of estrogen DNA-adducts in human breast tumor tissue and healthy tissue by combined nano LC–ESI-MS–MS*, J. Am. Soc. Mass Spectrom., 14 (2003) 482.

80.   L. Debrauwer, E. Rathahao, I. Jouanin, A. Paris, G. Clodic, H. Molines, O. Convert, F. Fournier, J.C. Tabet, *Investigation of the regio- and stereo-selectivity of deoxyguanosine linkage to deuterated 2-hydroxyestradiol by using LC–ESI-ion trap MS*, J. Am. Soc. Mass Spectrom., 14 (2003) 364.

INDEX

Printed in the United States
by Baker & Taylor Publisher Services